5. JUNAID KABIR
CIVIL - II

514

747 7373

170
50

320

Calculus and Analytic Geometry

third edition

Library of Congress Cataloging in Publication Data

FISHER, ROBERT CHARLES, (date)
  Calculus and analytic geometry.

    1. Calculus. 2. Geometry, Analytic. I. Ziebur,
Allen D., joint author. II. Title.
QA303.F57 1975   515'.15   74-13053
ISBN 0-13-112227-4

CALCULUS AND ANALYTIC GEOMETRY
third edition
Robert C. Fisher / Allen D. Ziebur

© 1975, 1965, 1961, 1960, 1959
by Prentice-Hall, Inc., Englewood Cliffs, N. J.

10 9 8 7 6 5 4 3 2 1

Printed in the United States of America

PRENTICE-HALL INTERNATIONAL, INC., *London*
PRENTICE-HALL OF AUSTRALIA, PTY. LTD., *Sydney*
PRENTICE-HALL OF CANADA, LTD., *Toronto*
PRENTICE-HALL OF INDIA PRIVATE LIMITED, *New Delhi*
PRENTICE-HALL OF JAPAN, INC., *Tokyo*

Contents

2

THE DERIVATIVE 59

3

APPLICATIONS OF THE DERIVATIVE 120

4

THE CONICS 154

5

THE INTEGRAL 192

6

EXPONENTIAL, LOGARITHMIC, INVERSE TRIGONOMETRIC, AND HYPERBOLIC FUNCTIONS 265

7

TECHNIQUES OF INTEGRATION 328

14

MULTIPLE INTEGRALS AND LINE INTEGRALS 687

Preface

Our reworking of the second edition has completely changed the face of the book. Topics have been rearranged so that the first ten chapters cover the calculus of functions of one variable, leaving multivariate calculus to the last four chapters (the typical third semester). But these changes are perhaps not as important as some of the less obvious ones. The changes that really count are those that make the material more accessible to the student.

The second edition was written at a time when the theoretical aspects of calculus were stressed in the beginning course. Now the pendulum is swinging the other way, and details of the theory are being de-emphasized. In organizing this edition, we first arranged the material so that the subject was developed step by step in a logical manner, proofs and all. Then we deleted most of the technical details of the proofs, leaving behind their basic ideas in the form of geometric or heuristic arguments. In this way, we hope to have provided an honest exposition of the theory without burying it in fine points that only confuse most students. In particular, the succession of theorems and examples on limits in Chapter 1 forms a coherent sequence of theorems and lemmas, so arranged that all the basic results are essentially corollaries of Theorems 10-2 and 11-1 and Examples 9-4 and 9-5.

Much, if not most, calculus is learned by working problems, so we have devoted special attention to the problem sets at the end of each section. They have been greatly enlarged, particularly by the addition of "routine" problems. Since one man's meat is another man's poison, it is impossible to list problems in order of their difficulty. Generally speaking, however, the earlier problems in our sets tend to be of a computational nature, while those containing words or involving theory occur later. A few of the more theoretical problems are intended primarily as "discussion questions." The student is not expected to be able to produce the required proofs without some help from the instructor. Extensive collections of review problems appear at the end of each chapter.

Problems taken from real life best convince the student that calculus is a worthwhile subject of study. But the more relevant the problem, the more background in physics, economics, biology, or whatever, is needed to understand it. We compromised by presenting a reasonably large array of "word problems," in which the words apply to situations that can be understood with only a superficial knowledge of geometry or the usual facts picked up while living in the twentieth century. We tried to select these problems so that they at least indicate the uses of calculus in various areas of human thought.

Although the book does not specifically call upon the electronic computer, the existence of these machines has had a considerable effect on our ordering and presentation of topics. For example, we have discussed Taylor's Formula much earlier than before because of its computational importance. And we stress the "definite," rather than the "indefinite" integral, not only because an integral is "definite" by definition, but also because "definite" integrals are what computers compute. The educational value of a calculus course is greatly enhanced by accompanying it with (or closely following it by) a "laboratory" in which such ideas as Newton's Method, Taylor's Formula, Simpson's Rule, and so on, are tried on a computer.

In the first chapter we discuss such ideas as functions, graphs, slopes of lines, and limits as a background to calculus proper. Chapter 2 contains definitions and techniques of differentiation, and Chapter 3 consists of applications of the derivative. In Chapter 4 we take up the conic sections, but students who already know this material could skip ahead to the introduction of the integral in Chapter 5.

The second semester might start with Chapters 6 and 7, which cover functions defined by integrals, some techniques of integration (antidifferentiation), and a brief introduction to differential equations. In Chapter 8 we are back to geometry, with polar coordinates and vectors. Of course, calculus plays a role in the study of polar and vector curves. Chapters 9 and 10 deal with the related topics of improper integrals and infinite series.

In Chapter 11 we look at other examples of vector spaces, R^2 and R^3. Here we learn about systems of linear equations and elementary operations with vectors and matrices. We put this knowledge to work in Chapter 12, where we study the geometry of three-space. Chapters 13 and 14 treat differentiation and integration of functions on R^n. Because this material (or at least its notation) is inherently complicated, our efforts were directed toward explaining it as simply as possible.

Proof by picture and analogy with the one-dimensional case take the place of rigorous mathematical arguments.

Although many people influenced this book in many ways, our special thanks are due to production editor Reynold Rieger for smoothly turning a large, complicated, and much-revised manuscript into the final book.

R.C.F.
A.D.Z.

Basic Concepts and the Limit

1

Scientists and engineers study *relations*. The manager of a municipal water plant, for example, is interested in the relation between the chlorine he adds to the city's water and its resultant potability. If he can express this relation in a quantitative way—the *number* of pounds of chlorine that reduces the bacterial count to a safe *number*—so much the better. Calculus is probably the most important mathematical tool available for studying quantitative relations.

We begin by considering such basic concepts as numbers, sets, functions, and graphs, concepts that are prerequisite for our study of calculus. Our presentation of these topics is somewhat brief, since it is mostly review for you. Nevertheless, experience shows that many students who take calculus have as much trouble with such items as absolute values, inequalities, and factoring, as they do with calculus itself. So it will be worthwhile for you to get this material under control right at the start. At the end of the chapter we introduce the central idea of calculus—the concept of a limit.

1. THE REAL NUMBERS

It should not surprise you if a mathematics book starts with a discussion of numbers—a great deal of mathematics is based on numbers. So let us summarize some of the facts about the real number system that we will need as we study calculus.

Each **real number** can be written as an unending decimal expression. Thus the real numbers π, 2, and $\frac{1}{3}$ can be written as $\pi = 3.14159\ldots$, $2 = 2.000\ldots$, and $\frac{1}{3} = .333\ldots$. Among the real numbers are the **integers**, $\ldots, -2, -1, 0, 1, 2, \ldots$, and the **rational numbers**, such as $-\frac{3}{2}, \frac{22}{7}, \frac{0}{3}$. A rational number can be written as a *ratio* of two integers. You are familiar with the rules of arithmetic of real numbers.

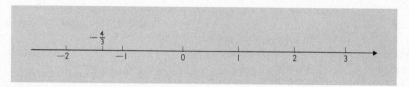

Figure 1-1

We represent the real numbers geometrically as points of the **number scale**. In order to construct the scale, we choose two points of a horizontal line and label the left-hand point 0 and the right-hand point 1. The first point is called the **origin** and the second is the **unit point**, and the distance between them is the unit of distance. Once the origin and the unit point are chosen, one can then find the points that represent other numbers. For example, since $-\frac{4}{3} = -1 - \frac{1}{3}$, the point representing $-\frac{4}{3}$ is one-third of the way from the point that represents -1 to the point that represents -2 (Fig. 1-1). Finding the point that represents an irrational number like π can get a little complicated, but it should be reasonable to you that *each point of the number scale is the graphical representation of a real number, and to each real number there corresponds a point of the number scale.*

Logically, a number and its corresponding point of the number scale should be distinguished as two different things. The number associated with a point is called the **coordinate** of the point. In practice, however, we often refer to the "point" 5, instead of using the cumbersome phrase "the point whose coordinate is 5."

When the number scale is displayed with the usual left-to-right orientation, smaller numbers lie to the left of larger ones. A formal definition of this order relation that does not rely on a picture reads as follows.

Definition 1-1. *The number a is* **less than** *b ($a < b$) and b is* **greater than** *a ($b > a$) if $b - a$ is a positive number. The symbols $<$ and $>$ are called* **inequality signs**.

Thus we see that the statements "a is positive" and "$a > 0$" say exactly the same thing. Notice how this formal definition is related to our geometric idea that smaller numbers lie to the left of larger ones. The number scale shows the point -5 to be to the left of -3, which is the geometric way of saying that $-5 < -3$. On the other hand, $-3 - (-5)$ is the positive number 2, which is also, according to Definition 1-1, a way of saying that $-5 < -3$. The notation $a < b < c$ means that $a < b$ *and* $b < c$. For obvious geometric reasons, we say that b is **between** a and c. Thus the number 4 is between 2 and 7, since $2 < 4 < 7$.

Because sums and products of positive numbers are positive, we have the following rules for operating with inequality symbols.

(1-1)   If $a < b$ and $b < c$, then $a < c$.

(1-2)   If $a < b$, then $a + c < b + c$ for any number c.

(1-3)   If $c > 0$ and $a < b$, then $ac < bc$.

(1-4)   If $c < 0$ and $a < b$, then $ac > bc$.

Let us show why Rule 1-3 is valid—and leave the verification of the other rules to you. In Rule 1-3 we assume that $a < b$; hence $b - a$ is positive. The product of the positive numbers $(b - a)$ and c is the positive number $(b - a)c = bc - ac$. But to say that this difference is positive is to say that $ac < bc$, which is exactly the conclusion of Rule 1-3.

The notation $a \leq b$ or $b \geq a$ means that a is a number that is either less than b or equal to b; that is, a is *not greater* than b. For example, $4 \leq 6$ and $6 \leq 6$, since neither 4 nor 6 is greater than 6. The basic rules that govern the use of the symbol $\leq$ are just like the rules we use with the symbol $<$.

The graphical representation of the system of real numbers as points of the number scale allows us to introduce the geometric concept of *distance* into our number system. The number of units between a point x of the number scale and the origin is called the **absolute value** of x, and it is denoted by the symbol $|x|$. Thus $|2| = 2$, $|-3| = 3$, $|\pi| = \pi$, and so on. The absolute value of a number gives only its distance from the origin, not its direction. If $|x| = 5$, for example, we know that x is either the number 5 or the number -5, since both these numbers, and no others, are 5 units from the origin.

A glance at the number scale shows that the number $|b - a|$ is the distance between the points a and b. Notice that $|b - a| = |a - b|$, so the distance between two points is obtained by subtracting their coordinates in either order and taking the absolute value of the difference. If we want to find the distance without using absolute values, we subtract the coordinate of the left point (smaller number) from the coordinate of the right point (larger number).

Example 1-1.   Verify that the points $a = -3$ and $b = 2$ are $|b - a|$ (or $|a - b|$) units apart.

Solution.   The points -3 and 2 are shown on the number scale in Fig. 1-2. It is obvious that they are 5 units apart, and $5 = |2 - (-3)| = |-3 - 2|$. Notice that we can also determine the distance 5 by subtracting the left number, -3, from the right number, 2.

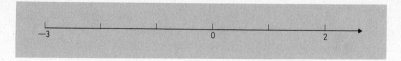

Figure 1-2

In words, the inequality

(1-5) $$|a - b| \le |a| + |b|$$

says that "the distance between the points a and b does not exceed the sum of the distances between a and the origin and b and the origin." That is, a and b are no farther apart than the sum of their distances to the origin. Substitute a few numbers in place of a and b to make yourself comfortable with this statement.

Now if in Inequality 1-5 we replace b with $-b$ and observe that $|-b| = |b|$, we obtain the inequality

(1-6) $$|a + b| \le |a| + |b|.$$

This result is called the **triangle inequality** for geometric reasons that we will discover when we take up vectors in Chapter 8. The important point to notice is that if you add two numbers and then take the absolute value of the sum, you don't necessarily end up with what you get by taking absolute values first and then adding. But you don't get anything larger.

PROBLEMS 1

1. What number is represented by the point of the number scale that is
   (a) $\frac{1}{3}$ of the way from 2 to 3?
   (b) $\frac{3}{4}$ of the way from -6 to -5?
   (c) $\frac{1}{5}$ of the way from -5 to 10?
   (d) midway between -3 and 4?

2. List all the integers between -2 and 3. Can you list all the real numbers between -2 and 3? All the rational numbers?

3. Replace each inequality with a simpler equivalent inequality.
   (a) $3x < 9$
   (b) $2x - 1 < 7$
   (c) $x < 4 - x$
   (d) $1 - x > 1 - 2x$
   (e) $3 < 4x - 1 < 15$
   (f) $x < 2x - 1 < 5$

4. What can you say about the number x if
   (a) $|x| = 4$?
   (b) $|x - 1| = 3$?
   (c) $|2x| = 2x$?
   (d) $|x - 1| = 1 - x$?

5. What can you say about the number x if
   (a) $|x| \le 4$?
   (b) $|x - 1| \ge 3$?
   (c) $|2x| \ge 2x$?
   (d) $|x - 1| \le 1 - x$?

6. Suppose that you know that $a < b$. Which of the following inequalities do you then know to be true?

    (a) $\dfrac{a}{3} < \dfrac{b}{3}$

    (b) $a + 5 \leq b + 4$

    (c) $a < \dfrac{a + b}{2} < b$

    (d) $a^2 < b^2$

    (e) $\sqrt{|a|} < \sqrt{|b|}$

    (f) $a^{-1} > b^{-1}$

7. If you know that $a^2 < b^2$, what can you conclude about the relation between a and b?

8. Express the following statements about points of a number scale in terms of inequality signs and absolute values.

    (a) The point x is more than 3 units to the right of the point y.
    (b) The point x is closer to the origin than is the point y.
    (c) The distance between the point x and the point 3 is less than $\frac{1}{2}$.
    (d) The points x and y are more than 5 units apart.

9. Find two rational numbers x and y such that

    (a) $x < \pi < y$ and $y - x < .01$.
    (b) $x < \sqrt{3} < y$ and $y - x < .1$.

10. Explain why the following statements are true.

    (a) $|x| = \sqrt{x^2}$
    (b) $|x|$ is the larger of the numbers x and $-x$.

11. If $a = .141414\ldots$, then $100a = 14.141414\ldots$, and $100a - a = 14.141414\ldots -$ $.141414\ldots$, or $99a = 14$. Thus $a = \frac{14}{99}$. Use a similar argument to write $b = .142142142\ldots$ and $c = 2.999\ldots$ as fractions.

12. Show that $\frac{1}{2}(a + b + |a - b|)$ is the larger of the numbers a and b and $\frac{1}{2}(a + b - |a - b|)$ is the smaller.

13. What can you say about the number x if

    (a) $\left| \dfrac{1}{x} - 2 \right| < 2$?

    (b) $\left| \dfrac{1}{x} - 2 \right| < 3$?

    (c) $\left| \dfrac{1}{x} - 2 \right| < 1$?

    (d) $|x^2 - 7| < 2$?

    (e) $(x^2 - 7) < 2$?

    (f) $|x(x - 7)| \leq x$?

14. In Inequality 1-6, replace a with $a - c$ and b with $c - b$ to show that, for any three numbers a, b, and c, $|a - b| \leq |a - c| + |b - c|$.

15. Replace a with $a - b$ in the triangle inequality to obtain the inequality $|a| - |b| \leq |a - b|$. Explain why it is true that $||a| - |b|| \leq |a - b|$ for any two numbers a and b.

16. Explain why Rules 1-1, 1-2, and 1-4 are valid.

We will often have occasion to discuss *sets* of mathematical objects, such as the set of positive integers or the set of points that make up some curve in the plane, so let us introduce some of the terminology and notation of set theory. The most straightforward way to specify a set is to list its **elements** or **members**. Thus {2, 4, 6, 8} is the set whose members are the four numbers 2, 4, 6, and 8. We enclosed the members of this set in braces { } to indicate that we think of the set as an entity in itself. In line with this idea, sets are often named by letters. For example, we might say, "let Q be the set of rational numbers."

Frequently we specify a set by *describing* rather than *listing* its members, perhaps because there are too many elements or because we don't know them. For instance, we can't list all the members of the set of rational numbers, and we don't want to write down all the integers between 1 and 1 billion. We denote this latter set by the symbols $\{n \mid 1 < n < 10^9$, and n is an integer$\}$. On the left of the vertical line appears a "typical" element of the set; on the right is the condition that this element be a member of the set. Thus the set $\{x \mid x^{12} - 17x^5 = 1492\}$ is the set of solutions of the equation $x^{12} - 17x^5 = 1492$. We don't list the members of this set because we don't know how to solve the equation. In more advanced math courses, however, we learn that the equation *has* solutions, and so there are some members of our set. Observe that the letter x plays no essential role in this notation. We could equally well write our set as $\{z \mid z^{12} - 17z^5 = 1492\}$. You will become familiar with this notation in its various forms as you see it used in the pages ahead.

When we specify a set by describing some property of its members, we have to be alert to the possibility that our set may be the **empty set** $\varnothing$, the set that has no members. Thus $\{x \mid x^2 + 1 = 0$, and x is a real number$\} = \varnothing$.

In order to indicate that an element a **is a member of** a set A, we write $a \in A$. We also say that a **belongs to** A or that a **is contained in** A. For example, $\frac{1}{2} \in Q$ (where Q is the set of rational numbers). To denote the fact that π is not a member of the set of rational numbers, we write $\pi \notin Q$. We will use R^1 to denote the set of real numbers. Thus the inclusion $x \in R^1$ is a symbolic way of saying that x is a real number. The letter R is to suggest the word "real," and the superscript 1 distinguishes R^1 from the sets R^2 and R^3, consisting of pairs and triples of real numbers, that we will introduce later.

If every member of a set A is also a member of a set B, we say that A **is a subset of** B or A **is contained in** B, and we write $A \subseteq B$. Notice that it is possible to have $A \subseteq B$ and $B \subseteq A$. These statements are both true if and only if the sets A and B contain the same elements; then we say that $A = B$. The **union** of two sets A and B is the set $A \cup B$ that we obtain by lumping together the elements of the sets A and B into a single set. For example, we can think of R^1 as the union of the set of rational numbers and the set of irrational numbers. The **intersection** of two sets A and B is the set $A \cap B$ whose members are the elements that the sets A and B have in common. Thus $\{1, 2, 3\} \cap \{2, 4, 6\} = \{2\}$.

In calculus, when we speak of a set of numbers, we mean a set of real numbers, and the sets of numbers we use most often are the *intervals*. Suppose that a and b are numbers, with $a < b$. Then the **open interval** (a, b) is the set of all numbers between a and b. For example, the open interval $(1, 5)$ consists of all the numbers between 1 and 5, so that $\pi \in (1, 5)$ and $2 \in (1, 5)$, but $0 \notin (1, 5)$ and $1 \notin (1, 5)$. The open interval (a, b) does not contain its **endpoints** a and b. If we adjoin them to (a, b), we obtain the **closed interval** $[a, b]$. Notice that we use *parentheses to denote an open interval and brackets to denote a closed interval*. In particular, a set consisting of a single point is a closed interval. Thus $[5, 5] = \{5\}$. We sometimes speak of "half-open" or "half-closed" intervals that contain one endpoint but not the other.

Example 2-1.   Find $(-1, 3) \cup [2, 5)$ and $(-1, 3) \cap [2, 5)$.

Solution.   From Fig. 2-1 we see that when we lump together the points of the intervals $(-1, 3)$ and $[2, 5)$, we obtain the interval $(-1, 5)$. Similarly, the intervals $(-1, 3)$ and $[2, 5)$ have the points of $[2, 3)$ in common. Therefore $(-1, 3) \cup [2, 5) = (-1, 5)$ and $(-1, 3) \cap [2, 5) = [2, 3)$.

Figure 2-1

We will consider the set of numbers that are greater than 3 as an interval whose left endpoint is 3 but which *does not have a right endpoint*. We employ our standard interval notation, but the symbol ∞ is used in place of the nonexistent right endpoint—$(3, \infty)$. The left parenthesis (indicates that 3 does not belong to our set (we are talking about numbers that are *greater than* 3). Since our interval does not have a right endpoint, it makes no sense to ask whether or not the right endpoint belongs to the interval. We will simply adopt the convention that with the symbol ∞ we can use either the bracket or the parenthesis, and either the word "closed" or the word "open." So we could equally well have denoted our interval as $(3, \infty]$. The symbol $-\infty$ is used in place of a nonexistent left endpoint. For example, $[-\infty, 7]$ is the set of numbers that are less than or equal to 7. Bear in mind that *the symbols $-\infty$ and ∞ do not stand for numbers*.

When we speak of an interval (a, b), you will have to infer from the context whether it is possible that a might be $-\infty$ or b might be ∞. If we refer specifically to the **finite interval** (a, b), we will mean that a and b stand for numbers, and the symbols $-\infty$ and ∞ are not allowed.

Example 2-2.   Write the set $\{x \mid x^2 - x - 2 > 0\}$ as an interval or a union of intervals.

Solution.   Since $x^2 - x - 2 = (x - 2)(x + 1)$, our given inequality may be written as $(x - 2)(x + 1) > 0$. A product is positive if and only if both factors have the same sign, so both numbers $(x - 2)$ and $(x + 1)$ are positive or both are negative. In the first case, we must have $x > 2$, and in the second case, we must have $x < -1$. Thus x must belong to the interval $(2, \infty)$ or to the interval $(-\infty, -1)$. In symbols,

$$\{x \mid x^2 - x - 2 > 0\} = (-\infty, -1) \cup (2, \infty).$$

Every set of real numbers is contained in the closed interval $[-\infty, \infty]$, but many sets are contained in smaller closed intervals. For example, the set of rational numbers with numerator 1 is contained in the closed interval $[-1, 1]$. This set is also contained in the closed intervals $[-7, 10]$ and $[-3, 2]$, and so on, but the *smallest closed interval* that contains the set of rational numbers with numerator 1 is the interval $[-1, 1]$. As another example, consider the set of all positive rational numbers whose squares are less than 2. This set is contained in the intervals $[-1, 5]$, $[0, 2]$, and so on. The *smallest* closed interval that contains this set is the interval $[0, \sqrt{2}]$. You might notice that neither 0 nor $\sqrt{2}$ belongs to the set of positive rational numbers with squares less than 2. But even though the endpoints of the interval $[0, \sqrt{2}]$ are not members of our given set, you cannot find a smaller closed interval that contains the set.

In the examples we have just discussed, it was easy to select the smallest closed interval that contains the given set. We are not always so fortunate. For instance, it is not an easy matter to find the smallest closed interval that contains the set $\{(1 + 1/n)^n \mid n$ a positive integer$\}$. Nevertheless, it is one of the fundamental properties of the system of real numbers that this set—and every other set of real numbers (except the empty set)—is contained in a smallest closed interval. Technically speaking, *the smallest closed interval that contains a set A is the closed interval that contains A and is contained in every closed interval that contains A.*

Property 2-1. (*The* **completeness** *of the real numbers.*) *Every nonempty set of real numbers is contained in a smallest closed interval.*

Table 2-1 gives some examples that illustrate the completeness property of the real number system. You should picture these examples (and others) on a number scale.

TABLE 2-1	The Set A	The Smallest Closed Interval Containing A
	$\{3\}$	$[3, 3]$
	$(-1, 5)$	$[-1, 5]$
	$(-1, 2) \cup [5, 10]$	$[-1, 10]$
	The set of positive integers.	$[1, \infty]$
	The set of rational numbers.	$[-\infty, \infty]$
	$\left\{ \dfrac{1}{n} \;\middle\vert\; n \text{ a positive integer} \right\}$	$[0, 1]$

PROBLEMS   2

1. On a number scale, make sketches of $A \cup B$ and $A \cap B$.

(a) $A = (-2, 3), B = (1, 4]$
(b) $A = [-2, 2], B = (1, 4)$
(c) $A = (2, 3), B = (3, 4)$
(d) $A = (2, 3], B = [3, 4)$
(e) $A = (0, \frac{22}{7}), B = (\pi, \infty)$
(f) $A = (0, 1), B = \varnothing$

2. Write the given set as an interval.

(a) $(-1, 4] \cup [0, 7)$
(b) $(-1, 4] \cap [0, 7)$
(c) $(-\infty, -1) \cup (-3, 7]$
(d) $(-\infty, -1) \cap (-3, 7]$
(e) $(-1, 2) \cup (-5, 8)$
(f) $(-1, 2) \cap (-5, 8)$

3. Express the solution set of the inequality as an interval or a union of intervals.

(a) $2x - 1 < 7$
(b) $1 - x > 5 - 3x$
(c) $4 < 2x < x + 7$
(d) $x + 7 < 2x < 4$

4. Express the solution set of the inequality as an interval or a union of intervals.

(a) $|x| > 1$ (b) $|x - 1| \le 2$ (c) $|x| < |2x|$ (d) $|x| \ge |2x|$

5. Express the solution set of the inequality as an interval or a union of intervals.

(a) $2x + 3 > 0$
(b) $2x - 3 \le 0$
(c) $x^2 - 5x > 0$
(d) $x^2 + 5x \le 0$
(e) $x^2 + x - 6 \ge 0$
(f) $x^2 + x - 6 < 0$
(g) $x^3 - 3x^2 + 2x > 0$
(h) $x^3 - 1 < 0$

6. Pick p and q so that the solution set of the inequality $x^2 + px + q < 0$ is the interval $(-4, 1)$.

7. (a) Explain why $[5, 13]$ is the solution set of the inequality $|x - 9| \le 4$.
(b) Explain why $[a, b]$ is the solution set of the inequality $|x - \frac{1}{2}(a + b)| \le \frac{1}{2}(b - a)$ if $[a, b]$ is any finite interval.

8. If $A \subseteq B$, explain why $A \cup B = B$ and $A \cap B = A$.

9. What can you conclude about a and b if you are told that $[a, b] \subseteq (a, b)$?

10. In each case, find the smallest closed interval that contains the specified set.

(a) $\{-1, 0, 1, 2\}$
(b) The set of prime numbers less than 100.
(c) The set of common logarithms of the positive numbers.
(d) The set of positive integral powers of $\frac{1}{2}$.
(e) $\{2^x \mid x \in R^1\}$
(f) $\{x^2 \mid x \in R^1\}$
(g) $\{.1, .11, .111, .1111, \dots\}$

11. If a is the smallest element of a set A of real numbers and b is the largest element, show that the smallest closed interval that contains A is the interval $[a, b]$. Does every set of real numbers contain a smallest element and a largest element?

12. If $A \subseteq R^1$ and $B \subseteq R^1$, define $A + B = \{x + y \mid x \in A, y \in B\}$. If $r \in R^1$, define $rA = \{rx \mid x \in A\}$.

(a) Find $3[-1, 2]$.
(b) Is it always true that $r[a, b] = [ra, rb]$?
(c) Find $[1, 2] + [3, 4]$.
(d) Find $(-\infty, 3) + (12, \infty)$.
(e) Show that $[-1, 0] + [0, 1] = [-1, 0] \cup [0, 1]$.
(f) Is it always true that $A + B = A \cup B$?

13.   Can you show that if $[a, b] \cap [c, d] \neq \varnothing$, then

$$[a, b] \cup [c, d] = [\tfrac{1}{2}(a + c - |a - c|), \tfrac{1}{2}(b + d + |b - d|)]$$

and

$$[a, b] \cap [c, d] = [\tfrac{1}{2}(a + c + |a - c|), \tfrac{1}{2}(b + d - |b - d|)]?$$

Try these formulas out with some specific numbers in place of a, b, c, and d.

3.   CARTESIAN COORDINATES AND THE DISTANCE FORMULA

Real numbers can be pictured as points of a line, and we will now move up to two dimensions. Here we use two number scales, meeting at right angles at their origins and with the positive direction upward on one scale and to the right on the other (see Fig. 3-1). The horizontal scale is called the **X-axis**, and the vertical scale is the **Y-axis**. Let P be any point of the plane and construct through P lines that are perpendicular to the axes. If x is the foot of the perpendicular to the X-axis and y is the corresponding point of the Y-axis, then the pair of numbers (x, y) is associated with P. Conversely, if we start with the pair (x, y) we can reverse this construction and end up with the point P. Thus with each point of the plane we associate a pair of numbers (x, y), and with each pair of numbers we associate a point P of the plane. The numbers x and y are the **coordinates** of P.

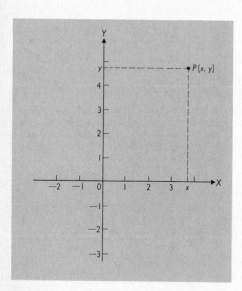

We have just described a **cartesian** coordinate system (named after the seventeenth-century French philosopher and mathematician René Descartes). By introducing coordinates, we establish a one-to-one correspondence between the points of a geometric plane and the set of ordered pairs of real numbers. We will denote this set of pairs by the symbol R^2; that is,

Figure 3-1

$$R^2 = \{(x, y) \mid x \text{ and } y \text{ are real numbers}\}.$$

As in the case of the number scale, we often ignore the logical distinction between a geometric point of a coordinate plane and its coordinates and simply speak of the "point (x, y)" instead of the "point whose coordinates are (x, y)." We will call the points $(x, 0)$ and $(0, y)$ the **projections** of (x, y) onto the X- and Y-axes; we obtain them by dropping perpendiculars from (x, y) to the coordinate axes.

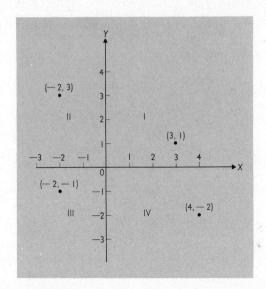

Figure 3-2

Figure 3-2 shows the points $(3, 1)$, $(-2, 3)$, $(-2, -1)$, and $(4, -2)$. Bear in mind that the first number of a coordinate pair is the X-coordinate and the second is the Y-coordinate. The two axes divide the coordinate plane into four **quadrants**, numbered I, II, III, and IV, as shown. We will agree that the axes do not belong to any quadrant.

The distance between points a and b of a number scale is $|a - b|$. We can also calculate the distance between two points of the coordinate plane.

Example 3-1. The point P_1 has coordinates $(-2, -1)$, and the point P_2 has coordinates $(2, 2)$. Find the distance $\overline{P_1P_2}$ between these points.

Solution. The points are plotted in Fig. 3-3. Let P_3 be the point $(2, -1)$, so that the points P_1, P_2, and P_3 are the vertices of a right triangle with the right angle at P_3. Since P_2 and P_3 lie in the same vertical line, you can easily see that they are 3 units apart. Similarly, $\overline{P_1P_3} = 4$. Now the Pythagorean Theorem tells us that

$$\overline{P_1P_2}^2 = \overline{P_1P_3}^2 + \overline{P_3P_2}^2 = 16 + 9 = 25,$$

and so $\overline{P_1P_2} = 5$.

We can use this same argument to develop a formula for the distance between any two points.

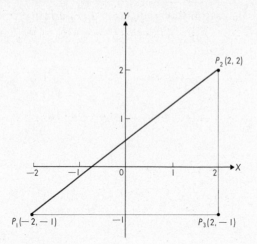

Figure 3-3

Theorem 3-1. *Let P_1 and P_2 with coordinates (x_1, y_1) and (x_2, y_2) be two points of the plane. Then the distance $\overline{P_1P_2}$ is given by the formula*

(3-1)
$$\overline{P_1P_2} = \sqrt{(x_2 - x_1)^2 + (y_2 - y_1)^2}.$$

PROOF

As in Example 3-1, we introduce the auxiliary point P_3 with coordinates (x_2, y_1) in such a way (see Fig. 3-4) that the points P_1, P_2, and P_3 form a right triangle,

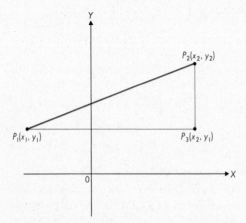

Figure 3-4

with P_3 the vertex of the right angle. The legs of this triangle are $\overline{P_1 P_3}$ and $\overline{P_2 P_3}$ units long, while the length of the hypotenuse, $\overline{P_1 P_2}$, is the distance we wish to find. Again, according to the Pythagorean Theorem,

$$(3\text{-}2) \qquad \overline{P_1 P_2}^2 = \overline{P_1 P_3}^2 + \overline{P_2 P_3}^2.$$

Since the points P_1 and P_3 have the same *Y*-coordinate, you can easily convince yourself that $\overline{P_1 P_3} = |x_2 - x_1|$, and so

$$\overline{P_1 P_3}^2 = |x_2 - x_1|^2 = (x_2 - x_1)^2.$$

Similarly,

$$\overline{P_2 P_3}^2 = |y_2 - y_1|^2 = (y_2 - y_1)^2.$$

We can therefore write Equation 3-2 as

$$\overline{P_1 P_2}^2 = (x_2 - x_1)^2 + (y_2 - y_1)^2,$$

which is equivalent to Equation 3-1. You will use the distance formula so often that you should memorize it.

Example 3-2. Find the distance between the points $(-3, 2)$ and $(2, -3)$.

Solution. It makes no difference which point is designated P_1. If the first point is labeled P_2 and the second P_1, the distance formula yields

$$\overline{P_1 P_2} = \sqrt{(-3 - 2)^2 + (2 + 3)^2} = 5\sqrt{2}.$$

Example 3-3. Find the distance between the point (x, y) and the origin.

Solution. Let P_1 be the point $(0, 0)$ and P_2 be the point (x, y) and apply the distance formula:

$$\sqrt{(x - 0)^2 + (y - 0)^2} = \sqrt{x^2 + y^2}.$$

If P_1, P_2, and P_3 are any three points, then

$$(3\text{-}3) \qquad \overline{P_1 P_3} \le \overline{P_1 P_2} + \overline{P_2 P_3},$$

and the equality sign holds if and only if the three points are collinear—that is, if they lie in a line. This property of distance is called the **triangle inequality** for reasons that will be obvious from a glance at Fig. 3-5. The truth of this property is apparent from the geometry of the situation, but we can prove it by using Formula 3-1 and considerable algebra.

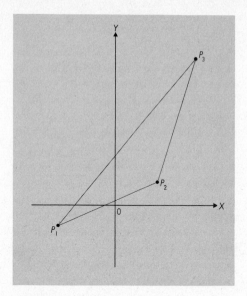

Figure 3-5

Example 3-4. Find the point Q that is $\frac{3}{4}$ of the way from the point $P(-4, -1)$ to the point $R(12, 11)$ along the segment PR.

Solution. Figure 3-6 illustrates the situation; we are to find the numbers x and y, the coordinates of Q. To find these two numbers, we might write the two equations $\overline{PQ} = \frac{3}{4}\overline{PR}$ and $\overline{QR} = \frac{1}{4}\overline{PR}$ in terms of x and y and solve. Although this method will work, it is easier to use a little geometry. If we introduce the auxiliary points $S(x, -1)$ and $T(12, -1)$ shown in Fig. 3-6, we obtain the similar triangles PSQ and PTR. Therefore

$$\frac{\overline{PS}}{\overline{PT}} = \frac{\overline{PQ}}{\overline{PR}} \quad \text{and} \quad \frac{\overline{QS}}{\overline{RT}} = \frac{\overline{PQ}}{\overline{PR}}.$$

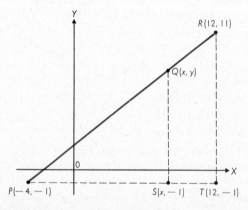

Figure 3-6

From our figure we see that $\overline{PS} = x + 4$, $\overline{PT} = 16$, $\overline{QS} = y + 1$, and $\overline{RT} = 12$, and it is a condition of the problem that $\overline{PQ}/\overline{PR} = \frac{3}{4}$. Hence

$$\frac{x + 4}{16} = \frac{3}{4} \quad \text{and} \quad \frac{y + 1}{12} = \frac{3}{4},$$

from which it follows that $x = 8$ and $y = 8$.

PROBLEMS 3

1. Plot the following points of R^2 in a coordinate plane and find the distance between them.
   (a) (1, 2) and (3, 7)
   (b) (2 −1) and (−2, 1)
   (c) (1, 1) and (−3, −2)
   (d) $(-\sqrt{2}, 2)$ and $(\sqrt{5}, 5)$
   (e) (cos 3, 0) and (0, sin 3)
   (f) $(1, \pi)$ and $\left(-1, \dfrac{1}{\pi}\right)$
   (g) (log 6, log 4) and (log 3, log 8)
   (h) $(\sqrt{7}, \sqrt{2})$ and $(-\sqrt{7}, -\sqrt{2})$

2. How far apart are the points
   (a) $(t, |t|)$ and $(-|t|, t)$?
   (b) (cos u, sin u) and (cos v, sin v)?
   (c) (5, tan t) and (4, 0)?
   (d) $(1 + t^2, t^3)$ and $(1 - t^2, t)$?

3. In what quadrant does the point $(-y, x)$ lie if the point (x, y) lies
   (a) in Quadrant I?
   (b) in Quadrant II?
   (c) in Quadrant III?
   (d) in Quadrant IV?

4. Plot every point (x, y) such that $x \in \{-1, 0, 1\}$ and $y \in \{2, 3, 4\}$. What is the radius of the smallest circular disk that contains this set?

5. Use the distance formula to determine if the triangle with vertices (1, 1), (3, 2), and (2, 12) is a right triangle.

6. Find the area of the triangular region whose vertices are the points (−1, 2), (3, 2), and (−4, 6).

7. Two vertices of a square are (−3, 2) and (−3, 6). Find three pairs of other possible vertices.

8. Use similar triangles, rather than the distance formula, to find
   (a) the point that is $\frac{2}{3}$ of the way from (−2, 1) to (4, 10).
   (b) a point that divides the segment with endpoints (−3, −1) and (7, 14) in the ratio $\frac{3}{2}$.
   (c) the midpoint of the segment with endpoints (x_1, y_1) and (x_2, y_2).

9. One of the points of a certain circle is the point $(-1, \sqrt{5})$, and its center is the origin. Show that $(\sqrt{2}, 2)$ is also a point of the circle. Can you find other points of this circle?

10. Find the point of the *X*-axis that is equidistant from the points (0, −2) and (6, 4).

11. Find the relation between the numbers x and y if the point (x, y) is equidistant from the origin and (1, 1).

12. Use Inequality 3-3 to determine whether or not the following points are collinear.
    (a) (4, −3), (−5, 4), and (0, 0)
    (b) (3, 2), $(-\frac{4}{3}, \frac{5}{9})$, and (6, 3)

13. What points of the X-axis are more than 5 units from the point $(2, 3)$?

14. Three vertices of a parallelogram are $(-1, -1)$, $(1, 4)$, and $(2, 1)$. Find all possibilities for the fourth vertex.

15. For two nonempty sets A and B of points of the plane, we think of all the possible distances between pairs of points, one of the points being a member of A and the other a member of B. This set of nonnegative numbers is contained in a smallest closed interval $[r, s]$, where $r \geq 0$, and we take the number r to be the distance between the sets A and B. Find the distance between the following pairs of sets.

   (a) $\{(0, y) \mid y \geq 1\}$ and the X-axis.
   (b) Quadrant I and Quadrant III.
   (c) The set of points that are less than 2 units from the point $(-4, 6)$ and the set of points that lie below the X-axis.

4. SUBSETS OF THE PLANE

Plane geometric figures, such as circles and lines, are simply sets of points, subsets of the plane. In order to specify a figure, therefore, we describe a collection of points. For instance, we think of a circle of radius 5 as the set of points that are 5 units from a given point, the center of the circle.

A coordinate system assigns pairs of numbers to points of the plane, so we can use numerical relations to specify plane point sets. For example, since Quadrant IV consists of points whose first coordinates are positive and whose second coordinates are negative, we can say that a point (x, y) belongs to Quadrant IV if and only if $y < 0 < x$. The term *analytic geometry* refers to such numerical descriptions of point sets, and as we go along you will see how the introduction of numbers into geometry enables us to use analytic techniques of algebra, calculus, and so on, to solve geometric problems. Conversely, we will often find it helpful to apply geometric arguments to analytic problems.

Example 4-1. Find a relation between the numbers x and y that specifies that a point (x, y) belongs to the union of Quadrants I and III.

Solution. The points of Quadrants I and III are completely characterized as those points whose coordinates have the same sign. Thus a point (x, y) belongs to (Quadrant I) $\cup$ (Quadrant III) if and only if $xy > 0$.

The set of points in a coordinate plane whose coordinates satisfy a numerical relation is called the **graph** of the relation. Thus the union of the first and third quadrants is the graph of the inequality $xy > 0$. Most of the relations whose graphs we will study will be *equations* in x and y. Two relations (in particular, two equations) are said to be **equivalent** if they have the same graph.

Example 4-2. Describe geometrically, and sketch, the graph of the equation $(x - 2)^2 + (y - 6)^2 = 25$.

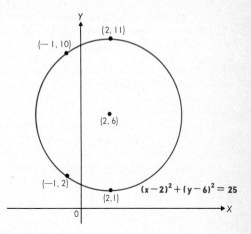

Figure 4-1

Solution. We can find any number of points of the graph of this equation simply by replacing x with a "suitable" number and solving the resulting equation for y. Thus if we replace x with 2, we obtain the equation $(y - 6)^2 = 25$, whose solutions are $y = 1$ and $y = 11$. The points $(2, 1)$ and $(2, 11)$ are then points of our graph. Similarly, we find the points $(-1, 2)$ and $(-1, 10)$ by letting $x = -1$ and solving for y, and so on. We could continue to calculate points indefinitely and plot them, as we plotted our known points in Fig. 4-1. The more points we plot, the more our figure will "take shape," but this method will never yield the complete graph of our equation. So let us turn to the equivalent equation

$$\sqrt{(x - 2)^2 + (y - 6)^2} = 5$$

for a geometric description of our graph. We recognize the left-hand side of this equation as the expression for the distance between the points (x, y) and $(2, 6)$. Thus, in words, the equation says, "The distance between the points (x, y) and $(2, 6)$ is 5." Consequently, a point (x, y) belongs to the graph of the equation $(x - 2)^2 + (y - 6)^2 = 25$ if and only if it is a point of the circle of radius 5 whose center is the point $(2, 6)$. We have sketched this circle in Fig. 4-1.

In Example 4-2 we started with an equation and found a geometric description of its graph. Now let us proceed in the opposite direction. We will start with a geometrically described point set in a coordinate plane and find an equation of which the set is the graph. We speak of such an equation as an *equation of the set*. In order to find an equation of a given set, we must translate into a numerical relation between x and y the geometric language that tells us that a point (x, y) belongs to the set.

Example 4-3. Find an equation of the perpendicular bisector of the segment that joins the points $(-5, 4)$ and $(1, -2)$.

Solution. The geometric description of our set leads to the sketch shown in Fig. 4-2. The given conditions tell us that a point (x, y) of our set is equidistant from the points $(-5, 4)$ and $(1, -2)$. We use the distance formula to express this statement as the equation

$$\sqrt{(x + 5)^2 + (y - 4)^2} = \sqrt{(x - 1)^2 + (y + 2)^2}.$$

17

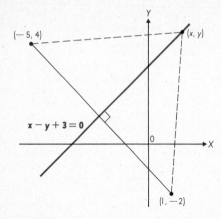

Figure 4-2

Conversely, each point whose coordinates satisfy this equation is equidistant from the points $(-5, 4)$ and $(1, -2)$ and therefore is a point of the perpendicular bisector of the segment that joins these points. This equation is perfectly correct, but we can write it in a simpler form. If we square both sides, multiply out the expressions in the parentheses, and collect terms, we obtain the equation $x - y + 3 = 0$, which is also an equation of the perpendicular bisector.

In the preceding example we found two equations of the same set of points in a coordinate plane. Instead of speaking of such equations as equivalent equations, we will often call them *two forms of the same equation.* Furthermore, in the same way that we identify a pair of coordinates (x, y) with the point it represents, we will talk about the equation of a point set in a coordinate plane as if the equation *were* the set. Thus we might speak of the perpendicular bisector $x - y + 3 = 0$.

In Example 4-2 we could quickly deduce that the graph of the given equation is a circle. It is important to be able to find equations of simple figures, such as lines and circles, and to know what the graphs of certain equations are. There are times, however, when we can't immediately tell what the graph of a given equation looks like. Then we plot as many points as practical, use various devices that we will learn as we go along (including some techniques of calculus), and fill in the rest of the graph "by eye."

Example 4-4. Sketch the graph of the equation $x^2 = y^3$.

Solution. We make the table that appears in Fig. 4-3 by making several replacements of x and solving for y in each case. The points of this table are then plotted and joined as shown in the figure. At this time, we can't be sure that the graph has a sharp point at the origin. Later we will use calculus to help us make such deci-

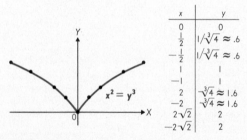

x	y
0	0
$\frac{1}{2}$	$1/\sqrt[3]{4} \approx .6$
$-\frac{1}{2}$	$1/\sqrt[3]{4} \approx .6$
1	1
-1	1
2	$\sqrt[3]{4} \approx 1.6$
-2	$\sqrt[3]{4} \approx 1.6$
$2\sqrt{2}$	2
$-2\sqrt{2}$	2

Figure 4-3

sions. The symmetry of the graph about the Y-axis tells us that the points of the graph can be paired, points of each pair having the same Y-coordinates and X-coordinates that differ only in sign, such as the pair $(1, 1)$ and $(-1, 1)$.

The **intersection** of the graphs of two relations is the intersection of two subsets of the plane and consists of those points whose coordinates satisfy both relations.

Example 4-5.   Find the intersection of the circle that we plotted in Example 4-2 and the line that we plotted in Example 4-3.

Solution.   We must find all pairs of numbers that satisfy the two equations $(x - 2)^2 + (y - 6)^2 = 25$ and $x - y + 3 = 0$. From the second of these equations, we find $y = x + 3$; and when we replace y with $x + 3$ in the first equation, we have

$$(x - 2)^2 + (x - 3)^2 = 25,$$
$$x^2 - 5x - 6 = 0,$$
$$(x + 1)(x - 6) = 0.$$

Therefore $x = -1$ or $x = 6$; and since $y = x + 3$, we see that the intersection is the two-point set $\{(-1, 2), (6, 9)\}$.

Graphs of the equations of calculus are "one-dimensional" figures, such as those shown in Figs. 4-1, 4-2, and 4-3. We call these graphs *curves*. Not all simple-looking equations in x and y, however, have graphs that are simple curves.

Example 4-6.   Sketch the graph of the equation $y + |y| = x + |x|$.

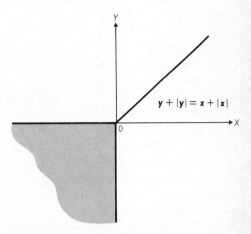

$y + |y| = x + |x|$

Solution.   To find the points of this graph, we check each quadrant separately. If $x \geq 0$ and $y \geq 0$, then $|x| = x$ and $|y| = y$, and our equation becomes $2y = 2x$; that is, $y = x$. As you can see, those points with equal coordinates [such as $(1, 1)$ and $(7, 7)$] lie in a line that bisects the first quadrant, as shown in Fig. 4-4. Each point of Quadrant II has the form $(-a, b)$, where a and b are positive, and for such a point the left-hand side of our equation becomes $b + b = 2b > 0$, while the right-hand side becomes $-a + a = 0$. Thus there are no points of our graph in Quadrant II. Similarly, there are no points of the graph in Quadrant IV. Every point of the form $(-a, -b)$, where $a > 0$ and $b > 0$, however, satisfies the equation, for in this case we have $-b + b =$

Figure 4-4

$-a + a$. Thus, as shown in Fig. 4-4, the graph of the equation $y + |y| = x + |x|$ is the union of Quadrant III, the negative X-axis, the negative Y-axis, and the half-line that contains the origin and bisects Quadrant I.

PROBLEMS   4

1. Make a table containing at least five points of the graph of the equation. Then sketch the graph.

    (a) $y = x$ (b) $y = -x$
    (c) $y = 2x + 3$ (d) $x = 2y + 3$
    (e) $3x + 2y = 5$ (f) $3y + 2x = 5$

2. Express these geometric statements as numerical relations.

    (a) The distance between the points (x, y) and $(-1, 3)$ is 3.
    (b) The distance between the points (x, y) and $(5, -2)$ is less than 2.
    (c) The distance between the point (x, y) and the X-axis is 7.
    (d) The distance between the point (x, y) and the Y-axis is greater than 5.
    (e) The point (x, y) belongs to one of the coordinate axes.
    (f) The point (x, y) is the origin of the coordinate system.

3. Describe the graph of the given relation.

    (a) $x > 0$ (b) $y < 1$ (c) $xy = 0$
    (d) $x^2 + y^2 = 0$ (e) $x^2 + y^2 < 4$ (f) $x^2 + y^2 > 4$

4. Find relations between x and y specifying that a point (x, y) belongs to the following sets.

    (a) The set of points that are more than 3 units from the point $(1, -2)$.
    (b) The set of points that are more than 4 units below the X-axis.
    (c) (Quadrant I) $\cup$ (Quadrant III).
    (d) The Y-axis.
    (e) The set of points that are between 2 and 3 units from the origin.
    (f) (The X-axis) $\cap$ (The Y-axis).

5. Sketch the graph of the equation.

    (a) $y = x^2$ (b) $x = y^2$
    (c) $y^2 = x^2$ (d) $y^2 = -x^2$
    (e) $(x - 3)^2 + (y - 2)^2 = 9$ (f) $(x + 3)^2 + (y + 2)^2 = 9$

6. By plotting points and using your judgment, sketch as best you can the graph of the equation.

    (a) $y = |x|$ (b) $|y| = x$ (c) $y = x|x|$
    (d) $y = x + |x|$ (e) $y = \sqrt{x}$ (f) $y = \sqrt{|x|}$

7. Find equations of the following sets of points.

    (a) The circle of radius 2 whose center is the point $(-1, -3)$.
    (b) The perpendicular bisector of the segment that joins the points $(-1, 3)$ and $(3, 9)$.
    (c) The points that are twice as far from the point $(2, -1)$ as they are from the point $(1, -2)$.
    (d) The points that are equidistant from the X- and Y-axes.
    (e) The line that is parallel to the Y-axis and contains the point (π, π).
    (f) The points whose Y-coordinates are 3 units greater than their X-coordinates.

8. Find the intersection of the graphs of the given pairs of equations.

    (a) $y = x^2$ and $x = y^2$
    (b) $y - 3x = 1$ and $x + 3y = 1$
    (c) $y = \sin x$ and $y = \tan x$
    (d) $(x - 2)^2 + (y + 1)^2 = 4$ and $y = x - 5$
    (e) $y = x^2$ and $2y - x = 1$
    (f) $x^2 - y^2 = 3$ and $x^2 + y^2 = 5$

9. Sketch the point sets that are defined by the following relations.

    (a) $|y| < |x|$ (b) $|y| + y < 2x$
    (c) $(|x| + x)(|y| + y) = 0$ (d) $|x| - x + |x - y| = 0$
    (e) $(x^2 + y^2 - 4)(x^2 + y^2 - 9) = 0$ (f) $x^2 > y^2$

10. Suppose that $a(x, y)$ and $b(x, y)$ are expressions in x and y. Explain why the graph of the equation $a(x, y)b(x, y) = 0$ is the union and the graph of the equation $|a(x, y)| + |b(x, y)| = 0$ is the intersection of the graphs of the equations $a(x, y) = 0$ and $b(x, y) = 0$.

11. Show that the union of the graphs of the equations

    $$y = \tfrac{3}{4}|x| + \sqrt{1 - x^2} \quad \text{and} \quad y = \tfrac{3}{4}|x| - \sqrt{1 - x^2}$$

    is the curve $25x^2 - 24|x|y + 16y^2 = 16$.

5. FUNCTIONS

A simple physical example illustrates the practical origins of the abstract mathematical idea of a function. In elementary physics we learn that if we apply E volts across a 5-ohm resistance, then a current of $\frac{1}{5}E$ amperes flows through the resistor (Ohm's Law). Thus with each number (of volts) there is paired a number (of amperes), and the collection of these number pairs is a *function*. Since one of the chief goals of any science is the development of laws or rules that pair with a given quantity some other quantity, we see that functions play a central role in scientific investigations. Calculus was developed to give us a better understanding of these quantitative relations. Because calculus is a study of functions, we must gain an understanding of the function concept before we can begin to talk about calculus itself.

The notion of a function involves three things: (1) a set D, called the *domain* of the function, (2) a set R, called the *range* of the function, and (3) a rule that assigns to each element $x \in D$ an element $y \in R$. The set of pairs $\{(x, y) \mid x \in D$ and y is the corresponding element of $R\}$ is the function. We generally use letters, such as f, g, F, G, and ϕ to name functions. Thus, in a particular context, f might be the name of the function whose domain is the set $\{1, 2, 3, 4\}$, whose range is the set $\{4, 5, 6\}$, and whose rule of correspondence that pairs elements of the range with elements of the domain is given by Table 5-1.

TABLE 5-1	Element of the domain of f	Corresponding element of the range of f
	1	5
	2	4
	3	4
	4	6

If x is an element of the domain of a function f, then the corresponding element of the range is denoted by $f(x)$. Thus in the present example, $f(1) = 5$, $f(2) = 4$, $f(3) = 4$, and $f(4) = 6$. The symbol $\textbf{f(x)}$ is read "f of x," and it is called the **value** of f at x. You can think of the equation $f(1) = 5$ as a symbolic abbreviation for the sentence, "In our function f, the element that corresponds to 1 is 5." Notice that f and $f(x)$ are two quite different things; f is the name of the function, whereas $f(x)$ is an element of its range. The function f is a set of pairs; in this example, $f = \{(1, 5), (2, 4), (3, 4), (4, 6)\}$. Notice, in particular, that $f(x)$ *does not* mean f times x.

Although the general definition of a function imposes no restriction on the kinds of objects that make up the sets D and R (points, angles, numbers, and so on, are perfectly acceptable), it is a fact that in most of the functions we meet in our early study of calculus, the sets D and R are sets of numbers. It will make our discussion of functions more direct and more concrete if we restrict our attention to such functions for the present. You will find it easy to make any necessary adjustments in terminology when we deal with other types of functions later. For the moment, then, we will consider a function to be a collection of pairs of real numbers. Not every such collection is a function, however. To each number x in the domain of a function there corresponds *just one number y*; that is, the first member of each number pair determines the second. Thus a function cannot contain two number pairs with the same first member, and we have the following formal definition.

Definition   5-1.   *A set of pairs of real numbers is called a* **function** *if it does not contain two pairs with the same first member. The set of all the first members of the pairs that constitute a function is the* **domain** *of the function, and the set of second members is the* **range**.

The functions of calculus consist of infinitely many pairs of numbers. Since we can't list all of them, we provide a rule or procedure for constructing them. Generally this rule consists of a mathematical formula for calculating the second member of the pair when the first member is given. Thus we specify functions by statements like the following ones.

(1)   Let f be the function that is defined by the equation $y = \sqrt{25 - x^2}$.

(2)   Let f be defined by the equation $f(x) = \sqrt{25 - x^2}$.

(3)   Let $f(x) = \sqrt{25 - x^2}$.

The first statement says that if (x, y) is one of the pairs that make up f, then $y = \sqrt{25 - x^2}$. The other two statements say the same thing in different notation; if $(x, f(x)) \in f$, then $f(x) = \sqrt{25 - x^2}$. In any case, our function f is composed of pairs whose general form is $(x, \sqrt{25 - x^2})$.

The domain of f is the set of first members of these pairs, the set of numbers we can substitute for x. When a function is defined in terms of a formula, we take it for granted that its domain consists of those numbers for which the formula makes sense—that is, those real numbers for which the formula yields real numbers. In our present example the formula $\sqrt{25 - x^2}$ makes sense whenever x is replaced by a number whose absolute value does not exceed 5. Hence the domain of f is understood to be the interval $[-5, 5]$. When the numbers of this interval are substituted in the given formula, they yield the numbers of the interval $[0, 5]$. This set is therefore the range of f.

Example 5-1.   For our present function f, find $6f(0) + f(-3) \cdot f(4)$. Compare the numbers $f(5 - r)$ and $f(5) - f(r)$ if r is a small positive number.

Solution.   The point of this example is to emphasize that $f(x)$ *is a number*. So our first task is to find the numbers $f(0)$, $f(-3)$, and $f(4)$, multiply the first by 6, and add the result to the product of the other two. By definition, $f(0) = \sqrt{25 - 0^2} = 5$, $f(-3) = \sqrt{25 - (-3)^2} = 4$, and $f(4) = \sqrt{25 - 4^2} = 3$. Therefore

$$6f(0) + f(-3) \cdot f(4) = 6 \cdot 5 + 4 \cdot 3 = 42.$$

In order to find $f(5 - r)$, we simply replace x with $5 - r$ in both sides of the defining equation $f(x) = \sqrt{25 - x^2}$:

$$f(5 - r) = \sqrt{25 - (5 - r)^2} = \sqrt{10r - r^2}.$$

On the other hand, $f(5) = 0$ and $f(r) = \sqrt{25 - r^2}$. Hence

$$f(5) - f(r) = -\sqrt{25 - r^2},$$

which is not the same as $f(5 - r)$.

The functions we have been discussing in this section are sets of pairs of numbers, and so we can picture them as sets of points in a coordinate plane. The point set that represents a function f is called the **graph** of f. A sketch of its graph can give you a "feel" for a function; whenever possible, you should have a picture of a function in your mind as you work with it.

Example 5-2.   Sketch the graph of the function f that is defined by the equation $y = x/|x|$.

Solution.   With each negative number our function f pairs the number -1, and with each positive number it pairs the number $+1$. When we plot these pairs in a coordinate plane, we obtain the graph shown in Fig. 5-1. Observe that the domain of a function is the projection of its graph onto the X-axis; here it is the entire axis except for the point 0. The range is the projection of the graph onto the Y-axis; here it is the two-point set $\{-1, 1\}$.

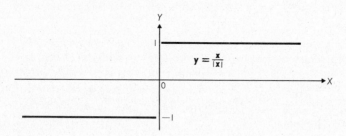

Figure 5-1

We will now give several illustrations of functions.

Illustration 5-1.   Suppose that c is a given real number and let $f(x) = c$. The domain of this function is R^1, and we see, for example, that $f(-1) = c$, $f(0) = c$, and $f(\pi) = c$. The range of f is the set $\{c\}$ that contains only the number c. This function is called the **constant function with value** c, and its graph for a positive number c is sketched in Fig. 5-2.

Illustration 5-2.   The equation $j(x) = x$ defines the **identity function** j. Both the domain and range of j are R^1, and $j(7) = 7, j(\sqrt{2}) = \sqrt{2}$, and so on. Thus $j = \{(x, x) \mid x \in R^1\}$.

Illustration 5-3.   The symbol $[\![x]\!]$ denotes the *greatest integer* n such that $n \le x$. To find $[\![x]\!]$, we therefore "round off" the decimal expression for x to the next lowest whole number. Thus $[\![\pi]\!] = 3$, $[\![\sqrt{2}]\!] = 1$, $[\![3]\!] = 3$, $[\![-3]\!] = -3$, $[\![-\tfrac{1}{2}]\!] = -1$, and $[\![-\pi]\!] = -4$. The equation $y = [\![x]\!]$ defines the **greatest integer function**, whose domain is the set R^1 of real numbers and whose range is the set of integers. Almost every time you state your age you are using the greatest integer function. You say that you are 18, for example, when you are actually $18\tfrac{1}{4}$, and $18 = [\![18\tfrac{1}{4}]\!]$. When pricing a certain item at \$4.95, a dealer hopes you will read the tag as $[\![4.95]\!] = 4$ dollars. If x is a given real number, then $[\![x]\!]$ can be determined graphically by locating x on the number scale and choosing the first integer at or to the left of this point. We have sketched the curve $y = [\![x]\!]$ in Fig. 5-3.

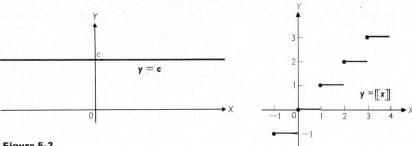

Figure 5-2

Figure 5-3

Illustration 5-4. The tin used in making a certain cylindrical can costs .012¢ per square inch for the side and .021¢ per square inch for the top or bottom. Suppose that we want to make a can with a capacity of 54π cubic inches. But the volume alone won't completely determine the dimensions of the can; it may be tall and thin or short and squat. We may select any positive number r and make the radius of the base of the can r inches. Since the volume is already fixed, the choice of r determines the height of the can, and so the dimensions are completely determined. The cost of the can depends on its dimensions and hence on our choice of the base radius r. Let us denote the cost, in cents, of a can of base radius r inches by c and express c in terms of r.

Clearly, $c =$ (area of side) $\times$ .012 + (area of ends) $\times$ .021. If the can is h inches high and has a base radius of r inches, then the area of the side is $2\pi rh$ square inches, and the area of the top (or bottom) is πr^2 square inches. Therefore

(5-1) $$c = (2\pi rh)(.012) + (2\pi r^2)(.021).$$

We have now expressed the cost of the can in terms of the dimensions r and h, but we wanted to express c in terms of r alone. Consequently, we write a formula for h in terms of r. Since the volume of the can is 54π cubic inches, $\pi r^2 h = 54\pi$ and $h = 54/r^2$. Now we replace h in Equation 5-1 with $54/r^2$ to obtain the equation

(5-2) $$c = \left(\frac{108\pi}{r}\right)(.012) + (2\pi r^2)(.021).$$

This equation defines a function. As far as the equation is concerned, the domain of the function could be the set of all nonzero numbers, but in terms of the original problem, we see that we must take $r > 0$. Thus the domain in which we are interested is the interval $(0, \infty)$. It is not so easy to find the range of this function; we will discuss that question in Section 22. A sketch of its graph appears in Fig. 23-1.

Illustration 5-5.   Let f be the function whose domain is the interval $(0, 1)$ and whose rule of correspondence is expressed as follows:

    (i) If we can write a decimal expression for x in which, after some point, all the digits are 0 (for example, $x = .135000\ldots$), then $f(x) = 0.$

    (ii) Otherwise $f(x) = 1.$

Thus we have $f(\frac{1}{4}) = 0$, $f(\frac{1}{3}) = 1$, $f(\frac{1}{2}) = 0$, and so on. Notice that if x is an irrational number, then $f(x) = 1$. Try to sketch the graph of this function.

When a function is defined by means of a formula, we view the formula as a "recipe" or "program" for generating the pairs of numbers that constitute the function. We may have to use mathematical tables (logarithms, square roots, trigonometric) and other aids in carrying out the directions of the program, but, at least in principle, we can use the formula to generate numbers. Today these programs addressed to people are augmented by programs addressed to machines. Instead of defining a function by means of a traditional-type formula, one could perhaps define it by giving a computer program for generating some of the desired pairs of numbers. A FORTRAN program serves just as well in defining a function as an algebraic formula does. In fact, it is often better, for machines can spew out pairs of numbers much faster than humans can.

PROBLEMS 5

1. If $f(x) = x^2 + 1$, find the following numbers.

    (a) $f(3 - 2)$ (b) $f(3) - f(2)$

    (c) $f(3 \cdot 2)$ (d) $f(3) \cdot f(2)$

    (e) $f(\frac{1}{4})$ (f) $1/f(4)$

    (g) $f(\sqrt{3})$ (h) $\sqrt{f(3)}$

    (i) $f(\log 2)$ (j) $\log f(2)$

    (k) $f(\tan 1)$ (l) $\tan f(1)$

2. If $f(x) = x^2 + 9$ and $g(x) = \sqrt{x + 4}$, find the following numbers.

    (a) $f(g(0))$ (b) $g(f(0))$

    (c) $g(5) - f(0)$ (d) $f(5) - g(0)$

    (e) $g(16 \cdot 2) - 16f(2)$ (f) $g(2 + 3) - f(2 - 3)$

    (g) $f(g(x))$ (h) $g(f(x))$

    (i) $f(0)^{g(0)}$ (j) $(-f(0))^{-g(0)}$

3. Determine the "understood" domains of the functions that are defined by the following equations.

    (a) $f(x) = \sqrt{x^{-1}}$ (b) $g(x) = \log (1 - x)^2$

    (c) $H(x) = 2 \log (1 - x)$ (d) $u(t) = \sqrt[3]{\dfrac{|t + 1|}{t + 1}}$

(e) $F(x) = \log \dfrac{x+1}{x-1}$

(f) $G(z) = \sqrt{\dfrac{z+1}{z-1}}$

(g) $p(x) = \dfrac{1}{[\![x]\!]}$

(h) $q(x) = \dfrac{[\![x]\!]}{x + |x|}$

4. Each of the following statements determines a function. In each case, give the function a letter as a name, determine its domain and range, find a formula that gives its rule of correspondence if you can, and find three points of its graph.

(a) To each positive integer let there correspond the next largest integer.
(b) To each nonnegative number let there correspond the sum of its square and its nonnegative square root.
(c) To each positive number let there correspond the sum of the number and its reciprocal.
(d) To each positive integer let there correspond the sum of its first and last digits.
(e) To each prime number let there correspond the number of times it appears when 1,000,000 is factored into a product of primes.

5. What is the difference between the function p that is defined by the equation $p(q) = q/6 + 6/q$ and the function q that is defined by the equation $q(p) = p/6 + 6/p$?

6. Find examples of constant functions for which the following statements are *not* true.

(a) $f(x^2) = f(x)^2$
(b) $f(|x|) = |f(x)|$
(c) $f(x + y) = f(x) + f(y)$
(d) $f(-x) = -f(x)$

(e) $f(-x) = f(x)$
(f) $f\left(\dfrac{1}{x}\right) = \dfrac{1}{f(x)}$

(g) $f([\![x]\!]) = [\![f(x)]\!]$

7. Find examples of constant functions for which the statements in the preceding question *are* true.

8. Let $f(k)$ denote the number of "n's" that occur in the kth word of the sentence in the preceding question. What is the domain of f? What is its range? Find $f(1) + f(4) - f(5)$.

9. The graph of a certain function g is the line segment that joins the points $(-3, 0)$ and $(0, 6)$. What is the domain of g? What is the range of g? Find $2g(-2) - 3g(-1)$.

10. Is the set $\{(x, y) \mid |y| = |x|\}$ a function?

11. In 1930 the postage on first-class mail was 2¢ per ounce or any fraction thereof. If it cost President Hoover $p(x)$ cents to mail a letter that weighed x ounces, what is the formula that expresses $p(x)$ in terms of x?

12. A box with a square base x inches by x inches has a volume of 25 cubic inches. Express the area of the surface of the box in terms of x.

13. A cubical box is constructed with wooden sides that cost 5¢ per square foot and with a cardboard top and bottom that cost 1¢ per square foot. Express the cost of the box in terms of the length of a side.

14. A conical cup is to be made of paper costing .01¢ per square inch and is to be fitted with a circular cardboard lid costing .1¢ per square inch. The volume of the container is to be 3π cubic inches. If the radius of the base of the cone is r inches and the cost is c cents, express c in terms of r. (The volume and lateral surface area of a cone are given by the formulas $V = \frac{1}{3}\pi r^2 h$, $S = \pi r \sqrt{r^2 + h^2}$.)

15. Sketch the graph of the function f.

    (a) $f(x) = -[\![x]\!]$ (b) $f(x) = [\![-x]\!]$

    (c) $f(x) = -[\![-x]\!]$ (d) $f(x) = [\![2x]\!]$

    (e) $f(x) = 2[\![x]\!]$ (f) $f(x) = [\![|x|]\!]$

    (g) $f(x) = |[\![x]\!]|$ (h) $f(x) = [\![x]\!]^{-1}$

6. THE TRIGONOMETRIC FUNCTIONS

Like the other functions of calculus, we treat the trigonometric functions as sets of pairs of numbers. Since you probably studied the sine, cosine, and tangent in geometric terms, we present here a brief review of the numerical aspects of trigonometry.

In order to define the cosine and sine functions numerically, we must provide a rule that pairs with a given *number t* a *number* cos t and a *number* sin t. This rule is based on the **unit circle**, the circle of radius 1 whose center is the origin shown in Fig. 6-1. To compute cos t and sin t when t is given, we start at the point $(1, 0)$ and proceed $|t|$ units along the circumference of the circle, counterclockwise if t is positive and clockwise if t is negative. Thus the number t determines a point, and *the X-coordinate of the point is the number* **cos t**, *while the Y-coordinate is* **sin t**. Notice that we simply assume that we know how to measure arclength along a circle; a discussion of this technical concept now would take us too far afield (see Section 71).

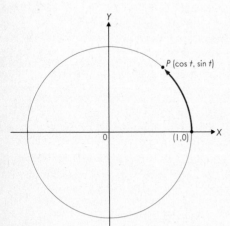

Figure 6-1

Example 6-1. Find the numbers cos π, sin $(-\frac{1}{2}\pi)$, and sin $\frac{3}{4}\pi$.

Solution. The circumference of the unit circle is 2π, so proceeding π units in the counterclockwise direction from the point $(1, 0)$ brings us to the point $(-1, 0)$. By definition, cos π is the X-coordinate of this point; that is, cos $\pi = -1$. Similarly, if we proceed $\frac{1}{2}\pi$ units in the clockwise direction, we arrive at the point $(0, -1)$. It follows that sin $(-\frac{1}{2}\pi) = -1$. If we proceed $\frac{3}{4}\pi$ units along the circle in the counterclockwise direction from $(1, 0)$, we will arrive at a point (x, y) midway along that portion of the circle that lies in the second quadrant (see Fig. 6-2). The second coordinate y is the number sin $\frac{3}{4}\pi$ we are seeking. The figure makes it clear that $x = -y$. Furthermore, since our point is one unit from the origin, $x^2 + y^2 = 1$. Thus $(-y)^2 + y^2 = 1$, or $y^2 = \frac{1}{2}$. Because y is positive in Quadrant II, we obtain $y = \sqrt{\frac{1}{2}}$, and so

$$\sin \tfrac{3}{4}\pi = \sqrt{\tfrac{1}{2}} = \tfrac{1}{2}\sqrt{2}.$$

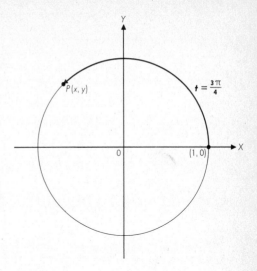

The other trigonometric functions—tan, cot, csc, and sec—are defined in terms of values of the cosine and sine functions by the equations

(6-1)
$$\tan t = \frac{\sin t}{\cos t}, \qquad \cot t = \frac{\cos t}{\tan t},$$

$$\csc t = \frac{1}{\sin t}, \qquad \sec t = \frac{1}{\cos t}.$$

Figure 6-2

The numerical sine and cosine functions are related to the same functions defined in terms of angles in a simple way. With an angle POQ in standard position (vertex the origin and initial side the positive X-axis, as in Fig. 6-3), we associate a number t, the directed distance along the unit circle from the initial side to the terminal side of the angle. This number t is the **radian measure** of $\angle POQ$. The definitions of the trigonometric functions are such that if T is any trigonometric function, then $T(\angle POQ) = T(t)$. Thus, for example, the sine of the *angle POQ* is the same as the sine of the *number t* that is the width

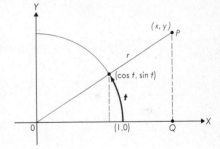

Figure 6-3

of $\angle POQ$ in radians. If (x, y) is a point of the terminal side of an angle of t radians, and if $r = \sqrt{x^2 + y^2}$, we can use similar triangles (see Fig. 6-3) to derive the fundamental equations

$$\sin t = \frac{y}{r}, \qquad \cos t = \frac{x}{r},$$

(6-2)
$$\tan t = \frac{y}{x}, \qquad \cot t = \frac{x}{y},$$

$$\csc t = \frac{r}{y}, \qquad \sec t = \frac{r}{x}.$$

A number of useful trigonometric identities and formulas are listed in Appendix A.

Example 6-2. Find the distance d between the points $(\cos t, \sin t)$ and $(1, 0)$ shown in Fig. 6-1.

Solution. The distance formula (Equation 3-1) gives us

$$d^2 = (\cos t - 1)^2 + \sin^2 t$$

$$= \cos^2 t - 2\cos t + 1 + \sin^2 t \qquad \text{(algebra)}$$

$$= 2 - 2\cos t \qquad \text{(Equation A-1 of the appendix)}$$

$$= 4\sin^2 \tfrac{1}{2}t \qquad \text{(Equation A-5).}$$

Therefore $\qquad\qquad\qquad d = \sqrt{4\sin^2 \tfrac{1}{2}t} = 2|\sin \tfrac{1}{2}t|.$

Figures 6-4, 6-5, and 6-6 show the graphs of the sine, cosine, and tangent functions. You should be able to sketch these graphs without referring to the book.

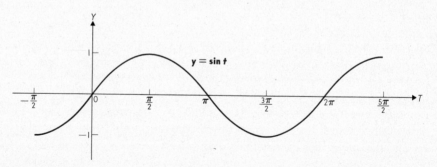

Figure 6-4

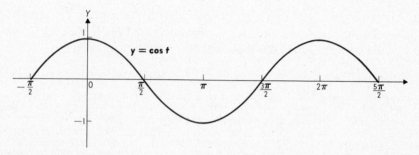

Figure 6-5

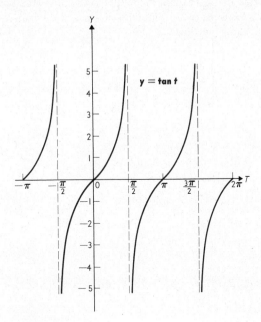

Figure 6-6

PROBLEMS 6

1. Use a sketch of the unit circle to estimate each of the following numbers.

   (a) $\sin 1$ (b) $\cos 1$ (c) $\cot 1.5$ (d) $\sin \frac{1}{2}$

2. Use a sketch of the unit circle to determine which of the following inequalities are true.

   (a) $\sin 2 < \sin 3$ (b) $\cos 2 < \cos 3$
   (c) $\cos 3 < \sin 3$ (d) $\cos 3 < \sin 2$
   (e) $\sin 3 < \cos 2$ (f) $|\cos 3| < |\cos 2|$

3. Use a sketch of the unit circle to find the following numbers.

   (a) $\sin \frac{1}{4}\pi$ (b) $\cos \frac{1}{4}\pi$ (c) $\tan \frac{1}{4}\pi$
   (d) $\cos \frac{1}{3}\pi$ (e) $\sin \frac{1}{3}\pi$ (f) $\sin \frac{1}{6}\pi$

4. The symbol $\sin 2°$ stands for the sine of an angle that measures $2°$. Use a sketch of the unit circle to determine which of the following inequalities are true.

   (a) $\sin 2 < \sin 2°$ (b) $\cos 2 < \cos 2°$
   (c) $\tan 2 < \tan 2°$ (d) $\cos 3 < \cos 3°$

5. Which of the following numbers are positive?

   (a) $\sin 2$ (b) $\tan 3$
   (c) $\cos 1.5$ (d) $\csc 4$
   (e) $\cot 5$ (f) $\sec 6$
   (g) $\cos 820°$ (h) $\sin (-460°)$

6. Find the following numbers.

   (a) $\sec -\frac{31}{4}\pi$ (b) $\cos \frac{31}{6}\pi$ (c) $\csc -\frac{31}{4}\pi$
   (d) $\sin \frac{19}{3}\pi$ (e) $\cot -\frac{31}{4}\pi$ (f) $\tan \frac{22}{3}\pi$

7. Show that $\sqrt{1 + \sin 2x} = |\sin x + \cos x|$.

8. Simplify.

   (a) $(\sin x + \cos x)^2 - \sin 2x$ (b) $\dfrac{\sin^2 2t}{(1 + \cos 2t)^2} + 1$

   (c) $\sin^2 2t - \cos^2 2t$ (d) $(\sin 2t)(\cos 2t)$

   (e) $\dfrac{2 \sin^2 t - 1}{\sin t \cos t} + \cot 2t$ (f) $\dfrac{\sec t + \tan t}{\cos t - \tan t - \sec t}$

   (g) $\dfrac{\sin t \cot t (\sec t - 1)}{1 - \cos t}$ (h) $\dfrac{\sin x}{\sec x + 1} + \dfrac{\sin x}{\sec x - 1}$

9. Conclude from geometric considerations that $|\sin t| \le |t|$. Hence conclude that $1 - t^2 \le \cos^2 t \le 1$.

10. List all solutions of the equation that belong to the interval $[0, 2\pi]$.

    (a) $\sin t = \cos t$ (b) $\sin t = -\cos t$
    (c) $\tan t = \cot t$ (d) $\sec t = \csc t$

11. From the unit circle it is easy to see that $\cos (-t) = \cos t$ for each t. Find similar relations for $\sin (-t)$ and $\tan (-t)$.

12. Describe the curve to which (x, y) belongs if t is a member of the interval $[0, \pi]$.

    (a) $x = \cos t, y = \sin t$ (b) $x = \sin t, y = \cos t$
    (c) $x = t, y = \sin 2t$ (d) $x = t, y = 2 \cos t$
    (e) $x = t, y = \frac{1}{2} \tan 2t$ (f) $x = |\cos t|, y = |\sin t|$

13. If a central angle in a circle of radius r is t radians wide, then $s = rt$, where s is the directed length of the intercepted arc. A circle with a radius of 2 contains a central angle that intercepts an arc 5 units long. How wide is the angle in radians? In degrees?

14. The radius of the front wheel of a child's tricycle is 10 inches and the rear wheels each have a radius of 6 inches. The pedals are fastened to the front wheel by arms that are 7 inches long. How far does a pedal travel when the rear wheels make 1 revolution?

15. Sketch the graphs of the following equations.

    (a) $\sin \pi x \sin \pi y = 0$ (b) $|\sin \pi x| + |\sin \pi y| = 0$
    (c) $[\![\sin \pi x \sin \pi y]\!] = 0$ (d) $[\![\sin \pi x]\!] + [\![\sin \pi y]\!] = 0$

16. Prove that these trigonometric "identities" are in fact false.

    (a) $\sin(u + v) = \sin u + \sin v$ (b) $\cos(u + v) = \cos u + \cos v$
    (c) $\sin uv = \sin u \sin v$ (d) $\cos uv = \cos u \cos v$

If $f(x) = mx + b$, where m and b are given numbers, then f is a **linear function**, so named because its graph is a line. Let us verify this assertion for a specific example.

Example 7-1. Suppose that $f(x) = 2x - 3$. Choose any three points of the graph of f and show that they lie in a line.

Solution. We were asked to choose *any* three points, so let us arbitrarily take $x = 0$, $x = 1$, and $x = 2$ to find the three points $P_1(0, -3)$, $P_2(1, -1)$, and $P_3(2, 1)$ shown in Fig. 7-1. The points will lie in a line if the distance $\overline{P_1P_3}$ equals the sum of the distances $\overline{P_1P_2}$ and $\overline{P_2P_3}$. Now

$$\overline{P_1P_3} = \sqrt{2^2 + 4^2} = \sqrt{20} = 2\sqrt{5}, \qquad \overline{P_1P_2} = \sqrt{1^2 + 2^2} = \sqrt{5},$$

$$\overline{P_2P_3} = \sqrt{1^2 + 2^2} = \sqrt{5},$$

and therefore $\overline{P_1P_3} = \overline{P_1P_2} + \overline{P_2P_3}$.

Precisely the same argument, with letters x_1, x_2, and x_3 replacing the specific numbers 0, 1, and 2, shows that the graph of the general equation $y = mx + b$ is a line. We ask you to carry out the details in Problem 7-10, and when you have done so, you will have proved the following theorem.

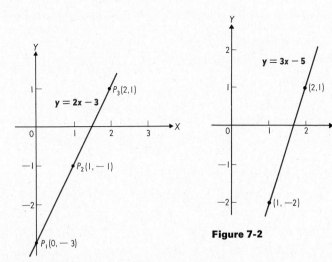

Figure 7-1

Figure 7-2

Theorem   7-1.   *The graph of a linear function is a line.*

Example 7-2.   Sketch the curve $y = 3x - 5$.

Solution.   The graph in question is a line, and so it is only necessary to find two of its points in order to draw it. Two points are $(2, 1)$ and $(1, -2)$, and the graph is shown in Fig. 7-2.

Example 7-3.   What can we say about a linear function f if $f(3) = f(1) + f(2)$?

Solution.   Since f is a linear function, there must be numbers m and b such that $f(x) = mx + b$. Therefore $f(3) = 3m + b$, $f(1) = m + b$, and $f(2) = 2m + b$. The equation $f(3) = f(1) + f(2)$ is valid only if $3m + b = (m + b) + (2m + b)$; that is, $3m + b = 3m + 2b$. But this equation says that $b = 2b$, and hence $b = 0$. So we know that the defining formula for f is $f(x) = mx$, but we don't have enough information to determine the number m.

Example 7-4.   If the temperature of a body is C degrees Celsius, and if the corresponding Fahrenheit reading is F degrees, then the relation between these numbers is a linear one. Find it.

Solution.   The statement that the relation is linear tells us that $F = mC + b$, and our problem is to determine the numbers m and b. Now, $F = 32$ when $C = 0$, and $F = 212$ when $C = 100$; and therefore $32 = b$ and $212 = 100m + b$. Thus $212 = 100m + 32$, and we see that $m = \frac{9}{5}$. The formula connecting the two units of temperature measurement is therefore $F = \frac{9}{5}C + 32$.

It is the choice of m and b in the defining equation $f(x) = mx + b$ that distinguishes one linear function from another, so let us discover what geometric significance these numbers have. The graph of f is the line $y = mx + b$, and we see immediately that $(0, b)$ is a point of this line. Thus the number b is the Y-coordinate of the intersection of the line and the Y-axis. We call b the **Y-intercept** of the line.

In interpreting the number m geometrically, we choose two points (x_1, y_1) and (x_2, y_2) of our line. These coordinate pairs must satisfy the equation of the line:

$$y_1 = mx_1 + b \quad \text{and} \quad y_2 = mx_2 + b.$$

When we solve these equations for m, we find that

$$(7\text{-}1) \qquad\qquad\qquad m = \frac{y_2 - y_1}{x_2 - x_1}.$$

This equation tells us that the number m in the equation $y = mx + b$ is the ratio of the difference of the Y-coordinates to the difference of the X-coordinates of any two points of the graph. We call m the **slope** of the line [or of any line *segment* that contains the points (x_1, y_1) and (x_2, y_2)]. Equation 7-1 is meaningless if these points lie in a line that is parallel to the Y-axis. We simply say that such lines *have no slope*. (This statement does not say that such lines have slope 0. What lines do have slope 0?)

Figure 7-3 shows that if we move along our line from the point (x_1, y_1) to the point (x_2, y_2), then we move $y_2 - y_1$ units in the Y-direction and $x_2 - x_1$, units in the X-direction. Therefore the quotient $(y_2 - y_1)/(x_2 - x_1)$ (that is, the slope) is the number of units moved in the Y-direction for each unit moved in the X-direction, or, as we say, the *rate* at which the line rises (or falls). A good part of calculus consists of studying the rates at which the graphs of other functions rise or fall. If $m > 0$, the graph of our linear function "rises to the right," as in Fig. 7-2. If $m < 0$, it falls to the right.

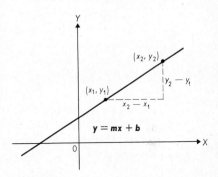

Figure 7-3

A linear function for which $f(x) = mx$ (that is, $b = 0$) is said to be **homogeneous**. Its graph is a line that contains the origin. These functions are used in many scientific applications: distance = rate × time (constant rate), work = force × distance (constant force), force = mass × acceleration (constant mass), and so on. The key properties of homogeneous linear functions are given in the following theorem.

Theorem   7-2.   *If f is a homogeneous linear function and u and v are any numbers, then*

(7-2) $$f(uv) = uf(v)$$

and

(7-3) $$f(u + v) = f(u) + f(v).$$

PROOF

We are told that f is a homogeneous linear function—in other words, that there is a number m such that for each number x, $f(x) = mx$. Therefore

$$f(uv) = m \cdot (uv) = u \cdot (mv) = uf(v).$$

Similarly,

$$f(u + v) = m \cdot (u + v) = mu + mv = f(u) + f(v).$$

If we replace u in Equation 7-2 with 2, it becomes $f(2v) = 2f(v)$. This equation tells us that doubling a number in the domain of f doubles the corresponding number in the range. Equation 7-2 simply says that this statement applies to any multiple of a number in the domain of f. In the problems we will ask you to show that a homogeneous linear function is the only function for which this statement is true.

Equation 7-3 says that if we add two numbers in the domain of f and then find the corresponding number in the range, or if we find the corresponding numbers first and then add, we end up with the same number. It is a common mistake to think that *all* functions have this property. Under stress, many of us will equate $(a + b)^2$ and $a^2 + b^2$, or $\sin(u + v)$ and $\sin u + \sin v$. But we know that these numbers are not, in general, equal; that is, the square function and the sine function do not have the property expressed by Equation 7-3. The homogeneous linear function is the only elementary function that has this property; you will study mathematics for some time before you meet another example of such a function.

PROBLEMS 7

1. Determine the formula for $f(x)$ if f is linear and its graph contains the given points.

    (a) (1, 1) and (2, 4)   (b) (0, 1) and (1, 0)
    (c) (1, 0) and (-2, π)   (d) (0, 0) and (π, $\frac{22}{7}$)
    (e) (0, log 5) and (log 25, log 125) (f) ($-\sin^2 3$, 0) and ($\cos^2 3$, 2)

2. Sketch the graphs of the lines with the given slope and Y-intercept.

    (a) slope 4, Y-intercept -7 (b) slope -7, Y-intercept 4
    (c) slope 0, Y-intercept -2 (d) slope -2, Y-intercept 0
    (e) slope 2, Y-intercept $\frac{1}{3}$ (f) slope $\frac{1}{2}$, Y-intercept 3

3. What is the intersection of the graphs of the linear functions f and g, where $f(x) = 2x + 3$ and $g(x) = \frac{1}{2}(x + 3)$? Check your result with a sketch.

4. What can you say about a linear function f if

    (a) $2f(x) = f(2x)$ for every number x?
    (b) $f(x + 1) = f(x) + 1$ for every number x?
    (c) $f(3) = 2$?
    (d) $f(3) = 2$ and $f(-1) = 1$?
    (e) $f(2x + 1) = f(2x) + 1$ for every number x?
    (f) $f(2) = -4$ and $f(-1) = -f(2)$?
    (g) $|f(x)| = f(|x|)$ for every number x?
    (h) $[\![f(x)]\!] = f([\![x]\!])$ for every number x?

5. Let $f(x) = 3x - 5$.

    (a) In what point does the graph of f intersect the Y-axis?
    (b) In what point does the graph of f intersect the X-axis?
    (c) How long is the segment of the graph of f that lies in Quadrant IV?
    (d) Sketch the curve $y = f(-x)$.

6. What is the linear function f for which the statement $\log 100x^3 = f(\log x)$ is true for each positive number x?

7. Suppose that $y = mx + b$. In what units are m and b expressed if x and y are expressed in the following units?
   (a) x in hours, y in miles
   (b) x in fortnights, y in leagues
   (c) x in students, y in books
   (d) x in men, y in women

8. If the number of calories of heat required to change 1 gram of solid ice at 0°C to water at T°C is denoted by Q, then the relation between Q and T is linear when $0 \le T \le 100$. If $Q = 90$ when $T = 10$, and $Q = 150$ when $T = 70$, how much heat is required to transform ice into 0° water?

9. (a) Find a pair of linear functions f and g such that $f(g(x)) = g(f(x))$.
   (b) What can you say about a linear function f if $f(f(x)) = x$?

10. If $P_1(x_1, mx_1 + b)$, $P_2(x_2, mx_2 + b)$, and $P_3(x_3, mx_3 + b)$, where we assume that $x_1 < x_2 < x_3$, are three points of the curve $y = mx + b$, show that $\overline{P_1P_3} = \overline{P_1P_2} + \overline{P_2P_3}$ and hence that the graph is a line. (*Hint:* A short calculation shows that $\overline{P_1P_3} = (x_3 - x_1)\sqrt{1 + m^2}$, and you get similar formulas for $\overline{P_1P_2}$ and $\overline{P_2P_3}$.)

11. Suppose that f is a function such that Equation 7-2 holds for each pair of real numbers u and v. Show that if we let $m = f(1)$, then $f(x) = mx$ for each real number x.

12. Show that h is a linear function if $h(x) = f(g(x))$, where f and g are linear.

13. Suppose that f is a function such that Equation 7-3 holds for each pair of numbers u and v. Let $u = 0$ and $v = 0$ to show that $f(0) = 0$. Then let $v = -u$ to show that $f(-u) = -f(u)$. Now let us write $m = f(1)$ and let $u = 1$ and $v = 1$ to see that $f(2) = m \cdot 2$. Continue in this way to show that $f(3) = m \cdot 3$, $f(4) = m \cdot 4$, and so on. Can you use mathematical induction to show that $f(x) = mx$ for each positive integer x? Once this step has been taken, it is not hard to show that $f(x) = mx$ for each *rational* number x.

8.   LINES

The graph of the equation $y = mx + b$ is a line. Now suppose that we are given a line; must its equation have the form $y = mx + b$? The answer is *yes*, if the line is not parallel to the Y-axis.

Theorem   8-1.   *A line not parallel to the Y-axis is the graph of an equation* $y = mx + b$.

PROOF

Given a line that is not parallel to the Y-axis, we must show that there are numbers m and b such that this line is the graph of the equation $y = mx + b$. Since our line is not parallel to the Y-axis, it intersects that axis in some point

$(0, b)$ and the line $x = 1$ in some point $(1, c)$. We now have a number b, and we will let $m = c - b$. It is our contention that the given line is the curve $y = mx + b$. This curve certainly is a line, and it will be the given line if it contains two of its points. But we chose m and b precisely so that the line $y = mx + b$ contains the points $(0, b)$ and $(1, c)$ of the line we started with. Thus the given line is the graph of the equation $y = mx + b$.

Equations of the form $y = mx + b$ do not represent lines parallel to the *Y*-axis. Such lines, it is clear, have equations of the form $x = a$. An equation that covers both cases is

$$(8\text{-}1) \qquad\qquad Ax + By + C = 0,$$

where A and B are not both 0. For if $B = 0$, Equation 8-1 can be written as $x = a$, with $a = -C/A$, whereas if $B \neq 0$, Equation 8-1 becomes $y = mx + b$, with $m = -A/B$ and $b = -C/B$. Since every equation of the form of Equation 8-1 represents a line and every line is represented by such an equation, we can call Equation 8-1 the **general linear equation.**

Example 8-1.   Find the slope and the *Y*-intercept of the line $2x + 3y + 4 = 0$.

Solution.   If we write this equation as $y = mx + b = -\frac{2}{3}x - \frac{4}{3}$, we can read off the slope $m = -\frac{2}{3}$ and the *Y*-intercept $b = -\frac{4}{3}$.

The slope of a line gives its direction, so we use slopes to determine if two lines are parallel.

Theorem 8-2.   *Two lines $y = m_1x + b_1$ and $y = m_2x + b_2$ are parallel if and only if $m_1 = m_2$.*

PROOF

Algebraically, our problem is to find the condition under which the system of equations $y = m_1x + b_1$ and $y = m_2x + b_2$ does not have *just one* solution (for then the lines intersect). To solve this system, we write $m_1x + b_1 = m_2x + b_2$; that is, $(m_1 - m_2)x = b_2 - b_1$. There is just one solution of this last equation if and only if $m_1 \neq m_2$. Otherwise either there are no solutions ($m_1 = m_2$ and $b_1 \neq b_2$, in which case our lines are parallel) or infinitely many solutions ($m_1 = m_2$ and $b_1 = b_2$, in which case the equations are identical and represent the same line).

The line $y = mx$ has *Y*-intercept 0 and is parallel to the line $y = mx + b$ (Fig. 8-1). It intersects the unit circle in a point $(\cos \alpha, \sin \alpha)$, and since this point

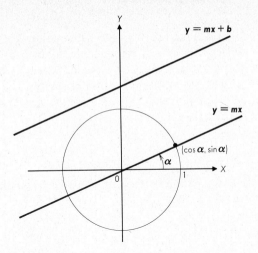

Figure 8-1

belongs to the curve $y = mx$, we have $\sin \alpha = m \cos \alpha$. Thus $m = \tan \alpha$. We call α the **angle of inclination** of the line $y = mx$, or of any parallel line $y = mx + b$, and so we see that the slope of a line is the tangent of its angle of inclination.

We can also use slopes to determine when two lines are perpendicular.

Theorem 8-3. *Two lines $y = m_1x + b_1$ and $y = m_2x + b_2$ are perpendicular if and only if $m_1m_2 = -1$.*

PROOF

Figure 8-2 shows the two lines intersecting in the point $P_0(x_0, y_0)$. If we move along the line $y = m_1x + b_1$ one unit to the right, then we move m_1 units in the Y-direction and arrive at the point $P_1(x_0 + 1, y_0 + m_1)$. Similarly, if we move along the line $y = m_2x + b_2$ one unit to the right, we arrive at the point $P_2(x_0 + 1, y_0 + m_2)$. The two lines are perpendicular if and only if the triangle $P_0P_1P_2$ is a right triangle with the right angle at P_0—that is, if

$$(8\text{-}2) \qquad \overline{P_1P_2}^2 = \overline{P_0P_1}^2 + \overline{P_0P_2}^2.$$

From the figure or from the distance formula, it is clear that

$$\overline{P_1P_2}^2 = [(y_0 + m_2) - (y_0 + m_1)]^2 = (m_2 - m_1)^2.$$

The distance formula gives us

$$\overline{P_0P_1}^2 = [(x_0 + 1) - x_0]^2 + [(y_0 + m_1) - y_0]^2 = 1 + m_1^2,$$
$$\overline{P_0P_2}^2 = [(x_0 + 1) - x_0]^2 + [(y_0 + m_2) - y_0]^2 = 1 + m_2^2,$$

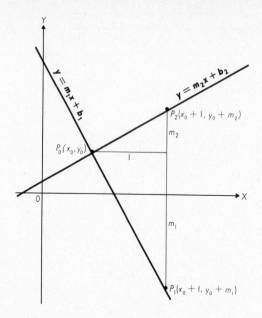

$y = m_1x + b_1$

$y = m_2x + b_2$

$P_2(x_0 + 1, y_0 + m_2)$

m_2

$P_0(x_0, y_0)$

1

m_1

$P_1(x_0 + 1, y_0 + m_1)$

Figure 8-2

so Equation 8-2 becomes

$$(m_2 - m_1)^2 = 1 + m_1^2 + 1 + m_2^2.$$

On simplifying this equation, we find that $-2m_2m_1 = 2$; that is, $m_1m_2 = -1$.

Example 8-2. Find the equation of the line that is perpendicular to $3x - y + 5 = 0$ at the point $(-1, 2)$.

Solution. The equation that we seek has the form $y = mx + b$; our task is to find the numbers m and b. By writing the original equation as $y = 3x + 5$, we see that the given line has slope 3. The desired line has slope m, so Theorem 8-3 tells us that $3m = -1$; that is, $m = -\frac{1}{3}$. Thus the equation we are looking for has the form $y = -\frac{1}{3}x + b$. This line must contain the point $(-1, 2)$, and so we find b by solving the equation $2 = -\frac{1}{3}(-1) + b$. Hence $b = \frac{5}{3}$, and we finally obtain the equation $y = -\frac{1}{3}x + \frac{5}{3}$. If we prefer, we can write this solution of our problem as $x + 3y - 5 = 0$.

Because it is written in terms of the slope m and Y-intercept b, the equation $y = mx + b$ is called the **slope-intercept** form of the equation of a line. Also useful is the **point-slope form**

(8-3) $$y - y_1 = m(x - x_1).$$

This equation explicitly shows the slope m and one of the points (x_1, y_1) of the line.

Example 8-3.   Find the line with slope π that contains the point (1, 2).

Solution.   Using Equation 8-3 with $m = \pi$, $x_1 = 1$, and $y_1 = 2$, we find that $y - 2 = \pi(x - 1)$; that is, $y = \pi x + (2 - \pi)$. Notice that the *Y*-intercept is $2 - \pi$. Sketch this line.

PROBLEMS 8

1. Find the slope and *Y*-intercept of the line.

    (a) $2x + 4y + 1 = 0$ (b) $x + 4y + \pi = 0$

    (c) $2y = 4x + 7$ (d) $2y = 5 - 4x$

2. Find the line of slope m that contains the listed point. What is the *Y*-intercept?

    (a) $m = 4$, (1, 3)   (b) $m = -7$, $(1, -7)$

    (c) $m = \frac{22}{7}$, $(1, \pi)$ (d) $m = \sqrt{2}$, (1, 1.414)

3. Find the line that contains the given points.

    (a) (1, 2) and (3, 4)   (b) $(-2, 1)$ and (2, 3)

    (c) (4, 1) and $(\pi, 1)$ (d) $(-1, 5)$ and $(\pi, 2)$

4. Determine whether or not the points are collinear.

    (a) $(4, -3)$, $(-5, 4)$, (0, 0)   (b) (3, 2), $(-\frac{4}{3}, \frac{5}{9})$, (6, 3)

5. Find the line that contains the point $(-1, 2)$ and is parallel to the line $3x + 5y + 4 = 0$. Replace the word "parallel" with "perpendicular" and do the problem.

6. Use the idea of slope to determine if the triangle with vertices (1, 1), (3, 2), and $(-5, 17)$ is a right triangle.

7. A circle has a radius of 2 units, and its center is the origin. Find the line that is tangent to the circle at the point $(1, \sqrt{3})$.

8. Find k so that

    (a) the lines $2x - y = 1$ and $ky + x = 2$ are parallel.

    (b) the line containing (1, k) and $(-1, 2)$ is parallel to $y = 5x$.

    (c) the lines $y = kx$ and $x = ky$ are perpendicular.

    (d) the triangle with vertices (0, 0), (1, 2), and (k, 5) is a right triangle.

9. (a) Convince yourself that the line whose *Y*-intercept is b ($\neq 0$) and whose *X*-intercept is a ($\neq 0$) is

$$\frac{x}{a} + \frac{y}{b} = 1.$$

    This equation is called the **intercept form** of the equation of a line.

    (b) What line contains the points (7, 0) and (0, 11)?

10. A circle is tangent to the line $y = \frac{3}{4}x + 5$ at the point (0, 5) and also tangent to the line $y = -4$. There are two possible choices for the center of such a circle. Find them.

11. A line with negative slope contains the point (4, 0) and is tangent to the unit circle. Find its slope.

12. Find the line that is tangent to the circle of radius 2 whose center is the point $(-1, 0)$, if the point of tangency belongs to (a) the positive *X*-axis or (b) the positive *Y*-axis.

13. Explain why Theorem 8-3 does not apply to the case of a horizontal line and a vertical line.

14. A "natural" way to find the perpendicular distance between the point $P(7, 2)$ and the line $L : 3y = 4x + 3$ is to find the equation of the line N that contains P and is perpendicular to L. Then the distance between P and the intersection of N and L is the desired distance. Find it.

15. Sketch the graphs of the following relations.

   (a) $|3x - 4y| = 5$ (b) $|3x - 4y| \leq 5$
   (c) $[\![3x - 4y]\!] = 5$ (d) $[\![3x - 4y]\!] \leq 5$

16. Given a triangle ABC, choose a coordinate plane so that the vertices are $A(-a, 0)$, $B(0, b)$, and $C(c, 0)$. For any point $P_1(x_1, y_1)$ of AB, determine the point Q_1 of BC and then the point R_1 of CA, such that P_1Q_1 is parallel to AC and Q_1R_1 is parallel to AB. Find the coordinates of the point P_2 of AB (in terms of x_1 and y_1) such that R_1P_2 is parallel to BC. If the foregoing procedure is repeated, starting with P_2, to obtain a point P_3 of AB, what are the coordinates of P_3?

9. THE LIMIT CONCEPT

In sketching the graph of a function f, we can plot only a few points of the form $(x, f(x))$. These plotted points suggest where the other points "should" lie. The idea that certain points of a graph determine where the other points "should" be is related to the limit concept, the basis of the whole subject of calculus. The study of limits can become quite complicated, but from examples like the ones that follow, you can obtain a working knowledge of what the idea of a limit is all about.

Our first example is typical of the limits found in calculus. We want to add a "missing point" to the graph of a function.

Example 9-1. If $s(x) = \dfrac{\sin x}{x}$, sketch the part of the graph of s that lies above the interval $[-\pi, \pi]$ of the X-axis.

Solution. When the 10 points of the table in Fig. 9-1 are plotted in the coordinate plane, they suggest reasonably well the curve we have drawn. But observe that this

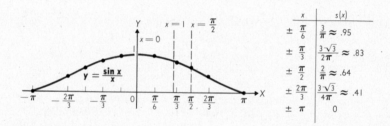

x	$s(x)$
$\pm \dfrac{\pi}{6}$	$\dfrac{3}{\pi} \approx .95$
$\pm \dfrac{\pi}{3}$	$\dfrac{3\sqrt{3}}{2\pi} \approx .83$
$\pm \dfrac{\pi}{2}$	$\dfrac{2}{\pi} \approx .64$
$\pm \dfrac{2\pi}{3}$	$\dfrac{3\sqrt{3}}{4\pi} \approx .41$
$\pm \pi$	0

Figure 9-1

curve has a "hole" in it. The domain of s does not include 0 (we can't divide by 0), so there is no point of its graph above this number.

The obvious candidate for a point to be added to the graph of the function s of our last example is (0, 1). In other words, the points we have sketched tell us that the graph of s "should" intersect the line $x = 0$ in the point (0, 1). Although (0, 1) is not actually a point of the graph, it is clearly the natural point to use in filling the gap. The number 1 is not the *value* of s at 0 (there is no number $s(0)$); it is the *limit of s at* 0.

More generally, let f be a given function and c a given number, and suppose that the graph of f "should" intersect the line $x = c$ in a certain point. The Y-coordinate of this point is called the **limit of f at c**, or the **limit of $f(x)$ as x approaches c**, and we denote it by the symbol $\lim_{x \to c} f(x)$. In geometric terms, then, *the number* $\lim_{x \to c} f(x)$ *is the Y-coordinate of the point in which the graph of f should intersect the line $x = c$.* Of course, this definition is not mathematically precise because of the vagueness of the word "should". However, it does convey the essential idea of what a limit is, and Fig. 9-1 makes it perfectly clear that $\lim_{x \to 0} (\sin x)/x = 1$. Notice that we sketched the curve $y = (\sin x)/x$ by plotting only a few points and joining them "by eye." In doing so, we are tacitly assuming that $\lim_{x \to 1} (\sin x)/x = \sin 1$, $\lim_{x \to \frac{1}{2}} (\sin x)/x = 2 \sin \frac{1}{2}$, and so on.

Example 9-2.   Determine graphically (if you can) the numbers

$$\lim_{x \to 0} [\![\cos x]\!], \ \lim_{x \to 1} [\![\cos x]\!], \ \text{and} \ \lim_{x \to \frac{1}{2}\pi} [\![\cos x]\!].$$

Solution.   Don't be alarmed by the symbols $[\![\cos x]\!]$. To draw the curve $y = [\![\cos x]\!]$, we simply sketch (in black) the cosine curve (Fig. 9-2) and observe that $[\![\cos x]\!] = -1$ if $-1 \le \cos x < 0$, $[\![\cos x]\!] = 0$ if $0 \le \cos x < 1$, and $[\![1]\!] = 1$ to get the desired graph in color.

From the figure it is clear that the point (0, 0) "should" belong to the graph (even though (0, 1) does belong instead), and hence $\lim_{x \to 0} [\![\cos x]\!] = 0$.
It is also apparent that the point (1, 0) "should" belong to the graph (and it does), so $\lim_{x \to 1} [\![\cos x]\!] = 0$.
It is not obvious what point with X-coordinate $\frac{1}{2}\pi$ "should" belong to the graph. True, the point $(\frac{1}{2}\pi, 0)$ *does* belong, but if we look only at $\{x > \frac{1}{2}\pi\}$, it would appear that $(\frac{1}{2}\pi, -1)$ would be a better candidate. On the other hand, if we restrict our

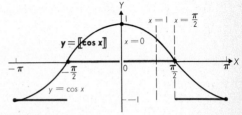

Figure 9-2

attention to $\{x < \frac{1}{2}\pi\}$, the point $(\frac{1}{2}\pi, 0)$ seems reasonable. We resolve the dilemma very simply; we say that $\lim_{x \to \frac{1}{2}\pi} [\![\cos x]\!]$ *does not exist.*

We don't need a picture in order to find limits. The number $\lim_{x \to c} f(x)$ is determined by values of f at numbers that are close to, *but not equal to, c*. Therefore we can put our idea of what a limit is in words that do not refer to a figure: $\lim_{x \to c} f(x)$ *is the number that is approximated by $f(x)$ when x is close to, but not equal to, c*.

Example 9-3. Find $\lim_{x \to 2} \dfrac{x^3 - 8}{x - 2}$.

Solution. We can't set $x = 2$ in the given quotient; that would be dividing by 0. However, when $x \neq 2$, $(x^3 - 8)/(x - 2) = x^2 + 2x + 4$, and when x is close to 2, this number is obviously close to 12. (For example, when $x = 2.1$, our quotient is 12.61.) Therefore $\lim_{x \to 2} (x^3 - 8)/(x - 2) = 12$.

Figure 9-2 shows that the point $(0, 0)$ *should* belong to the curve $y = [\![\cos x]\!]$, whereas the point $(0, 1)$ *does* belong. On the other hand, the point $(1, 0)$ both *should* and *does* belong to the graph. Thus if we let $g(x) = [\![\cos x]\!]$, then the *limit* of g at 1 is the *value* of g at 1, but the limit of g at 0 is *not* the value of g at 0. We say that g is *continuous at* 1 and *discontinuous at* 0 in accordance with the following definition.

Definition 9-1. *The function f is* **continuous** *at c if*

$$\lim_{x \to c} f(x) = f(c).$$

Example 9-4. Is a linear function continuous at each real number c?

Solution. Certainly. Numerically, we are asking "Is $mc + b$ the number that is approximated by $mx + b$ when x is close to c?" The answer clearly is *yes*. Geometrically, we are saying, "If the point $(c, mc + b)$ is removed from the line $y = mx + b$, is it obvious where to put it back?" Again the answer is *yes*.

Example 9-5. Is the function u continuous at 1 if $u(x) = 1/x$?

Solution. Here the value of u at 1 is the number $1/1 = 1$, so we are to show that $\lim_{x \to 1} 1/x = 1$. Thus we are asking whether or not 1 is the number that is approximated by $1/x$ when x is close to 1. The answer is certainly *yes*, but to make it even more obvious, we could rephrase the question: Is 0 the number that is approximated by $1/x - 1$ or $(1 - x)/x$ when x is close to 1? From the latter formula we see that $1/x$ will differ but little from 1 when x differs but little from 1.

Let us return to Fig. 9-2 to introduce one more limit concept. We pointed out that if x is close to, *but greater than*, $\frac{1}{2}\pi$, then $[\![\cos x]\!]$ is close to -1. If x is close to, *but less than*, $\frac{1}{2}\pi$, then $[\![\cos x]\!]$ is close to 0. In symbols, we express these two statements as

$$(9\text{-}1) \qquad \lim_{x \downarrow \frac{1}{2}\pi} [\![\cos x]\!] = -1 \qquad \text{and} \qquad \lim_{x \uparrow \frac{1}{2}\pi} [\![\cos x]\!] = 0.$$

If $g(x) = [\![\cos x]\!]$, these equations say that the *limit of g from the right* at $\frac{1}{2}\pi$ is -1, and the *limit of g from the left* at $\frac{1}{2}\pi$ is 0. Notice that Fig. 9-1 shows us that

$$\lim_{x \uparrow 0} \frac{\sin x}{x} = \lim_{x \downarrow 0} \frac{\sin x}{x} = \lim_{x \to 0} \frac{\sin x}{x} = 1.$$

This example illustrates the fact that the limit of a function g at c exists if and only if the limits from the right and from the left at c exist and are equal. Sometimes it is easier to examine the limits from the right and left separately. If these limits turn out to be different, then the limit does not exist. For example, Equations 9-1 confirm the fact that $\lim_{x \to \frac{1}{2}\pi} [\![\cos x]\!]$ does not exist.

PROBLEMS   9

1. Find the elements of the domain (if there are any) at which the functions whose graphs are shown in the following figures are not continuous. Find the limits from the left and from the right at these numbers (if the limits exist).

    (a) Fig. 4-3   (b) Fig. 5-1   (c) Fig. 5-2
    (d) Fig. 5-3   (e) Fig. 6-5   (f) Fig. 6-6

2. By thinking of $\lim_{x \to c} f(x)$ as "the number that is approximated by $f(x)$ when x is close to, but not equal to, c," find the following limits.

    (a) $\lim_{x \to -1} (3x - 2)$

    (b) $\lim_{x \to \frac{1}{2}} (4x + 5)$

    (c) $\lim_{x \to \frac{1}{3}\pi} \sin x$

    (d) $\lim_{x \to \frac{1}{3}} \cos \pi x$

    (e) $\lim_{x \to 2} \dfrac{x^2 - 4}{x - 2}$

    (f) $\lim_{x \to 2} \dfrac{x - 2}{x^2 - 4}$

    (g) $\lim_{x \to -1} \dfrac{x^3 + 1}{x + 1}$

    (h) $\lim_{x \to -1} \dfrac{x + 1}{x^3 + 1}$

    (i) $\lim_{x \to 9} \dfrac{x - 9}{\sqrt{x} - 3}$

    (j) $\lim_{x \to 9} \dfrac{\sqrt{x} - 3}{x - 9}$

3. Find the following limits.

    (a) $\lim_{x \to -1} \dfrac{1 - |x|}{1 + x}$

    (b) $\lim_{x \to 1} \dfrac{|1 - x|}{1 + x}$

    (c) $\lim_{x \to 0} x^0$

    (d) $\lim_{x \to 1} 0^x$

    (e) $\lim_{x \to -5} \tan(x + |x|)$

    (f) $\lim_{x \to 5} (\tan x + |\tan x|)$

4. Find by inspection (using your best judgment) $\lim\limits_{x \to c} f(x)$, and decide whether or not f is continuous at c.

(a) $f(x) = x^2 - 1, c = 1$ (b) $f(x) = x + 1/x, c = \frac{1}{2}$

(c) $f(x) = x|x|, c = 0$ (d) $f(x) = x + |x|, c = 0$

(e) $f(x) = [\![\sin x]\!], c = \frac{1}{2}\pi$ (f) $f(x) = [\![\sin x]\!], c = -\frac{1}{2}\pi$

5. Find the following limits.

(a) $\lim\limits_{x \downarrow 0} x/|x|$ (b) $\lim\limits_{x \uparrow 0} x/|x|$

(c) $\lim\limits_{x \downarrow 0} 0^x$ (d) $\lim\limits_{x \uparrow 0} (x + |x|) \csc x$

(e) $\lim\limits_{x \downarrow 0} [\![x]\!] \cot x$ (f) $\lim\limits_{x \uparrow \frac{1}{2}} [\![4x - [\![x]\!]]\!]$

6. Sketch the graph of f and see if you can determine $\lim\limits_{x \to 2} f(x)$.

(a) $f(x) = (3x^2 - 3x - 6)/(x - 2)$ (b) $f(x) = (x^2 - 2x)/(x - 2)$

(c) $f(x) = (x^2 - 4)/|x - 2|$ (d) $f(x) = |x^2 - 4|/(x - 2)$

7. Sketch the graph of f and see if you can determine $\lim\limits_{x \to 1} f(x)$.

(a) $f(x) = (x^3 - x^2)/(x - 1)$ (b) $f(x) = (x^3 - 2x^2 + x)/(x - 1)$

(c) $f(x) = (x^3 - 1)/(x - 1)$ (d) $f(x) = ((x - 1)^3 + 1)/(x - 1)$

8. Let $h(x) = [\![\sin \pi x]\!]$ and sketch the part of the graph of h that projects onto the interval $(-2, 2)$ of the X-axis.

(a) At what points of our interval is h discontinuous?

(b) Find the limit of h at $\frac{1}{2}$ and the value of h at $\frac{1}{2}$.

(c) What can you say about the limit of h at 0 and the value of h at 0?

(d) What is $\lim\limits_{x \uparrow 0} h(x)^2$?

(e) What is $(\lim\limits_{x \uparrow 0} h(x))^2$?

9. Let $f(x)$ be the distance between x and the nearest integer. For example, $f(\pi) = \pi - 3 = .14159\ldots, f(-2\frac{1}{4}) = \frac{1}{4}, f(3) = 0$, and so on.

(a) Sketch the graph of f and convince yourself that f is continuous at each real number.

(b) Find a formula for $f(x)$.

10. In Problem 5-11 we discussed the "postage function." Is it continuous? Can you think of other discontinuous functions that arise in real life? Continuous functions?

11. The function of Illustration 5-5 is discontinuous at every real number. How would you try to explain this statement?

10. SOME THEOREMS ON LIMITS AND CONTINUITY

From a geometric point of view, the ideas we presented in the last section are reasonably transparent. But mathematics does not consist of pointing to a picture and saying, "Don't you see?" A truly mathematical development of calculus requires a mathematical definition of the limit. Putting our geometric ideas about limits into precise mathematical terms is really not difficult, but the results *are* complicated. We are going to spare you these complications by continuing to use geometric and informal language in presenting the limit theorems that follow. Still, we have organized the material in such a way that if we had started with a more formal approach, this sequence of theorems and examples would produce a rigorous theory of limits.

A function is *continuous in an interval* if it is continuous at each number c of the interval. Every point of its graph is where it should be; there are no breaks or jumps. Thus Fig. 10-1 shows the graph of a function that is continuous in the interval $[a, b]$. Now look at the arc of the graph of f that projects onto $[a, b]$. If we project this arc onto the Y-axis, we obtain the **image** of $[a, b]$ under f, which we write as $f([a, b])$. Thus if x is a point of $[a, b]$, then $f(x)$ is the image of x under f, and $f([a, b])$ is the collection of all such numbers as x ranges over $[a, b]$. In our present instance, the image of the closed finite

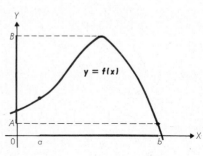

Figure 10-1

interval $[a, b]$ is itself a closed finite interval $[A, B]$. This statement is the most important fact that we need to know about continuous functions.

Theorem   10-1.   *If f is continuous in the closed finite interval $[a, b]$, then there are numbers A and B such that $f([a, b]) = [A, B]$.*

Theorem 10-1 is easier to state than to prove. Its proof depends on basic properties of the real number system (in particular, the Completeness Property 2-1), and, of course, we must use a formal definition of the limit. Every advanced calculus book contains a proof, although these key results are usually stated as several theorems.

Theorem 10-1 is more important for theory than for practice. For example, a practical person says, "How do I solve this equation?" The mathematician asks, "Does the equation *have* a solution?" And he uses Theorem 10-1 to answer that question, as the following example shows. We will see many other applications of Theorem 10-1 to the theoretical side of calculus as we go along.

Example 10-1.   Show that the equation $x^7 - 4x^3 + 1 = 0$ has a solution in the interval $[0, 1]$.

Solution.   We will soon see that if $f(x) = x^7 - 4x^3 + 1$, then f is continuous in any interval, in particular, in $[0, 1]$. Therefore Theorem 10-1 tells us that $f([0, 1])$ is a closed interval. Since $f(0) = 1$ and $f(1) = 1 - 4 + 1 = -2$, the numbers 1 and -2 belong to this image interval. If two numbers belong to an interval, then any number between them belongs to the interval, and so 0 belongs to the interval $f([0, 1])$. But this statement simply means that there is some number z in the interval $[0, 1]$ such that $f(z) = 0$, which is what we were to show. (How do you know that $z \neq 0$ and $z \neq 1$?)

If (r, s) is an open interval that contains a number c, we say that (r, s) is a **neighborhood** of c. For example, $(1, 5)$ is a neighborhood of π. In dealing with a limit at c, we are not required to consider the number c itself, so we may be working in a **punctured neighborhood** of c, the set we obtain by deleting c from our neighborhood (r, s). We denote this punctured neighborhood by $(r, s)_c$, so that $(r, s)_c = (r, c) \cup (c, s)$.

We use neighborhoods of a given number to measure the closeness of other numbers to it. Thus a number M approximates L closely if M belongs to a narrow neighborhood of L. In this way we can make precise the words "close to" and "is approximated by" in our informal definition of the limit and thereby produce a mathematically precise definition.

Definition   10-1.   *The equation* $\lim\limits_{x \to c} f(x) = L$ *means that each neighborhood of L contains the image under f of a punctured neighborhood of c.*

This definition (or one of its equivalent wordings) must be the basis for proofs of limit theorems. For the most part, the formality of such proofs tends to obscure the basic ideas behind them, so we will omit them. But here is an example.

Example 10-2.   If $\lim\limits_{x \to c} f(x) = L$, where $L > 0$, show that there is a neighborhood (r, s) of c such that $f(x) > 0$ for each number x in the punctured neighborhood $(r, s)_c$.

Solution.   Since L is positive, it belongs to the open interval $(0, \infty)$. This interval is therefore a neighborhood of L, so Definition 10-1 says that there must be a punctured neighborhood $(r, s)_c$ whose image under f belongs to $(0, \infty)$. In other words, if x belongs to $(r, s)_c$, then $f(x)$ belongs to $(0, \infty)$, as we were to show.

The use of neighborhoods also makes precise what we mean when we say that L is *the* number that is approximated by values of f, rather than *a* number.

Example 10-3.   Show that if $M \neq L$, then there is a neighborhood of L that does not contain M.

Solution.   We will simply write down such a neighborhood, for example,

$$(L - \tfrac{1}{4}|L - M|, \, L + \tfrac{1}{4}|L - M|).$$

The midpoint of this interval is L, so it surely contains L. But it can't contain M, because it is only $\tfrac{1}{2}|L - M|$ units long, and L and M are $|L - M|$ units apart. We usually word the statement of this example in the following form: *if M belongs to every neighborhood of L, then M = L.*

Our next theorem isn't so important in itself, but it leads to important results. And its truth is obvious, once the essential idea of a limit is grasped. It is known as the "squeeze theorem."

Theorem   10-2.   *Suppose that the number $g(x)$ is between the numbers $f(x)$ and $h(x)$ for each number x in some punctured neighborhood of c. If $\lim_{x \to c} f(x) = L$ and $\lim_{x \to c} h(x) = L$, then $\lim_{x \to c} g(x) = L$.*

This theorem simply says that if L is the number that is approximated by both $f(x)$ and $h(x)$ when x is close to c, and if $g(x)$ is between $f(x)$ and $h(x)$, then L is also the number that is approximated by $g(x)$, as shown in Fig. 10-2.

We can use Theorem 10-2 and the fact that linear functions are continuous to show that many other functions are continuous.

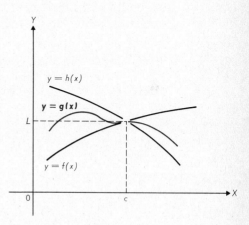

Example 10-4.   Show that if $g(x) = x^2$, then g is continuous at 0.

Solution.   We have to show that the limit of g at 0 is $g(0) = 0$; that is,

$$\lim_{x \to 0} x^2 = 0.$$

Figure 10-2

Observe that if x is close to 0, then x^2 is between $-x$ and x (for example, $(-\frac{1}{2})^2$ is between $\frac{1}{2}$ and $-\frac{1}{2}$). In Example 9-4 we stated that linear functions are continuous, and therefore $\lim_{x \to 0} -x = 0$ and $\lim_{x \to 0} x = 0$. So the squeeze theorem tells us that $-x$ and x squeeze x^2 to 0.

PROBLEMS   10

1.   Each function is continuous in the interval $[0, 1]$, so Theorem 10-1 asserts that $f([0, 1])$ is a closed interval. Find it.
     (a) $f(x) = 2x - 1$ (b) $f(x) = 1 - 2x$
     (c) $f(x) = (2x - 1)^2$ (d) $f(x) = |2x - 1|$
     (e) $f(x) = \sin \pi x$ (f) $f(x) = \cos x$
     (g) $f(x) = [\![\frac{1}{4}x + \pi]\!]$ (h) $f(x) = [\![\frac{1}{4}x]\!] + \pi$

2.   Sketch the graph of f if $f(x) = |x|$.
     (a) How does your figure show that f is continuous at 0?
     (b) Since the graph is squeezed between the lines $y = x$ and $y = -x$, explain how the results of Example 9-4 and Theorem 10-2 tell us that f is continuous at 0.

3.   Sketch the graph of f if $f(x) = [\![x]\!]$.
     (a) Is f continuous in the interval $(0, 1)$? the interval $[0, 1]$?
     (b) What is $f((0, 1))$? $f([0, 1])$?

4.   Convince yourself that the number $\sin x$ is between $-x$ and x for each real number x. Explain how this statement, the fact that linear functions are continuous, and the squeeze theorem tell us that the sine function is continuous at 0.

5.   (a) Show that for $x > 0$, $\sqrt{x}$ is between x and $1/x$.
     (b) How can we use Examples 9-4, 9-5, and the squeeze theorem to find $\lim\limits_{x \to 1} \sqrt{x}$?
     (c) What is $\lim\limits_{x \to 1} \sqrt[n]{x}$, where n is any positive integer?

6.   Show that if $x > 0$, then $1 \leq \sqrt{1 + x^2} \leq 1 + x$. What do these inequalities tell us about $\lim\limits_{x \downarrow 0} \sqrt{1 + x^2}$?

7.   In addition to the neighborhoods defined in the text, we say that an open interval (r, s) is a **right neighborhood** of r and a **left neighborhood** of s. Sketch the given neighborhoods of 3 on a number scale. Which are punctured neighborhoods, which are right neighborhoods, and which are left neighborhoods?

     (a) $(1/\pi, \pi)$ (b) $(1/\pi, 3) \cup (3, \pi)$
     (c) $\{x \mid |x - 3| < 1\}$ (d) $\{x \mid 0 < |x - 3| < 1\}$
     (e) $(\log 100, \log 1000)$ (f) $(\sin \pi, 6 \sin \frac{1}{6}\pi)$
     (g) $([\![\pi]\!], \pi)$ (h) $\{x \mid [\![3 - x]\!] = 0\}$

8.   How does the result of Example 10-2 tell us that the statement

     $$\lim_{x \to 0} (x^2 + 5)10^{\tan x} \cos (1/x) = .5$$

     is surely false? Could the limit be $-.5$?

9.   (a) Convince yourself that $\lim\limits_{x \to 0} x \sin 1/x = 0$.
     (b) Convince yourself that $\lim\limits_{x \to 0} \sin 1/x$ does not exist.

10. If $f(x) = 1/(1 + x^2)$, then it is true that f is continuous in R^1. How do your answers to the following questions square with Theorem 10-1?
     (a) Find the closed interval $f([-1, 1])$.
     (b) Is $f([-\infty, \infty])$ a closed interval?
     (c) Is $f((-1, 1))$ a closed interval?
     (d) Is $f((-1, 1))$ an open interval?

11. Draw the graphs of two arbitrary continuous functions f and g (two unbroken wavy curves) for x in the interval $(0, 3)$. For each x in this interval, let $h(x)$ be the larger of the numbers $f(x)$ and $g(x)$. Draw the curve $y = h(x)$, and explain how it shows that h is continuous. Is k continuous if $k(x)$ is the smaller of the numbers $f(x)$ and $g(x)$?

12. In each of these examples, $\lim\limits_{x \to 0} f(x) = 1$. Since $(.9, 1.1)$ is a neighborhood of 1, Definition 10-1 says that there are punctured neighborhoods of 0 whose images under f belong to $(.9, 1.1)$. Find such a punctured neighborhood in each case. (Notice that there are many possibilities.)

     (a) $f(x) = 1$ (b) $f(x) = x + 1$ (c) $f(x) = x^2 + 1$
     (d) $f(x) = 1 - x$ (e) $f(x) = x/x$ (f) $f(x) = (x^2 + x)/x$
     (g) $f(x) = \cos x$ (h) $f(x) = 1 - \sin x$

The order in which one performs a sequence of mathematical operations may or may not affect the final outcome. For example, if we add two numbers and double the result or double the numbers and add the results, we end up with the same number. But (see Inequality 1-6) if we add two numbers and compute the absolute value of the sum or compute the absolute values first and then add, we may well end up with different numbers.

One of the nicest features of the operation of taking a limit is that, for most practical purposes, it commutes with the other operations of mathematics. Thus we can add and then take limits, or take limits and then add, and so on. This statement is subject to some restrictions. For instance, in Example 9-1 we saw that $\lim_{x \to 0} (\sin x)/x = 1$. Here we are dividing $\sin x$ by x and then computing the limit. If we compute limits first ($\lim_{x \to 0} \sin x = 0$ and $\lim_{x \to 0} x = 0$), we can't divide the results because the divisor would be 0. But aside from this rather natural restriction against dividing by 0, examples of the failure of the limit operator to commute with other mathematical operators seem rather contrived. Thus in Example 9-2 we saw that $\lim_{x \to 0} [\![\cos x]\!] \neq [\![\cos \lim_{x \to 0} x]\!]$, but one isn't likely to run into either of these limits in the normal course of business.

The first limit theorem that we will introduce in this section deals with "composite" functions. Thus the equation $y = g(f(x))$ defines a function that is composed of given functions f and g. We call it the **composite function** formed by the composition of f by g. For example, the equation $y = \cos x^2$ is the rule of correspondence for the composition of the square function by the cosine function. Making up new functions by composing old ones is a normal pastime in calculus, and often it is important that we be able to use limits of the components to find a limit of the composite.

A function g is continuous at L if

(1) $g(L)$ is the number that is approximated by $g(u)$ when u is close to L.

Suppose that it is also true that $\lim_{x \to c} f(x) = L$. This equation says that

(2) $f(x)$ is close to L when x is close to, but not equal to, c.

Now if we substitute $f(x)$ from statement (2) for u in statement (1), the resulting statement becomes "$g(L)$ is the number that is approximated by $g(f(x))$ when x is close to, but not equal to, c." In symbols,

$$\lim_{x \to c} g(f(x)) = g(L).$$

These informal remarks can be converted into a rigorous proof of one of the most basic limit theorems. It says that we can calculate $g(f(x))$ and then take the limit, or take the limit of $f(x)$ first and then calculate its image under g; we end up with the same number in either case.

Theorem   11-1.   *If g is continuous at L, and if $\lim\limits_{x \to c} f(x) = L$, then*

$$\lim_{x \to c} g(f(x)) = g(L).$$

In other words,

$$\lim_{x \to c} g(f(x)) = g(\lim_{x \to c} f(x)).$$

Example 11-1.   What is $\lim\limits_{x \to 1} (x - 1)^2$?

Solution.   In this example, we can think of $g(x)$ as x^2, $f(x)$ as $x - 1$, $c = 1$, and $L = 0$. We know (Example 10-4) that g is continuous at 0 and (Example 9-4) that $\lim\limits_{x \to 1} f(x) = 0$. Hence the conditions of Theorem 11-1 are satisfied, and

$$\lim_{x \to 1} (x - 1)^2 = [\lim_{x \to 1} (x - 1)]^2 = 0^2 = 0.$$

To find the limit that we asked for in the previous example, we called upon a lot of machinery, in particular, Example 9-4 and Theorems 10-2 and 11-1. In a formal development of the theory of limits along the lines we are sketching, these results (together with Example 9-5) would have to be proved in terms of our precise definition of the limit. But once they are established, all of our further results can be proved in terms of them. Our next theorem is an example of such a result. We are not going to present a proof, since it is long and not particularly illuminating. Besides, you will no doubt be ready to believe the facts without proof, since they are quite apparent from our informal descriptions of the limit.

Theorem   11-2.   *If $\lim\limits_{x \to c} f(x) = L$ and $\lim\limits_{x \to c} g(x) = M$, then $\lim\limits_{x \to c} [f(x) + g(x)] = L + M$ and $\lim\limits_{x \to c} [f(x)g(x)] = LM$; that is,*

$$\lim_{x \to c} [f(x) + g(x)] = \lim_{x \to c} f(x) + \lim_{x \to c} g(x)$$

and

$$\lim_{x \to c} [f(x)g(x)] = \lim_{x \to c} f(x) \lim_{x \to c} g(x).$$

With the natural restriction that we are not allowed to divide by 0, there is a quotient rule for limits, too.

Theorem 11-3. *If* $\lim\limits_{x \to c} f(x) = L$ *and* $\lim\limits_{x \to c} g(x) = M$, *where* $M \neq 0$, *then* $\lim\limits_{x \to c} [f(x)/g(x)] = L/M$; *that is,*

$$\lim_{x \to c} [f(x)/g(x)] = \lim_{x \to c} f(x)/\lim_{x \to c} g(x).$$

Example 11-2.   Find $\lim\limits_{x \to 1} \dfrac{x^2 + 3x - 4}{3x + 2}$.

Solution.   According to our sum, product, and quotient rules (Theorems 11-2 and 11-3), this limit is the number

$$\frac{\lim\limits_{x \to 1} (x^2 + 3x - 4)}{\lim\limits_{x \to 1} (3x + 2)} = \frac{\lim\limits_{x \to 1} x \cdot \lim\limits_{x \to 1} x + \lim\limits_{x \to 1} (3x - 4)}{\lim\limits_{x \to 1} (3x + 2)}.$$

So we have reduced our problem to that of finding limits of linear expressions. But in Example 9-4 we stated that linear functions are continuous. In particular, this statement says that $\lim\limits_{x \to 1} (mx + b) = m + b$. When we replace x with 1 in our last quotient, we see that our desired limit is the number

$$\frac{1 \cdot 1 + (3 \cdot 1 - 4)}{3 \cdot 1 + 2} = \tfrac{0}{5} = 0.$$

If $f(x) = a_n x^n + a_{n-1} x^{n-1} + \cdots + a_1 x + a_0$, where $a_0, \ldots, a_n$ are given numbers, then f is called a **polynomial function**. For example, f is a polynomial function if $f(x) = 3x^2 - \tfrac{1}{2}x + 4$. Since the only mathematical operations that are involved in computing values of polynomial functions are the familiar ones of addition, subtraction, multiplication, and division, we naturally feel quite comfortable with such functions. And it adds to our feeling of comfort to find that polynomial functions are well-behaved as far as continuity is concerned.

Example 11-3.   Show that polynomial functions are continuous.

Solution.   To show that a polynomial function is continuous at a number c, we must show that

$$\lim_{x \to c} (a_n x^n + a_{n-1} x^{n-1} + \cdots + a_1 x + a_0) = a_n c^n + a_{n-1} c^{n-1} + \cdots + a_1 c + a_0.$$

This equation follows from our sum and product rules, together with the continuity of the identity and constant functions, for we have

$$\lim_{x \to c} (a_n x^n + a_{n-1} x^{n-1} + \cdots + a_1 x + a_0)$$

$$= \lim_{x \to c} a_n \cdot \lim_{x \to c} x \cdots \lim_{x \to c} x + \cdots + \lim_{x \to c} a_1 \cdot \lim_{x \to c} x + \lim_{x \to c} a_0$$

$$= a_n c^n + a_{n-1} c^{n-1} + \cdots + a_1 c + a_0.$$

Example 11-4. If $r(x) = p(x)/q(x)$, where p and q are polynomial functions, then r is a **rational function.** Discuss the continuity of rational functions.

Solution. According to Theorem 11-3, $\lim_{x \to c} r(x) = \lim_{x \to c} p(x)/\lim_{x \to c} q(x)$ if $\lim_{x \to c} q(x) \neq 0$, and according to Example 11-3, $\lim_{x \to c} p(x) = p(c)$ and $\lim_{x \to c} q(x) = q(c)$. Hence $\lim_{x \to c} r(x) = p(c)/q(c) = r(c)$, provided that $q(c) \neq 0$. In other words, r is continuous except at the solutions of the equation $q(x) = 0$.

PROBLEMS   11

1. Find the indicated limit and justify your reasoning by referring to the appropriate limit theorem.

   (a) $\lim_{x \to 1} (x^2 - 5x + 2)$

   (b) $\lim_{x \to 2} \dfrac{x^3 - 7}{x + 1}$

   (c) $\lim_{x \to 1} (2x - 1)(x + 1)$

   (d) $\lim_{x \to 2} \dfrac{x^2 + 1}{x + 1}$

   (e) $\lim_{x \to 1} \dfrac{x^3 + 2x^2 - x - 2}{x - 1}$

   (f) $\lim_{x \to 1} \dfrac{x - 1}{x^3 + 2x^2 - x - 2}$

2. For $x \neq -1$, $f(x)$ is defined by the given formula. What is $f(-1)$ if f is continuous at -1?

   (a) $f(x) = \dfrac{x^3 + 3}{x^2 + 3}$

   (b) $f(x) = (x^3 + 1)(x + 1)$

   (c) $f(x) = \dfrac{x^3 + 1}{x + 1}$

   (d) $f(x) = \dfrac{x + 1}{x^3 + 1}$

   (e) $f(x) = \dfrac{x^3 + 1}{|x| + 1}$

   (f) $f(x) = \left| \dfrac{x^3 + 1}{x + 1} \right|$

3. Assuming that the various functions involved are continuous, what are the following limits?

   (a) $\lim_{x \to 0} \log \cos x$

   (b) $\lim_{x \to 100} \cos (\pi \log x)$

   (c) $\lim_{x \to 0} 2^{\sin x}$

   (d) $\lim_{x \to 0} \sin 2^x$

   (e) $\lim_{x \to \pi} \tan [\![x]\!]$

   (f) $\lim_{x \to \pi} [\![\tan x]\!]$

4. Assuming (correctly) that the sine function is continuous at every real number, what is $\lim_{x \downarrow 0} \sin [\![x]\!]$? $\lim_{x \uparrow 0} \sin [\![x]\!]$?

5. (a) Show that the composition of two linear functions is also linear.
   (b) Find linear functions f and g so that the composition of f by g is not the same as the composition of g by f.

6. Discuss the relation between the composition of f by g and the composition of g by f.

   (a) $f(x) = x^2, g(x) = \sqrt{x}$ (b) $f(x) = g(x) = 3/x$
   (c) $f(x) = 2x, g(x) = \sin x$ (d) $f(x) = |x|, g(x) = x^4 + 1$
   (e) $f(x) = x - |x|, g(x) = x^2$ (f) $f(x) = [\![x]\!], g(x) = \sin \pi x$

7. In these quotients what happens if we divide first and then compute the limit or compute limits first and then divide?

   (a) $\lim_{x \to 3} \dfrac{x^2 - 9}{x - 3}$ (b) $\lim_{x \to 3} \dfrac{x^2 - 9}{x + 3}$ (c) $\lim_{x \to 2} \dfrac{x^4 - 16}{x - 2}$ (d) $\lim_{x \to 2} \dfrac{x^4 - 16}{x + 2}$

   (e) $\lim_{x \to 1} \dfrac{x^5 - 1}{x - 1}$ (f) $\lim_{x \to 1} \dfrac{x^6 - 1}{x^2 - 1}$ (g) $\lim_{x \to 3} \dfrac{x^5 - 3^5}{x - 3}$ (h) $\lim_{x \to 3} \dfrac{x^5 + 3^5}{x + 3}$

8. What is $\lim_{x \to 0} (1 - \cos^2 x)x^{-2}$? (*Hint:* What is the most basic trigonometric identity you know?)

9. In which of these examples is it true that $\lim_{x \to 0} |f(x)| = |\lim_{x \to 0} f(x)|$?

   (a) $f(x) = x/x$ (b) $f(x) = x/|x|$
   (c) $f(x) = x^3/|x|$ (d) $f(x) = (\sin x)/|x|$

10. (a) Can you find f, g, and c such that $\lim_{x \to c} f(x)$ and $\lim_{x \to c} g(x)$ do not exist, but $\lim_{x \to c} [f(x) + g(x)]$ does exist?

    (b) Can you find f, g, and c such that $\lim_{x \to c} f(x)$ does not exist, but $\lim_{x \to c} g(x)$ and $\lim_{x \to c} [f(x) + g(x)]$ do exist?

    (c) Can you find f, g, and c such that $\lim_{x \to c} f(x)$ and $\lim_{x \to c} g(x)$ do not exist, but $\lim_{x \to c} [f(x)g(x)]$ does exist?

    (d) Can you find f, g, and c such that $\lim_{x \to c} f(x)$ does not exist, but $\lim_{x \to c} g(x)$ and $\lim_{x \to c} [f(x)g(x)]$ do exist?

The contents of this introductory chapter are summed up in the section titles. Calculus is a study of *functions*, and functions are *sets* of pairs of *numbers*. Geometrically, these pairs of numbers are pictured as points of the *coordinate plane*, the collection of points being the *graph* of the function. We will find that much of calculus consists of approximating a curve by its tangent *lines*; the idea of *slope* is at the heart of the matter. Finally, the basic concept that pervades all of calculus is that of the *limit*. If there is a common thread to our presentation of these ideas, it is that we can view much of this material in terms of pictures. Draw many of them as you work through these review problems.

For example, Fig. I-1 illustrates the concept of continuity. By definition, f is continuous at p if its

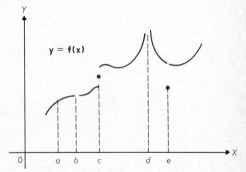

Figure I-1

limit at p is its value at p. Thus f is continuous at a, since $\lim\limits_{x \to a} f(x) = f(a)$. The "hole" in the graph above b indicates that b is not in the domain of f. In other words, f has no value at b, so it makes no sense to ask whether or not its limit at b is its value at b; f is not continuous at b. The same remarks apply to d. Both c and e are in the domain of f, so $f(c)$ and $f(e)$ are perfectly definite numbers. But the value $f(c)$ is not the limit of f at c; in fact, $\lim\limits_{x \to c} f(x)$ does not exist. And while the limit of f at e does exist, it doesn't equal the value of f at e, and so f is not continuous at e. Thus our figure indicates that f is discontinuous (not continuous) at b, c, d, and e, but it is continuous at the other points of its domain.

In terms of the theory of calculus, perhaps the most important equation to keep in mind is

$$\lim_{x \to c} \left[mf(x) + ng(x) \right] = m \lim_{x \to c} f(x) + n \lim_{x \to c} g(x)$$

This equation says that $\lim\limits_{x \to c}$ is a **linear operator**; it doesn't matter whether we form the linear combination $mf(x) + ng(x)$ and then apply the operator or apply the operator to $f(x)$ and $g(x)$ separately and then form the linear combination. Notice that if f is a homogeneous linear function, then (Theorem 7-2)

$$f(mu + nv) = mf(u) + nf(v),$$

so the operator $\lim\limits_{x \to c}$ has the basic property of these simple functions. We will find that other important operators of calculus, because they are defined in terms of limits, inherit this linearity property.

Review Problems, Chapter One

1. Is it true that the sum of two rational numbers is a rational number? Is the sum of two irrational numbers necessarily an irrational number? Can the sum of a rational number and an irrational number be rational? Can the product of a rational number and an irrational number be rational?

2. How would you try to convince someone that every open interval contains a rational number? An irrational number?

3. Show that $x \in [a, b)$ if and only if $[\![(x - a)/(b - a)]\!] = 0$.

4. Let f be a function whose range is a subset of R^1 and whose domain is R^1. Which of the following statements are surely true?

   (a) $f(3 - 2) = f(3) - f(2)$ (b) $f(3 - 2) = f(1)$
   (c) $f(\sqrt{2}) = \sqrt{f(2)}$ (d) $f(|-3|) = f(3)$
   (e) $f(2)f(3) = f(6)$ (f) $f(x^2) = f(x)f(x)$

5. The graph of a function f is shown in Fig. I-2. Use this figure to answer the following questions as best you can.

    (a) What is the domain of f?   (b) What is the range of f?
    (c) Find $f(\pi)$ and $f(3)$.   (d) If $f(x) = \frac{1}{2}$, what is x?
    (e) Find $f(f(4))$.   (f) Find $\lim\limits_{x\uparrow 3} f(x) + \lim\limits_{x\downarrow 3} f(x)$.

    (g) If $f(f(x)) = 2$, what is x?   (h) Find $f([4, 6])$.

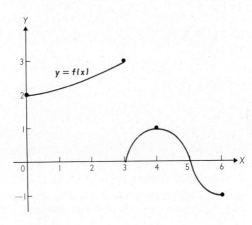

Figure I-2

6. Define a function by using the sets $\{1, 2, 3\}$ and $\{4, 5, 6, 7\}$ as the domain and range (not necessarily in that order).

7. If $f \subseteq g$, we say that g is an **extension** of f and f is a **restriction** of g. Show that $f \subseteq g$ for the following function pairs.

    (a) $f(x) = x/x$, $g(x) = 1$ (b) $f(x) = \sqrt{x}$, $g(x) = \sqrt[4]{x^2}$
    (c) $f(x) = 2(\sqrt{x})^2$, $g(x) = x + |x|$ (d) $f(x) = 2/(x + |x|)$, $g(x) = 1/x$

8. Six cannibals have a missionary stretched out between the points $(-1, -2)$ and $(5, 10)$. If they plan to divide him equally, where do they make their cuts?

9. Solve for x.

    (a) $\sin [\sin (\sin x)] = 0$ (b) $\cos (\cos x) = 0$
    (c) $\sin [\![x]\!] = [\![\sin x]\!]$ (d) $\cos [\![x]\!] = [\![\cos x]\!]$

10. Show that the area of the region enclosed by a regular polygon with n sides each of length a is $\frac{1}{4}na^2 \cot (\pi/n)$.

11. Show that $\cos 2t = \cos^4 t - \sin^4 t$ for each $t \in R^1$.

12. Sketch the graphs of the following equations.

    (a) $(\sin \pi x)(\sin \pi y) = 0$ (b) $|\sin \pi x| + |\sin \pi y| = 0$
    (c) $|\sin \pi y| - \sin \pi y + |\sin \pi x| - \sin \pi x = 0$

13. Let f and g be linear functions. Show that the graphs of the composition of f by g and the composition of g by f are parallel lines.

14. Closely pursued by a bull, Farmer Jones is at the point $(1, 4)$ and making for the fence $2y - 4x + 16 = 0$. For what point is he heading?

15. King Arthur's mead cost 12 florins per firkin for the first 10 firkins and 8 florins for each additional firkin.
   (a) Show that x firkins cost $c(x) = 10x + 20 - 2|x - 10|$ florins.
   (b) Sketch the graph of c.
   (c) Is c a continuous function?
   (d) The graph of c consists of line segments. Calculate their slopes and decide what they represent in terms of the price of mead.

16. The center of Paul Bunyon's buzz saw is the origin, and its radius is 2 yards. A tooth breaks at the point $(\frac{6}{5}, \frac{8}{5})$ and flies off on a tangent toward the blue ox Babe grazing at the point $(-5, 6)$. Try to find out graphically and analytically if Paul's supper menu lists ox steak.

17. Find the following limits.

   (a) $\lim\limits_{x \to 0} \dfrac{x}{|x|} \sin x$

   (b) $\lim\limits_{x \uparrow 0} \dfrac{x}{|x|} \cos x$

   (c) $\lim\limits_{x \downarrow 0} \dfrac{x}{|x|} (3x - 2)$

   (d) $\lim\limits_{x \to 0} \cos \pi \dfrac{x}{|x|}$

18. Suppose that $f(x) = x^2$, and let $q(z) = [f(z) - f(1)]/(z - 1)$.
   (a) What is the domain of q?
   (b) Sketch the curve $y = q(z)$ in a ZY-coordinate system.
   (c) What is $\lim\limits_{z \to 1} q(z)$? What is $\lim\limits_{z \to 2} q(z)$?

19. Let $f(x)$ denote the distance between x and the nearest even integer. Thus $f(17) = 1$, $f(31\frac{1}{4}) = \frac{3}{4}$, and so on.
   (a) Sketch the graph of f.
   (b) Do you think f is continuous?
   (c) Convince yourself that $f(x) = |x - 2[\![\frac{1}{2}(x + 1)]\!]|$.

20. The equation $\cos \pi[\![x]\!] = 2[\![\sin \pi x]\!] + 1$ isn't true for *each* $x \in (-\infty, \infty)$. How nearly true is it?

The Derivative 2

We now introduce one of the basic concepts of calculus, the derivative. A derivative is a rate, and we can gain some insight into its nature by considering the following two problems.

Problem 1. A bottle is dropped into a stream that is flowing 3 miles per hour. Plot a graph showing how the location of the bottle depends on the time it has been drifting.

Problem 2. A bottle is dropped from rest in a vacuum. Plot a graph showing how the location of the bottle depends on the time it has been falling.

In order to solve these problems, we use a little elementary science. If the bottle in Problem 1 drifts s miles in t hours, then the formula *distance = rate × time* gives us the equation $s = 3t$. Its graph (Fig. II-1) is a line with slope 3, and we observe that *the slope of the line is the velocity of the bottle.*

For Problem 2, we recall that the relation between the number of units s of distance that a body falls from rest in t units of time is expressed by the equation

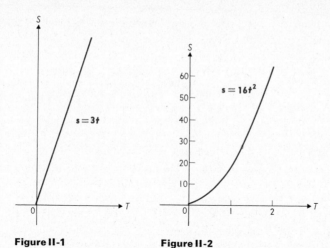

Figure II-1   **Figure II-2**

$s = \frac{1}{2}gt^2$, where g is a number denoting the acceleration due to gravity. If we measure distance in feet and time in seconds, then $g = 32$ (approximately), and the relation between s and t is $s = 16t^2$. The graph of this equation is shown in Fig. II-2.

One obvious difference between our graphs is that one is a line and the other isn't. The fact that the slope of the graph (line) in Fig. II-1 represents the velocity of the bottle in Problem 1 leads us to ask if the slope of the graph in Fig. II-2 represents the velocity of the bottle in Problem 2. But then we are faced with the question "What do we mean by the 'slope of the graph'?" since we have, so far, only talked about slope in connection with lines. In this chapter we will answer this last question, and we will also learn how to calculate the slopes of the graphs of many familiar functions.

Don't think that the idea of the slope of a graph is too unimportant to require your full attention. This concept has any number of applications, in all fields of thought. In particular, it does give us the velocity of our falling bottle.

12. A GEOMETRIC APPROACH TO THE DERIVATIVE

We will now give a geometric definition of the slope of the graph of a function f, postponing until the next section a translation of these geometric ideas into the formal terminology of calculus. If you understand what we are saying here, you will be well on your way toward understanding what a derivative is.

To a point of the graph of f we wish to assign a number that we call the slope of the graph *at the point*. Naturally, if the graph of f is a line, our assignment should yield the slope of the line. So the slope of the graph of a linear function at one point is the same as its slope at any other point. For other graphs, however, the slope may be different at different points.

Since we have already defined the slope of a line, we might try to obtain the slope of a general graph at a given point by drawing the line that best "fits" the graph at the point and then assigning the slope of this line as the slope of the graph at the point. The line that best "fits" the graph at a point is called the **tangent line** to the graph at the point. *The slope of the graph at a point is the slope of this tangent line.*

At a point of discontinuity (for example, a break), a curve will not have a tangent line, and even a continuous curve may lack tangent lines at some points (for example, sharp corners). But if for a given number x in the domain of f we can draw a tangent line at the point $(x, f(x))$, then we can associate with x a number, the slope of the graph at this point. This association produces a new function with the following rule of correspondence: to the number x there corresponds the slope of the graph at the point $(x, f(x))$. This new function is *derived* from f. We call it the **derived function** of f and designate it by the symbol f'. The value of the derived function at x is therefore denoted by $f'(x)$. Thus $f'(x)$ is the slope of the graph of f at the point $(x, f(x))$. The number $f'(x)$ is the **derivative** of $f(x)$.

A couple of examples will clarify what we are discussing.

Example 12-1.   Find $f'(x)$ if $f(x) = \sqrt{4 - x^2}$.

Solution. We first sketch the curve $y = f(x)$; that is, $y = \sqrt{4 - x^2}$. If (x, y) is a point of this graph, then x belongs to the interval $[-2, 2]$, $y \geq 0$, and $\sqrt{x^2 + y^2} = 2$. Thus each point of the graph is 2 units from the origin and lies above the X-axis. The graph of f is therefore the upper semicircle whose radius is 2 and whose center is the origin shown in Fig. 12-1.

We can use this graph to find values of the derived function f'. For example, our definition of f' tells us that the number $f'(0)$ is the slope of the tangent line at the point $(0, f(0)) = (0, 2)$. Obviously this tangent line is horizontal, and so its slope is 0. Therefore $f'(0) = 0$. When $x = 2$ or $x = -2$, the tangent line to the graph

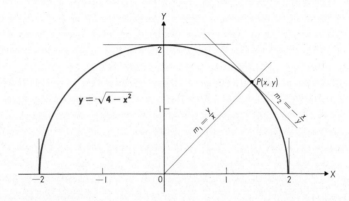

Figure 12-1

is vertical, and the slope of such a line is undefined. Thus the numbers 2 and -2 are not in the domain of the derived function f' although they are in the domain of f.

We can even find a formula that gives us the number $f'(x)$ for any number x in the interval $(-2, 2)$. Suppose that (x, y) is a point of our graph. We know that the tangent line to a circle at a given point is perpendicular to the radius at that point; hence the slope of the tangent line is the negative reciprocal of the slope of the radius (Theorem 8-3). The slope of the radius to the point (x, y) is y/x (why?), and so we see that the slope of the tangent line is $-x/y$. Therefore $f'(x) = -x/y$. Since $y = \sqrt{4 - x^2}$, we have derived the formula

$$f'(x) = \frac{-x}{\sqrt{4 - x^2}}.$$

Notice that this formula gives us the value $f'(0) = 0$ that we found earlier.

Example 12-2. If $f(x) = |x|$, find the derivative $f'(x)$.

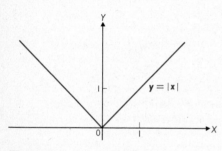

Figure 12-2

Solution. The graph of f is shown in Fig. 12-2. It is clear that for $x > 0$, the tangent line to the graph of f at the point $(x, |x|)$ is simply the line $y = x$. This line has slope 1. If $x < 0$, the tangent line is $y = -x$, whose slope is -1. At the point $(0, 0)$ there is no line that we consider to be a tangent line. It follows that the domain of f' is the set $(-\infty, 0) \cup (0, \infty)$ and that

$$f'(x) = 1 \text{ if } x > 0 \quad \text{and} \quad f'(x) = -1 \text{ if } x < 0.$$

Thus we have found that $f'(x) = \dfrac{x}{|x|}$.

Figure 12-3 illustrates a number of these ideas. It shows the graph of a certain function f. From it we see that the domain of f is the interval $[-1, 8]$, and we can read off some values; for example, $f(-1) = 1$, $f(1) = 3$, $f(5) = 5$, and so on. We have sketched "tangent lines" at various points of the graph. The derivative $f'(x)$ at such a point is the slope of the tangent line. We read from the figure the following approximate values: $f'(2) = 0$, $f'(4) = 1$, $f'(5) = 0$, and $f'(6) = -1$. When $x = 7$, the graph has a vertical tangent; so since a vertical line has no slope assigned to it, we see that the number 7 is not in the domain of f'. Finally, there is no line that we consider to be a tangent line either at the point $(1, 3)$ or at the point $(3, 3)$. The numbers 1 and 3 are not in the domain of the derived function.

Example 12-3. If $f(x) = \sin x$, sketch an approximation to the graph of the derived function f' for the interval $[0, 1.6]$.

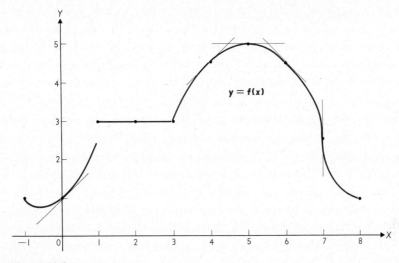

Figure 12-3

Solution. In Fig. 12-4 we have sketched the curve $y = \sin x$ in black, plotting the points in the accompanying table. The line segments joining these points are called "chords" of the graph, and because the points are quite close together, it appears that the slopes of the chords should approximate slopes of tangent lines. Thus as an approximation to the value of the derived function at one of our plotted points we take the slope of the chord that joins that point to the next point in the table.

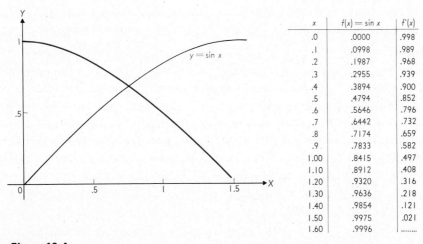

x	$f(x) = \sin x$	$f'(x)$
.0	.0000	.998
.1	.0998	.989
.2	.1987	.968
.3	.2955	.939
.4	.3894	.900
.5	.4794	.852
.6	.5646	.796
.7	.6442	.732
.8	.7174	.659
.9	.7833	.582
1.00	.8415	.497
1.10	.8912	.408
1.20	.9320	.316
1.30	.9636	.218
1.40	.9854	.121
1.50	.9975	.021
1.60	.9996	....

Figure 12-4

In this way, we see that the derivative at 0 is approximated by the number

$$\frac{.0998 - 0}{.1 - 0} = .998;$$

$f'(.1)$ is approximately

$$\frac{.1987 - .0998}{.2 - .1} = .989,$$

and so on. These approximate values appear in the third column of the table, from which we plotted the colored curve, the approximate graph of the derived function f'.

All sorts of notations are used in connection with the derivative. We shall use the "prime" notation for now and introduce others later. If f is a function and x is a number in its domain, then we use either the symbol $f(x)$ or simply the letter y to represent the value of f at x. Similarly, the derived function of f is a function f', and the value of f' at the point x is denoted either by $f'(x)$ or y'. The number y' is called the *derivative* of y. Thus the result of Example 12-1 can be written

$$\text{If} \quad y = \sqrt{4 - x^2}, \quad \text{then} \quad y' = \frac{-x}{\sqrt{4 - x^2}}.$$

PROBLEMS 12

1. If $f(x)$ is given by one of the following equations, the graph of f will be found on some earlier page of the book. Use the graph to estimate the required number.
   (a) $f(x) = \sin x$; find $f'(\frac{3}{2}\pi)$
   (b) $f(x) = \cos x$; find $f'(0)$
   (c) $f(x) = \sin x$; find $f'(\pi)$
   (d) $f(x) = \cos x$; find $f'(\pi)$
   (e) $f(x) = \sin x$; find $f'(\frac{1}{4}\pi) + f'(\frac{3}{4}\pi)$
   (f) $f(x) = \tan x$; find $f'(\frac{1}{4}\pi) - f'(\frac{3}{4}\pi)$

2. If the tangent line to the graph of f at a point is the line that "most nearly coincides" with the graph, what is the tangent line to the graph of the given function at the point with the given X-coordinate?
   (a) $f(x) = \sin x$, $x = 0$
   (b) $f(x) = \tan x$, $x = 0$
   (c) $f(x) = x + |x|$, $x = 3$
   (d) $f(x) = \sin x$, $x = \pi$

3. Find the number y'.
   (a) $y = \frac{1}{2}(x - 1)$
   (b) $3y = x - 4$
   (c) $2y = 4x - 7$
   (d) $1 - y = 2 - x$

4. Sketch the graphs of f and f'.
   (a) $f(x) = 2(x - 1)$
   (b) $f(x) = 2(1 - x)$
   (c) $f(x) = x + |1 - x|$
   (d) $f(x) = |x| + |1 - x|$

5. Use the method of Example 12-1 to find the formula for $f'(x)$.
   (a) $f(x) = \sqrt{9 - x^2}$
   (b) $f(x) = -\sqrt{25 - x^2}$

6. What is the derived function f' if f is a constant function?

7. If $f(x) = \cos x$, use a graph to solve the equation $f'(x) = 0$.

8. Sketch the curve $y = g(x)$, where $g(x) = x\sqrt{1 - x^2}/|x|$. What is the domain of g'? Calculate $g'(\frac{1}{2})$ and $g'(-\frac{1}{2}\sqrt{3})$. Find a formula for $g'(x)$ if $x \in (0, 1)$. Find a formula for $g'(x)$ if $x \in (-1, 0)$.

9. The equation $f(x) = 6 - \sqrt{21 + 4x - x^2}$ defines a function whose graph is a subset of the curve $(x - 2)^2 + (y - 6)^2 = 25$. Find a formula for $f'(x)$.

10. Let $f(x) = |x| + |x - 1| + |x - 2| + |x - 3|$. What numbers are not in the domain of the derived function f'?

11. Use graphs to find the solution sets of the inequality $f'(x) > 0$.
    (a) $f(x) = \sin x$ (b) $f(x) = \cos x$
    (c) $f(x) = \tan x$ (d) $f(x) = \cot x$

12. Give a geometric explanation of why the quotient $[f(1.1) - f(1)]/(1.1 - 1)$ is an approximation to the number $f'(1)$. Does a "typical" figure suggest that $[f(1.1) - f(.9)]/(1.1 - .9)$ might be an even better approximation? Calculate these quotients and compare with the given true value.
    (a) $f(x) = \sin x [f'(x) = \cos x]$ (b) $f(x) = \cos x [f'(x) = -\sin x]$
    (c) $f(x) = \tan x [f'(x) = \sec^2 x]$ (d) $f(x) = x^2 [f'(x) = 2x]$

13. Suppose that $f(x) = 3[\![x]\!]x + 4$.
    (a) Find $f'(\frac{1}{2})$, $f'(\pi)$, and $f'(-\sqrt{2})$.
    (b) What is the domain of f'?
    (c) Find a formula for $f'(x)$ that is valid for each x in the domain of f'.

14. Let $f(x)$ be the distance between x and the nearest prime number $(2, 3, 5, 7, \ldots)$. Thus $f(0) = 2$, $f(\pi) = .1415\ldots$, and so on. Sketch the graph of the equation $y = f(x)$ for x in the interval $[0, 5]$ and use this graph to help you calculate $f'(\sqrt{2})$, $f'(2.7)$, and $f'(\pi)$.

15. The graph of a certain function f whose domain is the interval $[1, \infty)$ consists of the line segments that successively join the points $(1, 1)$, $(2, \frac{1}{2})$, $(3, \frac{1}{3})$, ... Find $f'(\frac{1000}{3})$ and $f'(978\pi)$. Find a formula for $f'(x)$ if x is not an integer.

13. THE DEFINITION OF THE DERIVATIVE

We will now be a little more precise about the meaning of the derivative of $f(x)$. In order to calculate $f'(1)$ by the methods of the preceding section, we sketch the graph of f and then draw the tangent line at the point where $x = 1$. The slope of this tangent line is $f'(1)$. This device of using graphs and sketching in tangent lines "by eye" is useful for introducing the notion of the derivative, but it is not an exact description of a mathematical concept.

For a clue to a "numerical" approach to the derivative, we return to Example 12-3. There we approximate the derivative at a point by finding the slope of the "chord" joining that point and a nearby point. This chord is not the tangent line at the point, but its slope is an *approximation* to the slope of the tangent line.

Notice that we assume that there is a tangent line, that it has a slope, and it is this perfectly definite number that we approximate.

Let us apply this idea to find $f'(1)$ when $f(x) = x^2 + 1$. We will find the number $f'(1)$ indirectly, by calculating approximations of it and then deciding what number is approximated. Figure 13-1 shows the graph of f.

The number $f'(1)$ [the slope of the tangent line at the point $(1, 2)$] is approximated by the slope of the chord joining $(1, 2)$ and a second point P of the curve $y = x^2 + 1$. If we want a "good" approximation, we should choose P close to our given point $(1, 2)$. Therefore we find it convenient to denote the X-coordinate of P by $1 + h$, a number that will be close to 1 when h is close to 0. Then the Y-coordinate of P is $(1 + h)^2 + 1 = h^2 + 2h + 2$, and so P is the point $(1 + h, h^2 + 2h + 2)$. The number h may be positive or negative; we have shown h as a positive number in Fig. 13-1. The slope of the chord that joins our points $(1, 2)$ and $(1 + h, h^2 + 2h + 2)$ is

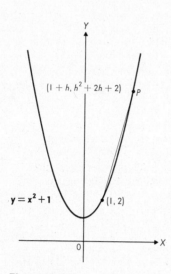

Figure 13-1

$$(13\text{-}1) \qquad \frac{(h^2 + 2h + 2) - 2}{(1 + h) - 1} = \frac{h^2 + 2h}{h}.$$

Formula 13-1 yields an approximation of the number $f'(1)$ for each choice of h different from zero. For example, we get the approximations 3, 1, and $\frac{5}{2}$ by taking $h = 1$, -1, and $\frac{1}{2}$. We get "good" approximations to $f'(1)$ when we choose h close to 0. We cannot choose h *equal to* 0, for then Formula 13-1 is meaningless. Thus we can say that $f'(1)$ is the number that is approximated by the expression $(2h + h^2)/h$ when h is close to, but not equal to, 0. But this sentence is our informal definition of the limit, and hence it appears that

$$f'(1) = \lim_{h \to 0} \frac{2h + h^2}{h}.$$

In calculating this limit, we write $(2h + h^2)/h = 2 + h$ (since $h \neq 0$), from which it follows that $f'(1) = 2$. Therefore we consider the line with slope 2 that contains the point $(1, 2)$ as the tangent line to our graph at that point. This line is the line $y = 2x$. We have sketched it in Fig. 13-2, and you will undoubtedly agree that the name "tangent line" is appropriate.

Let us continue to suppose that $f(x) = x^2 + 1$. For any $x \in R^1$, we can calculate $f'(x)$ by the same procedure we just used to determine $f'(1)$. We consider (see Fig. 13-3) the chord joining two points $(x, x^2 + 1)$ and $(x + h, (x + h)^2 + 1)$. Its slope is

$$\frac{(x + h)^2 + 1 - (x^2 + 1)}{h} = \frac{2xh + h^2}{h}.$$

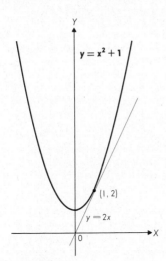

Figure 13-2

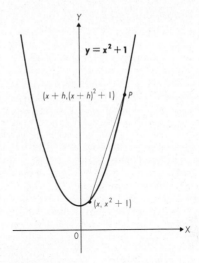

Figure 13-3

Geometrically, the slope $f'(x)$ of the tangent line should be the number that is approximated by this expression when h is close to, but not equal to, 0. Therefore we define $f'(x)$ to be the number

$$\lim_{h \to 0} \frac{2xh + h^2}{h} = \lim_{h \to 0} (2x + h) = 2x.$$

If you read the last sentence carefully, you noticed that we *defined* $f'(x)$ to be $2x$. We have reversed the position we took in Section 12. There we said that in order to calculate a derivative, we should *first* draw a tangent line and *then* calculate its slope. Here we calculated a derivative without drawing a tangent line, a procedure that eliminates the guesswork of fitting a tangent line to the curve "by eye."

We can now formulate a definition of the derivative in general terms. Let f be a function and suppose that x belongs to an open interval in its domain. We find the number $f'(x)$ exactly as we did in the last example. We compute the slope of the chord that joins the point $(x, f(x))$ to a second point $(x + h, f(x + h))$, see Fig. 13-4, by means of the formula

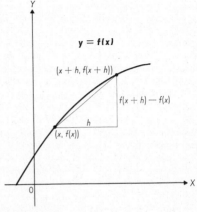

(13-2) $$\frac{f(x + h) - f(x)}{h}.$$

Then we define the derivative $f'(x)$ to be the limit as h approaches 0 of the **Difference Quotient** 13-2 (if this limit exists, of course).

Figure 13-4

Definition 13-1. *If the following limit exists, then*

$$f'(x) = \lim_{h \to 0} \frac{f(x + h) - f(x)}{h}.$$

This limit need not exist at every (or, indeed, any) number in the domain of a given function *f*. If it does exist at a particular number *x*, then we say that *f* is **differentiable** at *x*. The process of finding the derivative $f'(x)$ is called **differentiation**. Notice that we are assuming that the domain of *f* contains some neighborhood of *x*; otherwise we couldn't even set up the difference quotient, much less find its limit.

Example 13-1. Find $f'(x)$ if $f(x) = 5x^2 - 3x$.

Solution. The difference quotient for *f* at *x* is

$$\frac{f(x + h) - f(x)}{h} = \frac{5(x + h)^2 - 3(x + h) - (5x^2 - 3x)}{h}$$

$$= \frac{5(x^2 + 2xh + h^2) - 3x - 3h - 5x^2 + 3x}{h}$$

$$= \frac{10xh + 5h^2 - 3h}{h}$$

$$= 10x - 3 + 5h.$$

Therefore according to Definition 13-1,

$$f'(x) = \lim_{h \to 0} \frac{f(x + h) - f(x)}{h}$$

$$= \lim_{h \to 0} (10x - 3 + 5h)$$

$$= 10x - 3.$$

It sometimes simplifies our algebraic manipulations if we let $z = x + h$ (and hence $h = z - x$) in the Difference Quotient 13-2 and thus obtain the expression

$$\frac{f(z) - f(x)}{z - x}.$$

Then the limit that defines $f'(x)$ is written

$$f'(x) = \lim_{z \to x} \frac{f(z) - f(x)}{z - x}.$$

Example 13-2.   Find $f'(x)$ if $f(x) = 1/x$.

Solution.   The difference quotient for f at x is

$$\frac{f(z) - f(x)}{z - x} = \frac{1/z - 1/x}{z - x} = \frac{(x - z)/xz}{z - x} = -\frac{1}{xz}.$$

Therefore $f'(x) = \lim_{z \to x} \frac{f(z) - f(x)}{z - x} = \lim_{z \to x} -\frac{1}{xz} = -\frac{1}{x^2}.$

We started our discussion of the derivative $f'(x)$ by regarding it as the slope of a tangent line to the graph of f. Definition 13-1, however, enables us to talk about $f'(x)$ without first introducing the tangent line, so now we will define the tangent line in terms of the derivative. The geometric discussion that led us to calculate $f'(x)$ as the limit of the Difference Quotient 13-2 makes the following definition a natural one.

Definition 13-2.   *If a function f is differentiable at x, then the* **tangent line** *(or, simply,* **tangent***) to the graph of f at the point $(x, f(x))$ is the line that contains $(x, f(x))$ and has slope $f'(x)$.*

Example 13-3.   Find the tangent to the graph of the function f in the preceding example at the point $(1, 1)$.

Solution.   In Example 13-2 we found that $f'(1) = -1$. Therefore our desired tangent line has slope -1, and it contains the point $(1, 1)$. Hence its equation is $y = -x + 2$.

When we find $f'(x)$ as the limit of a difference quotient, we must use the limit theorems that we developed in Chapter 1. For instance, in Example 13-2 we know that $\lim_{z \to x} -1/xz = -1/x^2$ because rational functions are continuous. Since the limit of the denominator of a difference quotient is zero, we cannot find the limit of the quotient by dividing the limit of the numerator by the limit of the denominator. We must first "manipulate" the quotient into a form to which the limit rules apply.

Example 13-4.   Find $f'(x)$ if $f(x) = \sqrt{x}$.

Solution.   Our difference quotient is

$$\frac{f(z) - f(x)}{z - x} = \frac{\sqrt{z} - \sqrt{x}}{z - x}.$$

Experience shows that the proper approach now is to "rationalize the numerator":

$$\frac{f(z) - f(x)}{z - x} = \frac{\sqrt{z} - \sqrt{x}}{z - x} \cdot \frac{\sqrt{z} + \sqrt{x}}{\sqrt{z} + \sqrt{x}} = \frac{1}{\sqrt{z} + \sqrt{x}}.$$

It is easy to show that the square root function is continuous, and so we have

$$\lim_{z \to x} \frac{1}{\sqrt{z} + \sqrt{x}} = \frac{1}{\sqrt{x} + \sqrt{x}} = \frac{1}{2\sqrt{x}} = f'(x).$$

We have been finding *specific* differentiation rules, formulas that apply to particular functions. We can also use Definition 13-1 to derive *general* differentiation rules, theorems that apply to any differentiable function.

Theorem 13-1. *Suppose that f is differentiable at x and that g(x) = mf(x), where m is a given number. Then g is differentiable at x, and*

(13-3) $$g'(x) = mf'(x).$$

PROOF

$$g'(x) = \lim_{h \to 0} \frac{mf(x + h) - mf(x)}{h} \qquad \text{(Definition 13-1)}$$

$$= \lim_{h \to 0} m \cdot \frac{f(x + h) - f(x)}{h} \qquad \text{(Algebra)}$$

$$= m \lim_{h \to 0} \frac{f(x + h) - f(x)}{h} \qquad \text{(Theorem 11-2)}$$

$$= mf'(x) \qquad \text{(Definition 13-1).}$$

For example, we have already shown that the derivative of x^{-1} is $-x^{-2}$. Theorem 13-1 now tells us that the derivative of $10x^{-1}$ is $-10x^{-2}$, the derivative of $1492x^{-1}$ is $-1492x^{-2}$, and so on.

A differentiable function is quite "well-behaved." In particular, we will now show that a differentiable function is continuous. Notice that this theorem does *not* say that a continuous function must be differentiable.

Theorem 13-2. *If f is differentiable at x, then f is continuous at x.*

PROOF

According to Definition 9-1, f is continuous at x, provided that $\lim\limits_{z \to x} f(z) = f(x)$; that is, $\lim\limits_{z \to x} [f(z) - f(x)] = 0$. If $z \neq x$, we have

$$f(z) - f(x) = \frac{f(z) - f(x)}{z - x} (z - x).$$

Since $\lim\limits_{z \to x} \dfrac{f(z) - f(x)}{z - x} = f'(x)$ by hypothesis, and $\lim\limits_{z \to x} (z - x) = 0$, we may use our Product Rule for limits (Theorem 11-2) to see that $\lim\limits_{z \to x} [f(z) - f(x)] = f'(x) \cdot 0 = 0$, and our theorem is proved.

If we are dealing with a function defined on an interval that contains an endpoint, we may wish to speak of its continuity or differentiability at the endpoint. Then we must consider the limit that appears in the definition of continuity or differentiability as a limit from the right or from the left. For example, if the domain of f is a closed finite interval $[a, b]$, then we say that f is continuous at a if $\lim\limits_{x \downarrow a} f(x) = f(a)$ and f is differentiable at b if $\lim\limits_{z \uparrow b} \dfrac{f(z) - f(b)}{z - b}$ exists. Here f is *continuous from the right* at a and *differentiable from the left* at b. We may also use the modifying phrases "from the left" or "from the right" at points other than endpoints. For instance, the function discussed in Example 9-2 is continuous and differentiable from the left at $\frac{1}{2}\pi$, but it is not continuous from the right at that point.

PROBLEMS   13

1. In each of the following examples, set up and simplify the difference quotient $[f(z) - f(x)]/(z - x)$. Then find the limit as z approaches x, thus obtaining $f'(x)$. Finally, find the equation of the tangent line to the graph of f at the point $(2, f(2))$.

    (a) $f(x) = 2x - 3$ (b) $f(x) = 3 - 2x$
    (c) $f(x) = x^2 - 4$ (d) $f(x) = 4 - x^2$
    (e) $f(x) = x^3$ (f) $f(x) = x^3 - 12x$

2. In each of the following examples, set up and simplify (for h near 0) the difference quotient $[f(2 + h) - f(2)]/h$. Then calculate the limit as h approaches 0, thus obtaining $f'(2)$.

    (a) $f(x) = 3x^2$ (b) $f(x) = 2/x$
    (c) $f(x) = x^3$ (d) $f(x) = x^4$
    (e) $f(x) = |x + 1|$ (f) $f(x) = |x + 1|^{-1}$

3. If $f(x) = mx + b$, find $f'(x)$ by using Definition 13-1. What is the derived function of a constant function?

4.   Find y'.

(a) $y = 2\sqrt{x}$

(b) $y = \sqrt{2x}$

(c) $y = \sqrt{x + 2}$

(d) $y = \sqrt{x} + \sqrt{2}$

5.   Set up the difference quotient for each of the following functions and note that its limit is not obvious.

(a) $f(x) = \sin x$

(b) $f(x) = \sec x$

(c) $f(x) = \log x$

(d) $f(x) = 2^x$

6.   Let $f(x) = \sin x$ and set up the difference quotient for f at 0. In Section 9 we used a graphical argument to find the limit of this quotient; look up our result and hence find $f'(0)$. Does this number seem to be the slope of the sine curve at the origin?

7.   Set up and simplify the difference quotient $[f(z) - f(2)]/(z - 2)$ for z near 2. Then find the limit as z approaches 2, thus obtaining $f'(2)$. What is the equation of the tangent line to the graph of f at the point $(2, f(2))$?

(a) $f(x) = x|x - 1|$

(b) $f(x) = x + |x - 1|$

(c) $f(x) = (x - 2)|x - 2|$

(d) $f(x) = (x - 2) \sin x$

(e) $f(x) = x[\![x - .1]\!]$

(f) $f(x) = |x - 2|^3 + 4$

8.   If $f(x) = [\![x]\!]x^3$, find and simplify the difference quotient $[f(h) - f(0)]/h$ when h is small and positive. What is $\lim_{h \downarrow 0} [f(h) - f(0)]/h$? Find and simplify the difference quotient when h is small and negative. What is $\lim_{h \uparrow 0} [f(h) - f(0)]/h$? What is $f'(0)$?

9.   Set up the difference quotient $[f(h) - f(0)]/h$ for $f(x) = |x|$. What do you conclude about the existence of $f'(0)$? Could you have predicted this result from the graph of f?

10.   If $g(x) = f(x) + x^2$, how are $g'(x)$ and $f'(x)$ related?

11.   A 5-place table of common logarithms tells us that $\log 1.001 = .00043$. Use this fact to answer the following questions.

(a) If $f(x) = \log x$, find approximations to $f'(1)$, $f'(10)$, $f'(100)$, and $f'(\frac{1}{10})$.

(b) If $g(x) = 10^x$, find an approximation to $g'(0)$.

(c) Use the result of part (b) to find an approximation to $g'(x)$.

12.   If $f(x) = x^2 \sin (1/x)$ when $x \neq 0$ and $f(0) = 0$, show that $f'(0) = 0$.

13.   Suppose that f and g are differentiable at a number c and that $f(c) = g(c)$. Furthermore, suppose that $f(x) \leq g(x)$ if $x > c$ and $f(x) \geq g(x)$ if $x < c$. [For example, we might have $c = 0$, $f(x) = \sin x$, and $g(x) = \tan x$.] What is the relation between $f'(c)$ and $g'(c)$?

14.   What is the converse of Theorem 13-2? Is it true?

14.   RATE OF CHANGE AND THE SUM RULE

We interpret the slope m of the line $y = mx + b$ as the rate at which the line rises. Since the tangent line to a point of the graph of a function f is the line that best "fits" the graph at the point, it is natural to say that the slope of the tangent line measures the rate at which the graph rises. In other words, we use the derivative $f'(x)$ as a measure of the rate at which the values of f change. In this connection, it is customary to use a new notation for derivatives. If $y = f(x)$, then in addition to the notations y' and $f'(x)$ for the derivative of $f(x)$, we also write $\dfrac{dy}{dx}$. We call $\dfrac{dy}{dx}$ *the derivative of y with respect to x.* Thus

$$\frac{dy}{dx} = f'(x) = y'.$$

The derivative symbol $\dfrac{dy}{dx}$ looks like the quotient of a number dy by a number dx. Mathematicians do give meanings to the separate symbols dy and dx; in fact, they use them in several ways. But we are not going to separate dy and dx until Chapter 7. So think of $\dfrac{dy}{dx}$ as an entity. It is *a number*, the slope of the curve $y = f(x)$ at the point (x, y). We will find, however, that writing this number in quotient form helps us remember some of the basic theorems about derivatives that we will develop shortly.

The notation $\dfrac{dy}{dx}$ is not as arbitrary as we make it appear. It stems naturally from a different (and slightly more complicated) way of writing the difference quotient than the ones we use. This fact is not important to us. All we have to know is that *if* $y = f(x)$, *then (by definition)*

$$\frac{dy}{dx} = \lim_{h \to 0} \frac{f(x + h) - f(x)}{h} = \lim_{z \to x} \frac{f(z) - f(x)}{z - x}.$$

To see how the number $\dfrac{dy}{dx}$ measures rate of change, consider the concept of velocity. Suppose a car leaves New York for Boston at noon and t hours later is s miles north of the City. If $s = 100$ when $t = 2$, the car has *averaged* 50 miles per hour, but certainly the speedometer has not registered 50 at all times during that part of the trip. What the speedometer measures at a given instant is a *limit* of average velocities.

The relation between s and t defines a function f so that $s = f(t)$. The statement that $s = 100$ when $t = 2$ can therefore be written as $f(2) = 100$. Suppose that $f(2.25) = 115$; that is, the car is 115 miles north of New York $2\frac{1}{4}$ hours after noon. Then the average velocity during the first 15 minutes after 2 o'clock is $15/.25 = 60$. The numerator of this fraction represents the number of miles the car has gone and the denominator is the time of travel. We can therefore write this quotient as the difference quotient

$$\frac{f(2.25) - f(2)}{2.25 - 2}.$$

More generally, if t is a given number and z is another number, then the difference quotient

(14-1)
$$\frac{f(z) - f(t)}{z - t}$$

is the average velocity during the time interval between t and z. We cannot set $z = t$, for then the average velocity is not defined. But as in the problem of finding the slope of a curve, it is perfectly clear that the velocity *at time t* should be the number that is approximated by Difference Quotient 14-1 when z is near t. Therefore we *define* the **velocity** at t to be the limit of the difference quotient:

$$v = \frac{ds}{dt} = f'(t).$$

In order to distinguish this velocity from the average velocity, we call $\frac{ds}{dt}$ the **instantaneous velocity**.

Example 14-1. We noticed in the introduction to this chapter that if a body falls from rest in a vacuum s feet in t seconds, then s and t are related by the equation $s = 16t^2$. How fast is the body falling at the end of 5 seconds? Express the velocity of the body in terms of its displacement from the starting point.

Solution. We first have to calculate $f'(5)$, where $f(t) = 16t^2$. You did many examples of this sort in the last section, and you may easily verify that here $v = f'(t) = 32t$. Therefore $f'(5) = 32 \cdot 5 = 160$, and so the body is falling at the rate of 160 feet per second at the end of 5 seconds.

Since $s = 16t^2$, we see that $t = \frac{1}{4}\sqrt{s}$, and so the equation $v = 32t$ can be written as $v = 8\sqrt{s}$. For example, when the body is 100 feet below its starting point it is falling at the rate of $8\sqrt{100} = 80$ feet per second.

In Example 14-1 we measured distance in feet, and time in seconds; hence the velocity is measured in feet per second. If distance had been measured in kilometers and time in hours, the velocity would be measured in kilometers per hour and so on.

We use the concept of rate even though distance and time are not involved. For example, if y ergs of work are required to move a certain object x centimeters, then $\frac{dy}{dx}$ is the number of ergs per centimeter (or dynes) of force applied at a given point during the motion. Similarly, if a house can be heated to a temperature of T degrees by burning p pounds of coal, then the number $\frac{dT}{dp}$ measures the rate of change of temperature in degrees per pound.

An equation that involves derivatives is called a **differential equation**. For instance, in Example 14-1 we found that $s = 16t^2$ *satisfies* the differential equations $\dfrac{ds}{dt} = 32t$ and $\dfrac{ds}{dt} = 8\sqrt{s}$. Because a derivative is a rate, a differential equation is an equation that involves rates, and such equations play important roles in science. We shall mention differential equations from time to time, but a thorough study of them would require much more space than is available here.

Example 14-2. After t years from some initial date, the population of a certain country is P individuals, and it is increasing at the rate of 2% per year. Find a differential equation that is satisfied by P.

Solution. Our hypotheses tell us that the rate of change of the population at any particular time is $\frac{2}{100}$ of the population at that time, and so we have the equation $\dfrac{dP}{dt} = .02P$. Now the question arises, "Can we find a P that satisfies this equation, and, if so, how?" We will have to defer the answer until Section 46 when we will have had quite a bit more calculus.

It is convenient to think of $\dfrac{d}{dx}$ as a **differential operator** that we apply to values of differentiable functions. As we proceed, we will see how operations with $\dfrac{d}{dx}$ are related to other operations of mathematics, such as addition and multiplication. Thus Theorem 13-1 tells us that

$$(14\text{-}2) \qquad \frac{d}{dx}\, mf(x) = m\frac{d}{dx}\, f(x);$$

that is, we can multiply by a number m and then differentiate, or differentiate and then multiply by m, and we end up with the same value. Similarly, the differential operator $\dfrac{d}{dx}$ is *distributive* with respect to addition.

Theorem 14-1. *If f and g are differentiable at x, then*

$$(14\text{-}3) \qquad \frac{d}{dx}\big[f(x) + g(x)\big] = \frac{d}{dx}f(x) + \frac{d}{dx}g(x).$$

It is easy to extend this rule to the sum of any number of terms.

PROOF

We go back to the definition of the derivative and use our limit theorems:

$$\frac{d}{dx}[f(x) + g(x)] = \lim_{z \to x} \frac{[f(z) + g(z)] - [f(x) + g(x)]}{z - x} \quad \text{(Definition 13-1)}$$

$$= \lim_{z \to x} \left[\frac{f(z) - f(x)}{z - x} + \frac{g(z) - g(x)}{z - x} \right] \quad \text{(Algebra)}$$

$$= \lim_{z \to x} \frac{f(z) - f(x)}{z - x} + \lim_{z \to x} \frac{g(z) - g(x)}{z - x} \quad \text{(Theorem 11-2)}$$

$$= \frac{d}{dx} f(x) + \frac{d}{dx} g(x) \quad \text{(Definition 13-1).}$$

Equations 14-2 and 14-3 can be combined: *if f and g are differentiable functions and m and n are numbers, then*

$$(14\text{-}4) \qquad \frac{d}{dx}[mf(x) + ng(x)] = m\frac{d}{dx} f(x) + n\frac{d}{dx} g(x).$$

This equation says that $\dfrac{d}{dx}$ is a **linear operator**.

Example 14-3. Find $\dfrac{d}{dx}(3/x + 8\sqrt{x})$.

Solution. $\dfrac{d}{dx}(3/x + 8\sqrt{x}) = 3\dfrac{d}{dx}(1/x) + 8\dfrac{d}{dx}\sqrt{x} \quad$ (Equation 14-4)

$$= -3/x^2 + 4/\sqrt{x} \quad \text{(Examples 13-2 and 13-4).}$$

PROBLEMS 14

1. In the last few sections we have developed the differentiation formulas

$$\frac{d}{dx}(x^2 + 1) = 2x, \qquad \frac{d}{dx} x^{-1} = -x^{-2}, \qquad \frac{d}{dx}\sqrt{x} = \frac{1}{2\sqrt{x}},$$

and we know that $\dfrac{d}{dx}(mx + b) = m$ (this equation simply says that m is the slope of the line $y = mx + b$). Use these formulas and Equation 14-4 to find the following derivatives.

(a) $\dfrac{d}{dx} x^2$

(b) $\dfrac{d}{dx} (x^2 + 2)$

(c) $\dfrac{d}{dx} (5x - 2 + 3/x)$

(d) $\dfrac{d}{dx} (3x^2 - 4x + 5 + 6\sqrt{x})$

(e) $\dfrac{d}{dx} (x + 2)^2$

(f) $\dfrac{d}{dx} (\sqrt{x} + 3)^2$

(g) $\dfrac{d}{dx} (\sqrt{x} + 1/\sqrt{x})^2$

(h) $\dfrac{d}{dx} \left(\dfrac{x}{x + 1}\right)^{-1}$

2. Compare (see the preceding problem).

(a) $\dfrac{d}{dx} x^2$ and $\left(\dfrac{dx}{dx}\right)^2$

(b) $\dfrac{d}{dx} (1/x)$ and $1 \Big/ \dfrac{dx}{dx}$

(c) $\dfrac{d}{dx} \sqrt{x}$ and $\sqrt{\dfrac{dx}{dx}}$

(d) $\dfrac{d}{dx} (mx + b)$ and $m \dfrac{dx}{dx} + b$

3. Find the following derivatives.

(a) $\dfrac{d}{du} (3u - 7)$

(b) $\dfrac{d}{dv} 2(4v + 5)$

(c) $\dfrac{d}{dr} (3r)^2$

(d) $\dfrac{d}{ds} (3 + s)^2$

(e) $\dfrac{d}{dt} \sqrt{16t}$

(f) $\dfrac{d}{dz} (3z)^{-1}$

4. The motion of a particle along a line is described by the given equation. Find $v = \dfrac{ds}{dt}$ when $t = 1$.
(a) $s = t^2 - t$ (b) $s = (2 - t)^2$
(c) $s = t^3 - 3t^2$ (d) $s = (1 - t)^2$

5. A block slides down an inclined plane. At the end of t seconds, the block is s feet from its starting point. Find the average velocity of the block for the first 2 seconds and its instantaneous velocity at $t = 2$.
(a) $s = 3 + 2t$ (b) $s = 3t^2$
(c) $s = t^2 + t$ (d) $s = t^2 + \sqrt{t}$

6. How are $\dfrac{dy}{dx}$ and $\dfrac{dx}{dy}$ related if x and y are related by the given equation?

(a) $x + y = 0$ (b) $3x + 4y + 5 = 0$
(c) $xy = 1$ (d) $y = x^2$

7. A penny expands when heated. Find the expression for the rate of change of the area of one side with respect to the radius.

8. At t minutes after noon, the radius of a melting snowball is $60 - 3t$ inches. How fast is the volume changing at 12:10?

9. A machine rolls a wheel along a number scale marked off in centimeters in such

a way that the coordinate of the center of the wheel t seconds after the start of the motion is given by the equation $x = t + 1/t$.

(a) Find its velocity when $t = \frac{1}{2}$, $t = 1$, and $t = 2$.

(b) Which way is it moving at those times?

(c) If the wheel should break off the machine at one of those times and roll friction-lessly along the scale, where would it be 2 seconds later?

10.   Would you expect $\dfrac{dy}{dx}$ to be positive or negative if

(a) y is the number of miles traveled and x is the number of gallons of gas used?

(b) y is the number of miles per gallon of gas that the car gets at a speed of x miles per hour?

(c) y is the number of gallons of gas in the tank and x is the number of miles traveled?

(d) y is the number of miles traveled and x is the number of gallons of gas in the tank?

11.   Boyle's Law says that the volume V of a given quantity of gas is inversely propor-tional to its pressure P. Show that V satisfies the differential equation $\dfrac{dV}{dP} = -V/P$.

(*Hint:* By hypothesis, $V = k/P$, where k is a number that is independent of P. Now find $\dfrac{dV}{dP}$ and eliminate k.)

12.   A mass hanging from a spring bobs up and down in such a way that it is s inches from the equilibrium position t seconds after some initial moment, where $s = \sin t$.

(a) Use the table in Fig. 12-4 to estimate the velocity of the mass when $t = 0$, $t = .5$, and $t = 1$.

(b) When does the mass stop?

(c) Is it moving toward or away from the equilibrium position when $t = 2$?

(d) When is the velocity greatest?

13.   Show that $y = \sqrt{x}$ satisfies each of the differential equations (i) $y' = 1/2\sqrt{x}$, (ii) $y' = 1/2y$, and (iii) $y' = y/2x$. Show that $y = \sqrt{x} + 2$ satisfies (i), but not (ii) and (iii); $y = \sqrt{x} + 2$ satisfies (ii), but not (i) and (iii); and $y = 2\sqrt{x}$ satisfies (iii), but not (i) and (ii).

14.   Use the sine curve to graph as best you can the equation $y = \left[\!\left[\dfrac{d}{dx} \sin x \right]\!\right]$. What does the curve $y = \dfrac{d}{dx} \left[\!\left[\sin x \right]\!\right]$ look like?

15.   THE POWER FORMULA

The routine of setting up a difference quotient and then finding its limit each time we wish to calculate a derivative soon becomes tiresome, so it is worthwhile to establish (and memorize) some differentiation rules. These rules are of two types—specific and general. The specific rules are formulas for derivatives of certain basic functions. We use the general rules to find derivatives for functions that are built up out of basic functions. We have already established one of the basic general differentiation rules, the fact that $\dfrac{d}{dx}$ is a linear operator (Equation 14-4), and now we will begin to prove one of the most important specific differentiation formulas, the **Power Formula**.

Theorem 15-1. *If r is any real number, then*

(15-1) $$\frac{d}{dx} x^r = rx^{r-1}.$$

This theorem is easy to state and to use, but its proof is complicated because the definition of x^r depends on the "kind" of number r is. If r is a positive integer, we define x^r in one way, whereas if r is a fraction, we define it differently, and so on. For instance, you know that $x^3 = x \cdot x \cdot x$, $x^{1/3} = \sqrt[3]{x}$, $x^{-3} = 1/x^3$, $x^0 = 1$, and $x^{4/3} = \sqrt[3]{x^4} = (\sqrt[3]{x})^4$. When we prove Theorem 15-1, we must consider each of these possibilities. Furthermore, the theorem is also true when r is an irrational number, such as π. A definition of x^π is quite complicated, and we won't get around to it until Section 48. So our proof of Theorem 15-1 will be given in steps as we treat the different possibilities for r at different points in the book. At the end of this section we tell how to prove the theorem in case r is a positive integer and in case r is the reciprocal of a positive integer. Now let us give a few examples to show how we use it in those two cases.

Example 15-1. Find the equation of the tangent line to the graph of f at the point $(-1, f(-1))$ if $f(x) = x^4$.

Solution. The slope of the desired tangent is $f'(-1)$. Rule 15-1 tells us that

$$f'(x) = \frac{d}{dx} x^4 = 4x^3,$$

and so $$f'(-1) = -4.$$

Since our tangent line contains the point $(-1, f(-1)) = (-1, (-1)^4) = (-1, 1)$ and has slope -4, its equation is $y - 1 = -4(x + 1)$.

Example 15-2. Show that the tangent to the curve $y = x^5$ at the point $(2, 32)$ is parallel to the tangent to the curve $y = x^{80}$ at the point $(1, 1)$.

Solution. If $y = x^5$, then $\dfrac{dy}{dx} = 5x^4$, and if $y = x^{80}$, then $\dfrac{dy}{dx} = 80x^{79}$. Thus the slope of the tangent to the curve $y = x^5$ at a point (x, y) is $5x^4$, and so at the point (2, 32) the slope is $5 \cdot 2^4 = 80$. Similarly, the slope of the tangent to the curve $y = x^{80}$ at the point (1, 1) is $80 \cdot 1^{79} = 80$. Since the two tangents have the same slope, 80, they are parallel (Theorem 8-2).

The line that contains a point P of a given curve and is perpendicular to the tangent line at P is called the **normal line** (or **normal**) to the curve at P.

Example 15-3. Find the equation of the normal line to the curve $y = \sqrt{x}$ at (9, 3).

Solution. Since $\sqrt{x} = x^{1/2}$, we have $y = x^{1/2}$, so we may use Equation 15-1 to calculate the slope of the tangent:

$$\frac{dy}{dx} = \frac{d}{dx}\, x^{1/2} = \tfrac{1}{2}x^{1/2-1} = \tfrac{1}{2}x^{-1/2} = \frac{1}{2\sqrt{x}}.$$

When $x = 9$, the slope of the tangent is $\tfrac{1}{6}$. Thus the slope of the normal at the point (9, 3) is -6, and so its equation is $y - 3 = -6(x - 9)$.

Although we stated Rule 15-1 in terms of the letters x and r, the formula applies if other letters are used. For example,

$$\frac{d}{dy}\, y^s = \frac{dy^s}{dy} = sy^{s-1},$$

$$\frac{d}{du}\, u^7 = \frac{du^7}{du} = 7u^6,$$

$$\frac{d}{dt}\, \sqrt[5]{t} = \frac{dt^{1/5}}{dt} = \tfrac{1}{5}t^{-4/5} = 1/5\sqrt[5]{t^4}.$$

Example 15-4. Suppose that a body is s feet north of its starting point t seconds after it starts to move, and $s = t^2$. When does its velocity reach 10 feet per second?

Solution. We use Rule 15-1 to find $v = \dfrac{ds}{dt} = \dfrac{dt^2}{dt} = 2t$. Now we must find the number t such that $2t = 10$; clearly, $t = 5$. So 5 seconds after it starts, the body is traveling 10 feet per second.

In this section we will only discuss the proof of Theorem 15-1 in case r is a positive integer n or $r = 1/n$. In fact, let us be specific and suppose $n = 4$. Exactly the same argument works for $n = 44, n = 1492$, and so on. Our first step in finding

$\dfrac{dx^4}{dx}$ is to set up the difference quotient $\dfrac{f(z) - f(x)}{z - x} = \dfrac{z^4 - x^4}{z - x}$. Long division shows that

$$(15\text{-}2) \qquad \frac{z^4 - x^4}{z - x} = z^3 + z^2x + zx^2 + x^3,$$

so $\lim\limits_{z \to x} \dfrac{z^4 - x^4}{z - x} = x^3 + x^3 + x^3 + x^3 = 4x^3$. This equation tells us that $\dfrac{dx^4}{dx} = 4x^3$, which is Equation 15-1 with r replaced by 4.

Suppose now that $r = \frac{1}{4}$. Equation 15-2 is valid for any two unequal numbers z and x, so let us replace z with $z^{1/4}$ and x with $x^{1/4}$ and invert both sides to obtain the equation

$$\frac{z^{1/4} - x^{1/4}}{z - x} = \frac{1}{z^{3/4} + z^{2/4}x^{1/4} + z^{1/4}x^{2/4} + x^{3/4}}.$$

Hence $\lim\limits_{z \to x} \dfrac{z^{1/4} - x^{1/4}}{z - x} = \dfrac{1}{4x^{3/4}} = \frac{1}{4}x^{-3/4}$. This equation says that $\dfrac{dx^{1/4}}{dx} = \frac{1}{4}x^{1/4-1}$, which is Equation 15-1 with r replaced by $\frac{1}{4}$.

We will now feel free to use Equation 15-1 in case n is a positive integer or the reciprocal of a positive integer, but we will refrain from using it in other cases until we have verified it.

PROBLEMS 15

1. Find y'.

(a) $y = x^7 + x^3$ (b) $y = \sqrt[4]{x}$ (c) $y = 3\sqrt[5]{t}$

(d) $y = 2x^6 - x^3$ (e) $y = (u^3)^2$ (f) $y = v^3v^2$

(g) $y = x^3 + \sqrt[3]{x}$ (h) $y = (x^{1/5})^{10}$ (i) $y = (3x - 1)^2$

(j) $y = (x^{1/10})^5$ (k) $y = \sqrt{\sqrt{\sqrt{x}}}$ (l) $y = \left(\dfrac{|x| + x}{2}\right)^6$

2. Find the equation of the tangent and of the normal at the point $(1, 1)$.

(a) $y = \sqrt[3]{x}$ (b) $y = x^3$ (c) $y = x^n$ (d) $y = x^{1/n}$

3. A line parallel to the Y-axis intersects the given curves in points with the same X-coordinate. What is this X-coordinate if tangent lines drawn at the points are parallel?

(a) $y = x^2$ and $y = x^3$ (b) $y = \sqrt{x}$ and $y = \sqrt[3]{x}$

(c) $y = \sqrt{x}$ and $y = x^2$ (d) $y = x^n$ and $y = x^{1/n}$

4. What is the radius of the circle whose center is a point of the X-axis and that is tangent to the curve $y = x^{1/4}$ at the point $(16, 2)$?

5. A particle starts at the origin and moves along the curve $y = x^r$ in the direction of increasing x. When its X-coordinate is 8, it flies off on a tangent and hits a man standing at the point (32, 4). What is r?

6. (a) Does $\dfrac{d}{dx}(x^3x^5) = \dfrac{dx^3}{dx}\dfrac{dx^5}{dx}$?

   (b) Does $\dfrac{d}{dx}(x^7/x^2) = \dfrac{dx^7}{dx}\bigg/\dfrac{dx^2}{dx}$?

   (c) Does $\left(\dfrac{dx^3}{dx}\right)^5 = \left(\dfrac{dx^5}{dx}\right)^3$?

7. Prove Theorem 15-1 for the case $r = 0$.

8. Use Example 13-2 to show that Theorem 15-1 is true if $r = -1$.

9. Show that $y = x^r$ satisfies the differential equation $\dfrac{dy}{dx} = ry/x$.

10. Prove that if n is an odd positive integer, the lines tangent to the curve $y = x^n$ at the points (1, 1) and $(-1, -1)$ are parallel.

11. Prove that if n is an even positive integer, the line tangent to the curve $y = \sqrt[n]{x}$ at the point (1, 1) is perpendicular to the line tangent to the curve $y = x^n$ at the point $(-1, 1)$.

12. Suppose that $y = x^n$, where n is a positive integer. What is the relation between $\dfrac{dy}{dx}$ and $\dfrac{dx}{dy}$?

13. Show that the tangent to the curve $y = x^r$ at the point (1, 1) intersects the Y-axis in a point that is $r - 1$ units away from the origin.

14. Show that the Y-intercept of the tangent to the curve $y = x^r$ at the point (a, b) is $(1 - r)a^r$.

15. (a) Suppose that the tangent to the curve $y = x^2$ at the point (a, b) intersects the Y-axis in the point P and that the normal to the curve intersects the Y-axis in the point Q. Show that the coordinates of P are $(0, -b)$ and that the distance between P and Q is $2b + \frac{1}{2}$.

    (b) How far apart are P and Q if we consider the curve $y = x^r$?

16. Write out the proof of the Power Formula (Equation 15-1) in case $r = 6$ and in case $r = \frac{1}{6}$.

16. THE CHAIN RULE

We build new functions from old ones by composition, and derivatives of the resulting functions can be constructed from derivatives of the components. To see how, let us look at linear functions. Suppose that $f(x) = mx + b$ and $g(x) = nx + c$. Then the equation

$$y = f(g(x)) = m(nx + c) + b = mnx + mc + b$$

defines the composition of g by f. We observe that this composite function is also linear and that its slope is mn, the product of the slopes of the functions from which it is composed. One of the most important theorems in calculus is that this result holds in general: *to find a derivative of a composite function, multiply derivatives of its components.*

Writing this result in symbols, we suppose that f and g are given differentiable functions and let $u = g(x)$. Then the equation $y = f(u)$ defines the composition of g by f. We are saying that the derivative $\dfrac{d}{dx} f(u)$ of this composite function is simply the product of the derivatives $\dfrac{d}{du} f(u)$ and $\dfrac{du}{dx}$ of its components. The result is known as the **Chain Rule**.

Theorem   16-1.   *If f and g are differentiable functions, and $u = g(x)$, then*

(16-1)
$$\frac{d}{dx} f(u) = \frac{d}{du} f(u) \frac{du}{dx}.$$

Let us go back to our lines to illustrate this theorem. We have $u = nx + c$ and $f(u) = mu + b = mnx + mc + b$. Therefore $\dfrac{d}{dx} f(u) = mn$, $\dfrac{d}{du} f(u) = m$, and $\dfrac{du}{dx} = n$, so we see that Equation 16-1 is valid in this case. We will discuss the proof of this theorem at the end of the section, after we have seen it used in a few examples.

Example 16-1.   Use the Chain Rule to find $\dfrac{d}{dx} (x^{1/3})^6$.

Solution.   Since $(x^{1/3})^6 = x^2$, we don't need the Chain Rule to see that its derivative is $2x$. But let's go through the details to show that the Chain Rule also gives us this answer:

$$\frac{d}{dx} (x^{1/3})^6 = \frac{d}{dx} u^6 \qquad \text{(Let } u = x^{1/3})$$

$$= \frac{du^6}{du} \frac{du}{dx} \qquad \text{(Equation 16-1 with } f(u) = u^6)$$

$$= 6u^5 \frac{du}{dx} \qquad \left(\text{Power Formula to find } \frac{du^6}{du}\right)$$

$$= 6(x^{1/3})^5 \frac{dx^{1/3}}{dx} \qquad (u = x^{1/3})$$

$$= 6x^{5/3}\tfrac{1}{3}x^{-2/3} \qquad \left(\text{Power Formula to find } \frac{dx^{1/3}}{dx}\right)$$

$$= 2x \qquad \text{(Algebra)}.$$

Of course, we don't use the Chain Rule to find derivatives that we could have found more easily by other means. We use it to extend specific differentiation formulas by "building them into the Chain Rule." For example, the one specific differentiation formula we now know is the Power Formula $\dfrac{du^r}{du} = ru^{r-1}$. Therefore if we replace $f(u)$ with u^r in Equation 16-1, we obtain the equation

$$(16\text{-}2) \qquad\qquad \frac{du^r}{dx} = ru^{r-1}\frac{du}{dx}.$$

This equation does allow us to calculate derivatives that we couldn't calculate before, as we will now see.

Since we have verified the Power Formula only if r is a positive integer or the reciprocal of a positive integer, it looks as if we can apply Equation 16-2 only to these cases. But actually, we can pull ourselves up by the bootstraps and use Equation 16-2 for these special exponents to extend the validity of the Power Formula to all rational exponents.

Example 16-2.   Find $\dfrac{dx^{13/4}}{dx}$.

Solution.   Because $x^{13/4} = (x^{13})^{1/4}$ and we have already verified the Power Formula for $r = \frac{1}{4}$, we will set $u = x^{13}$ and $r = \frac{1}{4}$ in Equation 16-2:

$$\frac{dx^{13/4}}{dx} = \tfrac{1}{4}(x^{13})^{-3/4}\frac{dx^{13}}{dx}.$$

But we know that $\dfrac{dx^{13}}{dx} = 13x^{12}$, so

$$\frac{dx^{13/4}}{dx} = \tfrac{1}{4}(x^{13})^{-3/4}\cdot 13x^{12} = \tfrac{13}{4}x^{-39/4}x^{48/4} = \tfrac{13}{4}x^{9/4}.$$

This equation is simply the Power Formula $\dfrac{dx^r}{dx} = rx^{r-1}$ with r replaced by $\frac{13}{4}$.

Exactly the same argument verifies the Power Formula when r is replaced by any positive fraction p/q.

We can also use the Chain Rule to show that the Power Formula is valid for negative exponents.

Example 16-3.   Find $\dfrac{dx^{-s}}{dx}$, where s is a positive rational number.

Solution. We are to find $\dfrac{d}{dx} x^{-s} = \dfrac{d}{dx} (1/x)^s$, and we will therefore replace r with s and u with $1/x$ in Equation 16-2:

$$\frac{d}{dx} (1/x)^s = s(1/x)^{s-1} \frac{d}{dx} (1/x).$$

So now all we need to know is $\dfrac{d}{dx} (1/x)$, which we found to be $-1/x^2$ in Example 13-2. Therefore

$$\frac{dx^{-s}}{dx} = s(1/x)^{s-1}(-1/x^2) = -s(1/x)^{s+1} = -sx^{-s-1},$$

which is simply the Power Formula with r replaced by the negative rational exponent $-s$.

We have now shown how to verify the Power Formula for all rational exponents, and we will use it freely for such powers. For irrational exponents (such as π), we will have to wait until Section 48.

Example 16-4. Find $\dfrac{d}{dx} \sqrt{4x^3}$ in two ways.

Solution. We can replace u with $4x^3$ and r with $\frac{1}{2}$ in Equation 16-2:

$$\frac{d}{dx} \sqrt{4x^3} = \tfrac{1}{2}(4x^3)^{-1/2} \frac{d}{dx} 4x^3.$$

Since $\dfrac{d}{dx} 4x^3 = 4 \dfrac{dx^3}{dx} = 4 \cdot 3x^2$, we have

$$\frac{d}{dx} \sqrt{4x^3} = \tfrac{1}{2}(4x^3)^{-1/2} 12x^2 = \tfrac{1}{4}x^{-3/2} 12x^2 = 3x^{1/2} = 3\sqrt{x}.$$

On the other hand, we could first write $\sqrt{4x^3} = 2x^{3/2}$, and therefore

$$\frac{d}{dx} \sqrt{4x^3} = \frac{d}{dx} 2x^{3/2} = 2 \frac{d}{dx} x^{3/2} = 2 \cdot \tfrac{3}{2} x^{1/2} = 3\sqrt{x}.$$

As our examples illustrate, using Equation 16-2 is simply a matter of substitution. This remark applies to the many similar differentiation formulas we will shortly develop, and you should get the technique down pat right at the start. You won't, of course, explicitly write out the justification for every step each time you use a differentiation formula, but you should be able to.

Example 16-5. Find $\dfrac{d}{dx}\sqrt{2x^3 - 4x}$.

Solution. Since $\sqrt{2x^3 - 4x} = (2x^3 - 4x)^{1/2}$,

$$\frac{d}{dx}\sqrt{2x^3 - 4x} = \tfrac{1}{2}(2x^3 - 4x)^{-1/2}\frac{d}{dx}(2x^3 - 4x) \qquad \begin{array}{l}\text{(Set } r = \tfrac{1}{2} \text{ and } u = 2x^3 - \\ 4x \text{ in Equation 16-2)}\end{array}$$

$$= \tfrac{1}{2}(2x^3 - 4x)^{-1/2}\left(2\frac{dx^3}{dx} - 4\frac{dx}{dx}\right) \qquad \left(\text{Linearity of } \frac{d}{dx}\right)$$

$$= \tfrac{1}{2}(2x^3 - 4x)^{-1/2}(6x^2 - 4) \qquad \text{(Equation 15-1)}$$

$$= \frac{3x^2 - 2}{\sqrt{2x^3 - 4x}} \qquad \text{(Algebra)}.$$

Example 16-6. Show that

$$(16\text{-}3) \qquad \frac{d}{dx}[f(x) + g(x)]^2 = 2[f(x) + g(x)]\left[\frac{d}{dx}f(x) + \frac{d}{dx}g(x)\right].$$

Solution. Set $r = 2$ and $u = f(x) + g(x)$ in Equation 16-2 and use the Sum Rule:

$$\frac{d}{dx}[f(x) + g(x)]^2 = 2[f(x) + g(x)]\frac{d}{dx}[f(x) + g(x)]$$

$$= 2[f(x) + g(x)]\left[\frac{d}{dx}f(x) + \frac{d}{dx}g(x)\right].$$

Example 16-7. When heated, a dime expands radially at the rate of .0002 centimeters per minute. How fast is its area increasing when its radius is 1 centimeter?

Solution. The area of a circular disk is given by the formula $A = \pi r^2$, and hence

$$\frac{dA}{dt} = \frac{d}{dt}\pi r^2 = \pi\frac{dr^2}{dt} \qquad \left(\text{Linearity of } \frac{d}{dt}\right)$$

$$= 2\pi r\frac{dr}{dt} \qquad \text{(Equation 16-2)}.$$

We want the value of $\dfrac{dA}{dt}$ when $r = 1$ and $\dfrac{dr}{dt} = .0002$, so

$$\frac{dA}{dt} = 2\pi \times 1 \times .0002 = .0004\pi \text{ square centimeters per minute.}$$

We can write the Chain Rule Equation in many different ways. For example, if we replace $f(u)$ in Equation 16-1 with y, we obtain an especially attractive form: *if $y = f(u)$, where $u = g(x)$, then*

$$(16\text{-}4) \qquad\qquad \frac{dy}{dx} = \frac{dy}{du}\frac{du}{dx}.$$

This formula is a good reason for writing derivatives as quotients. Anyone who can do the simplest arithmetic with fractions will agree that it is true; merely cancel the du's. The only drawback to that argument is that, as far as we are concerned, the derivatives $\dfrac{dy}{du}$ and $\dfrac{du}{dx}$ are "inseparable;" dy, du, and dx alone don't mean anything. So some other way must be found to prove the Chain Rule.

Since $\dfrac{d}{du}f(u) = f'(u)$ and $u = g(x)$, we can also write Equation 16-1 as

$$(16\text{-}5) \qquad\qquad \frac{d}{dx}f(g(x)) = f'(g(x))g'(x).$$

It is this form that is usually used in proofs of the Chain Rule. The essential idea of such a proof is the observation that difference quotients approximate derivatives:

$$\frac{f(w) - f(u)}{w - u} \approx f'(u) \qquad \text{and} \qquad \frac{g(z) - g(x)}{z - x} \approx g'(x).$$

Let us write these approximations as

$$f(w) - f(u) \approx f'(u)(w - u) \qquad \text{and} \qquad g(z) - g(x) \approx g'(x)(z - x).$$

Now replace w and u in the first approximation with $g(z)$ and $g(x)$ and use the second approximation:

$$f(g(z)) - f(g(x)) \approx f'(g(x))(g(z) - g(x)) \approx f'(g(x))g'(x)(z - x).$$

This final approximation says that

$$\frac{f(g(z)) - f(g(x))}{z - x} \approx f'(g(x))g'(x),$$

which suggests that $\displaystyle\lim_{z \to x}\frac{f(g(z)) - f(g(x))}{z - x} = f'(g(x))g'(x)$. This last equation is Equation 16-5.

Of course, a rigorous proof requires us to replace these "approximations" with limits, using theorems from Chapter 1 and earlier sections of this chapter. Carrying out the details of such a proof sheds more heat than light on the problem, however.

PROBLEMS 16

1. Find $\dfrac{dy}{dx}$.

    (a) $y = 4x^{7/2}$ (b) $y = 1/4x^{7/2}$ (c) $y = (x^{7/2})^4$

    (d) $y = (4/x)^{7/2}$ (e) $y = (x + 4)^{7/2}$ (f) $y = (4x + 7)^{1/2}$

2. Find the tangent line to the graph of f at the point $(1, f(1))$.

    (a) $f(x) = -9x^{1/3}$ (b) $f(x) = 9x^{-1/3}$ (c) $f(x) = \sqrt{16x}$

    (d) $f(x) = \sqrt{16/x}$ (e) $f(x) = \sqrt{x + 8}$ (f) $f(x) = 1/(x + 8)$

3. Find v.

    (a) $s = 16t^2$ (b) $s = 16\sqrt{t}$ (c) $s = (16t)^2$

    (d) $s = \sqrt{16t}$ (e) $s = 16^2$ (f) $s = (-8t)^{2/3}$

4. Find the following derivatives.

    (a) $\dfrac{d}{dx}\sqrt{2x + 9}$ (b) $\dfrac{d}{dx}(2\sqrt{x} + 9)$ (c) $\dfrac{d}{dx}(27x + 8)^{-2/3}$

    (d) $\dfrac{d}{dx}(27x^{-2/3} + 8)$ (e) $\dfrac{d}{dx}\sqrt{3x^2 - 2}$ (f) $\dfrac{d}{dx}(1/(3x^2 - 2)^2)$

    (g) $\dfrac{d}{dx}(3x^2 + 2)^{1/2}$ (h) $\dfrac{d}{dx}\sqrt{2x^4 + x^2 - 1}$

5. Find $f'(1)$.

    (a) $f(x) = \sqrt{2x + 7}$ (b) $f(x) = \sqrt{1/(2x + 7)}$

    (c) $f(x) = (7 - 3x)^{3/2}$ (d) $f(x) = 1/(7 - 3x)^{3/2}$

    (e) $f(x) = [\![3 \sin x]\!]x^{-2/3}$ (f) $f(x) = x^{[\![3 \sin x]\!]}$

6. (a) Use the equation $|x| = \sqrt{x^2}$ and the Chain Rule to obtain the differentiation formula $\dfrac{d}{dx}|x| = x/|x|$ for $x \neq 0$ (see Example 12-2).

    (b) Incorporate this differentiation formula into the Chain Rule to obtain the equation

$$\frac{d}{dx}|u| = \frac{u}{|u|}\frac{du}{dx} \qquad (u \neq 0).$$

    (c) Suppose that f is differentiable on a certain interval. At what points of the interval could $\dfrac{d}{dx}|f(x)|$ fail to exist?

    (d) Explain why $\left|\dfrac{d}{dx}|f(x)|\right| = \left|\dfrac{d}{dx}f(x)\right|$ when all derivatives involved exist.

7. In Example 16-2, set $r = 13$ and $u = x^{1/4}$ to get the same result we obtained with a different substitution.

8. After t seconds, the radius of an expanding soap bubble is r inches. Express the time rate of change of the volume of the bubble in terms of $\dfrac{dr}{dt}$, the time rate of change of the radius.

9. A pebble dropped in a lake creates an expanding ripple whose radius is $g(t)$ feet t seconds later. Find a formula for the rate at which the area of the region enclosed by the ripple is growing.

10. Which of the following are correct statements of the Chain Rule Equation?

(a) $\dfrac{dx}{dy} = \dfrac{dx}{du}\dfrac{du}{dy}$ (b) $\dfrac{dy}{du} = \dfrac{dy}{dx}\dfrac{du}{dx}$ (c) $\dfrac{du}{dx} = \dfrac{dv}{dx}\dfrac{du}{dv}$

11. A point P moves along the curve $y = x^2$ so that the time rate of change of the X-coordinate of P is v. What is the time rate of change of the Y-coordinate?

12. (a) If $f(x) = x^r$ and $g(x) = x^s$, show that

$$\frac{d}{dx}\left[f(x)g(x)\right] = f(x)\frac{d}{dx}g(x) + g(x)\frac{d}{dx}f(x).$$

(b) Prove that the equation

$$\frac{d}{dx}\left[f(x)g(x)\right] = \frac{d}{dx}f(x)\frac{d}{dx}g(x)$$

is *not* one of the theorems of calculus.

13. Use Equation 16-2 and the sine curve to solve the equation $\dfrac{d}{dx}\sin^{13} x = 0$.

14. A function s is *even* if $s(-x) = s(x)$ and a function q is *odd* if $q(-x) = -q(x)$. Show that the derived function of an even function is odd and the derived function of an odd function is even.

15. A function f is **periodic** if there is a number p such that for each x, $f(x + p) = f(x)$. For example, the sine function is periodic, since $\sin(x + 2\pi) = \sin x$. Show that the derived function of a differentiable periodic function is periodic.

16. Suppose that $p(x) = f(g(x))$, where f is an arbitrary differentiable function and g is as given. Show that $p'(0) = 0$. (Use graphs.)

(a) $g(x) = \sqrt{4 - x^2}$ (b) $g(x) = \sec x$
(c) g is any even differentiable function.

17. What is the difference between $\dfrac{d}{dx}f(3x + 4)$ and $f'(3x + 4)$? Between $\dfrac{d}{dx}f(x + 4)$ and $f'(x + 4)$?

17. DERIVATIVES OF PRODUCTS AND QUOTIENTS

The rule for differentiating products is more complicated than the rule for differentiating sums. *It is not true, in general, that the derivative of a product is the product of the derivatives of its factors.* Instead, we have the following **Product Rule**.

Theorem 17-1. *If f and g are differentiable at x, then*

$$(17\text{-}1) \qquad \frac{d}{dx}\left[f(x)g(x)\right] = f(x)\frac{d}{dx}g(x) + g(x)\frac{d}{dx}f(x).$$

PROOF

In order to apply previously derived results, we write

$$f(x)g(x) = \tfrac{1}{2}[f(x) + g(x)]^2 - \tfrac{1}{2}f(x)^2 - \tfrac{1}{2}g(x)^2.$$

Therefore, because $\dfrac{d}{dx}$ is a linear operator,

$$\frac{d}{dx}[f(x)g(x)] = \tfrac{1}{2}\frac{d}{dx}[f(x) + g(x)]^2 - \tfrac{1}{2}\frac{d}{dx}f(x)^2 - \tfrac{1}{2}\frac{d}{dx}g(x)^2.$$

We have already calculated the first of the derivatives on the right-hand side of this equation (Equation 16-3), and Equation 16-2 immediately gives us

$$\frac{d}{dx}f(x)^2 = 2f(x)\frac{d}{dx}f(x) \qquad \text{and} \qquad \frac{d}{dx}g(x)^2 = 2g(x)\frac{d}{dx}g(x).$$

Therefore

$$\frac{d}{dx}[f(x)g(x)] = [f(x) + g(x)]\left[\frac{d}{dx}f(x) + \frac{d}{dx}g(x)\right]$$

$$- f(x)\frac{d}{dx}f(x) - g(x)\frac{d}{dx}g(x),$$

which you can quickly reduce to Equation 17-1.

Example 17-1. Find $\dfrac{d}{dx}[x(4x + 3)]$.

Solution. The easiest approach is to perform the indicated multiplication and then differentiate:

$$\frac{d}{dx}[x(4x + 3)] = \frac{d}{dx}(4x^2 + 3x) = 4\frac{dx^2}{dx} + 3\frac{dx}{dx} = 8x + 3.$$

But we could use Theorem 17-1 to write

$$\frac{d}{dx}[x(4x + 3)] = x\frac{d}{dx}(4x + 3) + (4x + 3)\frac{dx}{dx} = x \cdot 4 + (4x + 3) \cdot 1 = 8x + 3.$$

Example 17-2. Extend the Product Rule to three factors.

Solution. We can think of the product $f(x)g(x)h(x)$ as the product of the two factors $f(x)$ and $g(x)h(x)$. Therefore Theorem 17-1 applies, and we have

$$\frac{d}{dx}\left[f(x)g(x)h(x)\right] = f(x)\frac{d}{dx}\left[g(x)h(x)\right] + g(x)h(x)\frac{d}{dx}f(x).$$

Now we use Theorem 17-1 again, this time to show that

$$\frac{d}{dx}\left[g(x)h(x)\right] = g(x)\frac{d}{dx}h(x) + h(x)\frac{d}{dx}g(x),$$

and so we see that

$$\frac{d}{dx}\left[f(x)g(x)h(x)\right] = f(x)\cdot g(x)\cdot\frac{d}{dx}h(x)$$

$$+ f(x)\cdot\frac{d}{dx}g(x)\cdot h(x) + \frac{d}{dx}f(x)\cdot g(x)\cdot h(x).$$

The same ideas extend the rule to products containing any number of factors.

As with multiplication, the order in which we apply the operations of division and differentiation is important. If we divide and then differentiate, or differentiate and then divide, we will usually get different results.

Theorem   17-2. *If f and g are differentiable at x, and if $g(x) \neq 0$, then*

$$(17\text{-}2) \qquad \frac{d}{dx}\left(\frac{f(x)}{g(x)}\right) = \frac{g(x)\dfrac{d}{dx}f(x) - f(x)\dfrac{d}{dx}g(x)}{g(x)^2}.$$

PROOF

We use the Product Rule and the Power Formula:

$$\frac{d}{dx}\left(\frac{f(x)}{g(x)}\right) = \frac{d}{dx}[f(x)g(x)^{-1}] \qquad\qquad \text{(Algebra)}$$

$$= f(x)\frac{d}{dx}g(x)^{-1} + g(x)^{-1}\frac{d}{dx}f(x) \qquad \text{(Product Rule)}$$

$$= -f(x)g(x)^{-2}\frac{d}{dx}g(x) + g(x)^{-1}\frac{d}{dx}f(x) \qquad \begin{array}{l}\text{(Power Formula}\\\text{with } r = -1),\end{array}$$

and simple algebra reduces this equation to Equation 17-2.

Example 17-3.   Find y' if $y = (x - 1)/\sqrt{x + 1}$.

Solution.

$$\frac{d}{dx} \frac{x - 1}{\sqrt{x + 1}} = \frac{\sqrt{x + 1} \dfrac{d}{dx} (x - 1) - (x - 1) \dfrac{d}{dx} \sqrt{x + 1}}{x + 1} \qquad \text{(Quotient Rule)}$$

$$= \frac{\sqrt{x + 1} \cdot 1 - (x - 1)/2\sqrt{x + 1}}{x + 1} \qquad \left(\begin{array}{c} \text{Power Rule to} \\ \text{find } \dfrac{d}{dx} \sqrt{x + 1} \end{array} \right)$$

$$= \frac{x + 3}{2(x + 1)^{3/2}} \qquad \text{(Algebra)}.$$

Let us summarize the differentiation rules we have developed so far. If f and g are differentiable functions and $u = f(x)$ and $v = g(x)$, then for any given numbers m and n,

$$(17\text{-}3) \qquad \frac{d}{dx} (mu + nv) = m \frac{du}{dx} + n \frac{dv}{dx} \qquad \left(\frac{d}{dx} \text{ is a linear operator} \right).$$

We have also found that

$$(17\text{-}4) \qquad \frac{d}{dx} (uv) = u \frac{dv}{dx} + v \frac{du}{dx} \qquad \text{(Product Rule)}$$

and

$$(17\text{-}5) \qquad \frac{d}{dx} \left(\frac{u}{v} \right) = \frac{v \dfrac{du}{dx} - u \dfrac{dv}{dx}}{v^2} \qquad \text{(Quotient Rule)}.$$

PROBLEMS   17

1.   Find and simplify $\dfrac{dy}{dx}$.

(a) $y = \sqrt{4x - 9}$

(b) $y = 2\sqrt{x} - 3$

(c) $y = \sqrt{x + 3} - \sqrt{x} - \sqrt{3}$

(d) $y = (2x + 3)(3x + 2)$

(e) $y = (x + 3)^2(x + 2)^3$

(f) $h = \dfrac{x - 1}{x + 1}$

(g) $y = x\sqrt{x + 3}$

(h) $y = \dfrac{x}{\sqrt{x + 3}}$

2. Find $f'(2)$.

(a) $f(x) = \sqrt{2x} + \dfrac{1}{\sqrt{2x}}$

(b) $f(x) = \sqrt[3]{3x + 2}$

(c) $f(x) = (x - 2)^5(x - 5)^2$

(d) $f(x) = \dfrac{\sqrt{2x}}{x - 1}$

3. At what points of the following curves are the tangent lines perpendicular to the line $4y + x - 8 = 0$?

(a) $y = x^2 - 4x + 1$

(b) $y = \dfrac{x^2 - 3}{x}$

(c) $y = 3 + 2\sqrt{x}$

(d) $y = \dfrac{2 + x}{2 - x}$

4. At what point of the curve $y = x^3 + x$ is the slope least?

5. What is the rate of change of volume with respect to the radius of

(a) a sphere?
(b) a circular cylinder with fixed height?
(c) a cone with fixed height?

6. (a) Suppose that r is a rational number and $f(x) = x^{-r}$ and $g(x) = (x + r)^r$. Show that $\dfrac{d}{dx}[f(x)g(x)] = \dfrac{d}{dx}f(x)\dfrac{d}{dx}g(x)$. Can you find other pairs of functions for which this equation holds?

(b) Let r be a rational number and $f(x) = (1 + r/x)^r$ and $g(x) = x^{-r}$. Show that $\dfrac{d}{dx}[f(x)/g(x)] = \dfrac{d}{dx}f(x) \Big/ \dfrac{d}{dx}g(x)$. Can you find other pairs of functions for which this equation holds?

7. Find y'.

(a) $y = x|x|$

(b) $y = x + |x|$

(c) $y = (x - 1)^4(x + 2)^5(x - 3)^6$

(d) $y = \dfrac{(x - 1)^4(x - 3)^6}{(x + 2)^5}$

8. Find $f'(-2)$.

(a) $f(x) = |x| + x^2$

(b) $f(x) = [\![x + \tfrac{1}{2}]\!] + \sqrt[5]{16x}$

(c) $f(x) = [\![x + \tfrac{1}{2}]\!]\sqrt[5]{16x}$

(d) $f(x) = (|x| + x)\sqrt[5]{16x}$

(e) $f(x) = (|x| - x)\sqrt[5]{16x}$

(f) $f(x) = [\![\sin x]\!]\sqrt[5]{16x}$

9. A point moves along a number scale in accordance with the equation $s = \sqrt{t}/(t + 1)$ for $t \geq 0$.

(a) At what times is the point moving to the right (s is increasing)?
(b) At what times is it moving to the left?
(c) When is it farthest from the origin?
(d) What is the greatest distance from the origin that the point reaches?

10. Write a product rule for a product of four factors; that is, find a formula for $\dfrac{d}{dx}[a(x)b(x)c(x)d(x)]$. What is the rule for any number of factors?

11.   Suppose that we are told that sin and cos are differentiable functions.

(a) Use the identity $\cos^2 x + \sin^2 x = 1$ to show that

$$\cos x \, \frac{d \cos x}{dx} + \sin x \, \frac{d \sin x}{dx} = 0.$$

(b) What is $\dfrac{d \cos x}{dx}$ when $x = 0$? Does this result seem reasonable in terms of the graph of cos?

(c) Use the identity $\tan x = \sin x / \cos x$ and the rule for differentiating a quotient to show that $\dfrac{d \tan x}{dx} = \dfrac{d \sin x}{dx}$ when $x = 0$.

(d) What is $\dfrac{d \sec x}{dx}$ when $x = 0$?

18. DERIVATIVES OF THE TRIGONOMETRIC FUNCTIONS

Let us go back to finding specific differentiation formulas, this time for the trigonometric functions. To find $\dfrac{d \sin x}{dx}$, we set up the difference quotient

$$\frac{\sin (x + h) - \sin x}{h}$$

and compute its limit as h approaches 0. Since this limit is far from obvious, we first have to rewrite the quotient. Experience shows that it is a good idea to replace $\sin (x + h)$ with $\sin x \cos h + \cos x \sin h$ (see Formula A-2 in the appendix) and thus reduce the difference quotient to

$$\frac{\sin x \cos h + \cos x \sin h - \sin x}{h} = \cos x \, \frac{\sin h}{h} + \sin x \, \frac{\cos h - 1}{h}.$$

Therefore, according to the definition of the derivative and our theorems on limits,

$$(18\text{-}1) \qquad \frac{d \sin x}{dx} = \cos x \lim_{h \to 0} \frac{\sin h}{h} + \sin x \lim_{h \to 0} \frac{\cos h - 1}{h}.$$

Our problem of calculating $\dfrac{d \sin x}{dx}$ is now reduced to that of finding the two limits that appear in Equation 18-1. Although replacing one problem with two may not strike you as very efficient, it turns out that these two limits can be evaluated much more simply than the limit of our original difference quotient.

Observe that according to the definition of the derivative, *these limits are the derivatives of sin x and cos x when x = 0.* (Can you use this observation to guess what they are?) Our next two examples are devoted to finding them.

Example 18-1.   Find $\lim\limits_{h \to 0}$ (sin h)/h.

Solution.   In Fig. 9-1 we sketched the curve $y = (\sin x)/x$, from which it appeared that $\lim\limits_{h \to 0}$ (sin h)/h = 1. An analytic proof of this result requires more of a discussion of what we mean by the length of a circular arc than we care to undertake here, but we can show the basic idea by referring to a picture. In Fig. 18-1 we have sketched the unit circle, showing the points $P(\cos h, -\sin h)$ and $Q(\cos h, \sin h)$ that correspond to a small positive number h. Thus the length of the *segment PQ* is 2 sin h units, while the *arc PQ* is $2h$ units long. Since (sin h)/h = (2 sin h)/$2h$, we therefore see that the quotient (sin h)/h is the ratio of the length of a chord of a circle to the length of the corresponding arc. (If h were negative, we would get exactly the same result.) It is a consequence of the definition of arclength that the limit of the ratio of chord length to arclength is 1, and hence $\lim\limits_{h \to 0}$ (sin h)/h = 1.

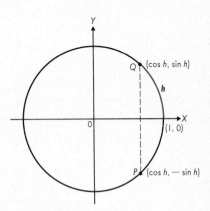

Figure 18-1

Example 18-2.   Find $\lim\limits_{h \to 0}$ (cos h − 1)/h.

Solution.   As we said, this number is simply the slope of the cosine curve at the point (0, 1), and it appears from Fig. 6-5 that this slope is 0. To find this limit by using the limit we found in Example 18-1, we write

$$\frac{\cos h - 1}{h} = \frac{\cos h - 1}{h} \cdot \frac{\cos h + 1}{\cos h + 1} = \frac{\cos^2 h - 1}{h(\cos h + 1)}$$

$$= \frac{-\sin^2 h}{h(\cos h + 1)} = -\left(\frac{\sin h}{h}\right)\left(\frac{\sin h}{\cos h + 1}\right).$$

In Example 18-1 we found that the first factor of this product approaches 1, and it is clear that the limit of the second factor is 0. Hence, $\lim\limits_{h \to 0}$ (cos h − 1)/h = 0.

The results of Examples 18-1 and 18-2 reduce Equation 18-1 to $\dfrac{d \sin x}{dx} =$ cos $x \cdot 1 +$ sin $x \cdot 0$; that is,

(18-2) $$\frac{d \sin x}{dx} = \cos x.$$

We used a geometric argument to obtain a rough sketch of the curve $y = \dfrac{d \sin x}{dx}$ in Fig. 12-4. Now that we have a differentiation formula for the sine function, you might reexamine that graph to see if our approximation agrees with Equation 18-2.

As we said when we introduced it, the Chain Rule greatly extends the applicability of specific differentiation formulas. In order to incorporate our newly derived specific Differentiation Formula 18-2 into the Chain Rule, we simply replace $f(u)$ with $\sin u$ in Equation 16-1:

$$\frac{d \sin u}{dx} = \frac{d \sin u}{du} \frac{du}{dx}.$$

Equation 18-2 (with x replaced by u, of course) tells us that $\dfrac{d \sin u}{du} = \cos u$, so we have the equation

$$(18\text{-}3) \qquad \qquad \frac{d \sin u}{dx} = \cos u \, \frac{du}{dx}.$$

The remaining theory and examples of this section illustrate how powerful this extension of Equation 18-2 is.

Example 18-3.   Find $\dfrac{d}{dx} \sin x^2$ and $\dfrac{d}{dx} \sin^2 x$.

Solution.   We find $\dfrac{d}{dx} \sin x^2$ by replacing u with x^2 in Equation 18-3:

$$\frac{d}{dx} \sin x^2 = \cos x^2 \, \frac{dx^2}{dx} = 2x \cos x^2.$$

On the other hand, $\sin^2 x$ means $(\sin x)^2$, and so we use the Power Formula to find $\dfrac{d}{dx} \sin^2 x$. Thus in Equation 16-2, with $r = 2$, we replace u with $\sin x$, and then use Equation 18-2 to calculate $\dfrac{d \sin x}{dx}$:

$$\frac{d}{dx} \sin^2 x = 2 \sin x \, \frac{d}{dx} \sin x = 2 \sin x \cos x = \sin 2x.$$

Instead of going back to first principles, we can use the rules already developed to find derivatives for the other trigonometric functions. For example, since $\cos x = \sin (\frac{1}{2}\pi - x)$, we have $\dfrac{d \cos x}{dx} = \dfrac{d}{dx} \sin (\frac{1}{2}\pi - x)$, and we find this last derivative by replacing u with $\frac{1}{2}\pi - x$ in Equation 18-3:

$$\frac{d \cos x}{dx} = \frac{d}{dx} \sin (\tfrac{1}{2}\pi - x) = \cos (\tfrac{1}{2}\pi - x) \frac{d}{dx} (\tfrac{1}{2}\pi - x) = -\cos (\tfrac{1}{2}\pi - x).$$

But $\cos (\frac{1}{2}\pi - x) = \sin x$, and so

$$(18\text{-}4) \qquad\qquad \frac{d \cos x}{dx} = -\sin x.$$

We find the derivative of $\tan x$ by applying the Quotient Rule to the right side of the identity $\tan x = (\sin x)/\cos x$ and using Equations 18-2 and 18-4:

$$\frac{d \tan x}{dx} = \frac{d}{dx} \frac{\sin x}{\cos x} = \frac{\cos x \dfrac{d \sin x}{dx} - \sin x \dfrac{d \cos x}{dx}}{\cos^2 x}$$

$$= \frac{\cos^2 x + \sin^2 x}{\cos^2 x} = \frac{1}{\cos^2 x} = \sec^2 x.$$

Therefore we have the differentiation formula

$$(18\text{-}5) \qquad\qquad \frac{d \tan x}{dx} = \sec^2 x.$$

We find $\dfrac{d \sec x}{dx}$ by writing $\sec x = (\cos x)^{-1}$ and applying the Power Formula:

$$\frac{d}{dx} (\cos x)^{-1} = -(\cos x)^{-2} \frac{d \cos x}{dx} = \frac{\sin x}{\cos^2 x} = \frac{1}{\cos x} \frac{\sin x}{\cos x} = \sec x \tan x.$$

Similar tricks yield derivatives for the other trigonometric functions:

$$(18\text{-}6) \qquad\qquad \frac{d \cot x}{dx} = -\csc^2 x,$$

$$(18\text{-}7) \qquad\qquad \frac{d \sec x}{dx} = \sec x \tan x,$$

$$(18\text{-}8) \qquad\qquad \frac{d \csc x}{dx} = -\csc x \cot x.$$

And, of course, we immediately incorporate these formulas into the Chain Rule Equation $\dfrac{df(u)}{dx} = \dfrac{df(u)}{du} \dfrac{du}{dx}$:

(18-9)
$$\frac{d \cos u}{dx} = -\sin u \frac{du}{dx},$$

(18-10)
$$\frac{d \tan u}{dx} = \sec^2 u \frac{du}{dx},$$

(18-11)
$$\frac{d \cot u}{dx} = -\csc^2 u \frac{du}{dx},$$

(18-12)
$$\frac{d \sec u}{dx} = \sec u \tan u \frac{du}{dx},$$

(18-13)
$$\frac{d \csc u}{dx} = -\csc u \cot u \frac{du}{dx}.$$

Perhaps you are beginning to wonder just how many differentiation formulas we are going to develop (and ask you to memorize). The answer appears in Table IV in the back of the book, where we have listed all the specific differentiation formulas that we will introduce.

Example 18-4. If $f(x) = \csc(1/x)$, find $f'(2/\pi)$.

Solution. According to Equation 18-13 with $u = 1/x$,

$$\frac{d}{dx} \csc \frac{1}{x} = -\csc \frac{1}{x} \cot \frac{1}{x} \frac{d}{dx} \frac{1}{x} = -\csc \frac{1}{x} \cot \frac{1}{x} \left(-\frac{1}{x^2}\right).$$

Therefore $f'(x) = (1/x^2) \csc(1/x) \cot(1/x)$, and so $f'(2/\pi) = (\tfrac{1}{2}\pi)^2 \csc \tfrac{1}{2}\pi \cot \tfrac{1}{2}\pi = 0$.

Example 18-5. Find the points of the curve $y = x + \cot x$ at which the tangent line is horizontal.

Solution. With the help of our differentiation formulas, we see that

$$y' = 1 - \csc^2 x = -\cot^2 x.$$

The slope of the tangent line is y', so the tangent line is horizontal at points where $y' = 0$. To find those points, we must solve the equation

$$-\cot^2 x = 0.$$

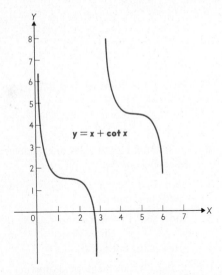

Figure 18-2

The solutions of this equation have the form $x = \frac{1}{2}\pi + k\pi$, where k can be any integer—positive, negative, or 0. Thus the points where the tangent line is horizontal form the set

$$\{(x, x) \mid x = \tfrac{1}{2}\pi + k\pi, \ k = 0, \ \pm 1, \ \pm 2, \dots \}.$$

The curve $y = x + \cot x$ is illustrated in Fig. 18-2. Does the figure agree with the fact that y' is never positive?

PROBLEMS 18

1. Find $\dfrac{dy}{dx}$.

(a) $y = \sqrt{\sin x} - \sin \sqrt{x}$ (b) $y = \tan x^2 - \tan^2 x$

(c) $y = \sec 2x - 2 \sec x$ (d) $y = \cot (x + 2) - (\cot x + \cot 2)$

(e) $y = \cos 2x - (\cos 2)(\cos x)$ (f) $y = \csc (3/x) - (\csc 3)/\csc x$

2. Find y'.

(a) $y = (\sin x)/x$ (b) $y = \cos (\sin x)$

(c) $y = \sin^2 3x$ (d) $y = \cos^3 (x - x^2)$

(e) $y = \sin^2 (\cos^2 x^2)$ (f) $y = \sqrt[3]{\sec^2 2x}$

3. Find the equations of the tangent line and the normal line to the sine curve at the following points.

(a) $(0, 0)$ (b) $(\tfrac{1}{4}\pi, \tfrac{1}{2}\sqrt{2})$ (c) $(\tfrac{1}{2}\pi, 1)$ (d) $(\pi, 0)$

4. Choose r so that $f'(0) = 3$.

   (a) $f(x) = \sin rx$ (b) $f(x) = \tan rx$

   (c) $f(x) = r \cos \sqrt{x + 1}$ (d) $f(x) = r \sin \sqrt{x + 4}$

   (e) $f(x) = \sqrt{\cos (rx + 1)}$ (f) $f(x) = \sqrt{\sin (rx + 1)}$

5. (a) Find a point of the sine curve at which the tangent is parallel to the line $x + 2y + 3 = 0$.

   (b) Find a point of the sine curve at which the normal is parallel to the line $2x + y + 3 = 0$.

6. Differentiate both sides of the identity $\sin 2x = 2 \sin x \cos x$ to obtain an identity involving $\cos 2x$.

7. Show that there are no points of the curve $y = \sec x - \tan x$ at which the tangent line is horizontal.

8. Show that $y = \tan x$ satisfies the differential equations (i) $y' \sin 2x = 2y$ and (ii) $y' = 1 + y^2$. Show that $y = 3 \tan x$ satisfies (i) but not (ii) and $y = \tan (x + 3)$ satisfies (ii) but not (i).

9. If the displacement of a moving body is related to the time by the equation $s = A \sin (bt + c)$, where A, B, and c are given numbers, then the motion of the body is termed *simple harmonic motion*. Find an expression for the velocity in simple harmonic motion.

10. Let $\sin \theta°$ and $\cos \theta°$ denote the sine and cosine of an angle of θ degrees. Show that $\dfrac{d}{d\theta} \sin \theta° = \dfrac{\pi}{180} \cos \theta°$.

11. Derive Formulas 18-6 and 18-8.

12. A point P moves along the unit circle at the rate of 1 rpm. Let Q denote the point in which the tangent to the circle at P intersects the X-axis. How fast is Q moving along the X-axis when it is 2 units to the right of the origin?

13. Let $f(x) = \cos x$ and $g(x) = \frac{1}{2} \sin 2x$. Find the angles between the tangents to the graphs of f and g at their points of intersection.

14. Let $f(x) = x^2 \sin (1/x)$ if $x \neq 0$ and $f(0) = 0$. In Problem 13-12 we asked you to show that f is differentiable at 0. Show that f' is not continuous at 0.

19. THE THEOREM OF THE MEAN

Because the theorems we discuss in this section are geometrically obvious, you will have no trouble accepting their truth. Their proofs, however, are based on one of the deeper theorems of analysis, Theorem 10-1. This theorem says that the continuous image of a finite closed interval $[a, b]$ is a finite closed interval $[A, B]$. The numbers A and B are the least and greatest values taken by the function in $[a, b]$. The function f whose graph appears in Fig. 19-1 takes its greatest value in $[a, b]$ at a number m in the *open* interval (a, b). It is geometrically apparent that the tangent to the graph at the maximum point $(m, f(m))$ has slope

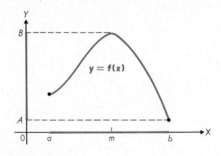

Figure 19-1

0; that is, $f'(m) = 0$. This fact is important enough to state as a theorem, for which we outline a proof in Problem 19-12.

Theorem 19-1. *Suppose that the greatest (or least) value that a function f takes in an open interval (a, b) is the number $f(m)$, and suppose that f is differentiable at m. Then $f'(m) = 0$.*

We use this result to prove the extremely versatile **Rolle's Theorem.**

Theorem 19-2. *If f is differentiable in the open interval (a, b) and continuous in the closed interval $[a, b]$, and if $f(a) = f(b)$, then there is at least one number $m \in (a, b)$ such that $f'(m) = 0$.*

PROOF

According to Theorem 19-1, it is sufficient for us to show that *either* the greatest or the least value that f takes in the closed interval $[a, b]$ is taken at a number m in the open interval (a, b)—that is, not at an endpoint. But, clearly, if *both* the least and the greatest values of f are taken at the endpoints a and b, the condition $f(a) = f(b)$ requires that $f(x) = f(a)$ for each $x \in (a, b)$, and so f takes its greatest (and least) value in (a, b), too.

If you sketch the graphs of a few functions that satisfy the hypotheses of Rolle's Theorem, you will see that it, too, is geometrically obvious. The theorem simply says that if the graph of such a function intersects a certain horizontal line when $x = a$, and again when $x = b$, then at some number between a and b the tangent line to the graph is horizontal. For the function f whose graph is shown

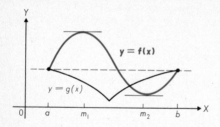

Figure 19-2

in Fig. 19-2, there are two points at which the tangent line is horizontal, and hence we can choose either m_1 or m_2 as the number m of Rolle's Theorem. For other functions, there may be 5, 100, or even infinitely many choices for m. But if the conditions of Rolle's Theorem are satisfied, then there must be *at least one choice for m*. The function g whose graph is pictured in Fig. 19-2 does not satisfy the hypotheses of Rolle's Theorem, so it is not surprising that the conclusion does not hold either. Rolle's Theorem can be exploited in many clever ways.

Example 19-1. In Example 10-1 we showed that the equation $x^7 - 4x^3 + 1 = 0$ has *at least* one solution in the interval $(0, 1)$. Can it have more?

Solution. As before, we let $f(x) = x^7 - 4x^3 + 1$. Suppose that there were two numbers r and s in the interval $(0, 1)$ such that $f(r) = 0$ and $f(s) = 0$. Then, according to Rolle's Theorem, there would be a number m between r and s—and hence in the interval $(0, 1)$—such that $f'(m) = 0$. Since $f'(x) = 7x^6 - 12x^2$, the equation $f'(m) = 0$ becomes $7m^6 = 12m^2$, which does not have a solution in the interval $(0, 1)$. Therefore our original equation cannot have more than one solution in the interval $(0, 1)$.

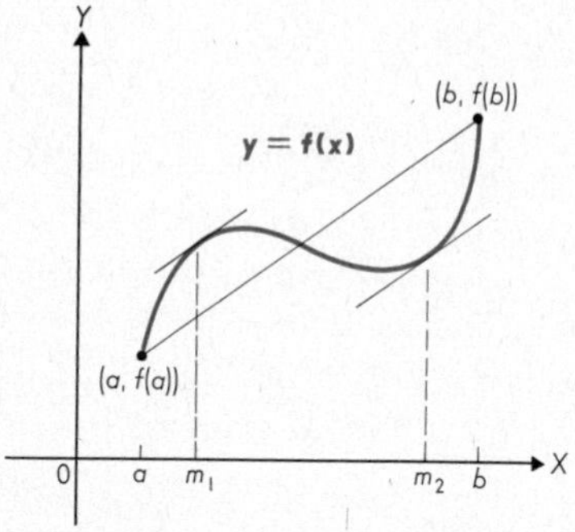

Figure 19-3 shows an arc of the graph of a function f with a chord joining the endpoints $(a, f(a))$ and $(b, f(b))$. At the points in which the lines $x = m_1$ and $x = m_2$ intersect the graph, the tangent lines are parallel to this chord. If you draw other smooth arcs, you will soon conclude that *every smooth arc must contain at least one point at which the tangent is parallel to the chord.* Analytically, we are saying that there must be a number m such that the slope $f'(m)$ of the tangent equals the slope $[f(b) - f(a)]/(b - a)$ of the chord. This result is formalized in the **Theorem of the Mean**.

Figure 19-3

Theorem 19-3. *If f is differentiable in the open interval (a, b) and continuous in the closed interval $[a, b]$, then there is a number $m \in (a, b)$ such that*

$$\frac{f(b) - f(a)}{b - a} = f'(m).$$

102

PROOF

We let

(19-2) $$E(x) = f(x) - f(a) - \frac{f(b) - f(a)}{b - a} (x - a)$$

and observe that $E(a) = E(b) = 0$. Therefore E satisfies the hypotheses of Rolle's Theorem, so there is a number $m \in (a, b)$ such that $E'(m) = 0$. Since $E'(x) = f'(x) - [f(b) - f(a)]/(b - a)$, the equation $E'(m) = 0$ is equivalent to Equation 19-1, and our proof is complete. Of course, the hard part of this proof consists of selecting the "right" function E to which to apply Rolle's Theorem.

Think of a point traveling along the curve shown in Fig. 19-4 from left to right. It goes up as its X-coordinate increases from 0 to 2 and then down as its X-coordinate increases from 2 to 5. We say that f is **increasing** in the interval $(0, 2)$

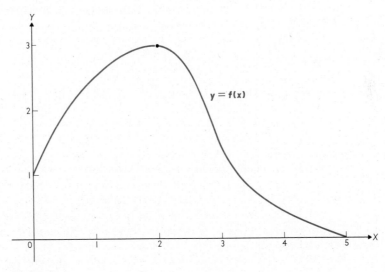

Figure 19-4

and **decreasing** in the interval $(2, 5)$. Formally, f increases in an interval I if for each pair of points a and b in I such that $a < b$, we have $f(a) < f(b)$. Less formally, *f is increasing if its graph "rises to the right."* It is decreasing if its graph "falls to the right." Notice that f is neither increasing nor decreasing in the entire interval $(0, 5)$ in Fig. 19-4.

A line rises to the right if its slope is positive, and it falls to the right if its slope is negative. This same rule applies to functions in general.

Theorem 19-4. *Suppose that f is continuous in an interval I. If $f'(x) > 0$ at every interior point (that is, not an endpoint) x of I, then f is increasing in I. If $f'(x) < 0$ at every interior point x, then f is decreasing in I.*

PROOF

We only write out the proof for the first case. Thus we are assuming that $f'(x) > 0$ for each interior point x of I, and we are to show that if a and b are two numbers in I such that $a < b$, then $f(a) < f(b)$. Now, according to the Theorem of the Mean, there is a number m between a and b such that

$$\frac{f(b) - f(a)}{b - a} = f'(m).$$

Since m is an interior point of I, our hypothesis tells us that $f'(m) > 0$. Thus $[f(b) - f(a)]/(b - a) > 0$, and since $b - a > 0$, we have $f(b) - f(a) > 0$; that is, $f(a) < f(b)$.

Example 19-2. Determine the intervals in which f is decreasing if

$$f(x) = 2x^3 + 3x^2 - 12x + 5.$$

Solution. Here $f'(x) = 6x^2 + 6x - 12 = 6(x - 1)(x + 2)$, so f is decreasing in intervals whose interior points satisfy the inequality

$$(x - 1)(x + 2) < 0.$$

The product $(x - 1)(x + 2)$ is negative where the factors $(x - 1)$ and $(x + 2)$ have opposite signs, and Fig. 19-5 shows that they do in the interval $(-2, 1)$. Therefore f is decreasing in any interval whose interior points belong to the interval $(-2, 1)$; the largest such interval is the closed interval $[-2, 1]$.

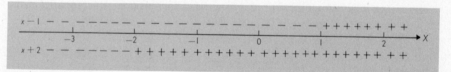

Figure 19-5

We list one more consequence of the Theorem of the Mean. A given differentiable function f gives rise to *just one* derived function f'. This statement does *not* say that two different functions cannot have the same derived function, however. For example, if $f(x) = x^3 + 3$ and $g(x) = x^3 - 5$, then f and g are different functions. But $f'(x) = g'(x) = 3x^2$, and so the derived functions f' and g' are the same. Nevertheless, if two functions have the same derived function, there is a definite relationship between them.

Theorem 19-5. *Let f and g be differentiable in some interval I and suppose that at each $x \in I$, $f'(x) = g'(x)$. Then there is a number C (independent of x) such that for each $x \in I$, $f(x) = g(x) + C$.*

PROOF

Let us write $F(x) = f(x) - g(x)$ and choose any number $a \in I$. Then according to the Theorem of the Mean, for each other number $x \in I$ there is a number m between a and x such that

$$\frac{F(x) - F(a)}{x - a} = F'(m).$$

Now $F'(m) = f'(m) - g'(m)$, and, by hypothesis, this number is 0. Therefore $F(x) - F(a) = 0$; that is, $F(x) = F(a)$. If we write $C = F(a)$ and recall that $F(x) = f(x) - g(x)$, we see that the equation $F(x) = F(a)$ asserts that $f(x) = g(x) + C$, as we were to prove.

Geometrically, the statement $f'(x) = g'(x)$ means that the graphs of f and g are "parallel." Any vertical line intersects the two graphs in points at which they have the same slope. Theorem 19-5 tells us that the two graphs therefore cut segments of equal lengths from such vertical lines (Fig. 19-6).

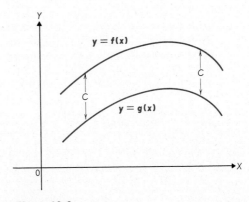

Figure 19-6

PROBLEMS 19

1. The maximum or minimum value of each of the following functions in the interval $[-1, 1]$ is $f(0)$. Make a sketch of the situation and see how Theorem 19-1 applies.

   (a) $f(x) = \cos x$ (b) $f(x) = \cos^2 \pi x$ (c) $f(x) = \sec x$

   (d) $f(x) = \tan^2 x$ (e) $f(x) = |x|$ (f) $f(x) = [\![x]\!]^2$

2. For each of the following functions, $f(0) = f(2)$. Is there a number in $(0, 2)$ at which f' takes the value 0?

   (a) $f(x) = x^2 - 2x + 4$

   (c) $f(x) = x^3 - 3x^2 + 2x + 6$

   (e) $f(x) = |x - 1|$

   (b) $f(x) = \sqrt{x^2 - 2x + 9}$

   (d) $f(x) = \sqrt{x^3 - 3x^2 + 2x + 9}$

   (f) $f(x) = \tan \pi x$

3. Find a number that can be chosen as a number m in the Theorem of the Mean for the given function and interval.

   (a) $f(x) = \sin x$, $[0, \pi]$

   (c) $f(x) = x^2$, $[0, 1]$

   (e) $f(x) = x^3 + 2x^2 + 1$, $[0, 3]$

   (b) $f(x) = x$, $[0, 1]$

   (d) $f(x) = x + 1/x$, $[1, 9]$

   (f) $f(x) = \sin x + \cos x$, $[0, 2]$

4. Find the intervals in which f is increasing and the intervals in which f is decreasing.

   (a) $f(x) = x^3 - 6x^2 + 9x + 3$

   (c) $f(x) = (x - 2)/(x + 1)^2$

   (e) $f(x) = x/(x^2 + 1)$

   (b) $f(x) = (x + 1)^2(x - 2)$

   (d) $f(x) = (x^2 + 1)/x$

   (f) $f(x) = \sqrt{1 - x}$

5. Find the subintervals of $[0, 2\pi]$ in which f is increasing.

   (a) $f(x) = \sin x + \cos x$

   (c) $f(x) = |\sin x| + |\cos x|$

   (b) $f(x) = \cos x + \sqrt{3} \sin x$

   (d) $f(x) = \sin x |\sin x|$

6. In each case, check that $f'(x) = g'(x)$ and find the relation between $f(x)$ and $g(x)$.

   (a) $f(x) = \sin^2 x$, $g(x) = -\frac{1}{2} \cos 2x$

   (b) $f(x) = (x - 2)^2$, $g(x) = x(x - 4)$

   (c) $f(x) = x/(1 + x)$, $g(x) = -1/(1 + x)$

   (d) $f(x) = x(1 + x + x^2)$, $g(x) = (1 - x^4)/(1 - x)$

   (e) $f(x) = \sec^2 x$, $g(x) = \tan^2 x$

   (f) $f(x) = \sin (x + 3)/\cos x$, $g(x) = \cos 3 \tan x$

7. Suppose that f and g are increasing in an interval I. (Sums and products of functions are defined in the "natural" way.)

   (a) Is it necessarily true that $f + g$ is increasing?

   (b) Is it necessarily true that fg is increasing?

   (c) Is it necessarily true that f^2 is increasing?

   (d) Is it necessarily true that f^3 is increasing?

8. (a) Suppose that $\dfrac{d}{dx} [10^x f(x)] = 0$ for each $x \in R^1$. If $f(0) = 3$, what is $f(x)$?

   (b) Suppose that $\dfrac{d}{dx} [10^x f(x)] \geq 0$ for each $x \geq 0$. If $f(0) = 3$, what can you say about $f(x)$ for $x \geq 0$?

9. The hypotheses of the Theorem of the Mean are not satisfied for the given function and interval. Explain why not. Does the conclusion hold?

   (a) $f(x) = |x|$, $[-1, 1]$

   (c) $f(x) = x - [\![x]\!]$, $[0, 1]$

   (b) $f(x) = \tan x$, $[0, \pi]$

   (d) $f(x) = [\![\cos x]\!]$, $[0, 2\pi]$

10. Show that

(a) if $f'(x)g(x) > -f(x)g'(x)$ for each x in an interval I, then the product of f and g is increasing in I.

(b) if $f'(x)g(x) > f(x)g'(x)$ for each x in an interval I, then the quotient of f by g is increasing in I.

11. Suppose that y satisfies the differential equation $xy' + y = 3x^2$ and that $y = 3$ when $x = 1$. Show that $\dfrac{d}{dx}(xy - x^3) = 0$ and hence deduce that $xy - x^3 = 2$; that is, $y = x^2 + 2/x$.

12. Fill in the details of the following outline of a proof of Theorem 19-1.

(a) If $f(m)$ is the largest value of f, show that

$$\frac{f(x) - f(m)}{x - m} \geq 0 \text{ if } x < m \quad \text{and} \quad \frac{f(x) - f(m)}{x - m} \leq 0 \text{ if } x > m.$$

(b) In the first of these inequalities, let $x \uparrow m$ to see that $f'(m) \geq 0$, and in the second, let $x \downarrow m$ to see that $f'(m) \leq 0$. Hence $f'(m) = 0$.

(c) How do we modify the proof if $f(m)$ is the least value of f?

13. (a) The converse of Theorem 19-4 says that if f is increasing in an interval I, then $f'(x) > 0$ at each interior point x of I. Is it true?

(b) Prove Theorem 19-4 if $f'(x) < 0$ for each interior point of I.

20. DERIVATIVES OF HIGHER ORDER

The derived function f' of a given differentiable function f may very well be a differentiable function itself. That is, we may be able to associate with it another function $(f')'$—the derived function of f'. This function is the **second derived function** of f and is denoted by f''. We have been calling $f'(x)$ the *derivative* of $f(x)$, but now we will call it the *first derivative* of $f(x)$ to distinguish it from the **second derivative** $f''(x)$.

We write $\dfrac{dy}{dx}$ for the first derivative of y with respect to x and $\dfrac{d^2y}{dx^2}$ for the second derivative. Here we are applying the differential operator $\dfrac{d}{dx}$ twice to y; that is, we are dealing with the square of $\dfrac{d}{dx}$. So it is natural to view it as $\dfrac{d^2}{dx^2}$, from which the notation $\dfrac{d^2y}{dx^2}$ follows. Similarly, $\dfrac{d^3y}{dx^3}$ is the third derivative of y with respect to x, and so on. We also use the simpler notation y'' and y''' to denote these derivatives.

Example 20-1. If $y = x^2 + \sin x$, find y''.

Solution. We calculate $y' = 2x + \cos x$ in the usual way. To find y'', we simply differentiate $2x + \cos x$. Thus $y'' = 2 - \sin x$.

After the third, we stop using primes for higher derivatives. We may, for example, use $f^{\text{iv}}(x)$ to denote the fourth derivative of $f(x)$, and $f^{(12)}(x)$ denotes the twelfth.

Example 20-2. Find $f^{(27)}(x)$ if $f(x) = \sin x$.

Solution. The first four of the indicated 27 differentiations,

$$f'(x) = \frac{d}{dx} \sin x = \cos x,$$

$$f''(x) = \frac{d}{dx} \cos x = -\sin x,$$

$$f'''(x) = \frac{d}{dx}(-\sin x) = -\cos x,$$

$$f^{\text{iv}}(x) = \frac{d}{dx}(-\cos x) = \sin x,$$

lead us back to our starting point. Hence we conclude that $f^{(8)}(x) = \sin x$, $f^{(12)}(x) = \sin x$, and so on, until finally, $f^{(24)}(x) = \sin x$. Now it is clear that $f^{(27)}(x) = f'''(x) = -\cos x$.

The first derivative measures rate of change. In particular, if a moving body is located at the point s of a number scale t time units after some initial instant, then $\frac{ds}{dt}$ measures the rate of change of position with respect to time; that is, $\frac{ds}{dt}$ measures the **velocity** of the moving point. So we write $v = \frac{ds}{dt}$. The second derivative of s with respect to t, $\frac{d^2s}{dt^2}$, is the number $\frac{d}{dt}\left(\frac{ds}{dt}\right) = \frac{dv}{dt}$. Thus $\frac{d^2s}{dt^2}$ *measures the rate of change of velocity with respect to time.* This rate is called the **acceleration** of the moving body. For example, in the formula $s = 16t^2$ (the relation between displacement measured in feet and time measured in seconds for a body falling from rest near the surface of the earth) we see that $v = \frac{ds}{dt} = 32t$ and $\frac{d^2s}{dt^2} = 32$. Thus the number 32 measures the acceleration of the body. Acceleration is measured in units of velocity per unit of time, so the acceleration of our falling body is 32 feet per second per second.

Example 20-3. If a body moves along a number scale in such a way that it is at the point s after t units of time have elapsed, and if $s = A \sin (at + b)$, we say the body is moving in **simple harmonic motion.** Show that the acceleration of the body is proportional to its position.

Solution. We are to show that the acceleration $\dfrac{d^2 s}{dt^2}$ and the position s are related by an equation $\dfrac{d^2 s}{dt^2} = ks$ for some number k. You may readily verify that $\dfrac{ds}{dt} = aA \cos (at + b)$ and $\dfrac{d^2 s}{dt^2} = -a^2 A \sin (at + b)$. Thus $\dfrac{d^2 s}{dt^2} = -a^2 s$, so the number $-a^2$ is our constant of proportionality, and our assertion is verified.

We have already pointed out that equations involving rates—that is, differential equations—are of great importance in science. A *differential equation of the first order* contains first, but not higher, derivatives; a *differential equation of the second order* contains second, but not higher, derivatives, and so on. For example, we have just seen that $s = A \sin (at + b)$ satisfies the differential equation of the second order $s'' + a^2 s = 0$.

Example 20-4. Show that $y = 1/x$ satisfies the differential equation of the first order (i) $xy' + y = 0$ and the differential equation of the second order (ii) $x^2 y'' + xy' - y = 0$. Show that $y = x$ satisfies (ii) but not (i).

Solution. It is a simple matter to make the required substitutions, and we leave it to you. As you can see, however, we would have much more of a problem on our hands if the question read, "Solve equations (i) and (ii) for y." We will learn how to solve equation (i) in Chapter 7.

Example 20-5. Show that if $f''(x) = 0$ for each $x \in R^1$, then f is a linear function.

Solution. It is easy to show that *if f is linear* (that is, $f(x) = mx + b$), then $y = mx + b$ satisfies the differential equation $y'' = 0$. We are asking the more difficult converse—namely, to show that the only possible solutions are determined by linear functions. We are to find numbers m and b such that $f(x) = mx + b$. The equation $f''(x) = 0$ says that $f'(x)$ and 0 have the same derivative, and so Theorem 19-5 tells us that there is a number m such that $f'(x) = 0 + m = m$. This equation, in turn, says that $f(x)$ and mx have the same derivative, so we can again apply Theorem 19-5 to see that there is a number b such that $f(x) = mx + b$, as we were to show.

PROBLEMS 20

1. Find y''.

(a) $y = \dfrac{1 - x}{1 + x}$ (b) $y = x + 1/x$ (c) $y = 2 \sin x$

(d) $y = \sin 2x$ (e) $y = \tan (1/x)$ (f) $y = 1/\tan x$

2. Find $f''(3)$.
    (a) $f(x) = x^3 - 2x^2 + 3x - 5$ (b) $f(x) = (x - 1)^4$
    (c) $f(x) = \sin(\pi x/9)$ (d) $f(x) = \cos(\pi/x)$
    (e) $f(x) = x|x|$ (f) $f(x) = x^3[\![x + \tfrac{1}{2}]\!] + x[\![x^3 + \tfrac{1}{2}]\!]$

3. (a) Show that for any pair of numbers a and b, $y = ax^2 + bx$ satisfies the differential equation $x^2y'' - 2xy' + 2y = 0$.
    (b) Choose numbers a and b so that y also satisfies the conditions $y = 3$ and $y' = 8$ when $x = 1$.

4. Show that $y = \sin kx$ satisfies the differential equation $y'' + k^2y = 0$. For what choices of k is it also true that $y = 0$ when $x = 0$ and $y = 0$ when $x = \pi$?

5. Show that $y = t \sin t$ satisfies the following differential equations.
    (a) $y'' + y = 2 \cos t$ (b) $y^{(4)} - y = -4 \cos t$

6. Determine k so that $y = \sec kx$ satisfies the differential equation $y'' + y = 2y^3$.

7. Let $f(x) = x^{13}$ and show that

    $$f^{(n)}(x) = \frac{13!}{(13 - n)!} x^{13-n} \text{ if } n \leq 13 \quad \text{and} \quad f^{(n)}(x) = 0 \text{ if } n > 13.$$

8. Verify the following equations.
    (a) $\cos^{(n)} x = \cos(x + \tfrac{1}{2}n\pi)$ (b) $\sin^{(n)} x = \sin(x + \tfrac{1}{2}n\pi)$

9. Show that $\dfrac{d^n}{dx^n} |f(x)| = \dfrac{f(x)}{|f(x)|} \dfrac{d^n}{dx^n} f(x)$.

10. Suppose that $f(a) = f(b) = f'(a) = f'(b) = 0$. Show that there are at least two solutions of the equation $f''(x) = 0$ in the interval (a, b). Illustrate that result if $f(x) = (x^2 - 1)^2$ and $[a, b] = [-1, 1]$.

11. (a) Show that $\dfrac{d^2y}{dx^2}$ and $\left(\dfrac{dy}{dx}\right)^2$ are not always the same.

    (b) Show that if $\dfrac{d^2y}{dx^2} = \left(\dfrac{dy}{dx}\right)^2$ (that is, $y'' = y'^2$), then $\dfrac{d}{dx}(x + 1/y') = 0$. What does this equation tell us about y'? Does it tell us anything about y?

12. Show that if $f'''(x) = 0$ for each $x \in R^1$, then f is a quadratic function; that is, show that there are numbers a, b, and c such that $f(x) = ax^2 + bx + c$.

13. Show that $(uv)'' = u''v + 2u'v' + uv''$ and $(uv)''' = u'''v + 3u''v' + 3u'v'' + uv'''$. Do these formulas remind you of the Binomial Theorem? If so, see if you can guess the formulas for $(uv)^{(4)}$, $(uv)^{(5)}$, and so on.

21. TAYLOR'S FORMULA

This section is devoted to an extension of the Theorem of the Mean, so let us recall the formula that expresses that theorem (Equation 19-1):

(21-1) $$f(b) = f(a) + f'(m)(b - a).$$

Here m belongs to the interval (a, b), and we are, of course, assuming that f is continuous in $[a, b]$ and differentiable in (a, b). If we assume that f' is continuous, too, and that b is close to a, then $f'(m)$ will be close to $f'(a)$, and Equation 21-1 gives us the approximation formula

(21-2) $$f(b) \approx f(a) + f'(a)(b - a).$$

Since the tangent line to the graph of f at the point $(a, f(a))$ contains this point and has slope $f'(a)$, its equation is $y = f(a) + f'(a)(x - a)$. Therefore the right-hand side of Approximation 21-2 is simply the Y-coordinate of the point in which the tangent line intersects the line $x = b$. In other words, we are approximating the value of f with the number we would get by following the tangent line rather than the graph of f itself (Fig. 21-1).

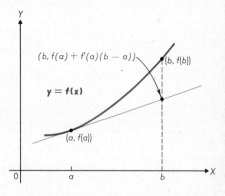

Figure 21-1

In order to see how good this approximation is, we use the idea behind our proof of the Theorem of the Mean and let

(21-3) $\quad E(x) = f(x) - f(a) - f'(a)(x - a)$

$\qquad - [f(b) - f(a) - f'(a)(b - a)]$

$\qquad \times (b - a)^{-2}(x - a)^2.$

Although this formula looks rather fearful, it is simply an expression of the form $E(x) = f(x) - A - Bx - Cx^2$, where A, B, and C are chosen so that

$$E(a) = E'(a) = E(b) = 0.$$

(You can verify these equations by substituting in Equation 21-3). Because $E(a)$ and $E(b)$ are both 0, we can apply Rolle's Theorem to see that there is a number m_1 between a and b such that $E'(m_1) = 0$. But $E'(a)$ is also 0, so we can apply Rolle's Theorem to E'. Therefore there is a number m between a and m_1 [and hence in (a, b)] such that $E''(m) = 0$. From Equation 21-3 we see that

$$E''(m) = f''(m) - 2[f(b) - f(a) - f'(a)(b - a)](b - a)^{-2},$$

so the equation $E''(m) = 0$ is equivalent to

(21-4) $\qquad f(b) = f(a) + f'(a)(b - a) + f''(m)\dfrac{(b - a)^2}{2}.$

Thus *the error that one makes by using Approximation 21-2 is $\frac{1}{2}f''(m)(b - a)^2$, where m is a number between a and b.*

Example 21-1. Find an approximate value of $\sqrt{105}$.

Solution. We will let $f(x) = \sqrt{x}$, $a = 100$, and $b = 105$. Since $f'(x) = 1/2\sqrt{x}$, Approximation 21-2 becomes

$$\sqrt{105} \approx 100 + \frac{1}{2\sqrt{100}}(105 - 100) = 10 + \frac{1}{4} = 10.25.$$

According to Equation 21-4, this result is in error by $\frac{25}{2}f''(m)$, where $m \in (100, 105)$. In our case, $\frac{25}{2}f''(m) = -\frac{25}{8}m^{-3/2}$, and (because $m > 100$) the absolute value of this number is less than $\frac{25}{8}100^{-3/2} = \frac{1}{320}$. We have therefore shown that $\sqrt{105} = 10.25$, correct to two decimal places.

Example 21-2. A cubical box with inside dimensions of 4 inches by 4 inches by 4 inches is made of lead $\frac{1}{32}$ inch thick. If lead weighs 5 pounds per cubic inch, approximately how much does the box weigh?

Solution. If we let $f(x) = x^3$, then the exterior of our box encloses a volume of $(4\frac{1}{16})^3 = f(4\frac{1}{16})$ cubic inches. On the other hand, the volume of the interior of the box is $4^3 = f(4)$ cubic inches. Therefore it requires

$$f(4\tfrac{1}{16}) - f(4)$$

cubic inches of lead to make our box. From Formula 21-2 (with $b = 4\frac{1}{16}$ and $a = 4$),

$$f(4\tfrac{1}{16}) - f(4) \approx f'(4)\tfrac{1}{16} = 3 \cdot 4^2 \cdot \tfrac{1}{16} = 3,$$

so our box is made out of about 3 cubic inches of lead and weighs approximately 15 pounds.

Equation 21-4 tells us that our approximation is in error by $\frac{1}{2}(\frac{1}{16})^2 f''(m) = \frac{3}{256}m$ cubic inches, where m is a number between 4 and $4\frac{1}{16}$. Our error is then surely less than $\frac{3}{256} \cdot 4\frac{1}{16} < .05$. Since we are dealing with a box whose lead content is about 3 cubic inches, our *relative error* is no worse than .05/3, which is less than two parts in a hundred.

We can think of Formula 21-4 as consisting of a linear approximation plus a quadratic error term. It can be generalized to a quadratic approximation plus a cubic error term, a cubic approximation plus a quartic error term, and so on. The result is known as **Taylor's Formula.**

Theorem 21-1. *Let n be an integer that is greater than or equal to 0 and suppose that $f^{(n)}$ is continuous in $[a, b]$ and differentiable in (a, b). Then there is a number $m \in (a, b)$ such that*

$$(21\text{-}5) \quad f(b) = f(a) + f'(a)(b - a) + \cdots + f^{(n)}(a) \frac{(b - a)^n}{n!}$$

$$+ f^{(n+1)}(m) \frac{(b - a)^{n+1}}{(n + 1)!} .$$

(By definition, $n! = 1 \cdot 2 \cdots n$.)

PROOF

(This proof is a direct generalization of our verification of Equation 21-4.) We let

$$(21\text{-}6) \quad E(x) = f(x) - f(a) - f'(a)(x - a) - \cdots - f^{(n)}(a) \frac{(x - a)^n}{n!}$$

$$- \left[f(b) - f(a) - \cdots - f^{(n)}(a) \frac{(b - a)^n}{n!} \right] (b - a)^{-n-1}(x - a)^{n+1}.$$

Then $E(a) = E'(a) = \cdots = E^{(n)}(a) = E(b) = 0$. Since $E(a)$ and $E(b)$ are both 0, Rolle's Theorem tells us that there is a number $m_1 \in (a, b)$ such that $E'(m_1) = 0$. Since $E'(a) = 0$, too, we may apply Rolle's Theorem to E' to see that there is a number $m_2 \in (a, m_1) \subseteq (a, b)$ such that $E''(m_2) = 0$. Now we apply Rolle's Theorem to E'', and so on. We finally see that there is a number $m \in (a, b)$ such that $E^{(n+1)}(m) = 0$. Simply by differentiating both sides of Equation 21-6 $n + 1$ times, replacing x with m, and setting the result equal to 0, we arrive at our desired Equation 21-5.

Taylor's Formula and our proof of it may look a little complicated, but *applying* the formula is easy. We proceed just as we do with Equation 21-4: *we decide what $f(x)$ is, what a, b, and n are, and simply substitute.* Notice that the Theorem of the Mean is the special case of Taylor's Formula in which $n = 0$, and Equation 21-4 is the case in which $n = 1$.

Example 21-3. Express the polynomial $f(x) = x^3 + 2x^2 - x + 5$ as a polynomial in $x - 2$.

Solution. We will set $n = 3$, $b = x$, and $a = 2$. Then $f(a) = f(2) = 19$, $f'(a) = f'(2) = 19$, $f''(a) = f''(2) = 16$, $f'''(2) = 6$, and Taylor's Formula reads

$$f(x) = 19 + 19(x - 2) + 16 \frac{(x - 2)^2}{2!} + 6 \frac{(x - 2)^3}{3!} + f^{(4)}(m) \frac{(x - 2)^4}{4!},$$

where m is some number (Theorem 21-1 doesn't tell us which) between 2 and x. Since $f(x)$ is a polynomial of degree 3, however, its fourth derivative $f^{(4)}(x)$ is 0 for each x. In particular, $f^{(4)}(m) = 0$, and so, after some simplification and re-arrangement, our expression for $f(x)$ in powers of $(x - 2)$ is

$$f(x) = (x - 2)^3 + 8(x - 2)^2 + 19(x - 2) + 19.$$

Since Taylor's Formula provides us with a *polynomial* approximation to values of a given function, the only mathematical operations we use when applying it are the elementary ones of addition, subtraction, multiplication, and division. If b is reasonably close to a and n is fairly large, we can often get very close approximations in this way, and, of course, it is easy to program a computer to do the arithmetic for us.

Example 21-4.   Show how Taylor's Formula could be used to compute the entries in a five-place table of values of the sine function.

Solution.   We set $f(x) = \sin x$, $a = 0$, and $n = 10$ in Taylor's Formula. Then $f(a) = \sin 0 = 0$, $f'(a) = \cos 0 = 1$, $f''(a) = -\sin 0 = 0$, $f'''(a) = -\cos 0 = -1, \ldots, f^{(10)}(a) = -\sin 0 = 0$, and $f^{(11)}(m) = -\cos m$. Thus for any number b, Taylor's Formula becomes

$$\sin b = 0 + 1(b - 0) + 0\,\frac{(b - 0)^2}{2!} + \cdots + 0\,\frac{(b - 0)^{10}}{10!} + (-\cos m)\,\frac{(b - 0)^{11}}{11!},$$

where m is a number between 0 and b. If we drop the last term and simplify, we obtain the approximation formula

(21-7)
$$\sin b \approx b - \frac{b^3}{3!} + \frac{b^5}{5!} - \frac{b^7}{7!} + \frac{b^9}{9!}.$$

The accuracy of this approximation can be judged by examining the size of the term that we dropped. In other words, the error we make by using the right-hand side of Formula 21-7 as an approximation to the left-hand side is $(-\cos m)b^{11}/11!$. All we know about m is that it is a number between 0 and b, but whatever it is, $|\cos m| \leq 1$, and so the absolute error is certainly no greater than $|b|^{11}/11!$. A table of trigonometric values need only contain values of the sine function that correspond to b in the interval $(0, \frac{1}{2}\pi)$, and therefore our absolute error will be less than $(\frac{1}{2}\pi)^{11}/11!$. We will leave it to you to show that this number is less than .000004. Hence Formula 21-7 can be used to compute values of the sine function for a five-place table, and the error (from rounding off) in any particular value would be, at worst, .00001.

Figure 21-2 shows output of a computer programmed to calculate with Formula 21-7. In the first column b varies from 0 to 1.6 (about $\frac{1}{2}\pi$) in steps of .2.

The second column shows the sum of the first three terms of Approximation 21-7, and the third column lists the sum of all five terms. The fourth column gives correct four-place values of the sine function. These figures should make you a believer in Taylor's Formula.

b	3 terms	5 terms	sin b
0.0	0.0000	0.0000	0.0000
0.2	0.1987	0.1987	0.1987
0.4	0.3894	0.3894	0.3894
0.6	0.5646	0.5646	0.5646
0.8	0.7174	0.7174	0.7174
1.0	0.8417	0.8415	0.8415
1.2	0.9327	0.9320	0.9320
1.4	0.9875	0.9855	0.9854
1.6	1.0047	0.9996	0.9996

Figure 21-2

PROBLEMS   21

1. Use Formula 21-2 to approximate the following numbers.

   (a) $\sqrt{26}$ (b) $\sqrt{120}$ (c) $\sqrt[5]{34}$ (d) $\sqrt[3]{9}$

   (e) $\dfrac{1}{\sqrt[3]{65}}$ (f) $\dfrac{1}{\sqrt[4]{85}}$ (g) sin 1°   (h) cos 1°

2. Use $\sqrt{2} \approx 1.4142$ and $\frac{1}{4}\pi \approx .7854$ to compute an approximation to sin .78 with the aid of Formula 21-2.

3. A washer with an outside radius of b inches and an inside radius of a inches has an area of A square inches. Let $f(r) = \pi r^2$ and use Formula 21-2 to find a formula for a number A_1 that is an approximation to A. From the formula for A, show that $A_1 < A$. How does this inequality follow from Equation 21-4?

4. Find a formula for the approximate volume of material in
   (a) a thin cylindrical shell h inches tall with an inside radius of r inches and a wall thickness of t inches.
   (b) a spherical shell of the same material with an inside radius of r inches.

5. A sum of P dollars invested at $x\%$ interest compounded annually is worth $P(1 + x/100)^n$ dollars at the end of n years. Show that this number is approximately $P(1 + nx/100)$, which is what the investment would be worth at simple interest of $x\%$.

6. Use Equation 21-4 to show that sin $(a + x) \approx$ sin $a + x$ cos a, with an error of no more than $\frac{1}{2}x^2$.

7. (a) What approximation formula do you get when you let $f(x) = \cos x$, $a = 0$, $n = 3$, and drop the error term in Taylor's Formula?
    (b) Use this formula to approximate cos 1. Is the result too large or too small?
    (c) Use Taylor's Formula to compute an error bound and see how it compares with the actual error (which you can find by consulting Table I).

8. Let $f(x) = \sqrt{x}$, $a = 100$, $b = 105$, $n = 2$, and drop the error term in Taylor's Formula. How does the resulting approximation of $\sqrt{105}$ compare with what we found in Example 21-1?

9. Express the polynomial $x^4 + 3x^2 + 2$ as a polynomial in $x - 1$. As a polynomial in $x + 1$.

10. Let $f(x) = (p + x)^5$, $a = 0$, $b = q$, and $n = 5$ in Equation 21-5. Compare the result with what you get when you expand $(p + q)^5$ by the Binomial Theorem.

11. (a) What approximation formula do we get when we let $f(x) = \sin x^2$, $a = 0$, $n = 4$, and drop the error term in Taylor's Formula?
    (b) What if we simply replace b with b^2 in Approximation 21-7?

12. Find an upper bound for the error incurred when using the approximation $\tan x \approx x$ for $x \in [0, .1]$.

13. Let $f(x) = x^3|x|$, $a = -1$, $b = 1$, and $n = 2$ in Taylor's Formula. What is m? What if we set $n = 3$?

14. In order to find an approximation to the number $f(b)g(b)$ in terms of the values of f, g, f', and g' at a, we could write $p(x) = f(x)g(x)$ and apply Formula 21-2 with f replaced by p. Or we could use Formula 21-2 to obtain approximations to $f(b)$ and $g(b)$ and then multiply the results. How do the answers obtained by these two methods differ?

15. It is clear that *if* $f(x)$ is a polynomial of degree n or less, then $f^{(n+1)}$ is the constant function with value 0. Show that the converse of this statement is also true.

It is convenient to make two lists of the differentiation formulas we have developed in this chapter. You should know these formulas by heart and be able to use them to differentiate combinations of algebraic and trigonometric expressions.

General Differentiation Formulas

(G1)
$$\frac{d}{dx} f(u) = \frac{df(u)}{du} \frac{du}{dx}$$

(G2)
$$\frac{d}{dx} [mf(x) + ng(x)] = m \frac{df(x)}{dx} + n \frac{dg(x)}{dx}$$

(G3)
$$\frac{d}{dx} [f(x)g(x)] = f(x) \frac{dg(x)}{dx} + g(x) \frac{df(x)}{dx}$$

(G4)
$$\frac{d}{dx} \frac{f(x)}{g(x)} = \frac{g(x) \dfrac{df(x)}{dx} - f(x) \dfrac{dg(x)}{dx}}{g(x)^2}$$

Specific Differentiation Formulas

(S1)
$$\frac{du^r}{dx} = ru^{r-1}\frac{du}{dx}$$

(S2)
$$\frac{d\sin u}{dx} = \cos u\,\frac{du}{dx}$$

(S3)
$$\frac{d\cos u}{dx} = -\sin u\,\frac{du}{dx}$$

(S4)
$$\frac{d\tan u}{dx} = \sec^2 u\,\frac{du}{dx}$$

(S5)
$$\frac{d\cot u}{dx} = -\csc^2 u\,\frac{du}{dx}$$

(S6)
$$\frac{d\sec u}{dx} = \sec u\tan u\,\frac{du}{dx}$$

(S7)
$$\frac{d\csc u}{dx} = -\csc u\cot u\,\frac{du}{dx}$$

You should also have a good geometric feeling for what a derivative is supposed to represent. A positive derivative denotes a rising curve, a negative derivative a falling curve, and so on. Bear in mind, too, that a derivative is a *rate*; it is this view of derivatives that is useful in other disciplines. It is hard to grasp the full significance of the Chain Rule (Equation G1) at the outset, and this is also true for the Theorem of the Mean,

$$f'(m) = \frac{f(b) - f(a)}{b - a}.$$

But at least this latter theorem is easy to interpret geometrically.

This chapter was mostly devoted to the "what's" and "how's" of differentiation. Some of the "why's" may become more apparent in the next chapter, when we take up applications.

Review Problems, Chapter Two

1. Find.

(a) $\dfrac{d}{dx}\sin^3 x$ (b) $\dfrac{d}{dx}\sin x^3$ (c) $\dfrac{d^3}{dx^3}\sin x$

(d) $\left(\dfrac{d\sin x}{dx}\right)^3$ (e) $\dfrac{d}{dx}(3\sin x)$ (f) $\dfrac{d}{dx}\sin 3x$

2. Solve the following inequalities.

(a) $\left(\dfrac{d}{dx}x^{10}\right)^2 \le \dfrac{d}{dx}(x^{10})^2$

(b) $\left(\dfrac{d}{dx}x^{-10}\right)^2 \le \dfrac{d}{dx}(x^{-10})^2$

(c) $\dfrac{d}{dx}\sin^2 x \le \sin\dfrac{dx^2}{dx}$

(d) $\dfrac{d}{dx}\cos^2 x \le \cos\dfrac{dx^2}{dx}$

3. Answer the following questions as best you can, referring to the graph of f that appears in Fig. II-3. The light lines represent tangent lines.

(a) Find $3f(7) - 7f(3)$.   (b) Find $3f'(7) - 7f'(3)$.

(c) Find $f''(2)/|f''(2)|$.   (d) Find $g'(1)$ if $g(x) = f(x)^2$.

(e) Find $h'(-1)$ if $h(x) = f(x^2)$.   (f) Find $q'(3)$ if $q(x) = f(x)/x$.

(g) What is the largest interval in which $f'(x) > 0$?

(h) What is the largest interval in which $f''(x) < 0$?

(i) Find $T'(1)$ if $T(x) = \tan f(x)$.   (j) Find $c'(1)$ if $c(x) = f(f(x))$.

(k) The Theorem of the Mean says that there is at least one point m in the interval $(-2, 1)$ so that $(f(1) - f(-2))/3 = f'(m)$. Actually, how many of these points are there?

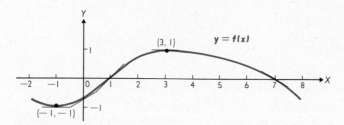

Figure II-3

4. Which of the following are necessarily the same?

(a) $\dfrac{d}{dx}f(x^2)$ and $f'(x^2)$

(b) $\dfrac{d}{dx}[f(x) + g(x)]$ and $f'(x) + g'(x)$

(c) $\dfrac{d}{dx}[f(x)g(x)]$ and $f'(x)g'(x)$

(d) $\dfrac{d}{dx}f(2)$ and $f\left(\dfrac{d2}{dx}\right)$

5. In Chapter 6 we are going to find that $\dfrac{d}{dx}2^x = .69 \cdot 2^x$ (the number .69 is only approximately correct, but it will suffice for the purposes of this problem). Use this fact and the Chain Rule to find $\dfrac{d}{dx}4^x$.

6. Is there a function f for which $\dfrac{d}{dx}[1/f(x)] = 1 \Big/ \dfrac{df(x)}{dx}$?

7. Suppose that for each $x \in R^1$, $f(x)g'(x) = f'(x)g(x)$ and that $g(x) \ne 0$. Furthermore, suppose that $f(0) = 5g(0)$. Then show that for each $x \in R^1$, $f(x) = 5g(x)$. (*Hint:* Consider the quotient $f(x)/g(x)$.)

8. If $f''(x) > 0$ for each x in an interval $[a, b]$, can you conclude that f is increasing in (a, b)? What if you also know that $f'(a) \ge 0$?

9. Think of the given expression as a difference quotient and find the required limit.

(a) $\lim\limits_{z \to 0} \dfrac{\sqrt{z + 4} - 2}{z}$

(b) $\lim\limits_{z \to 3} \dfrac{z^2 - 4z + 3}{z - 3}$

(c) $\lim\limits_{z \to 0} \dfrac{\cos z^2 - 1}{z}$

(d) $\lim\limits_{z \to 2} \dfrac{\cos z^2 - \cos 4}{z - 2}$

(e) $\lim\limits_{z \to \frac{1}{4}\pi} \dfrac{\sin 2z - 1}{z - \frac{1}{4}\pi}$

(f) $\lim\limits_{z \to 0} \dfrac{\sec^2 z - 1}{z}$

10. The Remainder Theorem of elementary algebra says that if $P(x)$ is a polynomial in x and r is any number, then there is a polynomial $Q(x)$ such that $P(x) = Q(x)(x - r) + P(r)$. What is $\lim\limits_{x \to r} Q(x)$?

11. Use elementary algebra to show that between each pair of numbers a and b, with $a < b$, there is a number m such that $m^2 = \frac{1}{3}(b^2 + ab + a^2)$ and then obtain the same result from the Theorem of the Mean. Which method is easier? Notice that a and b may have opposite signs. (*Hint:* Let $f(x) = x^3$.)

12. A sprinter covers $\frac{1}{2}(100 + t^2 - |t^2 - 100|)$ meters in t seconds. How fast is he going when $t = 5$? How fast is he going when he hits the tape at the end of 100 meters? What is the tape made of?

13. (a) If my car gets $f(x)$ miles per gallon at a speed of x (≥ 30) miles per hour, is $f'(x) > 0$ or $f'(x) < 0$?

(b) If your car hits $p(x)$ pedestrians after you have had x drinks, is $p'(x) > 0$ or $p'(x) < 0$?

(c) If your car is worth $d(x)$ dollars after being driven x miles, is $d'(x) > 0$ or $d'(x) < 0$? Is $d''(x) > 0$ or $d''(x) < 0$? What is $d''(100,000)$?

(d) If my car goes $m(x)$ miles on x gallons of gas and $m'(x) < 13$, am I happy or sad?

14. A ferris wheel with a radius of 16 feet revolves at 3 rpm. Assume that the sun is directly overhead and that its rays are parallel.

(a) How far is the shadow of a rider from the shadow of the center of the wheel t minutes after he passes the lowest point?

(b) How fast is his shadow moving when it is 8 feet from the shadow of the center?

(c) How fast is its velocity changing then?

(d) When is the rider's shadow moving the fastest? The slowest?

15. Find the intervals in which f is increasing if $f(x) = \sin(\cos x)$.

16. Show that if f is increasing in some interval I, then $-f$ and $1/f$ are decreasing in I.

17. Show that if f is increasing in some interval I and if g is increasing, then the composition of f by g (assuming it exists) is increasing in I.

18. If $f' = g'$, then Theorem 19-5 says that there is a number C such that $f(x) = g(x) + C$. Show that if $f'' = g''$, then there are numbers C and D such that $f(x) = g(x) + C + Dx$. In general, find the relation between $f(x)$ and $g(x)$ if $f^{(n)} = g^{(n)}$ for a given integer n.

19. Let $f(x)$ be the distance between x and the nearest perfect square. Thus $f(17) = 17 - 16 = 1$, $f(\pi) = 4 - \pi = .85840\ldots$, and so on. Find $f'(\pi)$ and $f'(1776)$. What is $f''(x)$? At what points do $f'(x)$ and $f''(x)$ fail to exist?

20. If $f(x) = x/10 + x^2 \sin(1/x)$ for $x \neq 0$, and $f(0) = 0$, then (see Problem 13-12) $f'(0) = \frac{1}{10} > 0$. Is there a positive number r such that f is increasing in $(0, r)$?

3 | Applications of the Derivative

Attempts to solve the problems that arise in natural science, from Newton's study of gravity to the present research in nuclear physics, have required the development of suitable mathematical tools. It is easy to see why calculus ranks high among them. In science we study relations; for example, we might be interested in the relation between the velocity of an earth satellite and its distance from the earth. We would ask such questions as "At what distances is the velocity increasing? At what distances is it decreasing? At what distances is the velocity greatest? At what distances is it least?" Calculus can help us answer such questions.

In this chapter we will look at some of the simpler applications of calculus to geometry, as well as to some problems in the "real world." Your courses in the natural and social sciences will present you with many meaningful problems to which calculus can provide the solutions.

22. CURVE SKETCHING: CONCAVITY, MAXIMA, AND MINIMA

When presented formally, some of the concepts of calculus can look rather forbidding. But we can usually bring them down to earth by looking at a picture. For example, the inequality $f'(x) > 0$ simply says that the curve $y = f(x)$ is rising, whereas $f'(x) < 0$ says it is falling. Now let us see what the sign of the second derivative can tell us.

Since $f''(x)$ is the derivative of $f'(x)$, it follows that if $f''(x) > 0$ in some interval, then f' is increasing in that interval. Figure 22-1 is designed to help us translate this statement into one about the graph of f. Suppose that P_1, P_2, and P_3 are the points of the graph of f that correspond to numbers x_1, x_2, and x_3 such that $x_1 < x_2 < x_3$. Then since f' is increasing, the tangent line at P_2 has a larger slope than the tangent line at P_1, and the tangent line at P_3 has a still larger slope. The segments we have drawn at P_1, P_2, and P_3 show this behavior. Since a curve is approximated in a neighborhood of one of its points by the tangent line at the point, Fig. 22-1 suggests that the graph of our function f may look like the one shown in Fig. 22-2.

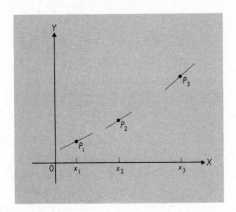

Figure 22-1

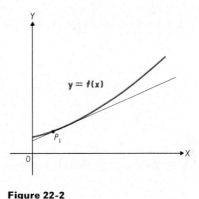

Figure 22-2

If we draw a tangent line to any point of the arc in Fig. 22-2, then the entire arc (with the obvious exception of the point of tangency) lies above it. Such an arc is said to be **concave up**. If an arc lies below every tangent line, it is **concave down**.

We are led to the arc shown in Fig. 22-2 by trying to find the geometric implications of the inequality $f''(x) > 0$. From our informal analysis, it looks as if the inequality implies that the graph of f is concave up, and the following theorem confirms this result.

Theorem   22-1.   *If $f''(x) > 0$ for every x in the interior of an interval I, then the graph of f is concave up in I. If $f''(x) < 0$ for every x in the interior of I, then the graph of f is concave down in I.*

PROOF

We will only write out the proof for the case in which $f''(x) > 0$. We are to show that if we draw the tangent line to a point $(a, f(a))$, where $a \in I$, then the part of the graph of f that projects onto I lies above this line. Therefore (see Fig. 21-1) if b is another number in I, we must show that the difference $f(b) - [f(a) + f'(a)(b - a)]$ is positive. But equation 21-4 tells us that this difference is the number $\frac{1}{2}f''(m)(b - a)^2$, which is positive, since $m \in (a, b)$ and we are assuming that f'' takes only positive values in I.

Example 22-1. Discuss the concavity of the curve $y = x^3 - 12x^2 + 7x - 4$.

Solution. Since $y'' = 6x - 24 = 6(x - 4)$, we see that $y'' > 0$ when $x > 4$ and $y'' < 0$ when $x < 4$. Therefore the part of our graph that projects onto the interval $(4, \infty)$ of the X-axis is concave up and the part that projects onto the interval $(-\infty, 4)$ is concave down.

Example 22-2. Sketch the graph of f if

$$f(x) = \frac{x^2}{8} - \frac{1}{x}.$$

Solution. For certain choices of x, we tabulated both

$$y = \frac{x^2}{8} - \frac{1}{x} \quad \text{and} \quad y' = \frac{x}{4} + \frac{1}{x^2}$$

in Fig. 22-3. In the left-hand coordinate system we plotted these points, and in the right-hand coordinate system we drew through each point (x, y) a short segment with slope y'. These segments are tangent to the graph of f.

Next we calculate

$$y'' = \frac{1}{4} - \frac{2}{x^3} = 2\left(\frac{1}{8} - \frac{1}{x^3}\right),$$

and hence $y'' < 0$ if $0 < x < 2$, whereas $y'' > 0$ if $x < 0$ or if $x > 2$. Thus our curve will be concave up when $x < 0$ or when $x > 2$, and it will be concave down when x is in the interval $(0, 2)$. So the curve lies above the tangent lines for points to the left of the Y-axis and for points to the right of the vertical line $x = 2$. The curve lies below the tangent lines for points between the Y-axis and the line $x = 2$. Figure 22-4 shows the graph of our function.

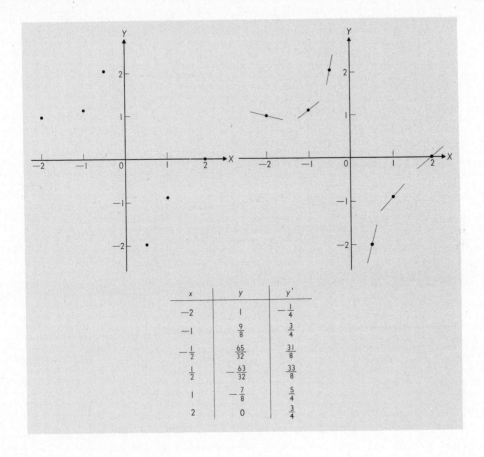

x	y	y'
-2	1	$-\frac{1}{4}$
-1	$\frac{9}{8}$	$\frac{3}{4}$
$-\frac{1}{2}$	$\frac{65}{32}$	$\frac{31}{8}$
$\frac{1}{2}$	$-\frac{63}{32}$	$\frac{33}{8}$
1	$-\frac{7}{8}$	$\frac{5}{4}$
2	0	$\frac{3}{4}$

The point labeled P of the curve in Fig. 22-4 is lower (its Y-coordinate is smaller) than every other point near it. We call such a point a **minimum point**. Similarly, a **maximum point** of a curve is the highest point in a neighborhood of it. Many authors use the term *relative* minimum (or *local* minimum) where we have simply said minimum point, because the Y-coordinate of a minimum point is only compared with the Y-coordinates of nearby points, not *all* Y-coordinates. For example, there are points of the graph in Fig. 22-4 that are lower than the minimum point P, so P is not an "absolute" minimum point. It is just the lowest point in its vicinity.

From Theorem 19-1 we know that: *if c belongs to an open interval in the domain of a differentiable function f, and if $(c, f(c))$ is a maximum or a minimum point of the graph of f, then $f'(c) = 0$.* This statement is illustrated in Fig. 22-4, where it is clear that the slope of the graph at the point P is 0. Thus in order to

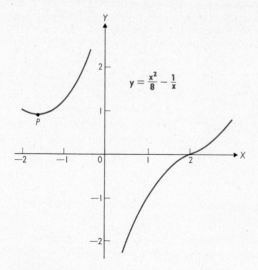

$$y = \frac{x^2}{8} - \frac{1}{x}$$

Figure 22-4

find the X-coordinate of P, we will solve the equation $f'(x) = 0$. Since $f'(x) = \frac{1}{4}x + x^{-2}$, this equation is

$$\frac{x}{4} + \frac{1}{x^2} = 0,$$

from which we find $x = -\sqrt[3]{4} = -1.6$. Now we calculate $f(-1.6) = [(-1.6)^2/8] + (1/1.6) = .95$, to see that P is the point $(-1.6, .95)$.

The equation $f'(c) = 0$ is neither a necessary nor a sufficient condition that the point $(c, f(c))$ be a maximum or a minimum point. Figure 22-5 illustrates

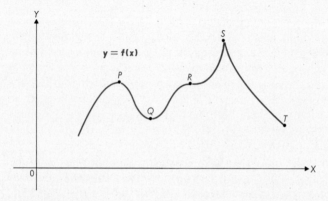

$y = f(x)$

Figure 22-5

various possibilities. At the maximum point P and the minimum point Q, the curve has a horizontal tangent. But the tangent is also horizontal at the point R, which is neither a maximum nor a minimum point. Furthermore, if we confine our search for maximum and minimum points to those points whose X-coordinates satisfy the equation $f'(x) = 0$, we will miss the maximum point S and the minimum point T. The graph doesn't have a slope at S (so it can't have slope 0 there), and the point T is an **endpoint**.

These remarks tell us that we are to look for maximum and minimum points of the graph of a function f among the points (x, y) such that

(1) $f'(x) = 0$,

(2) $f'(x)$ does not exist, or

(3) (x, y) is an endpoint of the graph.

We will call a point that satisfies one of these conditions a **critical point**. *A maximum or a minimum point must be a critical point, but critical points need not be maximum or minimum points.*

Example 22-3.   Find the maximum and minimum points and sketch the graph of f if $f(x) = 2\sqrt{x} - x$.

Solution.   The domain of f is the interval $[0, \infty)$. The derivative exists at all points in $(0, \infty)$, and

$$f'(x) = \frac{1}{\sqrt{x}} - 1.$$

Since the only solution of the equation $f'(x) = 0$ is 1, the critical points of our graph are the point $(1, 1)$ at which the tangent is horizontal and the endpoint $(0, 0)$. When we write $f(x) = \sqrt{x}\,(2 - \sqrt{x})$, it is clear that if x is slightly larger than 0, then $f(x) > 0$, and so $(0, 0)$ is a minimum point. On the other hand, $1 - f(x) = 1 - 2\sqrt{x} + x = (1 - \sqrt{x})^2$, which is positive if $x \neq 1$, and we see that $(1, 1)$ is a maximum point. Since $f''(x) = -x^{-3/2}$, a negative number, the curve is concave down. The graph of f appears in Fig. 22-6.

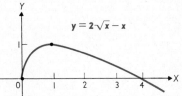

Figure 22-6

In order to find a rule for selecting maximum and minimum points from among the critical points of the graph of a function f, let us again refer to Fig. 22-5. At points to the left of the maximum points P and S, the derived function f' takes positive values (f is increasing), whereas for points to the right of P and S, f' takes negative values. The situation is reversed at the minimum point Q. In a neighborhood of a critical point such as R, the derivatives of $f(x)$ have the same signs. The following theorem, the **First Derivative Test**, tells this story in more formal terms. Its proof follows directly from Theorem 19-4.

Theorem 22-2.   *If f is continuous at c and is differentiable in some punctured neighborhood of c, then*

(i)   *f(c) is a maximum value of f if f'(x) is positive for each x in some left neighborhood of c and negative for each x in some right neighborhood.*

(ii)   *f(c) is a minimum value of f if f'(x) is negative for each x in some left neighborhood of c and positive for each x in some right neighborhood.*

(iii)   *f(c) is neither a maximum nor a minimum value of f if f'(x) has the same sign for each x in some punctured neighborhood of c.*

Example 22-4.   Apply the First Derivative Test to the function of Example 22-3.

Solution.   Let us first look at the critical number $c = 1$. If x is slightly less than 1, then $f'(x) = 1/\sqrt{x} - 1$ is positive, whereas if x is slightly greater than 1, $f'(x)$ is negative. Therefore statement (i) of our theorem applies, and it says that $f(1)$ is a maximum value of f. Furthermore, since $f'(x) > 0$ for each $x \in (0, 1)$, statement (ii) (modified for the endpoint case) tells us that $(0, 0)$ is a minimum point of the graph of f.

In applying the First Derivative Test to a critical point $(c, f(c))$, we must consider signs of values of f' in a neighborhood of c. Now we will introduce a test for maxima and minima that requires us to check the sign of only one number, the number $f''(c)$. This test is the **Second Derivative Test**, and you will notice that it applies only to those critical points at which the first derivative is 0, whereas the First Derivative Test applies to other types of critical points as well.

Theorem 22-3.   *Suppose that f is differentiable in an open interval that contains c and that $f'(c) = 0$. Then*

(i)   *f(c) is a maximum value if $f''(c) < 0$, and*

(ii)   *f(c) is a minimum value if $f''(c) > 0$.*

If $f''(c) = 0$, this test does not apply.

PROOF

We will base the proof of statement (i) on statement (i) of Theorem 22-2, which says that $f(c)$ is a maximum value if $f'(x)$ is positive for x slightly less than c and $f'(x)$ is negative for x slightly greater than c. We can combine these criteria by saying that $f'(x)/(x - c)$ is to be negative when x is near c. By hypothesis, $f'(c) = 0$, so this quotient can be written as

$$\frac{f'(x) - f'(c)}{x - c}.$$

This latter number is simply the difference quotient whose limit is $f''(c)$. Since $f''(c)$ is negative, it follows (see Example 10-2) that if x is sufficiently close to c, our quotient is also negative and statement (i) is proved. Statement (ii) follows from statement (ii) of Theorem 22-2 by similar reasoning.

Our discussion has shown that there are two steps in finding the maximum and minimum points of a given graph:

(1)   We first locate the critical points, and then

(2)   We test each critical point, perhaps by using the First or Second Derivative Test, to see which are maximum points, minimum points, or neither.

Example 22-5.   If $f(x) = \sin^{2/3} x$ in the interval $[0, 2\pi]$, test f for maximum and minimum points.

Solution.   If f is differentiable at x,

$$f'(x) = \tfrac{2}{3} \sin^{-1/3} x \cos x,$$

and f is not differentiable at the points for which $\sin x = 0$. Thus in this example we have critical points of all three kinds; $f'(x) = 0$ when $x = \tfrac{1}{2}\pi$ and $\tfrac{3}{2}\pi$, $f'(x)$ does not exist when $x = \pi$, and we must consider the endpoints $(0, 0)$ and $(2\pi, 0)$. The signs of $f'(x)$, for $x \in (0, 2\pi)$, are indicated in Fig. 22-7 [$*$ means that $f'(x)$ does not exist; 0 means that $f'(x) = 0$], and from this figure and Theorem 22-2 we see that $(0, 0)$, $(\pi, 0)$, and $(2\pi, 0)$ are minimum points, whereas $(\tfrac{1}{2}\pi, 1)$ and $(\tfrac{3}{2}\pi, 1)$ are maximum points. We find, furthermore, that $f''(x) = -\tfrac{2}{9} \sin^{-4/3} x \cos^2 x - \tfrac{2}{3} \sin^{2/3} x$. Therefore $f''(\tfrac{1}{2}\pi) = f''(\tfrac{3}{2}\pi) = -\tfrac{2}{3}$, and the fact that the second derivatives of f are negative at the points at which f' is 0 also shows, according to Theorem 22-3, that $f(\tfrac{1}{2}\pi)$ and $f(\tfrac{3}{2}\pi)$ are maximum values of f.

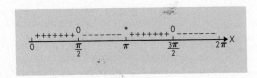

Figure 22-7

PROBLEMS 22

1. Let $y = x + 1/x$.

(a) Complete the following table.

x	-2	-1	$-\frac{1}{2}$	$\frac{1}{2}$	1	2
y						
y'						
y''						

(b) Discuss the concavity of the graph of the equation.

(c) Find the maximum and minimum points.

(d) Sketch the graph of the equation, showing the tangent lines at the points of the table, for x in the set $[-2, 0) \cup (0, 2]$.

2. Find the maximum and minimum points of the following curves.

(a) $y = x^3 - x^2 + 1$ (b) $y = x^2 - x^{-2}$

(c) $y = (x - 1)^{2/3}$ (d) $y = x(x + 1)^{-1}$

3. For each curve in the preceding problem, find the intervals in which it is concave up and sketch it.

4. Find all maximum and minimum values of f.

(a) $f(x) = 4x^3 + 9x^2 - 12x + 5$ (b) $f(x) = 3x^5 - 5x^3 - 8$

5. Sketch the curve, making use of information about maximum and minimum points and concavity.

(a) $y = (x^2 + 1)^{-1/2}$ (b) $y = \dfrac{x^{3/2}}{(2 - x)^{1/2}}$

(c) $y = x^2 + 2x^{-1}$ (d) $y = (x - 1)^3(x - 3)$

(e) $y = 3x^4 - 4x^3 + 1$ (f) $y = x(x - 1)^{1/2}$

6. Find the X-coordinates of the maximum and minimum points of the curve.

(a) $y = x^{1/2}(x^3 + 1)^{-1/2}$ (b) $y = x^{4/3} + 4x^{1/3}$

(c) $y = (\sin x + 1)^2$ (d) $y = \cos^2 x + 2 \cos x$

(e) $y = \sin 2x + 2 \cos x$ (f) $y = \cos 2x + 2 \sin x$

7. Show that the Second Derivative Test applies to the maximum points of the curve $y = |\cos x|$ but not to the minimum points.

8. Sketch the graphs of the following equations for x in the interval $[0, 2\pi]$, making use of your knowledge of maximum and minimum points, concavity, and so on.

(a) $y = \sin^{1/3} x$ (b) $y = \sin^{2/3} x$

(c) $y = \sin^4 x$ (d) $y = \sin^{4/3} x$

9. Find the maximum and minimum points of the graph of f.

(a) $f(x) = |x^2 - 1|$

(b) $f(x) = |\sin x|$

(c) $f(x) = \dfrac{x(x - 2)}{(x + 1)^2}$

(d) $f(x) = x^{3/2}(x - 1)^{2/3}$

(e) $f(x) = 1 - |\sin x|$

(f) $f(x) = \cos x + \sin x + |\cos x - \sin x|$

(g) $f(x) = \sqrt{x} \cos x$, $x \in [0, \frac{1}{2}\pi]$

(h) $f(x) = \sqrt{2 - x} - \sqrt{x + 1}$

10. If f is differentiable in a right and a left neighborhood of a point m at which it is continuous, and if the values of f'' are all positive in one of the two neighborhoods and negative in the other, then $(m, f(m))$ is a **point of inflection** of the graph of f. Find points of inflection for the following functions.

(a) $f(x) = x^3 + x - 1$

(b) $f(x) = \sqrt{x} + \dfrac{1}{\sqrt{x}}$

(c) $f(x) = \tan x$

(d) $f(x) = \tan x + \cot x$

(e) $f(x) = x|x|$

(f) $f(x) = \cos x + \sin x + |\cos x - \sin x|$

11. Find $f([a, b])$ for the following functions and intervals.

(a) $f(x) = x + \dfrac{1}{x}$, $[\frac{1}{2}, 3]$

(b) $f(x) = \dfrac{\sqrt{x}}{1 + x^2}$, $[0, \infty]$

12. What is the range of f if $f(x) = \sqrt{3 - x} - \sqrt{x + 2}$?

13. Discuss the problem of finding maximum and minimum values of f if $f(x) = (\sin x)/x$.

14. Suppose that $f(x) = ax + b/x$, where a and b are positive numbers. Show that the inequality $f(1) \geq f(m)$, where $f(m)$ is the minimum value of f in the interval $(0, \infty)$, states that the geometric mean of a and b cannot exceed their arithmetic mean.

15. If $f''(x) > 0$ for every number x in the interval $[a, b]$, then we know from Theorem 22-1 that the graph of f lies above any tangent line. Show that the graph lies below the chord that joins the points $(a, f(a))$ and $(b, f(b))$.

23. PROBLEMS INVOLVING MAXIMA AND MINIMA

Some of the most important problems in mathematics and its applications can be solved by finding maximum or minimum values of functions. Although the problems discussed in this section can hardly be classified as "important," they may indicate how calculus can be applied in real-life situations.

In Illustration 5-4 we showed that if a cylindrical tin can with a capacity of 54π cubic inches is constructed from material costing .012¢ per square inch for

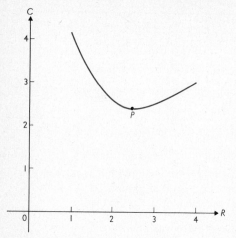

Figure 23-1

the side and .021¢ per square inch for the top and bottom, then the cost c in cents and the base radius r in inches are related by the equation

$$(23\text{-}1) \quad c = \frac{108\pi}{r}(.012) + 2\pi r^2 (.021).$$

For each choice of r, we obtain a number c. Thus, for example, if we select $r = 2$ we obtain $c = 2.57$, so a can with a 2-inch base radius costs 2.57 cents. A can with a base radius of 3 inches costs 2.54 cents, and so on. Since a different choice of a number r as radius may yield a different number c as cost, a natural question to ask is, "What choice of r results in the *smallest* number c?" That is, "What is the radius of the cheapest can?"

Our knowledge of the maximum and minimum points of graphs will help us answer this question. Figure 23-1 shows the graph of Equation 23-1 for $r > 0$. If (r, c) is a point of this graph, then a can with a radius of r inches costs c cents. The point labeled P has the smallest C-coordinate of any point of the graph. Therefore the R-coordinate of P represents the radius, and the C-coordinate of P represents the cost, of the cheapest can. So our question, "What is the radius of the cheapest can?" will be answered when we find the R-coordinate of the minimum point P.

It is clear from Fig. 23-1 that we can find this coordinate by solving the equation $\frac{dc}{dr} = 0$. From Equation 23-1 we find that

$$(23\text{-}2) \qquad \frac{dc}{dr} = -\frac{108\pi(.012)}{r^2} + 4\pi r(.021),$$

and so the equation $\frac{dc}{dr} = 0$ becomes

$$4\pi r(.021) - \frac{108\pi}{r^2}(.012) = 0.$$

Hence the radius of the cheapest can is

$$r = \sqrt[3]{\frac{108\pi(.012)}{4\pi(.021)}} = 2.49 \text{ inches.}$$

To find its cost, we simply replace r in Equation 23-1 with 2.49, and we find that $c = 2.45$. The cheapest can has a radius of 2.49 inches and costs 2.45 cents.

We only used the graph in Fig. 23-1 to convince ourselves that we could find the minimum point P by solving the equation $\frac{dc}{dr} = 0$. By relying on our previous theory, we could have dispensed with drawing the picture. The only critical points of the graph are those at which $\frac{dc}{dr} = 0$. If we have any doubts about whether such a critical point is a maximum point, a minimum point, or neither, we can consider the second derivative

$$\frac{d^2c}{dr^2} = 4\pi(.021) + (216\pi)(.012)/r^3.$$

Thus, $\frac{d^2c}{dr^2} > 0$ for every $r > 0$; the graph is concave up, and hence a point at which $\frac{dc}{dr} = 0$ is a minimum point. The First Derivative Test or the Second Derivative Test could also be used to verify that we have a minimum point.

In order to solve the "word problems" of this section, it will be worthwhile to follow these steps:

(1)   Read the problem carefully and make a sketch to illustrate it if you can.

(2)   Determine exactly what quantity you want to be a maximum or a minimum.

(3)   Express this quantity in terms of some one other quantity, thereby obtaining an equation of the form $y = f(x)$, $A = g(r)$, or the like.

(4)   Find the critical points of the graph of this equation and select those of the type you seek—maximum points or minimum points.

(5)   Reread the problem to see explicitly what question was asked and then see if you can answer it with the information you now have.

Example 23-1.   A farmer with 100 yards of fencing constructs a rectangular enclosure, using a stone wall as one side and fencing in the other three sides. What is the largest area he can enclose?

Solution.   We follow our five steps.

1. From its description, we make the drawing of the enclosure shown in Fig. 23-2.
2. We wish to find the maximum area.
3. If the side of the enclosure perpendicular to the wall is x yards long, then the side parallel to the wall is $100 - 2x$ yards long. So if the area is A square yards, then

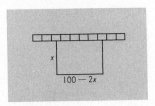

(23-3) $A = x(100 - 2x) = 100x - 2x^2.$ **Figure 23-2**

4. From the nature of the problem, x must satisfy the inequalities $0 < x < 50$ in order to determine a proper rectangle. So there are no endpoints to consider, and the X-coordinate of the critical point is found from the equation

$$\frac{dA}{dx} = 100 - 4x = 0.$$

Thus $x = 25$, and since $\dfrac{d^2A}{dx^2} = -4 < 0$, The Second Derivative Test tells us that we have found a maximum point.

5. Finally, we were asked to find the maximum area. We now know that the maximum area is obtained when $x = 25$, so we set $x = 25$ in Equation 23-3:

$$A_{max} = 25(100 - 50) = 1250 \text{ square yards.}$$

Example 23-2.   A man who can swim 20 feet per second and run 25 feet per second stands at the edge of a circular swimming pool that has a radius of 50 feet. He wishes to reach a point one-quarter of the way around the pool in the least possible time, and he plans to run along the edge of the pool for a way and then dive in and swim straight to his destination. What is the shortest possible time required for this maneuver?

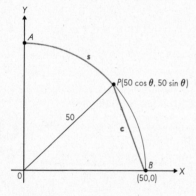

Figure 23-3

Solution.   Again we follow our five steps.

1. In Fig. 23-3 we have sketched a quarter circle that represents the perimeter of the pool. Our man will run along the circle from A to P, a distance of s feet, and then swim from P to B, a distance of c feet.
2. We wish to minimize the time it takes him to get from A to B.
3. The total number of seconds, t, that it takes our man to go from A to B is the sum of the time required to go from A to P and the time required to go from P to B. It takes $\frac{1}{25}s$ seconds to run from A to P and then $\frac{1}{20}c$ seconds to swim to B, and hence

(23-4) $$t = \tfrac{1}{25}s + \tfrac{1}{20}c.$$

We have now expressed t in terms of s and c; and in order to complete this step, we must express t in terms of *some one* quantity. So let us write both s and c in terms of the number θ, the width in radians of $\angle BOP$. The coordinates of P are $(50 \cos \theta, 50 \sin \theta)$, and so the Distance Formula gives the distance between P and the point $B(50, 0)$ as

$$c = \sqrt{(50 \cos \theta - 50)^2 + (50 \sin \theta)^2}$$

$$= 50\sqrt{\cos^2 \theta - 2 \cos \theta + 1 + \sin^2 \theta}$$

$$= 50\sqrt{2(1 - \cos \theta)}$$

$$= 50\sqrt{4 \sin^2 \tfrac{1}{2}\theta} = 100 \sin \tfrac{1}{2}\theta.$$

Furthermore, since $\angle POA$ is $\frac{1}{2}\pi - \theta$ radians wide, $s = 50(\frac{1}{2}\pi - \theta)$. Now we replace s and c in Equation 23-4 with their expressions in terms of θ, and we thus express t in terms of a single quantity:

(23-5)
$$t = \tfrac{1}{25}50(\tfrac{1}{2}\pi - \theta) + \tfrac{1}{20}100 \sin \tfrac{1}{2}\theta$$
$$= \pi - 2\theta + 5 \sin \tfrac{1}{2}\theta.$$

The statement of the problem tells us that $\theta \in [0, \frac{1}{2}\pi]$.

4. The critical points of the graph of Equation 23-5 are the endpoints—the points at which $\theta = 0$ and $\theta = \frac{1}{2}\pi$—and the points at which $\dfrac{dt}{d\theta} = 0$. We have

$$\frac{dt}{d\theta} = -2 + \tfrac{5}{2} \cos \tfrac{1}{2}\theta \quad \text{and} \quad \frac{d^2t}{d\theta^2} = -\tfrac{5}{4} \sin \tfrac{1}{2}\theta.$$

Thus we see that $\dfrac{d^2t}{d\theta^2} < 0$ for $\theta \in (0, \frac{1}{2}\pi)$, and so our curve is concave down. Hence the only minimum points are the endpoints of the arc.

5. From Equation 23-5 we find that $t = \pi$ when $\theta = 0$, and $t = \frac{5}{2}\sqrt{2}$ when $\theta = \frac{1}{2}\pi$. Since $\frac{5}{2}\sqrt{2} > \pi$, our man uses the least time when $\theta = 0$ (that is, he runs all the way and doesn't swim at all), and this minimum time is π seconds.

PROBLEMS 23

1. Show that the product of two positive numbers whose sum is 1 is a maximum when they are equal.

2. Show that the sum of two positive numbers whose product is 1 is a minimum when they are equal.

3. If x thousand sheep are in a certain pasture, each can eat $18 + 2/x^2$ pounds of grass per day (crowding results in less intake per animal). How many sheep should we put in the pasture if we want the least total amount of grass per day eaten?

4. (a) A bus company offers to charter a bus for a round trip excursion for x passengers at a fare of $15 - \frac{1}{4}x$ dollars per passenger. What number of passengers produces the greatest revenue, and how much does each passenger pay?
   (b) Suppose that the company sets the fare at \$10 less 25¢ for each passenger over 20.

5. Find the dimensions of the right triangle with a hypotenuse 10 units long whose area is a maximum by each of the following methods.
   (a) Suppose that one leg is x units long, express the area in terms of x, and maximize it.
   (b) Suppose that one of the acute angles is x radians wide, express the area in terms of x, and maximize it.

6. Equal squares are cut from each corner of a piece of tin measuring 6 inches by 6 inches, and the edges are then turned up to form a rectangular box with no lid. What is the volume of the largest such box?

7. A page is to contain 24 square inches of print. The margins at the top and bottom are each to be $1\frac{1}{2}$ inches. Each margin at the side is to be 1 inch. What are the dimensions of the smallest such page?

8. A wire 2 feet long is to be cut into two pieces. One piece is then used to form a circle, and the other piece is used to form a square. How much wire should be used for the square if the sum of the areas of the circular and square regions is to be a minimum?

9. If our man in Example 23-2 takes the longest possible time to get from A to B (even though he runs and swims at top speed), at what point does he dive in?

10. Suppose that it requires S square inches of tin to make a cylindrical can with a top and bottom. Show that for such a can to be of maximum volume, the diameter of the base must equal the altitude.

11. An E-volt battery has an internal resistance of r ohms. If a resistor of R ohms is connected across the battery terminals, then a current of I amps will flow and generate P watts of power, where

$$I = \frac{E}{R + r} \quad \text{and} \quad P = I^2 R.$$

For what resistance R will the power generated be a maximum?

12. An irrigation ditch is to have a cross section in the shape of an isosceles trapezoid, wider at the top than at the bottom, and with a horizontal base. The bottom and the equal sides of the trapezoid are each to be L feet long. Suppose that the side of the ditch makes an acute angle θ radians wide with the horizontal.
   (a) Show that the cross-sectional area of the ditch is given by the equation $A = L^2 \sin \theta (1 + \cos \theta)$.
   (b) For what angle will the carrying capacity of the ditch be a maximum?

13. A wall h feet high stands d feet away from a tall building. A ladder L feet long reaches from the ground outside the wall to the building. Suppose that the acute angle between the ladder and the horizontal is θ radians wide.
   (a) Show that if the ladder touches the top of the wall, $L = h \csc \theta + d \sec \theta$.
   (b) Find the shortest ladder that will reach the building if $h = 8$ and $d = 24$.

14. Two cottages are located on the shores of a circular lake that is 2 miles in diameter. Cottage A is one-fourth of the way around the lake from Cottage B. Where should a third cottage be located so that the sum of its distances from cottages A and B is a maximum?

15. An experiment is performed n times, and n numbers (measurements) $x_1, x_2, \ldots, x_n$ are obtained. In order to obtain an "average" number, we choose the number x that makes the sum of the squares of the differences between x and each x_i the least; that is, we choose the number x that makes $s = (x - x_1)^2 + (x - x_2)^2 + \cdots + (x - x_n)^2$ a minimum. Show that we get the usual arithmetic mean as our "average" number.

16.   While driving through the north woods, Bill runs out of gas. He can reach the nearest gas station either by walking 1 mile north and 10 miles east along paved roads or by walking through the woods in a northeasterly direction until he reaches the paved road and then proceeding east along the road to the station. Suppose that he can walk 5 miles per hour along the paved roads and 3 miles per hour through the woods. Describe the path he should follow to reach the station in the shortest time.

24.   LINEAR MOTION AND RELATED RATES

We want to recall and extend some of the ideas about rates that were introduced in Section 14, and we will start by illustrating them with an example. If an arrow shot straight up by Hiawatha at an initial velocity of 96 feet per second is s feet above its launching site after t seconds, then

$$(24\text{-}1) \qquad\qquad s = -16t^2 + 96t.$$

Of course, Hiawatha didn't know this formula. He had to wait for Newton to come along. We, too, have not yet had enough calculus to derive Equation 24-1 (we will do it in Section 60). For our purposes here, we can just accept it as correct.

We think of Hiawatha's arrow as traveling up a number scale whose origin is the launching site. The arrow is at the point s of the scale after t seconds. The velocity of the arrow is the rate of change of s with respect to t:

$$(24\text{-}2) \qquad\qquad v = \frac{ds}{dt} = -32t + 96.$$

Notice that $\frac{ds}{dt}$ may be positive or it may be negative   A positive value of $\frac{ds}{dt}$ (positive velocity) indicates that s is increasing, so the arrow is moving in the positive direction of the number scale (upward in this case). A negative value of v tells us that the arrow is coming down. The absolute value $|v|$ is called the **speed** of the arrow. Since we are measuring distance in feet and time in seconds, speed is measured in feet per second.

The number $a = \frac{dv}{dt}$ measures the rate of change of velocity with respect to time—that is, the acceleration of the arrow:

$$(24\text{-}3) \qquad\qquad a = \frac{dv}{dt} = -32$$

feet per second per second. Because the derivative $\frac{dv}{dt}$ is negative, we see that the velocity of the arrow decreases as time increases. It starts off at 96 feet per second in the positive direction and slows down to 0 feet per second. Then it begins to fall, *but its velocity continues to decrease* through negative values. The *speed* of the arrow increases as it falls.

Example 24-1. Find the maximum height, the total time of flight, and the maximum velocity of Hiawatha's arrow.

Solution. The arrow reaches its maximum height when $\dfrac{ds}{dt} = v = -32t + 96 = 0$ —that is, 3 seconds after launching. According to Equation 24-1, $s = -16 \cdot 3^2 + 96 \cdot 3 = 144$ feet at that moment. We will assume that an arrow shot straight up will return to its starting point (poor Hiawatha). Hence the duration of its flight is found by solving the equation $s = 0$; that is, $-16t^2 + 96t = 0$. There are two solutions of this equation: $t = 0$ and $t = 6$. When $t = 0$, the arrow is at its launching point at the start of the flight; and when $t = 6$, it is at its launching point after completing its flight. Thus its total time of flight is 6 seconds. The arrow's velocity decreases during its flight, so the maximum value of the velocity occurs when $t = 0$, and it is 96.

Our arrow problem is a typical example of motion of a body along a line. As the mathematical model of such motion we consider a point moving along a number scale (Fig. 24-1). Let s denote its coordinate t seconds after some initial

Figure 24-1

instant. The sign of s, of course, tells us on which side of the origin the moving point is located. It need not be true that $s = 0$ when $t = 0$. If the velocity of our moving point is v, its acceleration is a, and its speed is r, then, by definition

$$v = \frac{ds}{dt},$$

(24-4)
$$a = \frac{dv}{dt} = \frac{d^2s}{dt^2},$$

$$r = |v|.$$

Example 24-2. Discuss the motion of a point that moves along the number scale of Fig. 24-1 in accordance with the equation $s = t + 1/t$ for $t > 0$.

Solution. We have

$$s = t + t^{-1},$$

$$v = \frac{ds}{dt} = 1 - t^{-2},$$

$$a = \frac{d^2s}{dt^2} = 2t^{-3}.$$

For t close to 0, the number s is very large, and hence our point is very far from the origin. For t in the interval $(0, 1)$, v is negative, and hence s is decreasing. In other words, the point moves to the left. When $t = 1$, $v = 0$, and for $t > 1$, $v > 0$. Thus our point stops and reverses direction when $t = 1$. At this instant, $s = 2$, so the closest the point comes to the origin is 2 units to the right of it. For $t > 1$, the point moves to the right, and as time goes on it recedes indefinitely. The acceleration is always positive; that is, the velocity is continually increasing. Notice, however, that the velocity always remains less than 1.

Example 24-3. Show that a point moving along a line is speeding up in time intervals in which its velocity and acceleration have the same sign and slowing down in time intervals in which they have opposite signs.

Solution. We are to show that the speed r is increasing or decreasing in the indicated intervals, so we look at the sign of its derivative $\dfrac{dr}{dt}$. From the definitions of r and a,

$$\frac{dr}{dt} = \frac{d}{dt}|v| = \frac{v}{|v|}\frac{dv}{dt} = \frac{va}{|v|}.$$

This number will be positive (hence r is increasing) when v and a have the same sign and negative (hence r is decreasing) when they have opposite signs.

In Examples 24-1 and 24-2 we found the rate of change of s with respect to t in a straightforward way; we had an expression for s in terms of t, and we differentiated it. Sometimes, however, we want the rate of change of a quantity that we cannot, or at least find it inconvenient to, express in terms of time. Here is an example.

At the instant the radius of a melting spherical snowball is 4 inches, its volume is decreasing at the rate of 2 cubic inches per minute. How fast is the radius changing at that instant? In the notation of calculus, we are asked to find the number $\dfrac{dr}{dt}$ when $r = 4$, where we are supposing that the snowball has a radius of r inches t minutes after some initial time. If we could express r in terms of t, we would simply have a problem in differentiation. But we are not given enough information to write r in terms of t, so we must attack the problem differently.

What we are given is that the volume of the snowball is decreasing at the rate of 2 cubic inches per minute (at the instant we are interested in). If the volume is V cubic inches, this statement can be written as the equation $\dfrac{dV}{dt} = -2$ (the minus sign indicates that the volume is decreasing). Thus we are given $\dfrac{dV}{dt}$, and we are asked to find $\dfrac{dr}{dt}$. In order to solve the problem, we will find the relation between these two rates. We first find a relation between V and r and then differentiate to find out how the rates of change of V and r are related. Because the snowball is spherical, the relation between V and r is

$$V = \tfrac{4}{3}\pi r^3.$$

Now we differentiate with respect to t (using the Power Formula 16-2) to find the relation between $\dfrac{dV}{dt}$ and $\dfrac{dr}{dt}$:

$$\frac{dV}{dt} = 4\pi r^2 \frac{dr}{dt}.$$

Since we are to find $\dfrac{dr}{dt}$ when $r = 4$, we replace r with 4 and $\dfrac{dV}{dt}$ with -2 in this equation: $-2 = 64\pi \dfrac{dr}{dt}$. It follows that

$$\frac{dr}{dt} = -1/32\pi \text{ inches per minute when } r = 4.$$

In a typical problem in related rates, we are given one of two rates $\dfrac{dx}{dt}$ or $\dfrac{dy}{dt}$ (or two of three, $\dfrac{dx}{dt}$, $\dfrac{dy}{dt}$, or $\dfrac{dz}{dt}$, and so on) and are asked to find the other rate at some particular instant of time. Of course, we may use letters other than x and y as, for example, we used V and r in the snowball problem. We follow these steps:

(1) Find an equation that expresses y in terms of x; a figure is often helpful.

(2) Differentiate with respect to t to obtain the relation between the given and the desired rates.

(3) Substitute the data given in the problem and evaluate the unknown rate.

Example 24-4. A cameraman is televising a 100-yard dash from a position 10 yards from the track in line with the finish line (Fig. 24-2). When the runners are 10 yards from the finish line, his camera is turning at the rate of $\frac{3}{5}$ radians per second. How fast are the runners moving then?

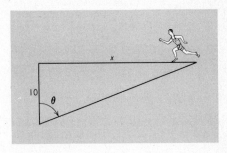

Solution. Figure 24-2 shows a runner x yards from the finish line, at which time the angle between the line joining the camera and the runners and the finish line extended to the camera is θ radians wide. We are told that $\dfrac{d\theta}{dt} = -\frac{3}{5}$ radians per second when

Figure 24-2

$x = 10$, and we follow our listed steps.
1. Figure 24-2 shows that $x = 10 \tan \theta$.
2. Therefore, according to the rule for differentiating $\tan \theta$ (Equation 18-10)

(24-5)
$$\frac{dx}{dt} = 10 \sec^2 \theta \, \frac{d\theta}{dt}.$$

3. We are interested in the speed of the runners when $x = 10$. At that point, $\theta = \frac{1}{4}\pi$, and hence $\sec \theta = \sqrt{2}$. So Equation 24-5 becomes

$$\frac{dx}{dt} = 10(\sqrt{2})^2(-\tfrac{3}{5}) = -12,$$

and we see that our runners are approaching the finish line at the rate of 12 yards per second.

Example 24-5. A freshman is drinking a soda through a straw from a conical cup 8 inches deep and 4 inches in diameter at the top. When the soda is 5 inches deep, he is drinking at the rate of 3 cubic inches per second. How fast is the level of the soda dropping?

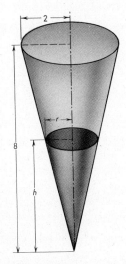

Solution. Suppose that t seconds after the freshman starts drinking, the soda is h inches deep and V cubic inches in volume. We are told that $\dfrac{dV}{dt} = -3$ when

$h = 5$, and we are asked to find $\dfrac{dh}{dt}$. We follow our three steps.
1. In Fig. 24-3 we have illustrated the situation at some instant of time. The formula for the volume of a cone gives us

Figure 24-3

(24-6) $V = \frac{1}{3}\pi r^2 h.$

Furthermore, from the similar triangles in the figure we see that $\dfrac{r}{2} = \dfrac{h}{8}$; that is,

(24-7)
$$r = \tfrac{1}{4}h.$$

We substitute this expression for r in Equation 24-6:

$$V = \tfrac{1}{3}\pi\tfrac{1}{16}h^2 h = \tfrac{1}{48}\pi h^3.$$

2. Therefore, from the Power Rule, $\dfrac{dV}{dt} = \tfrac{1}{16}\pi h^2 \dfrac{dh}{dt}$.

3. We substitute the given values $\dfrac{dV}{dt} = -3$ and $h = 5$, and we have $-3 = \tfrac{1}{16}\pi 5^2 \dfrac{dh}{dt}$, from which it follows that $\dfrac{dh}{dt} = -\dfrac{48}{25\pi}$. Thus the level of the soda is dropping at the rate of $\dfrac{48}{25\pi}$ inches per second.

We could have proceeded somewhat differently. If we apply the Product Rule to Equation 24-6, we find that

$$\frac{dV}{dt} = \tfrac{1}{3}\pi h \frac{dr^2}{dt} + \tfrac{1}{3}\pi r^2 \frac{dh}{dt} = \tfrac{2}{3}\pi r h \frac{dr}{dt} + \tfrac{1}{3}\pi r^2 \frac{dh}{dt}.$$

Also, Equation 24-7 gives us

$$\frac{dr}{dt} = \tfrac{1}{4}\frac{dh}{dt}.$$

Now we substitute our given values $\dfrac{dV}{dt} = -3$ and $h = 5$ in these equations and Equation 24-7 and solve for $\dfrac{dh}{dt}$. For our present problem, this second method is slightly more complicated than our first. There are problems, however, for which it is superior.

PROBLEMS 24

1. For the indicated time, find the position, velocity, speed, and acceleration of a point moving along a number scale in accordance with the given equation.

   (a) $s = t^3 + 2t^2$, $t = 2$ (b) $s = t + 2/t^2$, $t = \tfrac{1}{2}$
   (c) $s = \cos \tfrac{1}{3}\pi t$, $t = 1$ (d) $s = \tan \tfrac{1}{3}\pi t$, $t = 2$
   (e) $s = |t - 1| + |t - 3|^3$, $t = 2$ (f) $s = |\tfrac{1}{2} - \sin t|$, $t = \tfrac{4\,1}{4}\pi$

2. Discuss the motion of a point that moves along the number scale in accordance with the given equation.

   (a) $s = t^2 - 4t$, $t \in (0, \infty)$ (b) $s = 9t + \dfrac{4}{t}$, $t \in (0, \infty)$

   (c) $s = \sin t$, $t \in (0, \infty)$ (d) $s = 3 \cos 2t$, $t \in (0, \infty)$
   (e) $s = \tan t - 2t$, $t \in (0, \tfrac{1}{2}\pi)$ (f) $s = \sin t + \csc t$, $t \in (0, \pi)$

3. Discuss the motion of a point that moves along the number scale in accordance with the given equation.

   (a) $s = |t - 2|$, $t \in (0, 4)$ (b) $s = |t - 1| + 1 - t$, $t \in (0, \infty)$
   (c) $s = |\sin t|$, $t \in (0, \infty)$ (d) $s = |\cot t|$, $t \in (0, \pi)$

4. A point moves along the number scale in accordance with the equation $s = 3t^4 - 4t^3 + 2$ for $t > 0$. How close does it come to the origin? What is its minimum speed? What is its minimum velocity?

5. A projectile fired from a balloon is s feet above the ground t seconds after firing, and $s = -16t^2 + 96t + 256$.

   (a) How high was the balloon when the projectile was fired?
   (b) What is the maximum height reached by the projectile?
   (c) How fast is the projectile going when it hits the earth?

6. A rocket is s miles high t minutes after liftoff, and $s = 200 + 100 \sec (2t + \frac{2}{3}\pi)$.

   (a) How high does the rocket go?
   (b) How fast is the rocket going when it hits the earth?
   (c) How long does the flight last?
   (d) What is the maximum speed of the rocket?

7. A marble is dropped from a tall building, and T seconds later a second marble is dropped. What is the rate of change of distance between the marbles?

8. A 13-foot ladder leans against the wall of a house. Someone pulls the base of the ladder away from the house at the rate of 2 feet per second. How fast is the top of the ladder sliding down the wall when it is 5 feet from the ground?

9. A boat is moored to a dock by a 13-foot rope fastened 5 feet above the bow. A man pulls in the rope at the rate of 1 foot per second. What is the horizontal velocity of the boat at the instant he starts pulling?

10. A horizontal eaves trough 10 feet long has a triangular cross section 4 inches across the top and 4 inches deep. During a rainstorm, the water in the trough is rising at the rate of $\frac{1}{4}$ inch per minute when the depth is 2 inches. How fast is the volume of water in the trough increasing? After the rain has stopped, the water drains out of the trough at the rate of 40 cubic inches per minute. How fast is the surface of the water falling when it is 1 inch deep?

11. A man climbs a pole while holding one end of a rope that is wound around a spindle located 15 feet from the base of the pole. If he climbs at the rate of 1 foot per second, how fast is the rope coming off the spindle when he is 20 feet high?

12. An airplane flying at an altitude of 10,000 feet will pass directly over an observer on the ground. The observer notices that the angle between his line of sight to the plane and the horizontal is $\frac{1}{3}\pi$ radians wide and is increasing at the rate of .06 radian per second. How fast is the plane flying? How fast is the distance between the plane and the observer changing?

13. A 6-foot man walks at the rate of 5 feet per second directly away from a lamp 15 feet above the ground. Find the rate at which the end of his shadow is moving and the rate at which his shadow is lengthening when he is 20 feet away from the lamppost.

14. (a) A conical soda cup is 4 inches across the top and 6 inches deep. A soda jerk fills it at the rate of 2 cubic inches per second. What is the vertical velocity of a bubble on the top of the liquid when the soda is halfway to the top?

(b) What is the vertical velocity when $h = 3$ if the cup has a leak through which the soda is running out at the rate of 1 cubic inch per second as it is being filled?

15.   The radius of the base of a cone is increasing at the rate of 4 inches per minute, while the altitude is decreasing at the rate of 7 inches per minute. Is the volume of the cone increasing or decreasing at the instant the altitude and radius are equal?

16.   Show that the inequality $sv > pv$ indicates that our moving point is moving *away* from a point p of the number scale, whereas the inequality $sv < pv$ indicates that the moving point is moving *toward p*.

25. NEWTON'S METHOD

Solving equations is an important activity in mathematics and its applications. Some equations—linear and quadratic equations, for example—can be solved by straightforward methods. Others, such as

$$(25\text{-}1) \qquad\qquad \sin x - \frac{1}{x} = 0,$$

cannot be solved by simple formulas. If we write $f(x) = \sin x - 1/x$, Equation 25-1 reduces to $f(x) = 0$. The solutions of this equation are called **zeros** of the function f. A very useful technique for approximating a zero of a function is known as **Newton's Method**.

Newton's Method is best described in geometric language. A zero of f is the X-coordinate of a point in which the curve $y = f(x)$ cuts the X-axis. Thus r is a zero of the function whose graph appears in Fig. 25-1. We will suppose that we don't know r but that graphically or by using tables, we have found a number x_0 that approximates r. We wish to improve on this approximation. From Fig. 25-1 we see that the tangent line at $(x_0, f(x_0))$ intersects the X-axis in a point $(x_1, 0)$

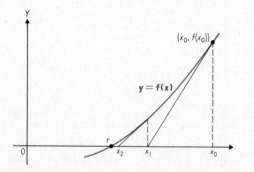

Figure 25-1

whose X-coordinate is a closer approximation of r than x_0 is. Since (as we mentioned at the beginning of Section 21) the equation of this tangent line is $y = f(x_0) + f'(x_0)(x - x_0)$, we see that

$$0 = f(x_0) + f'(x_0)(x_1 - x_0),$$

and hence

(25-2)
$$x_1 = x_0 - \frac{f(x_0)}{f'(x_0)}.$$

This equation gives us a rule that "transforms" an approximation x_0 into another approximation x_1. We can use the same rule to transform x_1 into a better approximation x_2; thus $x_2 = x_1 - f(x_1)/f'(x_1)$. The point $(x_2, 0)$, of course, is the point in which the tangent to the graph of f at $(x_1, f(x_1))$ intersects the X-axis, as shown in Fig. 25-1. We can repeat this procedure as often as we please and obtain a sequence of numbers $x_1, x_2, x_3, \ldots$, each of which is a closer approximation of r than its predecessor is.

We based our discussion of Newton's Method on Fig. 25-1, which shows a particular graph. Our arguments don't apply to graphs that are too "wiggly." In order to ensure that Newton's Method will always yield a sequence of numbers such that x_2 is a closer approximation of the desired zero than x_1 is, x_3 is still closer, and so on, we should stay in an interval in which $f'(x)f''(x) \neq 0$. Then the curve is either rising or falling and concave up (as in Fig. 25-1) or concave down. You might experiment with a few graphs to see how the sequence of numbers x_1, $x_2, \ldots$ "converges" toward r.

Example 25-1.   Use Newton's Method to approximate a solution of the equation $\sin x - 1/x = 0$ that is near $x_0 = \pi$.

Solution.   Here $f(x) = \sin x - 1/x$ and $x_0 = \pi \approx 3.142$. Equation 25-2 gives us the first approximation

$$x_1 = \pi - \frac{\sin \pi - (1/\pi)}{\cos \pi + (1/\pi)^2} \approx 2.789.$$

If we apply our transformation again, we get a second approximation,

$$x_2 = 2.789 - \frac{\sin 2.789 - (1/2.789)}{\cos 2.789 + (1/2.789)^2} \approx 2.776,$$

and additional approximations can be found in the same way. These computations are tedious to do by hand, but a computer can go through dozens of cycles of Newton's Method in the twinkling of an eye.

To get an idea of how the accuracy of our approximations $x_1, x_2, \ldots$ is improved at each step, let the numbers x_0 and r play the roles of a and b in Approximation Formula 21-2. The error in this approximation is the number

$$e(r) = f(r) - f(x_0) - f'(x_0)(r - x_0).$$

Since $f(r) = 0$, we can solve for r:

$$r = x_0 - \frac{f(x_0)}{f'(x_0)} - \frac{e(r)}{f'(x_0)}.$$

Now use Equation 25-2 to write this equation as $r = x_1 - e(r)/f'(x_0)$, and so

$$|r - x_1| = \frac{|e(r)|}{|f'(x_0)|}.$$

According to Equation 21-4, $e(r) = \frac{1}{2} f''(m)(r - x_0)^2$, where m is a number between x_0 and r. Therefore if we know that the values of f'' do not exceed some number M in the interval in which we are working, $|e(r)| \leq \frac{1}{2} M |r - x_0|^2$, and hence

$$|r - x_1| \leq \frac{M}{2|f'(x_0)|} |r - x_0|^2.$$

If we can now find a positive number m such that $|f'(x)| \geq m$ for every number x in our basic interval, then we will know that $|f'(x_0)| \geq m$, and therefore $M/2|f'(x_0)| \leq M/2m$. Thus we see that the absolute errors $|r - x_1|$ and $|r - x_0|$ are related by the inequality

(25-3)
$$|r - x_1| \leq \frac{M}{2m} |r - x_0|^2.$$

Clearly, this inequality applies to any two consecutive terms of the sequence $x_0, x_1, x_2, \ldots$, not just the first two.

Example 25-2.   To two decimal places, $\sqrt{45} \approx 6.71$. Use Newton's Method to find a closer approximation.

Solution.   The number $\sqrt{45}$ is a zero of the function defined by the equation $f(x) = x^2 - 45$. Here $f'(x) = 2x$, and if we use $x_0 = 6.71$, Equation 25-2 gives us

$$x_1 = 6.71 - \frac{(6.71)^2 - 45}{2(6.71)} \approx 6.70820.$$

Since $f''(x) = 2$, $M = 2$. Surely we need only consider numbers greater than $x = 6$, so that $f'(x) = 2x > 2 \cdot 6 = 12$. Therefore we can take $m = 12$, and Inequality 25-3 reads

$$|\sqrt{45} - x_1| \leq \tfrac{1}{12}|\sqrt{45} - 6.71|^2.$$

We said that 6.71 approximates $\sqrt{45}$ to two decimal places, and that statement means that

$$|\sqrt{45} - 6.71| \leq .005 = \tfrac{1}{2} \cdot 10^{-2}.$$

Therefore $$|\sqrt{45} - 6.71|^2 \leq \tfrac{1}{4} \cdot 10^{-4},$$

and we have

$$|\sqrt{45} - x_1| \leq \tfrac{1}{12} \cdot \tfrac{1}{4} \cdot 10^{-4} < \tfrac{1}{4} \cdot 10^{-5}.$$

We can conclude from this last inequality and our expression for x_1 that to five decimal places, $\sqrt{45} \approx 6.70820$.

PROBLEMS 25

1. Use one or two cycles of Newton's Method to approximate a solution near the given point.
   (a) $x^3 + x - 1 = 0$, $x_0 = 0$
   (b) $x^3 - 12x^2 + 45x - 35 = 0$, $x_0 = 1$
   (c) $x - \cos x = 0$, $x_0 = \tfrac{1}{4}\pi$
   (d) $1 - x - \sin x = 0$, $x_0 = \tfrac{1}{4}\pi$

2. Use Newton's Method to solve the equation $8x^3 + 8x - 5 = 0$, taking $x_0 = .5$ as your first approximation.

3. Use Newton's Method to approximate the solution of the equation in the indicated interval.
   (a) $x^3 + x = 1$, $[0, 1]$
   (b) $x^4 + x = 3$, $[1, 2]$

4. If $f(x) = x + 1 - \tan x$, you can easily convince yourself that f has a zero in the interval $(0, \tfrac{1}{2}\pi)$. What happens when you try to approximate this zero by Newton's Method, starting with $x_0 = 0$? With $x_0 = \tfrac{1}{4}\pi$?

5. Compare the results of approximating $\sqrt[3]{10}$ by the given method.
   (a) Let $f(x) = \sqrt[3]{x}$, $a = 8$, $b = 10$, and $n = 1$ in Taylor's Formula and discard the remainder. Let $n = 2$.
   (b) Use one cycle of Newton's Method with $f(x) = x^3 - 10$ and $x_0 = 2$. Use two cycles.

6. A limber stick 8 feet long is bent into a circular arc by tying the ends together with a piece of string 6 feet long. What is the radius of the arc?

7. Find the point of the curve $y = x^2$ that is closest to the point $(1, 0)$.

8. Apply Newton's Method twice to approximate a solution of the equation $x^2 - 25 = 0$, starting with $x_0 = 4$. What does Inequality 25-3 tell you about the ratio $|r - x_2|/|r - x_1|^2$? What is the actual ratio?

9. (a) Suppose that x_0 is an approximation of the number $\sqrt{a}$; that is, we are approximating a zero of $x^2 - a$. Show that Newton's Method leads to $\frac{1}{2}(x_0 + a/x_0)$ as the next approximation. In words, this rule reads, "To find a second approximation to the square root of a number a, divide a by the first approximation and average the resulting quotient and the first approximation." Apply this rule for a step or two to approximate a few square roots, such as $\sqrt{10}$, and see how well it works.

   (b) Sketch the curve $y = x^2 - a$ and conclude that if $x_0 > 0$, then $x_1 \geq \sqrt{a}$.

   (c) Verify the inequality in part (b) by showing that $x_1{}^2 - a = \frac{1}{4}(x_0 - a/x_0)^2$.

   (d) Convince yourself that (if $x_0 > 0$), then $x_2 \leq x_1$, $x_3 \leq x_2$, and so on.

10. Find a rule for approximating $\sqrt[n]{a}$, where n is an arbitrary positive integer.

11. Show that if f is a linear function that has exactly one zero, then Equation 25-2 leads to this zero, no matter what initial approximation x_0 we start with.

12. Use Theorem 10-1 to show that Equation 25-1 has a solution in the interval $(\frac{3}{4}\pi, \pi)$. Use Rolle's Theorem to show that there is only one solution in this interval.

13. Suppose that $f'(x)f''(x) > 0$ at each point of an interval I that contains a solution r of the equation $f(x) = 0$. By considering the graph of f, convince yourself that if x_0 is a point of I that is greater than r, then Newton's approximations line up like this: $x_0 \geq x_1 \geq x_2 \geq \cdots \geq r$. What if $x_0 < r$?

14. We say that a number A approximates a number N to **within p decimal places** if $|N - A| \leq \frac{1}{2}10^{-p}$ (Why?). Use Inequality 25-3 to show that if f is a function for which $M/m \leq 4$, and if x_0 approximates r to within p decimal places, then x_1 will approximate r to within $2p$ decimal places.

26. IMPLICIT DIFFERENTIATION

Suppose that the graph of a function—that is, the graph of an equation

$$(26\text{-}1) \qquad\qquad y = f(x),$$

is a subset of the graph of a more general equation

$$(26\text{-}2) \qquad\qquad A(x, y) = B(x, y).$$

For example, if $f(x) = \sqrt{4 - x^2}$, the graph of f (Fig. 12-1) is a subset of the circle

$$(26\text{-}3) \qquad\qquad x^2 + y^2 = 4.$$

To find the derivative of y at a point at which f is differentiable, we would naturally turn to Equation 26-1, which expresses y in terms of x. However, we can often find this derivative from Equation 26-2 instead. For example, if $y = f(x)$

satisfies Equation 26-3, we simply differentiate both sides of this latter equation with respect to x:

$$\frac{dx^2}{dx} + \frac{dy^2}{dx} = \frac{d4}{dx}.$$

Of course, $\frac{d}{dx} x^2 = 2x$ and $\frac{d4}{dx} = 0$. We find that $\frac{d}{dx} y^2 = 2y \frac{dy}{dx}$ by applying the Power Formula as built into the Chain Rule (Equation 16-2). Therefore

$$2x + 2y \frac{dy}{dx} = 0,$$

and so

$$\frac{dy}{dx} = -\frac{x}{y}.$$

This formula gives us the slopes of tangent lines to the curve $y = \sqrt{4 - x^2}$; in fact, it gives us the slope of the tangent line to any point of the circle $x^2 + y^2 = 4$.

Although we didn't use Equation 26-1 *explicitly* in our calculations, we did use it *implicitly* The equation $\frac{d}{dx} y^2 = 2y \frac{dy}{dx}$ makes no sense unless x and y are related by an equation like Equation 26-1. As we apply our general differentiation rules to the problems that follow, we will always assume that such a relation exists between y and x, although we won't always mention it. In the usual applications of this process of **implicit differentiation**, Equation 26-2 itself determines the relation between y and x. Thus in some neighborhood of a point (a, b) of the graph of Equation 26-2, we can (theoretically, at least) solve for y in terms of x and find Equation 26-1. We then say that our function f is *defined implicitly* by Equation 26-2. The question of when and how such an equation implicitly defines differentiable functions can safely be left to advanced calculus.

Example 26-1.   The point $(\sqrt{2}, \frac{1}{4}\pi)$ belongs to the curve $x \sin y = 2x^2 - 3$. Find the slope of the curve at this point.

Solution.   We differentiate both sides of our given equation with respect to x:

(26-4) $$\frac{d}{dx}(x \sin y) = \frac{d}{dx}(2x^2 - 3).$$

The right-hand side of this equation is $4x$, and

$$\frac{d}{dx}(x \sin y) = \sin y \frac{dx}{dx} + x \frac{d \sin y}{dx} \qquad \text{(Product Rule)}$$

$$= \sin y + x \cos y \frac{dy}{dx} \qquad \text{(Equation 18-3)}.$$

Equation 26-4 now becomes

$$\sin y + x \cos y \frac{dy}{dx} = 4x,$$

and we substitute the desired coordinates $(\sqrt{2}, \tfrac{1}{4}\pi)$:

$$\frac{1}{\sqrt{2}} + \frac{\sqrt{2}}{\sqrt{2}} \frac{dy}{dx} = 4\sqrt{2}.$$

Therefore
$$\frac{dy}{dx} = 4\sqrt{2} - 1/\sqrt{2} = 7/\sqrt{2}.$$

Example 26-2. At what points of the curve $\tfrac{1}{4}x^2 - y^2 = 1$ are the tangent lines parallel to the line $y = x + 3$?

Solution. Since the slope of the line $y = x + 3$ is 1, we are looking for those points of the graph at which the tangent lines have slope 1. We differentiate both sides of our given equation with respect to x:

$$\frac{d}{dx}(\tfrac{1}{4}x^2 - y^2) = \frac{d1}{dx},$$

$$\frac{d}{dx}\tfrac{1}{4}x^2 - \frac{d}{dx}y^2 = 0,$$

$$\tfrac{1}{2}x - 2y\frac{dx}{dy} = 0.$$

Therefore, the tangent line at a point (x, y) has slope

$$\frac{dy}{dx} = \frac{x}{4y},$$

so the requirement that the slope be 1 is expressed as the equation $x/4y = 1$, or

(26-5) $x = 4y.$

But (x, y) must also be a point of our curve; that is,

(26-6) $\dfrac{x^2}{4} - y^2 = 1.$

Equations 26-5 and 26-6 form a system of simultaneous equations whose solutions answer our original question. You may verify that these solutions are the points $(4/\sqrt{3}, 1/\sqrt{3})$ and $(-4/\sqrt{3}, -1/\sqrt{3})$.

Example 26-3. Sketch the curve $y^2 = 2x^2 - x^4 + 8$.

Solution. When $x = 0$, we have $y = \sqrt{8}$ or $y = -\sqrt{8}$. The X-coordinates of the points at which $y = 0$ satisfy the equation $2x^2 - x^4 + 8 = 0$. It is easy to solve this equation and find that $x = 2$ or $x = -2$ when $y = 0$. We find $\dfrac{dy}{dx}$ by implicit differentiation:

$$2y \frac{dy}{dx} = 4x - 4x^3;$$

and hence

$$\frac{dy}{dx} = \frac{2x(1 - x^2)}{y}.$$

Therefore, $\dfrac{dy}{dx} = 0$ when $x = 0, 1,$ or -1. These calculations are summarized in the table in Fig. 26-1, which also shows the desired graph.

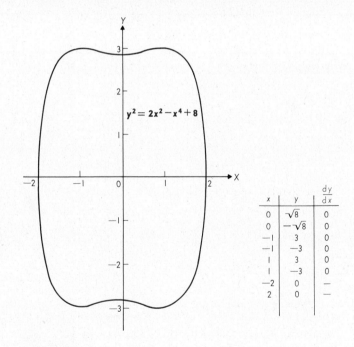

x	y	$\dfrac{dy}{dx}$
0	$\sqrt{8}$	0
0	$-\sqrt{8}$	0
-1	3	0
-1	-3	0
1	3	0
1	-3	0
-2	0	—
2	0	—

Figure 26-1

Example 26-4. When we replace the *parameter c* in the equation

(26-7) $$y^2 + (x - c)^2 = c^2$$

with numbers such as 1, π, and -3, we obtain various members of the *family of curves* whose "general member" is Equation 26-7. Show that if f is a differentiable function whose graph is a subset of any member of this family, then $y = f(x)$ satisfies the differential equation

(26-8) $$1 + y'^2 = \left(\frac{x}{y} + y'\right)^2.$$

Solution. Let c be any number and differentiate both sides of Equation 26-7:

$$\frac{d}{dx}\, y^2 + \frac{d}{dx}\, (x - c)^2 = \frac{d}{dx}\, c^2,$$

$$2y\, \frac{dy}{dx} + 2(x - c) = 0,$$

$$yy' + x - c = 0.$$

This last equation tells us that $x - c = -yy'$ and $c = x + yy'$, and when we substitute these expressions for $x - c$ and c in Equation 26-7 and simplify, we obtain Equation 26-8.

PROBLEMS 26

1. Use implicit differentiation to compute $\dfrac{dy}{dx}$.

    (a) $3y^5 + 5y^3 + 15y = 15x$ (b) $2x^2 - 5y^2 - 8 = 0$
    (c) $xy^3 - 2xy + y^2 = 7$ (d) $x^2y + y^3 - 4 = 0$
    (e) $\sqrt{1 + y} = x + y$ (f) $\sqrt{x^2 + y^2} = x + y$
    (g) $\sin(x + y) = \sin x + \sin y$ (h) $\sin xy = \sin x \sin y$

2. Find y' in two ways: (1) solve for y and then differentiate and (2) differentiate implicitly and then solve for y'.

    (a) $x^5 + y^5 = 1$ (b) $xy + 2x = 3y - 1$

    (c) $\dfrac{1}{x} + \dfrac{1}{y} = \dfrac{1}{3}$ (d) $\sqrt[3]{y} + \sqrt[3]{x} = 1$

3. Write the equation of the tangent to the curve at the indicated point.
    (a) $x^2y^2 = 9$, $(-1, 3)$ (b) $x^2 + xy + y^2 = 4$, $(2, -2)$
    (c) $2x^2 - xy + 3y^2 = 18$, $(3, 1)$ (d) $(y - x)^2 = 2x + 4$, $(6, 2)$

4. Find the equation of the tangent to the curve $x^my^n = a$ at the point (x_0, y_0).

5. Show that the graphs of the given pair of equations intersect at right angles.
    (a) $y^2 = 6x + 9$ and $y^2 = 9 - 6x$
    (b) $x^2 - y^2 = 5$ and $4x^2 + 9y^2 = 72$

6. Find the angle between the tangents to the curves $3y = 2x + x^4y^3$ and $2y + 3x + y^5 = x^3y$ at the origin.

7. Show that if $x^my^n = (x + y)^{m+n}$, then y satisfies the differential equation $xy' = y$.

8. Show that if $x^2 - 2y^2 = 4$, then (1) $y' = x/2y$, and hence $y'' = (y - xy')/2y^2$. Now substitute from Equation 1 to obtain $y'' = -1/y^3$. Use this method to compute y'' from the following equations.

(a) $x^2 + y^2 = 1$ (b) $x^3 + y^3 = 1$

(c) $y^2 = 4ax$ (d) $x^4 + 2x^2y^2 = \pi$

9. The curve $4x^2 + y^2 = 72$ is an *ellipse* (oval) and the point $(4, 4)$ is outside it. From this point, two tangent lines can be drawn to the ellipse. Find them.

10. Sketch the graph of the given equation. Use calculus to check the maximum and minimum points, concavity, and so on.

(a) $x^4 + y^4 = 1$

(b) $x^3 + y^3 = 1$

(c) $x^{2/3} + y^{2/3} = 1$

(d) $4x^2 - 4xy + y^2 - 10x + 5y + 6 = 0$

11. By "eliminating the parameter c" as we did in Example 26-4, show that $y = f(x)$ satisfies the given differential equation if the graph of f is a subset of one of the members of the given family.

(a) $y' = -\dfrac{x}{y}, \; x^2 + y^2 = c$

(b) $y' = -\sqrt{1 - y^2}/y, \; (x - c)^2 + y^2 = 1$

(c) $y' = \dfrac{1}{1 + x^2}, \; \tan(y - c) = x$

(d) $x^2 + 1/y'^2 = 1, \; \sin(y - c) = x$

12. Explain why the equation $2xy'' = 2y' + (5 + x^2)y'^3$ does or does not follow from the equation $\sin y = 5 + x^2$.

Your geometric feeling for the ideas of calculus should be such that the statements in the following table are second nature to you.

If for each x	the graph of f is
$f'(x) > 0$	rising
$f'(x) < 0$	falling
$f''(x) > 0$	concave up
$f''(x) < 0$	concave down

In addition, this table suggests to us all we need to know about the theory of maxima and minima.

By introducing "word problems" in connection with rates and maxima and minima, we tried to indicate how calculus might be useful in areas outside of mathematics. Often the hardest part of a practical problem is setting up a suitable mathematical model—changing words of economics, for example, into symbols of mathematics. Once a problem has been reduced to a mathematical equation, its solution may be quite mechanical.

Of the many techniques for solving equations, Newton's Method is particularly suitable for study in a calculus course. You should remember the graphical basis for Newton's Method of approximating a solution of the equation $f(x) = 0$, which quickly leads to the formula

$$x_1 = x_0 - f(x_0)/f'(x_0).$$

Notice that the topics of related rates and implicit differentiation are intimately connected with the Chain Rule. Since that rule is rather sophisticated for beginners, don't be alarmed if you have a slightly uneasy feeling about these topics. Most people don't feel at home with the Chain Rule until after a course in advanced calculus.

Review Problems, Chapter Three

1. Suppose that $f(x) = x\sqrt{|x|} - 1$. Show that f has only one maximum value and one minimum value and that these values are equal.

2. Suppose that $f(0) = 1$, $f(1) = 2$, $f(2) = 4$, and $f(3) = 3$. Furthermore, the signs of $f'(x)$ and $f''(x)$ are as shown in Fig. III-1. What does the graph of f look like?

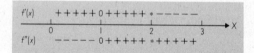

Figure III-1

3. Suppose that f is differentiable in an interval $[a, b]$ and let $g(x) = f(x)^2$. Show that critical points of f are critical points of g. Can g have a critical point that f doesn't have?

4. For x in the interval $[0, 2]$, sketch the graphs of the members of the family of curves $y = x + c \sin x$ for which $c = 2, 1, 0, -1$, and -2. Find maximum and minimum points, check the concavity, and so on. Show that regardless of the choice of c, y satisfies the differential equation $y' = 1 + (y - x) \cot x$.

5. If $f(x) = \cos 2x - 2 \cos x$, Theorem 10-1 tells us that $f([0, \pi])$ is a closed finite interval. What is it?

6. A point is moving along the number scale in accordance with the equation $s = |t^2 - 5| + \sqrt{t + 2}$ for t in the interval $(0, \infty)$ (distance measured in feet, time in seconds). When $t = 2$, where is the point, which way is it moving, and how fast (in feet per second) is it moving? At that instant, is it speeding up or slowing down? At what time is the point closest to the origin? At what time in the first second of motion is the point farthest from the origin? (You might want to use Newton's Method to calculate this number.)

7. A particle moves along a line in such a way that its acceleration is given by the equation $a = \sin t^2$. At what times $(t > 0)$ is the velocity a maximum?

8. The point $(1, 3)$ belongs to the curve $x^3 - 3xy + y^3 = 19$. Use implicit differentiation and Formula 21-2 to find an approximation to a number c such that $(1.1, c)$ also belongs to the curve.

9. If $g(x) = f(x)^2$, then a number r is a zero of g if it is a zero of f. Suppose that we start with a given initial approximation x_0 of r and apply Newton's Method to f and g, producing new approximations x_1 and x_1^*. How are these numbers related?

10. Suppose that you are told that there are no maximum or minimum points of the curve $y = ax^3 + bx^2 + cx + d$. What can you conclude about the numbers a, b, and c?

11. A rectangle is to have one side along the X-axis and two vertices on the circle with a radius of 5 whose center is the origin. What is the largest possible perimeter that such a rectangle can have?

12. A line segment is to join the point $(0, \frac{1}{2})$ to a first quadrant point P that belongs to the curve $y = 1/x^2$. Locate P so that the segment is as short as possible.

13. A cylindrical tin can, closed at the top and bottom, is to be made to contain a volume of 24 cubic inches. No waste is involved in cutting the tin for the vertical side of the can, but each circular end piece is to be stamped from a square, and the "corners" of the square are wasted. Find the radius of the most economical can that can be manufactured in this way.

14. (a) A large cylindrical glass of radius R inches is partly filled with water. A second cylindrical glass of radius $\frac{1}{2}R$ inches is pushed down into the large glass at the rate of K inches per second. How fast is the level of the water rising when the bottom of the smaller glass is 1 inch below the surface of the water?
    (b) Replace the smaller glass with a conical paper cup whose radius and height are both $\frac{1}{2}R$ inches and work the problem.

15. Suppose that we mail a rectangular package with one square side that measures x by x inches and whose other dimension is such that the package barely meets postal requirements that the sum of its length and girth not exceed 72 inches.
    (a) Show that its volume is given by the formula $V = \frac{1}{4}x^2(216 - 11x + |5x - 72|)$.
    (b) Find the dimensions of the largest such package.
    (c) Sketch the graph of the equation of part (a), taking note of such things as concavity.

16. Suppose that $f'(a) = f''(a) = \cdots = f^{(n)}(a) = 0$ and that $f^{(n+1)}(x) > 0$ for x in a neighborhood of a. Then if b is any point of this neighborhood, Taylor's Formula tells us that

$$f(b) - f(a) = f^{(n+1)}(m)\,\frac{(b - a)^{n+1}}{(n + 1)!},$$

where m is a number between a and b. Use this equation to show that $f(a)$ is a minimum value of f if n is odd. Use this result to show that if $f(x) = x^3 \sin x$, then f has a minimum value at 0.

17. Show that the curve $y = 3 \sin 2x + 5 \cos 2x$ is concave up when y is negative and concave down when y is positive.

18. Discuss the concavity of the curve $y = x^{p/q}$, where p and q are integers with no common factor.

4 | The Conics

The fundamental idea of analytic geometry is that a point set in the coordinate plane can be thought of as a subset of R^2, the set of pairs of real numbers, and so it can be specified by giving a numerical relation that is satisfied by the coordinates of its points. Thus a numerical relation $Ax + By + C = 0$ represents a geometric line, and so on. Now we are ready for a systematic study of more complicated plane point sets and their numerical representations. In this chapter we will take up the circle, the ellipse, the hyperbola, and the parabola. First we will treat these curves individually; then in Section 31 we will show how the last three, which are collectively known as *conics*, are related.

The material of this chapter is not really part of calculus, and if you have seen it before, you should go ahead to Chapter 5 now. Then this chapter will serve as a reference to remind you of forgotten terminology and formulas.

27. THE CIRCLE. TRANSLATION OF AXES

Geometrically speaking, a circle is a point set in the plane, each of whose members is the same distance from a given point called the center of the circle. We wish to translate this geometric statement into analytic language, so let us suppose that we are talking about a circle whose radius is a given positive number r and whose center is a point (h, k) of the coordinate plane. If (x, y) is a point of our circle, then its distance from (h, k) is the number

$$\sqrt{(x - h)^2 + (y - k)^2};$$

and hence the statement that this distance is r can be expressed as the equation $\sqrt{(x - h)^2 + (y - k)^2} = r$. This equation will look a little nicer if we square both sides, and when we do we get the following result: *the equation of the circle with a radius of r and with the point (h, k) as center is*

(27-1) $$(x - h)^2 + (y - k)^2 = r^2.$$

For example, the equation of the circle with a radius of 3 and $(0, -1)$ as center is $x^2 + (y + 1)^2 = 9$; that is, $x^2 + y^2 + 2y - 8 = 0$.

Example 27-1.   Find the center and radius of the circle

$$x^2 - 4x + y^2 + 8y - 5 = 0.$$

Solution.   We can find the radius and the coordinates of the center by inspection (and, incidentally, verify the fact that the graph of the given equation *is* a circle) if we write the equation in the form of Equation 27-1. To do so, we "complete the square":

$$(x^2 - 4x \quad) + (y^2 + 8y \quad) = 5,$$
$$(x^2 - 4x + 4) + (y^2 + 8y + 16) = 5 + 4 + 16,$$
$$(x - 2)^2 + (y + 4)^2 = 25.$$

From our final form of the given equation, we see that its graph is the circle with a radius of 5 and with the point $(2, -4)$ as its center.

Example 27-2.   Find the center of the circle that contains the three points $(0, 0)$, $(3, 3)$, and $(2, 4)$.

Solution.   If we denote the coordinates of the center of our circle by (h, k) and its radius by r, then its equation will be Equation 27-1; we are to find the numbers h and k. Since the three given points are to be points of the circle, the coordinates of each must satisfy this equation. So we substitute the coordinates of our three points in Equation 27-1 to obtain the system of equations

$$(0 - h)^2 + (0 - k)^2 = r^2$$
$$(3 - h)^2 + (3 - k)^2 = r^2$$
$$(2 - h)^2 + (4 - k)^2 = r^2.$$

After some simplification, this system takes the form

$$h^2 + k^2 = r^2$$

$$18 - 6h - 6k + h^2 + k^2 = r^2$$

$$20 - 4h - 8k + h^2 + k^2 = r^2.$$

Now we subtract the first equation from the second and the third and obtain a system of two equations in the two unknowns h and k:

$$18 - 6h - 6k = 0$$

$$20 - 4h - 8k = 0.$$

These equations immediately yield $h = 1$ and $k = 2$, so the center of our circle is the point (1, 2).

For the circle with a radius of r and with the origin as center, Equation 27-1 reduces to

(27-2) $$x^2 + y^2 = r^2.$$

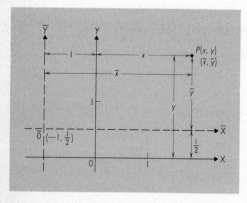

Figure 27-1

This equation is "simpler" than Equation 27-1. When dealing with geometric problems in terms of coordinates, it is wise to choose the coordinate axes so as to make your equations as simple as possible. Sometimes it is even convenient to change coordinate axes in order to simplify things. One method of changing the coordinate axes is to perform a *translation of axes*.

Let us suppose that we have chosen a set of coordinate axes, shown in Fig. 27-1 as solid lines. We have labeled these axes the X-axis and the Y-axis. We now select another set of coordinate axes, shown in Fig. 27-1 as dashed lines, and label them the $\overline{X}$-axis and the $\overline{Y}$-axis. The $\overline{X}$-axis is parallel to the X-axis and $\frac{1}{2}$ unit above it. The $\overline{Y}$-axis is parallel to the Y-axis and 1 unit to the left of it. We say that we have obtained the $\overline{X}\overline{Y}$-axes by a *translation*. Any point P in the plane can be described by either its XY-coordinates (x, y) or its $\overline{X}\overline{Y}$-coordinates $(\bar{x}, \bar{y})$. From Fig. 27-1 we see that the relation between the numbers x and $\bar{x}$ is $\bar{x} = x + 1$ and between the numbers y and $\bar{y}$ is $\bar{y} = y - \frac{1}{2}$. Let us write these equations as

(27-3) $$\bar{x} = x - (-1)$$
$$\bar{y} = y - \tfrac{1}{2}.$$

Now observe that the origin of the $\overline{X}\overline{Y}$-coordinate system has $\overline{X}\overline{Y}$-coordinates $(0, 0)$ and XY-coordinates $(-1, \frac{1}{2})$. Thus Equations 27-3 tell us that we obtain the $\overline{X}\overline{Y}$-coordinates of a point P by subtracting the XY-coordinates of the origin of the $\overline{X}\overline{Y}$-system from the XY-coordinates of P.

Let us generalize the problem we have just been looking at and consider a coordinate change in which we obtain new coordinates $(\bar{x}, \bar{y})$ for a point with original coordinates (x, y) by using the formulas

(27-4)
$$\bar{x} = x - h$$
$$\bar{y} = y - k,$$

where h and k are given numbers. It is clear from Equations 27-4 that the origin of the $\overline{X}\overline{Y}$-coordinate system—that is, the point whose $\overline{X}\overline{Y}$-coordinates are $(0, 0)$— is the point whose XY-coordinates are (h, k). It is just as easy to locate the new coordinate axes. The $\overline{Y}$-axis is the set of points for which $\bar{x} = 0$. According to Equations 27-4, this equation says that $x = h$. In other words, *the line $x = h$ is the $\overline{Y}$-axis.* Similarly, *the line $y = k$ is the $\overline{X}$-axis.* Thus we see that the coordinate change described by Equations 27-4 produces $\overline{X}$- and $\overline{Y}$-axes that are parallel to the X- and Y-axes, have the same directions, and intersect in the point whose XY-coordinates are (h, k). Equations 27-4 are called the **transformation of coordinate equations** for a **translation** of axes.

Example 27-3.   Find a translation of axes that simplifies the equation of the circle in Example 27-1.

Solution.   By completing the square, we expressed the given circle as

$$(x - 2)^2 + (y + 4)^2 = 25.$$

This equation makes the required translation obvious:

$$\bar{x} = x - 2 \quad \text{and} \quad \bar{y} = y + 4.$$

Our next example illustrates another technique for finding a new coordinate system, with respect to which a given equation takes a simpler form. We could have used this method in our preceding example, too.

Example 27-4.   Find a translation of axes that simplifies the equation

$$xy + 3x - 4y = 5.$$

Solution. We will replace x with $\bar{x} + h$ and y with $\bar{y} + k$ in the given equation and then choose h and k so that the resulting equation is simpler than the original. When we make the replacement, our equation becomes

$$(\bar{x} + h)(\bar{y} + k) + 3(\bar{x} + h) - 4(\bar{y} + k) = 5,$$

$$\bar{x}\bar{y} + (k + 3)\bar{x} + (h - 4)\bar{y} + hk + 3h - 4k = 5.$$

Now if we set $h = 4$ and $k = -3$, this equation takes the simple form $\bar{x}\bar{y} - 12 + 12 + 12 = 5$, or $\bar{y} = -7/\bar{x}$. An easy way to sketch the graph of the original equation, therefore, is to draw an $\overline{X}\overline{Y}$-coordinate system whose origin is the point $(4, -3)$ of the XY-system. Then, relative to this new system, sketch the graph of the equation $\bar{y} = -7/\bar{x}$. This example is illustrated in Fig. 27-2.

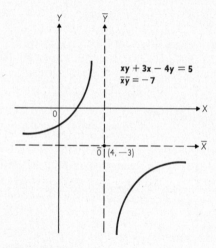

$$xy + 3x - 4y = 5$$
$$\bar{x}\bar{y} = -7$$

Figure 27-2

PROBLEMS 27

1. Find the equations of the circles with the given points as centers and the given numbers as radii.
   (a) $(2, -3)$, 5
   (b) $(-2, -2)$, 1
   (c) $(\cos 3, \sin 3)$, 1
   (d) $(\log \frac{1}{3}, \log 3)$, $\log 100$

2. Complete the square to find the center and radius of each of the following circles.
   (a) $x^2 + y^2 + 4x - 2y - 4 = 0$
   (b) $x^2 + y^2 - 4y = 0$
   (c) $x^2 + y^2 - x - 2y = 0$
   (d) $36x^2 + 36y^2 - 24x + 36y - 131 = 0$

3. Find the equation of the circle C in the following cases.
   (a) $(2, -3)$ is the center of C and $(1, 1)$ is one of its points.
   (b) $(0, 0)$, $(0, 2)$, and $(2, 0)$ are points of C.

(c) $(0, -1)$, $(2, 1)$, and $(-2, 5)$ are points of C.

(d) C is tangent to the X- and Y-axes, lies in Quadrant II, and has a radius of 5.

(e) The center of C is the point $(8, 14)$ and C just touches the circle

$$x^2 + y^2 + 2x - 4y - 20 = 0.$$

(f) The center of C is the point $(2, 6)$ and C just touches the circle

$$(x + 1)^2 + (y - 2)^2 = 225.$$

(g) The center of C is a point of the Y-axis, its radius is 5, and it contains the point $(4, 2)$.

(h) C lies in Quadrant I, has a radius of 1, and is tangent to the X-axis and the line $y = x$.

4. Describe the following sets geometrically.

(a) $\{(x, y) \mid x^2 + y^2 \geq 9\}$ (b) $\{(x, y) \mid x^2 + y^2 - 4x + 2y \leq 4\}$

(c) $\{(x, y) \mid 4 \leq x^2 + y^2 \leq 25\}$ (d) $\{(x, y) \mid |x^2 + y^2 - 4x + 2y| < 4\}$

5. We say that the equation $x^2 + y^2 = c^2$ determines a *family* of curves, each member of the family being obtained by choosing a different value of c. Here our family consists of circles whose centers are the origin. The members of the family that we obtain by setting $c = 1, 2,$ or 3 have radii 1, 2, or 3. Describe the following families of curves and sketch several members of each family.

(a) $(x - c)^2 + y^2 = 9$ (b) $(x + c)^2 + (y + 2c)^2 = 25$

(c) $x^2 + y^2 - 2cx = 0$ (d) $x^2 + y^2 + 2cx + 2cy = 0$

6. Suppose that we construct an $\bar{X}\bar{Y}$-coordinate system from an XY-coordinate system by translating the axes so that the origin is translated to the point $(-2, 3)$. First consider the following pairs of numbers as XY-coordinates of points and find the $\bar{X}\bar{Y}$-coordinates of the points; then consider the pairs as $\bar{X}\bar{Y}$-coordinates of points and find the XY-coordinates of the points.

(a) $(2, -5)$ (b) $(-2, 5)$ (c) $(-5, 2)$ (d) $(5, -2)$

(e) $(-\frac{7}{5}, \frac{5}{7})$ (f) $(\frac{3}{2}, -\frac{3}{2})$ (g) $(\log 3, \log 2)$ (h) $(\cos 1, \sin 1)$

7. Find translations of axes such that the origin of the new axis system is the point of intersection of the given pair of lines. What are the equations of the lines in the $\bar{X}\bar{Y}$-system?

(a) $y = 2x - 3$ (b) $y = 2$
    $x = 2$ $x - 2y = 1$

(c) $x - 2y = 4$ (d) $2x + y = 0$
    $3x + y = 5$ $x - 3y = 7$

8. Find transformation of coordinates equations under which a sine curve in the XY-system becomes a cosine curve in the $\bar{X}\bar{Y}$-system.

9. (Do these problems in two ways. First, complete the square and make the translation that the resulting equation suggests. Then replace x with $\bar{x} + h$ and y with $\bar{y} + k$, and choose h and k so that the resulting equation has the desired form.) Find a translation of axes that reduces the equation

(a) $2x^2 + 2y^2 - 8x + 4y + 2 = 0$ to the form $\bar{x}^2 + \bar{y}^2 = r^2$.

(b) $x^2 + 4y^2 + 2x - 16y + 13 = 0$ to the form $\bar{x}^2/a^2 + \bar{y}^2/b^2 = 1$.

(c) $9x^2 - 4y^2 - 36x + 8y - 4 = 0$ to the form $\bar{x}^2/a^2 - \bar{y}^2/b^2 = 1$.

(d) $y = 3x^2 + 6x + 2$ to the form $\bar{y} = c\bar{x}^2$.

10. For each choice of t, the pair of equations $x = 5 \cos t$ and $y = 5 \sin t$ gives us a pair of numbers (x, y). For example, if we choose $t = 0$, we get the point $(5, 0)$. For $t = \frac{1}{2}\pi$, we have $(0, 5)$, and so on. Show that the set of all such points is a circle. Describe geometrically the set

$$\{(x, y) \mid x = -3 \sin t, \ y = 3 \cos t, \ t \in R^1\}.$$

11. A point moves so that it is always three times as far from the point $(4, -3)$ as it is from the point $(0, 1)$. Describe the path of the moving point.

12. The equation $x^2 + y^2 + ax + by + c = 0$ does not represent a circle for every choice of a, b, and c (for example, not when $a = b = 0$ and $c = 1$). What relation among the coefficients a, b, and c is necessary in order that the graph should be a circle?

13. Explain why the following relations are analytic representations of the circular disk of radius r whose center is the origin. What would the relations be if the center of the disk were the point (h, k)?

    (a) $x^2 + y^2 < r^2$ $\qquad\qquad$ (b) $[\![(x^2 + y^2)r^{-2}]\!] = 0$

14. Sketch the graphs of the following equations.

    (a) $x^2 + 2xy + y^2 = 36$ $\qquad$ (b) $(x - |x|)^2 + 4y^2 = 4$
    (c) $(|x| + x)^2 + (|y| + y)^2 = 36$ $\qquad$ (d) $x^2 + y^2 + 2|x| - 4y = 0$
    (e) $[\![x]\!]^2 + [\![y]\!]^2 = 13$ $\qquad$ (f) $[\![x^2 + y^2]\!] = 13$

15. Show that if the curve $y = f(x)$ is an arc of the circle $x^2 + y^2 = r^2$, then y satisfies the differential equation $y' = -x/y$. Conversely, if $y' = -x/y$, then show that $\frac{d}{dx}(x^2 + y^2) = 0$ and hence that $x^2 + y^2 = r^2$ for some number r that is independent of x.

28. THE ELLIPSE. SYMMETRY

A circle can be thought of as a special case of a curve known as an **ellipse**. To describe a circle, we start with a given point and a given positive number, and we say that the circle is the set of points whose distance from the given point is the given number. To describe an ellipse, we start with *two* points and a positive number, and we say that *the ellipse is the set of points, the average of whose distances from the two given points is the given number.* In set notation, if F_1 and F_2 are the two given points and a is the given positive number, then the ellipse is the set of points $\{P \mid \frac{1}{2}(\overline{PF_1} + \overline{PF_2}) = a\}$. The points F_1 and F_2 are called the **foci** (singular, *focus*) of the ellipse. If the two foci coincide, the ellipse becomes a circle.

To translate our geometric description of an ellipse into analytic terms, we introduce a coordinate system into the plane, and it will simplify our formulas if we set up the coordinate system so that our foci are the points $(-c, 0)$ and $(c, 0)$, as shown in Fig. 28-1. For each point (x, y) of the ellipse, the average of its distances r_1 and r_2 from the foci is the given number a; that is, $\frac{1}{2}(r_1 + r_2) = a$, or

(28-1) $$r_1 + r_2 = 2a.$$

The foci are $2c$ units apart, so since the sum of the lengths of two sides of a triangle is greater than the length of the third side, $r_1 + r_2 > 2c$. When we compare this inequality with Equation 28-1, we see that $a > c$.

A simple way to construct our ellipse is to stick thumbtacks into each focus and then place a loop of string $2a + 2c$ units long around the tacks. Now put your pencil point inside the loop of string and draw it taut, thus obtaining the triangle shown in Fig. 28-1. As you move your pencil, keeping the string taut, you will trace out the ellipse.

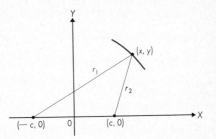

Figure 28-1

To find the equation of our ellipse, we use the distance formula to express r_1 and r_2:

$$r_1 = \sqrt{(x + c)^2 + y^2} \quad \text{and} \quad r_2 = \sqrt{(x - c)^2 + y^2}.$$

When we substitute these numbers in Equation 28-1, we find that a point (x, y) is one of the points of our ellipse if and only if its coordinates satisfy the equation

(28-2) $\qquad \sqrt{(x + c)^2 + y^2} + \sqrt{(x - c)^2 + y^2} = 2a.$

We could consider Equation 28-2 as the equation of our ellipse, but a little algebra enables us to reduce this equation to a much simpler form. To eliminate radicals, we transpose the term $\sqrt{(x - c)^2 + y^2}$ and square, thereby obtaining

$$(x + c)^2 + y^2 = 4a^2 - 4a\sqrt{(x - c)^2 + y^2} + (x - c)^2 + y^2.$$

When we simplify the last equation, we have $a\sqrt{(x - c)^2 + y^2} = a^2 - cx$, so we square again to get

$$a^2[(x - c)^2 + y^2] = a^4 - 2a^2cx + c^2x^2.$$

Simplification now yields

$$(a^2 - c^2)x^2 + a^2y^2 = a^2(a^2 - c^2).$$

We noticed that $a > c$, so $a^2 - c^2 > 0$. We can therefore introduce the positive number $b = \sqrt{a^2 - c^2}$, and our equation reduces to

$$b^2x^2 + a^2y^2 = a^2b^2.$$

Now we divide both sides of this equation by a^2b^2 to get the final form of the equation of our ellipse:

(28-3) $\qquad \dfrac{x^2}{a^2} + \dfrac{y^2}{b^2} = 1.$

The graph of this equation is shown in Fig. 28-2.

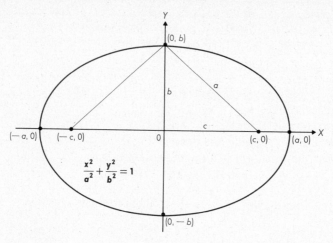

$$\frac{x^2}{a^2} + \frac{y^2}{b^2} = 1$$

Figure 28-2

An ellipse is a highly symmetrical figure, and this fact is very useful in constructing its graph. Before discussing the symmetry of the ellipse in particular, let us say a few words about the symmetry of graphs in general.

A plane point set S is **symmetric with respect to a line** L if for each point $P \in S$ there is a point $Q \in S$ such that the line L is the perpendicular bisector of the segment PQ. In particular, a set S of the coordinate plane is symmetric with respect to the Y-axis if whenever $(x, y) \in S$, we also have $(-x, y) \in S$, because, as you can see, the Y-axis is the perpendicular bisector of the segment that joins the points (x, y) and $(-x, y)$. Thus *the graph of a relation is symmetric with respect to the Y-axis if whenever the pair (x, y) satisfies the relation, the pair $(-x, y)$ also satisfies it.* A similar statement guarantees symmetry with respect to the X-axis. A moment's thought will also convince you that the points (a, b) and (b, a) are symmetric with respect to the line $y = x$. Thus *the graph of a relation is symmetric with respect to the line that bisects Quadrant I if whenever the pair (x, y) satisfies the relation, the pair (y, x) also satisfies it.*

A set S is **symmetric with respect to a point** Q if for each point $P \in S$ there is a point $R \in S$ such that the point Q is the midpoint of the segment PR. In particular, a set S of the coordinate plane is symmetric with respect to the origin if whenever $(x, y) \in S$, we also have $(-x, -y) \in S$, because the origin is clearly the midpoint of the segment that joins the points (x, y) and $(-x, -y)$. Thus *the graph of a relation is symmetric with respect to the origin if whenever the pair (x, y) satisfies the relation, the pair $(-x, -y)$ also satisfies it.*

We see immediately that if the pair (x, y) satisfies Equation 28-3, then the pairs $(-x, y)$, $(x, -y)$, and $(-x, -y)$ do, too, so our ellipse is symmetric with respect to the Y-axis, the X-axis, and the origin. The point of symmetry of an ellipse (here the origin) is called the **center** of the ellipse. An ellipse has two lines of symmetry, and the four points of intersection of the ellipse with its lines of symmetry are the **vertices** (singular, *vertex*) of the ellipse. The vertices of Ellipse 28-3 are the points $(a, 0)$, $(-a, 0)$, $(0, b)$, and $(0, -b)$. A line segment that contains the center of an ellipse and whose terminal points are points of the ellipse is called a **diameter** of the ellipse. The longest diameter of an ellipse is its **major diameter**, and the shortest diameter is its **minor diameter**. Figure 28-2 shows that the major diameter of our ellipse joins the vertices $(a, 0)$ and $(-a, 0)$ and is $2a$ units long, whereas the minor diameter joins the vertices $(0, b)$ and $(0, -b)$ and is $2b$ units long. Notice that the foci are points of the major diameter.

The graph of Equation 28-3 is also an ellipse if $b > a$. In that case, the roles of b and a are interchanged. The foci are the points $(0, -c)$ and $(0, c)$, where $c = \sqrt{b^2 - a^2}$, of the Y-axis. The major diameter is now the segment terminating in the points $(0, -b)$ and $(0, b)$, while the minor diameter terminates in $(-a, 0)$ and $(a, 0)$. The average of the distances of a point of this ellipse from the two foci is b rather than a.

Example 28-1.   Find the equation of the ellipse whose foci are the points $(0, 4)$ and $(0, -4)$ and that has $(0, 5)$ and $(0, -5)$ as vertices.

Solution.   The desired equation takes the form of Equation 28-3; we are to find the numbers a and b. Because the foci are points of the Y-axis, $b > a$. Here $b = 5$ and $c = 4$, and so $a = \sqrt{25 - 16} = 3$. The equation of our ellipse is

$$\frac{x^2}{9} + \frac{y^2}{25} = 1,$$

and a sketch is shown in Fig. 28-3.

Equation 28-3 represents an ellipse whose center is the origin. Now let us suppose that we have an ellipse whose center is the point (h, k). We still assume that the major and minor diameters of the ellipse are parallel to the coordinate axes. If we choose a translated $\overline{X}\,\overline{Y}$-coordinate system whose origin is the point (h, k), then the equation of the ellipse relative to this new coordinate system takes the form of Equation 28-3; that is,

$$\frac{\bar{x}^2}{a^2} + \frac{\bar{y}^2}{b^2} = 1.$$

Now we use the translation equations $\bar{x} = x - h$ and $\bar{y} = y - k$ to find the equation of our ellipse in the XY-coordinate system:

(28-4) $$\frac{(x - h)^2}{a^2} + \frac{(y - k)^2}{b^2} = 1.$$

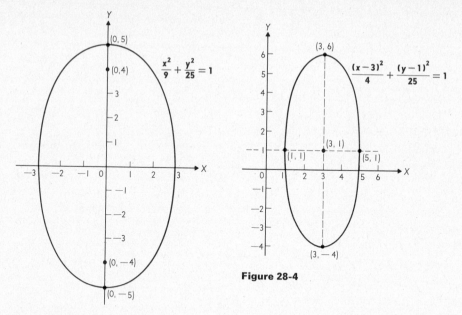

Figure 28-3

Figure 28-4

Example 28-2. Write the equation of the ellipse whose vertices are the points (1, 1), (5, 1), (3, 6), and (3, −4), and sketch its graph.

Solution. We first plot the given vertices (Fig. 28-4). These four points are the ends of the major and minor diameters of the ellipse. The diameter that is parallel to the *Y*-axis is 10 units long, and the diameter that is parallel to the *X*-axis is 4 units long. Thus $a = 2$ and $b = 5$. The center of an ellipse is the point in which its major and minor diameters intersect. In this case, the center is the point (3, 1). So $h = 3$ and $k = 1$, and Equation 28-4 becomes

$$\frac{(x-3)^2}{4} + \frac{(y-1)^2}{25} = 1.$$

Example 28-3. Reduce the equation $9x^2 + 4y^2 + 18x - 16y - 11 = 0$ to a simpler form by a translation of axes. What can you say about its graph?

Solution. We first rewrite our equation by completing the square:

$$9x^2 + 4y^2 + 18x - 16y - 11 = 0,$$

$$9(x^2 + 2x \qquad) + 4(y^2 - 4y \qquad) - 11 = 0,$$

$$9(x^2 + 2x + 1) + 4(y^2 - 4y + 4) - 11 = 9 + 16,$$

$$9(x + 1)^2 + 4(y - 2)^2 = 36.$$

So if we set $\bar{x} = x + 1$ and $\bar{y} = y - 2$, our equation becomes $9\bar{x}^2 + 4\bar{y}^2 = 36$, which reduces to the standard form (relative to the $\overline{XY}$-coordinate system) of the equation of an ellipse:

$$\frac{\bar{x}^2}{4} + \frac{\bar{y}^2}{9} = 1.$$

Here $a = 2$, $b = 3$, and $c = \sqrt{b^2 - a^2} = \sqrt{5}$. The coordinates of the vertices of our ellipse in the $\overline{XY}$-coordinate system are $(-2, 0)$, $(2, 0)$, $(0, -3)$, and $(0, 3)$, and its foci are the points with $\overline{XY}$-coordinates $(0, -\sqrt{5})$ and $(0, \sqrt{5})$. The XY- and the $\overline{XY}$-coordinates are related by the equations $x = \bar{x} - 1$ and $y = \bar{y} + 2$. Therefore the XY-coordinates of the vertices are $(-3, 2)$, $(1, 2)$, $(-1, -1)$, and $(-1, 5)$. The foci are the points $(-1, 2 - \sqrt{5})$ and $(-1, 2 + \sqrt{5})$.

Example 28-4.   At what point of the first quadrant is the tangent to the ellipse

$$\frac{x^2}{16} + \frac{y^2}{9} = 1$$

parallel to the chord that joins the vertices $(4, 0)$ and $(0, 3)$?

Solution.   To find the slope of the tangent at a point (x, y), we use implicit differentiation. Thus we differentiate both sides of the given equation with respect to x to obtain the equation

$$\frac{2x}{16} + \frac{2yy'}{9} = 0,$$

from which we see that

$$y' = \frac{-9x}{16y}.$$

The slope of the chord that joins the vertices $(4, 0)$ and $(0, 3)$ is $-\frac{3}{4}$. Since the tangent line is to be parallel to this chord, their slopes must be equal. Therefore the coordinates (x, y) of our desired point satisfy the equation

$$\frac{-9x}{16y} = \frac{-3}{4};$$

that is,

$$y = \frac{3}{4}x.$$

The numbers x and y also satisfy the equation of the ellipse, so

$$\frac{x^2}{16} + \frac{(3x/4)^2}{9} = 1,$$

and hence $x^2 = 8$. Thus (since (x, y) is in the first quadrant) $x = 2\sqrt{2}$, and $y = \frac{3}{4}x = \frac{3}{2}\sqrt{2}$.

PROBLEMS   28

1. Find the equation of, and then sketch, the ellipse whose
   (a) center is $(0, 0)$, with major diameter a part of the X-axis and 8 units long and minor diameter 6 units long.
   (b) center is $(0, 0)$, with major diameter a part of the Y-axis and 10 units long and minor diameter 2 units long.
   (c) center is $(2, -1)$, with major diameter parallel to the X-axis and 6 units long and minor diameter 4 units long.
   (d) center is $(2, -1)$, with major diameter parallel to the Y-axis and 12 units long and minor diameter 10 units long.

2. Find the equations of the ellipses whose foci are the given points and whose lengths of major diameters are the given numbers.
   (a) $(-1, 0)$, $(1, 0)$, 6   (b) $(0, -2)$, $(0, 2)$, 8
   (c) $(0, 4)$, $(2, 4)$, 4   (d) $(1, 0)$, $(1, 8)$, 16

3. Find the equations of the ellipses whose vertices are the given points.
   (a) $(-5, 0)$, $(5, 0)$, $(0, -3)$, $(0, 3)$ (b) $(2, 1)$, $(2, 5)$, $(-1, 3)$, $(5, 3)$
   (c) $(0, -1)$, $(10, -1)$, $(5, -3)$, $(5, 1)$ (d) $(1, -2)$, $(1, 6)$, $(-1, 2)$, $(3, 2)$

4. Find the foci of each of the following ellipses.
   (a) $9x^2 + 25y^2 = 225$ (b) $5x^2 + y^2 = 5$
   (c) $x^2 + 2y^2 + 6x + 7 = 0$ (d) $3x^2 + 7y^2 + 6x - 28y + 10 = 0$

5. Find the equation of the ellipse whose
   (a) minor diameter is 4 units long and whose major diameter ends with the points $(-1, 2)$ and $(9, 2)$.
   (b) major diameter ends with the points $(-4, -1)$ and $(2, -1)$ and whose foci are 4 units apart.
   (c) foci are the points $(1, 4)$ and $(1, 0)$ and whose minor diameter is 1 unit long.
   (d) major diameter ends with the points $(0, 1)$ and $(6, 1)$ and that contains the point $(3, 0)$.

6. Discuss the symmetry of the graphs of the following equations.
   (a) $3x^2 + 4y^2 = 5$ (b) $3x^2 - 4y^2 = 5$
   (c) $3|x| + 4|y| = 5$ (d) $3 \cos x + 4 \cos y = 5$
   (e) $3x^2 + 4y = 5$ (f) $3x^2 + 4 \tan y = 5$

7. (a) A point moves so that the line segments joining it to two given points are always perpendicular. Describe the path of the point.
   (b) Now suppose that the point moves so that the product of the slopes of the line segments joining it to the two given points is $-k^2$, where k is a given number. Describe the path of the point.

8. Use calculus to discuss the concavity of an ellipse.

9. An ellipse is to be constructed on an $8\frac{1}{2}$ inch by 11 inch sheet of paper by the string and thumbtack method so that its major diameter is parallel to the 11-inch edge. Where should we place the tacks and how long a piece of string should we use in order to draw the largest possible ellipse on the paper?

10. Give a geometric description of the set of points

$$\{(x, y) \mid x = 5 \cos t, \, y = 3 \sin t, \, 0 \le t \le \pi\}.$$

11. Discuss the graphs of the following equations.

(a) $\sin \pi\sqrt{x^2 + 4y^2} = 0$
(b) $4x^2 + 9y^2 - 8|x| - 36|y| + 4 = 0$
(c) $x(x + |x|) + 4y(y - |y|) = 8$
(d) $[\![x^2/25 + y^2/16]\!] = 0$

12. Circumscribe a circle of radius a about an ellipse whose major diameter is $2a$ units long and whose minor diameter is $2b$ units long. Now draw a line perpendicular to the major diameter of the ellipse and suppose that the segment of the line the ellipse cuts off is L units long and that the (longer) segment cut off by the circle is M units long. Show that $L/M = b/a$. Can you use this result to convince yourself that the ratio of the area of the elliptical region to the circular region is also b/a? What, therefore, is the area of the elliptical disk?

13. The segment that an ellipse cuts from a line that contains a focus and is perpendicular to the major diameter is called a **latus rectum** of the ellipse. Show that the latus rectum of Ellipse 28-3 (with $a > b$) is $2b^2/a$ units long. Find the slope of the tangent to this ellipse at the point of the first quadrant that is an endpoint of a latus rectum. Show that this tangent cuts off a segment of the positive Y-axis that is half as long as the major diameter of the ellipse.

14. A rectangle is to be inscribed in Ellipse 28-3 so that its sides are parallel to the major and minor diameters. Show that the largest area that such a rectangle can have is $2ab$.

29. THE HYPERBOLA

As with the ellipse, the geometric definition of the hyperbola starts with two given points F_1 and F_2 (again called the **foci**) and a given positive number a. An ellipse is the set of points, the *sum* of whose distances from the foci is $2a$, and *a* **hyperbola** *is the set of points, the differences of whose distances from the foci is $2a$.* In set notation, if F_1 and F_2 are the foci and a is a positive number, the hyperbola is the set of points $\{P \mid |\overline{PF_1} - \overline{PF_2}| = 2a\}$. In Fig. 29-1 we have chosen our coordinate system so that the foci of the hyperbola are the points $(-c, 0)$ and $(c, 0)$. A point (x, y) at distances of r_1 and r_2 from the foci is a point of our hyperbola if and only if

(29-1) $$|r_1 - r_2| = 2a.$$

The sides of the triangle whose vertices are the points $(-c, 0)$, $(c, 0)$, and (x, y) have lengths of $2c$, r_1, and r_2. Since the difference in the lengths of two sides of a triangle is less than the length of the third side, $|r_1 - r_2| < 2c$. When we compare this inequality with Equation 29-1, we see that for a hyperbola, $a < c$.

Just as in the case of the ellipse, we can construct an arc of a hyperbola by using a pencil, some string, and two thumbtacks. The tacks are placed at the foci, and the pencil is tied in the middle of the string. Then the string is looped around the tacks, as shown in Fig. 29-2 (the pencil point is at P). We keep the string taut

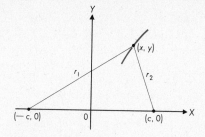

Figure 29-1

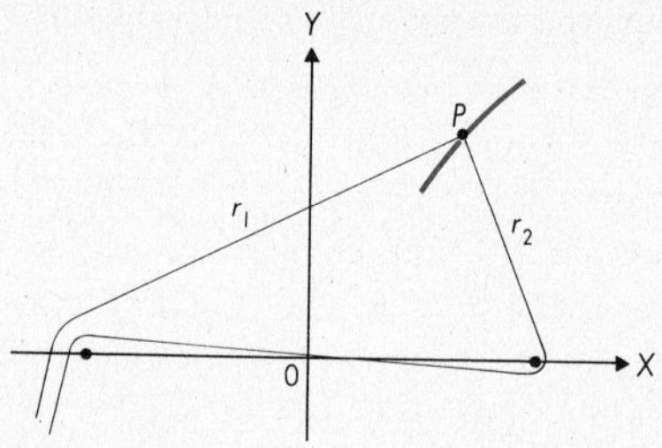

Figure 29-2

and move the pencil, paying out both strands of the string together. Thus the difference $r_1 - r_2$ is the same for any two points of the arc traced out by the pencil, so the arc must be part of a hyperbola.

To find the equation that tells us when a point (x, y) belongs to our hyperbola, we need only make the substitutions

$$r_1 = \sqrt{(x + c)^2 + y^2} \quad \text{and} \quad r_2 = \sqrt{(x - c)^2 + y^2}$$

in Equation 29-1. But the resulting equation is so complicated that it is to our advantage to rearrange Equation 29-1 a little before we make the substitution. We first observe that Equation 29-1 is equivalent to the equation $(r_1 - r_2)^2 = 4a^2$. Because r_1^2 and r_2^2 are "nicer" expressions than r_1 and r_2, we write our last equation as

$$- 2r_1r_2 = 4a^2 - r_1^2 - r_2^2$$

and square both sides. We leave it to you to show that the resulting equation can be written as

$$8a^2(r_1^2 + r_2^2) = (r_1^2 - r_2^2)^2 + 16a^4.$$

Now we substitute for r_1 and r_2:

$$8a^2[(x + c)^2 + y^2 + (x - c)^2 + y^2] = [(x + c)^2 - (x - c)^2]^2 + 16a^4.$$

When we simplify, this equation becomes

(29-2) $$(c^2 - a^2)x^2 - a^2y^2 = a^2(c^2 - a^2).$$

Since $c > a$, the equation $b = \sqrt{c^2 - a^2}$ defines a positive number b, in terms of which we can write Equation 29-2 as $b^2x^2 - a^2y^2 = a^2b^2$; that is,

(29-3) $$\frac{x^2}{a^2} - \frac{y^2}{b^2} = 1$$

is the equation of our hyperbola.

Replacing x with $-x$ and y with $-y$, separately or together, in Equation 29-3 does not alter the equation, so we see that our hyperbola is symmetric with respect to the Y-axis, the X-axis, and the origin. The point of symmetry of a hyperbola (in this case, the origin) is the **center** of the hyperbola. From Equation 29-3 we see that our hyperbola intersects one of its lines of symmetry, the X-axis, in the points $(-a, 0)$ and $(a, 0)$. These points are the **vertices** of the hyperbola. The line segment joining the two vertices is the **transverse diameter**. Since there is no real number y that satisfies Equation 29-3 when $x = 0$, we see that our hyperbola does not intersect its other line of symmetry, the Y-axis. Figure 29-3 shows the graph of Equation 29-3.

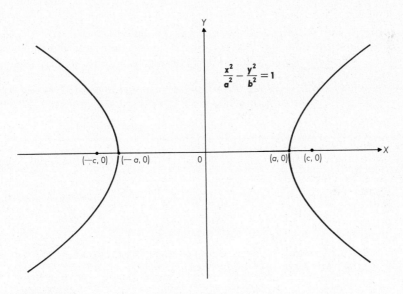

Figure 29-3

For the moment, let us consider only points in the first quadrant and solve Equation 29-3 for y to obtain $y = (b/a)\sqrt{x^2 - a^2}$, or equivalently,

$$(29\text{-}4) \qquad\qquad y = \frac{bx}{a}\sqrt{1 - \frac{a^2}{x^2}}.$$

Since $\sqrt{1 - a^2/x^2} < 1$, we see that $y < bx/a$. Hence the arc of the hyperbola that lies in the first quadrant is below the line $y = (b/a)x$ (Fig. 29-4). But if x is very large, $\sqrt{1 - a^2/x^2}$ is close to 1, and Equation 29-4 suggests that the equation of our hyperbola does not differ a great deal from the linear equation $y = (b/a)x$.

To verify that this statement is indeed correct, let x be a large number and suppose that (x, y_1) is the corresponding point of the hyperbola. We wish to show

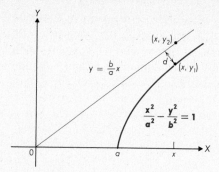

Figure 29-4

that the (perpendicular) distance d between this point and our line is small. Now if (x, y_2) is the point of the line that corresponds to x, it is clear (Fig. 29-4) that

$$d < y_2 - y_1.$$

By definition, $y_2 = (b/a)x$, and according to Equation 29-4,

$$y_1 = \frac{b}{a}\sqrt{x^2 - a^2}.$$

Hence

$$y_2 - y_1 = \frac{b}{a}(x - \sqrt{x^2 - a^2})$$

$$= \frac{b}{a}(x - \sqrt{x^2 - a^2})\left[\frac{x + \sqrt{x^2 - a^2}}{x + \sqrt{x^2 - a^2}}\right]$$

$$= \frac{b}{a}\left[\frac{x^2 - (x^2 - a^2)}{x + \sqrt{x^2 - a^2}}\right]$$

$$= \frac{ab}{x + \sqrt{x^2 - a^2}} < \frac{ab}{x}.$$

Consequently, $d < ab/x$, and we see that d is small when x is large. Thus for points in the first quadrant far to the right on our graph, the hyperbola practically coincides with the line $y = (b/a)x$. This line is called an **asymptote** of the hyperbola. It is not hard to see that for points in the third quadrant remote from the origin, the hyperbola also practically coincides with the line $y = (b/a)x$, whereas for points in the second and fourth quadrants, the line $y = -(b/a)x$ is an asymptote of the hyperbola.

The rectangle whose sides are parallel to the axes and contain the points $(-a, 0)$, $(a, 0)$, $(0, -b)$, and $(0, b)$ is called the **auxiliary rectangle** of our hyperbola.

The asymptotes of the hyperbola are extensions of the diagonals of this rectangle. Our hyperbola, with the asymptotes drawn in, is the curve on the left in Fig. 29-5. If the auxiliary rectangle is a square, then the hyperbola is an **equilateral hyperbola**.

 If we choose the coordinate axes so that the foci of a hyperbola are the points $(0, c)$ and $(0, -c)$, then its vertices are points of the Y-axis. We use $2b$ (rather than

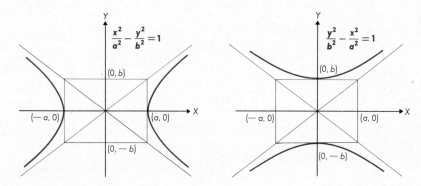

Figure 29-5

$2a$) to denote the difference of the focal distances of a point of this hyperbola, and we obtain the equation

(29-5)
$$\frac{y^2}{b^2} - \frac{x^2}{a^2} = 1,$$

where again $a^2 + b^2 = c^2$. The points $(0, -b)$ and $(0, b)$ are the vertices of this hyperbola. We also see that there is no real number x for which $y = 0$; that is, the curve does not intersect the X-axis. The graph of Equation 29-5 is shown on the right in Fig. 29-5. The asymptotes of both the hyperbolas in Fig. 29-5 are the same lines, and we say that these hyperbolas form a pair of **conjugate hyperbolas**.

Example 29-1. Find the equation of the hyperbola whose foci are the points $(5, 0)$ and $(-5, 0)$ and whose vertices are $(4, 0)$ and $(-4, 0)$.

Solution. We obtain the equation of this hyperbola by setting $c = 5$, $a = 4$, and $b = \sqrt{c^2 - a^2} = 3$ in Equation 29-3:

$$\frac{x^2}{16} - \frac{y^2}{9} = 1.$$

Its asymptotes are the lines $y = \frac{3}{4}x$ and $y = -\frac{3}{4}x$, and it looks like the hyperbola on the left in Fig. 29-5.

Now suppose that we have a hyperbola whose center is the point (h, k) and whose transverse diameter is parallel to the X-axis. If we choose a translated $\overline{X}\,\overline{Y}$-coordinate system whose origin is the point (h, k), then the equation of the hyperbola relative to this coordinate system is

$$\frac{\bar{x}^2}{a^2} - \frac{\bar{y}^2}{b^2} = 1.$$

Since $\bar{x} = x - h$ and $\bar{y} = y - k$, the equation of the hyperbola in the XY-coordinate system is

(29-6)
$$\frac{(x - h)^2}{a^2} - \frac{(y - k)^2}{b^2} = 1.$$

In a similar manner, the equation of a hyperbola whose center is the point (h, k) and whose transverse diameter is parallel to the Y-axis is

(29-7)
$$\frac{(y - k)^2}{b^2} - \frac{(x - h)^2}{a^2} = 1.$$

Example 29-2. The foci of an equilateral hyperbola are the points $(-3, 4)$ and $(-3, -2)$. Find the equation of the hyperbola and sketch its graph.

Solution. We are to replace the letters a, b, h, and k in Equation 29-7 with appropriate numbers. The center of the hyperbola is midway between the two foci, so it is the point $(-3, 1)$. Hence $h = -3$ and $k = 1$ (Fig. 29-6). Since the hyperbola is equilateral, $a = b$. Now c is the distance between the center of the hyperbola

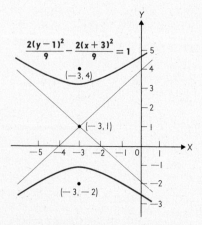

Figure 29-6

and a focus, so in the present case, $c = 3$. Thus the equation $c^2 = a^2 + b^2$ becomes $9 = 2a^2$, and therefore $a^2 = b^2 = \frac{9}{2}$. The equation of our hyperbola is

$$\frac{2(y-1)^2}{9} - \frac{2(x+3)^2}{9} = 1;$$

its graph is shown in Fig. 29-6.

PROBLEMS   29

1. Find the foci, vertices, and asymptotes of the following hyperbolas and sketch them.
   (a) $x^2 - 4y^2 = 4$ (b) $4x^2 - y^2 = 4$
   (c) $x^2 - 4y^2 = -4$ (d) $4x^2 - y^2 = -4$

   (e) $\dfrac{(x+1)^2}{16} - \dfrac{(y-2)^2}{9} = 1$ (f) $4x^2 - y^2 = 1$

   (g) $\dfrac{(y+2)^2}{9} - \dfrac{(x-1)^2}{16} = 1$ (h) $y^2 - 9x^2 = 1$

2. Find the equation of the hyperbola whose foci and vertices are the given points.
   (a) foci: $(-2, 0)$, $(2, 0)$, vertices: $(-1, 0)$, $(1, 0)$
   (b) foci: $(0, -3)$, $(0, 3)$, vertices: $(0, -2)$, $(0, 2)$
   (c) foci: $(0, 1)$, $(6, 1)$, vertices: $(1, 1)$, $(5, 1)$
   (d) foci: $(0, -1)$, $(0, 5)$, vertices: $(0, 0)$, $(0, 4)$

3. Find the foci of each of the following hyperbolas.
   (a) $9x^2 - 16y^2 + 36x + 32y - 124 = 0$
   (b) $16x^2 - 9y^2 + 64x + 18y - 89 = 0$
   (c) $9x^2 - 16y^2 + 36x + 32y + 164 = 0$
   (d) $16x^2 - 9y^2 + 64x + 18y + 199 = 0$

4. Find the equation of, and then sketch, the hyperbola whose
   (a) center is the origin, transverse diameter is part of the X-axis and is 12 units long, and whose foci are 16 units apart.
   (b) transverse diameter ends with the points $(0, 0)$ and $(0, 8)$ and whose foci are 10 units apart.
   (c) vertices are the points $(1, 0)$ and $(1, 10)$ and that contains the point $(4, 15)$.
   (d) foci are the points $(-2, -2)$ and $(-2, 8)$ and that has $(-2, 0)$ as one vertex.

5. Find the equation of the hyperbola whose asymptotes are the given lines and that contains the given point.
   (a) $y = 2x$ and $y = -2x$, $(1, 1)$ (b) $y = 2x$ and $y = -2x$, $(0, 1)$
   (c) $y = x + 1$ and $y = -x + 3$, $(2, 4)$
   (d) $y = x + 1$ and $y = -x + 1$, $(0, 0)$

6. (a) Find the equation of the hyperbola whose foci are vertices of the ellipse $11x^2 + 7y^2 = 77$ and whose vertices are the foci of this ellipse.
   (b) Find the equation of the ellipse whose vertices are the foci of the hyperbola $11x^2 - 7y^2 = 77$ and whose foci are the vertices of this hyperbola.

7. Are there points of a hyperbola at which the tangent line is parallel to an asymptote? Are there tangent lines that contain the center?

8. Find the tangent and normal lines to the hyperbola at the given point.
   (a) $8x^2 - 6y^2 = 48$, $(3, 2)$ (b) $(y - 1)^2 - 12(x + 2)^2 = 24$, $(-1, 7)$

9. Use calculus to discuss the concavity of a hyperbola.

10. A point moves so that the product of the slopes of the line segments that join it to two given points is k^2. Describe the curve traced out by the point.

11. A *convex* set in the plane is one that contains the line segment joining each pair of its points. Explain why the set

$$\{(x, y) \mid x \geq 0, 9x^2 - 16y^2 \geq 144\}$$

   is convex.

12. A man standing 2200 feet from the base of a cliff fires a rifle, and another man hears the echo 3 seconds after he hears the sound of the shot itself. Where is the second man standing? (Sound travels at 1100 feet per second.)

13. Describe the graph of the equation

$$\frac{x^4}{a^4} - \frac{y^4}{b^4} + \frac{2y^2}{b^2} = 1.$$

14. Sketch the graphs of the following equations.

   (a) $\left| \dfrac{x^2}{4} - y^2 \right| = 1$ (b) $\sin \pi \sqrt{x^2 - 4y^2} = 0$

   (c) $\dfrac{x|x|}{4} + y|y| = 1$ (d) $[\![x^2/4 - y^2]\!] = 0$

15. The segment cut by a hyperbola from a line that contains a focus and is perpendicular to the transverse diameter is called a **latus rectum** of the hyperbola. Show that a latus rectum of Hyperbola 29-3 is $2b^2/a$ units long. Find the slope of the tangent at the point of the first quadrant that is an endpoint of a latus rectum. Show that this tangent line cuts off a segment of the negative Y-axis that is half as long as the transverse diameter.

16. Consider the tangent at a point (h, k) of the hyperbola $x^2/a^2 - y^2/b^2 = 1$. Show that it intersects the asymptotes in the points $(h + ak/b, k + bh/a)$ and $(h - ak/b, k - bh/a)$. Use this result to prove that if the tangent to a hyperbola at a point A intersects the asymptotes in the points P and Q then A is the midpoint of the line segment PQ.

30. THE PARABOLA

For our geometric descriptions of the ellipse and the hyperbola, we started with two given points and a given positive number. In the case of the **parabola**, we start with a given point and a given line. The point is the **focus** of the parabola, the line is its **directrix**, and the *parabola is the set of points, each of which is equidistant from the focus and the directrix.*

To translate this geometric description of a parabola into an analytic one, let us choose our coordinate axes so that the focus is the point $(c, 0)$ and the directrix is the line $x = -c$. In Fig. 30-1 we have shown the focus and directrix if $c > 0$, but our discussion is valid if c is either positive or negative. The (perpendicular) distance between a point (x, y) and the directrix is $|x + c|$. The distance formula tells us that the distance between the points (x, y) and $(c, 0)$ is $\sqrt{(x - c)^2 + y^2}$. By definition, (x, y) is a point of our parabola if and only if these two numbers are equal; that is,

$$\sqrt{(x - c)^2 + y^2} = |x + c|.$$

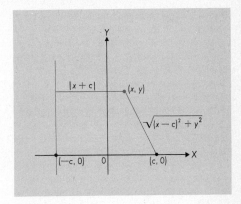

Figure 30-1

When we square both sides of this equation and simplify, we get the equation

(30-1) $$y^2 = 4cx.$$

Since we can replace y with $-y$ without changing Equation 30-1, we see that our parabola is symmetric with respect to the X-axis. The line of symmetry of a parabola (in this case, the X-axis) is called the **axis** of the parabola. The axis contains the focus and is perpendicular to the directrix. The point of intersection of a parabola with its axis (in this case, the origin) is the **vertex** of the parabola. The **focal length** of the parabola is the distance between the vertex and the focus (in this case, $|c|$).

We shall now draw the parabola $y^2 = 4cx$. Its "orientation" depends on the sign of c, so let us suppose for the moment that $c > 0$. Since the curve is symmetric about the X-axis, we need only plot those points for which $y \geq 0$, for we can obtain the remainder of the graph by reflecting this part about the X-axis. So we can suppose that

$$y = 2\sqrt{cx}.$$

Then it is easy to calculate that

$$y' = \sqrt{\frac{c}{x}} \quad \text{and} \quad y'' = -\frac{1}{2x}\sqrt{\frac{c}{x}}.$$

The expression for y' tells us that the slope is always positive (except where it is undefined at the origin), so the curve rises to the right. On the other hand, $y'' < 0$, so our parabola is concave down. Its graph is shown in Fig. 30-2. If $c < 0$, our parabola opens to the left, as shown in Fig. 30-3.

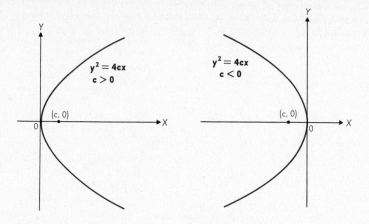

Figure 30-2

Figure 30-3

If we choose the coordinate axes so that the focus of the parabola is the point $(0, c)$ of the Y-axis and the directrix is the line $y = -c$ parallel to the X-axis, then the vertex is again the origin. The roles of x and y, however, are now interchanged, so the line of symmetry is the Y-axis and the equation of the parabola is

$$(30\text{-}2) \qquad\qquad x^2 = 4cy.$$

If $c > 0$, the parabola opens up, as shown on the left in Fig. 30-4. If $c < 0$, it opens down, as shown on the right.

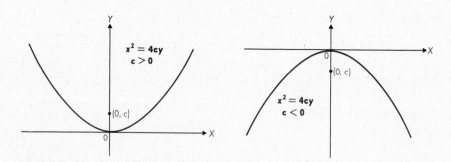

Figure 30-4

Example 30-1. The vertex of a parabola is the origin, and it is symmetric with respect to the X-axis. The slope of the parabola is -1 where it intersects the line $y = 6$. What is its equation?

Solution. The equation of a parabola that is symmetric with respect to the X-axis and has the origin as vertex has the form $y^2 = 4cx$. We must determine the number c. We use implicit differentiation to find that $2yy' = 4c$, and when we substitute $y = 6$ and $y' = -1$, we find that $c = -3$. The equation of our parabola is therefore $y^2 = -12x$.

The equation

(30-3) $$(y - k)^2 = 4c(x - h)$$

is also an equation of a parabola. If we use the translation equations $\bar{y} = y - k$ and $\bar{x} = x - h$ to introduce an $\overline{X}\overline{Y}$-coordinate system, our equation becomes $\bar{y}^2 = 4c\bar{x}$, which tells us that the vertex of the parabola is the origin of the $\overline{X}\overline{Y}$-system (that is, the point (h, k) of the XY-system) and the directrix is the line $\bar{x} = -c$ (that is, the line $x = h - c$ in the XY-system). Figure 30-5 shows what the parabola looks like if $c > 0$. Similarly, the equation

(30-4) $$(x - h)^2 = 4c(y - k)$$

represents a parabola whose vertex is the point (h, k) and whose directrix is the line $y = k - c$ (see Fig. 30-6 for the case $c > 0$).

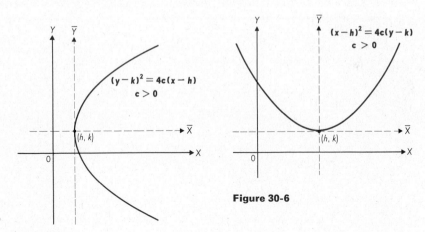

Figure 30-6

Figure 30-5

Example 30-2. Write the equation of the parabola whose directrix is the line $y = -5$ and whose focus is the point $(4, -1)$.

Solution. Since the directrix is horizontal, the axis of our parabola is vertical. The vertex is midway between the focus and the point in which the axis intersects the directrix. In this case, the vertex is the point $(4, -3)$. Clearly (see Fig. 30-7), the focal length is $c = 2$. Thus we obtain the equation of our parabola from Equation 30-4 by taking $h = 4$, $k = -3$, and $c = 2$; that is,

$$(x - 4)^2 = 8(y + 3).$$

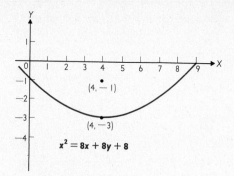

Figure 30-7

This equation can also be written

$$x^2 = 8x + 8y + 8.$$

Example 30-3. Show that the graph of the equation $y = ax^2 + bx + c$ is a parabola if $a \neq 0$.

Solution. We replace x with $\bar{x} + h$ and y with $\bar{y} + k$, where h and k are numbers to be determined, and we obtain the equation

$$\bar{y} + k = a(\bar{x} + h)^2 + b(\bar{x} + h) + c$$

$$= a\bar{x}^2 + (2ah + b)\bar{x} + ah^2 + bh + c.$$

Now if we let $h = -b/2a$ and then $k = ah^2 + bh + c = (4ac - b^2)/4a$, we obtain the standard form of the equation of a parabola,

$$\bar{x}^2 = \frac{1}{a}\bar{y}.$$

If $a > 0$, the parabola opens up; if $a < 0$, the parabola opens down. It would be a good exercise for you to obtain these translation equations by completing the square.

PROBLEMS 30

1. Find the focus and directrix of, and sketch, each of the following parabolas.
   (a) $y^2 = 16x$
   (b) $x^2 = -8y$
   (c) $(x - 2)^2 = 24(y + 3)$
   (d) $y^2 = 4(x - 2)$
   (e) $x^2 + 2x + 4y + 5 = 0$
   (f) $y^2 - 8x + 2y + 1 = 0$

2. Find the equation of the parabola whose focus is the given point and whose directrix is the given line.

   (a) $(4, 0)$, $x = -4$ (b) $(0, 6)$, $y = 0$
   (c) $(1, 2)$, $x = -5$ (d) $(-3, 2)$, $y = -6$

3. Find the equation of the parabola whose

   (a) focus is the point $(2, 4)$ and whose vertex is the point $(2, 0)$.
   (b) directrix is the Y-axis and whose vertex is the point $(4, 2)$.
   (c) axis is parallel to the Y-axis, whose vertex is the point $(-1, 3)$, and that contains the point $(1, 1)$.
   (d) focus is the point $(-2, 2)$, that contains the point $(-2, 4)$, and whose axis is parallel to a coordinate axis.

4. The segment that a parabola cuts off the line that contains the focus and is perpendicular to the axis is called the **latus rectum** of the parabola.

   (a) A parabola has a focal length of c. Show that the length of its latus rectum is $4c$.
   (b) Determine the length of the latus rectum of each of the parabolas in Problem 30-2.
   (c) Find the equation of the circle that contains the vertex and the two endpoints of the latus rectum of the parabola $y^2 = 4cx$.

5. Find the slopes of the tangents to the parabola $y^2 = 4cx$ at the endpoints of the latus rectum. What is the angle between these tangents? Show that the tangents intersect on the directrix of the parabola.

6. The axis of the parabola $y = ax^2 + bx + c$ is parallel to the Y-axis. Therefore its vertex is a maximum or a minimum point. Use derivatives to find the vertex and compare your result with what we found in Example 30-3. Use the second derivative to discuss the concavity of this parabola.

7. In the course of a yacht race, a boat is required to pass between a buoy and a straight shore 200 yards away. Assuming that the skipper stays as far as possible from both buoy and shore, describe his path as he rounds the buoy.

8. Discuss the graphs of the following equations.

   (a) $y^2 = 8|x|$ (b) $y|y| = 8x$
   (c) $y^2 = 4(x + |x|)$ (d) $(y + |y|)^2 = 16x$

9. Describe the following sets of pairs of numbers geometrically.

   (a) $\{(x, y) \mid y^2 < 8x\}$
   (b) $\{(x, y) \mid y > 4x^2 + 24x + 38\}$
   (c) $\{(x, y) \mid x = \sin t,\ y = \sin^2 t,\ t \in R^1\}$
   (d) $\{(x, y) \mid x = \sin t,\ y = \cos^2 t,\ t \in R^1\}$

10. From the geometric definition of a parabola, find the directrix of the "tilted" parabola whose vertex is the origin and whose focus is the point $(1, 1)$. Can you find the equation of the parabola?

11. Find the equations of the tangents and normals to each of the following parabolas at the given points.

   (a) $y^2 = 9x$, $(1, 3)$ (b) $x^2 = 4(y - 1)$, $(-2, 2)$
   (c) $(y + 1)^2 = 8(x - 4)$, $(4, -1)$ (d) $(y + 1)^2 = 8(x - 4)$, $(6, -5)$

12.   Let (x_1, y_1) and (x_2, y_2) be any two points of the parabola $x^2 = 4cy$. Show that the chord joining them is parallel to the tangent to the parabola at the point whose X-coordinate is the midpoint of the interval $[x_1, x_2]$.

13.   The vertex of the parabola $x^2 = 8y$ is the center of an ellipse, and the focus of the parabola is an endpoint of the minor diameter of the ellipse. The parabola and ellipse intersect at right angles. Find the equation of the ellipse.

14.   Show that if the curve $y = f(x)$ is a subset of the parabola $y^2 = 4cx$, then y satisfies the differential equation $2xy' = y$. Conversely, show that if y satisfies the differential equation, then $\dfrac{d}{dx}(y^2/x) = 0$, so $y^2 = 4cx$, where c is a number that is independent of x.

31.   CONICS

We have defined a parabola as the set of points, each of which is equidistant from a given point (the focus) and a given line (the directrix). Another way to word this definition is to say that the ratio of the distance between a point of our parabola and the focus to the distance between the point and the directrix is 1 for each point of the parabola.

Ellipses and hyperbolas have a similar property. Associated with each focus of an ellipse or a hyperbola is a line (called a *directrix*) such that the ratio of the distance between a point of the curve and the focus to the distance between the point and the directrix is a number e that is the same for every point of the curve.

Let us look at a hyperbola to see exactly what we mean. In Fig. 31-1 we have drawn the hyperbola

(31-1) $$\frac{x^2}{a^2} - \frac{y^2}{b^2} = 1.$$

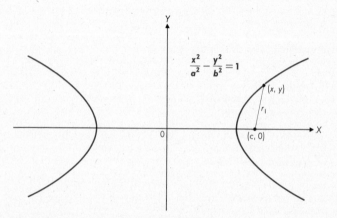

Figure 31-1

We have also shown the focus $(c, 0)$, where

(31-2) $$c = \sqrt{a^2 + b^2}.$$

Now we will compute the number r_1^2, where r_1 is the distance between a point (x, y) of our hyperbola and the focus $(c, 0)$. (Even though we have shown (x, y) as a point of the "right branch" in our figure, it could be a point of either branch of the hyperbola.) According to the distance formula,

$$r_1^2 = (x - c)^2 + y^2 = x^2 - 2cx + c^2 + y^2.$$

In this equation we may replace y^2 with $(b^2x^2/a^2) - b^2$ (from Equation 31-1) and get

$$r_1^2 = x^2 - 2cx + c^2 + \frac{b^2x^2}{a^2} - b^2$$

$$= \left(\frac{a^2 + b^2}{a^2}\right) x^2 - 2cx + c^2 - b^2.$$

Equation 31-2 tells us that $a^2 + b^2 = c^2$ and $c^2 - b^2 = a^2$, so

$$r_1^2 = \frac{c^2}{a^2} x^2 - 2cx + a^2$$

$$= \frac{c^2}{a^2}\left(x^2 - \frac{2a^2}{c} x + \frac{a^2}{c^2}\right)$$

$$= \frac{c^2}{a^2}\left(x - \frac{a^2}{c}\right)^2.$$

Therefore we see that

(31-3) $$r_1 = \frac{c}{a}\left|x - \frac{a^2}{c}\right|.$$

The number $|x - a^2/c|$ is the distance between the point (x, y) and the line parallel to the Y-axis and a^2/c units to the right of it (Fig. 31-2). We have labeled this distance as r_2, and so Equation 31-3 can be written as

(31-4) $$\frac{r_1}{r_2} = \frac{c}{a}.$$

Associated with the focus $(c, 0)$ we have found a line, $x = a^2/c$, such that the ratio of the distance between a point of our hyperbola and the focus to the distance between the point and the line is a number, c/a, that is independent of the

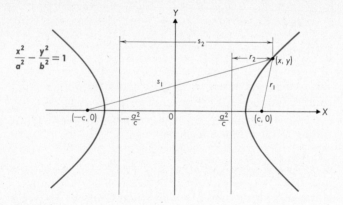

$$\frac{x^2}{a^2} - \frac{y^2}{b^2} = 1$$

Figure 31-2

choice of the point of the hyperbola. It is easy to show that the line a^2/c units to the left of the Y-axis is associated with the focus $(-c, 0)$ in the same way. That is, if s_1 is the distance between the point (x, y) and the focus $(-c, 0)$, and s_2 is the distance between the point and the associated line, then the ratio s_1/s_2 is again the number c/a. The lines $x = a^2/c$ and $x = -a^2/c$ are called the **directrices** of the hyperbola, and the ratio $e = c/a$ is the **eccentricity**.

The same reasoning leads to similar results for an ellipse. Let us consider the ellipse

(31-5) $$\frac{x^2}{a^2} + \frac{y^2}{b^2} = 1,$$

where $a > b$, that is shown in Fig. 31-3. Associated with the focus $(c, 0)$ is the directrix parallel to the Y-axis and a^2/c units to the right of it. The directrix a^2/c units to the left of the Y-axis is associated with the focus $(-c, 0)$. In the problems

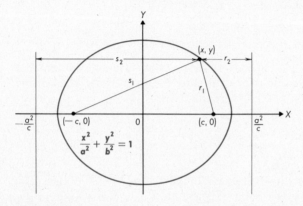

Figure 31-3

at the end of the section, we ask you to show that if (x, y) is any point of our ellipse, then the ratio r_1/r_2 of the distance between (x, y) and $(c, 0)$ to the distance between (x, y) and the associated directrix is the eccentricity $e = c/a$. The ratio s_1/s_2 of the distances to the other focus and directrix is also e.

We have shown that the ellipse, parabola, and hyperbola all share a common "ratio" property. Associated with a focus of one of these curves is a line, called a directrix, such that the ratio of the distance between a point of the curve and the focus to the distance between the point and the directrix is a number e (the eccentricity) that is independent of the choice of the point of the curve. For our ellipse and hyperbola, the number e is given by the formula $e = c/a$. Since $c < a$ for an ellipse and $c > a$ for a hyperbola, we see that *the eccentricity of an ellipse is a number that is less than 1, the eccentricity of a hyperbola is greater than 1, and the eccentricity of a parabola is equal to 1.*

Ellipses, parabolas, and hyperbolas make up the family of curves known as the *conics*. The members of this family can be geometrically described as follows: *A **conic** is determined by a given point F (a **focus**), a given line d not containing F (the **directrix** associated with F), and a number e (the **eccentricity**). A point P is a point of the conic if and only if the ratio of the distance $\overline{FP}$ to the distance between P and the line d is the number e.* The name "conic" comes from another geometric description of these curves. They are the curves that result when an ordinary cone in three-dimensional space is cut by a plane. Since we are not venturing into three dimensions, we will not prove this assertion.

A conic is an ellipse, parabola, or hyperbola, depending on whether $e < 1$, $e = 1$, or $e > 1$. For our Hyperbola 31-1 and our Ellipse 31-5 (in which $a > b$), the directrices are the lines

$$(31\text{-}6) \qquad x = -\frac{a^2}{c} \quad \text{and} \quad x = \frac{a^2}{c},$$

and the eccentricity is the number

$$(31\text{-}7) \qquad e = \frac{c}{a}.$$

If a hyperbola or ellipse has a different orientation, these formulas must be modified accordingly. For example, if $b > a$ in Equation 31-5 of the ellipse, then $e = c/b$, and the directrices are the lines $y = -b^2/c$ and $y = b^2/c$.

Example 31-1.   Find the equation of the conic that has an eccentricity of 2 and that has the point $(3, 0)$ as a focus for which the corresponding directrix is the Y-axis.

Solution.   In Fig. 31-4 we show the typical point (x, y) of our conic and also the focus $(3, 0)$. The geometric description of a conic tells us that $r_1/r_2 = 2$; that is,

$$r_1^2 = 4r_2^2.$$

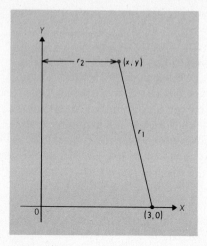

Figure 31-4

Since $r_1^2 = (x - 3)^2 + y^2$ and $r_2^2 = x^2$, the equation of our conic is

$$(x - 3)^2 + y^2 = 4x^2.$$

This equation can be written as $3x^2 - y^2 + 6x - 9 = 0$ or, after completing the square,

$$\frac{(x + 1)^2}{4} - \frac{y^2}{12} = 1.$$

In this form, it is plain that the equation represents a hyperbola, but we knew from the start that our conic is a hyperbola, since its eccentricity was greater than 1.

Example 31-2. Find the eccentricity and locate the directrices of the conic $9x^2 + 25y^2 = 1$.

Solution. If we write this equation as

$$\frac{x^2}{\frac{1}{9}} + \frac{y^2}{\frac{1}{25}} = 1,$$

we recognize it as the standard form of the equation of an ellipse for which $a^2 = \frac{1}{9}$ and $b^2 = \frac{1}{25}$. Hence

$$c = \sqrt{a^2 - b^2} = \sqrt{\frac{1}{9} - \frac{1}{25}} = \sqrt{\frac{16}{225}} = \frac{4}{15}.$$

Equation 31-7 gives the eccentricity of our ellipse as $e = c/a = \frac{4}{15}/\frac{1}{3} = \frac{4}{5}$, and we see from Equations 31-6 that the directrices are the lines $x = -\frac{5}{12}$ and $x = \frac{5}{12}$.

The foci of some conics play the role of focal points in a physical sense. In order to demonstrate the focusing property of the parabola, we find it convenient to choose the coordinate axes so that we are dealing with $y^2 = 4cx$, where $c > 0$. Figure 31-5 shows the upper half of the parabola; the point $F(c, 0)$ is its focus. If the figure represents the upper half of the cross section of a parabolic reflector, then we will show that a ray of light parallel to the X-axis will strike a point P of the parabola and be reflected to the focus F. The tangent to the parabola at P is the line TP, and it is a law of optics that the reflected beam and the incident beam will make equal angles with this line. From elementary geometry we see that $\angle PTF$ is the acute angle between the incident beam and the tangent line. The reflected beam will travel the path PF provided that $\angle FPT = \angle PTF$, and in order for these angles to be equal, the triangle TFP must be isosceles, with $\overline{PF} = \overline{TF}$. It is this equation that we will now verify.

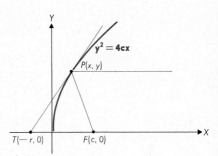

Figure 31-5

If the coordinates of P are (x, y), then the distance formula yields

$$\overline{PF} = \sqrt{(x - c)^2 + y^2}$$

$$= \sqrt{x^2 - 2cx + c^2 + y^2}.$$

Since P is a point of our parabola, $y^2 = 4cx$, and hence we may replace y^2 with $4cx$ in the last radical:

$$(31\text{-}8) \qquad \overline{PF} = \sqrt{x^2 + 2cx + c^2} = x + c.$$

Calculating the distance $\overline{TF}$ is somewhat more complicated. From Fig. 31-5 we see that $\overline{TF} = c + r$, where r is the positive number such that the tangent line at P intersects the X-axis in the point $T(-r, 0)$. To find this number r, we first note that the slope of the segment TP is $y/(x + r)$. The slope of this segment may also be found from the equation $y^2 = 4cx$ by implicit differentiation: $2yy' = 4c$. Hence $y' = 2c/y$, and we equate our two expressions for the slope of TP:

$$\frac{y}{x + r} = \frac{2c}{y}.$$

Then

$$x + r = \frac{y^2}{2c} = \frac{4cx}{2c} = 2x,$$

and so $r = x$. It follows that

$$\overline{TF} = c + r = c + x.$$

When we compare this equation with Equation 31-8, we see that $\overline{PF} = \overline{TF}$, and our argument is complete.

We won't go into quite so much detail with regard to the focusing property of the ellipse. Figure 31-6 shows an ellipse

$$\frac{x^2}{a^2} + \frac{y^2}{b^2} = 1,$$

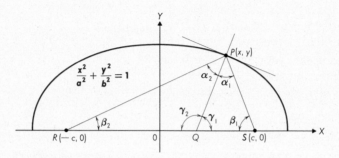

Figure 31-6

with $a > b$. We have drawn a tangent line at an arbitrary point P. Now we assert that the segments SP and RP (S and R are the foci) make equal angles with the tangent line at P (or, equivalently, with the normal line QP). Thus if our figure is the cross section of an elliptical reflector, we are saying that a ray of light emanating from the focus S will be reflected at P to the other focus R.

To prove this assertion, we must show that the angles labeled α_1 and α_2 are equal. These angles will be equal if $\beta_1 + \gamma_1 = \beta_2 + \gamma_2$, or equivalently here,

$$\tan(\beta_1 + \gamma_1) = \tan(\beta_2 + \gamma_2).$$

We can verify this last equation by first expanding the two sides,

$$\frac{\tan \beta_1 + \tan \gamma_1}{1 - \tan \beta_1 \tan \gamma_1} = \frac{\tan \beta_2 + \tan \gamma_2}{1 - \tan \beta_2 \tan \gamma_2},$$

and then calculating these tangents in terms of the coordinates of P. For example, $\tan \beta_2 = y/(x + c)$, and $\tan \gamma_1$ times the slope of the ellipse at P is -1. We will leave the details of verifying this equation to you.

PROBLEMS   31

1. Find the eccentricity and directrices of each of the following conics.

    (a) $3x^2 + 4y^2 = 12$ (b) $y^2 = 12x$ (c) $9x^2 - 16y^2 = 144$
    (d) $12x^2 + y = 3$ (e) $x^2 - y^2 = 1$ (f) $5x^2 + 9y^2 = 45$
    (g) $16y^2 - 9x^2 = 144$ (h) $9x^2 + 5y^2 = 45$

2. Use our geometric description to find the conic with the given focus, corresponding directrix, and eccentricity.

    (a) focus $(2, 0)$, directrix $x = -2$, $e = 2$
    (b) focus $(2, 0)$, directrix $x = -2$, $e = 1$
    (c) focus $(2, 0)$, directrix $x = -2$, $e = \frac{1}{2}$
    (d) focus $(0, 4)$, directrix $y = 0$, $e = 3$
    (e) focus $(1, 0)$, directrix $x = 2$, $e = 1$
    (f) focus $(1, 0)$, directrix $x = 4$, $e = \frac{1}{2}$
    (g) focus $(1, 2)$, directrix $y = -2$, $e = \frac{1}{3}$
    (h) focus $(0, 4)$, directrix $y = 1$, $e = 2$

3. Find the eccentricity and directrices of each of the following conics.

    (a) $3x^2 + 4y^2 - 6x + 16y + 7 = 0$
    (b) $9x^2 - 16y^2 + 18x - 32y = 151$
    (c) $x^2 - y^2 = 2x$
    (d) $5x^2 + 9y^2 - 20x - 36y + 11 = 0$

4. Find the eccentricity and the directrices of the hyperbola $a^2y^2 - b^2x^2 = a^2b^2$.

5. (a) Tell how the shape of an ellipse changes as the eccentricity varies from near 0 to near 1.
    (b) Tell how the shape of a hyperbola changes as the eccentricity varies from near 1 to a very large number.

6. Suppose that you are told that the Y-axis is a directrix of an equilateral hyperbola and that the corresponding focus is a point of the X-axis that is 3 units away from this directrix. Can you find the equation of the hyperbola?

7. What is the eccentricity of an ellipse in which the distance between its foci is one-half the distance between its directrices?

8. A vertex of a certain conic is the origin, and the nearest focus is the point $(3, 4)$. What is the corresponding directrix if

    (a) the conic is a parabola?
    (b) the conic is a hyperbola of eccentricity 2?
    (c) the conic is an ellipse of eccentricity $\frac{1}{2}$?

9. Mr. Jones owns a small rectangular woodlot 200 yards wide. One side is bounded by a very noisy highway, and right in the middle of the opposite edge is a particularly foul-smelling foundry. Describe the path Mr. Jones takes on his Sunday morning walks if

    (a) he dislikes the highway and foundry equally.
    (b) the wind is such that the smell of the foundry is twice as annoying as the sound of the highway.
    (c) the wind is such that the smell of the foundry is half as annoying as the sound of the highway.

10.   Find the eccentricity of the following conics.

(a) $x^2 - \dfrac{y^2}{\tan^2 c} = 1$

(b) $x^2 + \dfrac{y^2}{\cos^2 c} = 1$

(c) $\dfrac{x^2}{\cos^2 c} - \dfrac{y^2}{\sin^2 c} = 1$

(d) $\dfrac{x^2}{\cos^2 c} - \dfrac{y^2}{\sin^2 c} = -1$

11.   What is the eccentricity of an equilateral hyperbola?

12.   The vertex of the parabola $x^2 = 4cy$ is the center of an ellipse. The focus is an endpoint of the minor diameter of the ellipse, and the parabola and ellipse intersect at right angles. Find the eccentricity of the ellipse. (See Problem 30-13.)

13.   Suppose that the major diameter of an ellipse is $2S$ units long and its minor diameter is $2s$ units long. Show that its eccentricity satisfies the equation $e^2 = (S^2 - s^2)/S^2$.

14.   Let r be the ratio of the length of the minor diameter of an ellipse to the length of its major diameter.

(a) Find r for the ellipse

$$\frac{x^2}{16} + \frac{y^2}{9} = 1.$$

(b) Find r for the ellipse

$$\frac{(x+3)^2}{25} + \frac{(y-2)^2}{4} = 1.$$

(c) For the ellipses of parts (a) and (b), show that $e^2 + r^2 = 1$ (where e is the eccentricity of the ellipse).

(d) Show that $e^2 + r^2 = 1$ for any ellipse.

15.   Use algebra to show that the distance d between a point (x, y) of the ellipse $x^2/a^2 + y^2/b^2 = 1$ (with $a > b$) and the focus $(c, 0)$ is given by the equation

$$d = e \left| x - \frac{a^2}{c} \right|.$$

The very definition of a circle tells us that its equation is

$$(x - h)^2 + (y - k)^2 = r^2,$$

and this equation, in turn, helps us remember the translation equations

$$\bar{x} = x - h \quad \text{and} \quad \bar{y} = y - k.$$

But it is virtually impossible to remember the various formulas that are associated with the conics, so we will tabulate the ones that apply to a conic with its "major axis" along the X-axis.

Curve	Ellipse	Hyperbola	Parabola
Equation	$\dfrac{x^2}{a^2} + \dfrac{y^2}{b^2} = 1$	$\dfrac{x^2}{a^2} - \dfrac{y^2}{b^2} = 1$	$y^2 = 4cx$
Center	$(0, 0)$	$(0, 0)$	—
Foci	$(\pm\sqrt{a^2 - b^2}, 0)$	$(\pm\sqrt{a^2 + b^2}, 0)$	$(c, 0)$
Vertices	$(\pm a, 0), (0, \pm b)$	$(\pm a, 0)$	$(0, 0)$
Directrices	$x = \pm a^2/\sqrt{a^2 - b^2}$	$x = \pm a^2/\sqrt{a^2 + b^2}$	$x = -c$
Eccentricity	$\sqrt{a^2 - b^2}/a$	$\sqrt{a^2 + b^2}/a$	1
Asymptotes	—	$b^2x^2 = a^2y^2$	—
Graph	Fig. 28-2	Fig. 29-3	Fig. 30-2

Review Problems, Chapter Four

1. Some people say that a circle is a conic and some say that it isn't. Give a reason for each point of view.

2. Geometrically, it is obvious that the parabola $y^2 = 4cx$ and the ellipse $x^2/a^2 + y^2/b^2 = 1$ must intersect, whatever the choice of a, b, and c. Explain this fact algebraically.

3. You are on the seventh floor of College Hall, which is burning briskly. The firemen have a circular net of radius 8 centered at $(-2, -3)$. You jump for the point $(2, 4)$. What are your chances of graduating?

4. Jones' head has the equation $(x - 2)^2 + (y + 1)^2 = 4$. An axe is swung along the line $x + 2y = 3$. Is Jones singular or plural?

5. Is it true that if the graph of a relation is symmetric with respect to both coordinate axes, then it is also symmetric with respect to the origin?

6. Sketch several members (different choices of c) of each of the following families of curves.

    (a) $\dfrac{x^2}{9} + \dfrac{y^2}{c^2} = 1$ (b) $\dfrac{x^2}{c^2} - \dfrac{y^2}{4} = 1$

    (c) $\dfrac{(x - c)^2}{9} + \dfrac{y^2}{4} = 1$ (d) $x^2 - y^2 = c$

    (e) $(y - c)^2 = 12x$ (f) $y^2 = 12(x - c)$

7. Starting at the point $(r, 0)$ of the circle $x^2 + y^2 = r^2$, a man walks s units around the circumference in a counterclockwise direction. Express the coordinates of his stopping point in terms of r and s.

8. If the circle $x^2 + y^2 = 4y$ is rolled to the right exactly one revolution along the X-axis, what is its new center?

9. In his barnyard, Farmer Jones put a salt block and a watering trough 50 feet apart. Then he built a fence so that from any point inside it his cow could walk to the salt block, then to the watering trough, and back to her starting point without traveling more than 150 feet. Describe the fence.

10. The Union admiral attacking Magnolia Bay in 1865 knows that each of the two forts (200 yards apart) guarding the harbor is down to one cannonball, and between them they have only enough powder to shoot these cannonballs a total of 600 yards. He doesn't know how the forts have divided up their powder. Where is he in danger of being hit by both forts? Where is he safe from being hit by either? And where is it possible to get hit once but not twice?

11. An artificial earth satellite moves in an elliptical orbit with the center of the earth as one focus. The minimum distance from the center of the earth to the satellite's path is d_1 miles and the maximum distance is d_2 miles. Find the formula for the eccentricity of the elliptical path in terms of d_1 and d_2.

12. What conditions must a, b, c, d, and e satisfy in order that the curve $ax^2 + by^2 + 2cx + 2dy + e = 0$ be an ellipse? A hyperbola? A parabola?

13. Show that if the curve $y = f(x)$ is a subset of the graph of one of Equations 30-1, 29-3, 29-5, or 28-3, then $y^3 y''$ is independent of x.

14. Consider a hyperbola whose transverse diameter lies along the X-axis. Show that the slope of a tangent at an endpoint of a latus rectum is either e or $-e$.

15. Find the slope of the ellipse $b^2 x^2 + a^2 y^2 = a^2 b^2$ (where $a > b$) at the point of the first quadrant that is an endpoint of a latus rectum. How is this slope related to the eccentricity of the ellipse?

16. A man walks directly from the point $(4, 0)$ to a point of the curve $x^2/25 + y^2/9 = 1$ and then walks twice as fast directly to the point $(-4, 0)$. How far does he walk? To which point P should he go to complete his trip in minimum time?

17. Can you convince yourself that for first quadrant points far enough to the right of the Y-axis, the hyperbola $x^2/a^2 - y^2/b^2 = 1$ will be above the parabola $y^2 = 4cx$ ($c > 0$)?

18. One focus and one vertex of the hyperbola $b^2 x^2 - a^2 y^2 = a^2 b^2$ are points of the positive X-axis. Find the equation of the parabola with the same vertex and focus. Show that the parabola is "eaten by" the right-hand branch of the hyperbola.

19. Let $a > b$ so that one focus and one vertex of the ellipse $b^2 x^2 + a^2 y^2 = a^2 b^2$ are points of the positive X-axis. Find the equation of the parabola with the same vertex and focus. Show that the parabola "eats" the ellipse.

20. The equations of ellipses, hyperbolas, and parabolas in standard position can be written as

$$\frac{xx}{a^2} + \frac{yy}{b^2} = 1, \qquad \frac{xx}{a^2} - \frac{yy}{b^2} = 1, \qquad yy = 4c\,\frac{1}{2}(x + x),$$

and so on, where we replace x^2 and y^2 with xx and yy and replace x and y with $\frac{1}{2}(x + x)$ and $\frac{1}{2}(y + y)$. Show that the tangents to these curves at the point (x_1, y_1) are

$$\frac{xx_1}{a^2} + \frac{yy_1}{b^2} = 1, \qquad \frac{xx_1}{a^2} - \frac{yy_1}{b^2} = 1, \qquad yy_1 = 4c\,\frac{1}{2}(x + x_1),$$

and so on.

21. Discuss the following curves.

(a) $\dfrac{x^2}{9} + \dfrac{y|y|}{4} = 1$

(b) $\dfrac{x|x|}{9} + \dfrac{y|y|}{4} = 1$

(c) $(y + |y|)^2 + (x + |x|)^2 = 4$

(d) $(y + |y|)^2 - (x + |x|)^2 = 4$

(e) $y^2 = (|x| + x)^2 + 1$

(f) $y^2 = (|x| + x)^2$

5 | The Integral

The two most important concepts in calculus can be introduced by a study of geometric problems. We have seen how the concept of the *derivative* arose when we considered the problem of finding the tangent line to a curve at a point. In this chapter we shall see how the problem of finding the area of a plane region with a given curve as part of its boundary leads us to the second major concept of calculus—the *integral*.

Of course, we don't restrict our interpretation of the derivative to the geometric idea of slope. A derivative is a rate, and as such we can use it to discuss problems involving the velocity and acceleration of moving bodies, and so on. Similarly, although we shall introduce the integral as a solution to the problem of finding the areas of certain regions, we shall also see that other interpretations are possible. Our applications of integrals will range from finding the work required to pump out a tank of water to providing a new approach to the definition of logarithms.

32. APPROXIMATING AREAS

Before introducing the concept of the integral of a function over an interval, let us briefly recall how we introduced the concept of the derivative of a function at a point. In geometric terms, the derivative $f'(x)$ is the slope of the tangent line to the graph of f at the point $(x, f(x))$. Thus our first method of calculating $f'(x)$ was to draw "by eye" a line for which the name "tangent line" seemed appropriate and then measure its slope. This procedure suffers because the phrase "by eye" is too vague, so we turned to an analytic approach.

(1)   In order to get started, we assumed that there *is* a tangent line to the graph of the given function f at the given point.

(2)   Then we used geometric reasoning to convince ourselves that the slope of that line is approximated by some slopes that we can calculate analytically—namely, the slopes of certain chords. These slopes are calculated from our difference quotient $[f(z) - f(x)]/(z - x)$.

(3)   Finally, we examined these approximating slopes to see what number they approximate; that is, we found the limit of the difference quotient. This number we took to be the number $f'(x)$ that we were seeking.

Notice that from a strictly logical point of view we don't need steps (1) and (2). Their only purpose is to suggest why it might be useful to study the difference quotient. We could bypass these steps and simply define $f'(x)$ to be the number $\lim_{z \to x} [f(z) - f(x)]/(z - x)$. Thus we can formulate a precise definition of the derivative without first having a precise definition of a tangent line. In fact, we introduced our definition of the tangent line *after* we had defined $f'(x)$.

We will follow a similar sequence of steps as we introduce the integral of a function f over an interval $[a, b]$. As a first geometric approach to this concept, we will think of the integral as the area A of the region that is bounded by the graph of f, the X-axis, and the lines $x = a$ and $x = b$, the shaded region in Fig. 32-1. (For the moment, we will consider regions that lie wholly above the X-axis.) Therefore we could calculate the integral A by sketching the graph of f on graph paper and counting the squares that are contained in the shaded region. In general, the curve would cut some of the squares, so we would have to estimate "by eye" the fractions of squares that lie under the curve. Thus this graphical calculation of A suffers from difficulties like those we met when we tried to calculate $f'(x)$ by sketching a tangent line "by eye." To avoid these difficulties, we turn to an analytic approach.

(1)   In order to get started, we *assume* that there is a number A that we can say is the area of our region.

(2)   Then we use geometric reasoning to develop an analytic method of calculating approximations of this number A.

(3)   Finally, we examine these approximations to find the number that they approximate; that is, we find their limit.

As in the case of the derivative, the purpose of steps (1) and (2) is simply to suggest the analytic expression whose limit will be the integral. Thus we will be able to formulate a precise definition of the integral without first having a precise definition of the area of a plane region. We will have more to say about the connection between integrals and areas later; now we will concentrate on step (2), approximating the number that we think should measure the area of a region like the one shown in Fig. 32-1.

Figure 32-2 shows how we make the approximations. We have drawn some rectangles in the figure in such a way that the sum of their areas seems to approximate the area A we seek. To construct these approximating rectangles, we simply chose three points, x_1, x_2, and x_3 between a and b. In this way, the interval $[a, b]$

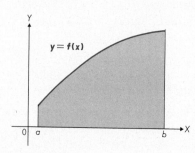

Figure 32-1   **Figure 32-2**

is divided into four smaller intervals called **subintervals** of $[a, b]$. In each subinterval we erected a vertical segment from the X-axis to the curve. The lengths of these segments are y_1, y_2, y_3, and y_4. The areas of our four rectangles are $y_1(x_1 - a)$, $y_2(x_2 - x_1)$, $y_3(x_3 - x_2)$, and $y_4(b - x_3)$. From the figure, it appears that the sum of these areas, the number

$$s = y_1(x_1 - a) + y_2(x_2 - x_1) + y_3(x_3 - x_2) + y_4(b - x_3),$$

approximates our desired area A.

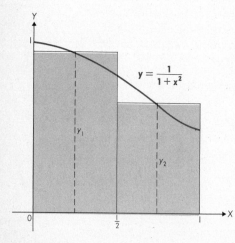

Figure 32-3

Example 32-1. Suppose that

$$f(x) = \frac{1}{1 + x^2}.$$

Approximate the area A of the region bounded by the graph of f, the X-axis, and the lines $x = 0$ and $x = 1$.

Solution. Figure 32-3 shows the graph of f in the interval $[0, 1]$. Let us divide this interval into the two subintervals $[0, \frac{1}{2}]$ and $[\frac{1}{2}, 1]$ and then construct rectangles with these one-half unit long subintervals as bases. For the altitudes of the rectangles, we will choose the values of f at the midpoints of the subintervals. Thus,

$$y_1 = f(\tfrac{1}{4}) = \frac{1}{1 + \frac{1}{16}} = \tfrac{16}{17}$$

and

$$y_2 = f(\tfrac{3}{4}) = \tfrac{16}{25}.$$

The area A is therefore approximated by the sum

$$s = \tfrac{16}{17} \cdot \tfrac{1}{2} + \tfrac{16}{25} \cdot \tfrac{1}{2} = .79 \ldots .$$

Later we shall find that $A = \tfrac{1}{4}\pi \approx .78$, so our approximation is quite close.

Now let us go into a little more detail on how we calculate the sums that approximate the area of a region like the one illustrated in Fig. 32-1. First we divide our basic interval $[a, b]$ into a number of subintervals. To form n sub-intervals, we choose $n - 1$ points $x_1, x_2, \ldots, x_{n-1}$ lying between a and b such that $a < x_1 < x_2 < \cdots < x_{n-1} < b$. These points determine a **partition** of (that is, a "division of") the interval $[a, b]$ into the subintervals $[a, x_1]$, $[x_1, x_2], \ldots,$ $[x_{n-1}, b]$. In order to simplify notation, we write $x_0 = a$ and $x_n = b$, and then we can talk about a subinterval $[x_{i-1}, x_i]$, where i is one of the numbers $1, 2, \ldots, n$. In Example 32-1, we chose $n = 2$, $x_0 = 0$, $x_1 = \tfrac{1}{2}$, and $x_2 = 1$; these division points determined the subintervals $[0, \tfrac{1}{2}]$, and $[\tfrac{1}{2}, 1]$.

Now we construct a rectangle on each of the subintervals in such a way that the sum of the areas of the rectangles approximates the area of the region under the graph of f. It seems reasonable to choose as the altitude of each rectangle the value of f at a point of the subinterval that forms the base of the rectangle. And that is exactly what we do. In each subinterval $[x_{i-1}, x_i]$, we choose a number x_i^*. (In Example 32-1, for instance, we chose $x_1^* = \tfrac{1}{4}$ in the subinterval $[0, \tfrac{1}{2}]$ and $x_2^* = \tfrac{3}{4}$ in the subinterval $[\tfrac{1}{2}, 1]$.) Then we erect a rectangle whose altitude is $f(x_i^*)$ and whose base is the subinterval $[x_{i-1}, x_i]$; the area of this rectangle is $f(x_i^*)(x_i - x_{i-1})$. The sum of the areas of all the rectangles will approximate the area A. Thus if $n = 4$, for example, we will have the **approximating sum**

$$(32\text{-}1) \qquad s = f(x_1^*)(x_1 - x_0) + f(x_2^*)(x_2 - x_1)$$
$$+ f(x_3^*)(x_3 - x_2) + f(x_4^*)(x_4 - x_3).$$

If we had partitioned the interval $[a, b]$ into 104 subintervals, our number s would have been the sum of 104 terms. To condense our formulas for approximating sums to a manageable length when n is a large number, we must introduce some notation. The sum in Equation 32-1 may be obtained by carrying out the following directions: "Replace the letter i in the expression $f(x_i^*)(x_i - x_{i-1})$ with each of the numbers $1, 2, 3$, and 4 in turn and add up the terms that result." Such directions occur so often in mathematics that we have a special symbol for them. If $P(i)$ is some mathematical expression and n is a positive integer, then the symbol

$$\sum_{i=1}^{n} P(i)$$

means, "Successively replace the letter i in the expression $P(i)$ with the numbers 1, 2, 3, ..., n and add up the resulting terms." The symbol $\sum$ is the Greek letter **sigma**, and it is used to suggest "sum," since we are to add a number of terms. Thus

$$\sum_{i=1}^{6} i^2 = 1^2 + 2^2 + 3^2 + 4^2 + 5^2 + 6^2 = 91;$$

$$\sum_{i=1}^{5} 2^i = 2^1 + 2^2 + 2^3 + 2^4 + 2^5 = 62.$$

In this $\sum$-notation, Equation 32-1 can be written as

$$s = \sum_{i=1}^{4} f(x_i^*)(x_i - x_{i-1}).$$

The letter i is called the **index of summation**, and in the preceding sum the numbers 1 and 4 are the **limits** of i. Other letters may be used as indices of summation, and the lower limit of the index need not be 1. For instance,

$$\sum_{r=3}^{5} \log r = \log 3 + \log 4 + \log 5 = \log 60,$$

and so on. In the $\sum$-notation, an approximating sum based on a partition of the interval into n subintervals takes the form

(32-2) $$s = \sum_{i=1}^{n} f(x_i^*)(x_i - x_{i-1}).$$

The length $x_i - x_{i-1}$ of the subinterval $[x_{i-1}, x_i]$ occurs so often in our formulas that it is worthwhile to introduce an abbreviation for it,

$$\Delta x_i = x_i - x_{i-1}.$$

The symbol Δ is the Greek letter **delta**, and it is supposed to suggest the word "difference." Equation 32-2 for an approximating sum now becomes

(32-3) $$s = \sum_{i=1}^{n} f(x_i^*) \, \Delta x_i.$$

Example 32-2. Approximate the area of the region between the curve $y = \sin x$ and the interval $[0, \frac{1}{2}\pi]$ of the X-axis (Fig. 32-4).

Solution. The statement of the problem allows us considerable freedom to choose the number of intervals into which the basic interval $[0, \frac{1}{2}\pi]$ is to be partitioned and the points that determine the partition. Suppose that we decide to partition the interval into $n = 2$ subintervals by selecting the point $x_1 = \frac{1}{4}\pi$. Therefore $x_0 = 0$, $x_1 = \frac{1}{4}\pi$, and $x_2 = \frac{1}{2}\pi$, and so $\Delta x_1 = \frac{1}{4}\pi - 0$, and $\Delta x_2 = \frac{1}{2}\pi - \frac{1}{4}\pi$. Now let us arbitrarily select the point $x_1^* = \frac{1}{6}\pi$ in the interval $[0, \frac{1}{4}\pi]$ and the point $x_2^* = \frac{1}{3}\pi$ in the interval $[\frac{1}{4}\pi, \frac{1}{2}\pi]$. Thus Equation 32-3 becomes in this case

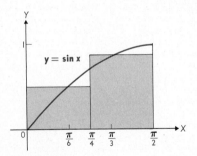

Figure 32-4

$$s = (\sin \tfrac{1}{6}\pi) \cdot \tfrac{1}{4}\pi + (\sin \tfrac{1}{3}\pi) \cdot \tfrac{1}{4}\pi$$
$$= \tfrac{1}{2} \cdot \tfrac{1}{4}\pi + \tfrac{1}{2}\sqrt{3} \cdot \tfrac{1}{4}\pi \approx 1.1.$$

It appears from Fig. 32-4 that our approximation is probably too large. Later we will find that half the area bounded by an arch of the sine curve and the X-axis is 1 square unit.

In both Example 32-1 and Example 32-2 we partitioned the interval into subintervals of equal length. Thus in Example 32-1, $\Delta x_1 = \Delta x_2 = \frac{1}{2}$; and in Example 32-2, $\Delta x_1 = \Delta x_2 = \frac{1}{4}\pi$. Choosing equal subintervals may make it easier to calculate an approximating sum, but it is not necessary to make such a choice. In fact, in some cases it is not even desirable. Here is an example of such a case.

Example 32-3. Approximate the area of the region under the graph of f in the interval $[0, 4]$ if $f(x) = 2x^4/(1 + x^4)$.

Solution. Figure 32-5 shows the graph of f for the interval $[0, 4]$. This graph suggests that we partition the basic interval into subintervals that are shortest when the graph is steepest. So let us partition the interval $[0, 4]$ by means of the points $0, \frac{1}{2}, \frac{3}{4}, 1, \frac{5}{4}, \frac{3}{2}, 2$, and 4. If we then pick the number x_i^* to be the midpoint of the subinterval $[x_{i-1}, x_i]$, our approximating sum is

$$s = \tfrac{1}{2}f(\tfrac{1}{4}) + \tfrac{1}{4}f(\tfrac{5}{8}) + \tfrac{1}{4}f(\tfrac{7}{8}) + \tfrac{1}{4}f(\tfrac{9}{8}) + \tfrac{1}{4}f(\tfrac{11}{8}) + \tfrac{1}{2}f(\tfrac{7}{4}) + 2f(3) \approx 5.81.$$

If we use the techniques of Section 59, we can show that the actual area of our region is 5.79 square units.

Example 32-4. Find $\displaystyle\sum_{i=1}^{n} f(x_i^*) \, \Delta x_i$ if $n = 5$, $x_i = 2 + .4i$ for $i = 0, 1, 2, 3, 4, 5$, $x_i^* = \frac{1}{2}(x_{i-1} + x_i)$, and $f(x) = 5x - 10$.

Solution. When we successively replace i with 0, 1, 2, 3, 4, and 5 in the equation $x_i = 2 + .4i$, we find that our basic interval is partitioned by the points

$$2, 2.4, 2.8, 3.2, 3.6, 4.$$

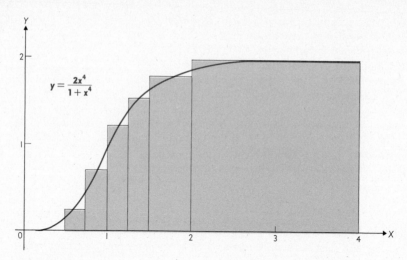

$$y = \frac{2x^4}{1+x^4}$$

Figure 32-5

The first of these numbers is the left endpoint and the last is the right endpoint, so we are working in the interval $[2, 4]$. Each of our subintervals is .4 units long; that is, $\Delta x_i = .4$ for each index i. The equation $x_i^* = \frac{1}{2}(x_{i-1} + x_i)$ simply says that x_i^* is the midpoint of the interval $[x_{i-1}, x_i]$, and hence $x_1^* = 2.2$, $x_2^* = 2.6$, $x_3^* = 3.0$, $x_4^* = 3.4$, and $x_5^* = 3.8$. Therefore, $f(x_1^*) = 5 \times 2.2 - 10 = 1$, $f(x_2^*) = 3$, $f(x_3^*) = 5$, $f(x_4^*) = 7$, and $f(x_5^*) = 9$, and so

$$\sum_{i=1}^{n} f(x_i^*) \Delta x_i = f(x_1^*) \times .4 + f(x_2^*) \times .4 + f(x_3^*) \times .4$$
$$+ f(x_4^*) \times .4 + f(x_5^*) \times .4$$
$$= (1 + 3 + 5 + 7 + 9) \times .4 = 25 \times .4 = 10.$$

How well does this number approximate the area of the region "between" the curve $y = 5x - 10$ and the interval $[2, 4]$ of the X-axis?

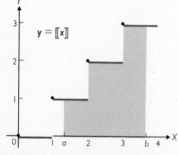

$$y = [\![x]\!]$$

Figure 32-6

You should notice that Formula 32-3 applies to discontinuous functions as well as to continuous ones. If the function f has some discontinuities in the interval $[a, b]$ (for instance, the greatest integer function as shown in Fig. 32-6), it is not strictly correct to talk about the region that is bounded by the graph of f, the X-axis, and the lines $x = a$ and $x = b$, but we will do so, and you will know what we mean (the shaded region).

PROBLEMS 32

1. Find the points that partition the following intervals into five equal subintervals.

   (a) $[2, 12]$ (b) $[-5, 10]$ (c) $[-3, 5]$ (d) $[\log 3, \log 96]$

2. Approximate the area of the region that is bounded by the graph of f, the X-axis, and the lines $x = 2$ and $x = 10$ by partitioning the interval $[2, 10]$ into four equal subintervals and using rectangles based on these subintervals, with altitudes equal to the values of f at the midpoints of the subintervals. Use a figure to judge if your approximation is too large or too small.

   (a) $f(x) = 3x + 1$ (b) $f(x) = 20 - 2x$
   (c) $f(x) = 3 + |x - 6|$ (d) $f(x) = 4 - |x - 6|$
   (e) $f(x) = [\![x]\!]$ (f) $f(x) = [\![\frac{1}{2}x]\!]$

3. Let $f(x) = x^2$ and consider the region between the graph of f and the interval $[1, 9]$.

   (a) What is the largest approximating sum you can obtain using four equal subintervals? Eight equal subintervals?
   (b) What is the smallest approximating sum you can get using four equal subintervals? Eight equal subintervals?
   (c) What approximating sum do you get using four equal subintervals and letting each x_i^* be the midpoint of a subinterval? What if you use eight equal subintervals?

4. Use the ideas in this section to approximate the area of the region under the graph of f and above the indicated interval.

   (a) $f(x) = x^2$, $[0, 2]$ (b) $f(x) = 1 - x^2$, $[-1, 1]$
   (c) $f(x) = \sin^2 x$, $[0, \frac{1}{2}\pi]$ (d) $f(x) = |\sin x|$, $[0, 2\pi]$

5. Suppose that the interval $[-1, 1]$ is partitioned into four equal subintervals. What is the largest and the smallest (if there are such) sums that we can get from Equation 32-1 (by various choices of x_1^*, x_2^*, x_3^*, and x_4^*) in the following cases?

   (a) $f(x) = 2x + 1$ (b) $f(x) = 3 - 2x$ (c) $f(x) = |x|$
   (d) $f(x) = 2 - |x|$ (e) $f(x) = 2 + [\![x]\!]$ (f) $f(x) = 2 - [\![x]\!]$

6. Since the area of the circular disk $x^2 + y^2 \leq 1$ is π square units, the area of the region in the first quadrant bounded by the X-axis, the Y-axis, and the curve $y = \sqrt{1 - x^2}$ is $\frac{1}{4}\pi$ square units. Partition the interval $[0, 1]$ into four equal subintervals and erect altitudes at the midpoints of the subintervals so as to obtain an approximating sum of four terms that approximates the area of this region, and hence obtain an approximation of the number π.

7. Find the following sums.

   (a) $\displaystyle\sum_{i=2}^{5} (1 + i)^2$ (b) $\displaystyle\sum_{i=2}^{5} (1 + i^2)$

   (c) $\displaystyle\sum_{i=1}^{4} \sin (\pi/i)$ (d) $\displaystyle\sum_{i=1}^{25} \cos \pi i$

   (e) $\displaystyle\sum_{i=1}^{5} [F(i) - F(i - 1)]$ (f) $\displaystyle\sum_{i=1}^{100} [F(i) - F(i - 1)]$

8. Find $\displaystyle\sum_{i=1}^{n} f(x_i^*)\,\Delta x_i$.

   (a) $n = 10$, $x_i = 2 + i$ for $i = 0, 1, \ldots, 10$, $x_i^* = x_{i-1}$, and $f(x) = 3x + 1$
   (b) $n = 10$, $x_i = 2 + i$ for $i = 0, 1, \ldots, 10$, $x_i^* = x_i$, and $f(x) = 3x + 1$
   (c) $n = 3$, $x_0 = 0$, $x_i = 2^i$ for $i = 1, 2, 3$, $x_i^* = \frac{1}{2}(x_i + x_{i-1})$, and $f(x) = x^2$
   (d) $n = 3$, $x_0 = 0$, $x_i = 2^i$ for $i = 1, 2, 3$, $x_i^* = \sqrt{(x_i x_{i-1})}$, and $f(x) = x^2$

9. Suppose that for each $x \in [a, b]$, the number $f(x)$ belongs to the interval $[A, B]$. Show that each sum that we can obtain from Equation 32-3 belongs to the interval $[A(b - a), B(b - a)]$.

10. Suppose that we are given two functions f and g, an interval $[a, b]$, and a number c. Let $h(x) = f(x) + g(x)$ and $k(x) = cf(x)$. Now choose a particular partition of $[a, b]$ and a set of points $\{x_1^*, x_2^*, \ldots, x_n^*\}$, and denote by F, G, H, and K the sums that we obtain from Equation 32-3 for the functions f, g, h, and k.

   (a) Show that $H = F + G$.
   (b) Show that $K = cF$.
   (c) If we further assume that $f(x) \le g(x)$ for each $x \in [a, b]$, show that $F \le G$.

11. Suppose that f is an increasing function in the interval $[a, b]$. Let s and t be two sums that are given by Equation 32-3 for the same partition of $[a, b]$ but using different choices of the points $x_1^*, x_2^*, \ldots, x_n^*$. Show that $|s - t| \le [f(b) - f(a)]u$, where u is the length of the longest subinterval of our partition of $[a, b]$.

33. THE INTEGRAL

In the preceding section we used sums of the form

$$(33\text{-}1) \qquad\qquad s = \sum_{i=1}^{n} f(x_i^*)\,\Delta x_i$$

to approximate the area A of the region bounded by the graph of a function f that takes positive values, the X-axis, and the lines $x = a$ and $x = b$. In looking at the figures in that section, it appears that we can get a close approximation of A by choosing the partition points $x_1, x_2, \ldots, x_{n-1}$ to be close together. Then the rectangles that approximate the region under the graph of f will be narrow, and we feel that the narrower the rectangles, the more nearly they "fill up" the region. Thus a "fine" partition of the interval $[a, b]$ should lead to a close approximation of the area, regardless of our choice of the points $x_1^*, x_2^*, \ldots, x_n^*$ at which we compute the functional values $f(x_1^*), f(x_2^*), \ldots, f(x_n^*)$ that serve as the altitudes of our approximating rectangles.

We measure the "fineness" of a partition by means of a number called the **norm** of the partition, which is the largest of the numbers Δx_i, $i = 1, 2, \ldots, n$. Thus the norm of a partition of the interval $[a, b]$ is the length of the longest subinterval into which it is divided. For example, the norm of the partition in

Example 32-3 is 2, and the norm of the partition in Example 32-2 is $\frac{1}{4}\pi$. If the norm of a partition is small, then *all* the subintervals into which the basic interval is divided are short. Intuitively speaking, the area of the region under the graph of f is approximated by Sum 33-1, and a close approximation can be ensured by using a partition with a small norm when we form this approximating sum.

To a calculus student, the words "close approximation" should suggest that we are really dealing with a limit. If u is the norm of the partition $[a, b]$, then the area A we are talking about is the number $\lim_{u \downarrow 0} s$. This limit differs slightly from the limit concept introduced in Section 10. Our number s is not determined merely by specifying the norm u of the partition that we use when calculating it. We may choose many different partitions with the same norm, and even after the partition is fixed, the choice of the number x_i^* in each subinterval must still be made. In general, we will not get the same sum s for each choice. Thus for a given norm u, Equation 33-1 yields a whole *set* $S(u)$ of numbers. The area of our region is given by the equation

$$(33\text{-}2) \qquad\qquad A = \lim_{u \downarrow 0} S(u).$$

This limit is the number that is approximated by the members of the set $S(u)$ when u is small. A precise definition of the limit for our set-valued function S is completely analogous to the precise definition of the limit for a numerical-valued function that we set forth in Section 10. Every open interval that contains A contains all the members of $S(u)$ (all the sums formed by using Equation 33-1) for u in some open interval whose left endpoint is 0.

Now we can give our mathematical definition of the integral.

Definition 33-1. *Let f be a function whose domain contains an interval $[a, b]$ and suppose that $S(u)$ is the set of all sums given by Equation 33-1 when $[a, b]$ is partitioned so that the maximum distance between successive division points is u. Then if $\lim_{u \downarrow 0} S(u)$ exists, we say that f is* **integrable** *on the interval $[a, b]$. The limit is called the* **integral** *of f over $[a, b]$, and we denote this number by the symbol*

$$\int_a^b f(x)\,dx.$$

Definition 33-1 is rather complicated reading, and it makes the idea of the integral clearer if we write the equation that defines it as

$$\int_a^b f(x)\,dx = \lim_{u \downarrow 0} \sum_{i=1}^{n} f(x_i^*)\,\Delta x_i.$$

Stripped of technical fine points, this equation simply says that the integral is the number that is approximated by all the sums of the form $\sum_{i=1}^{n} f(x_i^*) \, \Delta x_i$ in which the first step of the calculation is to divide the basic interval $[a, b]$ into short subintervals.

Before we say more about the meaning of the integral, let us discuss the symbol $\int_a^b f(x) \, dx$. The **integral sign** $\int$ is an elongated S to suggest that the integral is the limit of "sums." The numbers a and b are called the lower and upper **limits of integration,** but here the word "limit" is used in a new sense. The limits of integration are analogous to the limits of the index in a summation—for example, the numbers 1 and n in Equation 33-1. There, the index of summation i can be replaced by another letter without altering the sum. The analogous letter x in the symbol $\int_a^b f(x) \, dx$ is called the **variable of integration,** and it can be replaced by other letters without altering the integral. Thus

$$\int_a^b f(x) \, dx = \int_a^b f(t) \, dt = \int_a^b f(z) \, dz.$$

The "dx" or "dt" or "dz" tells us what the variable of integration is. For example,

$$\int_0^1 x^2 t \, dx \quad \text{and} \quad \int_0^1 x^2 t \, dt$$

are different integrals, limits of the sums

$$\sum_{i=1}^{n} x_i^{*2} t \, \Delta x_i \quad \text{and} \quad \sum_{i=1}^{n} x^2 t_i^* \, \Delta t_i.$$

The fact that the notation for derivatives $\left(\dfrac{dy}{dx}\right)$ and integrals both use the same symbols is not a coincidence. We will soon see that differentiation and integration are closely related indeed.

Finally, we refer to $f(x)$ as the **integrand** of $\int_a^b f(x) \, dx$. Thus $\sqrt{x}$ is the integrand of $\int_2^3 \sqrt{x} \, dx$.

In the geometric discussion suggesting that the sums given by Equation 33-1 are approximations to the area of the region between the graph of f and the interval $[a, b]$ of the X-axis, we assumed that f does not take negative values. But we don't need this restriction to form these sums and hence construct our set-valued function S and take its limit. In other words, Definition 33-1 permits us to talk about $\int_a^b f(x)\,dx$ even though the graph of f crosses the X-axis between a and b. We will interpret such integrals in terms of area in Section 39.

To help fix the definition of the integral in our minds, we now state a simple but useful theorem.

Theorem 33-1. *If f is integrable on an interval $[a, b]$ and if $f(x) \geq 0$ for each point $x \in [a, b]$, then $\int_a^b f(x)\,dx \geq 0$.*

The strategy for proving this theorem is simple. The integral $\int_a^b f(x)\,dx$ is approximated by sums formed from Equation 33-1. Since f does not take negative values in $[a, b]$, each such sum is greater than or equal to 0. Therefore these sums cannot approximate a negative number, which is what the theorem says.

Definition 33-1 deals with an integral $\int_a^b f(x)\,dx$ in which the upper limit b is larger than the lower limit a. In order to complete our definition of the integral, we add a further definition.

Definition 33-2. *For any function f whose domain contains a point a,*

$$\int_a^a f(x)\,dx = 0.$$

If $a < b$ and if f is integrable on the interval $[a, b]$, then

$$\int_b^a f(x)\,dx = -\int_a^b f(x)\,dx.$$

Our formal definition of the integral is logically independent of the concept of the area of a plane region. Nevertheless, the discussion that led to this formal definition makes it clear that if the mathematical concept of area is to agree with our commonsense notions about area, then the area of the region between the graph of a function f that takes nonnegative values and an interval $[a, b]$ of the X-axis must be the number $\int_a^b f(x)\,dx$.

One way to ensure this result is to *define* the area of the region to be the integral. This approach, however, merely gives us the areas of geometric figures having the form of "rectangles with one curved side." It is not easy to extend this definition to make it cover other geometric shapes; consequently, most mathematicians prefer a definition of area that applies directly to more complicated regions.

We will have to leave a full explanation of the mathematical definition of area to some later course. Here we will only say that one aspect of finding the area of a plane figure consists of "packing" it with simple figures whose areas we can compute (such as rectangles) and then calculating a limit of the sums of these areas. This process is closely related to the way we might "pack" one of our regions with rectangles before taking the limit that is the integral, so it turns out that the integral really is the area (in the sense in which mathematicians use the term) of our region. Furthermore, the mathematical definition of area has all the properties you would expect; that is, the area of a rectangular region is the product of its base and its altitude, the area of a circular disk of radius r is πr^2, the area of the union of two sets that have no common points is the sum of their areas, and so on. Because integrals do give areas, we will use areas to calculate integrals and integrals to calculate areas.

Example 33-1.   Suppose that a is a positive number. Evaluate the integral

$$\int_0^a \sqrt{a^2 - x^2} \, dx.$$

Solution.   The curve $y = \sqrt{a^2 - x^2}$ is a semicircle with a radius of a units whose center is the origin. This semicircle and the interval $[-a, a]$ of the X-axis bound a region with an area of $\frac{1}{2}\pi a^2$ square units. We are interested only in the half of this region that lies above the interval $[0, a]$, so our integral is only one-half of this number:

(33-3) $\int_0^a \sqrt{a^2 - x^2} \, dx = \frac{1}{4}\pi a^2.$

PROBLEMS   33

1. Use the geometric interpretation of $\int_a^b f(x) \, dx$ as an area to evaluate the following integrals.

(a) $\int_1^2 (2x - 1) \, dx$ (b) $\int_{-1}^3 (2x + 3) \, dx$

(c) $\int_{-3}^5 |x| \, dx$ (d) $\int_{-1}^1 (1 - |x|) \, dx$

(e) $\int_0^{5/2} [\![x]\!] \, dx$ (f) $\int_{-1}^3 ([\![x]\!] + 1) \, dx$

(g) $\int_{-1}^2 (|x| + x) \, dx$ (h) $\int_{-1}^2 (|x| - x) \, dx$

(i) $\int_0^4 (|x - 1| + |x - 2|) \, dx$ (j) $\int_{-1}^3 (x + 1 + 3|1 - x|) \, dx$

2. Observe that we can construct the curve $y = 1 + f(x)$ by sliding the graph of f up one unit. Use this fact to calculate the following integrals.

(a) $\int_{-1}^{1} (1 + \sqrt{1 - x^2})\, dx$ (b) $\int_{-1}^{1} (1 - \sqrt{1 - x^2})\, dx$

(c) $\int_{0}^{2\pi} (1 + [\![\cos x]\!])\, dx$ (d) $\int_{0}^{2\pi} (1 - [\![\cos x]\!])\, dx$

3. Show that none of the following integrals is negative.

(a) $\int_{0}^{3} (4 + 3x - x^2)\, dx$ (b) $\int_{-8}^{-4} (x - 2)(x + 1)^{-1}\, dx$

(c) $\int_{-1}^{0} (z^3 - z^2 - 2z)\, dz$ (d) $\int_{-1}^{0} (u^3 - u)\, du$

(e) $\int_{1/2}^{2} \left(u + \dfrac{1}{u} - 2\right) du$ (f) $\int_{-2}^{-1/2} -\left(u + \dfrac{1}{u} + 2\right) du$

4. Evaluate the following integrals.

(a) $\int_{6}^{3} 12\, dx$ (b) $\int_{-5}^{-10} 15\, dx$ (c) $\int_{5}^{1} 3x\, dx$

(d) $\int_{-1}^{-5} -3x\, dx$ (e) $\int_{12}^{12} \sec x^3\, dx$ (f) $\int_{\sqrt{\pi}}^{\sqrt{\pi}} \log x\, dx$

5. Evaluate the following integrals.

(a) $\int_{1}^{9} (\log 100x - \log x)\, dx$ (b) $\int_{-1}^{2} (1 + x/|x|)\, dx$

(c) $\int_{-1}^{1} (\sec^2 x - \tan^2 x)\, dx$ (d) $\int_{0}^{3} \tfrac{1}{3}\sqrt{81 - (3x)^2}\, dx$

6. Find $\int_{2}^{9} f(x)\, dx$ if $f(x)$ is the distance between x and the nearest prime number.

7. By an argument that is similar to our justification of Theorem 33-1, show that if $f(x) \le 0$ for each point $x \in [a, b]$, then $\int_{a}^{b} f(x)\, dx \le 0$.

8. Suppose that f has positive values in the interval $[1, 10]$.

(a) Use the area interpretation of $\int_{1}^{10} f(x)\, dx$ to conclude that if f is increasing in $[1, 10]$, then

$$\sum_{k=1}^{9} f(k) \le \int_{1}^{10} f(x)\, dx \le \sum_{k=2}^{10} f(k).$$

(b) Write a set of inequalities similar to those in part (a) if f is decreasing in the interval $[1, 10]$.

9. Suppose that f is the constant function with value c. Show that for each partition of an interval $[a, b]$ and each choice of points $x_1^*, x_2^*, \ldots, x_n^*$ from the partition, Equation 33-1 gives us $s = c(b - a)$. What is the set $S(u)$ for any $u \in (0, b - a]$? What is $\int_a^b f(x)\, dx$? (This simple problem is designed to help you get acquainted with our terminology.)

10. Show that if for each x in the interval $[a, b]$ we have $f(x) \in [A, B]$, then for each $u \in (0, b - a]$, $S(u) \subseteq [A(b - a), B(b - a)]$. If you also know that f is integrable on $[a, b]$, show that $\int_a^b f(x)\, dx \in [A(b - a), B(b - a)]$.

11. Let f be the greatest integer function, $[a, b]$ the interval $[0, 1]$, and $S(1)$ the set of all sums formed by using Equation 33-1 when $[0, 1]$ is partitioned into "subintervals" the largest of which is 1 unit long. Describe this set.

12. Let f be defined, as in Illustration 5-5, by the statements (i) $f(x) = 0$ if there is a decimal expression for x that has all 0's after a certain stage and (ii) $f(x) = 1$ otherwise, and let $[a, b]$ be the interval $[0, 2]$. Show that $S(u) = [0, 2]$ if $u \in (0, 1]$. Is there a *number* A such that $\lim_{u \downarrow 0} S(u) = A$? Is f integrable on the interval $[0, 2]$?

34. GENERAL INTEGRAL THEOREMS

When we consider the integral of a function f on an interval $[a, b]$, we raise two questions:

(1) Is f integrable on $[a, b]$; that is, does the integral (a limit) exist?

(2) If so, what is the value of the integral; that is, what is the number

$$\int_a^b f(x)\, dx?$$

In this first course in calculus, we will concentrate most of our attention on the second question. Because of its difficulty, we must leave a full discussion of question (1) to a later course.

It is not hard to show that the answer to question (1) is *no* if f is *unbounded* in the interval $[a, b]$; that is, if the set $f([a, b])$ is not contained in some finite interval. For example, if $f(x) = \tan x$ for $x \in [0, \frac{1}{2}\pi)$ and $f(\frac{1}{2}\pi) = 1066$, then f is unbounded in the interval $[0, \frac{1}{2}\pi]$, and it is *not* integrable on that interval. But if f *is* bounded and "reasonably continuous" in an interval, then it is integrable. In particular, if there are only a finite number (maybe none, maybe 1 million) of points of an interval $[a, b]$ at which a given function f is discontinuous, and if f is bounded in $[a, b]$, then $\int_a^b f(x)\, dx$ exists. These criteria cover all the functions you will meet in this course.

Example 34-1.   Explain how you know that the following integrals exist:

$$\int_0^\pi \sin x \, dx, \quad \int_{-10}^{17} [\![x]\!] \, dx, \quad \text{and} \quad \int_{-3}^3 \frac{\sin x}{x} \, dx. \quad \text{What about} \quad \int_0^1 \frac{1}{x} \, dx?$$

Solution.   The sine function is continuous in the interval $[0, \pi]$, and it is certainly bounded there. Thus it is integrable. The greatest integer function is bounded in the interval $[-10, 17]$ and has only a finite number (27) of points of discontinuity there. Consequently, it, too, is integrable. The equations $s(x) = (\sin x)/x$ and $q(x) = 1/x$ define functions whose domains do not contain the point 0. Strictly speaking, therefore, we should not talk about integrals of these functions over any interval that contains 0 unless we first extend their definitions so that their domains do contain 0. If we choose $s(0)$ to be 1, then s is continuous at 0 and at every other point of the X-axis. If we take $s(0)$ to be a number other than 1, then s will not be continuous at 0, but it will be continuous everywhere else. In either case, s has only a finite number (0 or 1) of points of discontinuity. It is also bounded in the interval $[-3, 3]$, and hence it is integrable on this interval. (Incidentally, the value of the integral is the same, whatever number we choose as $s(0)$.)

No matter what number we choose as $q(0)$, the function q will be discontinuous at 0. But it will be continuous at every other point, and so it has only a finite number (1) of points of discontinuity. Nevertheless, since q is unbounded in the interval $[0, 1]$, the integral $\int_0^1 \frac{1}{x} \, dx$ does not exist.

We have repeatedly emphasized the linearity of the differential operator $\frac{d}{dx}$; if f and g are differentiable functions and m and n are numbers, then $\frac{d}{dx}[mf(x) + ng(x)] = m\frac{d}{dx}f(x) + n\frac{d}{dx}g(x)$. This linearity property is inherited from the limit operator, and since the integral is a limit, it is linear, too.

Theorem   34-1.   *If f and g are integrable on an interval $[a, b]$ and m and n are numbers, then*

$$\int_a^b [mf(x) + ng(x)] \, dx = m\int_a^b f(x) \, dx + n\int_a^b g(x) \, dx.$$

This theorem follows from the fact that limits of sums of set-valued functions are sums of limits, and so on. Conceptually, a proof is not difficult, but we will omit one because plowing through its technical complexities would not add to your understanding of the theorem.

Example 34-2.   Find $\int_0^2 (3[\![x]\!] - 2\sqrt{4 - x^2}) \, dx$.

Solution.   We have

$$\int_0^2 (3[\![x]\!] - 2\sqrt{4 - x^2}) \, dx$$

$$= 3 \int_0^2 [\![x]\!] \, dx - 2 \int_0^2 \sqrt{4 - x^2} \, dx \qquad \text{(Theorem 34-1)}$$

$$= 3 \cdot 1 - 2\pi \qquad\qquad \text{(Evaluate the}$$
$$\qquad\qquad\qquad\qquad\qquad\qquad \text{integrals geometrically.)}$$

$$= 3 - 2\pi \approx -3.28.$$

Example 34-3.   If $f(x) \leq g(x)$ for each $x \in [a, b]$, show that

$$\int_a^b f(x) \, dx \leq \int_a^b g(x) \, dx.$$

Solution.   Since $g(x) - f(x) \geq 0$, Theorem 33-1 tells us that

(34-1)
$$\int_a^b [g(x) - f(x)] \, dx \geq 0.$$

We use Theorem 34-1 to rewrite this integral as

$$\int_a^b [g(x) - f(x)] \, dx = \int_a^b g(x) \, dx + (-1) \int_a^b f(x) \, dx,$$

and hence Inequality 34-1 becomes $0 \leq \int_a^b g(x) \, dx - \int_a^b f(x) \, dx$, which is simply another form of the inequality we were to verify.

Geometrically, we think of the integral $\int_a^b f(x) \, dx$ of the function whose graph is shown in Fig. 34-1 as the area of the shaded region. From the figure it seems reasonable to suppose that there is a point $m \in [a, b]$ such that the rectangle of altitude $f(m)$ whose base is the interval $[a, b]$ (and hence is $b - a$ units long) has the same area as our shaded region. Thus

$$(b - a)f(m) = \int_a^b f(x) \, dx.$$

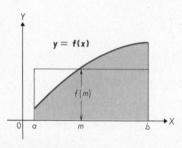

Figure 34-1

The number $f(m)$ is called the **average value** of f in the interval $[a, b]$. Our equation is geometrically obvious, and we will dispense with a formal proof.

Theorem 34-2. *If f is continuous in the interval* $[a, b]$, *then there is a point* $m \in [a, b]$ *such that*

(34-2)
$$f(m) = \frac{1}{b - a} \int_a^b f(x) \, dx.$$

Example 34-4. If $f(x) = \sqrt{9 - x^2}$, what is the average value of f in the interval $[-3, 3]$?

Solution. We leave it to you to show (geometrically) that $\int_{-3}^3 \sqrt{9 - x^2} \, dx = \frac{9}{2}\pi$. In this example, $b - a = 3 - (-3) = 6$, and so Equation 34-2 becomes

$$f(m) = \tfrac{1}{6} \cdot \tfrac{9}{2}\pi = \tfrac{3}{4}\pi.$$

Thus the average value of f is about 2.38. Since $f(m) = \sqrt{9 - m^2}$, we see that $9 - m^2 = \frac{9}{16}\pi^2$, so $m^2 = \frac{9}{16}(16 - \pi^2)$. Therefore f takes its average value at $m = \frac{3}{4}\sqrt{16 - \pi^2}$ and at $-\frac{3}{4}\sqrt{16 - \pi^2}$.

The final theorem of this section is also suggested by our geometric interpretation of the integral. Because the union of the regions R_1 and R_2 of Fig. 34-2 is the entire shaded region, it appears that the integral of f over the interval $[a, c]$ is the sum of the integrals of f over $[a, b]$ and $[b, c]$.

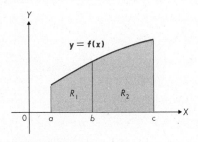

Figure 34-2

Theorem 34-3. *Let a, b, and c be three points of an interval that is contained in the domain of a function f. Then*

(34-3)
$$\int_a^c f(x) \, dx = \int_a^b f(x) \, dx + \int_b^c f(x) \, dx.$$

We omit a proof of this theorem, which simply amounts to translating our geometric remarks into relations between limits of set-valued functions. Our remarks preceding the theorem only apply if $a < b < c$, whereas the theorem is true regardless of the order of these numbers. For example, suppose that $b < c < a$. Then geometry (and a rigorous proof based on the definition of the integral) only tells us that

$$(34\text{-}4) \qquad \int_b^a f(x)\, dx = \int_b^c f(x)\, dx + \int_c^a f(x)\, dx.$$

To show that this equation is equivalent to our desired equation, we use Definition 33-2 to write

$$\int_b^a f(x)\, dx \quad \text{as} \quad -\int_a^b f(x)\, dx \quad \text{and} \quad \int_c^a f(x)\, dx \quad \text{as} \quad -\int_a^c f(x)\, dx.$$

These substitutions transform Equation 34-4 into Equation 34-3, and it is clear that this argument will work for any ordering of the numbers a, b, and c.

Example 34-5.   Explain why $\displaystyle\int_{-\pi}^{\pi} \sin x\, dx = 0$.

Solution.   According to Equation 34-3,

$$(34\text{-}5) \qquad \int_{-\pi}^{\pi} \sin x\, dx = \int_{-\pi}^{0} \sin x\, dx + \int_{0}^{\pi} \sin x\, dx.$$

Let us look at the number $\displaystyle\int_{-\pi}^{0} \sin x\, dx$ from a geometric point of view. We first use Theorem 34-1 to write this integral as $-\displaystyle\int_{-\pi}^{0} (-\sin x)\, dx$. Then we consider the integral $\displaystyle\int_{-\pi}^{0} (-\sin x)\, dx$ to be the area of the region bounded by the X-axis

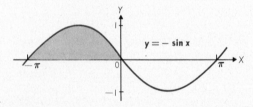

$y = -\sin x$

Figure 34-3

and the curve $y = -\sin x$ that is shaded in Fig. 34-3. Since this region is bounded by one arch of the sine curve, its area is $\int_0^\pi \sin x\, dx$. Thus we see that

$$\int_{-\pi}^0 (-\sin x)\, dx = \int_0^\pi \sin x\, dx,$$

and hence Equation 34-5 becomes

$$\int_{-\pi}^\pi \sin x\, dx = -\int_0^\pi \sin x\, dx + \int_0^\pi \sin x\, dx = 0.$$

PROBLEMS 34

1. Determine which of the following integrals exist. If the integral exists, use general integral theorems and geometric arguments to evaluate it.

   (a) $\displaystyle\int_0^\pi \sec x\, dx$ (b) $\displaystyle\int_0^\pi \csc x\, dx$ (c) $\displaystyle\int_{-1}^1 x^{-3}\, dx$

   (d) $\displaystyle\int_0^2 (x-1)^{-2}\, dx$ (e) $\displaystyle\int_{-2}^3 \left(2 + \frac{x}{|x|}\right) dx$ (f) $\displaystyle\int_{-2}^3 \left(2 - \frac{x}{|x|}\right) dx$

   (g) $\displaystyle\int_0^2 \frac{x^2-1}{x-1}\, dx$ (h) $\displaystyle\int_{-2}^0 \frac{1-x^2}{1+x}\, dx$

2. What general integral theorem tells us that $\displaystyle\int_0^4 (3x+2)\, dx = 3\int_0^4 x\, dx + 2\int_0^4 1\, dx$? Use the geometric interpretation of the integral to evaluate the three integrals in this equation and thus show that it is correct.

3. Use general integration theorems and geometric arguments to evaluate the following integrals.

   (a) $\displaystyle\int_0^2 (4 + \sqrt{4-x^2})\, dx$ (b) $\displaystyle\int_0^2 (4 - \sqrt{4-x^2})\, dx$

   (c) $\displaystyle\int_0^3 (2x + 1 + 4\sqrt{9-x^2})\, dx$ (d) $\displaystyle\int_{-1}^1 \left(4 + \frac{x}{|x|}\sqrt{1-x^2}\right) dx$

4. How does Theorem 34-1 tell us that $\displaystyle\int_a^b 0\, dx = 0$?

5. Let $f(x) = \sqrt{16-x^2}$. What is the average value of f in the interval $[-4, 4]$? At what points does f take its average value?

6. Let $f(x) = mx + b$. What is the average value of f in the interval $[c, d]$? At what point does f take this average value?

7. What is the average value of the sine function in the interval $[-\pi, \pi]$ (see Example 34-5)?

8. Evaluate the integral $\displaystyle\int_0^2 (x - [\![x^2]\!]) \, dx$.

9. Verify the following inequalities.

   (a) $\displaystyle\int_0^1 x^2 \, dx \leq \int_0^1 x \, dx$

   (b) $\displaystyle\int_0^1 x^4 \, dx \leq \int_0^1 x^2 \, dx$

   (c) $\displaystyle\int_0^1 (x^2 - 2x - 3) \, dx \leq \int_0^{-1} (x^2 - 2x - 3) \, dt$

   (d) $\displaystyle\int_0^{\pi/4} \sec x \, dx \geq \int_0^{-\pi/4} \sec x \, dx$

10. If f and g are integrable on the interval $[a, b]$, is it necessarily true that

    (a) $\displaystyle\int_a^b f(x)g(x) \, dx = \int_a^b f(x) \, dx \int_a^b g(x) \, dx$?

    (b) $\displaystyle\int_a^b [f(x) - g(x)] \, dx = \int_a^b f(x) \, dx - \int_a^b g(x) \, dx$?

    (c) $\displaystyle\int_a^b \frac{f(x)}{g(x)} \, dx = \frac{\displaystyle\int_a^b f(x) \, dx}{\displaystyle\int_a^b g(x) \, dx}$?

11. Give an example, using a discontinuous function, of course, for which Equation 34-2 does not hold. Find a discontinuous function for which it does hold.

12. Extend Theorem 34-1 to show that

$$\int_a^b [f(x) + g(x) + h(x)] \, dx = \int_a^b f(x) \, dx + \int_a^b g(x) \, dx + \int_a^b h(x) \, dx.$$

13. Use Example 34-3 to show that if f is integrable on $[a, b]$, then $\left| \displaystyle\int_a^b f(x) \, dx \right| \leq \displaystyle\int_a^b |f(x)| \, dx$. Hence conclude that if $|f(x)| \leq M$ for each point $x \in [a, b]$, then $\left| \displaystyle\int_a^b f(x) \, dx \right| \leq M(b - a)$.

35. INTEGRALS OF SOME PARTICULAR FUNCTIONS

We don't go back to the definition of differentiation each time we want a specific derivative; we refer to a differentiation formula. Similarly, we don't calculate specific integrals from their definition. We use integration formulas instead. The three formulas that we will now develop are important in themselves, and they also form the basis for numerical methods of integration (introduced in the next section) by which a computer can evaluate much more complicated integrals.

Figure 35-1 shows the curves $y = 1$, $y = x$, and $y = x^2$ for x in an interval $[0, h]$. The first graph tells us that

(35-1)
$$\int_0^h 1 \cdot dx = h,$$

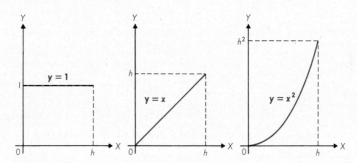

Figure 35-1

since the integral is simply the area of a 1 by h rectangular region. The second graph says that

(35-2)
$$\int_0^h x \, dx = \frac{h^2}{2},$$

since the integral is one-half the area of an h by h rectangular region. The third graph shows that the integral $\int_0^h x^2 \, dx$ is some fraction (less than half) of the area of an h^2 by h rectangular region. The simplest guess would be

(35-3)
$$\int_0^h x^2 \, dx = \frac{h^3}{3},$$

but we write this equation with considerably less confidence than we wrote the other two. Nevertheless, it is correct, as we shall now see.

A "proper" way to evaluate an integral should rely on our basic Definition 33-1, rather than a picture. In Problem 33-9 we gave you some hints for using the definition to verify Equation 35-1. Now we will introduce a theorem that helps us verify many other integration formulas, including Equations 35-2 and 35-3.

The integral $\int_a^b f(x)\,dx$ is the number I that is approximated by sums of the form $\sum_{i=1}^n f(x_i^*)\,\Delta x_i$ that are constructed using "fine" partitions of the interval $[a, b]$. Each partition gives rise to many such sums, but they are all close to I. Frequently we can show that there is a number A such that for each partition, one of our approximating sums will be A. The definition of the limit of a set-valued function says that every neighborhood of I will have to contain A, and so A must be I (Example 10-3); that is, $A = \int_a^b f(x)\,dx$. We state this result as a theorem.

Theorem   35-1.   *Suppose that the function f is integrable on the interval $[a, b]$. If there is a number A such that for every partition of $[a, b]$ we can choose the numbers $x_1^*, x_2^*, \ldots, x_n^*$ so that the approximating sum $\sum_{i=1}^n f(x_i^*)\,\Delta x_i$ is equal to A, then $\int_a^b f(x)\,dx = A$.*

Example 35-1.   Verify Equation 35-2.

Solution.   (In these verifications we will assume that h is positive, but the same sort of arguments apply when h is negative.)   According to Theorem 35-1, we need only show that for each partition of the interval $[0, h]$, there is an approximating sum $A = \frac{1}{2}h^2$. So suppose that we have a partition determined by the points $x_0 = 0$, $x_1, x_2, \ldots, x_{n-1}, x_n = h$. Every approximating sum will take the form

$$s = \sum_{i=1}^n f(x_i^*)\,\Delta x_i = \sum_{i=1}^n x_i^*(x_i - x_{i-1}),$$

and we must show that a suitable choice of the numbers $x_1^*, x_2^*, \ldots, x_n^*$ will give us $s = \frac{1}{2}h^2$. In this example the "suitable choice" consists of taking x_i^* to be the midpoint of the interval $[x_{i-1}, x_i]$; that is, $x_i^* = \frac{1}{2}(x_i + x_{i-1})$. For then

$$s = \sum_{i=1}^n \tfrac{1}{2}(x_i + x_{i-1})(x_i - x_{i-1}) = \sum_{i=1}^n (\tfrac{1}{2}x_i^2 - \tfrac{1}{2}x_{i-1}^2)$$
$$= (\tfrac{1}{2}x_1^2 - \tfrac{1}{2}x_0^2) + (\tfrac{1}{2}x_2^2 - \tfrac{1}{2}x_1^2) + \cdots + (\tfrac{1}{2}x_n^2 - \tfrac{1}{2}x_{n-1}^2)$$
$$= \tfrac{1}{2}x_n^2 - \tfrac{1}{2}x_0^2 = \tfrac{1}{2}h^2 - \tfrac{1}{2}0^2 = \tfrac{1}{2}h^2.$$

Example 35-2.   Verify Equation 35-3.

Solution.   For each partition of the interval $[0, h]$, our approximating sums take the form $s = \sum_{i=1}^n x_i^{*2}\,\Delta x_i$; and Theorem 35-1 tells us that we can verify Equation 35-3 by showing that there are numbers $x_1^*, x_2^*, \ldots, x_n^*$ such that this sum is $\frac{1}{3}h^3$.

Just as in the last example, for each i we will pick x_i^* such that

(35-4)
$$x_i^{*2} \, \Delta x_i = \tfrac{1}{3}x_i^3 - \tfrac{1}{3}x_{i-1}^3,$$

and then we will have the approximating sum

$$\sum_{i=1}^{n} x_i^{*2} \, \Delta x_i = \sum_{i=1}^{n} (\tfrac{1}{3}x_i^3 - \tfrac{1}{3}x_{i-1}^3)$$

$$= (\tfrac{1}{3}x_1^3 - \tfrac{1}{3}x_0^3) + (\tfrac{1}{3}x_2^3 - \tfrac{1}{3}x_1^3) + \cdots + (\tfrac{1}{3}x_n^3 - \tfrac{1}{3}x_{n-1}^3)$$

$$= \tfrac{1}{3}x_n^3 - \tfrac{1}{3}x_0^3 = \tfrac{1}{3}h^3 - \tfrac{1}{3}0^3 = \tfrac{1}{3}h^3.$$

To complete our verification of Equation 35-3, we still have to show that we can indeed pick x_i^* so that Equation 35-4 holds. But solving that equation for x_i^* is simply a matter of algebra. Since $\Delta x_i = x_i - x_{i-1}$, Equation 35-4 is equivalent to the equation

$$x_i^{*2} = \tfrac{1}{3}\frac{x_i^3 - x_{i-1}^3}{x_i - x_{i-1}} = \tfrac{1}{3}(x_i^2 + x_i x_{i-1} + x_{i-1}^2),$$

and therefore we take

$$x_i^* = \sqrt{\tfrac{1}{3}(x_i^2 + x_i x_{i-1} + x_{i-1}^2)}.$$

It is easy to show that $x_i^* \in [x_{i-1}, x_i]$, so we are done.

We can use our general integration formulas to broaden Equation 35-1. Thus

(35-5)
$$\int_a^b c \, dx = c(b - a),$$

where a, b, and c are any given numbers, since

$$\int_a^b c \, dx = c \left(\int_a^0 1 \cdot dx + \int_0^b 1 \cdot dx \right) \qquad \text{(Theorems 34-1 and 34-3)}$$

$$= c \left(\int_0^b 1 \cdot dx - \int_0^a 1 \cdot dx \right) \qquad \text{(Definition 33-2)}$$

$$= c(b - a) \qquad \text{(Equation 35-1)}.$$

Similar reasoning generalizes Equations 35-2 and 35-3 to

(35-6)
$$\int_a^b x \, dx = \frac{b^2}{2} - \frac{a^2}{2}$$

and

(35-7)
$$\int_a^b x^2 \, dx = \frac{b^3}{3} - \frac{a^3}{3}.$$

Together the *specific* Integration Formulas 35-4, 35-5, and 35-6 and the *general* integration formula of Theorem 34-1 allow us to integrate any quadratic expression.

Example 35-3. Evaluate $\int_2^3 (6x^2 + 2x - 5)\, dx$.

Solution.

$$\int_2^3 (6x^2 + 2x - 5)\, dx$$

$$= 6 \int_2^3 x^2\, dx + 2 \int_2^3 x\, dx + \int_2^3 (-5)\, dx \qquad \text{(Theorem 34-1)}$$

$$= 6 \left(\frac{3^3}{3} - \frac{2^3}{3} \right) + 2 \left(\frac{3^2}{2} - \frac{2^2}{2} \right) - 5(3 - 2) \qquad \text{(Equations 35-5 to 35-7)}$$

$$= 38.$$

Example 35-4. Evaluate $\int_0^3 |x - 1|(x - 1)\, dx$.

Solution. We can't use Formulas 35-5 to 35-7 until we get rid of the absolute value signs:

$$\int_0^3 |x - 1|(x - 1)\, dx$$

$$= \int_0^1 |x - 1|(x - 1)\, dx + \int_1^3 |x - 1|(x - 1)\, dx \qquad \text{(Theorem 34-3)}$$

$$= \int_0^1 -(x - 1)^2\, dx + \int_1^3 (x - 1)^2\, dx \qquad \begin{array}{l} (|x - 1| \\ \quad = -(x - 1) \\ \text{if } x \le 1) \end{array}$$

$$= \int_0^1 (-x^2 + 2x - 1)\, dx + \int_1^3 (x^2 - 2x + 1)\, dx \qquad \text{(Algebra)}$$

$$= -\int_0^1 x^2\, dx + 2 \int_0^1 x\, dx + \int_0^1 (-1)\, dx$$

$$\quad + \int_1^3 x^2\, dx - 2 \int_1^3 x\, dx + \int_1^3 1 \cdot dx \qquad \text{(Theorem 34-1)}$$

$$= -\frac{1}{3} + 2 \cdot \frac{1}{2} - 1 + \frac{3^3}{3} - \frac{1}{3} - 2 \left(\frac{3^2}{2} - \frac{1}{2} \right) + 3 - 1 \qquad \begin{array}{l} \text{(Equations} \\ \text{35-5 to 35-7)} \end{array}$$

$$= \frac{7}{3}.$$

The technique we used in the last two examples obviously works in general, so let us use it to develop a formula for the integral of a quadratic function that will be useful in the next section.

Example 35-5.   Show that

$$(35\text{-}8) \qquad \int_a^b (Ax^2 + Bx + C)\, dx = \frac{b-a}{6}\left[g(a) + 4g\left(\frac{a+b}{2}\right) + g(b)\right],$$

where $g(x) = Ax^2 + Bx + C$.

Solution.   As in the last two examples,

$$(35\text{-}9) \qquad \int_a^b (Ax^2 + Bx + C)\, dx$$

$$= A\int_a^b x^2\, dx + B\int_a^b x\, dx + \int_a^b C\, dx$$

$$= A\left(\frac{b^3}{3} - \frac{a^3}{3}\right) + B\left(\frac{b^2}{2} - \frac{a^2}{2}\right) + C(b - a).$$

Now we show that the right-hand sides of Equations 35-8 and 35-9 are the same. Thus since

$$4g\left(\frac{a+b}{2}\right) = 4A\left(\frac{a+b}{2}\right)^2 + 4B\left(\frac{a+b}{2}\right) + 4C$$

$$= A(b^2 + 2ab + a^2) + 2B(b + a) + 4C,$$

you can easily verify that

$$g(a) + 4g\left(\frac{a+b}{2}\right) + g(b) = 2A(b^2 + ab + a^2) + 3B(b + a) + 6C.$$

So

$$\frac{b-a}{6}\left[g(a) + 4g\left(\frac{a+b}{2}\right) + g(b)\right] = \frac{A(b^3 - a^3)}{3} + \frac{B(b^2 - a^2)}{2} + C(b - a),$$

and therefore Equation 35-9 can be expressed in the form of Equation 35-8.

PROBLEMS   35

1.   Evaluate the following integrals.

(a) $\displaystyle\int_{-1}^{7} x\, dx$

(b) $\displaystyle\int_{-3}^{-1} t\, dt$

(c) $\displaystyle\int_{-2}^{4} z^2 \, dz$

(d) $\displaystyle\int_{-3}^{-5} x^2 \, dx$

(e) $\displaystyle\int_{0}^{2} (x - 4x^2) \, dx$

(f) $\displaystyle\int_{1}^{4} (3t^2 - 6t) \, dt$

(g) $\displaystyle\int_{-1}^{0} (1 - r^2) \, dr$

(h) $\displaystyle\int_{6}^{6} (s^2 + 6s - 7) \, ds$

2. Suppose that $mx + b > 0$ for $x \in [c, d]$. Use the area interpretation of the integral to compute $\displaystyle\int_{c}^{d} (mx + b) \, dx$ and show that you get the same result when you evaluate the integral by using our integration formulas.

3. Compute the following numbers.

(a) $\displaystyle\int_{-1}^{3} (t + 1)^2 \, dt$

(b) $\displaystyle\int_{-2}^{3} (x + 2)(x - 3) \, dx$

(c) $\displaystyle\int_{0}^{3} \frac{x^3 - 8}{x - 2} \, dx$

(d) $\displaystyle\int_{-3}^{1} \frac{x^3 + 8}{x + 2} \, dx$

(e) $\displaystyle\int_{-a}^{a} [(x + a)^3 - (x - a)^3] \, dx$

(f) $\displaystyle\int_{-a}^{a} (x + a)^3 \, dx - \int_{-a}^{a} (x^3 + a^3) \, dx$

4. Evaluate the following integrals.

(a) $\displaystyle\int_{-2}^{3} |x|(3x + 2) \, dx$

(b) $\displaystyle\int_{-2}^{1} \frac{4x^3 - x^2}{|x|} \, dx$

(c) $\displaystyle\int_{0}^{4} \frac{x^3 - 2x^2}{|x - 2|} \, dx$

(d) $\displaystyle\int_{0}^{4} |x^2 - 1| \, dx$

(e) $\displaystyle\int_{-1}^{2} [\![x]\!](x^2 - 4x) \, dx$

(f) $\displaystyle\int_{-1}^{2} [\![x + 1]\!](x^2 - 4x) \, dx$

5. Show that

$$\int_{a}^{b} (x - c)^2 \, dx = \frac{(b - c)^3}{3} - \frac{(a - c)^3}{3}.$$

6. Verify Formula 35-8 if $a = 0$, $b = 2$, $A = 3$, $B = 4$, and $C = 5$.

7. Find the average value of f on the interval $[1, 3]$ if $f(x) = 3x^2 + 2x + 1$. At what point of the given interval does f take its average value?

8. What is the average height of a castle door that is bounded by the parabola $y = 8x(2 - x)$ and the X-axis?

9. Evaluate $\displaystyle\int_{a}^{b} x^{-2} \, dx$, where $0 < a < b$, by choosing $x_i^* = \sqrt{x_i x_{i-1}}$ and using the method of Example 35-2.

10. Follow the method of Example 35-2 and show that

$$\int_{a}^{b} x^3 \, dx = \frac{b^4}{4} - \frac{a^4}{4}.$$

(We end up choosing $x_i^* = \sqrt[3]{\frac{1}{4}(x_i^3 + x_i^2 x_{i-1} + x_i x_{i-1}^2 + x_{i-1}^3)}$.)

11. Use the formula we developed in the preceding problem to evaluate the integral.

(a) $\displaystyle\int_1^6 x^3 \, dx$

(b) $\displaystyle\int_{-1}^0 t^3 \, dt$

(c) $\displaystyle\int_{-1}^2 (4x^3 - 2x + 1) \, dx$

(d) $\displaystyle\int_{-1}^2 |x(x - 1)(x - 2)| \, dx$

(e) $\displaystyle\int_{-1}^2 \frac{x^4}{|x|} \, dx$

(f) $\displaystyle\int_{-1}^2 (4|x|^3 - 2|x| + 1) \, dx$

12. What is the average depth of a river whose cross section is the part of the third quadrant that is bounded by the X-axis and the curve $y = 10x(2 + x - x^2)$?

13. (a) Show that for each pair of numbers a and b,

$$\int_a^b |x| \, dx = \frac{|b|b - |a|a}{2}.$$

(b) Use part (a) and Problem 34-13 to show that $|b^2 - a^2| \leq |b|b - |a|a$ if $a \leq b$.

(c) Show that for each pair of numbers a and b,

$$\int_a^b |x|^3 \, dx = \frac{|b|^3 b - |a|^3 a}{4}.$$

(d) Use part (c) and Problem 34-13 to show that $|b^4 - a^4| \leq |b|^3 b - |a|^3 a$ if $a \leq b$.

14. In Example 35-5 we showed that

$$\int_a^b g(x) \, dx = \frac{b - a}{6}\left[g(a) + 4g\left(\frac{a + b}{2}\right) + g(b)\right]$$

when $g(x) = Ax^2 + Bx + C$. Show that this equation is also valid when $g(x) = Ax^3 + Bx^2 + Cx + D$. (You might want to use the formula we developed in Problem 35-10.)

15. Derive Equations 35-6 and 35-7 from Equations 35-2 and 35-3 in the same way we derived Equation 35-5 from Equation 35-1.

36. NUMERICAL INTEGRATION

An integral of a function f over an interval $[a, b]$ is a *number*. For certain simple functions, we have formulas (Equations 35-5, 35-6, and 35-7, for example) with which we can compute this number very easily. Soon we will have many more such formulas. These formulas are especially suitable for hand calculation, so a student who does his homework with pencil and paper may come away from a calculus course believing that they are the *only* way to evaluate integrals. Not so.

In the first place, no matter how many formulas we develop, they cannot apply to all possible integrals. Secondly, although our numerical methods may look complicated to you, they (and some of their exotic descendants) look rather simple to a modern computer. If you get the chance to try some numerical integration on a machine, you will be surprised and pleased to see how quickly it evaluates complicated-looking integrals. Any "reasonable" function is theoretically integrable. The computer allows us to say that a function that is theoretically integrable is practically integrable.

We will use an example to introduce our first method of numerical integration.

Example 36-1. Find an approximation to $\int_0^4 \sqrt{x}\, dx$.

Solution. Figure 36-1 shows the curve $y = \sqrt{x}$ for $x \in [0, 4]$. The number $\int_0^4 \sqrt{x}\, dx$ that we seek is the area of the region that lies between this arc and the X-axis. As an approximation to this number, we will find the sum of the areas of

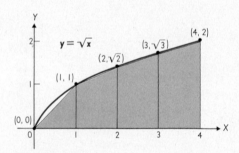

Figure 36-1

the shaded trapezoidal regions that are shown in the figure. The parallel sides of the four trapezoids have lengths 0 and 1, 1 and $\sqrt{2}$, $\sqrt{2}$ and $\sqrt{3}$, and $\sqrt{3}$ and 2, and each trapezoid has an altitude (the distance between the parallel sides) of 1. Since the area A of a trapezoid with altitude h and parallel sides a and b is $A = \frac{1}{2}(a + b)h$, we find that the area of the entire shaded region is

$$\tfrac{1}{2}(0 + 1)\cdot 1 + \tfrac{1}{2}(1 + \sqrt{2})\cdot 1 + \tfrac{1}{2}(\sqrt{2} + \sqrt{3})\cdot 1 + \tfrac{1}{2}(\sqrt{3} + 2)\cdot 1$$

$$\approx \tfrac{1}{2}[(0 + 1) + (1 + 1.4) + (1.4 + 1.7) + (1.7 + 2)] = 5.1.$$

Therefore $\int_0^4 \sqrt{x}\, dx \approx 5.1$. From the figure it appears that this number is too small, and we will soon find (Example 37-1) that the actual value of the integral is $\frac{16}{3} \approx 5.3$. It is apparent that we could get a more accurate approximation with our "trapezoidal rule" by choosing trapezoids of smaller altitudes, such as $\frac{1}{2}$, $\frac{1}{4}$, and .1.

The way we approximated the integral in the last example suggests a general method of approximating an integral $\int_a^b f(x)\,dx$. We pick an integer n and partition the interval $[a, b]$ into n subintervals of equal length. Each of these subintervals will therefore be $h = (b - a)/n$ units long, and the points that determine our partition will be

$$x_0 = a, \quad x_1 = a + h, \quad x_2 = a + 2h, \ldots, x_n = a + nh = b.$$

Now we join the successive points $(x_0, y_0), (x_1, y_1), \ldots, (x_n, y_n)$, where $y_i = f(x_i)$, to obtain a polygonal (broken line) arc $y = p(x)$ that approximates the graph of f (Fig. 36-2). Because this arc approximates the graph of f, it seems reasonable to suppose that the integral $\int_a^b p(x)\,dx$ approximates the given integral $\int_a^b f(x)\,dx$. So now we will evaluate this integral of p, first writing it as a sum:

$$(36\text{-}1) \quad \int_a^b p(x)\,dx = \int_{x_0}^{x_1} p(x)\,dx + \int_{x_1}^{x_2} p(x)\,dx + \cdots + \int_{x_{n-1}}^{x_n} p(x)\,dx.$$

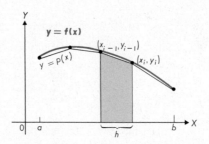

Figure 36-2

The integrand of a typical term $\int_{x_{i-1}}^{x_i} p(x)\,dx$ of this sum is linear, so it has the form $p(x) = m_i x + b_i$. Therefore

$$\int_{x_{i-1}}^{x_i} p(x)\,dx = \int_{x_{i-1}}^{x_i} (m_i x + b_i)\,dx$$

$$= m_i\left(\frac{x_i^2}{2} - \frac{x_{i-1}^2}{2}\right) + b_i(x_i - x_{i-1}) \quad \text{(Which of our integration formulas did we use?)}$$

$$= \tfrac{1}{2}(m_i x_i + m_i x_{i-1} + 2b_i)(x_i - x_{i-1}) \quad \text{(Algebra)}.$$

We recall that $x_i - x_{i-1} = h$, and $y_i = m_i x_i + b_i$ and $y_{i-1} = m_i x_{i-1} + b_i$, so this equation takes the simple form

$$\int_{x_{i-1}}^{x_i} p(x)\, dx = \frac{y_i + y_{i-1}}{2}\, h.$$

When we substitute these expressions (for $i = 1, 2, \ldots, n$) in the right-hand side of Equation 36-1, we obtain the value of $\int_a^b p(x)\, dx$:

$$\frac{y_0 + y_1}{2} h + \frac{y_1 + y_2}{2} h + \cdots + \frac{y_{n-2} + y_{n-1}}{2} h + \frac{y_{n-1} + y_n}{2} h$$

$$= \left(\frac{y_0}{2} + \frac{y_1}{2} + \frac{y_1}{2} + \frac{y_2}{2} + \cdots + \frac{y_{n-2}}{2} + \frac{y_{n-1}}{2} + \frac{y_{n-1}}{2} + \frac{y_n}{2} \right) h$$

$$= \left(\frac{y_0}{2} + y_1 + y_2 + \cdots + y_{n-1} + \frac{y_n}{2} \right) h.$$

We are assuming that this number approximates our given integral, so we have the following numerical integration rule, whose name stems from the type of geometric argument we used in Example 36-1.

THE TRAPEZOIDAL RULE. Let n be a positive integer and take $h = (b - a)/n$. If we write $y_i = f(a + ih)$ for $i = 0, 1, 2, \ldots, n$, then

$$(36\text{-}2) \qquad \int_a^b f(x)\, dx \approx \left(\frac{y_0}{2} + y_1 + y_2 + \cdots + y_{n-1} + \frac{y_n}{2} \right) h.$$

Example 36-2. Use the Trapezoidal Rule with $n = 4$ to approximate the integral

$$\int_1^2 \frac{1}{x}\, dx.$$

Solution. We divide the basic interval $[1, 2]$ into four parts, each with a length of $h = (2 - 1)/4 = \frac{1}{4}$, by means of the points $x_0 = 1$, $x_1 = \frac{5}{4}$, $x_2 = \frac{3}{2}$, $x_3 = \frac{7}{4}$, and $x_4 = 2$. Since $y_i = 1/x_i$, then $y_0 = 1$, $y_1 = \frac{4}{5}$, $y_2 = \frac{2}{3}$, $y_3 = \frac{4}{7}$, and $y_4 = \frac{1}{2}$. Thus according to Formula 36-2,

$$\int_1^2 \frac{1}{x}\, dx \approx \left(\frac{1}{2} + \frac{4}{5} + \frac{2}{3} + \frac{4}{7} + \frac{1}{4} \right) \frac{1}{4}$$

$$= \tfrac{1171}{1680} \approx .697.$$

The actual value of this integral, correct to three decimal places, is .693.

In order to approximate an integral by means of the Trapezoidal Rule, we divide the interval of integration into equal subintervals and replace our given function with a function that is linear over each of the subintervals. Often we get an even closer approximation of the integral by replacing the given function with a function that is quadratic over certain subintervals of the interval of integration.

In approximating the number $\int_a^b f(x)\, dx$ by this method, we divide the interval $[a, b]$ into an *even* number n of equal subintervals, each with a length of $h = (b - a)/n$ (Fig. 36-3). Suppose that the points of subdivision are $x_0 = a$, $x_1 = a + h$, $x_2 = a + 2h$, and so on, and that we denote the corresponding function values by y_0, y_1, y_2, and so on. It is a matter of simple algebra to find numbers A, B, and C such that the graph of the equation $y = Ax^2 + Bx + C$ contains the points (x_0, y_0), (x_1, y_1), and (x_2, y_2). Since this graph is a parabola, we say that we have "passed a parabola through the points." Through the points (x_2, y_2), (x_3, y_3), and (x_4, y_4) we pass another parabola whose equation is $y = Dx^2 + Ex + F$, and so on, for each successive group of three points.

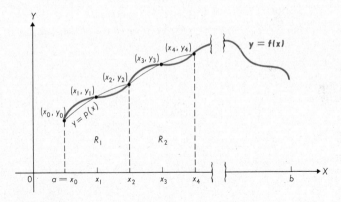

Figure 36-3

Together these parabolic arcs form a curve $y = p(x)$, and because this curve approximates the graph of f, it seems reasonable to suppose that the integrals $\int_a^b f(x)\, dx$ and $\int_a^b p(x)\, dx$ are also approximately equal. Again we evaluate our approximating integral by writing it as a sum, but not the same sum that we used in Equation 36-1; here we cover two subintervals at a time:

$$(36\text{-}3) \quad \int_a^b p(x)\, dx = \int_{x_0}^{x_2} p(x)\, dx + \int_{x_2}^{x_4} p(x)\, dx + \cdots + \int_{x_{n-2}}^{x_n} p(x)\, dx.$$

To evaluate the integrals on the right-hand side of this equation, we replace $p(x)$ in each integral with the appropriate quadratic expression. Thus we replace the first integral with $\displaystyle\int_{x_0}^{x_2} (Ax^2 + Bx + C)\, dx$, the second with $\displaystyle\int_{x_2}^{x_4} (Dx^2 + Ex + F)\, dx$, and so on.

Now we evaluate the integrals of these quadratic expressions. From Formula 35-8 we see that

(36-4) $\displaystyle\int_{x_0}^{x_2} (Ax^2 + Bx + C)\, dx$

$$= \frac{x_2 - x_0}{6}\left[p(x_0) + 4p\left(\frac{x_0 + x_2}{2}\right) + p(x_2)\right],$$

where $p(x) = Ax^2 + Bx + C$. We are supposing that $x_2 - x_0 = 2h$, and $(x_0 + x_2)/2 = x_1$, so the right-hand side of Equation 36-4 can be written as

$$\frac{h}{3}\left[p(x_0) + 4p(x_1) + p(x_2)\right].$$

Furthermore, since the graph of p contains the three points (x_0, y_0), (x_1, y_1), and (x_2, y_2), we have $p(x_0) = y_0, p(x_1) = y_1$, and $p(x_2) = y_2$. Thus Equation 36-4 can be written as

$$\int_{x_0}^{x_2} (Ax^2 + Bx + C)\, dx = \frac{h}{3}(y_0 + 4y_1 + y_2).$$

In exactly the same way, we find that

$$\int_{x_2}^{x_4} (Dx^2 + Ex + F)\, dx = \frac{h}{3}(y_2 + 4y_3 + y_4),$$

and so on. We now replace the terms in the right-hand side of Equation 36-3 with these expressions, and we obtain

$$\frac{h}{3}(y_0 + 4y_1 + y_2) + \frac{h}{3}(y_2 + 4y_3 + y_4) + \cdots + \frac{h}{3}(y_{n-2} + 4y_{n-1} + y_n)$$

$$= \frac{h}{3}(y_0 + 4y_1 + 2y_2 + 4y_3 + \cdots + 2y_{n-2} + 4y_{n-1} + y_n).$$

Our result is another method for calculating integrals numerically.

SIMPSON'S PARABOLIC RULE. Let n be a positive even integer and take $h = (b - a)/n$. Then

$$(36\text{-}5) \quad \int_a^b f(x)\, dx \approx \frac{h}{3}\, (y_0 + 4y_1 + 2y_2 + 4y_3 + \cdots$$
$$+ 2y_{n-2} + 4y_{n-1} + y_n),$$

where $y_i = f(a + ih)$, $i = 0, 1, \ldots, n$. Notice that inside the parentheses the numbers y_0 and y_n are multiplied by 1, whereas the other y's with even subscripts are multiplied by 2. All the y's with odd subscripts are multiplied by 4.

Example 36-3. Use Simpson's Rule with $n = 4$ to approximate $\int_1^2 \frac{1}{x}\, dx$.

Solution. We use the numbers y_0, y_1, y_2, y_3, and y_4 that we calculated in Example 36-2, together with Formula 36-5, to get

$$\int_1^2 \frac{1}{x}\, dx \approx \frac{\frac{1}{4}}{3} \left[1 + 4 \cdot \left(\frac{4}{5}\right) + 2 \cdot \left(\frac{2}{3}\right) + 4 \cdot \left(\frac{4}{7}\right) + \frac{1}{2} \right] = \frac{1747}{2520} \approx .6933.$$

The value of the integral, correct to four decimal places, is .6931. Our new approximation is somewhat more accurate than the result we obtained with the Trapezoidal Rule in Example 36-2.

As with any approximation method, we would naturally like to know how accurate the Trapezoidal Rule and Simpson's Parabolic Rule are. We won't go into the details, which are more tedious than difficult, but by methods very similar to those used in Section 21 we can prove the following theorem.

Theorem 36-1. *Let M and N be numbers such that for each $x \in [a, b]$, we have the inequalities $|f''(x)| \le M$ and $|f^{(4)}(x)| \le N$. If the Trapezoidal Rule gives the number T, and Simpson's Rule gives the number S as approximations to $\int_a^b f(x)\, dx$, then*

$$(36\text{-}6) \qquad \left| \int_a^b f(x)\, dx - T \right| \le \frac{b - a}{12} M h^2$$

and

$$(36\text{-}7) \qquad \left| \int_a^b f(x)\, dx - S \right| \le \frac{b - a}{180} N h^4.$$

Example 36-4.   Use the error bounds of Theorem 36-1 to test the accuracy of the approximations we found in Examples 36-2 and 36-3.

Solution.   Here $f(x) = x^{-1}$, so $f''(x) = 2x^{-3}$ and $f^{(4)}(x) = 24x^{-5}$. Then $|f''(x)| \leq 2$ and $|f^{(4)}(x)| \leq 24$ in the interval $[1, 2]$, so we set $M = 2$ and $N = 24$ in Formulas 36-6 and 36-7. In both our examples, $h = \frac{1}{4}$ and $b - a = 1$. Thus we see that the error we made in Example 36-2 was no greater than $\frac{1}{12} \cdot 2 \cdot (\frac{1}{4})^2 = \frac{1}{96} \approx .01$, whereas the error we made in Example 36-3 was no greater than $\frac{1}{180} \cdot 24 \cdot (\frac{1}{4})^4 = \frac{1}{1920} \approx .0005$.

PROBLEMS   36

1. Compare your results when you compute $\int_0^2 (1 + 2x - x^2)\, dx$ using
    (a) integration formulas.
    (b) the Trapezoidal Rule with $n = 4$.
    Sketch a figure showing the curve and the trapezoids involved.

2. Compute $\int_1^2 x^2\, dx$ exactly and by the Trapezoidal Rule with $n = 4$. What percentage error do you get with the numerical integration?

3. Compute an approximation to the integral, using the Trapezoidal Rule with the given number n.

    (a) $\int_0^1 (1 + x^2)^{-1}\, dx, \quad n = 2$ (b) $\int_1^2 x^{-2}\, dx, \quad n = 2$

    (c) $\int_{-2}^2 2^{-x}\, dx, \quad n = 4$ (d) $\int_{-2}^2 2^{-|x|}\, dx, \quad n = 4$

    (e) $\int_0^\pi \sin x\, dx, \quad n = 4$ (f) $\int_0^\pi \cos x\, dx, \quad n = 4$

4. Use Simpson's Parabolic Rule with $n = 4$ to approximate the integrals in parts (c), (d), (e), and (f) of the preceding problem.

5. Use Simpson's Parabolic Rule with $n = 4$, plus the tables in the back of the book, to approximate $\int_0^{.4} \sin x\, dx$. Use Formula 36-7 to show that your calculation is as exact as the tables will allow. Use the same technique to approximate $\int_0^{.4} \frac{\sin x}{x}\, dx$.

6. (a) Use the geometric interpretation to find $\int_{-1}^3 |x|\, dx$.
    (b) Use Simpson's Rule with $n = 4$ to approximate the integral.
    (c) Use Simpson's Rule with $n = 8$ to approximate the integral.
    (d) Explain the results of parts (b) and (c) geometrically.

7. Show that the parabola that contains the points (x_0, y_0), (x_1, y_1), and (x_2, y_2) is

$$y = \frac{y_0(x - x_1)(x - x_2)}{(x_0 - x_1)(x_0 - x_2)} + \frac{y_1(x - x_0)(x - x_2)}{(x_1 - x_0)(x_1 - x_2)} + \frac{y_2(x - x_0)(x - x_1)}{(x_2 - x_0)(x_2 - x_1)}.$$

8. Does Simpson's Rule always give a better approximation than the Trapezoidal Rule?

9. Suppose that $f(x) > 0$ for each $x \in [a, b]$. Can you convince yourself that if $f''(x) > 0$, then the Trapezoidal Rule gives too large a number as an approximation to $\int_a^b f(x) \, dx$, whereas if $f''(x) < 0$, the approximate value is too small?

10. Suppose that $f(x)$ is a polynomial of degree 3 or less. What can you conclude from Formula 36-7 about the accuracy of the Simpson Rule approximation?

11. The **Midpoint Rule** approximation formula is

$$\int_a^b f(x) \, dx \approx h \sum_{i=1}^n f(m_i),$$

where $m_i = \frac{1}{2}(x_{i-1} + x_i)$.

(a) Geometrically interpret the right-hand side as a sum of areas of rectangles.

(b) Can you give an alternative geometric interpretation of the right-hand side as a sum of areas of trapezoids, the top ends of which are tangent to the graph of f at the points $(m_i, f(m_i))$?

(c) Compare the approximations to $\int_0^{.4} \frac{\sin x}{x} \, dx$ that you get by using the Trapezoidal Rule and the Midpoint Rule with $n = 2$.

12. From our geometric interpretation of the integral, we know that $\int_0^1 \sqrt{1 - x^2} \, dx = \frac{1}{4}\pi$. Find approximate values of π by using this equation, $n = 4$, and

(a) the Trapezoidal Rule.

(b) Simpson's Parabolic Rule.

(c) the Midpoint Rule (see the preceding problem).

13. Can you choose a value of n for which the Trapezoidal Rule gives the exact value of the integral?

(a) $\int_0^4 [\![x]\!] \, dx$
(b) $\int_{-1}^{-1+\sqrt{2}} |x| \, dx$

37. THE FUNDAMENTAL THEOREM OF CALCULUS

So far we have kept the concepts of the derivative and the integral separate. In Chapter 2 we found out what a derivative is and how to calculate it. In the early sections of this chapter we learned what an integral is and acquired some techniques for calculating integrals. Nothing we have said up to now suggests that the processes of differentiation and integration are not independent. But they aren't. In a sense, they are inverses of each other.

In order to get a hint about the relation between differentiation and integration, we will express the specific integration Formulas 35-6 and 35-7 in a slightly new notation. If F is a given function and a and b are numbers in its domain, we will write

$$F(x)\Big|_a^b = F(b) - F(a).$$

For example, $x^3\Big|_1^2 = 2^3 - 1^3 = 8 - 1 = 7$. This notation allows us to write our specific integration formulas in the compact form

(37-1)
$$\int_a^b x^r\,dx = \frac{x^{r+1}}{r+1}\Big|_a^b.$$

Thus Equation 35-7 is obtained from Equation 37-1 by replacing r with 2:

$$\int_a^b x^2\,dx = \frac{x^3}{3}\Big|_a^b = \frac{b^3}{3} - \frac{a^3}{3}.$$

We get Equation 35-6 by replacing r with 1, and we will shortly find that Equation 37-1 is correct for every choice of r (except $r = -1$).

Because its derivative is x^r, we call $\dfrac{x^{r+1}}{r+1}$ an *antiderivative* of x^r. In general, if $\dfrac{d}{dx}F(x) = f(x)$, then $F(x)$ is an **antiderivative** of $f(x)$ [of course, $f(x)$ is the derivative of $F(x)$]. One of the milestones of mathematics was the discovery that if x^r in Equation 37-1 is replaced by a general $f(x)$ and the corresponding antiderivative $\dfrac{x^{r+1}}{r+1}$ is replaced by an antiderivative $F(x)$ of $f(x)$, then the equation remains valid. This result is called the **Fundamental Theorem of Calculus**.

Theorem 37-1. *If a function f is integrable on the interval $[a, b]$ and if F is a function that is continuous in $[a, b]$ and is such that $\dfrac{d}{dx}F(x) = f(x)$ for each $x \in (a, b)$, then*

(37-2)
$$\int_a^b f(x)\,dx = F(b) - F(a).$$

PROOF

Our proof of the Fundamental Theorem is like the argument we used to establish Equation 37-1. According to Theorem 35-1, the number $F(b) - F(a)$ is the integral of f over $[a, b]$ if for *each* partition of this interval there is an approximating sum such that

(37-3)
$$\sum_{i=1}^n f(x_i^*)\,\Delta x_i = F(b) - F(a).$$

When we verifed Equation 37-1 (for $r = 1$ and 2) we used algebraic arguments to find numbers $x_1^*, x_2^*, \ldots, x_n^*$ for which Equation 37-3 is valid. Now we will use the Theorem of the Mean (Theorem 19-3). We are assuming that F is continuous in every interval $[x_{i-1}, x_i]$ and that $F'(x)$ is defined at each point in the open interval (x_{i-1}, x_i). Therefore there is a number m_i in the interval (x_{i-1}, x_i) such that

(37-4) $$F(x_i) - F(x_{i-1}) = F'(m_i)(x_i - x_{i-1}).$$

But $F'(m_i) = f(m_i)$, since the main hypothesis of Theorem 37-1 is that $F'(x) = f(x)$. Also, $x_i - x_{i-1} = \Delta x_i$, so Equation 37-4 can be written as

$$F(x_i) - F(x_{i-1}) = f(m_i)\,\Delta x_i.$$

Therefore, if we choose $x_i^* = m_i$, the left-hand side of Equation 37-3 becomes

$$
\begin{aligned}
\sum_{i=1}^{n} f(m_i)\,\Delta x_i &= \sum_{i=1}^{n} [F(x_i) - F\,x_{i-1})] \\
&= [F(x_1) - F(x_0)] + [F(x_2) - F(x_1)] + \cdots + [F(x_n) - F(x_{n-1})] \\
&= F(x_n) - F(x_0) = F(b) - F(a).
\end{aligned}
$$

We have found numbers $x_1^*, x_2^*, \ldots, x_n^*$ for which Equation 37-3 is valid and have thereby completed the proof of the Fundamental Theorem of Calculus.

We can make the connection between differentiation and integration seem more direct if we write Equation 37-2 in a different form. We are supposing that $\dfrac{d}{dx} F(x) = f(x)$, and hence Equation 37-2 becomes

$$\int_a^b \frac{d}{dx} F(x)\,dx = F(x)\Big|_a^b.$$

This equation shows us how integration "undoes" the operation of differentiation. When you remember that we introduced the process of differentiation in terms of finding a tangent line to a curve and the process of integration in terms of the apparently unrelated problem of finding the area of a region under a curve, you will probably agree that the Fundamental Theorem is not an obvious result! The Fundamental Theorem tells us that we can reduce integration to subtraction if we know our rules of differentiation backward as well as forward. For to evaluate $\int_a^b f(x)\,dx$, we need only find an antiderivative $F(x)$ of $f(x)$ and then compute the number $F(b) - F(a)$. Of course, it is not always easy to solve the equation $\dfrac{d}{dx} F(x) = f(x)$ for $F(x)$, but in many cases our familiarity with the rules of differentiation lets us solve this equation by inspection.

We have only verified Equation 37-1 in case $r = 1$ or $r = 2$ ($r = 3$ in Problem 35-10). But with the aid of the Fundamental Theorem it is easy to show that it is true if r is any rational number other than -1. We need merely note that $\dfrac{x^{r+1}}{r+1}$ is an antiderivative of x^r for such an r; that is,

$$\frac{d}{dx}\left(\frac{x^{r+1}}{r+1}\right) = \left(\frac{r+1}{r+1}\right)x^r = x^r.$$

When we see how easily the Fundamental Theorem yields Equation 37-1, which includes the formulas we so laboriously derived in Section 35, we begin to see what a valuable tool it is.

Example 37-1.   Evaluate the integral $\displaystyle\int_0^4 \sqrt{x}\, dx$ that we approximated in Example 36-1.

Solution.   Our given integral can be written as $\displaystyle\int_0^4 x^{1/2}\, dx$, so we apply Formula 37-1 with $r = \frac{1}{2}$:

$$\int_0^4 x^{1/2}\, dx = \frac{x^{3/2}}{\frac{3}{2}}\bigg|_0^4 = \frac{2}{3}(4^{3/2} - 0^{3/2}) = \frac{2}{3}\cdot 8 = \frac{16}{3}.$$

Example 37-2.   How do we use the Fundamental Theorem to evaluate $\displaystyle\int_{-1}^8 \frac{1}{\sqrt[3]{t^5}}\, dt$?

Solution.   We don't! You might be tempted to use Equation 37-1 with $r = -\frac{5}{3}$ to obtain the answer

$$\frac{t^{-(5/3)+1}}{-\frac{5}{3}+1}\bigg|_{-1}^8 = -\frac{3}{2} t^{-2/3}\bigg|_{-1}^8 = -\frac{3}{2}\left(\frac{1}{4} - 1\right) = \frac{9}{8}.$$

But the integrand is not bounded in the interval $[-1, 8]$, so our integral does not exist. Obviously, the Fundamental Theorem does not apply.

We can evaluate many integrals by combining Equation 37-1 with general integration rules like Theorem 34-1.

Example 37-3.   Evaluate the integral $\displaystyle\int_1^2 \left(2z^3 - \frac{3}{z^2}\right) dz.$

Solution.

$$\int_1^2 \left(2z^3 - \frac{3}{z^2}\right) dz = \int_1^2 (2z^3 - 3z^{-2}) \, dz$$

$$= 2 \int_1^2 z^3 \, dz - 3 \int_1^2 z^{-2} \, dz$$

$$= 2 \left(\frac{z^4}{4}\right)\Big|_1^2 - 3 \left(\frac{z^{-1}}{-1}\right)\Big|_1^2$$

$$= 2 \left(\frac{16}{4} - \frac{1}{4}\right) - 3 \left(\frac{2^{-1}}{-1} - \frac{1}{-1}\right) = 6.$$

PROBLEMS 37

1. Compute.

    (a) $\displaystyle\int_0^2 10x^4 \, dx$

    (b) $\displaystyle\int_0^1 \sqrt[4]{t^3} \, dt$

    (c) $\displaystyle\int_{-4}^{-2} \frac{1}{t^2} \, dt$

    (d) $\displaystyle\int_0^{-1} \sqrt[3]{z} \, dz$

    (e) $\displaystyle\int_1^4 \frac{1}{\sqrt{z}} \, dz$

    (f) $\displaystyle\int_1^8 \frac{1}{3\sqrt[3]{x^2}} \, dx$

2. Compute.

    (a) $\displaystyle\int_0^1 (x^4 - \sqrt{x}) \, dx$

    (b) $\displaystyle\int_0^a (a^2 x - x^3) \, dx$

    (c) $\displaystyle\int_{-1}^3 (3 + 2x - x^2) \, dx$

    (d) $\displaystyle\int_1^4 (\sqrt[3]{x^2} - \sqrt{x^3}) \, dx$

    (e) $\displaystyle\int_{-2}^2 x(x^2 + 3x) \, dx$

    (f) $\displaystyle\int_0^{-1} x(\sqrt[3]{x} + x^3) \, dx$

    (g) $\displaystyle\int_1^4 (\sqrt{t} + 2)(2t - 1) \, dt$

    (h) $\displaystyle\int_0^8 (2 - \sqrt[3]{x})^2 \, dx$

3. Compute.

    (a) $\displaystyle\int_0^2 (3x)^2 \, dx$

    (b) $\displaystyle\int_{1/2}^2 \frac{x^2 + 1}{x^2} \, dx$

    (c) $\displaystyle\int_{-8}^{-1} \left(\sqrt[3]{8x} + \frac{1}{\sqrt[3]{8x}}\right)^2 \, dx$

    (d) $\displaystyle\int_1^{16} \frac{5\sqrt[4]{x^3} - 3x}{\sqrt{x}} \, dx$

    (e) $\displaystyle\int_1^4 x^{2^3} \, dx$

    (f) $\displaystyle\int_1^{a^{16}} \sqrt{x\sqrt{x\sqrt{x}}} \, dx$

4. Compute.

(a) $\displaystyle\int_0^4 |4 - x^2|\, dx$ (b) $\displaystyle\int_0^2 x|1 - x|\, dx$

(c) $\displaystyle\int_{-1}^1 \sqrt{|x| + x}\, dx$ (d) $\displaystyle\int_{-1}^5 \sqrt[3]{4|x| + x)}\, dx$

5. Show that if p and q are positive integers, then $\displaystyle\int_0^1 (x^{p/q} + x^{q/p})\, dx = 1$.

6. Find the area of the region that is above the X-axis and below the curve.
    (a) $y = 16 - x^2$ (b) $y = 2x^2 - x^3$
    (c) $y = 4a^3x - 6a^2x^2 + 4ax^3 - x^4$ (d) $y = \sqrt{x} + 2 - x - |\sqrt{x} + x - 2|$

7. Find the average value of f in the interval $[1, 4]$ if $f(x) = 3\sqrt{x} + (2/\sqrt{x})$. Find a point at which f takes this value.

8. Find a positive number x such that

    $$\int_0^x (2 - t + t^2)\, dt = \tfrac{14}{3}.$$

    Interpret this problem geometrically.

9. Use an argument that is part geometric and part based on integration formulas to find $\displaystyle\int_{-1}^4 \sqrt{|x|}\, dx$.

10. Are there any positive rational numbers r and s such that

    $$\int_0^1 x^r\, dx \cdot \int_0^1 x^s\, dx = \int_0^1 x^r \cdot x^s\, dx?$$

11. Compute.

(a) $\displaystyle\int_0^1 \frac{d}{dx} \sqrt{x^2 + 8}\, dx$ (b) $\displaystyle\int_{-1}^1 \frac{d}{dt} \sin t^2\, dt$

(c) $\displaystyle\int_0^{\pi/4} \frac{d}{dt} \tan t\, dt$ (d) $\displaystyle\int_2^8 \frac{d}{dx} \log x\, dx$

12. If the position of a point along the number scale is given by the equation $s = f(t)$, then its *average velocity* in a time interval $[t_1, t_2]$ is $[f(t_2) - f(t_1)]/(t_2 - t_1)$. In the terminology of Theorem 34-2, the *average of the velocity* v is the number

    $$\frac{1}{t_2 - t_1} \int_{t_1}^{t_2} v\, dt.$$ Show that these two average velocities are the same.

13. (a) Show that for a nonnegative number r,

    $$\int_a^b |x|^r\, dx = \frac{b|b|^r - a|a|^r}{r + 1}.$$

    (b) Show that this equation is valid for any rational $r \neq -1$ if a and b have the same sign.
    (c) Does the equation hold for any negative number r if a and b have opposite sign?

The Fundamental Theorem of Calculus tells us that

$$(38\text{-}1) \qquad \int_a^b f(x)\, dx = F(x)\Big|_a^b$$

whenever f is integrable, F is continuous in $[a, b]$, and $\dfrac{d}{dx} F(x) = f(x)$ for each $x \in (a, b)$. For example,

$$\int_a^b 6x^5\, dx = x^6\Big|_a^b ,$$

since $\dfrac{d}{dx} x^6 = 6x^5$. Here the limits of integration a and b are arbitrary. They could be the numbers 2 and 3, $-\pi$ and π, or any other pair of real numbers. Thus

$$\int_2^3 6x^5\, dx = x^6\Big|_2^3 = 665, \qquad \int_{-\pi}^{\pi} 6x^5\, dx = x^6\Big|_{-\pi}^{\pi} = 0,$$

and so on.

If, as in the preceding example, Equation 38-1 is independent of the limits a and b, we drop them and simply write

$$(38\text{-}2) \qquad \int f(x)\, dx = F(x).$$

Thus *Integration Formula 38-2 is merely an abbreviated form of Equation* 38-1. *When we say that Formula 38-2 holds in an interval I, we mean that Equation* 38-1 *is valid for each pair of numbers a and b of I.* We get Equation 38-1 from Formula 38-2 by inserting the limits a and b.

The Fundamental Theorem says that if $\dfrac{d}{dx} F(x) = f(x)$ for each x in an interval I, then Equation 38-1 does hold for each pair of numbers in I, and hence Formula 38-2 is valid in I. Therefore with each of our specific differentiation formulas, we can now associate an integration formula. For example, because

$$\frac{d}{dx} \frac{x^{r+1}}{r+1} = x^r,$$

we have the integration formula that we introduced in the preceding section,

$$(38\text{-}3) \qquad \int x^r\, dx = \frac{x^{r+1}}{r+1} \qquad (r \text{ a rational number, not equal to } -1).$$

Our differentiation formulas for the trigonometric functions lead to the following integration formulas:

$$(38\text{-}4) \qquad\qquad \int \sin x \; dx = -\cos x,$$

$$(38\text{-}5) \qquad\qquad \int \cos x \; dx = \sin x,$$

$$(38\text{-}6) \qquad\qquad \int \sec^2 x \; dx = \tan x,$$

$$(38\text{-}7) \qquad\qquad \int \csc^2 x \; dx = -\cot x,$$

$$(38\text{-}8) \qquad\qquad \int \sec x \tan x \; dx = \sec x,$$

$$(38\text{-}9) \qquad\qquad \int \csc x \cot x \; dx = -\csc x.$$

To verify one of these formulas, we need only show that the expression on the right is an antiderivative of the integrand. For example, Formula 38-8 is valid because $\dfrac{d}{dx} \sec x = \sec x \tan x$. You should memorize these integration formulas and realize that they are nothing but our old differentiation formulas written in a new form.

Example 38-1. Find the area of the region bounded by an arch of the sine curve and the X-axis.

Solution. A glance at the curve $y = \sin x$ shows us that the area we seek is given by the integral $\displaystyle\int_0^\pi \sin x \; dx$. This integral is evaluated by inserting the limits 0 and π into Formula 38-4:

$$\int_0^\pi \sin x \; dx = -\cos x \Big|_0^\pi = (-\cos \pi) - (-\cos 0) = 2.$$

(In Example 32-2 we used rectangles to find an approximation of the area of one-half of this region.)

In addition to the integration formulas listed above, which deal with specific functions, there are a number of general integration rules, two of which we already know (Theorems 34-1 and 34-3):

$$(38\text{-}10) \qquad \int_a^b [mf(x) + ng(x)] \, dx = m \int_a^b f(x) \, dx + n \int_a^b g(x) \, dx,$$

$$(38\text{-}11) \qquad \int_a^c f(x) \, dx = \int_a^b f(x) \, dx + \int_b^c f(x) \, dx.$$

In these equations f and g are assumed to be integrable functions and m and n are real numbers.

The best way to become familiar with integration formulas is to use them.

Example 38-2. Evaluate the integral $\int_0^1 (8t^3 - \sec^2 t) \, dt$.

Solution.

$$\int_0^1 (8t^3 - \sec^2 t) \, dt = 8 \int_0^1 t^3 \, dt - \int_0^1 \sec^2 t \, dt \qquad \text{(Equation 38-10)}$$

$$= 8 \frac{t^4}{4} \bigg|_0^1 - \tan t \bigg|_0^1 \qquad \text{(Formulas 38-3 and 38-6)}$$

$$= 8(\tfrac{1}{4} - 0) - (\tan 1 - \tan 0)$$

$$= 2 - \tan 1.$$

From Table I we find that $\tan 1 = 1.557$, so

$$\int_0^1 (8t^3 - \sec^2 t) \, dt = .443.$$

Example 38-3. Evaluate the integral $\int_{-\pi}^{2\pi} \sin |x| \, dx$.

Solution. None of the Formulas 38-3 to 38-9 applies directly to this case. However, since $|x| = -x$ when $x \in [-\pi, 0]$ and $|x| = x$ when $x \in [0, 2\pi]$, we can use these formulas if we first apply Equation 38-11:

$$\int_{-\pi}^{2\pi} \sin |x| \, dx = \int_{-\pi}^0 \sin |x| \, dx + \int_0^{2\pi} \sin |x| \, dx \qquad \text{(Equation 38-11)}$$

$$= \int_{-\pi}^0 \sin (-x) \, dx + \int_0^{2\pi} \sin x \, dx \qquad \text{(Definition of } |x|\text{)}$$

$$= -\int_{-\pi}^0 \sin x \, dx + \int_0^{2\pi} \sin x \, dx \qquad \begin{array}{l}(\sin (-x) = -\sin x \text{ and} \\ \text{Equation 38-10)}\end{array}$$

$$= \cos x \bigg|_{-\pi}^0 - \cos x \bigg|_0^{2\pi} = 2 \qquad \text{(Formula 38-4)}.$$

You will often find the following generalizations of Formulas 38-4 and 38-5 useful. If a and b are any numbers ($a \neq 0$), then

$$(38\text{-}12) \qquad \int \cos{(ax + b)} \, dx = \frac{1}{a} \sin{(ax + b)}$$

and

$$(38\text{-}13) \qquad \int \sin{(ax + b)} \, dx = -\frac{1}{a} \cos{(ax + b)}.$$

We verify these integration formulas by using our rules of differentiation to show that

$$\frac{d}{dx}\left[\frac{1}{a} \sin{(ax + b)} \right] = \cos{(ax + b)},$$

and

$$\frac{d}{dx}\left[-\frac{1}{a} \cos{(ax + b)} \right] = \sin{(ax + b)}.$$

It is not hard to write similar generalizations of the other formulas on our list.

Finally, notice that the Fundamental Theorem tells how to evaluate an integral of a function f by using *any* antiderivative of $f(x)$. We get the value of the integral no matter which antiderivative of $f(x)$ we choose. For example, if we evaluate the integral $\int_a^b \cos x \, dx$ by choosing the antiderivative $\sin x$, we obtain the number $\sin b - \sin a$. On the other hand, since $\frac{d}{dx}(\sin x + 7) = \cos x$, we see that $\sin x + 7$ is also an antiderivative of $\cos x$, and hence

$$\int_a^b \cos x \, dx = (\sin x + 7)\Big|_a^b = (\sin b + 7) - (\sin a + 7)$$

$$= \sin b - \sin a.$$

Since it makes no difference which of the antiderivatives of $f(x)$ we use when applying the Fundamental Theorem, we normally choose the "simplest" anti-derivative when stating an integration formula. Thus we write $\int \cos x \, dx = \sin x$, for example, although it would be equally correct to write $\int \cos x \, dx = \sin x + 7$.

We cannot infer from these two formulas that $\sin x$ and $\sin x + 7$ are equal, of course. These integration formulas are simply abbreviated forms of the equations

$$\int_a^b \cos x \, dx = \sin x \Big|_a^b \quad \text{and} \quad \int_a^b \cos x \, dx = (\sin x + 7)\Big|_a^b \, ;$$

it is the numbers $\sin x \Big|_a^b$ and $(\sin x + 7)\Big|_a^b$ that are the same.

PROBLEMS   38

1. Use integration formulas from this section to evaluate the following integrals.

(a) $\displaystyle\int_0^{\pi/2} (\sin x + 2\cos x)\, dx$

(b) $\displaystyle\int_0^{\pi} (x + \sin x)\, dx$

(c) $\displaystyle\int_{-\pi/4}^{0} \sec^2 x\, dx$

(d) $\displaystyle\int_{\pi/4}^{\pi/2} (2x + \csc^2 x)\, dx$

(e) $\displaystyle\int_{-\pi/4}^{\pi/4} \sec t \tan t\, dt$

(f) $\displaystyle\int_{\pi/4}^{3\pi/4} \csc z \cot z\, dz$

2. Use integration formulas from this section to evaluate the following integrals.

(a) $\displaystyle\int_0^{1} (4x + \sin x)\, dx$

(b) $\displaystyle\int_{-1}^{0} (3x^2 - \cos x)\, dx$

(c) $\displaystyle\int_{-\pi/2}^{\pi/2} (\sin 2x - 2\sin x)\, dx$

(d) $\displaystyle\int_{-\pi/2}^{\pi/2} (\cos 2x - 2\cos x)\, dx$

(e) $\displaystyle\int_{\pi/4}^{3\pi/4} \csc^2 z \cos z$

(f) $\displaystyle\int_{-\pi/4}^{\pi/4} \sec^2 t \sin t\, dt$

(g) $\displaystyle\int_1^{2} [\cos (x - 1) - \sin 1]\, dx$

(h) $\displaystyle\int_1^{2} [\sin (x - 1) - \cos 1]\, dx$

3. Find the area of the region bounded by the X-axis and an arch of the given curve.
    (a) $y = 2\cos x$
    (b) $y = \cos 2x$
    (c) $y = \sin 3x$
    (d) $y = \sin (3x - 12)$

4. Evaluate the following integrals.

(a) $\displaystyle\int_{-\pi}^{\pi} \sin (|x| + x)\, dx$

(b) $\displaystyle\int_{-\pi}^{\pi} \cos (|x| + x)\, dx$

(c) $\displaystyle\int_{-\pi}^{\pi} (|\sin x| + \sin x)\, dx$

(d) $\displaystyle\int_{-\pi}^{\pi} (|\cos x| + \cos x)\, dx$

(e) $\displaystyle\int_{-\pi}^{\pi} (\sin |x| + |\sin x|)\, dx$

(f) $\displaystyle\int_{-\pi}^{\pi} (\cos |x| + |\cos x|)\, dx$

5. Determine the positive number c such that the first quadrant region that is bounded by the curve $y = \cos x$, the X-axis, the Y-axis, and the line $x = c$ has the same area as the first quadrant region that is bounded by the curve $y = 2\sin 2x$, the X-axis, and the line $x = c$.

6. Verify the following integration formulas.

(a) $\displaystyle\int \sin^2 x \cos x\, dx = \tfrac{1}{3}\sin^3 x$

(b) $\displaystyle\int \cos^3 x \sin x\, dx = -\tfrac{1}{4}\cos^4 x$

(c) $\displaystyle\int \sec^2 (ax + b)\, dx = \frac{1}{a}\tan (ax + b)$

(d) $\int \csc^2 (ax + b) \, dx = -\dfrac{1}{a} \cot (ax + b)$

(e) $\int \sqrt{ax + b} \, dx = \dfrac{2}{3a} (ax + b)^{3/2}$ (f) $\int \dfrac{dx}{\sqrt{ax + b}} = \dfrac{2}{a} \sqrt{ax + b}$

(g) $\int \dfrac{\sin x}{\sqrt{1 - \cos x}} \, dx = 2\sqrt{1 - \cos x}$ (h) $\int \dfrac{\csc x - \cot x}{\sin x} \, dx = \csc x - \cot x$

7. Find a number r that makes the given integration formula correct.

(a) $\int x(x^2 + 3)^3 \, dx = r(x^2 + 3)^4$ (b) $\int x^2(x^3 + 3)^3 \, dx = r(x^3 + 3)^4$

(c) $\int \dfrac{x}{\sqrt{x^2 + 3}} \, dx = r\sqrt{x^2 + 3}$ (d) $\int x\sqrt{3 - x^2} \, dx = r(3 - x^2)^{3/2}$

(e) $\int x^2 \cos x^3 \, dx = r \sin x^3$ (f) $\int x \sin x^2 \, dx = r \cos x^2$

(g) $\int \sin x \cos^3 x \, dx = -\dfrac{1}{r} \cos^r x$ (h) $\int \cos x \sin^3 x \, dx = \dfrac{1}{r} \sin^r x$

8. Prove that there is a point $x \in [0, \pi]$ at which the value of the sine function is equal to the average value of the sine function in the interval $[0, x]$. Use Newton's Method to find this point (approximately).

9. How could you prove that the formula $\int \sin^3 x \, dx = \frac{1}{4} \sin^4 x$ is false? (*Hint:* Insert a pair (a, b) of limits that lead to an absurd "equation.")

10. Suppose that a and b are given numbers, with $a \neq 0$. Show that if $F'(x) = f(x)$ for each $x \in R^1$, then $\int f(ax + b) \, dx = \dfrac{1}{a} F(ax + b)$.

11. Can you convince yourself that $\displaystyle\int_a^b \dfrac{x}{|x|} \, dx = |x| \,\Big|_a^b$ for every pair of numbers a and b and hence that the integration formula $\displaystyle\int \dfrac{x}{|x|} \, dx = |x|$ is valid in R^1? Is the equation $\dfrac{d}{dx} |x| = \dfrac{x}{|x|}$ true for every number x?

12. By definition, the fearsome formula $\int [\![x]\!] \, dx = x[\![x]\!] - \frac{1}{2}[\![x]\!][\![x + 1]\!]$ means that

$$\int_a^b [\![x]\!] \, dx = (x[\![x]\!] - \tfrac{1}{2}[\![x]\!][\![x + 1]\!]) \,\Big|_a^b$$

is valid for each pair of numbers (a, b). Use a geometric argument to show that this equation is true for the following pairs.

(a) $(1, 2)$ (b) $(-1, 1)$ (c) $(0, 4)$ (d) $(-4, 0)$ (e) $(-3, \sqrt{2})$

(f) $(-2, \sqrt{3})$

In order to fix the idea of the integral in your mind, we are now going to take up some applications. Although you can find applications of the integral in almost any scientific subject you choose to study, we will mostly restrict ourselves to geometric problems. By doing so, we avoid the necessity of introducing specialized terminology from physics, economics, and the like. But even in our relatively simple geometric problems, we will be basing our discussion on intuitive ideas rather than on strict mathematical definitions. As we said before, a beginning calculus course is no place for a digression into the fine points of what a mathematician means by area and volume. As far as these concepts are concerned, we ask you to proceed on the principle that "What seems reasonable, is reasonable."

If $f(x) \geq 0$ for each $x \in [a, b]$, then the awesome-looking equation $R = \{(x, y) \mid a \leq x \leq b, 0 \leq y \leq f(x)\}$ simply says that R is the region that is bounded by the graph of f, the X-axis, and the lines $x = a$ and $x = b$. The area of R is the number

(39-1) $$A = \int_a^b f(x) \, dx.$$

In the graphical argument that makes Equation 39-1 seem reasonable, we divide the basic interval $[a, b]$ into a set of subintervals of lengths $\Delta x_1, \Delta x_2, \ldots, \Delta x_n$ such as the "typical subinterval" shown in Fig. 39-1. Then we choose a point x_i^* in our typical subinterval and drop a perpendicular from the curve to the X-axis at this point. Let us denote the length of this segment by h_i. Next we draw the rectangle whose base is our subinterval and whose altitude is this perpendicular. The area of this rectangle is $h_i \, \Delta x_i$. Since $h_i = f(x_i^*)$, the formula for the area of

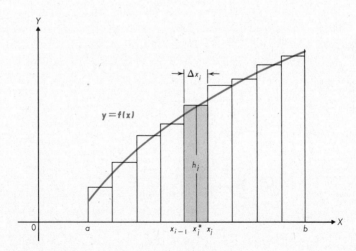

Figure 39-1

our typical rectangle can be written as $f(x_i^*)\,\Delta x_i$. The sum of the areas of all the rectangles,

$$\sum_{i=1}^{n} f(x_i^*)\,\Delta x_i,$$

approximates the area of our region R. This sum also approximates the integral

$$\int_a^b f(x)\,dx,$$

which we therefore take to be the area of R.

Whenever we use integrals to compute areas, we shall follow the sequence of steps that we used in the example above:

(1) Draw a figure that shows the region whose area we are to find.

(2) In the figure, show a typical subinterval with a length of Δx_i on the X-axis and a point x_i^* in this subinterval.

(3) Find the segment with a length of h_i that is perpendicular to the X-axis at x_i^* and that serves as the altitude of our typical rectangle; then draw in the typical rectangle.

(4) Express h_i in terms of x_i^* and hence derive an expression for the area $h_i\,\Delta x_i$ of our typical rectangle.

(5) Form the sum of these areas and then write down the integral that this sum suggests.

(6) Evaluate this integral and take it to be the area of the region.

Let us see how these steps apply to a region like the one shown in Fig. 39-2. Here we are considering a function f that takes only negative values in $[a, b]$, and we wish to find the area of the region R that is bounded by the graph of f, the X-axis, and the lines $x = a$, $x = b$. Figure 39-2 shows how we have carried out steps (1) to (3) above. Clearly, the altitude h_i of our typical rectangle is the number $h_i = -f(x_i^*)$. Thus our approximating sum is

$$\sum_{i=1}^{n} -f(x_i^*)\,\Delta x_i,$$

and this sum suggests that the area A of our region is given by the equation

$$(39\text{-}2) \quad A = \int_a^b -f(x)\,dx = -\int_a^b f(x)\,dx.$$

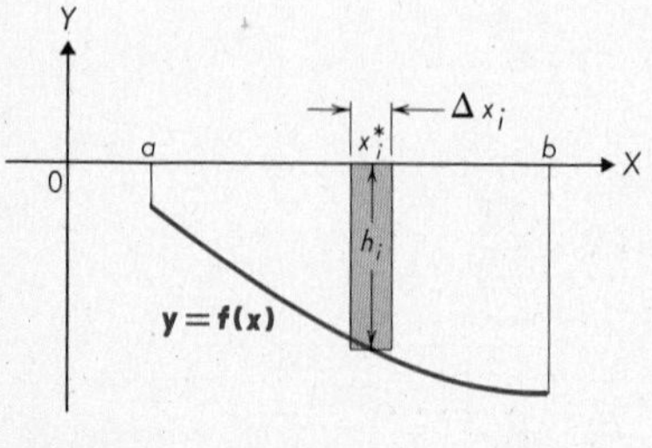

Figure 39-2

Equation 39-1 applies when f is a function whose graph lies above the X-axis, and Equation 39-2 applies when f is a function whose graph lies below the X-axis. The equation

(39-3)
$$A = \int_a^b |f(x)| \, dx$$

covers both cases. Furthermore, Equation 39-3 gives us the area of the region bounded by the graph of a function f, the X-axis, and the lines $x = a$ and $x = b$ even when this region lies partly above and partly below the X-axis. Thus if f is the function whose graph is shown in Fig. 39-3, then

$$\int_a^c |f(x)| \, dx = \int_a^b |f(x)| \, dx + \int_b^c |f(x)| \, dx.$$

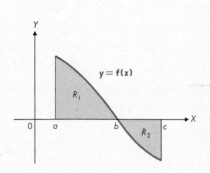

The first integral on the right-hand side of this equation gives us the area of the region R_1; the second integral gives us the area of R_2, and so their sum is the area of $R_1 \cup R_2$.

Now let us give a graphical interpretation of the integral $\int_a^c f(x) \, dx$ for the function whose graph appears in Fig. 39-3. We have

$$\int_a^c f(x) \, dx = \int_a^b f(x) \, dx + \int_b^c f(x) \, dx.$$

Figure 39-3

The first integral on the right-hand side of this equation is the area of the region R_1. The second integral is the *negative* of the area of R_2, since $f(x) \leq 0$ if $x \in [b, c]$. Therefore the integral $\int_a^c f(x) \, dx$ represents the *difference* in the areas of the regions R_1 and R_2. In general, the integral $\int_a^b f(x) \, dx$ represents the number of square units by which the part of the region bounded by the graph of f, the X-axis, and the lines $x = a$ and $x = b$ that lies above the X-axis exceeds the part that lies below the X-axis. In other words, $\int_a^b f(x) \, dx$ represents a "net" area.

Example 39-1. Compute the area of the region between the X-axis and the graph of f if $f(x) = \frac{1}{4}(x^3 + x - 2)$ in the interval $[0, 2]$.

Solution. The graph of f is shown in Fig. 39-4. Because the curve lies below the X-axis for x in the interval $(0, 1)$, the area A_1 of the region R_1 is

$$A_1 = \int_0^1 |f(x)| \, dx = -\frac{1}{4} \int_0^1 (x^3 + x - 2) \, dx = \frac{5}{16}.$$

The area of the region R_2 is

$$A_2 = \int_1^2 |f(x)|\, dx = \tfrac{1}{4} \int_1^2 (x^3 + x - 2)\, dx = \tfrac{13}{16}.$$

Hence the area of the shaded region, $R_1 \cup R_2$, is the number

$$A_1 + A_2 = \tfrac{5}{16} + \tfrac{13}{16} = \tfrac{9}{8}.$$

Notice that

$$\tfrac{1}{4} \int_0^2 (x^3 + x - 2)\, dx = \tfrac{1}{2}.$$

This number is the amount by which A_2 exceeds A_1; that is,

$$A_2 - A_1 = \tfrac{13}{16} - \tfrac{5}{16} = \tfrac{1}{2}.$$

Instead of Equation 39-3, we frequently write the formula for the area of the region between the curve $y = f(x)$ and the interval $[a, b]$ of the X-axis as

$$(39\text{-}4) \qquad\qquad A = \int_a^b |y|\, dx,$$

or if $f(x) \geq 0$ in the interval $[a, b]$, simply as

$$(39\text{-}5) \qquad\qquad A = \int_a^b y\, dx.$$

We understand, of course, that we must replace y with $f(x)$ when we evaluate the integral.

Example 39-2. Find the area of the region interior to an ellipse whose diameters have lengths of $2a$ and $2b$.

Solution. Figure 39-5 shows one quarter of our ellipse; so Formula 39-5 gives us

$$\tfrac{1}{4}A = \int_0^a y\, dx.$$

Since the equation of our ellipse is $x^2/a^2 + y^2/b^2 = 1$, we have $y = (b/a)\sqrt{a^2 - x^2}$. Hence

$$A = \frac{4b}{a} \int_0^a \sqrt{a^2 - x^2}\, dx.$$

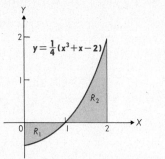

Figure 39-4

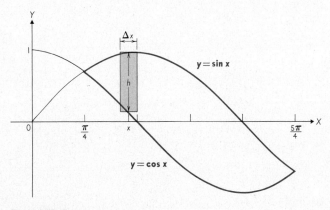

Figure 39-5

We have already seen (Equation 33-3) that $\int_0^a \sqrt{a^2 - x^2}\, dx = \tfrac{1}{4}\pi a^2$, and therefore the area of the region bounded by the ellipse is $(4b/a)\tfrac{1}{4}\pi a^2$; that is,

$$A = \pi ab.$$

When we use our six steps to find an area by integration, we ordinarily use a simplified notation in which we write Δx rather than Δx_i for the length of a typical subinterval, x rather than x_i^* for a point in the subinterval, and h rather than h_i for the altitude of a typical rectangle. The next example illustrates this simplified notation.

Example 39-3. Find the area of the region between the sine and cosine curves in the interval $[\tfrac{1}{4}\pi, \tfrac{5}{4}\pi]$.

Solution. Figure 39-6 shows the region whose area we are to find. A typical rectangle whose altitude is h and base is Δx is also shown. Now we must express h

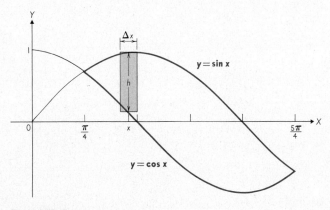

Figure 39-6

in terms of x and substitute the result in the formula $h \, \Delta x$ for the area of the typical rectangle. Since the upper boundary of our region is an arc of the sine curve and the lower boundary is an arc of the cosine curve, $h = \sin x - \cos x$. Therefore the area of our typical rectangle is $(\sin x - \cos x) \, \Delta x$, and a sum of such terms, indicated as

$$\sum (\sin x - \cos x) \, \Delta x,$$

approximates the area of our region. This indicated sum is the simplified notation for the sum

$$\sum_{i=1}^{n} (\sin x_i^* - \cos x_i^*) \, \Delta x_i,$$

and this latter sum suggests that the area we seek is given by the equation

$$A = \int_{\pi/4}^{5\pi/4} (\sin x - \cos x) \, dx = -\cos x \Big|_{\pi/4}^{5\pi/4} - \sin x \Big|_{\pi/4}^{5\pi/4} = 2\sqrt{2}.$$

Example 39-4. Find the area A of the region bounded by the Y-axis and the curve $y^2 - 4y + 2x = 0$.

Solution. The region whose area we seek appears on the left-hand side of Fig. 39-7. A typical rectangle is also shown, and now we must express the altitude h of this rectangle in terms of x. In this case, $h = y_2 - y_1$, where y_2 is the larger and y_1 is the smaller solution of the equation $y^2 - 4y + 2x = 0$. Thus

$$y_2 = \frac{4 + \sqrt{16 - 8x}}{2} = 2 + \sqrt{4 - 2x},$$

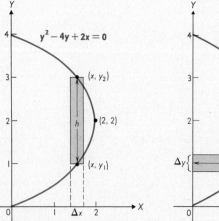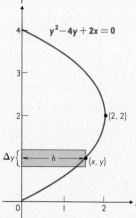

Figure 39-7

and

$$y_1 = 2 - \sqrt{4 - 2x}.$$

Therefore $h = y_2 - y_1 = 2\sqrt{4 - 2x}$, so the area of our typical rectangle is $2\sqrt{4 - 2x}\, \Delta x$. The indicated sum $\sum 2\sqrt{4 - 2x}\, \Delta x$ suggests that the area of our region is given by the equation

$$A = 2 \int_0^2 \sqrt{4 - 2x}\; dx.$$

In part (e) of Problem 38-6, we have an integration formula that applies in this case; if you use it, you will find that $A = \frac{16}{3}$.

Instead of using "vertical rectangles" to approximate our region, it is simpler to use "horizontal" ones, such as that shown on the right-hand side of Fig. 39-7. In this case, we are subdividing an interval of the Y-axis rather than an interval of the X-axis. Consequently, we must express the altitude h of our typical rectangle in terms of y. It is clear that h is the X-coordinate of the point of the curve $y^2 - 4y + 2x = 0$ that corresponds to y, so we find h by solving for x:

$$h = x = 2y - \tfrac{1}{2}y^2.$$

Therefore the area of our typical rectangle is

$$h\, \Delta y = (2y - \tfrac{1}{2}y^2)\, \Delta y,$$

and a sum of such terms,

$$\sum (2y - \tfrac{1}{2}y^2)\, \Delta y,$$

approximates the area of our region. This sum suggests that the area A we seek is

$$A = \int_0^4 \left(2y - \frac{1}{2}y^2\right) dy = y^2 \Big|_0^4 - \frac{y^3}{6}\Big|_0^4 = \frac{16}{3}.$$

PROBLEMS   39

1. Compute the area of the region between the curve $y = f(x)$ and the X-axis in the given interval.
    (a) $f(x) = \sin x$, $[0, 2\pi]$ (b) $f(x) = \cos x$, $[0, \pi]$
    (c) $f(x) = x^3$, $[-1, 1]$ (d) $f(x) = 3x^5 - x$, $[-2, 2]$
    (e) $f(x) = [\![x]\!]$, $[-1, 2]$ (f) $f(x) = 1 - |x|$, $[-2, 2]$
2. Compute the area of the region bounded by the X-axis and the parabola.
    (a) $y = 4 - x^2$ (b) $y = 2 + x - x^2$
3. Find the area of the region bounded by the Y-axis and the curve.
    (a) $x + y^2 + 3 = 4y$ (b) $x + y^2 = 2y$

4. Find the area of the region bounded by the given curves.
   (a) $y = x^2$ and $y = 4x - x^2$ (b) $y = 4 - x^2$ and $y = 8 - 2x^2$
   (c) $y = x^3 + x^2$ and $y = x^3 + 1$ (d) $x + 2y = 3$ and $x = y^2 - 3y + 1$

5. Find the area of the region between the line $y = 1 - x$ and the parabola $y^2 = x$.

6. Find the area of the region bounded by the given curves. Draw pictures!
   (a) $y = \frac{1}{4}x^2$ and $y = \frac{1}{2}x + 2$
   (b) $y^2 = 6x$ and $x^2 = 6y$
   (c) $y^2 = 2x$ and $x - y = 4$
   (d) $\sqrt{x} + \sqrt{y} = \sqrt{a}$ and the coordinate axes
   (e) $x = y^3 - 4y$, $x = 4 - y^2$
   (f) $y = x^3$, $y = (2 - x)^2$, and the X-axis

7. Three bounded regions in the first quadrant have boundaries made up of parts of all three of the curves $y^2 = x$, $y = (x - 2)^2$, and $y^2 = 4 - x$. Find the area of the largest such region.

8. Use geometric reasoning to evaluate the following integrals.

   (a) $\displaystyle\int_0^8 (-1)^{[\![x]\!]} [\![x]\!]\, dx$ (b) $\displaystyle\int_0^{-5} (-1)^{[\![x]\!]} [\![x]\!]\, dx$

9. Through opposite corners of a rectangle, draw an arc of a parabola whose vertex is one of the corner points and whose axis lies along one side of your rectangle. Show that you have formed two regions such that the area of one is twice the area of the other.

10. Draw the rectangle formed by the two coordinate axes and the lines $x = a$ and $y = b$. The curve $y = bx^n/a^n$ (n a positive integer) divides the rectangle into two regions. Show that the area of one region is n times the area of the other region.

11. Sketch the region in the first quadrant between the circles $x^2 + y^2 = a^2$ and $x^2 + y^2 = b^2$. Draw an arc of the ellipse $x^2/a^2 + y^2/b^2 = 1$. Show that your arc divides the region between the two circles into two regions whose areas are in the ratio a/b.

12. Let k be a positive integer and let $A(k)$ denote the area of the region bounded by the curves $y = x^k$ and $y = x^{k+1}$. Find an expression for $\displaystyle\sum_{k=1}^{n} A(k)$. What number does this sum approximate if n is very large? Interpret your result geometrically.

13. Let $pn(x)$ denote the prime number that is nearest to x (if there are two equally near, choose the smaller). Determine m such that the region bounded by the X-axis, the line $x = 10$, and the line $y = mx$ has the same area as the region bounded by the X- and Y-axes, the line $x = 10$, and the curve $y = pn(x)$.

14. If f is integrable on the interval $[a, b]$, what is the area of the region
$$\left\{ (x, y) \mid x \in [a, b], y \in \left[\frac{f(x) - |f(x)|}{2}, \frac{f(x) + |f(x)|}{2} \right] \right\}?$$

40. VOLUME OF A SOLID OF REVOLUTION

We have been using our knowledge of calculus to compute the areas of various plane regions that are somewhat more complicated than the triangles, trapezoids, and other figures whose areas you learned to measure in plane geometry.

In your elementary geometry course you were told formulas that give the volume of various solid figures, such as spheres, cylinders, and cones. Now we will find that calculus can help us determine these formulas and the volumes of other solids as well. We first consider the problem of computing the volume of a solid obtained by rotating a plane region about a line, a **solid of revolution**.

Let f be a function whose domain contains the interval $[a, b]$. Figure 40-1 shows the solid of revolution that is generated by rotating the region bounded by the

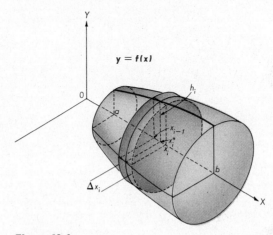

Figure 40-1

graph of f, the X-axis, and the lines $x = a$ and $x = b$ about the X-axis. In order to find the volume of this solid, we proceed as we did in our area problems. We partition the interval $[a, b]$ into n subintervals. Then in each subinterval $[x_{i-1}, x_i]$ we choose a point x_i^* and erect a rectangle with an altitude of $h_i = |f(x_i^*)|$, as shown in Fig. 40-1. When we rotate this "typical rectangle" about the X-axis, we obtain a circular disk with a radius of h_i and a thickness of $\Delta x_i = x_i - x_{i-1}$. The volume of this disk is therefore $\pi h_i^2 \, \Delta x_i = \pi f(x_i^*)^2 \, \Delta x_i$, and the sum of these volumes,

$$\sum_{i=1}^{n} \pi f(x_i^*)^2 \, \Delta x_i,$$

approximates the volume of our solid of revolution. Now we find the limit of this sum as the norm of the partition of the interval $[a, b]$ approaches zero (if this limit exists), and we take this limit to be the volume of our solid of revolution. If f is an integrable function, then the limit of this sum will be (as the form of its terms suggests) $\int_a^b \pi f(x)^2 \, dx$. So we take the volume of our solid of revolution to be the number

(40-1) $$V = \pi \int_a^b f(x)^2 \, dx.$$

Instead of Equation 40-1, we frequently write the formula for the volume of the solid obtained by rotating the curve $y = f(x)$ about the X-axis as

(40-2)
$$V = \pi \int_a^b y^2 \, dx.$$

Of course, before we evaluate the integral in Formula 40-2, we must replace y with $f(x)$.

Example 40-1. Use Equation 40-2 to compute the volume of a spherical ball whose radius is r.

Solution. The curve $y = \sqrt{r^2 - x^2}$ is a semicircle with a radius of r and with its diameter along the X-axis. We obtain our sphere by rotating this arc about the X-axis. Now we make use of Equation 40-2 with $y = \sqrt{r^2 - x^2}$, $a = -r$, and $b = r$ to obtain

$$V = \pi \int_{-r}^r y^2 \, dx = \pi \int_{-r}^r (r^2 - x^2) \, dx = \pi \left(r^2 x \Big|_{-r}^r - \tfrac{1}{3}x^3 \Big|_{-r}^r \right) = \tfrac{4}{3}\pi r^3.$$

Rather than rely on formulas like Equation 40-1, you should follow the steps we used in arriving at the formula:

(1)   Draw a figure that includes a sketch of a "typical rectangle."

(2)   Write a formula that expresses, in terms of x and Δx, the volume of the disk obtained by rotating this typical rectangle. (We usually use the simplified notation x and Δx instead of x_i^* and Δx_i.)

(3)   Take the volume of the solid of revolution to be the integral that is suggested by a sum of the volumes of these disks.

(4)   Evaluate the integral.

Example 40-2. Find the volume of a solid right-circular cone with an altitude of h and a base radius of r.

Solution. The surface of the cone can be obtained by rotating the line segment joining the origin and the point (h, r) about the X-axis (Fig. 40-2). The altitude of our illustrated rectangle is y, and its base is Δx. Therefore the volume of the disk that it generates is $\pi y^2 \, \Delta x$. In order to write this formula in terms of x and Δx, we must express y in terms of x. The numbers x and y are related by the equation of the line that contains the point (h, r) and the origin. Thus $y = (r/h)x$, so the volume of our disk is $\pi(r/h)^2 x^2 \, \Delta x$. The sum of these volumes has the form

$$\sum \pi \left(\frac{r}{h}\right)^2 x^2 \, \Delta x,$$

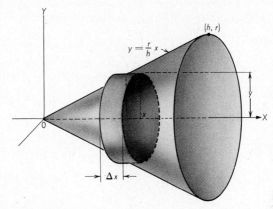

Figure 40-2

and the volume of our cone is given by the integral suggested by this sum:

$$V = \pi \frac{r^2}{h^2} \int_0^h x^2 \, dx = \pi \left(\frac{r^2}{h^2}\right)\left(\frac{h^3}{3}\right) = \tfrac{1}{3}\pi r^2 h.$$

No essentially new ideas are involved when we consider solids of revolution obtained by rotating plane regions about the Y-axis.

Example 40-3. The region bounded by the Y-axis, the line $y = 1$, and the curve $y = \sqrt{x}$ (Fig. 40-3) is rotated about the Y-axis. What is the volume of the solid that is generated?

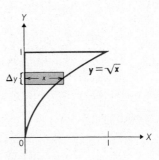

Figure 40-3

Solution. A "typical rectangle" is shown in Fig. 40-3. The volume of the disk generated by this rectangle is $\pi x^2 \, \Delta y$. Here we will express this volume in terms of y and Δy, and to do so, we must express x in terms of y. The numbers x and y are related by the equation of our curve; that is, $y = \sqrt{x}$. Therefore $x^2 = y^4$, so the volume of our typical disk is $\pi y^4 \, \Delta y$. A sum of such volumes has the form

$$\sum \pi y^4 \, \Delta y,$$

which suggests that the volume of our solid of revolution is:

$$V = \pi \int_0^1 y^4 \, dy = \tfrac{1}{5}\pi y^5 \Big|_0^1 = \tfrac{1}{5}\pi.$$

The next example is somewhat more complicated than the preceding one, but we analyze it in much the same way.

Example 40-4. Find the volume of the solid obtained by rotating the region described in Example 40-3 about the *X*-axis.

Solution. When we rotate the typical rectangle shown in Fig. 40-4, it generates a solid "washer," not a disk. An end view of half this washer is shown on the right in Fig. 40-4. From this figure we see that the area of the base of the washer is

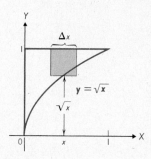
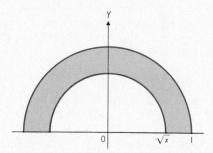

Figure 40-4

$\pi \cdot 1^2 - \pi(\sqrt{x})^2 \doteq \pi(1 - x)$. Since the washer is Δx units thick, its volume is $\pi(1 - x) \Delta x$. A sum of such terms, indicated by $\sum \pi(1 - x) \Delta x$, approximates the volume of the solid under consideration. The volume is given by the integral suggested by this sum:

$$V = \pi \int_0^1 (1 - x)\, dx = \pi \left(x \Big|_0^1 - \tfrac{1}{2}x^2 \Big|_0^1 \right) = \tfrac{1}{2}\pi.$$

Example 40-5. Find the volume obtained by rotating the region of the last two examples about the line $y = 1$.

Solution. When we rotate the typical rectangle shown in Fig. 40-4 about the line $y = 1$, it generates a disk with a radius of $(1 - \sqrt{x})$ and a thickness of Δx. Therefore its volume is $\pi(1 - \sqrt{x})^2 \Delta x$, and a sum of terms of the form $\sum \pi(1 - \sqrt{x})^2 \Delta x$ approximates the volume of our solid. The volume is the integral suggested by this sum:

$$V = \pi \int_0^1 (1 - \sqrt{x})^2\, dx = \pi \int_0^1 (1 - 2\sqrt{x} + x)\, dx$$

$$= \pi \int_0^1 (1 - 2x^{1/2} + x)\, dx$$

$$= \pi \left(x \Big|_0^1 - \tfrac{4}{3}x^{3/2} \Big|_0^1 + \tfrac{1}{2}x^2 \Big|_0^1 \right) = \tfrac{1}{6}\pi.$$

PROBLEMS   40

1.   Find the volume of the solid generated by rotating about the X-axis the region bounded by the given curves.

   (a) $y = x^3$, $x = 2$, and the X-axis
   (b) $y = -4x^2 + 8x - 1$, $x = 1$, and the coordinate axes
   (c) $\sqrt{x} + \sqrt{y} = \sqrt{a}$ and the coordinate axes
   (d) $y = x^{2/3}$, $x = -1$, $x = 1$, and the X-axis
   (e) $y = \sec x$, $x = 0$, $x = \frac{1}{3}\pi$, and the X-axis
   (f) $y = \csc x$, $x = \frac{1}{3}\pi$, $x = \frac{1}{2}\pi$, and the X-axis

2.   Find the volume of the solid generated by rotating about the X-axis the region bounded by the given curves.

   (a) $y = 2 - x$ and the coordinate axes
   (b) $y = 4 - x^2$ and the X-axis
   (c) $y = 1$ and $y = 2 - x^2$
   (d) $y = a$ and $x^2 = 4ay$

3.   Find the volume of the solid obtained by rotating about the Y-axis the region bounded by the given curves.

   (a) $y = x^3$, $y = 8$, and the Y-axis
   (b) $y = x^3$, $x = 2$, and the X-axis
   (c) $2y^2 = x^3$ and $x = 2$
   (d) $y = 1 + x^2$, $x = 2$, and the coordinate axes

4.   Find the volume of the solid generated by rotating about the Y-axis the region bounded by the given curves.

   (a) $y = x$ and $y = x^2$ (b) $y = 4$ and $y = x^2$

5.   Find the volume of the solid obtained by rotating about the line $x = 4$ the region bounded by the given curves.

   (a) $y = x$, $y = 4 - x$, and $x = 4$ (b) $y = x^{3/2}$, $y = 0$, and $x = 4$
   (c) $y = x^{3/2}$, $x = 0$, and $y = 8$ (d) $y = x$ and $y = \frac{1}{2}x^2$

6.   Find the volume of the solid generated by rotating one arch of the sine curve about the X-axis. [Use the identity $\sin^2 x = \frac{1}{2}(1 - \cos 2x)$.]

7.   Find the volume of the solid generated by rotating about the X-axis the region in the first quadrant that is bounded by the Y-axis and the sine and cosine curves. (Use the identity $\cos^2 x - \sin^2 x = \cos 2x$.)

8.   Find the volume of the solid ellipsoid generated by rotating the region bounded by the ellipse $b^2x^2 + a^2y^2 = a^2b^2$ about the X-axis.

9.   Derive the formula for the volume of a solid truncated cone whose height is h and whose two radii are a and b.

10.   Find the volume of the solid generated by rotating the region bounded by the line $x = a$ and the parabola $y^2 = 4ax$ about the line $x = 2a$.

11.   Let a be a positive number and let n be a positive integer. The curve $y = x^n$ divides the rectangle formed by the coordinate axes and the lines $x = a$ and $y = a^n$ into two regions. Two solids are obtained by rotating these regions about the X-axis. What is the ratio of their volumes? What is the ratio of the volumes of the solids obtained by rotating the two regions about the Y-axis?

12. Find the volume of the *torus* obtained by rotating the circle $x^2 + (y - b)^2 = a^2$ about the X-axis (assume $0 < a < b$).

13. The region common to the two ellipses $a^2x^2 + b^2y^2 = a^2b^2$ and $b^2x^2 + a^2y^2 = a^2b^2$ is rotated about the X-axis. Find the volume of the resulting solid.

14. (a) Show that the volume of a right circular cylindrical shell of altitude h, inner radius x_1, and outer radius x_2 is $2\pi \bar{x}_2 h \, \Delta x_2$, where $\bar{x}_2 = \frac{1}{2}(x_1 + x_2)$ and $\Delta x_2 = x_2 - x_1$.

    (b) Let f be a function that is integrable on an interval $[a, b]$ that lies to the right of the origin ($a \geq 0$), and suppose that $f(x) \geq 0$ for each $x \in [a, b]$. We are going to find a formula for the volume of the solid that is obtained by rotating the region $\{(x, y) \mid 0 \leq a \leq x \leq b, \, 0 \leq y \leq f(x)\}$ about the Y-axis. We partition the interval $[a, b]$ into n subintervals, and on each subinterval $[x_{i-1}, x_i]$ we erect a rectangle with an altitude of $f(\bar{x}_i)$. When a typical rectangle is rotated about the Y-axis, a cylindrical shell is formed. Use part (a) to show that it is reasonable to expect that the volume of the solid is given by the formula

$$V = 2\pi \int_a^b xf(x) \, dx = 2\pi \int_a^b xy \, dx.$$

15. Use the formula in the preceding problem to compute the volume of the solid obtained by rotating the given region about the Y-axis.

    (a) $\{(x, y) \mid 0 \leq x \leq 1, \, 0 \leq y \leq x^2\}$
    (b) $\{(x, y) \mid 0 \leq x \leq 4, \, 0 \leq y \leq (x - 2)^2\}$
    (c) $\{(x, y) \mid 0 \leq x \leq \pi, \, 0 \leq y \leq (\sin x)/x\}$
    (d) $\{(x, y) \mid 0 \leq x \leq 3, \, 0 \leq y \leq [\![2 \sin x]\!] x^2\}$

41. VOLUMES BY SLICING

We can generalize the methods of the last section so that they apply to solids other than solids of revolution. The integral $\pi \int_a^b f(x)^2 \, dx$ that is the volume of the solid of revolution shown in Fig. 40-1 is the limit of sums of the form

$$\sum \pi f(x)^2 \, \Delta x.$$

In each term of this sum, the number $\pi f(x)^2$ is the area of the base of a cylindrical disk "sliced" from our solid. If we denote this area by $A(x)$, then our sum takes the form $\sum A(x) \, \Delta x$. In the case of a solid of revolution, the cross-sectional area $A(x)$ is easy to calculate because the cross sections are circles. But even when the cross sections are not circles, we can still find the volume of a solid by slicing if we can calculate the cross-sectional area $A(x)$.

In Fig. 41-1 we have shown a solid whose volume we wish to determine. This solid projects onto an interval $[a, b]$ of a conveniently chosen X-axis. We partition this interval into subintervals—a typical subinterval having a length of

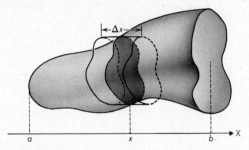

Figure 41-1

Δx. Through a point x in this subinterval we pass a plane perpendicular to the X-axis. This plane intersects our solid in the shaded region, whose area we will denote by $A(x)$. The volume of the cylindrical "slice" whose thickness is Δx and that has the shaded region as a cross section is $A(x) \Delta x$. A sum of such volumes approximates the volume of our solid and is indicated by

$$\sum A(x)\, \Delta x.$$

We take the integral that is suggested by this sum to *be* the volume:

(41-1)
$$V = \int_a^b A(x)\, dx.$$

In order to apply Equation 41-1 to find the volume of a given figure, we need only find the expression $A(x)$ for the cross-sectional area and then evaluate the integral, as in the following examples.

Example 41-1. The plans for a wave guide antenna are shown in Fig. 41-2. Each cross section in a plane perpendicular to the central axis of the wave guide (here the X-axis) is an ellipse whose major diameter is twice as long as its minor diameter. The upper boundary of the widest longitudinal cross section is the parabola $y = \frac{1}{3}x^2 + 1$. The entire antenna is 3 feet long. Find the volume of the region enclosed by the wave guide.

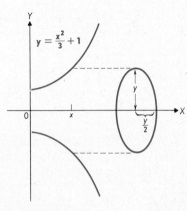

Solution. In order to use Equation 41-1 to find the volume, we must find the area $A(x)$ of a cross section cut from the antenna by a plane perpendicular to the axis at a point x. This cross section is the elliptical region shown in Fig. 41-2. In Example 39-2 we found that an elliptical region with diameters of $2a$ and $2b$

Figure 41-2

253

has an area of πab square units. Therefore the area of our illustrated cross section is $\frac{1}{2}\pi y^2$ square feet. Since $y = \frac{1}{3}x^2 + 1$,

$$A(x) = \frac{1}{2}\pi(\frac{1}{3}x^2 + 1)^2 = \pi(\frac{1}{18}x^4 + \frac{1}{3}x^2 + \frac{1}{2}).$$

Thus Formula 41-1 gives us

$$V = \pi \int_0^3 (\frac{1}{18}x^4 + \frac{1}{3}x^2 + \frac{1}{2})\, dx = \frac{36}{5}\pi \text{ cubic feet.}$$

Example 41-2. A regular pyramid is 100 feet high and has a base 100 feet square. If it is made of rock weighing 100 pounds per cubic foot, how much does it weigh?

Solution. We will first find the volume, in cubic feet, of the pyramid and then multiply this result by 100 to get its weight. Figure 41-3 shows the entire pyramid

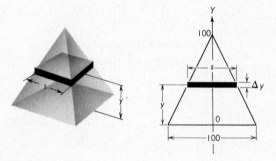

Figure 41-3

and also a triangular cross section. At a height of y feet above the base, we have shaded a "typical slice." If we denote the area of a face of this slice by $A(y)$, then the volume of our pyramid is

$$V = \int_0^{100} A(y)\, dy.$$

So in order to find V, we must find an expression for $A(y)$. A face of our typical slice is obviously square; let s denote the length of a side. Then from similar triangles in Fig. 41-3 we see that

$$\frac{s}{100 - y} = \frac{100}{100}.$$

Thus $s = 100 - y$ and $A(y) = (100 - y)^2$. And so

$$V = \int_0^{100} (100 - y)^2 \, dy = \int_0^{100} (100^2 - 200y + y^2) \, dy = \frac{100^3}{3}.$$

The weight W of the pyramid is 100 times this amount; therefore

$$W = \frac{100^4}{3} = \frac{10^8}{3}.$$

The pyramid weighs $10^8/3$ pounds or nearly 17,000 tons.

The next example tests your ability to visualize three-dimensional figures.

Example 41-3. The base of a certain solid is a circular disk with a 2-inch radius. Cross sections perpendicular to one of the diameters of the disk are square. Find the volume of the solid.

Solution. On the left side of Fig. 41-4 we have shown a top view of our solid with the mentioned diagonal lying along the X-axis. On the right of the figure is the view we would see if we cut the solid at the point x and looked at it down the

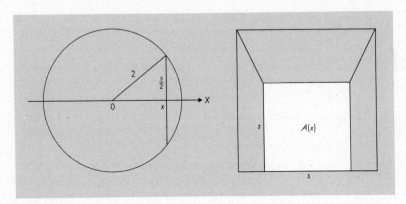

Figure 41-4

X-axis. Our cross section is an s by s square, and we will be able to find its area $A(x)$ as soon as we find s in terms of x. From the right triangle of hypotenuse 2 and legs $\frac{1}{2}s$ and x that we see inside the circle, we obtain the relation $4 = (\frac{1}{2}s)^2 + x^2$. Hence $s^2 = 16 - 4x^2$. But $A(x) = s^2$, so

$$V = \int_{-2}^{2} A(x) \, dx = \int_{-2}^{2} (16 - 4x^2) \, dx$$

$$= (16x - \tfrac{4}{3}x^3) \Big|_{-2}^{2} = \tfrac{128}{3} \text{ cubic inches.}$$

PROBLEMS 41

1. A four-sided solid is formed by sawing a corner off a rectangular block. The base of the figure is a right triangle whose legs are 3 and 4 inches long, and the altitude of the figure is 5 inches. Use Equation 41-1 to find its volume.

2. Use Equation 41-1 to compute the volume of a section taken from a spherical orange with a radius of r inches that contains k identical sections. Now find the answer without using calculus.

3. A cylindrical hole with a radius of 1 inch is drilled along the diameter of a solid metal sphere whose radius is 4 inches. Find the volume of the part of the sphere that remains.

4. Find the volume of a solid whose base is the region bounded by the X-axis and the lines $y = 2x$ and $x = 2$ and such that each cross section perpendicular to the Y-axis is a triangle of altitude 3.

5. A wedge is cut from a solid right-circular cylinder of radius 3 inches by passing a plane through a diameter of the base. Find the volume of the wedge if the plane is inclined at a 30° angle to the base.

6. A solid is constructed with a circular base of radius 4 and such that every cross section perpendicular to a certain diameter of the base is bounded by an equilateral triangle. Find the volume of the solid.

7. A solid is constructed with a circular base of radius 2 and such that every cross section perpendicular to a certain diameter of the base is an isosceles right triangle with its hypotenuse in the base plane. Find the volume of the solid.

8. A solid is constructed with a circular base of radius 1 and such that every cross section perpendicular to a certain diameter of the base is an isosceles triangle whose altitude is 2. Find the volume of the solid.

9. The base of a solid is the region bounded by the line $x = 1$ and the parabola $y^2 = 4x$, and each cross section perpendicular to the X-axis is an equilateral triangle. Find the volume of the solid.

10. The base of a solid is the region bounded by the hyperbola $16x^2 - 9y^2 = 144$ and the line $x = 6$. Find the volume of the solid if each cross section perpendicular to the X-axis is (a) an equilateral triangle, (b) a square.

11. A solid is constructed with a base that is bounded by an ellipse whose major diameter is $2a$ units long and whose minor diameter is $2b$ units long such that every cross section perpendicular to the major diameter is a square. Find the volume of the solid.

12. Calculate the volume of a solid ellipsoid such that three mutually perpendicular cross sections are bounded by concentric ellipses whose diameters are $2a$, $2b$, and $2c$ units long.

13. When two pieces of 1-inch quarter round molding are cut so as to fit together in the corner of a room, how much material is cut away from each piece?

14. The axes of two right-circular cylinders with equal radii of r inches intersect at right angles. Find the volume of the region common to both cylinders.

The formula *area = length × width* gives the area of a rectangle, but in order to find areas of more general regions, we must turn to integration. This phenomenon is quite widespread. The generalization of the formula *distance = rate × time* is the statement *distance is the integral of velocity*, and so on. In this section we are going to generalize the formula

(42-1)   *work = force × distance.*

Suppose that one end of a spring is fixed at a point to the left of the origin of a number scale and that the free end is at the origin when the spring is in its natural (unstretched) position. We will denote by $f(x)$ the force that is necessary to hold the free end at the point x (see Fig. 42-1). Thus $f(0) = 0$, for example. We want to find out how much work is required to pull the free end from the point a to the point b.

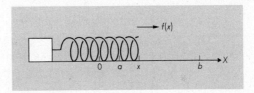

Figure 42-1

The end obviously moves $b - a$ units, but the force varies from point to point during the course of the motion, so we can't use Equation 42-1 over the whole interval. We will apply it a step at a time in order to obtain an approximation and take the limit of such approximations to be the actual work.

Specifically, let x be a point of the interval $[a, b]$ and let Δx be a small number. As the end of the spring moves from x to $x + \Delta x$, let us neglect the variation in the force and simply assume that its value throughout the short interval $[x, x + \Delta x]$ is its value $f(x)$ at the beginning. Then Equation 42-1 gives an approximation to the amount of work that is performed as the free end of the spring is pulled from x to $x + \Delta x$:

$$f(x)\, \Delta x.$$

A sum of such quantities, $\sum f(x)\, \Delta x$, approximates the work required to pull the free end from a to b. It would appear that the smaller Δx is, the better our approximation will be. So it would seem that the actual amount W of work is expressed by the integral

(42-2) $$W = \int_a^b f(x)\, dx.$$

Hooke's Law tells us that the force $f(x)$ required to hold the end of the spring x units from the origin is

$$f(x) = kx,$$

where k is a number, called the **spring constant**, that depends on the spring. Therefore the work that we do when we stretch the free end of our spring from a to b is

$$W = \int_a^b kx\ dx = \tfrac{1}{2}k(b^2 - a^2).$$

Example 42-1. If it requires a force of 10 pounds to hold a certain spring when it is stretched 2 inches, how much work is performed in pulling the free end 6 inches from its natural position?

Solution. Here $a = 0$ and $b = 6$, so $W = \int_0^6 kx\ dx = 18k$. Now we must find the spring constant k. We know that $f(2) = 10$, and therefore the equation $f(x) = kx$ tells us that $2k = 10$. Thus $k = 5$, so $W = 18 \cdot 5 = 90$ inch pounds $= 7.5$ foot pounds.

Of course, our derivation of Equation 42-2 is not very rigorous. It does suggest that the amount of work required to pull the end of our spring from a to b is the limit of sums of the work required to go from a to b in steps, just as the equation $area = \int_a^b f(x)\ dx$ suggests that area is the limit of sums of areas of rectangles. In the latter case, we stated that the definition of the area of a plane region is such that this equation is correct. In the case of work, things are simpler. We consider Equation 42-2 to be the *definition* of the work that is done as an object moves from a to b, at each point x being subjected to a force $f(x)$. Thus *work is the integral of force*. If the force function f is a constant function with value c, then the amount of work that is done as the force is applied to a body that moves from a to b is given by the formula

$$W = \int_a^b c\ dx = c(b - a).$$

This equation is our old formula *work = force × distance*, and so we see that the integral definition of work agrees with the elementary definition when both apply.

The next example is another typical work problem that we can solve with calculus.

Example 42-2. A vat has the shape of a paraboloid of revolution. It is 2 feet high, and the radius of its top is 1 foot. It is filled with water that weighs 62.5 pounds per cubic foot. How much work is required to pump the contents of the vat through a nozzle 4 feet above the top?

Solution. Figure 42-2 shows our vat and its exhaust pipe. We have drawn in a coordinate system to help us formulate our problem mathematically. We can picture an expanding piston (here shown y feet from the bottom of the vat) pushing the water out. The piston moves from $y = 0$ to $y = 2$, so if the force on it is $f(y)$ pounds, then the work required to empty the vat is

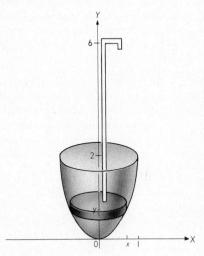

$$\int_0^2 f(y)\, dy$$

foot pounds. By definition, pressure is force per unit area, so the force on our piston is

$$f(y) = \pi x^2 p(y)$$

pounds, where $p(y)$ is the pressure in pounds per square foot and πx^2 is the area of the piston (its radius is x feet). The water above our piston is $6 - y$ feet deep, and we are assuming that it weighs 62.5 pounds per cubic foot, so the pressure is $p(y) = 62.5(6 - y)$. Therefore

Figure 42-2

$$f(y) = 62.5\pi x^2 (6 - y).$$

Before we can compute its integral, we must express $f(y)$ entirely in terms of y, so let us now replace x^2 with an expression in y. We chose the coordinate system of Fig. 42-2 so that our parabola would be in standard position, and hence its equation is $x^2 = 4cy$, where c is a number that we will now determine. Since the vat is 2 feet tall and the radius of its top is 1 foot, we see that the point $(1, 2)$ belongs to the parabola. Then $1 = 4c \cdot 2$, so $4c = \frac{1}{2}$, and the equation of our parabola is $x^2 = \frac{1}{2}y$. Therefore $f(y) = \frac{1}{2} \cdot 62.5\pi y(6 - y)$, and so it requires

$$\frac{1}{2} \cdot 62.5\pi \int_0^2 y(6 - y)\, dy = \frac{875}{3}\pi$$

foot pounds of work to empty the vat.

PROBLEMS 42

1. A force of 50 pounds stretches a spring 2 inches beyond its natural length. How much work is required to stretch it those 2 inches?

2. A spring is stretched 1 inch beyond its natural length by a force of 100 pounds. How much work is done in stretching it 4 inches?

3. A spring whose natural length is 3 feet is stretched to $4\frac{1}{2}$ feet by a force of 25 pounds. How much work is required to stretch it from $4\frac{1}{2}$ to 6 feet?

4. Find the work required to pump out a cylindrical tank of diameter 6 feet and height 8 feet through a pipe at the top if the tank contains an oil weighing 50 pounds per cubic foot and (a) it is full, (b) it is half-full.

5. A cylindrical cistern full of water is to be emptied by a hand pump mounted on top of the cistern. How far down from the top of the cistern is the water level when half the work is done?

6. How much work is required to empty a tankful of water by pumping it over the rim if the tank is a hemisphere with a radius of 2 feet?

7. A tank in the shape of a paraboloid of revolution is 6 feet deep, has a radius on the top of 3 feet, and is full of water. How much work must be done to lower the water level 2 feet by pumping the water over the rim?

8. A conical cistern is 20 feet across the top, 15 feet deep, and is filled to within 5 feet of the top with water. Find the work done in pumping the water over the top to empty the tank.

9. How much work is required to construct the pyramid in Example 41-2?

10. A 1-ton elevator is lifted from the bottom of a 50-foot shaft by means of a cable that weighs 10 pounds per foot. Find the work done in lifting the elevator to the top of the shaft.

11. A leaky bucket is used to draw water from a well. The water level of the well is 10 feet below the surface of the ground, and the bucket leaks so that if it is raised at the rate of 1 foot per second, it is just half full when it reaches the top. Find the work done in drawing a bucket of water if the bucket weighs 1 pound and a bucket full of water weighs 60 pounds.

12. An empty tank that is a right-circular cone with its vertex down has a base radius of 3 feet, an altitude of 8 feet, and stands on a platform 30 feet above the ground. The tank is filled by pumping water from ground level through a pipe that enters the bottom of the tank. How much work is done?

13. The weight of a body varies inversely as the square of its distance from the center of the earth. A satellite weighs 10 tons on the surface of the earth. Neglecting the atmospheric resistance, compute the work done in propelling the satellite to a height of 1000 miles. (Assume that the radius of the earth is 4000 miles.)

14. A rocket is loaded with 1000 pounds of fuel at the surface of the earth. The fuel is burned at a steady rate with respect to distance and is entirely consumed when the rocket reaches a height of 10 miles. Neglecting the variation in the force of gravity, how much work is required just to lift the rocket's fuel?

This chapter was devoted to explaining *what* an integral is. By definition, an integral $\int_a^b f(x)\,dx$ is a number, which we picture geometrically as the area of the region between the curve $y = f(x)$ and the interval $[a, b]$. Analytically, the integral is a limit of approximating sums, so we can use it to "add up little slices of volume" or "little increments of work," and so on.

The general integral theorems

$$\int_a^b (mf(x) + ng(x))\,dx = m \int_a^b f(x)\,dx + n \int_a^b g(x)\,dx$$

and

$$\int_a^c f(x)\, dx = \int_a^b f(x)\, dx + \int_b^c f(x)\, dx$$

are ready consequences of the fact that lim is a linear operator. The expression

$$f(m) = \frac{1}{b-a} \int_a^b f(x)\, dx$$

for the average value of f on $[a, b]$ comes from the geometric interpretation.

Calculating an integral (a number) frequently requires such numerical methods as Simpson's Rule and such mechanical aids as an electronic computer. But in many special cases the Fundamental Theorem of Calculus

$$\int_a^b f(x)\, dx = F(b) - F(a)$$

reduces integration to subtraction.

Think of this chapter as an introduction to these ideas. A good part of the rest of the book is devoted to amplifying some of the things that were only touched upon here.

Review Problems, Chapter Five

1. Evaluate the following integrals.

(a) $\displaystyle\int_0^1 x^2 t\, dt - \int_0^1 x^2 t\, dx$

(b) $\displaystyle\int_1^2 (\sqrt{2x} + 1/\sqrt{2x})\, dx$

(c) $\displaystyle\int_{-1}^8 \sqrt[3]{|x|}\, dx$

(d) $\displaystyle\int_{-1}^1 \sqrt[3]{4(|x| + x)}\, dx$

(e) $\displaystyle\int_0^1 [\sin(x - 1) - \sin x - \sin 1]\, dx$

(f) $\displaystyle\int_{3\pi/4}^\pi [-2\cos x + \cos(-2x) + \cos^{-2} x]\, dx$

(g) $\displaystyle\int_{\pi/6}^{\pi/3} \left[\frac{d}{dx}\sin^{-2} x - \left(\frac{d}{dx}\sin x\right)^{-2}\right] dx$

(h) $\displaystyle\int_0^{2\pi} (|2\cos x + 1| + [\![2\cos x + 1]\!])\, dx$

2. Determine numbers p and q that make the following integration formulas correct.

   (a) $\displaystyle\int \frac{1}{(1 + x)^2}\, dx = p(1 + x)^q$ $\qquad$ (b) $\displaystyle\int \frac{\sin \sqrt{x}}{\sqrt{x}}\, dx = p \cos \sqrt{x}$

   (c) $\displaystyle\int \sqrt{8x}\, dx = px^q$

   (d) $\displaystyle\int (\sin x + \cos x)\, dx = p \sin x + q \cos x$

   (e) $\displaystyle\int x^{-2} \sec x^{-1} \tan x^{-1}\, dx = p \cos^q x^{-1}$

   (f) $\displaystyle\int \sin 2x\, dx = p \cos^q x$

3. Which member of the pair is the larger, do you think?

   (a) $\displaystyle\int_0^1 x^2\, dx$ or $\displaystyle\int_0^1 2^x\, dx$ $\qquad$ (b) $\displaystyle\int_0^\pi \sin x\, dx$ or $\displaystyle\int_0^\pi \sin^2 x\, dx$

   (c) $\displaystyle\int_0^{\pi/4} \tan x\, dx$ or $\displaystyle\int_0^{\pi/4} \sec x\, dx$ $\qquad$ (d) $\displaystyle\int_{-\pi/3}^{\pi/4} \tan x\, dx$ or 0

   (e) $\displaystyle\int_1^{10} \log x\, dx$ or 3 $\qquad$ (f) $\displaystyle\int_0^1 4^x\, dx$ or 2.5

4. Let R be the region between the parabolas $y = x^2$ and $y = \sqrt{8x}$.
   (a) Find the area of R.
   (b) Find the volume of the solid of revolution that we obtain when we rotate R about the X-axis.
   (c) Find the volume of the solid of revolution that we obtain when we rotate R about the Y-axis.

5. A gumdrop has the shape of a paraboloid of revolution 1 inch high and with base diameter of 1 inch. How much gum is in the drop?

6. A 5-pound bucket, originally containing 100 pounds of water, is hoisted at a steady rate from a 50-foot well. The bucket leaks so fast that it becomes empty just as it gets to the top. How much work is required to get the bucket to the surface?

7. An oil drum 4 feet long and 2 feet in diameter is lying on its side. If the oil is $1\frac{1}{2}$ feet deep at the deepest point, how many cubic feet of oil are in the drum? (Use Simpson's Parabolic Rule with $n = 4$ to evaluate the integral you get.)

8. Suppose that $f(t) = 2 + t - |t - 2|$. Sketch the graph of f to help you solve the following equations for x.

   (a) $\displaystyle\int_0^x f(t)\, dt = 1$ $\qquad$ (b) $\displaystyle\int_0^x f(t)\, dt = 8$

   (c) $\displaystyle\int_x^2 f(t)\, dt = 3$ $\qquad$ (d) $\displaystyle\int_x^2 f(t)\, dt = 4$

(e) $\int_{2}^{x} f(t)\, dt = 8$ (f) $\int_{2}^{x} f(t)\, dt = 0$

(g) $\int_{2}^{x^2} f(t)\, dt = 8$ (h) $\int_{x^2}^{2} f(t)\, dt = 0$

9. Show that

$$1 - x + x^2 - x^3 \le \frac{1}{1 + x} \le 1 - x + x^2 - x^3 + x^4$$

for each number x in the interval $[0, 1]$. What do these inequalities tell us about the number $\int_{0}^{1} \frac{1}{1 + x}\, dx$? Can you improve the accuracy of this approximation by increasing the number of terms in the expressions on the left and right sides of the inequalities? Use a geometric argument to convince yourself that $\int_{0}^{1} \frac{1}{1 + x}\, dx = \int_{1}^{2} \frac{1}{x}\, dx$, and then look at Example 36-3 to find an approximate value of this integral.

10. Use the inequality $[f(x) - g(x)]^2 \ge 0$, together with some general integral theorems, to show that if f and g are integrable on an interval $[a, b]$, then

$$\int_{a}^{b} f(x)^2\, dx + \int_{a}^{b} g(x)^2\, dx \ge 2 \int_{a}^{b} f(x) g(x)\, dx.$$

11. Suppose that $f''(x) > 0$ for each $x \in [a, b]$. Use some ideas from Section 21 to show that the Trapezoidal Rule always yields a number that is larger than $\int_{a}^{b} f(x)\, dx$. What about the Midpoint Rule (Problem 36-11)?

12. (a) Show that $\int_{a}^{b} [f(x) - u]^2\, dx$ is a minimum when the number u is the average value of f in the interval $[a, b]$.

(b) Use this result to show that for any function f that is integrable on $[a, b]$,

$$\left[\int_{a}^{b} f(x)\, dx \right]^2 \le (b - a) \int_{a}^{b} f(x)^2\, dx.$$

(c) Let S_1 be the solid we generate when we rotate the graph of f about the X-axis, and let S_2 be the solid we generate when we rotate the graph of the constant function whose value is the average value of f about the X-axis. Use the result of part (b) to find which of these solids has the larger volume.

13. The generalization of the equation *distance = rate × time* is *distance traveled in the time interval* $[a, b] = \int_{a}^{b} v(t)\, dt$, where $v(t)$ is the velocity at time t. For a freely falling body, $v(t) = 32t$. Show that it travels 720 feet during the fifth through ninth seconds of fall.

14. The equation *income = hourly wage × number of hours worked* generalizes to *income from working during the time interval* $[a, b] = \int_{a}^{b} w(t)\, dt$, where $w(t)$ is

the hourly wage t hours after starting work. For example, the blacksmith Thor was paid $3 + (t - 8)/|t - 8|$ talers per hour. Discuss his wage scale and figure his paycheck for a 12-hour day.

15. The equation *revenue = selling price × number of items sold* generalizes to *revenue from the sale of the a*th *through b*th *item* $= \displaystyle\int_{a}^{b} p(x)\, dx$, where $p(x)$ is the selling price of the xth item. Why would you expect $p'(x)$ to be nonpositive? Suppose that the Indians who sold Manhattan Island charged 2 pounds of gunpowder per acre for the first 100 acres and then reduced their price 1 ounce of powder per acre for each 10 acres sold beyond 100. Show that $p(x) = 32 - \frac{1}{10}(x - 100)$ ounces for $x \geq 100$ and that 125 acres went for about 248 pounds of powder.

Exponential, Logarithmic, Inverse Trigonometric, and Hyperbolic Functions

6

This chapter is pretty much on the theoretical side, which is another way of saying that it gets down to fundamentals. Thus a function *is* a set of pairs of numbers. We choose a number x and pair with it a number y calculated by some rule or other. In this chapter that rule will be "evaluate the following integral."

The first members of the number pairs that make up a function constitute its domain and the second members give its range. If the roles of these sets are reversed—domain becomes range and range becomes domain—we have the *inverse* of the function we started with. We are going to use integration to construct a new logarithmic function and then look at its inverse and the inverses of the trigonometric functions.

Ideas we pick up along the way, as well as a number we already know, go into the making of the hyperbolic functions. In a sense, these latter are hybrids between exponential and trigonometric functions.

43. FUNCTIONS DEFINED BY INTEGRALS

The fundamental process of differentiation produces a derived function f' from a given function f. Now we are going to use the fundamental process of integration to construct a function F from a given function f. Here is an example of what we have in mind.

Example 43-1.   If

$$F(x) = \int_0^x |\sin t|\, dt,$$

find $F(0)$, $F(\tfrac{1}{2}\pi)$, $F(\tfrac{3}{2}\pi)$, and $F(27\pi)$.

Solution.   Let us interpret the functional values of F graphically. In Fig. 43-1 we have drawn the curve $y = |\sin t|$. Graphically speaking, $F(x)$ is the area of the shaded region between this curve and the interval $[0, x]$ of the T-axis. Thus

$$F(0) = \int_0^0 |\sin t|\, dt = 0$$

and

$$F(\tfrac{1}{2}\pi) = \int_0^{\pi/2} |\sin t|\, dt = \int_0^{\pi/2} \sin t\, dt = 1.$$

To calculate $F(\tfrac{3}{2}\pi)$ and $F(27\pi)$, we see from the graph that

$$F(\tfrac{3}{2}\pi) = \int_0^{3\pi/2} |\sin t|\, dt = 3 \int_0^{\pi/2} |\sin t|\, dt = 3F(\tfrac{1}{2}\pi) = 3$$

and

$$F(27\pi) = \int_0^{27\pi} |\sin t|\, dt = 54 \int_0^{\pi/2} |\sin t|\, dt = 54F(\tfrac{1}{2}\pi) = 54.$$

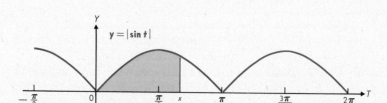

Figure 43-1

In general, if f is a function that is integrable on an interval that contains a point c, then we can define a function F in this interval by means of the equation

(43-1) $$F(x) = \int_c^x f(t)\, dt.$$

We may be forced to evaluate functional values like $F(3)$ and $F(5)$ by approximation methods, such as Simpson's Rule or the Trapezoidal Rule. But Equation 43-1 does assign a number $F(x)$ to each number x in the given interval, and this correspondence defines a function, even though it may be a function whose values are hard to find.

Our function F is determined by the function f and the number c. We are free to choose whatever letters we wish to indicate the variable of integration and to denote a number in the domain of F. To avoid confusion, however, the same letter should not be used to indicate both. Thus, for example, the equations

$$F(u) = \int_0^u |\sin v|\, dv, \qquad F(t) = \int_0^t |\sin u|\, du, \quad \text{and} \quad F(t) = \int_0^t |\sin x|\, dx$$

all define the function F that we discussed in Example 43-1.

To get an idea of the properties of functions that are defined by integrals, let us turn to another example.

Example 43-2. Let f be the greatest integer function (that is, $f(u) = [\![u]\!]$) and let $F(x) = \int_0^x f(u)\, du$. Sketch the graphs of the functions f and F for the interval $[0, 2]$.

Solution.   The graph of the greatest integer function for the interval $[0, 2]$ is shown on the left in Fig. 43-2. The number $F(x)$ is the area of the region under this graph and above the interval $[0, x]$ of the horizontal axis. It is clear that if $0 \le x \le 1$, the area is zero, so that $F(x) = 0$ for $x \in [0, 1]$. If $1 < x \le 2$, $F(x)$ is the area of the shaded rectangular region in Fig. 43-2. This area is obviously $(x - 1) \cdot 1 = x - 1$ square units. Thus $F(x) = x - 1$ if $x \in (1, 2]$. The graph of F appears on the right-hand side of Fig. 43-2.

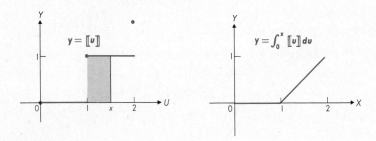

Figure 43-2

The graphs of the functions f and F in the preceding example illustrate two important facts. First, we see that f is not continuous at 1 and at 2; its graph has breaks at those points. On the other hand, the graph of F has no breaks; F is continuous in the interval $[0, 2]$. Thus the operation of integration produces continuous functions, even if we start with discontinuous ones.

Theorem   43-1.   *If f is integrable on an interval I that contains c and if $F(x) = \int_c^x f(t)\, dt$, then F is continuous in I.*

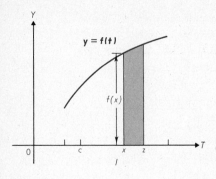

Figure 43-3

It is easy to see graphically why this theorem is true. In symbols, the statement that F is continuous at x is

$$(43\text{-}2) \qquad \lim_{z \to x} F(z) = F(x);$$

that is, $\lim_{z \to x} [F(z) - F(x)] = 0$. If you picture the region in Fig. 43-3 whose area is $F(z)$ and the region whose area is $F(x)$, you will see that the difference $F(z) - F(x)$ is the area of the shaded region (we are tacitly assuming that $z > x$ here; in the contrary case we end up with the same conclusions). Equation 43-2 simply says that this area shrinks to 0 as z approaches x.

Figure 43-2 also shows that the derivative $F'(x)$, the slope of the graph of F, is defined everywhere in the interval $(0, 2)$, except at the number 1. Furthermore, we see that $F'(x) = 0$ when x is between 0 and 1 and $F'(x) = 1$ when x is between 1 and 2. Thus $F'(x) = f(x)$ wherever f is continuous. If you were given the graph on the right-hand side of Fig. 43-2 and were asked to make a graph showing its slope, you would draw the graph on the left-hand side.

This behavior is typical; derivatives of functions defined by integrals are values of the integrand. Our next theorem is a precise statement of this basic fact of calculus.

Theorem 43-2. *Suppose that f is integrable on an interval I that contains a number c and that $F(x) = \displaystyle\int_c^x f(t) \, dt$ for each $x \in I$. If f is continuous at x, then F is differentiable at x, and $F'(x) = f(x)$.*

This theorem, too, has a proof which is motivated by geometric considerations. According to the definition of the derivative, we must show that

$$(43\text{-}3) \qquad \lim_{z \to x} \frac{F(z) - F(x)}{z - x} = f(x).$$

In rough terms, this limit equation says that the approximation

$$F(z) - F(x) \approx f(x)(z - x)$$

should be a good one when z is close to x. We have already observed that $F(z) - F(x)$ is the area of the shaded region in Fig. 43-3, and $f(x)(z - x)$ is the area of the rectangular region of altitude $f(x)$ and base $z - x$. The figure makes it clear that these two regions do have approximately the same area, and when these geometric ideas are translated into the formal terminology of calculus, Equation 43-3 is established.

The essential content of Theorem 43-2 can be summed up in the equation

(43-4) $\dfrac{d}{dx} \displaystyle\int_c^x f(t)\, dt = f(x)$ (if f is continuous at x).

This equation shows us how differentiation "undoes" integration, just as integration "undoes" differentiation. Theorem 43-2 is, in a sense, the converse of the Fundamental Theorem of Calculus (Theorem 37-1). They are very closely related, and there is really no reason to say that one is more "fundamental" than the other. *The Fundamental Theorem tells us that if we have an antiderivative $F(x)$ of $f(x)$, then we can evaluate integrals of f. Theorem 43-2 tells us that if we can evaluate integrals of f, then we can calculate an antiderivative $F(x)$ of $f(x)$.*

Example 43-3.   Find the number x for which the function F defined in the interval $[-1, 1]$ by the equation $F(x) = \displaystyle\int_{-1}^x \tan t\, dt$ takes its minimum value.

Solution.   According to Theorem 43-2, $F'(x) = \tan x$. We see that $F'(x) > 0$ if $x > 0$, and $F'(x) < 0$ if $x < 0$. Therefore, the point $(0, F(0))$ is a minimum point of the graph of F (Theorem 22-2). To find the minimum *value* of F (that is, the number $F(0) = \displaystyle\int_{-1}^0 \tan t\, dt$), we could use some numerical technique, such as Simpson's Rule. We would find that $F(0) \approx -.62$.

Equation 43-4 is a differentiation formula, and a mathematician's first response to a differentiation formula is to incorporate it into the Chain Rule Equation $\dfrac{d}{dx} F(u) = \dfrac{d}{du} F(u) \dfrac{du}{dx}$. Thus we replace $F(u)$ with $\displaystyle\int_c^u f(t)\, dt$ and observe from Equation 43-4 that $\dfrac{d}{du} \displaystyle\int_c^u f(t)\, dt = f(u)$, thereby obtaining the more general equation

(43-5) $\dfrac{d}{dx} \displaystyle\int_c^u f(t)\, dt = f(u) \dfrac{du}{dx}$.

Example 43-4.   Find $\dfrac{d}{dx} \displaystyle\int_3^{x^2} \cos t^2\, dt$.

Solution.

$$\dfrac{d}{dx} \int_3^{x^2} \cos t^2\, dt = \cos x^4 \dfrac{dx^2}{dx} \quad \text{(Equation 43-5 with } u = x^2 \text{ and } f(t) = \cos t^2)$$

$$= 2x \cos x^4 \quad \left(\dfrac{dx^2}{dx} = 2x\right).$$

PROBLEMS 43

1. (a) Sketch the curve $y = |t|$ for $t \in [-2, 2]$.

   (b) If $F(x) = \displaystyle\int_{-2}^{x} |t|\ dt$, find $F(-2)$, $F(-1)$, $F(0)$, $F(1)$, and $F(2)$.

   (c) Use a graphical or other argument to show that $F(x) = 2 - \frac{1}{2}x^2$ if $x \in [-2, 0]$ and $F(x) = 2 + \frac{1}{2}x^2$ if $x \in [0, 2]$.

   (d) Sketch the curve $y = F(x)$.

   (e) How do your curves illustrate Theorems 43-1 and 43-2?

2. (a) Sketch the curve $y = t/|t|$ for $t \in [-2, 0) \cup (0, 2]$.

   (b) If $F(x) = \displaystyle\int_{-2}^{x} \frac{t}{|t|}\ dt$, find the numbers $F(-2)$, $F(-1)$, $F(0)$, $F(1)$, and $F(2)$.

   (c) Sketch the curve $y = F(x)$.

   (d) How do your curves illustrate Theorems 43-1 and 43-2?

   (e) Find $F'(x)$ and $F''(x)$.

   (f) Find a simple formula for $F(x)$.

3. The equation $F(x) = \displaystyle\int_{-3}^{x} \sqrt{9 - t^2}\ dt$ defines a function F. Use the area interpretation of the integral to determine its domain and range. What is $F(0)$? $F'(0)$? $F''(0)$? Show that $F(x) = \frac{9}{2}\pi - F(-x)$ and make a crude sketch of F.

4. Let $F(x) = \displaystyle\int_{0}^{x} \sin^3 t\ dt$.

   (a) Use a graphical argument to find $F(2\pi)$.

   (b) For what numbers does F take maximum values? Minimum values?

   (c) Convince yourself that $0 \le F(x) \le 2$.

   (d) In what intervals is the graph of F concave up? Concave down?

5. Let $f(x)$ be the distance between x and the nearest prime number, for $x \in [0, 5]$. Sketch the graphs of $f, f',$ and F, where $F(x) = \displaystyle\int_{0}^{x} f(t)\ dt$. How does the concavity of the graph of F show up in the graph of f'?

6. Show that $\dfrac{d}{dx} \displaystyle\int_{x}^{c} f(t)\ dt = -f(x)$.

7. Find $\dfrac{dy}{dx}$.

   (a) $y = \displaystyle\int_{3}^{x} (5^t + t^5)\ dt$

   (b) $y = \displaystyle\int_{-7}^{x} \sqrt[4]{|\sin u|}\ du$

   (c) $y = \displaystyle\int_{x}^{3} (1 + r^2)^{-1}\ dr$

   (d) $y = \displaystyle\int_{x}^{2} (3z + 1)^{12}\ dz$

8. Use Equation 43-5 to find $\dfrac{dy}{dx}$.

   (a) $y = \displaystyle\int_{2}^{x^2} t^2\ dt$

   (b) $y = \displaystyle\int_{0}^{\sin x} 2^t\ dt$

   (c) $y = \displaystyle\int_{-x}^{0} f(t)\ dt$

   (d) $y = \displaystyle\int_{3x+2}^{x^2} \sin t^2\ dt$

9. Let x be a number in the interval $[-1, 1]$ and let $F(x)$ be twice the area of the region that is bounded by the semicircle $y = \sqrt{1 - x^2}$ and the line segments joining the origin to the points $(x, \sqrt{1 - x^2})$ and $(1, 0)$.

    (a) Show that $F(x) = x\sqrt{1 - x^2} + 2\displaystyle\int_x^1 \sqrt{1 - t^2}\, dt$.

    (b) Show that F is decreasing in the interval $[-1, 1]$.

    (c) Discuss the concavity of the graph of F and sketch it.

    (d) Let $S(x) = F(x) + F(-x)$. Show that $S'(x) = 0$ for $-1 < x < 1$, and thus S is a constant function. What is its value? Interpret this result geometrically.

10. Let x be a number in the interval $[1, \infty)$ and let $G(x)$ be twice the area of the region that is bounded by the hyperbola $x^2 - y^2 = 1$ and the line segments joining the origin to the points $(x, \sqrt{x^2 - 1})$ and $(1, 0)$. (Notice the analogy to the function F in the preceding problem.)

    (a) Show that $G(x) = x\sqrt{x^2 - 1} - 2\displaystyle\int_1^x \sqrt{t^2 - 1}\, dt$.

    (b) Show that G is increasing in the interval $[1, \infty)$.

    (c) Discuss the concavity of the graph of G.

11. For what numbers does F take maximum and minimum values if

$$F(x) = \int_0^x (2^t + 2^{t-1} - 3)\, dt\,?$$

12. Sketch the graph of the equation $y = \displaystyle\int_0^x (1 + 2[\![\cos t]\!])\, dt$ for $x \in [0, 2\pi]$. Find the maximum and minimum points of the graph.

13. Show that

$$\frac{d}{dx} \int_c^x f(t)\, dt = \int_c^x \frac{d}{dt} f(t)\, dt + f(c).$$

14. Let $F(x) = \displaystyle\int_{-x}^x f(t)\, dt$, where f is a continuous, odd function [that is, $f(-t) = -f(t)$]. Show that $F'(x) = 0$ for each number x. Then use Theorem 19-5 to see that $F(x) = 0$ for each x.

15. Let $F(x) = \displaystyle\int_{-x}^x f(t)\, dt$ and $G(x) = 2\displaystyle\int_0^x f(t)\, dt$, where f is a continuous, even function [that is, $f(-t) = f(t)$]. Show that $F'(x) = G'(x)$. Does it follow that $F(x) = G(x)$?

16. Suppose that the integration formula $\displaystyle\int f(x)\, dx = F(x)$ is valid in an interval I in which f is continuous. Show that $\dfrac{d}{dx} F(x) = f(x)$ for each $x \in I$. Could we make this assertion if we did not assume that f is continuous? (See Problem 38-11.)

44. THE FUNCTION ln

In the last section we discussed the general problem of defining functions by means of integrals; now we will turn our attention to an important specific example. Instead of denoting our new function by a single letter, such as F, we will use the two-lettered symbol ln and call the function, for the moment, the "ell-en" function. As we develop its properties, the function ln will look more and more familiar, and, in fact, it will turn out to be a logarithmic function.

Definition 44-1. *The function ln is defined by the equation*

$$(44\text{-}1) \qquad\qquad \ln x = \int_1^x \frac{1}{t}\, dt.$$

Since $1/t$ is unbounded in any interval that contains 0, we see that the integral that defines ln x exists only if x is a point of the interval $(0, \infty)$. *In other words, the domain of the function* ln *is the set of positive numbers.* We will find that its range is the set of all real numbers. Furthermore, it follows from Theorem 43-1 that ln is continuous at each point of the interval $(0, \infty)$.

Let us see what we can say about the values of ln. Since $\int_1^1 \frac{1}{t}\, dt = 0$, we have ln $1 = 0$. If $x > 1$, then $\int_1^x \frac{1}{t}\, dt > 0$, and the integral is the area of the region between the curve $y = 1/t$ and the interval $[1, x]$. On the other hand, if $x \in (0, 1)$, then $\int_1^x \frac{1}{t}\, dt = -\int_x^1 \frac{1}{t}\, dt$, and we see that the integral defining ln x is negative. This negative number is the negative of the area of the shaded region that is shown in Fig. 44-1. Thus our function ln has the following properties:

$$\ln x < 0 \quad \text{if} \quad x \in (0, 1),$$
$$(44\text{-}2) \quad \ln x = 0 \quad \text{if} \quad x = 1,$$
$$\ln x > 0 \quad \text{if} \quad x \in (1, \infty).$$

We could use Simpson's Parabolic Rule, the Trapezoidal Rule, or some other method of numerical integration to calculate values of ln. For example, we have already found in Example 36-3 that ln $2 \approx .6933$. The function ln is an important function, as we shall soon see, and its values have been tabulated (Table II).

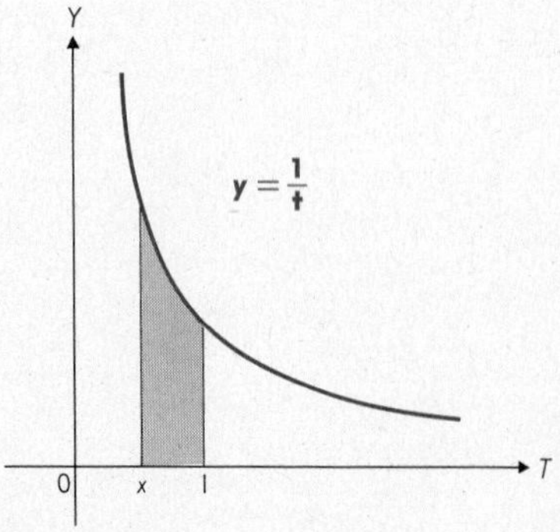

Figure 44-1

The equation $y = 1/x$ defines a function that is continuous in the interval $(0, \infty)$. Therefore, according to Theorem 43-2,

$$\frac{d}{dx} \int_1^x \frac{1}{t} \, dt = \frac{1}{x},$$

and so we have rather easily derived the differentiation formula

(44-3) $$\frac{d \ln x}{dx} = \frac{1}{x}.$$

It follows from Equation 44-3 that

$$\frac{d^2 \ln x}{dx^2} = \frac{dx^{-1}}{dx} = -x^{-2}.$$

Thus $\dfrac{d \ln x}{dx} > 0$ and $\dfrac{d^2 \ln x}{dx^2} < 0$, so ln is an *increasing* function and its graph is *concave down* in the interval $(0, \infty)$. Using this information and some values from Table II, we can plot the graph of ln shown in Fig. 44-2.

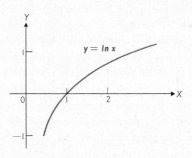

Figure 44-2

To make Formula 44-3 apply more generally, we incorporate it into the Chain Rule Equation $\dfrac{df(u)}{dx} = \dfrac{df(u)}{du} \dfrac{du}{dx}$. When we replace $f(u)$ with ln u, this equation becomes $\dfrac{d \ln u}{dx} = \dfrac{d \ln u}{du} \dfrac{du}{dx}$, and from Equation 44-3 we see that $\dfrac{d \ln u}{du} = \dfrac{1}{u}$. Therefore if $u = g(x)$, we have the differentiation formula

(44-4) $$\frac{d \ln u}{dx} = \frac{1}{u} \frac{du}{dx}.$$

Example 44-1. Let a be a positive number and r a rational number. Show that

(44-5)
$$\frac{d \ln ax}{dx} = \frac{1}{x}$$

and

(44-6)
$$\frac{d \ln x^r}{dx} = \frac{r}{x}.$$

Solution. To verify Equation 44-5, we simply replace u with ax in Equation 44-4:

$$\frac{d \ln ax}{dx} = \frac{1}{ax} \frac{dax}{dx} = \frac{1}{ax} \cdot a = \frac{1}{x}.$$

Similarly, to verify Equation 44-6, we replace u with x^r:

$$\frac{d \ln x^r}{dx} = \frac{1}{x^r} \frac{dx^r}{dx} = \frac{1}{x^r} rx^{r-1} = \frac{r}{x}.$$

We can use the results of the preceding example to show that our function ln behaves like a logarithmic function.

Theorem 44-1. *If a and b are positive numbers, and r is a rational number, then*

(44-7)
$$\ln ab = \ln a + \ln b$$

and

(44-8)
$$\ln a^r = r \ln a.$$

PROOF

According to Definition 44-1, $\ln b = \int_1^b \frac{1}{x} dx$. Equation 44-5 tells us that $\ln ax$ is an antiderivative of $\frac{1}{x}$, so we can use the Fundamental Theorem of Calculus to express this integral as

$$\int_1^b \frac{1}{x} dx = \ln ax \Big|_1^b = \ln ab - \ln a.$$

Thus we have shown that $\ln b = \ln ab - \ln a$, which is equivalent to Equation 44-7.

Similarly, to verify Equation 44-8, we use Definition 44-1 to write

$$r \ln a = r \int_1^a \frac{1}{x} \, dx = \int_1^a \frac{r}{x} \, dx.$$

Equation 44-6 tells us that $\ln x^r$ is an antiderivative of $\dfrac{r}{x}$, so we can use the Fundamental Theorem to express this integral as

$$\int_1^a \frac{r}{x} \, dx = \ln x^r \bigg|_1^a = \ln a^r - \ln 1.$$

But $\ln 1 = 0$, so our integral equals $\ln a^r$, and Equation 44-8 is established.

We can use Theorem 44-1 to find values of ln outside the range of Table II.

Example 44-2.   Find ln .045.

Solution.

$$\ln .045 = \ln (4.5)(10)^{-2} = \ln 4.5 + \ln 10^{-2} \qquad \text{(Equation 44-7)}$$
$$= \ln 4.5 - 2 \ln 10 \qquad \text{(Equation 44-8)}$$
$$= 1.5041 - 2(2.3026) \qquad \text{(Table II)}$$
$$= 1.5041 - 4.6052 = -3.1011.$$

Example 44-3.   Show that the range of ln is R^1.

Solution.   For each positive integer n, $\ln 2^{2n} = 2n \ln 2$ and $\ln 2^{-2n} = -2n \ln 2$; that is, $2n \ln 2$ and $-2n \ln 2$ belong to the range of ln. Therefore, according to Theorem 10-1, every point between them belongs to the range. So each interval $(-2n \ln 2, 2n \ln 2)$ is a subset of the range of ln. Although the approximation $\ln 2 \approx .697$ that we obtained in Example 36-2 is certainly not correct in the third decimal place, it is obviously true that $\ln 2 > \frac{1}{2}$. It follows that $(-n, n)$ is a subinterval of $(-2n \ln 2, 2n \ln 2)$ and hence is contained in the range of ln. Now we finally observe that if y is any real number, then we can choose n so large that $y \in (-n, n)$. Thus y belongs to the range of ln, and our assertion is verified.

PROBLEMS   44

1.   Compute ln 3 from the definition of ln, using $n = 4$ and
      (a) the Trapezoidal Rule.
      (b) Simpson's Parabolic Rule.
      (c) the Midpoint Rule (Problem 36-11).
      Check your results against Table II.

2. Use Theorem 44-1 and the values ln 2 = .69 and ln 3 = 1.10 to compute the following numbers.
   (a) ln 6 (b) ln ½ (c) ln 27 (d) ln 108 (e) ln $\sqrt[3]{12}$ (f) ln $\sqrt{6}$

3. Find $\dfrac{dy}{dx}$.
   (a) $y = \ln \sqrt{x} - \sqrt{\ln x}$ (b) $y = \ln 3x - 3 \ln x$
   (c) $y = \ln 3 - \ln x$ (d) $y = \ln (3 - x)$
   (e) $y = \ln \sin x - \sin \ln x$ (f) $y = \ln (1/x) - 1/\ln x$

4. Find $f'(x)$.
   (a) $f(x) = \ln (1/x^2)$ (b) $f(x) = (\ln x)^2$
   (c) $f(x) = \ln (\ln x)$ (d) $f(x) = \ln \sin x^2$
   (e) $f(x) = \ln \sin^2 x$ (f) $f(x) = \ln \sec^2 x$

5. Find $\dfrac{dy}{dx}$.
   (a) $y = \ln |x| - |\ln x|$ (b) $y = x \ln |x| - x$
   (c) $y = \ln |\sec x + \tan x|$ (d) $y = \ln (|\sec x| + |\tan x|)$

6. Use Theorem 44-1 to show that for positive numbers A and B, $\ln (A/B) = \ln A - \ln B$.

7. Use Table II, the rules of logarithms (Theorem 44-1), and linear interpolation to calculate the following numbers.
   (a) ln .541 (b) ln 541 (c) $\dfrac{\sqrt[3]{53.1}}{135\sqrt{.315}}$ (d) $\dfrac{1}{\sqrt{\ln 351}}$

8. Find the average value of the function q on the interval $[5, 10]$ if $q(x) = 1/x$.

9. Show that $y = 3 \cos \ln x - 2 \sin \ln x$ satisfies the differential equation $x^2 y'' + xy' + y = 0$. What are y and y' when $x = 1$?

10. If we use linear interpolation to find logarithms of numbers that lie between two numbers listed in Table II, do we get a number that is too large or too small? If we use Formula 21-2 as an interpolation formula, do we get a number that is too large or too small?

11. Use a graphical argument to show that $\ln \frac{1}{3} = -\ln 3$.

12. Let $G(x) = \displaystyle\int_1^x \ln t\, dt$ and use Fig. 44-2 to estimate $G(2)$ and $G(\frac{1}{2})$. Check your result by using a numerical integration technique and Table II.

13. In each case, determine how the functions defined by the following equations are related to the function ln.
    (a) $F(x) = \displaystyle\int_2^x \frac{1}{t}\, dt$ (b) $G(x) = \displaystyle\int_{1/x}^1 \frac{1}{t}\, dt$
    (c) $H(x) = \displaystyle\int_2^x \frac{1}{t-1}\, dt$ (d) $K(x) = \displaystyle\int_{x^4}^{x^5} \frac{1}{t}\, dt$.

14. Show graphically that (see Problem 33-8)
    $$\tfrac{1}{2} + \tfrac{1}{3} + \tfrac{1}{4} + \tfrac{1}{5} < \ln 5 < 1 + \tfrac{1}{2} + \tfrac{1}{3} + \tfrac{1}{4}.$$

15.   Find the following numbers, using Table II as necessary.

   (a) The area of the region that is bounded by the curves $y = 1/x$, $y = 0$, $x = -10$, and $x = -7$.

   (b) The volume of the solid that is obtained by rotating about the X-axis the region bounded by the curves $y = x^{-1/2}$, $x = 1$, $x = 4$, and $y = 0$.

16.   Show that if $x > -1$, then

$$\frac{x}{x+1} \leq \ln(1 + x) \leq x.$$

Use Table II to check this inequality when $x = .1$.

45.   INVERSE FUNCTIONS

The heart of a function is the rule that assigns to each number x in its domain exactly one number y. For example, the equation $y = \sqrt{x-1}$ defines a function whose domain is the interval $[1, \infty)$ and whose range is $[0, \infty)$. If we set $x = 5$, we obtain $y = 2$. For this function, we could also start by setting $y = 2$ and solve the equation $2 = \sqrt{x-1}$ to obtain $x = 5$. In fact, if y is any number of the interval $[0, \infty)$, we can solve the equation $y = \sqrt{x-1}$ for the number $x = y^2 + 1$ in the interval $[1, \infty)$. Thus by interchanging the roles of domain and range and using the rule of correspondence backward, we get a new function, the *inverse* of the one we started with. In this section we want to discuss the inverses of functions in general, and later we will apply our results to the inverses of specific functions, such as the function ln and the trigonometric functions.

Suppose that f is a function whose domain and range are sets of numbers. The equation $y = f(x)$ determines the number y that corresponds to a given number x, and now we ask whether it also determines x when y is given. If so, we have a rule that pairs with the numbers of the range of f the numbers of its domain. We therefore have all the ingredients of a function, and since we obtained this function from f by reading its rule of correspondence backward, we naturally call it the inverse of f, in accordance with the following definition.

Definition   45-1.   *If for each number y in the range of a function f there is exactly one number x such that $y = f(x)$, then f has an* **inverse** *function f^{-1}. The domain of f^{-1} is the range of f, and the range of f^{-1} is the domain of f. The number that corresponds to a given number y in the range of f is the number x that satisfies the equation $y = f(x)$. Thus*

$$x = f^{-1}(y) \quad \text{if and only if} \quad y = f(x).$$

If we view our given function f as a collection of pairs of numbers, we construct the set of pairs that make up f^{-1} simply by interchanging the members of each pair of f:

$$(45\text{-}1) \qquad f^{-1} = \{(y, x) \mid (x, y) \in f\}.$$

To say that for each number y in the range of f, the equation $y = f(x)$ has *just one* solution x means that two different pairs in the set f^{-1} do not have the same first member, which is our criterion that a subset of R^2 should constitute a function.

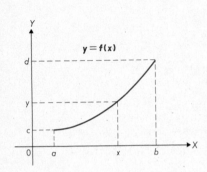

Figure 45-1

Not every function has an inverse. For example, the equation $y = (x - 1)^2$ pairs with the number $x = 5$ the number $y = 16$. But the "reverse" equation $16 = (x - 1)^2$ merely tells us that x could be 5 or it could be $- 3$.

The question of whether or not a given function has an inverse has a simple graphical answer. Figure 45-1 shows the graph of a function f whose domain is the interval (a, b) and whose range is (c, d). As with the graph of any function, a line parallel to the Y-axis and intersecting the X-axis in a point of the domain intersects the graph of f in just one point, which is a graphical expression of the statement that the number x determines the pair (x, y). For a function with an inverse, the number y also determines this pair. Thus lines parallel to the X-axis and intersecting the Y-axis in a point of the range also intersect the graph of f in just one point. *For a function without an inverse, there is at least one line parallel to the X-axis and intersecting the graph in more than one point.* Horizontal lines do not intersect the graph in Fig. 45-1 more than once, so our function f has an inverse. Notice that f is an increasing function in the interval (a, b). Every increasing function has an inverse. (Why?) So does every decreasing function.

Example 45-1. Find the equation that defines f^{-1} if $f(x) = 3x + 2$.

Solution. It is clear from its graph (a line with positive slope) that f is an increasing function and therefore has an inverse. To find the value of f^{-1} at a given number y, we must solve the equation $y = f(x)$ (that is, $y = 3x + 2$) for x. Thus we have $x = \frac{1}{3}y - \frac{2}{3}$, or, in other words, $f^{-1}(y) = \frac{1}{3}y - \frac{2}{3}$. Of course, the letter that we use to denote a number in the domain of the inverse function is of no importance whatsoever, so this last equation can be rewritten $f^{-1}(u) = \frac{1}{3}u - \frac{2}{3}$, or $f^{-1}(s) = \frac{1}{3}s - \frac{2}{3}$, or even $f^{-1}(x) = \frac{1}{3}x - \frac{2}{3}$, and it will still define the same function f^{-1}.

Example 45-2. Does the sine function have an inverse?

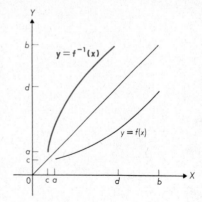

Figure 45-2

Solution. No. A horizontal line that intersects the interval $[-1, 1]$ of the Y-axis intersects the sine curve infinitely often, not just once. Later we will introduce a restriction of the sine function that does have an inverse.

Equation 45-1 says that (u, v) is a point of the graph of f^{-1} if and only if (v, u) is a point of the graph of f. It is easy to see that the points (u, v) and (v, u) are symmetric with respect to the line $y = x$, and so *the graph of f^{-1} is obtained by reflecting the graph of f about the line that bisects the first and third quadrants.* Figure 45-2 illustrates what we mean for the function whose graph appears in Fig. 45-1. Notice that if we reflect the graph of f^{-1} about the line $y = x$, we come back to the graph of f, which is a geometric way of saying that *the inverse of f^{-1} is the original function f.*

According to Definition 45-1, the equation $v = f(u)$ is equivalent to $u = f^{-1}(v)$. By replacing v in the second equation with $f(u)$ from the first, we obtain the identity

(45-2) $$u = f^{-1}(f(u)),$$

which is valid for each point u in the domain of f. The pair of equations $v = f(u)$ and $u = f^{-1}(v)$ also yields the identity

(45-3) $$v = f(f^{-1}(v)),$$

which is valid for each point v in the range of f (the domain of f^{-1}).

Example 45-3. Show how Equations 45-2 and 45-3 apply to the function of Example 45-1.

Solution. Since
$$f^{-1}(x) = \tfrac{1}{3}x - \tfrac{2}{3}$$
it follows that
$$f^{-1}(f(u)) = \tfrac{1}{3}f(u) - \tfrac{2}{3} = \tfrac{1}{3}(3u + 2) - \tfrac{2}{3} = u.$$
Similarly,
$$f(f^{-1}(v)) = 3f^{-1}(v) + 2 = 3(\tfrac{1}{3}v - \tfrac{2}{3}) + 2 = v.$$

We pointed out earlier that if a function f is continuous in an interval I, then its graph has no breaks. Reflecting a curve about a line cannot introduce breaks into a graph that doesn't already have them, so it seems reasonable to suppose that if f is continuous in I, then f^{-1} is continuous in $f(I)$. Similarly, if a graph is smooth enough so that we can draw a tangent at each point, then we should also be able to draw tangents to its reflection about a line. This argument suggests that the inverse of a differentiable function is also differentiable (except at points where $f'(x) = 0$). The geometric reasoning we have just used is not, of course, a mathematical proof of the conclusions we drew from it. It is relatively easy, but quite tedious, to turn this geometric argument into a proof that continuity of f implies continuity of f^{-1}. Once we establish continuity of f^{-1}, an argument very similar to the proof of the Chain Rule can be used to prove that f^{-1} is differentiable. Let us collect the essential results as a theorem whose precise proof we will omit.

Theorem   45-1. *If f is differentiable in an interval I and if $f'(x) \neq 0$ for each $x \in I$, then f^{-1} exists and is a differentiable function.*

Our theorem tells us that the derivative of $f^{-1}(x)$ exists; it doesn't tell us how to find it. Instead of calculating $\dfrac{dy}{dx}$ directly from the equation $y = f^{-1}(x)$, let us apply implicit differentiation to the equivalent equation $x = f(y)$. When we differentiate both sides of this equation with respect to x, using the Chain Rule, we obtain the equation

$$1 = \frac{df(y)}{dx} = \frac{df(y)}{dy}\frac{dy}{dx} = f'(y)\frac{dy}{dx}$$

Then we solve for $\dfrac{dy}{dx}$ and obtain an important result: *If $y = f^{-1}(x)$, then*

(45-4) $$\frac{dy}{dx} = \frac{1}{f'(y)}.$$

It is frequently helpful to have this equation written in different, but equivalent, forms. So let us replace y with $f^{-1}(x)$ to obtain the equation

(45-5) $$\frac{d}{dx}f^{-1}(x) = \frac{1}{f'(f^{-1}(x))}.$$

Also, since $x = f(y)$, we see that $\dfrac{dx}{dy} = f'(y)$, and Equation 45-4 can be written in the symmetric form

(45-6) $$\frac{dy}{dx} = 1 \left/ \frac{dx}{dy} \right. .$$

Of course, our practice of writing derivatives as quotients makes this equation look obvious.

Example 45-4.   Show that Equation 45-5 yields the correct answer when $f(x) = x^3$.

Solution.   The inverse of the cube function is the cube root function, so $f^{-1}(x) = x^{1/3}$. Therefore,

$$\frac{d}{dx} f^{-1}(x) = \frac{d}{dx} x^{1/3} = \tfrac{1}{3} x^{-2/3}.$$

On the other hand, $f'(x) = 3x^2$, and when we substitute in Equation 45-5, we obtain

$$\frac{d}{dx} f^{-1}(x) = \frac{1}{3(f^{-1}(x))^2} = \frac{1}{3(x^{1/3})^2} = \tfrac{1}{3} x^{-2/3},$$

as before.

Example 45-5.   Let $F(x) = \displaystyle\int_0^x \frac{1}{1 + t^2} \, dt$ and show that $y = F^{-1}(x)$ satisfies the differential equation $y' = 1 + y^2$.

Solution.   According to Theorem 43-2, $F'(x) = 1/(1 + x^2)$. Thus we see that F is an increasing function in any interval and therefore has an inverse. If we set $y = F^{-1}(x)$, we find from Equation 45-4 that

$$\frac{dy}{dx} = \frac{1}{F'(y)} = \frac{1}{1/(1 + y^2)} = 1 + y^2,$$

as we were to show.

PROBLEMS   45

1.   Find f^{-1} if it exists and verify Equations 45-2 and 45-3.
   (a) $f(x) = 2x - 1$ (b) $f(x) = 3 - 4x$
   (c) $f(x) = x^2 + 1$ (d) $f(x) = x^3 + 1$

2.   Find the equation that defines f^{-1} and show how Equations 45-2 and 45-3 apply.

   (a) $f(x) = x^3 - 1$ (b) $f(x) = \dfrac{2 - x}{3 + x}$

   (c) $f(x) = \dfrac{4 - 4x}{5x + 4}$ (d) $f(x) = \dfrac{8}{x^3 + 1}$

3.   Sketch the graphs of f and f^{-1} if $f(x) = x|x|$. Does $f'(x)$ exist for all x? Does $\dfrac{d}{dx} f^{-1}(x)$ exist for all x?

4.   In each of the following examples, show that $f = f^{-1}$.

   (a) $f(x) = x$ (b) $f(x) = b - x$ (c) $f(x) = \dfrac{b}{x}$

   (d) $f(x) = \dfrac{x + b}{cx - 1}$

5. Determine whether or not f has an inverse, and if it does, determine the domain of the inverse function.

   (a) $f(x) = x^3 + 2x - 1$ (b) $f(x) = x^3 - 2x - 1$

   (c) $f(x) = x^5 + 2x^3 - 23$ (d) $f(x) = x^4 - 2x^3 + 23$

   (e) $f(x) = 3|x| + x$ (f) $f(x) = \dfrac{x}{1 + |x|}$

   (g) $f(x) = \displaystyle\int_0^x (1 + \sin^2 t)\, dt$ (h) $f(x) = \displaystyle\int_0^x (1 + |\cos t|)\, dt$

6. Use implicit differentiation to find $\dfrac{dy}{dx}$ and $\dfrac{dx}{dy}$. How do your results square with Equation 45-6?

   (a) $x^2 + y^2 = 1$ (b) $xy^3 + x^2 = 1$

   (c) $\sin x^2 y + y^2 = x$ (d) $y^2 + \sqrt{xy} = \cos x^2$

7. Without using calculus, show that if we interchange x and y in the equation $y = mx + b$, we obtain the equation of a line whose slope is the reciprocal of the slope of our given line. How is this fact related to Equation 45-6?

8. Under what conditions on a, b, c, and d can you be sure that a function f defined by an equation of the form $f(x) = ax^3 + bx^2 + cx + d$ has an inverse?

9. Show that if $F(x) = \displaystyle\int_c^x f(t)\, dt$, then $y = F^{-1}(x)$ satisfies the differential equation $y' = 1/f(y)$ and the initial condition $y = c$ when $x = 0$. Use this fact to find y if it satisfies the given differential equation and the initial condition $y = 1$ when $x = 0$.

   (a) $y' = \frac{1}{3}y^{-2}$ (b) $y' = y^4$

10. (a) Let F be the function defined in Problem 43-9. Show that $y = F^{-1}(x)$ satisfies the differential equation $y'^2 + y^2 = 1$.

    (b) Let G be the function defined in Problem 43-10. Show that $y = G^{-1}(x)$ satisfies the differential equation $y'^2 - y^2 = -1$.

11. Suppose that f and g have inverses and that $p(x) = f(g(x))$. Convince yourself that $p^{-1}(x) = g^{-1}(f^{-1}(x))$.

12. Suppose that f^{-1} exists and that $F(x) = \displaystyle\int_0^x f(t)\, dt$. Show that

$$\frac{d}{dx}\left(xf^{-1}(x) - F(f^{-1}(x))\right) = f^{-1}(x).$$

13. The Chain Rule Equation reads $\dfrac{dz}{dx} = \dfrac{dz}{dy}\dfrac{dy}{dx}$. If we replace z with $\dfrac{dy}{dx}$, we have $\dfrac{d^2y}{dx^2} = \dfrac{d}{dy}\left(\dfrac{dy}{dx}\right)\dfrac{dy}{dx}$, and in this equation we may, according to Equation 45-6,

    replace $\dfrac{dy}{dx}$ with $1\left/\dfrac{dx}{dy}\right.$ and obtain $\dfrac{d^2y}{dx^2} = \dfrac{d}{dy}\left(1\left/\dfrac{dx}{dy}\right)\dfrac{dy}{dx}\right.$. Continue this computation and obtain the formula

$$\frac{d^2y}{dx^2} = -\frac{d^2x}{dy^2}\left/\left(\frac{dx}{dy}\right)^3\right..$$

14. Suppose that $F(x) = \displaystyle\int_0^x \dfrac{1}{\sqrt{1 - t^2}}\, dt$. Use the equation we developed in the problem above to show that $y = F^{-1}(x)$ satisfies the differential equation $y'' + y = 0$. (In Section 51 we will see that $F^{-1}(x) = \sin x$.)

Now we will apply our knowledge of inverse functions to the function ln that we defined in Section 44 by the equation

$$\ln x = \int_1^x \frac{1}{t}\, dt \qquad \text{for } x > 0.$$

Since $\dfrac{d}{dx} \ln x = 1/x > 0$, ln is an increasing function, and hence it has an inverse. We denote this inverse function by the three-lettered symbol **exp**. (The letters exp form an abbreviation of the word "exponent," and we will soon see why this choice is appropriate.) Thus

(46-1) $y = \ln x$ if and only if $x = \exp y.$

From our knowledge of inverse functions in general, we see that not only is exp the inverse of ln, but ln is also the inverse of exp.

We can find some values of exp by reading Table II "backward"—that is, from right to left. For example, since $\ln 8 = 2.0794$, $\exp 2.0794 = 8$. The function exp is so important, however, that it is tabulated directly. You will shortly find that Table III gives values of exp.

We need only reflect the graph of the function ln (the black curve) about the line that bisects the first and third quadrants to obtain the curve $y = \exp x$ that is shown in Fig. 46-1. From the graph of exp we see that

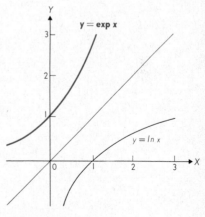

Figure 46-1

$$0 < \exp x < 1 \quad \text{if} \quad x < 0,$$

$$\exp x = 1 \quad \text{if} \quad x = 0,$$

$$\exp x > 1 \quad \text{if} \quad x > 0.$$

These relations correspond to Relations 44-2 for ln. Just as you should know how to sketch the graphs of the trigonometric functions, so should you know what the graphs of ln and exp look like.

We have already shown that ln has the basic logarithmic properties

(46-2) $\ln ab = \ln a + \ln b$

and

(46-3) $\ln a^r = r \ln a.$

Furthermore, since exp and ln are inverses of each other, we have (see Equations 45-2 and 45-3)

(46-4) $$\exp(\ln u) = u$$

for any positive number u and

(46-5) $$\ln(\exp v) = v$$

for any number v. These equations enable us to express the logarithmic properties of ln as exponential properties of exp.

Theorem   46-1. *If x and y are any two numbers and if r is a rational number, then*

(46-6) $$\exp(x + y) = (\exp x)(\exp y)$$

and

(46-7) $$\exp rx = (\exp x)^r.$$

PROOF

We will verify Equation 46-6 and leave the verification of Equation 46-7 for the problems:

$$\exp(x + y) = \exp\left[\ln(\exp x) + \ln(\exp y)\right] \qquad \text{(Eq. 46-5 applied twice, once with } v = x \text{ and once with } v = y)$$

$$= \exp\left[\ln(\exp x \cdot \exp y)\right] \qquad \text{(Eq. 46-2)}$$

$$= (\exp x)(\exp y) \qquad \text{(Eq. 46-4 with } u = \exp x \cdot \exp y).$$

Equations 46-6 and 46-7 tell us that our function exp shares certain properties with exponential functions. For instance, if $f(x) = 2^x$, then $f(x + y) = f(x) \cdot f(y)$ and $f(rx) = [f(x)]^r$ so that f and exp both satisfy the same functional equations (Equations 46-6 and 46-7). The fact that our function exp satisfies functional equations that are also satisfied by exponential functions makes us suspect that exp is itself an exponential function. We shall now see that it is.

If we set $x = 1$ in Equation 46-7, we obtain

(46-8) $$\exp r = (\exp 1)^r.$$

Therefore the value of exp at the rational number r is the rth power of the number exp 1. This number is so important in mathematics that we use a particular letter for it, just as we use the letter π for the number $3.14159\ldots$.

Definition   46-1.
$$\exp 1 = e.$$

From the graph in Fig. 46-1 it appears that the number e is slightly smaller than 3. It can be shown that to five decimal places,

$$e = 2.71828\ldots.$$

Now that we have agreed to write e in place of exp 1, Equation 46-8 becomes

(46-9) $\exp r = e^r.$

Thus the function exp and the exponential function with base e take the same values at the rational numbers, and we are left with the question of whether or not the numbers exp a and e^a are the same for each irrational number a, too. When attempting to answer this question, we discover, as we pointed out in Section 15, that we have never defined what we mean by an irrational power of a number. One definition suggests itself now. If we want e^a to be equal to exp a, we may *define e^a to be equal to exp a for each irrational number a.*

Then we see that exp is an exponential function whose domain is the set of all real numbers. For if x is any real number, then Equation 46-9 tells us that exp $x = e^x$ if x is a rational number, and our definition of irrational powers of e tells us that exp $x = e^x$ if x is irrational. Therefore the equation

(46-10) $\exp x = e^x$

is true for every real number x. This equation allows us to write Equations 46-6 and 46-7 as

$$e^{x+y} = e^x e^y \quad \text{and} \quad e^{rx} = (e^x)^r.$$

These equations are two of the basic laws of exponents, so our definition of irrational powers of e does not violate these two fundamental laws.

According to Equation 46-10, exp $y = e^y$, so Equations 46-1 can be written

$$y = \ln x \quad \text{if and only if} \quad x = e^y.$$

Let us recall the definition of the logarithm of a positive number x to a positive base $b \neq 1$:

$$y = \log_b x \quad \text{if and only if} \quad x = b^y.$$

This definition assumes that we know what the symbol b^y means for each real number y, irrational as well as rational. We will take up this point in Section 48. If we set $b = e$, however, we have just defined what we mean by the symbol e^y for every real number y, and so we have the statement

$$y = \log_e x \quad \text{if and only if} \quad x = e^y.$$

Thus we see that ln x and $\log_e x$ are the same number; namely, the solution of the equation $x = e^y$. Hence,

(46-11) $$\ln x = \log_e x.$$

From now on we will call **ln** the **logarithmic function** (with base e) and **exp** the **exponential function** (with base e). We shall continue to use the notation ln x, but many authors simply write log x for $\log_e x$. The number ln x is called the **natural logarithm** of x, and Table II is a table of natural logarithms. We will use the notation e^x more often than exp x. Table III contains some values of the exponential function with base e.

In Section 44 we learned how to differentiate the logarithm, and now we will use that differentiation formula to find $\dfrac{de^x}{dx}$. We differentiate both sides of the equation $x = \ln y$ with respect to x, using Equation 44-4 (with $u = y$) to find $\dfrac{d \ln y}{dx}$:

$$1 = \frac{d \ln y}{dx} = \frac{1}{y}\frac{dy}{dx}.$$

Therefore $\dfrac{dy}{dx} = y$, and since the equation $x = \ln y$ is equivalent to the equation $y = e^x$, we have derived the specific differentiation formula

(46-12) $$\frac{de^x}{dx} = e^x.$$

Formula 46-12 says that e^x is "immune" to differentiation. We incorporate this formula into the Chain Rule Equation $\dfrac{df(u)}{dx} = \dfrac{df(u)}{du}\dfrac{du}{dx}$ by setting $f(u) = e^u$ (where we suppose that $u = g(x)$): $\dfrac{de^u}{dx} = \dfrac{de^u}{du}\dfrac{du}{dx}$. Since $\dfrac{de^u}{du} = e^u$,

(46-13) $$\frac{de^u}{dx} = e^u \frac{du}{dx}.$$

Example 46-1. Find $\dfrac{d}{dx} e^{\sqrt{x}}$.

Solution. According to Equation 46-13, with $u = \sqrt{x}$,

$$\frac{d}{dx} e^{\sqrt{x}} = e^{\sqrt{x}} \frac{d}{dx} \sqrt{x} = \frac{e^{\sqrt{x}}}{2\sqrt{x}}.$$

In Example 14-2 we met the differential equation $\dfrac{dP}{dt} = .02P$ and were unable to solve it. Now we can see what the solution looks like. It is an exponential function.

Example 46-2. Show that $P = P_0 e^{.02t}$ satisfies the differential equation $\dfrac{dP}{dt} = .02P$ and the initial condition $P = P_0$ when $t = 0$.

Solution. We simply substitute:

$$\frac{dP}{dt} = \frac{d}{dt} P_0 e^{.02t} = P_0 e^{.02t} \frac{d}{dt} .02t$$

$$= .02 P_0 e^{.02t} = .02P.$$

When $t = 0$, $P = P_0 e^{.02 \times 0} = P_0 e^0 = P_0 \cdot 1 = P_0$, and our verification is complete.

Example 46-3. The curve $y = e^{-x^2}$ is important in probability theory. Discuss its concavity.

Solution. Since

$$y' = \frac{d}{dx} e^{-x^2} = e^{-x^2} \frac{d}{dx}(-x^2) = -2xe^{-x^2} \qquad \text{(Equation 46-13)},$$

$$y'' = \frac{d}{dx}(-2xe^{-x^2}) = -2e^{-x^2} - 2x\frac{d}{dx}e^{-x^2} \qquad \text{(Product Rule)}$$

$$= -2e^{-x^2} + 4x^2 e^{-x^2} \qquad \text{(Equation 46-13)}$$

$$= 2e^{-x^2}(2x^2 - 1) \qquad \text{(Algebra)}.$$

But $e^{-x^2} > 0$ for every x, and so $y'' < 0$ if $x \in (-\frac{1}{2}\sqrt{2}, \frac{1}{2}\sqrt{2})$ and $y'' > 0$ if $|x| > \frac{1}{2}\sqrt{2}$. In other words, our curve is concave down above the interval $(-\frac{1}{2}\sqrt{2}, \frac{1}{2}\sqrt{2})$, and concave up otherwise.

PROBLEMS 46

1. Find y'.
   (a) $y = \sin 2e^x$ (b) $y = \sin e^{2x}$
   (c) $y = \sin^2 e^x$ (d) $y = \sin(\exp x^2)$
   (e) $y = \exp(\sin x)$ (f) $y = \exp(\sin^2 x)$

2. Find $\dfrac{dy}{dx}$.

(a) $y = e^{x^2} - (e^x)^2$ (b) $y = e^{\sin x} \sin e^x$

(c) $y = \exp e^x$ (d) $y = \ln \sqrt{e^x} - \exp (\ln \sqrt{x})$

(e) $y = \sqrt{\ln e^x} - \ln e^{\sqrt{x}}$ (f) $x = \exp e^y$

3. Use the tables in the back of the book to find the following numbers.

(a) $e^{2.35}$ (b) $\sqrt{e^{\sin 1}}$ (c) $1/e^e$ (d) $(e^{\sin \frac{1}{4}}) \cos \frac{1}{2}$

4. Solve for x.

(a) $\displaystyle\int_1^x \frac{1}{t}\, dt = 1$ (b) $\displaystyle\int_{x^2}^{x^5} \frac{2}{t}\, dt = 54$

(c) $\displaystyle\int_5^{1/x} \frac{1}{t}\, dt = \ln 5$ (d) $\displaystyle\int_{e^{x^3}}^{e^{x^2}} \frac{1}{t}\, dt = 4$

5. Find the maximum and minimum points, investigate the concavity, and sketch the graph of f.

(a) $f(x) = x^2 e^{-x}$ (b) $f(x) = x \ln x$

6. (a) Show that $y = Ae^{4x} + Be^{-4x}$ satisfies the differential equation $y'' - 16y = 0$.

(b) Now choose A and B so that y also satisfies the initial conditions $y = 1$ and $y' = 0$ when $x = 0$.

7. Use implicit differentiation to find y' from the equation $e^x + e^y = e^{x+y}$.

8. Show that $y = e^{-x} \sin 2x$ satisfies the differential equation $y'' + 2y' + 5y = 0$.

9. (a) Show that if $y = Ae^{mx}$, then $\dfrac{d^n y}{dx^n} = m^n y$.

(b) Find m such that $y = Ae^{mx}$ satisfies the differential equation $2y'' + 3y' - 2y = 0$.

10. Let $f(x) = e^x - \ln x$. For what number x is the value of f a minimum? (Use Newton's Method to approximate the solution.)

11. We introduced the concepts of extension and restriction of functions in Review Problems I. Show that $f \subseteq g$ for each of the following pairs of functions.

(a) $f(x) = \exp (\ln x)$, $g(x) = \ln (\exp x)$

(b) $f(x) = 2 \ln x$, $g(x) = \ln x^2$

(c) $f(x) = \ln x$, $g(x) = \ln |x|$

(d) $f(x) = \ln x + \ln (x - 1)$, $g(x) = \ln (x^2 - x)$

12. Sketch the curves $y = e^{-x}$ and $y = \sin x$. Use them to graph the equation $y = e^{-x} \sin x$ for $x \geq 0$. Find the X-coordinates of the maximum points of this graph. Discuss its concavity.

13. Show that $\displaystyle\int_1^2 \frac{1}{t}\, dt < 1$ and explain why you can conclude from this inequality that $e > 2$.

14. Verify Equation 46-7.

15. Suppose that a certain population increases at the rate of 3% a year and reaches a figure of 1000 when $t = 0$. Use the ideas of Examples 14-2 and 46-2 to find a formula for the population P at the end of t years. Use the tables in the back of the book to find the population after 10 years, plus the number of years that it will take the population to double.

16.   Let $A(x) = \displaystyle\int_0^x p(t)\, dt$, and $B(x) = \displaystyle\int_0^x e^{A(t)} q(t)\, dt$.

(a) Show that for any choice of the number C, $y = Ce^{-A(x)}$ satisfies the differential
equation $y' + p(x)y = 0$. Find a solution of $y' - xy = 0$.

(b) Show that for any choice of the number C, $y = Ce^{-A(x)} + e^{-A(x)}B(x)$ satisfies
the differential equation $y' + p(x)y = q(x)$. Find a solution of $y' - y = e^{2x}$.

47. INTEGRATION FORMULAS
FOR EXPONENTS AND LOGARITHMS

In the preceding sections we developed the differentiation formulas

(47-1)
$$\frac{d \ln u}{dx} = \frac{1}{u}\frac{du}{dx}$$

(47-2)
$$\frac{d e^u}{dx} = e^u \frac{du}{dx}$$

Because of the Fundamental Theorem of Calculus, differentiation formulas are
always accompanied by integration formulas; these two give us the integration
formulas we will introduce now.

Our first is

(47-3)
$$\int e^{ax}\, dx = \frac{1}{a}e^{ax} \qquad (a \neq 0).$$

To verify this formula, we simply use Equation 47-2 to show that $\dfrac{d}{dx}\left(\dfrac{1}{a}e^{ax}\right) = e^{ax}$.

Example 47-1.   The region that is bounded by the curve $y = e^x$, the X-axis, the
Y-axis, and the line $x = 1$ is rotated about the X-axis. What is the volume of the
resulting solid of revolution?

Solution.   Our formula for the volume of a solid of revolution tells us that

$$V = \pi \int_0^1 y^2\, dx = \pi \int_0^1 (e^x)^2\, dx = \pi \int_0^1 e^{2x}\, dx.$$

Now we use Formula 47-3 with $a = 2$:

$$V = \pi \int_0^1 e^{2x}\, dx = \tfrac{1}{2}\pi e^{2x}\Big|_0^1 = \tfrac{1}{2}\pi(e^2 - 1) \approx 10.$$

Our next integration formula is

(47-4)
$$\int \frac{1}{x} \, dx = \ln |x|.$$

To verify this formula, we replace u with $|x|$ in Equation 47-1:

(47-5)
$$\frac{d}{dx} \ln |x| = \frac{1}{|x|} \frac{d}{dx} |x| = \frac{1}{|x|} \cdot \frac{x}{|x|} = \frac{x}{x^2} = \frac{1}{x}.$$

Example 47-2. Evaluate the integral $\displaystyle\int_{-8}^{-4} \frac{1}{x} \, dx$.

Solution. According to Formula 47-4,

$$\int_{-8}^{-4} \frac{1}{x} \, dx = \ln |x| \Big|_{-8}^{-4} = \ln |-4| - \ln |-8| = \ln 4 - \ln 8$$

$$= \ln \tfrac{4}{8} = \ln \tfrac{1}{2} = -\ln 2 = -.6931.$$

(Explain the geometric significance of the negative answer.)

Our final integration formula of this section is

(47-6)
$$\int \ln |x| \, dx = x \ln |x| - x.$$

To verify this formula, we must show that $\dfrac{d}{dx} (x \ln |x| - x) = \ln |x|$. According to the sum and product rules of differentiation,

$$\frac{d}{dx} (x \ln |x| - x) = x \frac{d}{dx} \ln |x| + \ln |x| \frac{dx}{dx} - \frac{dx}{dx}.$$

We have just seen that $\dfrac{d}{dx} \ln |x| = 1/x$, and $\dfrac{dx}{dx} = 1$, so

$$\frac{d}{dx} (x \ln |x| - x) = \frac{x}{x} + \ln |x| - 1 = \ln |x|,$$

which verifies Formula 47-6.

Usually Formulas 47-4 and 47-6 are applied in intervals in which x is positive, and in that case the absolute value signs can be deleted.

Example 47-3. Evaluate the integral $\int_1^2 \ln x^2 \, dx$.

Solution. We do not have an integration formula in which the integrand is $\ln x^2$. However, $\ln x^2 = 2 \ln x$, so

$$\int_1^2 \ln x^2 \, dx = 2 \int_1^2 \ln x \, dx = 2(x \ln x - x)\Big|_1^2$$

$$= 2[(2 \ln 2 - 2) - (1 \ln 1 - 1)]$$

$$= 2[2 \ln 2 - 1] \approx .77.$$

Before closing this section, we will squeeze one more formula from Equation 47-1, a limit relation that we will need in the next section.

Example 47-4. Show that $\displaystyle\lim_{h\to 0} \frac{\ln (1 + h)}{h} = 1$.

Solution. If we let $f(x) = \ln x$, then we see from Equation 47-1 that $f'(1) = 1$. But the definition of a derivative tells us that

$$f'(1) = \lim_{h\to 0} \frac{f(1 + h) - f(1)}{h} = \lim_{h\to 0} \frac{\ln (1 + h) - \ln 1}{h} = \lim_{h\to 0} \frac{\ln (1 + h)}{h},$$

and so our conclusion follows.

PROBLEMS 47

1. Evaluate the following integrals.

(a) $\displaystyle\int_0^1 e^{3x} \, dx$

(b) $\displaystyle\int_0^2 (e^x + 1)^2 \, dx$

(c) $\displaystyle\int_1^2 \frac{1 + x}{x^2} \, dx$

(d) $\displaystyle\int_{-1}^{-2} \frac{1 + x}{x^2} \, dx$

(e) $\displaystyle\int_1^e \ln x \, dx$

(f) $\displaystyle\int_{-1}^{-e} \ln |x| \, dx$

2. Find the volume of the solid obtained by rotating the region between the curve $y = e^x$ and the interval $[0, 1]$ about the line $y = 1$.

3. Evaluate the following integrals.

(a) $\displaystyle\int_{-8}^{-4} \frac{1}{|x|} \, dx$

(b) $\displaystyle\int_0^{\ln 2} e^{3t} \, dt$

(c) $\displaystyle\int_0^1 \frac{(e^{2x} + 1)^2}{e^x}\, dx$

(d) $\displaystyle\int_1^2 \frac{x^3 + x^2 + x + 1}{x^2}\, dx$

(e) $\displaystyle\int_1^2 e^{\ln 2x}\, dx$

(f) $\displaystyle\int_1^2 e^{-2\ln x}\, dx$

4. Verify the following integration formulas.

(a) $\displaystyle\int xe^{ax}\, dx = e^{ax}(ax - 1)/a^2$

(b) $\displaystyle\int (1 + e^x)^{-1}\, dx = -\ln (1 + e^{-x})$

(c) $\displaystyle\int \frac{xe^x}{(1 + x)^2}\, dx = \frac{e^x}{1 + x}$

(d) $\displaystyle\int \frac{(1 - n)}{x(\ln |x|)^n}\, dx = (\ln |x|)^{1-n}$

5. Compute the following derivatives and write the integration formulas that the results suggest.

(a) $\dfrac{d}{dx} [e^x(x - 1)]$

(b) $\dfrac{d}{dx} [-e^{-x}(2 + 2x + x^2)]$

(c) $\dfrac{d}{dx} [e^x/(1 + x)]$

(d) $\dfrac{d}{dx} (\tfrac{1}{2}x^2 \ln x - \tfrac{1}{4}x^2)$

6. Sketch the curve $y = \ln |x|$. Draw in a few tangent lines to verify graphically that $\dfrac{dy}{dx} = 1/x$.

7. Evaluate the following integrals.

(a) $\displaystyle\int_1^{e^3} \ln x\, dx$

(b) $\displaystyle\int_1^e \ln x^3\, dx$

(c) $\displaystyle\int_1^e \ln 3x\, dx$

(d) $\displaystyle\int_1^e \ln 3^x\, dx$

(e) $\displaystyle\int_1^e \ln \sqrt[3]{x}\, dx$

(f) $\displaystyle\int_1^e \ln \frac{3}{x}\, dx$

8. Use the equation $\ln ax = \ln a + \ln x$ to find an integration formula whose left-hand side is $\displaystyle\int \ln ax\, dx$.

9. Find the volume of the solid that is generated by rotating about the X-axis the region that is bounded by the following curves: $y = e^{-x}$, $y = 0$, $x = -\tfrac{1}{2}$, and $x = \tfrac{1}{2}$.

10. Show that $\displaystyle\int \ln |x|\, dx = x \ln (|x|/e)$.

11. Find the minimum value of $e^{5x} - 20e^{2x}$.

12. Find the maximum value of the product xe^{-x}.

13. Prove that the equation $e^x = 1 + x$ has only one real solution.

14. Use a method similar to that used in Example 47-4 to show that $\lim\limits_{h\to 0} (e^h - 1)/h = 1$.

15.   (a) Show that if $y > 0$ and n is a positive integer, then

$$n(1 - y^{-1/n}) \le \ln y \le n(y^{1/n} - 1).$$

(*Hint*: Consider the integrals $\displaystyle\int_1^y t^{-(1/n)-1}\, dt, \int_1^y t^{-1}\, dt,$ and $\displaystyle\int_1^y t^{(1/n)-1}\, dt$ in

case $0 < y < 1$ and in case $y \ge 1$.)

(b) Suppose that x is a given number and set $y = e^x$ in the inequalities of part (a). If we choose n to be a positive integer that is larger than $|x|$, show that these inequalities yield the inequalities

$$\left(1 + \frac{x}{n}\right)^n \le e^x \le \left(1 - \frac{x}{n}\right)^{-n}.$$

(c) Set $x = 1$ and $n = 5$ in the inequalities of part (b) to show that $2.4 \le e \le 3.2$.

16.   Show that if f' is an odd function, then $\dfrac{d}{dx} f(|x|) = f'(x)$.

48.   EXPONENTIAL AND LOGARITHMIC FUNCTIONS (WITH BASES OTHER THAN *e*)

If b is a positive number (for example, 10), we know what $b^2, b^{-3}, b^{5/3}, b^0$, and so on, mean. Specifically, we know what b^r represents if r is a rational number. In this section we will come to grips with the question, "What is the number b^a when a is an irrational number?" Since we have already decided on the answer when $b = e$ ($e^a = \exp a$), we will build our definition of b^a on this special case.

We know (Equation 46-4) that $b = e^{\ln b}$. Consequently, however we define b^a, we will have $b^a = (e^{\ln b})^a$. Whenever we introduce a new mathematical concept, we try to preserve familiar rules. In this case, we will try to preserve the law of exponents $(x^m)^n = x^{mn}$. In order for this rule to hold, we must have $(e^{\ln b})^a = e^{a \ln b}$, and so we make the following definition.

Definition   48-1.   *If b is a positive number and a is an irrational number, then*

(48-1) $$b^a = e^{a \ln b} = \exp\,(a \ln b).$$

For example, $\pi^{\sqrt{2}} = e^{\sqrt{2} \ln \pi}$. From our table of logarithms, we find $\ln \pi \approx 1.144$, so $\sqrt{2} \ln \pi \approx 1.62$. From Table III, $e^{1.62} \approx 5.05$. Thus $\pi^{\sqrt{2}} \approx 5.05$. If a is irrational, Equation 48-1 is true by definition. If a is rational, it follows from the formulas of Section 46; Equations 46-3 and 46-4 tell us that $e^{a \ln b} = e^{\ln b^a} = b^a$. Therefore Equation 48-1 is valid for any real number a.

We were forced to make Definition 48-1 to keep from violating *just one* rule of exponents, and now it is natural to ask whether, with this definition, all the rules of exponents and logarithms hold. The answer is *yes*, but, of course, it requires demonstration. For example, we must show that $\ln x^r = r \ln x$ when r is any real number. (We already know that this equation is valid if r is a *rational* number.) We have

$$\ln x^r = \ln \left[\exp \left(r \ln x \right) \right] \qquad \text{(Equation 48-1)}$$

$$= r \ln x \qquad \text{(Equation 46-5 with } v = r \ln x\text{)}.$$

Furthermore, we must show that $b^a \cdot b^c = b^{a+c}$, when a and c are any real numbers. Here we have

$$b^a \cdot b^c = (e^{a \ln b})(e^{c \ln b}) \qquad \text{(Equation 48-1)}$$

$$= e^{a \ln b + c \ln b} \qquad \text{(Equation 46-6)}$$

$$= e^{(a+c) \ln b}$$

$$= b^{a+c} \qquad \text{(Equation 48-1)}.$$

We will leave the proofs of the laws $(b^a)^c = b^{ac}$ and $(ab)^c = a^c b^c$ to you in the problems.

When we started the proof of the Power Formula $\dfrac{d}{dx} x^r = rx^{r-1}$ in Section 15, we pointed out that its details would vary, depending on the "kind" of number r is. We have already proved this formula in case r is a rational number, and now we have the tools at hand to prove it in case r is *any* real number, rational or irrational. According to Equation 48-1,

$$x^r = e^{r \ln x}.$$

Thus we can find the derivative of x^r by setting $u = r \ln x$ in the formula $\dfrac{d}{dx} e^u = e^u \dfrac{du}{dx}$:

$$\frac{d}{dx} e^{r \ln x} = e^{r \ln x} \frac{d}{dx} (r \ln x) = x^r \cdot \frac{r}{x} = rx^{r-1}.$$

Therefore the formula

$$\frac{d}{dx} x^r = rx^{r-1}$$

is valid for any real number.

We have just seen how to differentiate x to a real power. Now let us consider a case in which x is the exponent.

Example 48-1. Find $\dfrac{d}{dx}\,10^x$.

Solution. According to Equation 48-1, $10^x = e^{x\ln 10}$. Now we use the formula $\dfrac{d}{dx}\,e^u = e^u\,\dfrac{du}{dx}$ with $u = x\ln 10$:

$$\frac{d}{dx}\,e^{x\ln 10} = e^{x\ln 10}\,\frac{d}{dx}\,(x\ln 10) = (\ln 10)10^x.$$

Thus

$$\frac{d}{dx}\,10^x = (\ln 10)10^x \approx (2.303)10^x.$$

We can use the method of the preceding example to derive a more general differentiation formula. If b is any positive number,

$$\frac{d}{dx}\,b^x = \frac{d}{dx}\,e^{x\ln b} = e^{x\ln b}\,\frac{d}{dx}\,(x\ln b) = e^{x\ln b}\ln b = b^x\ln b.$$

Thus

(48-2) $$\frac{d}{dx}\,b^x = b^x\ln b.$$

We incorporate this specific differentiation formula into the Chain Rule Equation $\dfrac{df(u)}{dx} = \dfrac{df(u)}{du}\dfrac{du}{dx}$ by replacing $f(u)$ with b^u: $\dfrac{d}{dx}\,b^u = \dfrac{d}{du}\,b^u\,\dfrac{du}{dx}$. Equation 48-2 says that $\dfrac{d}{du}\,b^u = b^u\ln b$, and so

(48-3) $$\frac{d}{dx}\,b^u = b^u\ln b\,\frac{du}{dx}.$$

It follows from Equation 48-2 that if $b \neq 1$, then

$$\frac{d}{dx}\left(\frac{b^x}{\ln b}\right) = \frac{1}{\ln b}\frac{d}{dx}\,b^x = b^x,$$

so $b^x/\ln b$ is an antiderivative of b^x. Thus we have the integration formula

(48-4) $$\int b^x\,dx = \frac{b^x}{\ln b} \qquad (b \neq 1).$$

Example 48-2. The curve $y = 3^x$, the lines $x = -1$ and $x = 1$, and the X-axis enclose a region R. Find the volume of the solid obtained by rotating this region about the X-axis.

Solution. The volume we seek is given by the integral

$$V = \pi \int_{-1}^{1} y^2 \, dx = \pi \int_{-1}^{1} (3^x)^2 \, dx = \pi \int_{-1}^{1} 3^{2x} \, dx = \pi \int_{-1}^{1} 9^x \, dx.$$

Now Formula 48-4 (with $b = 9$) applies:

$$V = \frac{\pi}{\ln 9} 9^x \bigg|_{-1}^{1} = \frac{\pi}{\ln 9} \left(9 - \frac{1}{9}\right) \approx 13.$$

Let us now consider logarithmic functions with bases other than e. Suppose that b is a positive number and let $y = \log_b x$. We wish to find $\dfrac{dy}{dx}$. We first notice that, by definition,

$$y = \log_b x \quad \text{if and only if} \quad b^y = x.$$

From the equation $b^y = x$ it follows that $\ln b^y = \ln x$; that is, $y \ln b = \ln x$. Thus $y = \ln x / \ln b$, so that

$$(48\text{-}5) \qquad\qquad \log_b x = \frac{\ln x}{\ln b}.$$

From this equation we get

$$\frac{d}{dx} \log_b x = \frac{d}{dx}\left(\frac{\ln x}{\ln b}\right) = \frac{1}{\ln b}\frac{d}{dx} \ln x = \frac{1}{x \ln b}.$$

Since $1/\ln b = \log_b e$ (Equation 48-5 with $x = e$), we can write our differentiation formula as

$$(48\text{-}6) \qquad\qquad \frac{d}{dx} \log_b x = \frac{1}{x \ln b} = \frac{\log_b e}{x}.$$

When we incorporate this formula into the Chain Rule Equation, we obtain

$$(48\text{-}7) \qquad\qquad \frac{d}{dx} \log_b u = \frac{1}{u \ln b}\frac{du}{dx} = \frac{\log_b e}{u}\frac{du}{dx}.$$

Example 48-3.   Verify the integration formula

(48-8) $$\int \log_b x \, dx = x \log_b x/e.$$

Solution.   We need only show that $\log_b x$ is the derivative of $x \log_b x/e$:

$$\frac{d}{dx}(x \log_b x/e) = \frac{d}{dx}(x \log_b x - x \log_b e)$$

$$= \log_b x \frac{dx}{dx} + x \frac{d}{dx} \log_b x - \log_b e \frac{dx}{dx}$$

$$= \log_b x + (x \log_b e)/x - \log_b e = \log_b x.$$

Example 48-4.   Find $\dfrac{dy}{dx}$ if $y = x^{e^x}$.

Solution.   We can write

$$y = e^{e^x \ln x} = e^u, \text{ where } u = e^x \ln x,$$

and use the formula $\dfrac{d}{dx} e^u = e^u \dfrac{du}{dx}$. You may verify that

$$\frac{du}{dx} = e^x \ln x + e^x/x,$$

and so

$$\frac{dy}{dx} = e^u(e^x \ln x + e^x/x) = ye^x(\ln x + 1/x)$$

$$= x^{e^x}e^x(\ln x + 1/x).$$

Example 48-5.   Show that

(48-9) $$\lim_{h \to 0} (1 + h)^{1/h} = e.$$

Solution.   If we let $f(h) = (1 + h)^{1/h}$, then $\ln f(h) = \dfrac{1}{h} \ln (1 + h)$. We substitute this result in the identity $f(h) = \exp [\ln f(h)]$ and obtain the equation

$$f(h) = \exp \left[\frac{1}{h} \ln (1 + h) \right].$$

Thus

$$\lim_{h \to 0} f(h) = \lim_{h \to 0} \exp \left[\frac{1}{h} \ln (1 + h) \right]$$

$$= \exp \left[\lim_{h \to 0} \frac{\ln (1 + h)}{h} \right] \qquad \text{(exp is a continuous function)}$$

$$= \exp 1 \qquad \text{(Example 47-4)}$$

$$= e.$$

In many books you will find the number *e defined* by Equation 48-9.

PROBLEMS   48

1. Find $\dfrac{dy}{dx}$.

    (a) $y = 2 \cdot 3^x$ (b) $y = 3^{2x}$ (c) $y = 1/3^x$
    (d) $y = 3^{1/x}$ (e) $y = 3^{\sin x}$ (f) $y = \sin 3^x$
    (g) $y = x^\pi \pi^x$ (h) $y = 1/x^\pi \pi^x$

2. Find $\dfrac{dy}{dx}$.

    (a) $y = \log_3 x$ (b) $y = \log_3 x^3$ (c) $y = \log_3 (\ln x)$
    (d) $y = \ln (\log_3 x)$ (e) $y = \log_3 5^x$ (f) $y = \log_5 3^x$

3. (a) Find the largest value of $e^{-x} x^e$.
    (b) Set $x = \pi$ and conclude that $\pi^e < e^\pi$.
    (c) Use the tables in the back of the book to compute π^e and e^π.

4. Evaluate the following integrals.

    (a) $\displaystyle\int_0^1 e^\pi \, dx$ (b) $\displaystyle\int_0^1 e^{\pi x} \, dx$ (c) $\displaystyle\int_0^1 \pi^{ex} \, dx$ (d) $\displaystyle\int_0^1 x^{\pi e} \, dx$

    (e) $\displaystyle\int_{-1}^1 3^{-t} \, dt$ (f) $\displaystyle\int_0^2 (2^x - x^2) \, dx$

    (g) $\displaystyle\int_e^{10e} \log_{10} x \, dx$ (h) $\displaystyle\int_1^{10} \log_{10} \sqrt{xe} \, dx$

5. Find the smallest value of f.

    (a) $f(x) = 4x^\pi + x^{-\pi}$ (b) $f(x) = 4\pi^x + \pi^{-x}$

6. Find the average value of f on the given interval.

    (a) $f(x) = 2^x$ on $[0, 2]$ (b) $f(x) = \log_{10} x$ on $[\frac{1}{10}e, 100e]$

7. Suppose that $u = f(x)$ and $v = g(x)$, where $f(x) > 0$. By first writing $u^v = e^{v \ln u}$, show that

$$\frac{d}{dx} u^v = u^v \left[\frac{v}{u} \frac{du}{dx} + (\ln u) \frac{dv}{dx} \right].$$

Use this formula to find $\dfrac{dy}{dx}$.

   (a) $y = x^{\sqrt{x}}$ (b) $y = \sqrt{x^x}$ (c) $y = x^{\sin x}$ (d) $y = \sin^x x$

8. What is the minimum value of x^x (for $x > 0$)?

9. Sketch the graph of the equation $y = (1 + x)^{1/x}$.

10. Find $\dfrac{d}{dx} \log_x 3$.

11. Show that $\lim\limits_{h \to 0} (b^h - 1)/h = \ln b$. (*Hint*: Let $f(x) = b^x$ and write out the definition of $f'(0)$.)

12. (a) Show that $(ab)^c = a^c b^c$.
    (b) Show that $(b^c)^a = b^{ac}$.

49.   INVERSE TRIGONOMETRIC FUNCTIONS

In Example 45-2 we pointed out that the sine function does not have an inverse. In fact, no trigonometric function does. If f is a trigonometric function and y is a given number of its range, then to each solution of the equation $y = f(x)$ we can add $2\pi, 4\pi, -2\pi$, and so on, and obtain other solutions. Thus the equation $y = f(x)$ does not have *just one* solution for each choice of y in the range of f, and so f does not have an inverse. But we can construct **restrictions** (subsets) of the trigonometric functions that do have inverses.

Figure 49-1 shows the graph of the sine function in black. The colored curve is the graph of a restriction that obviously has an inverse. This restriction is the **principal sine function**, and we call it **Sin** (the capital letter S distinguishes it from

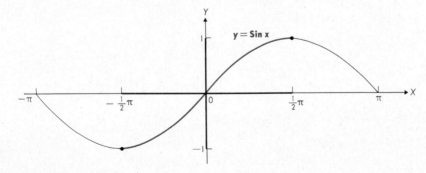

Figure 49-1

the ordinary sine function). The domain of the principal sine function is the interval $[-\frac{1}{2}\pi, \frac{1}{2}\pi]$, and it has the same rule of correspondence as the sine function. Thus Sin $\frac{1}{6}\pi = \sin \frac{1}{6}\pi = \frac{1}{2}$, but Sin $\frac{7}{6}\pi$ is a meaningless symbol, since $\frac{7}{6}\pi$ is not in the domain of the principal sine function.

The inverse of the Sine function is the **Arcsine function**. The range of the Sine function is the interval $[-1, 1]$, so this interval serves as the domain of the Arcsine function. The number that the Arcsine function associates with a given number y of its domain is commonly denoted either by Arcsin y or by Sin^{-1} y. If we think of our functions as subsets of R^2, then Sin $= \{(x, y) \mid x \in [-\frac{1}{2}\pi, \frac{1}{2}\pi], y = \sin x\}$. We obtain its inverse by interchanging the numbers in each pair; that is, Arcsin $= \{(y, x) \mid x \in [-\frac{1}{2}\pi, \frac{1}{2}\pi], y = \sin x\}$. Our formal definition of the Arcsine function states the conditions under which a pair belongs to this set.

Definition 49-1. *The equation*

$$x = \text{Arcsin } y = \text{Sin}^{-1} y$$

is equivalent to the two statements

(i) $y = \sin x$ *and* (ii) $x \in [-\frac{1}{2}\pi, \frac{1}{2}\pi]$.

Example 49-1. Find Arcsin $\frac{1}{2}$.

Solution. We know from our study of trigonometry that $\frac{1}{2} = \sin \frac{1}{6}\pi$. Furthermore, $\frac{1}{6}\pi \in [-\frac{1}{2}\pi, \frac{1}{2}\pi]$, so both conditions (i) and (ii) of Definition 49-1 are satisfied, and hence $\frac{1}{6}\pi = $ Arcsin $\frac{1}{2}$.

Example 49-2. Show that cos (Sin^{-1} x) $= \sqrt{1 - x^2}$ for each $x \in [-1, 1]$.

Solution. If we write $t = $ Sin^{-1} x, then according to Definition 49-1,

(i) $x = \sin t$ and (ii) $t \in [-\frac{1}{2}\pi, \frac{1}{2}\pi]$.

We are to find cos t. Now since $\cos^2 t + \sin^2 t = 1$ and $\sin t = x$, we have $\cos^2 t = 1 - x^2$. Therefore cos $t = \sqrt{1 - x^2}$ or cos $t = -\sqrt{1 - x^2}$. Because $t \in [-\frac{1}{2}\pi, \frac{1}{2}\pi]$, it follows that cos $t \geq 0$, and so cos $t = \sqrt{1 - x^2}$.

We can obtain the graph of the Arcsine function by reflecting the graph of the Sine function about the line $y = x$. The resulting curve $y = $ Sin^{-1} x is shown in Fig. 49-2.

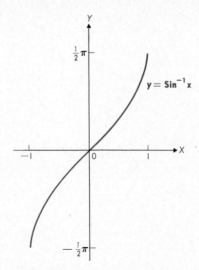

Figure 49-2

We treat the cosine function as we have just treated the sine function. We first define the **Cosine function** (again using a capital letter to distinguish this new function from the cosine function) as the function whose domain is the interval $[0, \pi]$ and whose rule of correspondence is expressed by the equation $y = \cos x$. The graph of the Cosine function is the colored curve in Fig. 49-3; it shows that Cos has an inverse whose domain is the interval $[-1, 1]$. This inverse is called the **Arccosine function**, and the number corresponding to a given number y in the domain of the Arccosine function is denoted either by Arccos y or by $\text{Cos}^{-1} y$. Since Cos $= \{(x, y) \mid x \in [0, \pi], y = \cos x\}$, we have Arccos $= \{(y, x) \mid x \in [0, \pi], y = \cos x\}$. Our formal definition of the Arccosine function states the conditions under which a pair belongs to this set.

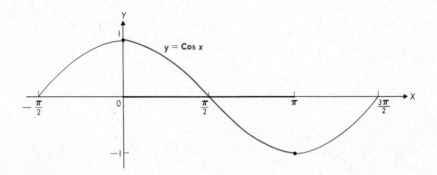

Figure 49-3

Definition 49-2. *The equation*

$$x = \text{Arccos } y = \text{Cos}^{-1} y$$

is equivalent to the two statements

(i) $y = \cos x$ *and* (ii) $x \in [0, \pi]$.

The curve $y = \text{Cos}^{-1} x$ is shown in Fig. 49-4.

Now let us turn to the tangent function. We first restrict its domain to the interval $(-\tfrac{1}{2}\pi, \tfrac{1}{2}\pi)$, thereby obtaining a new function called the **Tangent function.** Thus $\text{Tan} = \{(x, y) \mid x \in (-\tfrac{1}{2}\pi, \tfrac{1}{2}\pi), y = \tan x\}$. We have drawn the graph of this function in Fig. 49-5. It shows that Tan has an inverse whose *domain is the set of all real numbers.* This inverse function is called the **Arctangent function.** and the number that it associates with a given real number y is denoted by $\text{Arctan } y$ or by $\text{Tan}^{-1} y$. From our description of the Tangent function as a subset of R^2, we obtain the equation $\text{Arctan} = \{(y, x) \mid x \in (-\tfrac{1}{2}\pi, \tfrac{1}{2}\pi), y = \tan x\}$. Our formal definition of the Arctangent function states the conditions under which a pair belongs to this set.

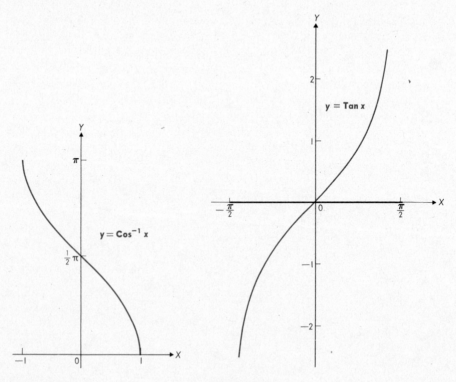

Figure 49-4   **Figure 49-5**

Definition   49-3.   *The equation*

$$x = \text{Arctan } y = \text{Tan}^{-1} y$$

is equivalent to the two statements

(i) $y = \tan x$ *and*   (ii) $x \in (-\tfrac{1}{2}\pi, \tfrac{1}{2}\pi)$.

We obtain the graph of the Arctangent function (Fig. 49-6) by reflecting the graph of the Tangent function about the line $y = x$. It is important that you know what this graph looks like.

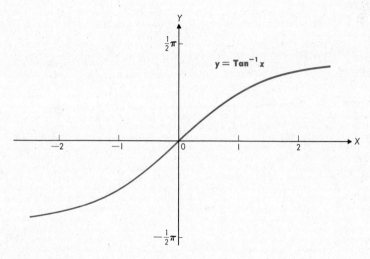

Figure 49-6

Example 49-3.   Calculate

$$\text{Tan}^{-1} x \Big|_{-\sqrt{3}/3}^{\sqrt{3}/3}.$$

Solution.

$$\text{Tan}^{-1} x \Big|_{-\sqrt{3}/3}^{\sqrt{3}/3} = \text{Tan}^{-1}(\sqrt{3}/3) - \text{Tan}^{-1}(-\sqrt{3}/3)$$

$$= \tfrac{1}{6}\pi - (-\tfrac{1}{6}\pi)$$

$$= \tfrac{1}{3}\pi.$$

PROBLEMS   49

1. Compute the following numbers.

(a) $\operatorname{Sin}^{-1} x \Big|_{1/2}^{1}$ (b) $\operatorname{Cos}^{-1} t \Big|_{1/2}^{1}$ (c) $\operatorname{Tan}^{-1} u \Big|_{-1}^{1}$

(d) $\operatorname{Sin}^{-1} x \Big|_{-\sqrt{2}/2}^{\sqrt{3}/2}$ (e) $\operatorname{Tan}^{-1} x \Big|_{.4228}^{.6841}$ (f) $\operatorname{Sin}^{-1} x \Big|_{.7833}^{.9854}$

(g) $\operatorname{Sin}^{-1} (\sin x) \Big|_{\pi/2}^{3\pi/2}$ (h) $\sin (\operatorname{Sin}^{-1} x) \Big|_{-2/3}^{1/3}$

2. Find the following numbers.
(a) $\cos (\operatorname{Sin}^{-1} .8)$ (b) $\sin (\operatorname{Arccos} \frac{12}{13})$ (c) $\sin (\operatorname{Cos}^{-1} \frac{3}{5})$
(d) $\cos (\operatorname{Tan}^{-1} 0)$ (e) $\sin (2 \operatorname{Cos}^{-1} \frac{1}{2})$ (f) $\cos (2 \operatorname{Sin}^{-1} \frac{1}{2})$

3. Use the graphs in this section to determine which number is larger if $x > 0$.
(a) x or $\operatorname{Sin} x$ (b) x or $\operatorname{Sin}^{-1} x$
(c) $\operatorname{Tan}^{-1} x$ or $\operatorname{Sin}^{-1} x$ (d) $\exp (\operatorname{Sin} x)$ or $\exp (\operatorname{Sin}^{-1} x)$

4. From the graphs in this section determine whether the following numbers are positive or negative.

(a) $\dfrac{d}{dx} \operatorname{Sin}^{-1} x$ (b) $\dfrac{d}{dx} \operatorname{Cos}^{-1} x$

(c) $\dfrac{d}{dx} \operatorname{Tan}^{-1} x$ (d) $\dfrac{d}{dx} (\operatorname{Cos}^{-1} x)^2$

(e) $\dfrac{d}{dx} (\operatorname{Sin}^{-1} x)^2$ for $x < 0$ (f) $\dfrac{d}{dx} (\operatorname{Tan}^{-1} x)^2$ for $x < 0$.

5. Verify the following identities.
(a) $\operatorname{Arctan} (-x) = -\operatorname{Arctan} x$ (b) $\operatorname{Arcsin} (-x) = -\operatorname{Arcsin} x$
(c) $\operatorname{Arccos} (-x) = \pi - \operatorname{Arccos} x$ (d) $\operatorname{Sin}^{-1} x + \operatorname{Cos}^{-1} x = \frac{1}{2}\pi$

6. Use the graphs of the inverse trigonometric functions to help you evaluate the following integrals.

(a) $\displaystyle\int_0^1 \operatorname{Sin}^{-1} x \, dx$ (b) $\displaystyle\int_0^1 \operatorname{Cos}^{-1} t \, dt$

(c) $\displaystyle\int_0^{1/2} \operatorname{Sin}^{-1} u \, du$ (d) $\displaystyle\int_{-1}^0 \operatorname{Cos}^{-1} u \, du$

7. Solve for t: $\operatorname{Arctan} \frac{1}{3} + \operatorname{Arctan} \frac{1}{2} = \operatorname{Arcsin} t$.
8. Find the point of intersection of the graphs of the Arccosine function and the Arctangent function.
9. Explain the difference, if any, between
(a) $\sin (\operatorname{Sin}^{-1} x)$ and $\operatorname{Sin}^{-1} (\sin x)$.
(b) $\operatorname{Tan}^{-1} (\operatorname{Tan} x)$ and $\operatorname{Tan} (\operatorname{Tan}^{-1} x)$.
(c) $\cos (\operatorname{Cos}^{-1} x)$ and $\operatorname{Cos}^{-1} (\cos x)$.

10. Sketch the following curves.
    (a) $y = \sin (\text{Sin}^{-1} x)$ (b) $y = \text{Sin}^{-1} (\sin x)$
    (c) $y = \text{Sin}^{-1} (\cos x)$ (d) $y = \cos (\text{Sin}^{-1} x)$
    (e) $y = \tan (\text{Tan}^{-1} x)$ (f) $y = \text{Tan}^{-1} (\tan x)$

11. Verify the following equalities. [We will need results (a) and (b) in the next section.]
    (a) $\sin (\text{Cos}^{-1} x) = \sqrt{1 - x^2}$ (b) $\sec^2 (\text{Tan}^{-1} x) = 1 + x^2$

    (c) $\text{Tan}^{-1} x = \text{Sin}^{-1} \dfrac{x}{\sqrt{1 + x^2}}$ (d) $\tan (\tfrac{1}{2} \text{Cos}^{-1} x) = \dfrac{\sqrt{1 - x}}{\sqrt{1 + x}}$

12. The **Arccotangent function** is defined as follows: $x = \text{Arccot } y = \text{Cot}^{-1} y$ if and only if (i) $y = \cot x$ and (ii) $x \in (0, \pi)$. Show that if $x > 0$, then $\text{Arctan } x = \text{Arccot } (1/x)$.

13. What are the following sets?
    (a) $\cos \cap \text{Cos}$ (b) $\sin \cup \text{Sin}$ (c) $\text{Sin} \cap \text{Cos}$
    (d) $\sin \cap \text{Cos}$ (e) $\tan \cap \text{Tan}$ (f) $\tan \cap \text{Tan}^{-1}$

14. Show that if x is not an odd multiple of $\tfrac{1}{2}\pi$, then

$$\text{Tan}^{-1} (\tan x) = x - \pi[\![\tfrac{1}{2} + (x/\pi)]\!].$$

What if x *is* an odd multiple of $\tfrac{1}{2}\pi$?

50. DIFFERENTIATION OF THE INVERSE TRIGONOMETRIC FUNCTIONS

In Section 45 we derived the natural-looking equation $\dfrac{dy}{dx} = 1 \bigg/ \dfrac{dx}{dy}$ for finding derivatives of inverse functions. We will use it now to find derivatives of the inverse trigonometric functions. Thus if $y = \text{Sin}^{-1} x$, then $x = \text{Sin } y$, and so $\dfrac{dx}{dy} = \cos y$. Therefore, $\dfrac{dy}{dx} = \dfrac{1}{\cos y}$ or (replacing y in terms of x) $\dfrac{d}{dx} \text{Sin}^{-1} x = \dfrac{1}{\cos (\text{Sin}^{-1} x)}$. In Example 49-2 we found that $\cos (\text{Sin}^{-1} x) = \sqrt{1 - x^2}$, so we have derived the differentiation formula

$$(50\text{-}1) \qquad \frac{d}{dx} \text{Sin}^{-1} x = \frac{1}{\sqrt{1 - x^2}}.$$

Let us incorporate this differentiation formula for the Arcsine function into the Chain Rule Equation $\dfrac{df(u)}{dx} = \dfrac{df(u)}{du} \dfrac{du}{dx}$. Since $\dfrac{d}{du} \text{Sin}^{-1} u = 1/\sqrt{1 - u^2}$, replacing $f(u)$ with $\text{Sin}^{-1} u$ gives us the differentiation formula

$$(50\text{-}2) \qquad \frac{d}{dx} \text{Sin}^{-1} u = \frac{1}{\sqrt{1 - u^2}} \frac{du}{dx}.$$

Example 50-1. Show that if a is any positive number,

$$(50\text{-}3) \qquad \frac{d}{dx} \operatorname{Sin}^{-1}(x/a) = 1/\sqrt{a^2 - x^2}.$$

Solution. If we replace u with x/a in Formula 50-2, we obtain

$$\frac{d}{dx} \operatorname{Sin}^{-1}(x/a) = \frac{1}{\sqrt{1 - (x/a)^2}} \frac{d}{dx}(x/a)$$

$$= 1/a\sqrt{1 - (x/a)^2} = 1/\sqrt{a^2 - x^2}.$$

At what point did we use the assumption that a is positive?

It is easy to see (we asked you to verify it in part (d) of Problem 49-5) that $\operatorname{Cos}^{-1} x + \operatorname{Sin}^{-1} x = \frac{1}{2}\pi$. Therefore,

$$\frac{d}{dx} \operatorname{Cos}^{-1} x = \frac{d}{dx}(\tfrac{1}{2}\pi - \operatorname{Sin}^{-1} x) = -\frac{d}{dx} \operatorname{Sin}^{-1} x,$$

and now a glance at Equation 50-1 shows us that

$$(50\text{-}4) \qquad \frac{d}{dx} \operatorname{Cos}^{-1} x = -\frac{1}{\sqrt{1 - x^2}}.$$

When we incorporate this formula into the Chain Rule Equation, we obtain the differentiation formula

$$(50\text{-}5) \qquad \frac{d}{dx} \operatorname{Cos}^{-1} u = -\frac{1}{\sqrt{1 - u^2}} \frac{du}{dx}.$$

Now suppose that $y = \operatorname{Tan}^{-1} x$; hence, $x = \tan y$. Therefore, $\dfrac{dx}{dy} = \sec^2 y$ and so the equation $\dfrac{dy}{dx} = 1 \Big/ \dfrac{dx}{dy}$ becomes $\dfrac{dy}{dx} = 1/\sec^2 y = 1/(1 + \tan^2 y)$. (We used the trigonometric identity $\sec^2 y = 1 + \tan^2 y$.) When we replace $\tan y$ with x, our differentiation formula for calculating the derivative of $\operatorname{Tan}^{-1} x$ becomes

$$(50\text{-}6) \qquad \frac{d}{dx} \operatorname{Tan}^{-1} x = \frac{1}{1 + x^2}.$$

Now we incorporate this formula into the Chain Rule Equation and obtain the differentiation formula

$$(50\text{-}7) \qquad \frac{d}{dx} \operatorname{Tan}^{-1} u = \frac{1}{1 + u^2} \frac{du}{dx}.$$

Example 50-2. Show that if $a \neq 0$, then

(50-8) $$\frac{d}{dx} \operatorname{Tan}^{-1}(x/a) = a/(a^2 + x^2).$$

Solution. When we replace u with x/a, Equation 50-7 becomes

$$\frac{d}{dx} \operatorname{Tan}^{-1}(x/a) = \frac{1/a}{1 + (x/a)^2} = a/(a^2 + x^2).$$

Example 50-3. Find $\dfrac{d}{dx} \operatorname{Tan}^{-1}(\cot x).$

Solution. If we replace u with $\cot x$, Equation 50-7 becomes

$$\frac{d}{dx} \operatorname{Tan}^{-1}(\cot x) = \frac{1}{1 + \cot^2 x} \frac{d \cot x}{dx} = \frac{-\csc^2 x}{1 + \cot^2 x}.$$

Now we use the trigonometric identity $1 + \cot^2 x = \csc^2 x$, and we see that

$$\frac{d}{dx} \operatorname{Tan}^{-1}(\cot x) = -1.$$

It would be a good test of your understanding of the trigonometric functions and their inverses to verify that the curve $y = \operatorname{Tan}^{-1}(\cot x)$ is the one shown in Fig. 50-1.

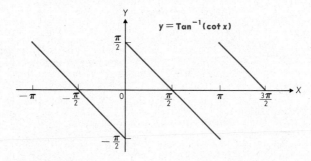

Figure 50-1

Example 50-4. A searchlight 40 feet from the base of a building makes a spot 30 feet up on the side of the building. Approximately how much must the search-light be rotated in order to raise the spot 1 foot?

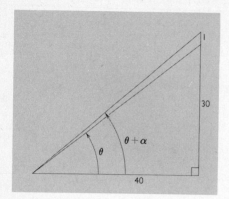

Figure 50-2

Solution. Figure 50-2 illustrates the situation; we are to find α. From elementary trigonometry we see that

$$\tan(\theta + \alpha) = \tfrac{31}{40}$$

and

$$\tan \theta = \tfrac{3}{4},$$

so

$$\alpha = (\theta + \alpha) - \theta = \text{Tan}^{-1} \tfrac{31}{40} - \text{Tan}^{-1} \tfrac{3}{4}.$$

Now we will set $f(x) = \text{Tan}^{-1}(x/40)$, and so our solution α can be expressed as

$$\alpha = f(31) - f(30).$$

Formula 21-2 gives us an approximate value of this difference,

$$f(31) - f(30) \approx f'(30)(31 - 30) = f'(30),$$

so let us calculate $f'(30)$. When we set $a = 40$ in Equation 50-8, we find that $f'(x) = 40/(40^2 + x^2)$, and hence $f'(30) = 40/(40^2 + 30^2) = .016$. Therefore the searchlight must be rotated upward approximately .016 radian, or .9°.

PROBLEMS 50

1. Find y'.

   (a) $y = \text{Sin}^{-1} \tfrac{1}{2}x$

   (b) $y = \text{Cos}^{-1} \left(1 - \dfrac{2}{x}\right)$

   (c) $y = \text{Sin}^{-1} e^x$

   (d) $y = \exp \text{Tan}^{-1} x$

   (e) $y = \text{Tan}^{-1} \dfrac{x - 1}{x + 1}$

   (f) $y = \text{Sin}^{-1} \dfrac{x}{x + 1}$

2. Find y'.

   (a) $y = \text{Arctan } x^2 - (\text{Arctan } x)^2$

   (b) $y = \sqrt{\text{Sin}^{-1} x} - \text{Sin}^{-1} \sqrt{x}$

   (c) $y = \ln \text{Cos}^{-1} x - \text{Cos}^{-1} \ln x$

   (d) $y = \text{Sin}^{-1} x - (\text{Sin } x)^{-1}$

   (e) $y = \text{Sin}^{-1} (\cos x) - \cos (\text{Sin}^{-1} x)$

   (f) $y = (\text{Tan}^{-1} x^{-1})^{-1}$

3. Find $\dfrac{d}{dx} [2x \text{ Tan}^{-1} 2x - \ln \sqrt{1 + 4x^2}]$.

4. Compute y' and sketch the graph of the equation.

   (a) $y = \text{Sin}^{-1} (\sin x)$

   (b) $y = \sin (\text{Sin}^{-1} x)$

   (c) $y = \text{Cos}^{-1} (\cos x)$

   (d) $y = \text{Tan}^{-1} (\tan x)$

5. Use implicit differentiation to calculate y' from the given equation.

   (a) $\text{Sin}^{-1} x + \text{Sin}^{-1} y = \tfrac{1}{2}\pi$

   (b) $\text{Tan}^{-1} x + \text{Tan}^{-1} y = \tfrac{1}{2}\pi$

   (c) $\text{Sin}^{-1} x + \text{Tan}^{-1} y = 0$

   (d) $\text{Tan}^{-1} (x + y) = x$

6. Let $y = \text{Sin}^{-1} (\cos x) + \text{Cos}^{-1} (\sin x)$. Find y' and sketch the graph of the equation for $x \in [0, 2\pi]$.

7. Show that $\dfrac{d}{dx} \sin \text{Cos}^{-1} x = \dfrac{d}{dx} \cos \text{Sin}^{-1} x$. Is it all right to cancel $\dfrac{d}{dx}$?

8. What is the relation between $\dfrac{d}{dx} \text{Arctan} (1/x)$ and $\dfrac{d}{dx} \text{Arctan } x$? Use this result to find the relation between $\text{Arctan} (1/x)$ and $\text{Arctan } x$. (You will want to look at the case $x > 0$ and the case $x < 0$ separately.)

9. Verify the following identities. Would they still be true if we "canceled" the symbol $\dfrac{d}{dx}$?

(a) $\dfrac{d}{dx} \text{Sin}^{-1} (-x) = -\dfrac{d}{dx} \text{Sin}^{-1} x$

(b) $\dfrac{d}{dx} \text{Cos}^{-1} (-x) = -\dfrac{d}{dx} \text{Cos}^{-1} x$

(c) $\dfrac{d}{dx} \text{Tan}^{-1} \left(\dfrac{2x}{1 - x^2}\right) = \dfrac{d}{dx} (2 \text{ Tan}^{-1} x)$

(d) $\dfrac{d}{dx} \text{Cos}^{-1} \dfrac{\sqrt{1 - x^2} - x\sqrt{3}}{2} = \dfrac{d}{dx} \text{Sin}^{-1} x$

10. A statue 7 feet high is placed on a pedestal so that the base of the statue is 9 feet above your eye level. How far back should you stand from the base of the statue in order to get the "best" view?

11. Two guy wires are to be fastened at the same point P, which is h feet up a pole. The wires and the pole are in the same vertical plane, and one wire is to be fastened to the ground at a point x feet from the base of the pole and the other to a point on the same side of the pole and $2x$ feet from its base. Determine x so that the angle between the wires at the point P is a maximum.

12. Find $\lim\limits_{h \to 0} (\text{Sin}^{-1} h)/h$. (Write out the definition of $f'(0)$ if $f(x) = \text{Sin}^{-1} x$.)

13. Let $F(x) = \displaystyle\int_0^x \text{Tan}^{-1} (\cot t) \, dt$ (see Fig. 50-1).

(a) Sketch the graph of F.   (b) Evaluate the integral $\displaystyle\int_{-10\pi}^{10\pi} F(x) \, dx$.

14. Use the graph of the Arcsine function to show that

$$\int_0^x \text{Sin}^{-1} t \, dt = x \text{ Sin}^{-1} x - \int_0^{\text{Sin}^{-1} x} \sin t \, dt$$

$$= x \text{ Sin}^{-1} x + \cos (\text{Sin}^{-1} x) - 1$$

$$= x \text{ Sin}^{-1} x + \sqrt{1 - x^2} - 1.$$

Verify this equation by differentiation.

15. Let F be the function defined in Problem 43-9. Prove that $F(x) = \text{Cos}^{-1} x$.

(*Hint*: First show that $F'(x) = \dfrac{d}{dx} \text{Cos}^{-1} x$.) Check your work in Problem 45-10 now!

51. INTEGRATION FORMULAS THAT INVOLVE INVERSE TRIGONOMETRIC FUNCTIONS

According to the Fundamental Theorem of Calculus, we can write an integration formula $\int f(x)\,dx = F(x)$ for each differentiation formula $\dfrac{d}{dx}F(x) = f(x)$. Thus the Differentiation Formulas 50-3 and 50-8 lead to the integration formulas

$$(51\text{-}1) \qquad \int \frac{1}{\sqrt{a^2 - x^2}}\,dx = \mathrm{Sin}^{-1}\frac{x}{a} \qquad (a > 0)$$

and

$$(51\text{-}2) \qquad \int \frac{1}{a^2 + x^2}\,dx = \frac{1}{a}\,\mathrm{Tan}^{-1}\frac{x}{a} \qquad (a \neq 0).$$

Notice that there is a factor $1/a$ on the right-hand side of Formula 51-2 that is not present in Formula 51-1 and that we require a to be positive in Formula 51-1 but not in Formula 51-2.

Example 51-1. Evaluate the integral

$$\int_{-3}^{3} \frac{1}{\sqrt{12 - x^2}}\,dx.$$

Solution. We may use Formula 51-1 with $a = \sqrt{12} = 2\sqrt{3}$ to find that

$$\int_{-3}^{3} \frac{1}{\sqrt{12 - x^2}}\,dx = \mathrm{Sin}^{-1}\frac{x}{2\sqrt{3}}\bigg|_{-3}^{3} = \mathrm{Sin}^{-1}\tfrac{1}{2}\sqrt{3} - \mathrm{Sin}^{-1}\left(-\tfrac{1}{2}\sqrt{3}\right)$$

$$= \tfrac{1}{3}\pi - \left(-\tfrac{1}{3}\pi\right) = \tfrac{2}{3}\pi.$$

Example 51-2. Use Simpson's Rule with $n = 4$ to find π from the equation

$$(51\text{-}3) \qquad \tfrac{1}{4}\pi = \int_{0}^{1} \frac{1}{1 + x^2}\,dx.$$

Solution. First, let us use Formula 51-2 to verify that Equation 51-3 is correct:

$$\int_{0}^{1} \frac{1}{1 + x^2}\,dx = \mathrm{Tan}^{-1}x\bigg|_{0}^{1} = \tfrac{1}{4}\pi - 0 = \tfrac{1}{4}\pi.$$

Now we will approximate the integral by Simpson's Rule. We partition the interval $[0, 1]$ into four subintervals by the points $x_0 = 0$, $x_1 = \tfrac{1}{4}$, $x_2 = \tfrac{1}{2}$, $x_3 = \tfrac{3}{4}$, and $x_4 = 1$. The corresponding numbers obtained from the equation $y = 1/(1 + x^2)$

are $y_0 = 1$, $y_1 = \frac{16}{17}$, $y_2 = \frac{4}{5}$, $y_3 = \frac{16}{25}$, and $y_4 = \frac{1}{2}$. In this case, $h = \frac{1}{4}$, and Simpson's Rule gives us

$$\tfrac{1}{4}\pi \approx \tfrac{1}{12}[1 + 4(\tfrac{16}{17}) + 2(\tfrac{4}{5}) + 4(\tfrac{16}{25}) + \tfrac{1}{2}] = .785392.$$

Hence $\pi \approx 3.14157$. Since the value of π, correct to five decimal places, is 3.14159, we see that our approximation is a good one.

Example 51-3. Verify the integration formula

(51-4) $\qquad \displaystyle\int \sqrt{a^2 - x^2}\; dx = \tfrac{1}{2}\left(x\sqrt{a^2 - x^2} + a^2\,\text{Sin}^{-1}\frac{x}{a}\right) \qquad (a > 0).$

Solution. To verify an integration formula, we must differentiate the expression on the right-hand side and obtain the integrand. We first differentiate $x\sqrt{a^2 - x^2}$:

$$\frac{d}{dx}\,(x\sqrt{a^2 - x^2}) = \sqrt{a^2 - x^2}\,\frac{dx}{dx} + x\,\frac{d}{dx}\,\sqrt{a^2 - x^2}$$

$$= \sqrt{a^2 - x^2} - x^2/\sqrt{a^2 - x^2}$$

$$= (a^2 - 2x^2)/\sqrt{a^2 - x^2}.$$

From Formula 50-3 we see that

$$\frac{d}{dx}\left(a^2\,\text{Sin}^{-1}\frac{x}{a}\right) = a^2/\sqrt{a^2 - x^2}.$$

Thus

$$\frac{d}{dx}\,\tfrac{1}{2}\left(x\sqrt{a^2 - x^2} + a^2\,\text{Sin}^{-1}\frac{x}{a}\right) = \frac{1}{2}\left(\frac{a^2 - 2x^2}{\sqrt{a^2 - x^2}} + \frac{a^2}{\sqrt{a^2 - x^2}}\right)$$

$$= \sqrt{a^2 - x^2},$$

and Formula 51-4 is verified.

Example 51-4. Derive a formula for the area of the smaller region cut from a circle whose radius is r by a chord s units from the center of the circle.

Solution. Let us choose our circle with the origin as its center so that its equation is $x^2 + y^2 = r^2$. Then we will choose our chord to lie along the line $x = s$, where $s > 0$ (see Fig. 51-1). Because of the symmetry of the circle, the area we are seeking is given by the equation

$$A = 2\int_s^r \sqrt{r^2 - x^2}\; dx.$$

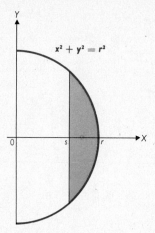

$x^2 + y^2 = r^2$

Figure 51-1

We use Formula 51-4 to evaluate the integral:

$$A = \left[x\sqrt{r^2 - x^2} + r^2 \text{Sin}^{-1}(x/r) \right]\Big|_s^r$$

$$= r^2 \text{Sin}^{-1} 1 - s\sqrt{r^2 - s^2} - r^2 \text{Sin}^{-1}(s/r)$$

$$= \tfrac{1}{2}\pi r^2 - s\sqrt{r^2 - s^2} - r^2 \text{Sin}^{-1}(s/r).$$

PROBLEMS 51

1. Evaluate the following integrals.

(a) $\displaystyle\int_0^{1/2} \frac{1}{\sqrt{1 - x^2}}\, dx$

(b) $\displaystyle\int_{-1/2}^{1/\sqrt{2}} \frac{1}{\sqrt{1 - x^2}}\, dx$

(c) $\displaystyle\int_{-1}^{0} \frac{1}{1 + x^2}\, dx$

(d) $\displaystyle\int_{-\sqrt{3}}^{1} \frac{1}{1 + x^2}\, dx$

(e) $\displaystyle\int_0^1 \frac{1}{\sqrt{36 - 9x^2}}\, dx$

(f) $\displaystyle\int_0^2 \frac{1}{36 + 9x^2}\, dx$

(g) $\displaystyle\int_0^{1/2} \frac{1}{1 + 4x^2}\, dx$

(h) $\displaystyle\int_0^{1/6} \frac{1}{\sqrt{1 - 9x^2}}\, dx$

2. Find the area of the region under the curve $y = 10/(9 + x^2)$ and above the interval $[-\sqrt{3}, \sqrt{3}]$.

3. Find the average value of f on the indicated interval.

(a) $f(x) = 3/\sqrt{12 - x^2}$, $[-3, 3]$
(b) $f(x) = 5/\sqrt{8 - x^2}$, $[0, 2]$
(c) $f(x) = 7/(3 + x^2)$, $[-1, 1]$
(d) $f(x) = 7/(3 + x^2)$, $[0, 3]$
(e) $f(x) = 1/(4 + x^2)$, $[-1, 1]$
(f) $f(x) = 1/(4 + x^2)$, $[0, 1]$

4. Evaluate the following integrals.

(a) $\displaystyle\int_0^1 \sqrt{4 - x^2}\,dx$

(b) $\displaystyle\int_0^3 \sqrt{36 - x^2}\,dx$

(c) $\displaystyle\int_0^{\sin 1} \sqrt{1 - x^2}\,dx$

(d) $\displaystyle\int_0^{\cos 1} \sqrt{1 - x^2}\,dx$

5. Find the area of the region that is bounded by the circle $x^2 + y^2 = 25$ and the lines $x = -3$ and $x = 4$.

6. Let x be a number in the interval $(0, \tfrac{1}{2}\pi)$ and compute.

(a) $\displaystyle\int_{\sin x}^{\cos x} \frac{1}{\sqrt{1 - t^2}}\,dt$

(b) $\displaystyle\int_{\cot x}^{\tan x} \frac{1}{1 + t^2}\,dt$

7. Show that for any numbers a and b the area of the region bounded by the curve $y = 1/(1 + x^2)$, the X-axis, and the lines $x = a$ and $x = b$ is less than π.

8. The region bounded by the curve $y = 1/\sqrt{1 + x^2}$, the line $x = 1$, and the co-ordinate axes is rotated about the X-axis. What is the volume of the resulting solid of revolution?

9. Find the area of the region interior to the ellipse $x^2/16 + y^2/9 = 1$ and to the right of the line $x = 2$.

10. If $f(x) = 1/\sqrt{1 - x^2}$ and b is a number between 0 and 1 but "close" to 1, then $f(b)$ is a large number. Show that the average value of f on the interval $[0, b]$ is a number close to $\tfrac{1}{2}\pi$.

11. Draw the curve $y = 1/\sqrt[4]{1 - x^2}$. The region bounded by this graph, the X-axis, and the lines $x = a$ and $x = b$, where $-1 < a < b < 1$, is rotated about the X-axis. Show that the volume of the resulting solid is less than π^2.

12. The region bounded by the curve $x\sqrt{4 + y^2} = 1$, the coordinate axes, and the line $y = 100$ is rotated about the Y-axis. Find the approximate volume of the resulting solid.

52. THE HYPERBOLIC FUNCTIONS

Certain combinations of e^x and e^{-x} occur so often in mathematical applications that they are used to define new functions. The **hyperbolic cosine** function and the **hyperbolic sine** function are defined by the equations

(52-1) $$\cosh x = \frac{e^x + e^{-x}}{2}$$

and

(52-2) $$\sinh x = \frac{e^x - e^{-x}}{2}.$$

These functions are called "hyperbolic" functions because their values are related to the coordinates of points of a hyperbola in somewhat the same way that the values of the trigonometric functions are related to the coordinates of points of a circle. We won't go into the details of this relationship, but the following example will give you the essential idea. (Also see Problems 43-9, 43-10, 50-15, and 53-10.)

Example 52-1. Show that for any number t,

(52-3) $$\cosh^2 t - \sinh^2 t = 1.$$

Solution. According to our definitions of the hyperbolic functions,

$$\cosh^2 t - \sinh^2 t = \left(\frac{e^t + e^{-t}}{2}\right)^2 - \left(\frac{e^t - e^{-t}}{2}\right)^2.$$

It is a matter of simple algebra to expand the terms on the right-hand side of this equation and see that the resulting number is 1. Equation 52-3 tells us that the point $(\cosh t, \sinh t)$ belongs to the hyperbola $x^2 - y^2 = 1$, just as the point $(\cos t, \sin t)$ belongs to the circle $x^2 + y^2 = 1$.

The remaining hyperbolic functions can be defined in terms of hyperbolic sines and cosines:

(52-4) $$\tanh x = \frac{\sinh x}{\cosh x} = \frac{e^x - e^{-x}}{e^x + e^{-x}},$$

(52-5) $$\coth x = \frac{1}{\tanh x} = \frac{e^x + e^{-x}}{e^x - e^{-x}},$$

(52-6) $$\operatorname{sech} x = \frac{1}{\cosh x} = \frac{2}{e^x + e^{-x}},$$

(52-7) $$\operatorname{csch} x = \frac{1}{\sinh x} = \frac{2}{e^x - e^{-x}}.$$

Figures 52-1, 52-2, and 52-3 show the graphs of the hyperbolic cosine, sine, and tangent functions. These functions are tabulated in Table III. It is clear from the defining equations that the entire interval $(-\infty, \infty)$ serves as domain for cosh, sinh, tanh, and sech, but that 0 does not belong to the domain of coth or csch. You might find it an interesting exercise to show that the range of cosh is the interval $[1, \infty)$, the range of sinh is $(-\infty, \infty)$, the range of tanh is $(-1, 1)$, the range of coth is $(-\infty, -1) \cup (1, \infty)$, the range of sech is $(0, 1]$, and the range of csch is $(-\infty, 0) \cup (0, \infty)$.

The values of the hyperbolic functions satisfy many identities similar to those satisfied by the trigonometric functions. In addition to Equation 52-3, which is a fundamental identity,

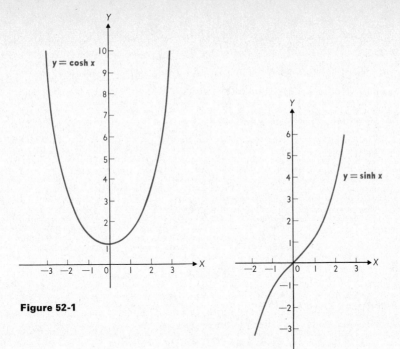

Figure 52-1

Figure 52-2

(52-8) $1 - \tanh^2 x = \operatorname{sech}^2 x,$

(52-9) $\coth^2 x - 1 = \operatorname{csch}^2 x,$

(52-10) $\sinh (u + v) = \sinh u \cosh v + \cosh u \sinh v,$

(52-11) $\cosh (u + v) = \cosh u \cosh v + \sinh u \sinh v.$

These identities can be verified by straightforward algebraic manipulation.

Example 52-2. Find expressions for $\sinh 2x$ and $\cosh 2x$.

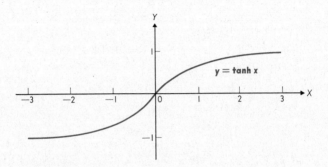

Figure 52-3

Solution. By setting $u = x$ and $v = x$ in Equations 52-10 and 52-11, we obtain

$$(52\text{-}12) \qquad\qquad \sinh 2x = 2 \sinh x \cosh x$$

and

$$(52\text{-}13) \qquad\qquad \cosh 2x = \cosh^2 x + \sinh^2 x$$

To find the derivatives of $\cosh x$, $\sinh x$, $\tanh x$, and so on, we utilize our defining equations. For example,

$$\frac{d}{dx} \cosh x = \frac{d}{dx} \tfrac{1}{2}(e^x + e^{-x}) = \frac{1}{2}\left(\frac{de^x}{dx} + \frac{de^{-x}}{dx}\right)$$

$$= \tfrac{1}{2}(e^x - e^{-x}) = \sinh x.$$

Thus we find that

$$(52\text{-}14) \qquad\qquad \frac{d \cosh x}{dx} = \sinh x.$$

When we incorporate this differentiation formula into the Chain Rule Equation $\dfrac{df(u)}{dx} = \dfrac{df(u)}{du}\dfrac{du}{dx}$, we obtain the formula

$$(52\text{-}15) \qquad\qquad \frac{d \cosh u}{dx} = \sinh u \, \frac{du}{dx}.$$

In the problems at the end of the section, we leave to you the verification of the differentiation formulas

$$(52\text{-}16) \qquad\qquad \frac{d \sinh u}{dx} = \cosh u \, \frac{du}{dx},$$

$$(52\text{-}17) \qquad\qquad \frac{d \tanh u}{dx} = \operatorname{sech}^2 u \, \frac{du}{dx},$$

$$(52\text{-}18) \qquad\qquad \frac{d \coth u}{dx} = -\operatorname{csch}^2 u \, \frac{du}{dx},$$

$$(52\text{-}19) \qquad\qquad \frac{d \operatorname{sech} u}{dx} = -\operatorname{sech} u \tanh u \, \frac{du}{dx},$$

$$(52\text{-}20) \qquad\qquad \frac{d \operatorname{csch} u}{dx} = -\operatorname{csch} u \coth u \, \frac{du}{dx}.$$

Example 52-3.   Compute $\dfrac{d}{dx} \ln (\cosh x)$.

Solution.   In the differentiation formula $\dfrac{d \ln u}{dx} = \dfrac{1}{u} \dfrac{du}{dx}$ we replace u with $\cosh x$ and use Equation 52-14:

$$\frac{d}{dx} \ln (\cosh x) = \frac{1}{\cosh x} \frac{d \cosh x}{dx} = \frac{\sinh x}{\cosh x}.$$

Thus we have derived the formula

(52-21)
$$\frac{d}{dx} \ln (\cosh x) = \tanh x.$$

Example 52-4.   Show that

$$y = A \sinh kx + B \cosh kx$$

satisfies the differential equation $y'' - k^2 y = 0$.

Solution.   From Formulas 52-15 and 52-16, with $u = kx$, we see that

$$y' = Ak \cosh kx + Bk \sinh kx$$

and

$$y'' = Ak^2 \sinh kx + Bk^2 \cosh kx.$$

Thus $y'' = k^2 y$; in other words, $y'' - k^2 y = 0$.

Example 52-5.   Show that the graph of the function defined by the equation

$$y = 8 \sinh x - 17 \cosh x$$

is concave down everywhere and find the maximum value of the function.

Solution.   We have
$$y' = 8 \cosh x - 17 \sinh x$$

and
$$y'' = 8 \sinh x - 17 \cosh x.$$

To show that our graph is concave down, we will show that $y'' < 0$—that is, that $8 \sinh x < 17 \cosh x$. This last inequality is equivalent to the inequality $\tanh x < \frac{17}{8}$, which is clearly true, since (see Fig. 52-3) $\tanh x \leq 1$. At the maximum point of the graph, $y' = 0$; that is,

$$8 \cosh x = 17 \sinh x.$$

This equation is equivalent to the equation tanh $x = \frac{8}{17}$, and we see from Fig. 52-3 that its solution is a number x_0 approximately equal to .5. The maximum value of our function is therefore

$$y_0 = 8 \sinh x_0 - 17 \cosh x_0,$$

and we will now proceed to calculate this number. First of all, tanh $x_0 = \frac{8}{17}$, and so it follows from Equations 52-6 and 52-8 that cosh $x_0 = \frac{17}{15}$. Now we have

$$y_0 = 8 \sinh x_0 - 17 \cosh x_0$$
$$= \cosh x_0 (8 \tanh x_0 - 17)$$
$$= \tfrac{17}{15}[8(\tfrac{8}{17}) - 17] = -15.$$

Corresponding to each of the differentiation formulas for the hyperbolic functions is an integration formula. Thus corresponding to Formula 52-14 is the integration formula

(52-22) $$\int \sinh x \, dx = \cosh x.$$

Similarly, corresponding to the other differentiation formulas (with $u = x$) are the integration formulas

(52-23) $$\int \cosh x \, dx = \sinh x,$$

(52-24) $$\int \operatorname{sech}^2 x \, dx = \tanh x,$$

(52-25) $$\int \operatorname{csch}^2 x \, dx = -\coth x,$$

(52-26) $$\int \operatorname{sech} x \tanh x \, dx = -\operatorname{sech} x,$$

(52-27) $$\int \operatorname{csch} x \coth x \, dx = -\operatorname{csch} x.$$

Example 52-6. Find the area of the region that is bounded by the coordinate axes, the curve $y = \cosh x$, and the line $x = 1$.

Solution. We are to find the area of the shaded region in Fig. 52-4, and so we must evaluate the integral $\int_0^1 \cosh x \, dx$. Using Integration Formula 52-23, we have

$$\int_0^1 \cosh x \, dx = \sinh x \Big|_0^1 = \sinh 1 - \sinh 0 = 1.1752.$$

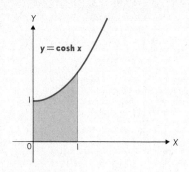

Figure 52-4

We can immediately state another integration formula from Differentiation Formula 52-21 that we found in Example 52-3:

(52-28)
$$\int \tanh x \ dx = \ln (\cosh x).$$

We will leave the verification of the following integration formula for the problems:

(52-29)
$$\int \coth x \ dx = \ln |\sinh x|.$$

PROBLEMS 52

1. Find y'.
   (a) $y = \cosh e^x - e^{\cosh x}$
   (b) $y = \sinh^2 x - \sinh x^2$
   (c) $y = \ln (\operatorname{csch} x) - \ln (\operatorname{sech} x)$
   (d) $y = \operatorname{csch} (\ln x) - \operatorname{sech} (\ln x)$
   (e) $y = \sin (\cosh x) - \cos (\sinh x)$
   (f) $y = \operatorname{sech} x \operatorname{csch} x$

2. Evaluate the following integrals, using tables as necessary.

   (a) $\displaystyle\int_0^1 3 \cosh x \ dx$

   (b) $\displaystyle\int_{-1}^2 \sinh x \ dx$

   (c) $\displaystyle\int_0^1 \tanh x \ dx$

   (d) $\displaystyle\int_{-3}^3 \tanh x \ dx$

   (e) $\displaystyle\int_0^{\ln 2} \cosh x \ dx$

   (f) $\displaystyle\int_{\ln 1/2}^{\ln 2} \sinh x \ dx$

3. Verify that the graphs of sinh and cosh are concave up in the first quadrant.

4. (a) Verify Formula 52-16.
   (b) Verify Formula 52-17.
   (c) Verify Formula 52-18.
   (d) Verify Formula 52-19.
   (e) Verify Formula 52-20.
   (f) Verify Formula 52-29.

5. Show that for each number u, $\cosh u + \sinh u = e^u$. Use this fact to verify the identity

$$(\cosh x + \sinh x)^n = \cosh nx + \sinh nx.$$

6. (a) Show that $y = e^{ax} \sinh bx$ satisfies the differential equation

$$y'' - 2ay' + (a^2 - b^2)y = 0.$$

(b) Show that $y = e^{ax} \cosh bx$ also satisfies it.

7. Show that $\mathrm{Tan}^{-1}(\sinh x) = \mathrm{Sin}^{-1}(\tanh x)$. (*Hint*: First show that their derivatives are equal.)

8. (a) Show that $\dfrac{d}{dx}(2 \mathrm{Tan}^{-1} e^x) = \dfrac{d}{dx} \mathrm{Tan}^{-1}(\sinh x)$.

(b) Is the equation we found in part (a) still valid if we "cancel" $\dfrac{d}{dx}$?

(c) State two integration formulas with sech x as the integrand.

9. Suppose that $F(x) = \displaystyle\int_0^x \operatorname{sech} t \, dt$. Use the inequality $\operatorname{sech} t \leq 2e^{-t}$ to show that $F(x) \leq 2(1 - e^{-x})$ if $x \geq 0$.

10. (a) If $a \neq 0$, show that $\displaystyle\int \sinh ax \, dx = (1/a) \cosh ax$ and $\displaystyle\int \cosh ax \, dx = (1/a) \sinh ax$.

(b) Use the results of part (a) and suitable identities to find integration formulas whose left-hand members are $\displaystyle\int \cosh^2 x \, dx$ and $\displaystyle\int \sinh x \cosh x \, dx$.

(c) Use the results of part (b) to find the volume of the solid that is generated by rotating the region of Example 52-6 about the X-axis.

11. Use algebra to verify the following identities.

(a) $\sinh (u - v) = \sinh u \cosh v - \cosh u \sinh v$

(b) $\cosh (u - v) = \cosh u \cosh v - \sinh u \sinh v$

(c) $\tanh (u + v) = (\tanh u + \tanh v)/(1 + \tanh u \tanh v)$

12. Show that for each pair of numbers a and b,

$$\int_a^b \cosh x \, dx = 2 \cosh \tfrac{1}{2}(b + a) \sinh \tfrac{1}{2}(b - a).$$

53. INVERSE HYPERBOLIC FUNCTIONS

You can see from Figs. 52-1, 52-2, and 52-3 that the hyperbolic sine and hyperbolic tangent functions have inverses but that the hyperbolic cosine function doesn't. Just as we did with the trigonometric functions, we choose a restriction of the hyperbolic cosine function that does have an inverse. Thus we define the **Hyperbolic Cosine** function as the function whose domain is the interval $[0, \infty)$ and whose rule of correspondence is $y = \cosh x$. This function has an inverse. Similarly, we choose a restriction of the hyperbolic secant function, the **Hyperbolic**

Secant function, that has an inverse. The inverse hyperbolic functions are thus defined by the equations

(53-1) $y = \sinh^{-1} x$ if and only if $x = \sinh y,$

(53-2) $y = \mathrm{Cosh}^{-1} x$ if and only if $x = \cosh y$ and $y \geq 0,$

(53-3) $y = \tanh^{-1} x$ if and only if $x = \tanh y,$

(53-4) $y = \coth^{-1} x$ if and only if $x = \coth y,$

(53-5) $y = \mathrm{Sech}^{-1} x$ if and only if $x = \mathrm{sech}\, y$ and $y \geq 0,$

(53-6) $y = \mathrm{csch}^{-1} x$ if and only if $x = \mathrm{csch}\, y.$

Since the hyperbolic functions are defined in terms of exponentials with base e, it is not surprising that we can express the values of the inverse hyperbolic functions in terms of logarithms with base e.

Example 53-1. Show that

(53-7) $$\tanh^{-1} x = \tfrac{1}{2} \ln \left(\frac{1 + x}{1 - x} \right), \qquad |x| < 1.$$

Solution. If we let $y = \tanh^{-1} x$, then

$$x = \tanh y = \frac{e^y - e^{-y}}{e^y + e^{-y}}.$$

Thus

$$xe^y + xe^{-y} = e^y - e^{-y},$$

and so

$$(x - 1)e^y = -(1 + x)e^{-y}.$$

When we multiply both sides of this equation by e^y, we obtain

$$(x - 1)e^{2y} = -(1 + x),$$

and hence

$$e^{2y} = (1 + x)/(1 - x).$$

Now we take the logarithm of both sides, and remember that $y = \tanh^{-1} x$, to obtain Equation 53-7.

By proceeding as we did in the foregoing example, you can verify that

(53-8) $$\sinh^{-1} x = \ln (x + \sqrt{x^2 + 1}),$$

(53-9) $$\mathrm{Cosh}^{-1} x = \ln (x + \sqrt{x^2 - 1}),$$

(53-10) $$\coth^{-1} x = \tfrac{1}{2} \ln \left(\frac{x + 1}{x - 1} \right), \qquad |x| > 1.$$

In order to find differentiation formulas for the inverse hyperbolic functions, we may either use the rules for finding derived functions of inverse functions that we developed in Section 45 or we may differentiate the logarithmic expressions for the inverse hyperbolic functions.

Example 53-2. Show in two ways that

(53-11) $$\frac{d}{dx}\sinh^{-1} x = \frac{1}{\sqrt{x^2 + 1}}.$$

Solution. Method 1.

If $y = \sinh^{-1} x$, then

$$\frac{d}{dx}\sinh^{-1} x = 1/\frac{d}{dy}\sinh y \qquad \text{(Equation 45-4)}$$

$$= 1/\cosh y \qquad \text{(Equation 52-16)}$$

$$= 1/\sqrt{\sinh^2 y + 1} \qquad \text{(Equation 52-3)}$$

$$= 1/\sqrt{x^2 + 1} \qquad (\sinh y = x).$$

Method 2.

$$\frac{d}{dx}\sinh^{-1} x = \frac{d}{dx}\ln (x + \sqrt{x^2 + 1}) \qquad \text{(Equation 53-8)}$$

$$= \frac{1}{x + \sqrt{x^2 + 1}}\frac{d}{dx}(x + \sqrt{x^2 + 1}) \qquad \text{(Equation 44-4)}$$

$$= \frac{1 + x/\sqrt{x^2 + 1}}{x + \sqrt{x^2 + 1}} = 1/\sqrt{x^2 + 1}.$$

When we incorporate Equation 53-11 into the Chain Rule Equation $\dfrac{df(u)}{dx} = \dfrac{df(u)}{du}\dfrac{du}{dx}$, we obtain

(53-12) $$\frac{d}{dx}\sinh^{-1} u = \frac{1}{\sqrt{u^2 + 1}}\frac{du}{dx}.$$

The differentiation formulas for the other inverse hyperbolic functions may be obtained in the same way:

(53-13) $$\frac{d}{dx}\text{Cosh}^{-1} u = \frac{1}{\sqrt{u^2 - 1}}\frac{du}{dx} \qquad (u > 1),$$

(53-14) $\dfrac{d}{dx}\tanh^{-1}u = \dfrac{1}{1-u^2}\dfrac{du}{dx}$ $(|u| < 1)$,

(53-15) $\dfrac{d}{dx}\coth^{-1}u = \dfrac{1}{1-u^2}\dfrac{du}{dx}$ $(|u| > 1)$,

(53-16) $\dfrac{d}{dx}\operatorname{Sech}^{-1}u = \dfrac{-1}{u\sqrt{1-u^2}}\dfrac{du}{dx}$ $(0 < u < 1)$,

(53-17) $\dfrac{d}{dx}\operatorname{csch}^{-1}u = \dfrac{-1}{|u|\sqrt{1+u^2}}\dfrac{du}{dx}$.

These differentiation formulas lead to some useful integration formulas. For example,

(53-18) $\displaystyle\int \dfrac{1}{\sqrt{x^2 + a^2}}\, dx = \sinh^{-1}\dfrac{x}{a}$ $(a > 0)$.

To verify this formula, we merely have to replace u with x/a in Equation 53-12:

$$\frac{d}{dx}\sinh^{-1}(x/a) = \frac{1}{\sqrt{(x/a)^2 + 1}}\frac{d}{dx}(x/a)$$

$$= \frac{1}{a\sqrt{(x/a)^2 + 1}} = \frac{1}{\sqrt{x^2 + a^2}}.$$

(Where did we use the fact that $a > 0$?)

In exactly the same way, we can use Formula 53-13 to verify the integration formula

(53-19) $\displaystyle\int \dfrac{1}{\sqrt{x^2 - a^2}}\, dx = \operatorname{Cosh}^{-1}\dfrac{x}{a}$ $(0 < a < x)$.

From Differentiation Formulas 53-14 and 53-15 come the integration formulas

(53-20) $\displaystyle\int \dfrac{1}{a^2 - x^2}\, dx = \begin{cases} a^{-1}\tanh^{-1}(x/a) & \text{if } |x| < |a|, \\ a^{-1}\coth^{-1}(x/a) & \text{if } |x| > |a|. \end{cases}$

Equation 53-7 shows us the relation between the inverse hyperbolic tangent and the natural logarithm, and we could use such equations, together with Equations 53-20, to obtain the integration formula

(53-21) $\displaystyle\int \dfrac{1}{a^2 - x^2}\, dx = \dfrac{1}{2a}\ln\left|\dfrac{a+x}{a-x}\right|$ $(a \neq 0)$.

PROBLEMS 53

1. Sketch the graphs of the following equations.

   (a) $y = \sinh^{-1} x$ (b) $y = \mathrm{Cosh}^{-1} x$

   (c) $y = \tanh^{-1} x$ (d) $y = \coth^{-1} x$

2. Use Tables II and III to verify the following statements.

   (a) $\sinh^{-1} \frac{3}{4} = \ln 2$ (b) $\tanh^{-1} .5 = \frac{1}{2} \ln [(1 + .5)/(1 - .5)]$

   (c) $\mathrm{Cosh}^{-1} 2 = \ln (2 + \sqrt{2^2 - 1})$ (d) $\coth^{-1} 3 = \frac{1}{2} \ln [(3 + 1)/(3 - 1)]$

3. Find y'.

   (a) $y = \sinh^{-1} x^2$ (b) $y = (\sinh^{-1} x)^2$

   (c) $y = (\sinh x)^{-2}$ (d) $y = \sinh x^{-2}$

4. Compute y'.

   (a) $y = \tanh^{-1} (\cos x)$ (b) $y = \mathrm{csch}^{-1} (\tan x)$

   (c) $y = \coth^{-1} (\cosh x)$ (d) $y = \sinh^{-1} (\tan x)$

   (e) $y = \mathrm{Cosh}^{-1} (\cos x)$ (f) $y = \mathrm{Cosh}^{-1} (\sec x)$

5. Evaluate the following integrals.

   (a) $\displaystyle\int_0^2 \frac{1}{\sqrt{x^2 + 4}}\, dx$ (b) $\displaystyle\int_0^1 \frac{1}{\sqrt{1 + x^2}}\, dx$

   (c) $\displaystyle\int_2^3 \frac{1}{\sqrt{x^2 - 1}}\, dx$ (d) $\displaystyle\int_5^7 \frac{1}{\sqrt{x^2 - 9}}\, dx$

   (e) $\displaystyle\int_{-1/2}^{1/2} \frac{1}{1 - x^2}\, dx$ (f) $\displaystyle\int_{-1}^1 \frac{1}{2 - x^2}\, dx$

   (g) $\displaystyle\int_4^5 \frac{1}{x^2 - 9}\, dx$ (h) $\displaystyle\int_{-1}^1 \frac{1}{\sqrt{2 - x^2}}\, dx$

6. Verify the following differentiation formulas and rewrite them as integration formulas.

   (a) $\dfrac{d}{dx} [x \sinh^{-1} x - \sqrt{1 + x^2}] = \sinh^{-1} x$

   (b) $\dfrac{d}{dx} [x \mathrm{Cosh}^{-1} x - \sqrt{x^2 - 1}] = \mathrm{Cosh}^{-1} x$

   (c) $\dfrac{d}{dx} [x \tanh^{-1} x + \ln \sqrt{1 - x^2}] = \tanh^{-1} x$

7. (a) How are f and g related if $f(x) = \sinh (\sinh^{-1} x)$ and $g(x) = \sinh^{-1} (\sinh x)$?

   (b) How are f and g related if $f(x) = \sin (\mathrm{Sin}^{-1} x)$ and $g(x) = \mathrm{Sin}^{-1} (\sin x)$?

8. Find the following numbers.

   (a) $\displaystyle\lim_{h \to 0} \frac{\sinh^{-1} h}{h}$ (b) $\displaystyle\lim_{z \to 3} \frac{\mathrm{Cosh}^{-1} z - \mathrm{Cosh}^{-1} 3}{z - 3}$

9. Verify the integration formulas.

(a) $\displaystyle\int \sqrt{x^2 + a^2}\, dx = \frac{1}{2}\left(x\sqrt{x^2 + a^2} + a^2 \sinh^{-1}\frac{x}{a}\right)$

(b) $\displaystyle\int \sqrt{x^2 - a^2}\, dx = \frac{1}{2}\left(x\sqrt{x^2 - a^2} - a^2 \mathrm{Cosh}^{-1}\frac{x}{a}\right),$ $(x > a).$

10. Let G be the function defined in Problem 43-10. Prove that $G(x) = \mathrm{Cosh}^{-1} x$. (*Hint*: See Problem 50-15.)

11. Show that if $b > 1$, then $\sinh^{-1}\sqrt{b^2 - 1} = \mathrm{Cosh}^{-1} b$.

12. Show that $\sinh^{-1}(\mathrm{Tan}\, x) = \tanh^{-1}(\mathrm{Sin}\, x)$. (You might start by comparing derivatives.)

13. Either use algebra, as we did in Example 53-1, or use calculus (Equations 53-12, 53-13, 53-15, and the differentiation rules for logarithms) to verify Formulas 53-8, 53-9, and 53-10.

The Fundamental Theorem of Calculus says that if $F'(x) = f(x)$, then $\displaystyle\int_a^b f(x)\, dx = F(x)\Big|_a^b$. In other words, if we can find antiderivatives, then we can integrate. Conversely, Theorem 43-2 tells us that $\displaystyle\int_a^x f(t)\, dt$ is an antiderivative of $f(x)$. So if we can integrate, then we can find antiderivatives. These points are a little subtle but extremely important, nevertheless.

Most of the time we find antiderivatives by using our memory or using tables. Thus our experience with differentiation immediately tells us that x^2 is an antiderivative of $2x$. But we could also rely on Theorem 43-2 to say, just as correctly, that $\displaystyle\int_0^{} 2t\, dt$ is an antiderivative of $2x$. Graphically, this latter number is the area of the shaded region in Fig. VI-1, which is $\frac{1}{2}x\cdot 2x = x^2$, as before. Of course, this example is too simple to show any reason for defining functions by integrals, but in this chapter we learned one very important function that is so defined, the logarithmic function. Many mathematicians prefer, on theoretical grounds, to define the inverse trigonometric functions by means of integrals (for example, $\mathrm{Sin}^{-1} x = \displaystyle\int_0^x \frac{1}{\sqrt{1 - t^2}}\, dt$) and then invert them to obtain the trigonometric functions. This definition avoids some touchy questions about arclength that we glossed over when we gave the standard definitions of the trigono-

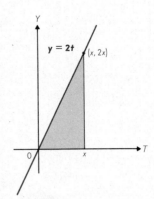

Figure VI-1

metric functions in Section 6. The introduction of the high-speed computer has increased the practicability of defining functions by means of integrals; numerical integration on a computer produces a table of values in no time at all.

In any case, you should know the formulas for derivatives and integrals of logarithmic and exponential functions and some trigonometric functions and their inverses (Tan^{-1} is a particularly important function). The hyperbolic functions are a little specialized, but they do occur in some applications of mathematics.

Review Problems, Chapter Six

1. Find y'.
   (a) $y = x^{e^e} + e^{x^e} + e^{e^x}$
   (b) $y = 3 \ln x + \ln 3x + \ln 3^x$
   (c) $y = \text{Sin}^{-1} x + \text{Sin } x^{-1} + (\text{Sin } x)^{-1}$
   (d) $y = (\text{Tan}^{-1} x)^2 + \text{Tan}^{-1} x^2 + (\text{Tan } x)^2$
   (e) $y = \cosh \ln x + \ln \cosh x$
   (f) $y = \sinh^{-1} \exp x + \exp^{-1} \sinh x$

2. Sketch the curves.
   (a) $\ln x + \ln y = 0$
   (b) $\ln x \ln y = 0$
   (c) $\ln x - \ln y = 0$

   (d) $\dfrac{\ln x}{\ln y} = 0$
   (e) $\ln xy = 0$
   (f) $\ln \dfrac{x}{y} = 0$

3. Show that $\displaystyle\int_a^b \frac{c}{x}\, dx = \ln \left(\frac{b}{a}\right)^c$ if a and b are positive numbers.

4. Solve for x.
   (a) $\ln x^2 = (\ln x)^2$
   (b) $e^{x^2} = (e^x)^2$

   (c) $\ln \dfrac{1}{x} = \dfrac{1}{\ln x}$
   (d) $e^{1/x} = \dfrac{1}{e^x}$

5. Show that $\dfrac{d}{dx} x^{\log 56} = \log 56 x^{\log 5 \cdot 6}$.

6. Verify the following integration formulas.

   (a) $\displaystyle\int e^{ax} \cos bx\, dx = \left(\frac{a \cos bx + b \sin bx}{a^2 + b^2}\right) e^{ax}$

   (b) $\displaystyle\int e^{ax} \sin bx\, dx = \left(\frac{a \sin bx - b \cos bx}{a^2 + b^2}\right) e^{ax}$

7. Sketch the curves.

   (a) $y = \displaystyle\int_1^{3x} \frac{1}{t}\, dt$
   (b) $y = \displaystyle\int_x^{3x} \frac{x}{t}\, dt$

8. How many solutions does the equation $e^x = ex$ have?

9. Sketch the curve $y = x^2 \ln 1/|x|$. Check for maximum and minimum points, concavity, and so on.

10. Evaluate the following integrals.

(a) $\displaystyle\int_{-b}^{b} \frac{1}{\sqrt{4b^2 - x^2}}\, dx$

(b) $\displaystyle\int_{-b}^{b} \frac{1}{\sqrt{4b^2 + x^2}}\, dx$

(c) $\displaystyle\int_{-b}^{b} \frac{1}{4b^2 + x^2}\, dx$

(d) $\displaystyle\int_{-b}^{b} \frac{1}{4b^2 - x^2}\, dx$

11. Show that $y = A \cosh 4x + B \sinh 4x$ satisfies the differential equation $y'' = 16y$ regardless of what A and B are. Find A and B so that $y = 3$ and $y' = -8$ when $x = 0$.

12. Evaluate the integral $\displaystyle\int_{1}^{e^{100}} [\![\ln x]\!]\, dx$.

13. Convince yourself that the equation $(\text{Tan } x)^{-1} = \text{Tan}^{-1}\, x$ has a positive solution r and a negative solution $-r$. Is $r < 1$ or $r > 1$?

14. (a) Sketch the curve $y = \text{Arccos}\,(\cos t)$ in the TY-plane.
(b) Use the geometric interpretation of the integral to find $F(\tfrac{1}{2}\pi)$, $F(\pi)$, and $F(-\tfrac{1}{2}\pi)$

if $F(x) = \displaystyle\int_{0}^{x} \text{Arccos}\,(\cos t)\, dt$.

(c) How do you know that F is an increasing function?
(d) Discuss the concavity of the graph of F.
(e) Sketch the graph of F.

15. Suppose that $0 < x < \tfrac{1}{2}\pi$, $y > 0$, and $\cos x \cosh y = 1$.
(a) Show that $y = \ln (\sec x + \tan x)$.
(b) Compute y' from the equation of part (a) and by implicit differentiation of the given equation. Reconcile your answers.

16. After t minutes of flight, the coordinates of an airplane are $(x, y) = (4 \cosh t, 3 \sinh t)$. Use Equation 52-3 to describe its path in the XY-plane.

17. If $F(x) = \displaystyle\int_{2}^{x} (|t - 2| + |t - 4|)\, dt$, find $F(6)$, $F'(6)$, $F''(6)$, and $F^{-1}(6)$.

18. Suppose that f is an even function $[f(-x) = f(x)]$ and g is odd $[g(-x) = -g(x)]$, and that $F(x) = \displaystyle\int_{0}^{x} f(t)\, dt$ and $G(x) = \displaystyle\int_{0}^{x} g(t)\, dt$. Can you convince yourself by graphical interpretation of integrals that F is odd and G is even?

19. Suppose that f is continuous in some interval I that contains a point c and let $F(x) = \displaystyle\int_{c}^{x} f(t)\, dt$. Show that replacing f with F in Equation 19-1 gives Equation 34-2.

20. Suppose that $u = f(x)$ and $v = g(x)$, where $f(x) > 1$ and $g(x) > 0$. Show that

$$\frac{d}{dx} \log_u v = \left(\frac{1}{v}\frac{dv}{dx} - \frac{1}{u}\frac{du}{dx}\log_u v\right)\log_u e.$$

21. Use the trigonometric identities $\sin(x - t) = \sin x \cos t - \cos x \sin t$ and $\cos(x - t) = \cos x \cos t + \sin x \sin t$ to show that $\dfrac{d}{dx}\displaystyle\int_{0}^{x} \sin(x - t)\sqrt{t}\, dt = \displaystyle\int_{0}^{x} \cos(x - t)\sqrt{t}\, dt$. What do we get if we differentiate again?

7 Techniques of Integration

We have developed a number of specific integration formulas, rules that tell us how to find integrals of polynomials, some trigonometric functions, logarithmic functions, and so forth. In order to utilize these formulas more fully, we will now introduce some additional general techniques of integration. Our sampling of integration techniques will be far from exhaustive. Not many years ago a large part of the standard calculus course was devoted to various integration tricks. Today we restrict our presentation of methods of integration to a few essentials, thereby leaving time for more important topics. You can look in almost any one of the old calculus books and find tricks that we have omitted.

54. USING INTEGRATION FORMULAS

An integral $\int_a^b f(x)\,dx$ is the limit of sums of the form $\sum_{i=1}^n f(x_i^*)\,\Delta x_i$. In some cases, we must use numerical methods, such as Simpson's Parabolic Rule or the Trapezoidal Rule, to evaluate an integral, but often an integration formula

$$(54\text{-}1) \qquad \int f(x)\,dx = F(x)$$

applies in some interval *I*. Formula 54-1 means that the integral $\int_a^b f(x)\,dx$ is the number $F(x)\Big|_a^b = F(b) - F(a)$ for each pair of numbers *a* and *b* of *I*. We also extend this convention to more general integration formulas, such as Formula 54-2 below. Thus a formula like

$$\int f(x)\,dx = F(x) + \int g(x)\,dx$$

is really an abbreviation of the equation

$$\int_a^b f(x)\,dx = F(x)\Big|_a^b + \int_a^b g(x)\,dx.$$

When you use Formula 54-1 to replace $\int f(x)\,dx$ *with F(x), you must realize that you are really replacing* $\int_a^b f(x)\,dx$ *with* $F(x)\Big|_a^b$.

The Fundamental Theorem of Calculus and its converse, Theorems 37-1 and 43-2, tell us that if *f* is continuous in an interval *I*, then Formula 54-1 is equivalent to saying that $F'(x) = f(x)$ for each $x \in I$. In other words, *F(x) is an antiderivative of f(x).*

We have listed a number of differentiation formulas in Table IV and integration formulas in Table V in the back of the book. The first 25 formulas of Table V are labeled Basic Integration Formulas. They are essentially the differentiation formulas of Table IV read from right to left. You should learn most of these basic formulas by heart. Table V is actually a very short list of integration formulas. Longer lists can be found in the standard collections of mathematical tables, and you can find extensive "tables of integrals" in any mathematics library. Since the most practical way to find an integral you don't know is to look it up in a table, you should learn how to use these tables. Notice that the tables use the abbreviation $\int \dfrac{du}{f(u)}$ for $\int \dfrac{1}{f(u)}\,du$.

We have simply set down a list of integration formulas in Table V, many of which you have not seen before. What assurance have we that these formulas are correct? Because of the relation between differentiation and integration formulas that we just mentioned, it is easy to answer this question for any of our integration formulas. *To verify a specific integration formula in Table V, we differentiate the right-hand side to obtain the integrand.*

Example 54-1.   Verify Integral Formula V-33.

Solution. We differentiate the right-hand side and obtain the integrand:

$$\frac{d}{du} \ln |\sec u| = \frac{1}{\sec u} \frac{d \sec u}{du} \quad \text{(Formula IV-10 with } \sec u \text{ in place of } u \text{ and } u \text{ in place of } x)$$

$$= \frac{\sec u \tan u}{\sec u} \quad \text{(Formula IV-7 with } u \text{ in place of } x)$$

$$= \tan u \quad \text{(Algebra).}$$

Example 54-2. Evaluate the integral $\displaystyle\int_1^{100} \frac{dx}{x(4x + 3)}$.

Solution. We set $a = 4$, $b = 3$, and $u = x$ in Formula V-28 to obtain the equation

$$\int_1^{100} \frac{dx}{x(4x + 3)} = \frac{1}{3} \ln \left| \frac{x}{4x + 3} \right| \Big|_1^{100}$$

$$= \tfrac{1}{3} (\ln \tfrac{100}{403} - \ln \tfrac{1}{7})$$

$$= \tfrac{1}{3} \ln \tfrac{700}{403} \approx .1826.$$

Some of the formulas in Table V are *reduction formulas.* They may not give us the answer we seek directly, but they help us proceed one step in the right direction.

Example 54-3. Find $\displaystyle\int \frac{x \, dx}{(x^2 - 2x + 2)^2}$. That is, find an integration formula whose left-hand side is the given expression.

Solution. Here we apply Formula V-49 to obtain the formula

$$(54\text{-}2) \quad \int \frac{x \, dx}{(x^2 - 2x + 2)^2} = \frac{-(4 - 2x)}{4(x^2 - 2x + 2)} - \left(-\frac{2}{4}\right) \int \frac{dx}{(x^2 - 2x + 2)}$$

$$= \frac{x - 2}{2(x^2 - 2x + 2)} + \frac{1}{2} \int \frac{dx}{x^2 - 2x + 2}.$$

Our reduction Formula V-49 has reduced the exponent of the term $(x^2 - 2x + 2)$ from 2 in our original integrand to 1 in the new integrand. Now we may use Formula V-46 to find that

$$\int \frac{dx}{x^2 - 2x + 2} = \frac{2}{\sqrt{4}} \operatorname{Tan}^{-1} \frac{2x - 2}{\sqrt{4}} = \operatorname{Tan}^{-1} (x - 1).$$

We substitute this result in Formula 54-2 and obtain the integration formula we were seeking:

$$\int \frac{x\,dx}{(x^2 - 2x + 2)^2} = \frac{x - 2}{2(x^2 - 2x + 2)} + \frac{1}{2}\,\mathrm{Tan}^{-1}(x - 1).$$

Example 54-4. Find $\int_0^\pi \cos^4 x\,dx$.

Solution. According to the second equation in Formula V-41 with $m = 0$ and $n = 4$,

$$\int_0^\pi \cos^4 x\,dx = \frac{\cos^3 x \sin x}{4}\Big|_0^\pi + \tfrac{3}{4}\int_0^\pi \cos^2 x\,dx = \tfrac{3}{4}\int_0^\pi \cos^2 x\,dx.$$

Now we use Formula V-41 again, this time with $m = 0$ and $n = 2$:

$$\int_0^\pi \cos^2 x\,dx = \frac{\cos x \sin x}{2}\Big|_0^\pi + \tfrac{1}{2}\int_0^\pi 1\,dx = 0 + \tfrac{1}{2}\pi.$$

It follows that

$$\int_0^\pi \cos^4 x\,dx = \tfrac{3}{4}\cdot\tfrac{1}{2}\pi = \tfrac{3}{8}\pi.$$

Theorem 34-1 says that

$$\int_a^b [mf(x) + ng(x)]\,dx = m\int_a^b f(x)\,dx + n\int_a^b g(x)\,dx.$$

Here we assume that m and n are given numbers and f and g are integrable functions. When we drop the limits of integration, we obtain the integration formula

$$(54\text{-}3) \quad \int [mf(x) + ng(x)]\,dx = m\int f(x)\,dx + n\int g(x)\,dx,$$

which we can use to reduce complicated integrands to simple ones. Thus in Section 58 we shall have to find integration formulas whose left-hand sides are of the form $\int \frac{Bx + C}{(x^2 + bx + c)^r}\,dx$, where B, C, b, and c are real numbers, with $b^2 - 4c < 0$, and r is a positive integer. According to Formula 54-3, this problem reduces to finding

$$\int \frac{x}{(x^2 + bx + c)^r}\,dx \quad \text{and} \quad \int \frac{1}{(x^2 + bx + c)^r}\,dx.$$

Example 54-3 illustrates how we use our integral tables to handle such problems.
Here is an example of how we might use Formula 54-3 to find an anti-derivative.

Example 54-5. Solve the equation $\dfrac{dy}{dx} = 3\cos x - 2xe^x$; that is, find an anti-derivative of $3\cos x - 2xe^x$.

Solution.

$$\int (3\cos x - 2xe^x)\,dx = 3\int \cos x\,dx - 2\int xe^x\,dx \qquad \text{(Formula 54-3)}$$

$$= 3\sin x - 2xe^x + 2\int e^x\,dx \qquad \text{(Formulas V-2 and V-43)}$$

$$= 3\sin x - 2xe^x + 2e^x \qquad \text{(Formula V-8)}.$$

Since f is continuous if $f(x) = 3\cos x - 2xe^x$, this integration formula tells us
that an antiderivative of $3\cos x - 2xe^x$ is $3\sin x + 2(1 - x)e^x$. It is always
wise to check such a result by differentiating, and when we check now we find that

$$\frac{d}{dx}[3\sin x + 2(1 - x)e^x] = 3\cos x - 2xe^x,$$

as it should.

PROBLEMS 54

1. Use Table V to find the following.

(a) $\displaystyle\int \frac{dx}{\sqrt{1 - 4x^2}}$

(b) $\displaystyle\int \frac{dx}{x^2 + 2x}$

(c) $\displaystyle\int \frac{x\,dx}{x^2 + 2x + 1}$

(d) $\displaystyle\int x^2 \ln x\,dx$

2. Evaluate.

(a) $\displaystyle\int_0^2 \frac{dx}{\sqrt{4 + x^2}}$

(b) $\displaystyle\int_0^2 \frac{dx}{4 + x^2}$

(c) $\displaystyle\int_0^1 \frac{dx}{4 - x^2}$

(d) $\displaystyle\int_0^1 \frac{dx}{\sqrt{5 - 4x^2}}$

3. Determine k so that the given integration formula is correct.

(a) $\displaystyle\int xe^{3x^2}\,dx = ke^{3x^2}$

(b) $\displaystyle\int \frac{dx}{x(\ln x)^{2/3}} = k\sqrt[3]{\ln x}$

(c) $\displaystyle\int \sin x \cos (\cos x)\, dx = k \sin (\cos x)$

(d) $\displaystyle\int \frac{2^{1/x}}{x^2}\, dx = k2^{1/x}$

4. Use Table V and Formula 54-3 to find the following.

(a) $\displaystyle\int (e^{x+2} + x^{e+2})\, dx$ (b) $\displaystyle\int \left(\ln w^4 + \ln \frac{4}{w}\right) dw$

5. Verify the following formulas from Table V.
   (a) V-26   (b) V-29   (c) V-30   (d) V-35   (e) V-37   (f) V-38

6. Use Table V to find an antiderivative of $f(x)$.
   (a) $f(x) = 1/x(3x - 4)^2$ (b) $f(x) = \tan^2 x$
   (c) $f(x) = x^2\sqrt{5 - x^2}$ (d) $f(x) = e^{3x+4} \sin (5x + \tfrac{1}{3}\pi)$
   (e) $f(x) = x^3 \ln x^x$ (f) $f(x) = (x^2 - 2x + 2)^{-2}$

7. Evaluate the following integrals.

(a) $\displaystyle\int_0^{\pi/2} \sin^8 x\, dx$ (b) $\displaystyle\int_0^{\pi/2} \sin^9 x\, dx$

(c) $\displaystyle\int_0^1 x^4 e^x\, dx$ (d) $\displaystyle\int_0^{\pi/2} x^9 \sin x\, dx$

8. Evaluate the following integrals with the aid of reduction formulas.

(a) $\displaystyle\int_0^2 \frac{dx}{(x^2 + 4)^3}$ (b) $\displaystyle\int_0^2 \frac{x\, dx}{(x^2 + 4)^3}$

(c) $\displaystyle\int_1^3 \frac{x\, dx}{(x^2 - 4x + 5)^2}$ (d) $\displaystyle\int_{-2}^{-1} \frac{dx}{(x^2 + 4x + 5)^2}$

9. Show that $\displaystyle\int \sec u\, du = -\ln |\sec u - \tan u|$. Does this formula agree with Formula V-35?

10. Find the area of the region between the curve $y = (1 + x^2)^{-1}$ and the interval $[-1/\sqrt{3}, 1/\sqrt{3}]$.

11. The region between the curve $y\sqrt{4x^2 + 9} = 12$ and the interval $[0, 2]$ is rotated about the X-axis. Find the volume of the resulting solid.

12. Find the area of the region between the line $y = 1$ and the curve $y\sqrt{25 - 16x^2} = 4$.

13. Find the area of the region bounded by the X-axis, the lines $x = 1$ and $x = 2$, and the curve $x^2 y + xy = 1$.

14. The region in the first quadrant bounded by the Y-axis, the lines $y = 1$ and $y = 4$, and the curve $yx^2(y + 1)^2 = 1$ is rotated about the Y-axis. Find the volume of the resulting solid.

15. If you know that the integration formula $\displaystyle\int f(x)\, dx = \int g(x)\, dx$ is valid in an interval I, can you necessarily conclude that $f(x) = g(x)$ for each point $x \in I$?

16. Show, taking care to use all symbols correctly, that the integration formula $\int f(x)\,dx = H(x) + \int g(x)\,dx$ is valid in an interval I if and only if the formula $\int [f(x) - g(x)]\,dx = H(x)$ is valid in I.

55. THE METHOD OF SUBSTITUTION

According to the Fundamental Theorem of Calculus, we can verify the integration formula

$$(55\text{-}1) \qquad \int 2x \cos x^2\,dx = \sin x^2$$

by showing that $\dfrac{d}{dx} \sin x^2 = 2x \cos x^2$. This latter equation, in turn, is verified by means of the Chain Rule. We let $u = x^2$ and then

$$\frac{d}{dx}\sin x^2 = \frac{d \sin u}{dx} = \frac{d \sin u}{du}\frac{du}{dx}$$

$$= \cos u \frac{du}{dx} = \cos x^2 \frac{d}{dx}x^2$$

$$= 2x \cos x^2.$$

But why merely *verify* an integration formula? We want to learn how to find one. For example, if we are given the left-hand side of Formula 55-1, how do we come up with the right-hand side? We apply the Chain Rule backward, as we will now illustrate.

We again let $u = x^2$. Then $\dfrac{du}{dx} = 2x$, and the integrand in Formula 55-1 becomes

$$(55\text{-}2) \qquad 2x \cos x^2 = \cos u \frac{du}{dx}.$$

We recognize the right-hand side of this equation as $\dfrac{d \sin u}{dx}$ (Equation 18-3), so $2x \cos x^2 = \dfrac{d}{dx}\sin x^2$. Once we have written an expression as a derivative, finding an antiderivative is like dating the War of 1812. An antiderivative of $\dfrac{d}{dx}\sin x^2$ is $\sin x^2$, and so we arrive at Formula 55-1.

Of course, one key step in our argument was observing that the right-hand side of Equation 55-2 is the derivative of sin u—that is, recognizing that sin u is an antiderivative of $\cos u \dfrac{du}{dx}$. In this case, recognition was relatively easy; we recalled one of our memorized differentiation formulas. Usually we refer to an integration formula, and in order to bring these integration formulas into the picture, we turn to differential notation, the practice of writing a derivative as a formal quotient.

Let us multiply both sides of Equation 55-2 by dx:

$$(55\text{-}3) \qquad 2x \sin x^2 \, dx = \cos u \, \frac{du}{dx} \, dx.$$

If we treat $\dfrac{du}{dx}$ as a real quotient, then $\dfrac{du}{dx} \, dx = du$, and this equation reduces to

$$2x \sin x^2 \, dx = \cos u \, du.$$

Now we introduce the integral sign:

$$\int 2x \sin x^2 \, dx = \int \cos u \, du.$$

Since Table V tells us that $\int \cos u \, du = \sin u$, and since $u = x^2$, we again come back to Formula 55-1.

Our description of how to change the variable of integration isn't mathematically precise. The heart of the matter is the idea that the expression $\dfrac{du}{dx}$ can be treated as an ordinary quotient in this setting. Since we have not given (and will not give) du and dx any status as numbers, such treatment may not seem justified. But it is convenient, and it will not lead us into error here. In the next section we will give a more rigorous justification of our method of substitution. It is, as we have tried to suggest by our example, a way of using the Chain Rule backward.

For the moment, let us simply follow these steps in finding $\int f(x) \, dx$, where f is a given continuous function:

(1) Choose a function h and set $u = h(x)$.

(2) Calculate $\dfrac{du}{dx}$ from this equation and write $du = \dfrac{du}{dx} \, dx$.

(3) Use the equations from steps (1) and (2) to write $f(x) \, dx = g(u) \, du$ and hence $\int f(x) \, dx = \int g(u) \, du.$

(4) Find $\int g(u)\,du$, perhaps by Table V.

(5) Now use the substitution equation $u = h(x)$ to express this latter antiderivative in terms of x.

Example 55-1. Find an antiderivative of $\sec^2 (x + 2)$.

Solution.   We follow our five steps:

   (1) Let $u = x + 2$.

   (2) Since $\dfrac{du}{dx} = \dfrac{d}{dx}(x + 2) = 1$, we write $du = 1 \cdot dx = dx$.

   (3) Hence $\int \sec^2 (x + 2)\,dx = \int \sec^2 u\,du$.

   (4) From Table V, $\int \sec^2 u\,du = \tan u$.

   (5) Since $u = x + 2$, our desired antiderivative is $\tan (x + 2)$.

This last example was particularly simple; we were merely illustrating the technique. Most of the time we won't be so explicit about listing our five steps, but we follow them nevertheless.

Example 55-2. Find $\displaystyle\int \frac{x\,dx}{\sqrt{x + 1}}$.

Solution.   We will indicate two different substitutions that could be used to solve this problem.

Substitution 1. Set $u = \sqrt{x + 1}$. Then $du = \dfrac{d}{dx}\sqrt{x + 1}\,dx = \dfrac{1}{2\sqrt{x + 1}}\,dx$, and $x = u^2 - 1$. Therefore

$$\int \frac{x\,dx}{\sqrt{x + 1}} = \int 2(u^2 - 1)\,du.$$

Now $\int 2(u^2 - 1)\,du = 2(\frac{1}{3}u^3 - u)$, and so

$$\int \frac{x\,dx}{\sqrt{x + 1}} = \tfrac{2}{3}(x + 1)^{3/2} - 2(x + 1)^{1/2}.$$

Substitution 2. Set $u = x + 1$. Then $du = dx$, $x = u - 1$, and we have

$$\int \frac{x\,dx}{\sqrt{x + 1}} = \int \frac{u - 1}{\sqrt{u}}\,du = \int (u^{1/2} - u^{-1/2})\,du.$$

Now $\int (u^{1/2} - u^{-1/2})\, du = \tfrac{2}{3}u^{3/2} - 2u^{1/2}$, and so

$$\int \frac{x\, dx}{\sqrt{x + 1}} = \tfrac{2}{3}(x + 1)^{3/2} - 2(x + 1)^{1/2}.$$

Our method of substitution is more of an art than a science. No magic rules exist to tell us what substitution to make so that $f(x)\, dx$ becomes $g(u)\, du$, with g being a function for which we know an integration formula. Since both u and $\dfrac{du}{dx}$ are involved in our calculations, we try to seek out certain expressions and their derivatives, such as $\sin x$ and $\cos x$, $\ln |x|$ and $1/x$. For example, to find $\int f(\sin x) \cos x\, dx$, we naturally think of the substitution $u = \sin x$. Then $du = \cos x\, dx$, and

$$\int f(\sin x) \cos x\, dx = \int f(u)\, du.$$

Example 55-3. Find $\displaystyle\int \frac{\cos x}{\sqrt{9 - \sin^2 x}}\, dx.$

Solution. We have just said that the substitution $u = \sin x$ produces the equation

$$\int \frac{\cos x}{\sqrt{9 - \sin^2 x}}\, dx = \int \frac{du}{\sqrt{9 - u^2}}.$$

From Formula V-12 we have $\displaystyle\int \frac{du}{\sqrt{9 - u^2}} = \operatorname{Sin}^{-1} \tfrac{1}{3}u$, and so

$$\int \frac{\cos x}{\sqrt{9 - \sin^2 x}}\, dx = \operatorname{Sin}^{-1} (\tfrac{1}{3} \sin x).$$

It sometimes helps us to decide on a particular substitution if we first perform certain algebraic or trigonometric manipulations on our given integrand. In the next section we will look at some trigonometric examples; here is a standard algebraic manipulation.

Example 55-4. Find $\displaystyle\int \frac{dx}{3x^2 - 12x + 15}.$

Solution. We factor and complete the square in the denominator of the integrand:

$$\int \frac{dx}{3x^2 - 12x + 15} = \frac{1}{3} \int \frac{dx}{(x-2)^2 + 1} .$$

This form suggests the substitution $u = x - 2$. Then $du = dx$, and so

$$\int \frac{dx}{3x^2 - 12x + 15} = \frac{1}{3} \int \frac{du}{u^2 + 1} .$$

Since $\int \frac{du}{u^2 + 1} = \text{Tan}^{-1} u$ and $u = x - 2$, we obtain the integration formula

$$\int \frac{dx}{3x^2 - 12x + 15} = \tfrac{1}{3} \text{Tan}^{-1} (x - 2).$$

PROBLEMS 55

1. Use the indicated substitution to find an integration formula with the given left-hand side.

(a) $\int 3x^2 e^{x^3} dx, u = x^3$

(b) $\int 3^{x^2} x \, dx, u = x^2$

(c) $\int x^2 \sin x^3 \, dx, u = x^3$

(d) $\int x \ln x^2 \, dx, u = x^2$

(e) $\int \frac{dx}{4 + 9x^2}, u = 3x$

(f) $\int \frac{x \, dx}{3 + 4x^2}, u = 3 + 4x^2$

(g) $\int \frac{(\ln x)^3}{x} dx, u = \ln x$

(h) $\int \sin^2 x \cos x \, dx, u = \sin x$

(i) $\int \tan x \, dx, u = \cos x$

(j) $\int e^x \cos e^x \, dx, u = e^x$

2. Use a substitution to find the following.

(a) $\int x \sin (x^2 + 1) \, dx$

(b) $\int x \csc^2 x^2 \, dx$

(c) $\int \frac{x}{(x^2 - 4)^{3/2}} dx$

(d) $\int x e^{x^2} dx$

(e) $\int \sin \tfrac{1}{2}x \sqrt{\cos \tfrac{1}{2}x} \, dx$

(f) $\int \cos 2x \sin^2 2x \, dx$

(g) $\displaystyle\int \cos^3 6x \sin 6x \, dx$

(h) $\displaystyle\int \sec^4 x \sin x \, dx$

(i) $\displaystyle\int \frac{dx}{e^{-x}\sqrt{e^{2x}+1}}$

(j) $\displaystyle\int \frac{dx}{e^x + 1}$

3. Find an antiderivative of each of the following.

(a) $e^{\sin x} \cos x$

(b) $\cos(3x - 4)$

(c) $\sec^2(2x + 3)$

(d) $x^{-1} \sin(\ln x)$

(e) $\dfrac{x^2}{\sqrt{x^3 - 5}}$

(f) $\dfrac{3x - 1}{3x^2 - 2x + 1}$

(g) $e^x \sec e^x \tan e^x$

(h) $\dfrac{e^x - e^{-x}}{e^x + e^{-x}}$

4. Find an integration formula with the given left-hand side.

(a) $\displaystyle\int \sec^3 x \tan x \, dx$

(b) $\displaystyle\int \frac{\cos 2x}{3 + 4\sin 2x} \, dx$

(c) $\displaystyle\int \frac{3 + 4\sin 2x}{\cos 2x} \, dx$

(d) $\displaystyle\int \ln e^{3x^2} \, dx$

(e) $\displaystyle\int \frac{e^{2x}}{e^x + 1} \, dx$

(f) $\displaystyle\int \exp(e^x + x) \, dx$

5. When finding integration formulas for the following, manipulate the integrand algebraically before deciding on your substitution.

(a) $\displaystyle\int \frac{dx}{x^2 + 4x + 13}$

(b) $\displaystyle\int \frac{dx}{\sqrt{6x - x^2 - 8}}$

(c) $\displaystyle\int \frac{dx}{\sqrt{8x - 4x^2}}$

(d) $\displaystyle\int \frac{(x + 2)\,dx}{x^2 - 6x + 10}$

6. What are the "natural" substitutions in the following cases? What do you get when you make this substitution?

(a) $\displaystyle\int e^x f(e^x) \, dx$

(b) $\displaystyle\int \frac{1}{x} f(\ln |x|) \, dx$

(c) $\displaystyle\int \frac{f(\sqrt{x})}{\sqrt{x}} \, dx$

(d) $\displaystyle\int \frac{f(1/x)}{x^2} \, dx$

(e) $\displaystyle\int x^{r-1} f(x^r) \, dx$

(f) $\displaystyle\int \frac{f(\operatorname{Tan}^{-1} x)}{1 + x^2} \, dx$

(g) $\displaystyle\int f(mx + b) \, dx$

(h) $\displaystyle\int g(x) f\left(\int_c^x g(t)\,dt\right) dx$

7. Use a substitution to find the following.

(a) $\displaystyle\int \frac{x}{\sqrt{1 - x^4}} \, dx$

(b) $\displaystyle\int \frac{x}{1 + x^4} \, dx$

(c) $\displaystyle\int 2\sin(\ln x) \, dx$

(d) $\displaystyle\int \frac{\sin 2x}{\sin^2 x + \sin x + 1} \, dx$

8. Find an antiderivative of $\sec^6 x$ by first writing

$$\sec^6 x = \sec^4 x \sec^2 x = (1 + \tan^2 x)^2 \sec^2 x$$

and then letting $u = \tan x$.

9. (a) Explain how the substitution $u^2 = ax^2 + b$ could be used to find

$$\int \frac{x^{2n-1}}{\sqrt{ax^2 + b}} \, dx,$$

where n is a positive integer.

(b) Show how the substitution $u = 1/x$ reduces the problem of finding

$$\int \frac{dx}{x^{2n}\sqrt{ax^2 + b}},$$

where n is a positive integer, to the problem of part (a).

56. CHANGING THE VARIABLE OF INTEGRATION

We now go into a little more detail on the theory of changing the variable of integration. As we said, this theory is based on the Chain Rule. The Fundamental Theorem of Calculus and its converse also play key roles. In order to evaluate the integral $\int_a^b f(x)\, dx$ of a given continuous function f, we start by making a substitution $u = h(x)$ such that $f(x)\, dx = g(u)\, du$ for some continuous function g. Since dx and du are not formally defined, this "equation" should be considered as representing the actual equation $f(x) = g(u) \dfrac{du}{dx}$ or

$$(56\text{-}1) \qquad\qquad f(x) = g[h(x)] \frac{dh(x)}{dx}.$$

Recall from Theorem 43-2 that if c is any number, then $\dfrac{d}{dv}\displaystyle\int_c^v g(u)\, du = g(v)$. When we incorporate this differentiation formula into the Chain Rule Equation $\dfrac{dG(v)}{dx} = \dfrac{dG(v)}{dv}\dfrac{dv}{dx}$ by replacing $G(v)$ with $\displaystyle\int_c^v g(u)\, du$, we see that

$$(56\text{-}2) \qquad\qquad \frac{d}{dx}\int_c^v g(u)\, du = g(v)\frac{dv}{dx}.$$

Now we replace v with $h(x)$:

$$\frac{d}{dx}\int_c^{h(x)} g(u)\, du = g[h(x)]\frac{dh(x)}{dx}.$$

In view of Equation 56-1, therefore, we have developed the differentiation formula

(56-3)
$$\frac{d}{dx} \int_c^{h(x)} g(u) \, du = f(x).$$

Now we apply the Fundamental Theorem of Calculus. Equation 56-3 says that $\int_c^{h(x)} g(u) \, du$ is an antiderivative of $f(x)$, and so the Fundamental Theorem tells us that

$$\int_a^b f(x) \, dx = \int_c^{h(x)} g(u) \, du \Big|_a^b$$

$$= \int_c^{h(b)} g(u) \, du - \int_c^{h(a)} g(u) \, du$$

$$= \int_{h(a)}^{h(b)} g(u) \, du;$$

that is,

(56-4)
$$\int_a^b f(x) \, dx = \int_{h(a)}^{h(b)} g(u) \, du.$$

Equation 56-4 is our "real" change-of-variable equation. The formula we discussed in the last section is obtained by dropping the limits of integration. Notice that when we do write the equation $\int f(x) \, dx = \int g(u) \, du$, a substitution $u = h(x)$ is always in the background.

We work with Equation 56-4 just about as we worked with the equation of the last section except that we use our substitution equation $u = h(x)$ to change the limits of integration as well as the variable of integration itself.

Example 56-1. Evaluate the integral $\int_3^5 x\sqrt{x^2 - 9} \, dx$.

Solution. (1) Set $u = x^2 - 9$.

(2) Then $\dfrac{du}{dx} = 2x$, and so $du = 2x \, dx$.

(3) From the equations in steps (1) and (2), we see that $x\sqrt{x^2 - 9} \, dx = \frac{1}{2}\sqrt{u} \, du$.

(4) Our new limits of integration are $3^2 - 9 = 0$ and $5^2 - 9 = 16$, and so

$$\int_3^5 x\sqrt{x^2 - 9} \, dx = \frac{1}{2} \int_0^{16} \sqrt{u} \, du.$$

(5) From our earlier work with integrals we find

$$\tfrac{1}{2} \int_0^{16} \sqrt{u}\ du = \tfrac{1}{3}u^{3/2} \Big|_0^{16} = \tfrac{64}{3}.$$

Example 56-2. Compute $\displaystyle\int_0^{\pi/6} \sin^2 x \cos x\ dx.$

Solution. We set $u = \sin x$, and so $du = \cos x\ dx$. Therefore $\sin^2 x \cos x\ dx = u^2\ du$. We see that $u = 0$ when $x = 0$ and $u = \tfrac{1}{2}$ when $x = \tfrac{1}{6}\pi$, so our new limits of integration are 0 and $\tfrac{1}{2}$, and

$$\int_0^{\pi/6} \sin^2 x \cos x\ dx = \int_0^{1/2} u^2\ du = \tfrac{1}{3}u^3 \Big|_0^{1/2} = \tfrac{1}{24}.$$

We already know that the letter we use as the variable of integration has nothing to do with the value of an integral. Thus, for example,

$$\int_a^b \sin x\ dx = \int_a^b \sin u\ du = \cos a - \cos b.$$

This equality agrees with what we have been saying in this section, because we can go from the integral on the left to the one on the right by means of the substitution $u = x$. But sometimes when we deal with several integrals at once, it is easy to lose sight of the basic fact that the value of an integral is independent of the letter used as the variable of integration. Our next example illustrates this kind of problem. Working out such a problem is a real test of your understanding of the nature of the variable of integration.

Example 56-3. Show that for any continuous function f,

$$\int_0^\pi \theta f(\sin \theta)\ d\theta = \tfrac{1}{2}\pi \int_0^\pi f(\sin \theta)\ d\theta.$$

Solution. If we make the substitution $\phi = \pi - \theta$, we have $d\phi = -d\theta$ and $\sin \theta = \sin \phi$. Furthermore, $\phi = \pi$ when $\theta = 0$, and $\phi = 0$ when $\theta = \pi$. Thus

$$\int_0^\pi \theta f(\sin \theta)\ d\theta = - \int_\pi^0 (\pi - \phi) f(\sin \phi)\ d\phi$$

$$= \pi \int_0^\pi f(\sin \phi)\ d\phi - \int_0^\pi \phi f(\sin \phi)\ d\phi.$$

Therefore

(56-5) $$\int_0^\pi \theta f(\sin \theta)\ d\theta + \int_0^\pi \phi f(\sin \phi)\ d\phi = \pi \int_0^\pi f(\sin \phi)\ d\phi.$$

Now remember that it makes no difference which letter we use as the variable of integration. In particular, we may replace all the ϕ's, letter by letter, with θ's. If we make this replacement, Equation 56-5 can be written

$$2 \int_0^\pi \theta f(\sin \theta) \, d\theta = \pi \int_0^\pi f(\sin \theta) \, d\theta,$$

and the desired equation follows immediately.

Finally, let us remark again on the nonspecific nature of the substitution $u = h(x)$. To evaluate $\int_a^b f(x) \, dx$, we try to find a substitution such that Equation 56-1 holds with a function g for which we know $\int g(u) \, du$ or can find it in a table. As our next example shows, there may be a number of satisfactory decompositions of $f(x)$. Or there may not be any.

Example 56-4. Suppose that $f(x) = e^x \tan e^x$. Find four pairs of functions g and h such that Equation 56-1 holds.

Choice	h(x)	g(u)
1	e^x	$\tan u$
2	$\sec e^x$	$\dfrac{1}{u}$
3	$\ln \sec e^x$	1
4	x	$e^u \tan e^u$

TABLE 56-1

Solution. We list four choices in Table 56-1. In Choice 1, $\dfrac{dh(x)}{dx} = e^x$, so

$$g[h(x)] \frac{dh(x)}{dx} = e^x \tan e^x = f(x),$$

and you may similarly verify that Equation 56-1 holds for the other choices. According to our theory, these choices reduce the problem of finding an integral of f to finding an integral of g. Therefore the "natural" Choice 1 is a good one because $\int \tan u \, du$ is listed in Table V. Choice 2 is much less natural, but it is certainly as good as or better than Choice 1, because the resulting integrand is so simple. Choice 3 is the best of all, in the sense that it results in the simplest possible $g(u)$, but it is most unlikely that you would hit on it as your first choice for $h(x)$.

Of course, it really makes no sense to say that one of the first three choices is better than another. We can use any one of them to evaluate integrals of f. Choice 4 is a correct, but trivial, choice. We can always satisfy Equation 56-1 with $h(x) = x$ and $g(u) = f(u)$, but then Equation 56-4 becomes $\int_a^b f(x)\,dx = \int_a^b f(u)\,du$—our new integral is the one we started with.

PROBLEMS 56

1. Use the indicated change of variable of integration to evaluate the integral.

(a) $\int_0^{\pi/6} \cos 2x\,dx,\ u = 2x$

(b) $\int_0^2 xe^{x^2}\,dx,\ u = x^2$

(c) $\int_0^1 \dfrac{x^2}{1 + x^3}\,dx,\ u = 1 + x^3$

(d) $\int_0^1 \dfrac{2x + 3}{x^2 + 3x + 2}\,dx,\ u = x^2 + 3x + 2$

(e) $\int_1^2 \dfrac{2x}{(3x^2 - 2)^2}\,dx,\ u = 3x^2 - 2$

(f) $\int_{-1}^1 (2x - 3)\sqrt{x^2 - 3x + 5}\,dx,\ u = x^2 - 3x + 5$

(g) $\int_0^{\pi/2} \cos x \sin(\sin x)\,dx,\ u = \sin x$

(h) $\int_0^{\pi/6} \dfrac{\cos x - \sin x}{\sin x + \cos x}\,dx,\ u = \sin x + \cos x$

(i) $\int_1^4 \dfrac{e^{\sqrt{x}}}{\sqrt{x}}\,dx,\ u = \sqrt{x}$

(j) $\int_0^{\pi/6} 4^{-\sin x} \cos x\,dx,\ u = -\sin x$

(k) $\int_0^1 \text{sech}^5 x \sinh x\,dx,\ u = \cosh x$

(l) $\int_1^2 \text{csch } x \cosh x\,dx,\ u = \sinh x$

2. Evaluate the integral $\int_{\sqrt{\pi/6}}^{\sqrt{\pi/2}} x \cot x^2\,dx$ by means of the following changes of variable of integration.
    (a) $u = x^2$ (b) $u = \sin x^2$ (c) $u = \ln \sin x^2$ (d) $u = \ln \sqrt{\sin x^2}$

3. By making a suitable change of variable of integration, each of the following integrals can be evaluated by using an integration formula for $c \int u^r \, du$, where c and r are appropriate numbers.

(a) $\displaystyle\int_3^4 x\sqrt{25 - x^2} \, dx$

(b) $\displaystyle\int_0^1 \frac{x + 4}{(x^2 + 8x + 1)^2} \, dx$

(c) $\displaystyle\int_0^{\ln 5} \tanh x \, dx$

(d) $\displaystyle\int_0^\pi \sin x \, (\cos^{13} x + 6 \cos^8 x - 2 \cos^5 x + 3) \, dx$

(e) $\displaystyle\int_1^e \frac{(\ln x)^5}{x} \, dx$

(f) $\displaystyle\int_0^{\sqrt{3}} \frac{(\text{Arctan } x)^3}{1 + x^2} \, dx$

4. Find $\displaystyle\int_{2\sqrt{2}}^4 \frac{dx}{x\sqrt{x^2 - 4}}$ in two ways. In one case, use the change of variable $u = \sqrt{x^2 - 4}$, and in the other set $v = 1/x$.

5. Evaluate the following integrals.

(a) $\displaystyle\int_{\ln 2}^{\ln 7} e^x e^{e^x} \, dx$

(b) $\displaystyle\int_0^2 \frac{x}{4 + x^2} \, dx$

(c) $\displaystyle\int_0^{\pi/2} \frac{\sin t}{1 + \cos^2 t} \, dt$

(d) $\displaystyle\int_0^{\pi/2} \frac{\cos t}{1 + \sin^2 t} \, dt$

(e) $\displaystyle\int_0^1 \sin \tfrac{1}{2}\pi(x + 3) \, dx$

(f) $\displaystyle\int_0^1 \frac{e^x}{1 + e^{2x}} \, dx$

(g) $\displaystyle\int_{\pi/4}^{\pi/2} (\cot x) \ln (\sin x) \, dx$

(h) $\displaystyle\int_0^1 \frac{x \ln \sqrt{x^2 + 1}}{\sqrt{x^2 + 1}} \, dx$

6. Find the area of the region enclosed by the curve $y^2 + x^4 = 4x^2$.

7. Show that the following equations hold for arbitrary continuous functions f and g.

(a) $\displaystyle\int_2^3 f(x + 3) \, dx = \int_5^6 f(x) \, dx$

(b) $\displaystyle\int_0^1 f(x)g(1 - x) \, dx = \int_0^1 f(1 - x)g(x) \, dx$

(c) $\displaystyle\int_a^b f(x) \, dx = \int_a^b f(a + b - x) \, dx$

(d) $\displaystyle\int_a^b f(x - a) \, dx = \int_a^b f(b - x) \, dx$

8. Use a change of variable of integration to show that $\displaystyle\int_{-a}^a f(x) \, dx$ is $2 \int_0^a f(x) \, dx$ if f is an even continuous function and 0 if f is an odd continuous function.

9. Evaluate $\displaystyle\int_0^{2\pi} |\cos nx|\ dx$, where n is a positive integer.

10. Sketch the curve $y = (\ln x)/\sqrt{x}$. The region bounded by this curve, the X-axis, and a vertical line through the maximum point of the curve is rotated about the X-axis. Find the volume of the solid that is generated.

11. Find the volume of the solid formed by rotating the first quadrant region that is bounded by the curve $y = x\sqrt[3]{1 - x^3}$ and the X-axis about the X-axis.

57. INTEGRANDS INVOLVING TRIGONOMETRIC FUNCTIONS

If an integrand contains values of the trigonometric functions, we can often use appropriate trigonometric identities to help us as we search for an integration formula. Some of the most useful trigonometric identities for this purpose are

$$\cos^2 x + \sin^2 x = 1,$$
$$(57\text{-}1)\qquad \tan^2 x = \sec^2 x - 1,$$
$$\cot^2 x = \csc^2 x - 1,$$

and

$$(57\text{-}2)\qquad \sin^2 x = \frac{1 - \cos 2x}{2}, \qquad \cos^2 x = \frac{1 + \cos 2x}{2}.$$

Example 57-1. Find an antiderivative of $\tan^2 x$.

Solution. When we use the second of Equations 57-1, we obtain the integration formula

$$\int \tan^2 x\ dx = \int (\sec^2 x - 1)\ dx = \tan x - x,$$

which tells us that an antiderivative of $\tan^2 x$ is $\tan x - x$.

We often use our trigonometric identities to write a given integrand in the form $f(\sin x) \cos x$ or $f(\cos x) \sin x$; the first form suggests the substitution $u = \sin x$, and the second the substitution $u = \cos x$.

Example 57-2. Without using Table V, evaluate the integral

$$\int_0^{\pi/2} \sin^4 x \cos^5 x\ dx.$$

Solution. In order to write $\sin^4 x \cos^5 x$ in the form $f(\sin x) \cos x$, we first express $\cos^4 x$ in terms of $\sin x$. Since $\cos^2 x = 1 - \sin^2 x$, we see that $\cos^4 x = (1 - \sin^2 x)^2$. Therefore

$$\int_0^{\pi/2} \sin^4 x \cos^5 x \, dx = \int_0^{\pi/2} \sin^4 x (1 - \sin^2 x)^2 \cdot \cos x \, dx.$$

Now we let $u = \sin x$. Then $du = \cos x \, dx$, and $u = 0$ when $x = 0$ and $u = 1$ when $x = \frac{1}{2}\pi$. Therefore our integral becomes

$$\int_0^1 u^4 (1 - u^2)^2 \, du = \int_0^1 (u^4 - 2u^6 + u^8) \, du = \tfrac{8}{315}.$$

Of course, there are many trigonometric identities in addition to Equations 57-1 and 57-2, and many of them are useful for finding integration formulas. We will mention just one more set of identities, the equations

$$2 \sin A \cos B = \sin (A + B) + \sin (A - B),$$

(57-3) $$2 \sin A \sin B = \cos (A - B) - \cos (A + B),$$

$$2 \cos A \cos B = \cos (A - B) + \cos (A + B).$$

Example 57-3. In certain mathematical applications we meet the expression

$$B(m, n) = \int_{-\pi}^{\pi} \sin mx \sin nx \, dx,$$

where m and n are positive integers. Evaluate $B(5, 3)$ and $B(5, 5)$.

Solution. From the second of Equations 57-3 we have

$$\sin mx \sin nx = \frac{\cos (m - n)x - \cos (m + n)x}{2},$$

and so

(57-4) $$B(m, n) = \tfrac{1}{2} \int_{-\pi}^{\pi} [\cos (m - n)x - \cos (m + n)x] \, dx.$$

Therefore

$$B(5, 3) = \tfrac{1}{2} \int_{-\pi}^{\pi} (\cos 2x - \cos 8x) \, dx$$

$$= \frac{1}{2} \left(\frac{\sin 2x}{2} - \frac{\sin 8x}{8} \right) \Bigg|_{-\pi}^{\pi}$$

$$= \frac{\sin 2\pi}{4} - \frac{\sin 8\pi}{16} + \frac{\sin 2\pi}{4} - \frac{\sin 8\pi}{16}$$

$$= 0.$$

Equation 57-4 also gives us

$$B(5, 5) = \tfrac{1}{2} \int_{-\pi}^{\pi} (1 - \cos 10x)\, dx$$

$$= \frac{1}{2} \left(x - \frac{\sin 10x}{10} \right) \Big|_{-\pi}^{\pi}$$

$$= \pi.$$

You should be able to show that $B(m, n) = 0$ if m and n are unequal positive integers, whereas $B(m, m) = \pi$ for every positive integer m.

Now let us look at some situations in which trigonometric methods can be applied, even though we start with algebraic, rather than trigonometric, integrands. The trick is to change the algebraic integrands to trigonometric ones by means of a substitution suggested by the Theorem of Pythagoras and our knowledge of right-triangle trigonometry. We apply this technique to integrands that contain powers of x and expressions of the form $(a^2 + x^2)^r$, $(a^2 - x^2)^r$, or $(x^2 - a^2)^r$. Figure 57-1 shows the right triangles that we draw in these cases. We use the Pythagorean Theorem to label the legs and hypotenuse of each triangle, the largest number being the length of the hypotenuse, of course. The first triangle suggests the substitution $u = \mathrm{Tan}^{-1}(x/a)$, the second $u = \mathrm{Sin}^{-1}(x/a)$, and the third $u = \mathrm{Sin}^{-1}(a/x)$.

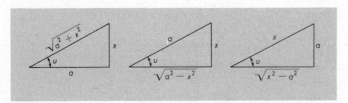

Figure 57-1

Example 57-4. In Example 51-3 we *verified* an integration formula whose left-hand side is $\int \sqrt{a^2 - x^2}\, dx$. How might we have arrived at this formula in the first place?

Solution. Since our integrand contains the expression $\sqrt{a^2 - x^2}$, we refer to the middle triangle in Fig. 57-1. This triangle suggests the substitution $u = \mathrm{Sin}^{-1}(x/a)$, from which we calculate $du = \dfrac{dx}{\sqrt{a^2 - x^2}}$. We could also use the substitution equation to establish that $\sqrt{a^2 - x^2} = a \cos u$, but it is easier to read this equation directly from the figure. Now we use these last two equations to write

$$\int \sqrt{a^2 - x^2} \, dx = \int a^2 \cos^2 u \, du$$

$$= \tfrac{1}{2}a^2 \int (1 + \cos 2u) \, du \qquad \text{(Equation 57-2)}$$

$$= \tfrac{1}{2}a^2(u + \tfrac{1}{2} \sin 2u)$$

$$= \tfrac{1}{2}a^2(u + \sin u \cos u).$$

To get our answer in terms of x, we recall that $u = \text{Sin}^{-1}(x/a)$ (so $\sin u = x/a$) and $\cos u = (1/a)\sqrt{a^2 - x^2}$. These substitutions give us Formula 51-4.

Example 57-5. Find $\displaystyle\int_0^{\sqrt{3}} \frac{dx}{(1 + x^2)^2}$.

Solution. Our integrand contains the expression $1 + x^2$, so we refer to the left-hand triangle in Fig. 57-1 (with $a = 1$) to see that $x = \tan u$ and $\sqrt{1 + x^2} = \sec u$. Hence $dx = \sec^2 u \, du$, and $u = 0$ when $x = 0$ and $u = \tfrac{1}{3}\pi$ when $x = \sqrt{3}$. Therefore

$$\int_0^{\sqrt{3}} \frac{dx}{(1 + x^2)^2} = \int_0^{\pi/3} \frac{\sec^2 u}{\sec^4 u} \, du = \int_0^{\pi/3} \cos^2 u \, du$$

$$= \tfrac{1}{2} \int_0^{\pi/3} (1 + \cos 2u) \, du = \tfrac{1}{2}(u + \tfrac{1}{2} \sin 2u)\Big|_0^{\pi/3}$$

$$= \tfrac{1}{2}(\tfrac{1}{3}\pi + \tfrac{1}{2} \sin \tfrac{2}{3}\pi) = \tfrac{1}{6}\pi + \tfrac{1}{8}\sqrt{3}.$$

PROBLEMS 57

1. Find the following.

(a) $\displaystyle\int \cot^2 x \, dx$

(b) $\displaystyle\int \sin x \cos^2 x \, dx$

(c) $\displaystyle\int \cos^3 x \sin^2 x \, dx$

(d) $\displaystyle\int \tan^2 x \sec^2 x \, dx$

(e) $\displaystyle\int \sin x \sqrt{\cos x} \, dx$

(f) $\displaystyle\int \sqrt[3]{\cos^2 x} \sin^5 x \, dx$

2. Evaluate the following integrals.

(a) $\displaystyle\int_0^1 \sin^2 \pi x \, dx$

(b) $\displaystyle\int_0^{1/4} \tan^2 \pi x \, dx$

(c) $\displaystyle\int_{-\pi/2}^{\pi/2} \cos^2 x \, dx$

(d) $\displaystyle\int_{\pi/8}^{\pi/4} \cot^2 2x \, dx$

3. By using Equation 57-2, we find that $\displaystyle\int \cos^2 x \, dx = \tfrac{1}{2}x + \tfrac{1}{4}\sin 2x$. Do we get the same result from Integration Formula V-41?

4. Find the following.

(a) $\displaystyle\int \sec^5 x \sin x \, dx$

(b) $\displaystyle\int \ln (\sin x) \cot x \, dx$

(c) $\displaystyle\int \tan^2 x \cos x \, dx$

(d) $\displaystyle\int \tan^4 x \, dx$

5. Use formulas from Table V to evaluate the following integrals.

(a) $\displaystyle\int_0^{\pi/2} \sin^8 x \, dx$

(b) $\displaystyle\int_0^{\pi/2} \sin^9 x \, dx$

(c) $\displaystyle\int_0^{\pi/2} \cos^8 x \, dx$

(d) $\displaystyle\int_0^{\pi/2} \cos^9 x \, dx$

(e) $\displaystyle\int_0^{\pi/2} \cos^8 x \sin^8 x \, dx$

(f) $\displaystyle\int_0^{\pi/2} \cos^8 x \sin^9 x \, dx$

(g) $\displaystyle\int_{-\pi}^{\pi} |\cos^3 t| \, dt$

(h) $\displaystyle\int_{-\pi}^{\pi} |\sin^5 t| \, dt$

6. (a) Show that $\displaystyle\int_{-\pi}^{\pi} \cos mx \sin nx \, dx = 0$ if m and n are integers.

(b) Set $A(m, n) = \displaystyle\int_{-\pi}^{\pi} \cos mx \cos nx \, dx$, where m and n are positive integers, and show that $A(m, n) = 0$ if $m \neq n$, while $A(m, m) = \pi$.

7. Find.

(a) $\displaystyle\int \frac{dx}{x^2\sqrt{x^2 + 25}}$

(b) $\displaystyle\int \frac{dx}{x^2\sqrt{9 - x^2}}$

(c) $\displaystyle\int \frac{dx}{x^2\sqrt{x^2 - 4}}$

(d) $\displaystyle\int \frac{dx}{(9 + x^2)^{3/2}}$

(e) $\displaystyle\int \frac{\sqrt{9 - x^2}}{x^2} \, dx$

(f) $\displaystyle\int \frac{\sqrt{x^2 - 9}}{x^2} \, dx$

(g) $\displaystyle\int \frac{\sqrt{x^2 - 9}}{x} \, dx$

(h) $\displaystyle\int \frac{\sqrt{9 - x^2}}{x} \, dx$

8. Evaluate the following integrals.

(a) $\displaystyle\int_0^2 \frac{dx}{(4 + x^2)^2}$

(b) $\displaystyle\int_{-5}^5 \frac{dx}{(25 + x^2)^{3/2}}$

(c) $\displaystyle\int_{\sqrt{3}}^3 \frac{dx}{x^2\sqrt{x^2 + 9}}$

(d) $\displaystyle\int_2^{2\sqrt{3}} \frac{dx}{x^2\sqrt{16 - x^2}}$

(e) $\displaystyle\int_{3/2}^3 \frac{\sqrt{9 - x^2}}{x^2}\, dx$

(f) $\displaystyle\int_3^6 \frac{\sqrt{x^2 - 9}}{x^2}\, dx$

9. Evaluate the integral $\displaystyle\int_0^{\pi/2} \sqrt{2 - \sin^2 x}\, \cos^3 x\, dx$.

10. One arch of the sine curve is rotated about the X-axis. What is the volume of the resulting solid?

11. The region bounded by the X-axis and the curves $y = \mathrm{Sin}^{-1}\, x$ and $y = \mathrm{Cos}^{-1}\, x$ is rotated about the Y-axis. Find the volume of the resulting solid.

58. INTEGRANDS THAT ARE RATIONAL EXPRESSIONS

So far the algebraic operations we have performed on a given integrand have involved nothing more complicated than completing the square. Now we will introduce some deeper algebraic theory that will enable us to handle a very broad class of algebraic integrands.

If n is a positive integer or 0 and if $a_0, a_1, \ldots, a_n$ are $n + 1$ real numbers, where $a_n \neq 0$, then the expression

$$P(x) = a_n x^n + a_{n-1} x^{n-1} + \cdots + a_0$$

is called a **polynomial** of **degree** n in x with real **coefficients** $a_0, a_1, \ldots, a_n$. A *ratio* of two polynomials is termed a **rational expression**. For example, $\dfrac{3x^2 + 2x - 1}{x - 2}$ and $\dfrac{1}{x^2 + 1}$ are rational expressions. In general, a rational expression takes the form

(58-1) $\qquad F(x) = \dfrac{a_n x^n + a_{n-1} x^{n-1} + \cdots + a_0}{b_m x^m + b_{m-1} x^{m-1} + \cdots + b_0} = \dfrac{P(x)}{D(x)}.$

Every rational expression can be written as a sum of terms of the forms

(58-2) $\qquad q_r x^r + q_{r-1} x^{r-1} + \cdots + q_0, \qquad \dfrac{A}{(x - a)^r}, \qquad \text{and} \qquad \dfrac{Bx + C}{(x^2 + bx + c)^r},$

where $b^2 - 4c < 0$. For example,

$$(58\text{-}3) \quad \frac{x^9 - 6x^8 + 19x^7 - 35x^6 + 34x^5 - 5x^4 - 30x^3 + 48x^2 - 39x + 12}{x^7 - 6x^6 + 17x^5 - 28x^4 + 28x^3 - 16x^2 + 4x}$$

$$= x^2 + 2 + \frac{3}{x} + \frac{2}{x - 1} - \frac{1}{(x - 1)^2}$$

$$+ \frac{1}{x^2 - 2x + 2} + \frac{-8x + 11}{(x^2 - 2x + 2)^2}.$$

How we arrive at such a decomposition of a rational expression is a question that we will discuss shortly. But once we get there, it is easy to find an antiderivative of the given expression. We can simply take a sum of antiderivatives of the individual terms in the decomposition.

It is not hard to find antiderivatives of Expressions 58-2. Polynomials present no problem, and we reduce $\displaystyle\int \frac{A}{(x - a)^r}\, dx$ to $\displaystyle A \int u^{-r}\, du$ by means of the substitution $u = x - a$. Expressions of the third type are most easily disposed of by referring to various formulas in Table V; see Example 54-3, for instance.

So now let us turn to the question of how to obtain the decomposition of a rational expression such as $F(x)$ of Equation 58-1. If the degree of the numerator is greater than or equal to the degree of the denominator (that is, $n \geq m$), then we can use long division to write

$$(58\text{-}4) \qquad F(x) = Q(x) + \frac{R(x)}{D(x)},$$

where $Q(x)$ is a polynomial $q_r x^r + \cdots + q_0$ and $R(x)$ is a polynomial whose degree is actually less than the degree of $D(x)$. The first step in finding the decomposition of a rational expression $F(x)$ is to write it in the form indicated in Equation 58-4.

Example 58-1. Perform the first step in the decomposition of $\dfrac{x^3 + 1}{x - 2}$ and find $\displaystyle\int \frac{x^3 + 1}{x - 2}\, dx.$

Solution. We use long division

$$
\begin{array}{r}
x^2 + 2x + 4 \\
x - 2 \overline{\smash{\big)}\, x^3 + 1} \\
\underline{x^3 - 2x^2} \\
2x^2 + 1 \\
\underline{2x^2 - 4x} \\
4x + 1 \\
\underline{4x - 8} \\
9
\end{array}
$$

to find the quotient $Q(x) = x^2 + 2x + 4$ and the remainder 9 that enable us to write

$$\frac{x^3 + 1}{x - 2} = x^2 + 2x + 4 + \frac{9}{x - 2}.$$

Notice that the degree of the remainder is 0, a number less than the degree (1) of the denominator $x - 2$. From this last equation we find that

$$\int \frac{x^3 + 1}{x - 2}\, dx = \int (x^2 + 2x + 4)\, dx + 9 \int \frac{dx}{x - 2}$$

$$= \tfrac{1}{3}x^3 + x^2 + 4x + 9 \ln |x - 2|.$$

After we have performed any necessary long division, we are left with the rational expression $R(x)/D(x)$ to decompose. The degree of $R(x)$ is less than the degree of $D(x)$. Now we factor the denominator $D(x)$ into a product of real factors. These factors can all be reduced to one of two forms: either $x - a$ or $x^2 + bx + c$, where $b^2 - 4c < 0$. (If $b^2 - 4c \geq 0$, we can factor $x^2 + bx + c$ into two real factors.) Some of the factors may be repeated, so our denominator will be a product of powers of such factors. [The so-called Fundamental Theorem of Algebra guarantees that such a factorization is (theoretically) possible.] For example, the denominator of the rational expression on the left-hand side of Equation 58-3 is the product

$$x(x - 1)^2(x^2 - 2x + 2)^2.$$

Each factor $(x - a)^r$ in the denominator leads to a sum of the form

$$\frac{A_1}{x - a} + \frac{A_2}{(x - a)^2} + \cdots + \frac{A_r}{(x - a)^r}$$

in the decomposition of $F(x)$. For example, the factor $(x - 1)^2$ in the denominator of the expression on the left-hand side of Equation 58-3 led to the sum $\dfrac{2}{x - 1} - \dfrac{1}{(x - 1)^2}$ in its decomposition. An example will show us how we find such sums.

Example 58-2.   Decompose the rational expression

$$\frac{x^2 - 7x + 8}{(x - 2)(x - 3)^2}.$$

Solution.   We notice first that the degree of the numerator (2) is less than the degree of the denominator (3), so we can omit the long division. The factor $x - 2$

in the denominator leads to a term $\dfrac{A}{x-2}$ and the factor $(x-3)^2$ leads to a sum $\dfrac{B}{x-3} + \dfrac{C}{(x-3)^2}$ in the decomposition. In other words,

(58-5)
$$\frac{x^2 - 7x + 8}{(x-2)(x-3)^2} = \frac{A}{x-2} + \frac{B}{x-3} + \frac{C}{(x-3)^2},$$

and now we have to determine the numbers A, B, and C. Our first step is to multiply both sides of Equation 58-5 by $(x-2)(x-3)^2$:

(58-6)
$$x^2 - 7x + 8 = A(x-3)^2 + B(x-2)(x-3) + C(x-2).$$

There are now two ways to proceed. We could multiply out the terms on the right in Equation 58-6 and obtain the equation

$$x^2 - 7x + 8 = (A+B)x^2 + (-6A - 5B + C)x + 9A + 6B - 2C.$$

Equating coefficients of corresponding powers of x yields the system of equations

$$A + B = 1$$

$$-6A - 5B + C = -7$$

$$9A + 6B - 2C = 8,$$

whose solution is $(A, B, C) = (-2, 3, -4)$.

We can arrive at the same numbers via a somewhat simpler calculation. Equation 58-6 is to be valid for each real number x. If we set, in turn, $x = 2$, 3, and 0, we get the system of equations

$$A = -2$$

$$C = -4$$

$$9A + 6B - 2C = 8.$$

The first two equations give us A and C, and then the third equation tells us that $B = 3$. Thus both methods give us the decomposition

$$\frac{x^2 - 7x + 8}{(x-2)(x-3)^2} = \frac{-2}{x-2} + \frac{3}{x-3} - \frac{4}{(x-3)^2}.$$

If a factor $(x^2 + bx + c)^r$ appears in the denominator of our rational expression, then in its decomposition we will have a sum of terms

$$\frac{B_1 x + C_1}{(x^2 + bx + c)} + \frac{B_2 x + C_2}{(x^2 + bx + c)^2} + \cdots + \frac{B_r x + C_r}{(x^2 + bx + c)^r}.$$

For example, the decomposition shown in Equation 58-3 contains the sum

$$\frac{1}{(x^2 - 2x + 2)} + \frac{-8x + 11}{(x^2 - 2x + 2)^2}.$$

Again an example will show us how we find such sums.

Example 58-3. Decompose the rational expression

$$\frac{2x^2 + 1}{(x^2 - x + 1)^2}.$$

Solution. To find numbers A, B, C, and D such that

$$\frac{2x^2 + 1}{(x^2 - x + 1)^2} = \frac{Ax + B}{(x^2 - x + 1)} + \frac{Cx + D}{(x^2 - x + 1)^2},$$

we multiply both sides of this equation by $(x^2 - x + 1)^2$:

$$2x^2 + 1 = (Ax + B)(x^2 - x + 1) + Cx + D$$
$$= Ax^3 + (-A + B)x^2 + (A - B + C)x + (B + D).$$

Now we equate the coefficients of corresponding powers of x:

$$A = 0$$
$$-A + B = 2$$
$$A - B + C = 0$$
$$B + D = 1.$$

When we solve this system of equations, we find $(B, C, D) = (2, 2, -1)$, and hence

$$\frac{2x^2 + 1}{(x^2 - x + 1)^2} = \frac{2}{x^2 - x + 1} + \frac{2x - 1}{(x^2 - x + 1)^2}.$$

Example 58-4. Find

$$\int \frac{6x^3 + 5x^2 + 21x + 12}{x(x + 1)(x^2 + 4)} \, dx.$$

Solution. We first write

$$\frac{6x^3 + 5x^2 + 21x + 12}{x(x + 1)(x^2 + 4)} = \frac{A}{x} + \frac{B}{x + 1} + \frac{Cx + D}{x^2 + 4}$$

and then clear of fractions:

$$6x^3 + 5x^2 + 21x + 12 = A(x + 1)(x^2 + 4) + Bx(x^2 + 4)$$

$$+ (Cx + D)x(x + 1)$$

$$= (A + B + C)x^3 + (A + D + C)x^2$$

$$+ (4A + 4B + D)x + 4A.$$

Therefore

$$A + B + C = 6$$

$$A + C + D = 5$$

$$4A + 4B + D = 21$$

$$4A = 12,$$

from which $(A, B, C, D) = (3, 2, 1, 1)$. Consequently,

$$\int \frac{6x^3 + 5x^2 + 21x + 12}{x(x + 1)(x^2 + 4)}\, dx = \int \frac{3}{x}\, dx + \int \frac{2}{x + 1}\, dx + \int \frac{x + 1}{x^2 + 4}\, dx,$$

and we leave it to you to use integration formulas to write

$$\int \frac{6x^3 + 5x^2 + 21x + 12}{x(x + 1)(x^2 + 4)}\, dx$$

$$= 3 \ln |x| + 2 \ln |x + 1| + \tfrac{1}{2} \ln (x^2 + 4) + \tfrac{1}{2} \text{ Arctan } \tfrac{1}{2}x$$

$$= \ln |x^3(x + 1)^2 \sqrt{x^2 + 4}| + \tfrac{1}{2} \text{ Arctan } \tfrac{1}{2}x.$$

PROBLEMS 58

1. Decompose the following rational expressions and find an antiderivative.

(a) $\dfrac{x^3}{x + 1}$

(b) $\dfrac{x + 1}{x^3}$

(c) $\dfrac{x^2}{x^2 - x}$

(d) $\dfrac{x^2 - 2x + 1}{x^2 - x}$

(e) $\dfrac{5x + 2}{x^2 + 4}$

(f) $\dfrac{x^2 + 2x - 5}{x^3 + 9}$

(g) $\dfrac{5x + 2}{x^2 - 4}$

(h) $\dfrac{x^2 + 2x - 5}{x^2 - 4}$

2. Find.

(a) $\int \dfrac{1}{(x + 1)^2}\, dx$

(b) $\int \dfrac{x}{(x + 1)^2}\, dx$

(c) $\int \dfrac{x^2}{(x + 1)^2}\, dx$

(d) $\int \dfrac{x^3}{(x + 1)^2}\, dx$

3. Use the methods of this section to find $\int \dfrac{x}{(x - 3)^2}\, dx$ and then use the substitution $u = x - 3$ to arrive at the same result.

4. Use the methods of this section to find the following.

(a) $\int \dfrac{x + 5}{(x + 2)(x + 3)}\, dx$

(b) $\int \dfrac{x - 5}{x^2 - 5x + 6}\, dx$

(c) $\int \dfrac{x^3}{(x + 1)^2}\, dx$

(d) $\int \dfrac{5x^2 - 3}{x^3 - x}\, dx$

(e) $\int \dfrac{x^2 + x + 4}{x^3 + 4x}\, dx$

(f) $\int \dfrac{x^3 - x + 4}{x^2 + 4x}\, dx$

5. Find.

(a) $\int \dfrac{3x^2 + 2x + 1}{(x + 1)(x^2 + x + 1)}\, dx$

(b) $\int \dfrac{3}{x^3 - 1}\, dx$

(c) $\int \dfrac{x^2 + x}{(x - 1)(x^2 + 1)}\, dx$

(d) $\int \dfrac{x^2 + x}{(x + 1)(x^2 - 1)}\, dx$

(e) $\int \dfrac{2}{x^4 - 1}\, dx$

(f) $\int \dfrac{32}{x^4 - 16}\, dx$

6. Evaluate the following integrals.

(a) $\int_1^2 \dfrac{x - 3}{x^3 + x^2}\, dx$

(b) $\int_1^3 \dfrac{2 - x^2}{x^3 + 3x^2 + 2x}\, dx$

(c) $\int_0^1 \dfrac{dx}{x^3 + 1}$

(d) $\int_{-1}^0 \dfrac{dx}{x^3 - 1}$

(e) $\int_3^4 \dfrac{5x^3 - 4x}{x^4 - 16}\, dx$

(f) $\int_3^4 \dfrac{x^3 - 2x^2 - 4x - 8}{x^4 - 16}\, dx$

7. You could use the methods of this section to find $\int \dfrac{x^2}{x^3 + 1}\, dx$. What is a better way?

8. Use the method of this section to "derive" the following formulas of Table V.

(a) V-27

(b) V-29

9. Use the method of this section to "derive" the formula

$$\int \frac{du}{u^2 - a^2} = \frac{1}{2a} \ln \left| \frac{u - a}{u + a} \right|.$$

(See Formulas V-22 and V-23.)

10.   One method of "deriving" the formula for $\int \csc u \, du$ is to proceed as follows. Let $u = 2v$ so that

$$\int \csc u \, du = \int \frac{du}{\sin u} = \int \frac{dv}{\sin v \cos v} = \int \frac{\cos v \, dv}{\sin v \cos^2 v} = \int \frac{\cos v \, dv}{\sin v (1 - \sin^2 v)}.$$

Now make the substitution $x = \sin v$ and use the method of this section to obtain Formula V-36.

11.   If we want to find $\int R(\cos x, \sin x) \, dx$, where $R(\cos x, \sin x)$ is made up of sums, products, and quotients of numbers and $\cos x$ and $\sin x$, we can use the substitution $u = \tan \frac{1}{2}x$ and then the methods described in this section. With this substitution,

$$du = \frac{1}{2} \sec^2 \tfrac{1}{2}x \, dx = \frac{dx}{2 \cos^2 \tfrac{1}{2}x} = \frac{dx}{1 + \cos x},$$

and now we must find a way of expressing $\cos x$ and $\sin x$ in terms of u. From Equation A-5 (Appendix A), with $t = \frac{1}{2}x$, we see that

$$\tan^2 \tfrac{1}{2}x = \frac{1 - \cos x}{1 + \cos x}.$$

Thus $u^2 = (1 - \cos x)/(1 + \cos x)$, from which we find that

$$\cos x = \frac{1 - u^2}{1 + u^2}.$$

From the third of Equations A-4, with $t = \frac{1}{2}x$, we have

$$\tan x = \frac{2 \tan \tfrac{1}{2}x}{1 - \tan^2 \tfrac{1}{2}x};$$

that is,

$$\frac{\sin x}{\cos x} = \frac{2u}{1 - u^2}.$$

Since we already know $\cos x$, it is a simple matter to solve this equation for $\sin x$. In summary, then,

$$\cos x = \frac{1 - u^2}{1 + u^2}, \quad \sin x = \frac{2u}{1 + u^2}, \quad \text{and} \quad dx = \frac{2}{1 + u^2} \, du.$$

Use this substitution procedure to find the following.

(a) $\displaystyle\int \frac{dx}{1 + \sin x}$

(b) $\displaystyle\int \frac{dx}{3 + 5 \sin x}$

(c) $\displaystyle\int \frac{dx}{5 + 3 \sin x}$

(d) $\displaystyle\int \frac{dx}{3 \sin x + 4 \cos x}$

Our method of substitution is a device for utilizing the Chain Rule to calculate integrals and find antiderivatives. Now we are going to put another general differentiation formula to work. The Product Rule of differentiation states that if the functions f and g are differentiable in an interval I, then

$$\frac{d}{dx}[f(x)g(x)] = f(x)g'(x) + g(x)f'(x)$$

for each $x \in I$. In other words, $f(x)g(x)$ is an antiderivative of $f(x)g'(x) + g(x)f'(x)$. Therefore (if f' and g' are integrable), the Fundamental Theorem of Calculus tells us that

$$\int_a^b [f(x)g'(x) + g(x)f'(x)]\,dx = f(x)g(x)\Big|_a^b$$

for an arbitrary pair of points a and b of I. We may rewrite this equation as

$$(59\text{-}1) \qquad \int_a^b f(x)g'(x)\,dx = f(x)g(x)\Big|_a^b - \int_a^b g(x)f'(x)\,dx.$$

In our abbreviated notation, we drop the limits of integration and simply write

$$(59\text{-}2) \qquad \int f(x)g'(x)\,dx = f(x)g(x) - \int g(x)f'(x)\,dx.$$

We will change the appearance, but not the content, of this last integration formula if we set $u = f(x)$ and $v = g(x)$. Then $f'(x) = \dfrac{du}{dx}$ and $g'(x) = \dfrac{dv}{dx}$, and our formula becomes

$$\int u \frac{dv}{dx}\,dx = uv - \int v \frac{du}{dx}\,dx.$$

Finally, we replace $\dfrac{dv}{dx}\,dx$ with dv and $\dfrac{du}{dx}\,dx$ with du:

$$(59\text{-}3) \qquad \int u\,dv = uv - \int v\,du.$$

When using Formula 59-3, we have to keep the substitution equations $u = f(x)$ and $v = g(x)$ in mind; the formula is really just a notational simplification of Formula 59-2.

The rule that is expressed by Equation 59-1 and Formulas 59-2 and 59-3 is known as the process of **integration by parts**. We use it to find antiderivatives and integrals that would otherwise be difficult to obtain. Basically the rule replaces the problem of finding $\int u \, dv$ with the problem of finding $\int v \, du$, and sometimes the second problem is easier to deal with than the first. For example, suppose that we let $u = x$ and $v = \sin x$ in Formula 59-3. Then $du = dx$ and $dv = \cos x \, dx$, and our formula becomes

$$\int x \cos x \, dx = x \sin x - \int \sin x \, dx = x \sin x + \cos x.$$

These formulas show that we transformed the problem of finding $\int x \cos dx$ into the easier problem of finding $\int \sin x \, dx$. The key step in applying the integration-by-parts technique is to choose u and v properly. Sometimes a little experimentation is necessary.

Example 59-1. Find $\int x \ln x \, dx$.

Solution. Here we let $u = \ln x$, and we will choose v so that $dv = x \, dx$. The simplest choice is $v = \frac{1}{2}x^2$. We see that $du = \frac{1}{x} \, dx$, and so Formula 59-3 reads

$$\int x \ln x \, dx = \frac{1}{2}x^2 \ln x - \int \frac{x^2}{2x} \, dx$$

$$= \frac{1}{2}x^2 \ln x - \frac{1}{2} \int x \, dx = \frac{1}{2}x^2 \ln x - \frac{1}{4}x^2.$$

It is easy to verify this result by differentiation.

In our usual applications, of course, we are not told what substitutions u and v to make. We are presented with $\int F(x) \, dx$, and it is up to us to write $F(x) \, dx$ as $u \, dv$. We are free to choose u arbitrarily. This choice determines dv, and then to finish filling out the formula, we must compute $du = \frac{du}{dx} \, dx$ and find v such that $\frac{dv}{dx} \, dx$ is our chosen dv. There is no set procedure for making the proper choice of u, but once we make our choice, the rest of the calculation is straightforward.

Example 59-2.   Find $\int x(x - 3)^5\, dx.$

Solution.   In this case, we will choose $u = x$, and so $dv = (x - 3)^5\, dx$. The first equation gives us $du = dx$, and now we must find v such that $\dfrac{dv}{dx} = (x - 3)^5$; that is, v must be an antiderivative of $(x - 3)^5$. In symbols we have

$$v = \int (x - 3)^5\, dx,$$

from which you can find, by substitution if necessary, that $\frac{1}{6}(x - 3)^6$ is a suitable choice for v. Now Formula 59-3 becomes

$$\int x(x - 3)^5\, dx = \tfrac{1}{6}x(x - 3)^6 - \tfrac{1}{6}\int (x - 3)^6\, dx$$

$$= \tfrac{1}{6}x(x - 3)^6 - \tfrac{1}{42}(x - 3)^7.$$

Example 59-3.   Find $\int x^2 e^x\, dx.$

Solution.   We let $u = x^2$ and hence $dv = e^x\, dx$. Then $du = 2x\, dx$, and $v = \int e^x\, dx = e^x$, so Formula 59-3 reads

(59-4) $$\int x^2 e^x\, dx = x^2 e^x - 2\int x e^x\, dx.$$

To finish our problem, we must find $\int x e^x\, dx$. What we have done so far is to reduce the problem of finding $\int x^2 e^x\, dx$ to the simpler problem of finding $\int x e^x\, dx$, which we do by using the integration-by-parts procedure a second time. Here we set $u = x$ and again take $v = e^x$. Then $du = dx$, and we have

$$\int x e^x\, dx = x e^x - \int e^x\, dx = x e^x - e^x.$$

When this result is substituted in Formula 59-4, we get

$$\int x^2 e^x\, dx = x^2 e^x - 2x e^x + 2e^x.$$

Before leaving this problem, let us see how we could have gone wrong in our choice of u. Suppose that we had originally let $u = e^x$ and $dv = x^2\, dx$. Then

$du = e^x \, dx$, $v = \int x^2 \, dx = \frac{1}{3}x^3$, and Formula 59-3 becomes

$$\int x^2 e^x \, dx = \frac{1}{3}x^3 e^x - \frac{1}{3} \int x^3 e^x \, dx.$$

Although this formula is correct, it leaves us with the problem of finding $\int x^3 e^x \, dx$, which is at least as difficult as the original problem. It is quite common to make such a false start, and when we do we just have to go back and start over.

Example 59-4. Find $\displaystyle\int_0^1 \mathrm{Tan}^{-1} x \, dx$.

Solution. Here let us set $f(x) = \mathrm{Tan}^{-1} x$ and $g'(x) = 1$. Then $f'(x) = 1/(1 + x^2)$, and we can take $g(x) = x$. Thus Equation 59-1 becomes

$$\int_0^1 \mathrm{Tan}^{-1} x \, dx = x \, \mathrm{Tan}^{-1} x \Big|_0^1 - \int_0^1 \frac{x}{1 + x^2} \, dx$$

$$= x \, \mathrm{Tan}^{-1} x \Big|_0^1 - \frac{1}{2} \ln(1 + x^2) \Big|_0^1$$

$$= \mathrm{Tan}^{-1} 1 - \frac{1}{2} \ln 2 \approx .44.$$

If you study mathematics further, or go on in fields in which mathematics is used, you will find that integration by parts is an important tool in theoretical investigations. The following example is a simplified version of a problem that arises in the study of differential equations.

Example 59-5. Suppose that u is a function such that $u(a) = u(b) = 0$. Show that

$$\int_a^b u(x)u''(x) \, dx = -\int_a^b [u'(x)]^2 \, dx.$$

Solution. We will let $f(x) = u(x)$ and $g(x) = u'(x)$ in Equation 59-1. Then

$$\int_a^b u(x)u''(x) \, dx = u(x)u'(x) \Big|_a^b - \int_a^b u'(x)u'(x) \, dx$$

$$= 0 - \int_a^b [u'(x)]^2 \, dx.$$

Our final example emphasizes once more that Integration Formula 59-3 is an abbreviation; the basic rule is Equation 59-1. We must insert the limits of integration to get the basic equation.

Example 59-6. Apply the integration-by-parts technique to $\int \frac{1}{x}\,dx$.

Solution. We will set $u = 1/x$, and so $dv = dx$. Then $du = -\frac{1}{x^2}\,dx$, and we may take $v = x$. Therefore Formula 59-3 becomes

$$\int \frac{1}{x}\,dx = 1 + \int \frac{1}{x}\,dx,$$

and it would seem to follow that $1 = 0$. Actually, however, our last formula is an abbreviation for the equation

$$\int_a^b \frac{1}{x}\,dx = 1\Big|_a^b + \int_a^b \frac{1}{x}\,dx,$$

and $1\Big|_a^b$ *is* 0, so there is no contradiction.

PROBLEMS   59

1. Use integration by parts to find the following.

(a) $\int x \sin x\,dx$

(b) $\int x \cos 2x\,dx$

(c) $\int xe^{2x}\,dx$

(d) $\int x \sec^2 2x\,dx$

(e) $\int x(x + 25)^{100}\,dx$

(f) $\int x\,10^x\,dx$

(g) $\int \mathrm{Sin}^{-1} x\,dx$

(h) $\int \mathrm{Tan}^{-1} 2x\,dx$

(i) $\int \sinh^{-1} x\,dx$

(j) $\int \mathrm{Cosh}^{-1} x\,dx$

2. Use integration by parts to evaluate the following integrals.

(a) $\int_0^2 x(x - 2)^4\,dx$

(b) $\int_0^1 xe^{3x}\,dx$

(c) $\int_0^{\pi/2} x \cos x\,dx$

(d) $\int_0^1 \mathrm{Cos}^{-1} x\,dx$

(e) $\int_1^2 (\ln x)^2\,dx$

(f) $\int_0^1 x\,\mathrm{Tan}^{-1} x\,dx$

3. It may be necessary to apply the integration-by-parts procedure more than once to find the following.

(a) $\int x^2 \cos x \, dx$

(b) $\int x^4 (\ln x)^2 \, dx$

(c) $\int x^3 \operatorname{Tan}^{-1} x \, dx$

(d) $\int_1^2 (x-1)^2 (x-2)^{10} \, dx$

(e) $\int_0^{\pi/2} x^2 \sin x \, dx$

(f) $\int_0^1 x \operatorname{Sin}^{-1} x \, dx$

4. Find.

(a) $\int x^{12} \ln x \, dx$

(b) $\int x \ln 12x \, dx$

(c) $\int x \ln x^{12} \, dx$

(d) $\int x^{12} \ln 12x^{12} \, dx$

5. Find $\int \ln x \, dx$ by taking $u = \ln x$ and $dv = dx$ and using integration by parts.

6. Find $\int_a^b e^{-x} \cos 2x \, dx$ by the following procedure. First, take $u = e^{-x}$ and $dv = \cos 2x \, dx$. In the new integral that results from using the integration-by-parts procedure, take $u = e^{-x}$ and $dv = \sin 2x \, dx$ to obtain the equation

$$\int_a^b e^{-x} \cos 2x \, dx = \tfrac{1}{2} e^{-x} \sin 2x \Big|_a^b - \tfrac{1}{4} e^{-x} \cos 2x \Big|_a^b - \tfrac{1}{4} \int_a^b e^{-x} \cos 2x \, dx.$$

Find $\int_a^b e^{-x} \cos 2x \, dx$ from this equation, and check your result with Table V.

7. Evaluate the following integrals.

(a) $\int_1^e x^2 \ln 2x \, dx$

(b) $\int_1^e x^2 (\ln x)^2 \, dx$

(c) $\int_1^{e^\pi} \sin (\ln x) \, dx$

(d) $\int_1^{e^\pi} \cos (\ln x) \, dx$

8. Find.

(a) $\int x^3 \cos x^2 \, dx$

(b) $\int x^5 e^{x^2} \, dx$

9. Find the volume of the solid obtained by rotating about the X-axis the region bounded by the X-axis and the curve $y = \sqrt[4]{x^2(1-x)}$.

10. Verify the formula

$$\int x f'(x) \, dx = x f(x) - \int f(x) \, dx.$$

What specific formulas do you obtain by replacing $f(x)$ with $\csc^2 x$, $\cosh x$, and e^x?

11. (a) Suppose that m is a positive integer and take $u = \sin^{m-1} x$ and $dv = \sin x \, dx$ in the formula for integration by parts to obtain the equation

(i) $\displaystyle\int_0^{\pi/2} \sin^m x \, dx = \left. -\cos x \, \sin^{m-1} x \right|_0^{\pi/2} + (m-1) \int_0^{\pi/2} \sin^{m-2} x \cos^2 x \, dx.$

(b) In Equation (i) replace $\cos^2 x$ by $1 - \sin^2 x$ and solve the resulting equation for

$\displaystyle\int_0^{\pi/2} \sin^m x \, dx.$

(c) Use the formula you obtained in part (b) to evaluate the integrals $\displaystyle\int_0^{\pi/2} \sin^5 x \, dx$

and $\displaystyle\int_0^{\pi/2} \sin^6 x \, dx.$

12. Use integration by parts to obtain the following formulas.

(a) $\displaystyle\int_0^a x^2 f'''(x) \, dx = a^2 f''(a) - 2a f'(a) + 2f(a) - 2f(0)$

(b) $\displaystyle\int_a^b f(x)g''(x) \, dx = \left. [f(x)g'(x) - f'(x)g(x)] \right|_a^b + \int_a^b f''(x)g(x) \, dx$

13. We can use the integration-by-parts technique to find the integral $\displaystyle\int_0^a (a^2 - t^2)^{3/2} \, dt$ as follows. First write

$$\int_0^a (a^2 - t^2)^{3/2} \, dt = a^2 \int_0^a \sqrt{a^2 - t^2} \, dt - \int_0^a t^2 \sqrt{a^2 - t^2} \, dt.$$

We know that the first integral on the right-hand side of this equation is $\frac{1}{4}\pi a^2$. Apply the integration-by-parts formula, with $u = t$, $dv = t\sqrt{a^2 - t^2} \, dt$, to the second integral and obtain an equation that you can solve for $\displaystyle\int_0^a (a^2 - t^2)^{3/2} \, dt$.

60. INITIAL VALUE PROBLEMS

To say that $F(x)$ is an antiderivative of $f(x)$ means that $y = F(x)$ satisfies the *differential equation*

(60-1) $y' = f(x).$

We solve this differential equation when we use the Fundamental Theorem of Calculus to evaluate the integral $\displaystyle\int_a^b f(x) \, dx$ as the difference $F(b) - F(a)$. For this purpose, any solution of Equation 60-1 will do. For example, to evaluate the integral $\displaystyle\int_1^2 3x^2 \, dx$, we could take $F(x)$ to be x^3, $x^3 + \pi$, $x^3 - 1492$, and so on, since $y = x^3$, $y = x^3 + \pi$, $y = x^3 - 1492$, and so on, all satisfy the differential equation

$$(60\text{-}2) \qquad\qquad y' = 3x^2.$$

In most applications of differential equations, however, it is not true that just any solution will do. Only a particular solution, one that also satisfies some given "initial conditions," will solve our problem. Thus in addition to Equation 60-1, we might be given two numbers x_0 and y_0 and be asked to find $F(x)$ such that $y = F(x)$ satisfies the differential equation *and* the initial condition $y = y_0$ when $x = x_0$. For example, $y = x^3 + 3$ satisfies Equation 60-2 and the initial condition $y = 3$ when $x = 0$. By adjoining an initial condition to Differential Equation 60-1, we obtain the **initial value problem**

$$(60\text{-}3) \qquad y' = f(x) \quad\text{and}\quad y = y_0 \quad\text{when } x = x_0.$$

This problem is a simple one; we are merely looking for a particular antiderivative of $f(x)$. But since we are going to reduce more complicated initial value problems to this special case, we will now write down a general formula to solve it.

Theorem   60-1.   *If f is continuous in an interval I that contains x_0, then*

$$y = \int_{x_0}^{x} f(t)\, dt + y_0$$

satisfies Initial Value Problem 60-3.

PROOF

The proof is simply a matter of verification. From Theorem 43-2 we see that $y' = f(x)$, so y satisfies the given differential equation. Furthermore, when we set $x = x_0$, we obtain $y = y_0$, and hence our initial condition is also satisfied.

Example 60-1.   At each point (x, y) of a certain graph, the slope is $\sec^2 x$. Furthermore, the graph contains the point $(\frac{1}{4}\pi, 3)$. Find the equation of the graph.

Solution.   Since the slope of the graph is y', we see that y must satisfy the differential equation $y' = \sec^2 x$. In addition, we are told that $y = 3$ when $x = \frac{1}{4}\pi$. Therefore we are to solve the initial value problem

$$y' = \sec^2 x \quad\text{and}\quad y = 3 \quad\text{when } x = \frac{1}{4}\pi.$$

According to Theorem 60-1, this problem is satisfied by

$$y = \int_{\pi/4}^{x} \sec^2 t \, dt + 3 = \tan t \Big|_{\pi/4}^{x} + 3 = \tan x + 2.$$

Example 60-2. The graph of a function f is shown in Fig. 60-1. Let y satisfy the problem $y' = f(x)$ and $y = 3$ when $x = 0$. Find y when $x = 2$, 3, and 6.

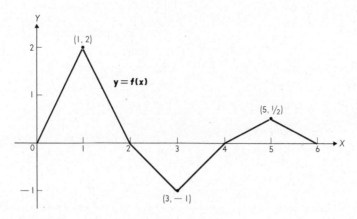

Figure 60-1

Solution. According to Theorem 60-1, our problem is satisfied by

$$y = \int_{0}^{x} f(t) \, dt + 3.$$

Therefore when $x = 2$, $y = \int_{0}^{2} f(t) \, dt + 3$. To evaluate the integral $\int_{0}^{2} f(t) \, dt$, we notice that it is simply the area of the triangle based on the interval $[0, 2]$ and with vertex $(1, 2)$. The base of this triangle is 2 units long, and its altitude is 2, so its area is 2 square units. Thus $\int_{0}^{2} f(t) \, dt = 2$, so $y = 2 + 3 = 5$ when $x = 2$. When $x = 3$,

$$y = \int_{0}^{3} f(t) \, dt + 3 = \int_{0}^{2} f(t) \, dt + \int_{2}^{3} f(t) \, dt + 3$$

$$= 2 + \int_{2}^{3} f(t) \, dt + 3 = 5 + \int_{2}^{3} f(t) \, dt.$$

To evaluate the integral $\int_{2}^{3} f(t) \, dt$, we see from Fig. 60-1 that it is simply the negative of the area of a triangle whose base and altitude are both 1. Therefore $\int_2^3 f(t)\,dt = -\frac{1}{2}$, so $y = 5 - \frac{1}{2} = \frac{9}{2}$ when $x = 3$. When $x = 6$,

$$y = \int_0^6 f(t)\,dt + 3 = \int_0^2 f(t)\,dt + \int_2^4 f(t)\,dt + \int_4^6 f(t)\,dt + 3.$$

We again evaluate our integrals by interpreting them as areas and get

$$y = 2 - 1 + \tfrac{1}{2} + 3 = \tfrac{9}{2}.$$

The following table lists the answers to our problem.

x	y
2	5
3	$\frac{9}{2}$
6	$\frac{9}{2}$

Recall that if a body is displaced s feet from the origin of a number scale t seconds after some initial instant, then the numbers $s' = v$ and $s'' = v' = a$ represent its velocity and acceleration. It is a physical fact that the acceleration of an unsupported body in a vacuum near the surface of the earth is approximately -32 feet per second per second, so its displacement s satisfies the differential equation $s'' = -32$. (Here we are measuring distance on a number scale that is perpendicular to the surface of the earth; the minus sign indicates that this number scale is pointed upward. Let us suppose that the origin of our number scale is on the surface of the earth so that the displacement s is the number of feet the body is above the surface.) If our body has a velocity of v_0 feet per second when $t = 0$ and is s_0 feet above the surface of the earth at that instant, we have two initial conditions to adjoin to our differential equation to produce the initial value problem

$$(60\text{-}4) \quad s'' = -32, \quad s = s_0 \text{ and } s' = v_0 \text{ when } t = 0.$$

(Because the highest derivative of s that appears in our differential equation is the second, the differential equation is said to be of the *second order*. As this problem illustrates, when we deal with a differential equation of the second order, we need two initial conditions to determine our solution; when our differential equation is of the third order, we need three initial conditions, and so on.) If we write $v = s'$, we can replace Initial Value Problem 60-4 with the two problems

$$(60\text{-}5) \qquad v' = -32 \text{ and } v = v_0 \text{ when } t = 0,$$

$$(60\text{-}6) \qquad s' = v \text{ and } s = s_0 \text{ when } t = 0.$$

These problems are in the form of Initial Value Problem 60-3, so we can solve them successively by means of Theorem 60-1. Thus

$$v = \int_0^t (-32)\, dz + v_0 = -32t + v_0.$$

Now we substitute this expression for v in Initial Value Problem 60-6 and use Theorem 60-1 to solve it:

$$(60\text{-}7) \quad s = \int_0^t (-32z + v_0)\, dz + s_0 = -16t^2 + v_0 t + s_0.$$

You probably recognize this equation from elementary physics. (We used it in Section 24.)

Example 60-3. You are at a hotel window 100 feet up with a paper bag full of water. A man is walking toward the spot directly under your window at the rate of 10 feet per second. If you throw the bag straight down when he is 20 feet from the spot, how fast should you throw it to hit him?

Solution. We may use Equation 60-7 to find the initial velocity v_0. We are told that the initial height is $s_0 = 100$. Furthermore, since the target will be in position when $t = 2$, we want $s = 0$ at that time (humanely, we aim the bag at the victim's feet). Therefore Equation 60-7 becomes

$$0 = -16 \cdot 2^2 + 2v_0 + 100.$$

From this equation we find that $v_0 = -18$. We must throw the bag with a speed of 18 feet per second (the minus sign indicates that we are to throw the bag downward).

Initial Value Problem 60-3 is a special case of the problem $y' = f(x, y)$ and $y = y_0$ when $x = x_0$. The more complicated $f(x, y)$ is, the harder the problem is to solve. Simple transformations reduce a couple of important initial value problems to the form of Problem 60-3. In the next section we will show you how to transform the *linear* differential equation $y' = p(x)y + q(x)$, and now we will look at the **separable** problem

$$(60\text{-}8) \qquad y' = \frac{f(x)}{g(y)} \quad \text{and} \quad y = y_0 \quad \text{when } x = x_0.$$

Since we are dividing by it, we assume that $g(y) \neq 0$ if y is in some neighborhood of y_0.

To solve Initial Value Problem 60-8, we let $z = \int_{y_0}^y g(t)\, dt$. Then according to Equation 56-2, $z' = \dfrac{d}{dx} \int_{y_0}^y g(t)\, dt = g(y)y'$. We replace y' with $f(x)/g(y)$ from our given differential equation, observe that $z = 0$ when $y = y_0$, and we see that z satisfies the initial value problem

$$z' = f(x) \quad \text{and} \quad z = 0 \quad \text{when } x = x_0.$$

Now we are back to Initial Value Problem 60-3, and Theorem 60-1 tells us that $z = \int_{x_0}^{x} f(t)\, dt$ satisfies it. When we equate our two expressions for z, we have the following theorem.

Theorem 60-2. *The solution of Initial Value Problem 60-8 is found by solving the equation*

$$(60\text{-}9) \qquad \int_{y_0}^{y} g(t)\, dt = \int_{x_0}^{x} f(t)\, dt$$

for y.

Differential notation helps us remember this theorem. We write Differential Equation 60-8 as

$$\frac{dy}{dx} = \frac{f(x)}{g(y)}$$

and formally multiply both sides by $g(y)\, dx$ to obtain

$$g(y)\, dy = f(x)\, dx.$$

Then we supply the symbols of integration $\int_{y_0}^{y}$ and $\int_{x_0}^{x}$ to the left- and right-hand sides (and change the dummy variable of integration to avoid confusion with the upper limits), and we have our solution, Equation 60-9.

Example 60-4. Solve the initial value problem $y' = \sqrt{\dfrac{1 - y^2}{1 - x^2}}$ and $y = 0$ when $x = \tfrac{1}{2}$.

Solution. We may write our differential equation as

$$\frac{dy}{dx} = \sqrt{\frac{1 - y^2}{1 - x^2}} = \frac{\sqrt{1 - y^2}}{\sqrt{1 - x^2}},$$

which suggests that

$$\frac{dy}{\sqrt{1 - y^2}} = \frac{dx}{\sqrt{1 - x^2}}.$$

Our initial condition $(x_0, y_0) = (\frac{1}{2}, 0)$ gives us the lower limits on the integrals in the equation

$$\int_0^y \frac{dt}{\sqrt{1 - t^2}} = \int_{1/2}^x \frac{dt}{\sqrt{1 - t^2}}$$

that we are to solve for y. This equation is simply (see Table V)

$$\mathrm{Sin}^{-1}\, t \Big|_0^y = \mathrm{Sin}^{-1}\, t \Big|_{1/2}^x \quad \text{or} \quad \mathrm{Sin}^{-1}\, y = \mathrm{Sin}^{-1}\, x - \tfrac{1}{6}\pi.$$

Hence

$$y = \mathrm{Sin}\,(\mathrm{Sin}^{-1}\, x - \tfrac{1}{6}\pi) = \tfrac{1}{2}(\sqrt{3}x - \sqrt{1 - x^2}).$$

Example 60-5.   A projectile is fired straight up from the surface of the earth. What must its initial velocity be so that it never comes back down?

Solution.   We will first express the given physical problem as an initial value problem. To obtain our differential equation, we equate two expressions for the acceleration of the projectile. First we use the Chain Rule to write the acceleration $\dfrac{dv}{dt}$ as $\dfrac{dv}{ds}\dfrac{ds}{dt} = v\dfrac{dv}{ds}$. Our second expression for the acceleration comes from Newton's Inverse Square Law of Gravitation, the physical law that tells how gravitational force varies with distance. In the units we are using, Newton's law states that when the projectile is s miles from the center of the earth, its acceleration is about $-95{,}000s^{-2}$ miles per second per second. We equate our two expressions for acceleration, adjoin the initial condition that $v = v_0$ when $s = 3960$ (the radius of the earth is about 3960 miles), and we obtain the initial value problem

$$\frac{dv}{ds} = -\frac{95{,}000}{s^2 v} \quad \text{and} \quad v = v_0 \quad \text{when } s = 3960.$$

Here, in the notation of Initial Value Problem 60-8, we have $g(v) = v$ and $f(s) = -95{,}000s^{-2}$. Therefore, according to Theorem 60-2, we are to solve the equation

$$\int_{v_0}^v t\, dt = -95{,}000 \int_{3960}^s t^{-2}\, dt$$

for v. After we integrate and simplify, this equation reduces to

$$v^2 = \frac{190{,}000}{s} + v_0^2 - \frac{4750}{99}.$$

We have expressed the square of our projectile's velocity in terms of its distance s from the center of the earth and its initial velocity v_0. If the projectile is to escape from the earth, its velocity can never become 0 (if it ever stops, it will fall back). Therefore no matter how large s gets, we must have $v^2 > 0$, and so we must choose

v_0 so that $v_0^2 - \frac{4750}{99} \geq 0$. The smallest possible such number is $v_0 = \sqrt{\frac{4750}{99}} \approx 7$. So we must fire our projectile with a speed of at least 7 miles per second or it will fall back.

PROBLEMS 60

1. Use Theorem 60-1 to find the equation of the curve containing the given point and having the given slope m.
   (a) $m = x \sin x^2$, $(\sqrt{\pi}, 2)$ (b) $m = e^{2x}$, $(\ln 2, 5)$
   (c) $m = 3/x$, $(1, 3)$ (d) $m = x/3$, $(1, 3)$

2. Use Theorem 60-1 to find the formula for $f(x)$.

   (a) $f'(x) = \dfrac{\cos (\ln x)}{x}$ and $f(1) = 2$ (b) $f'(x) = \ln x$ and $f(1) = 3$

   (c) $f'(x) = 2|x|$ and $f(-1) = 3$ (d) $f'(x) = x + |x|$ and $f(0) = 0$

3. Use Theorem 60-2 to find the equation of the curve containing the given point and having the given slope m.

   (a) $m = \dfrac{\sin x}{\cos y}$, $(0, 0)$ (b) $m = \dfrac{\cos x}{\sin y}$, $(0, \frac{1}{2}\pi)$

   (c) $m = \dfrac{1 + y^2}{1 + x^2}$, $(0, 0)$ (d) $m = \dfrac{1 + x^2}{1 + y^2}$, $(0, 0)$

   (e) $m = e^{x+y}$, $(0, 0)$ (f) $m = e^{x-y}$, $(0, 0)$
   (g) $m = 2x \cos^2 y$, $(0, \pi)$ (h) $m = 2y \cos^2 x$, $(0, \pi)$

4. Solve the following initial value problems.
   (a) $y'' = -\sin x$ and $y = 0$, $y' = 2$ when $x = 0$
   (b) $y'' = 3 \sin x + 2 \cos x$ and $y = 1$, $y' = -1$ when $x = 0$
   (c) $y'' = 6x^2 - 2$ and $y = 1$, $y' = 5$ when $x = 0$
   (d) $xy'' = 1$ and $y = 0$, $y' = 1$ when $x = 1$

5. Use Simpson's Parabolic Rule with four subdivisions to compute y when $x = 2$ if y satisfies the given initial value problem.
   (a) $y' = \sqrt{1 + x^3}$ and $y = 0$ when $x = 0$
   (b) $y' = e^{-y}\sqrt{1 + x^3}$ and $y = 0$ when $x = 0$

6. At every point of a certain curve, the tangent line is perpendicular to the segment that joins the point to the origin. The curve contains the point $(2, 3)$. What is its equation?

7. Find the equation of the curve that contains the points $(0, 4)$ and $(1, 8)$ and whose slope at a point is proportional to (a) the X-coordinate of the point, (b) the Y-coordinate of the point, (c) the product of the X- and Y-coordinates of the point. (*Hint:* The differential equation for part (a) is $y' = kx$, where k is a constant of proportionality to be determined. To this differential equation adjoin initial conditions determined by one of our given points, solve the resulting initial value problem, and then use the other point to determine k.)

8. At each point (x, y) of an arc connecting the points $(-1, 2)$ and $(4, 7)$, the slope is directly proportional to x and inversely proportional to y. What is the equation of the arc?

9. A ball is thrown directly upward with an initial velocity of v_0. One second later a second ball is thrown directly upward along the path of the first ball and with an initial velocity of $2v_0$. The second ball collides with the first ball just as the first ball reaches the top of its upward flight. Find v_0. (Neglect the radii of the balls.)

10. A ball is thrown directly upward with an initial velocity of v_0, and t_0 seconds later a second ball is thrown directly upward along the path of the first ball and with an initial velocity of w_0. Is it possible to choose t_0 and w_0 in such a way that the two balls collide just as both of them reach the top of their upward flight? (Neglect the radii of the balls.)

11. At liftoff a certain small rocket weighs 1500 pounds, of which 1000 pounds are fuel. The fuel burns at the rate of 10 pounds per second, producing a thrust of 2000 pounds. Show that at the end of t seconds the rocket is acted on by an upward force of 2000 pounds and a downward force of $1500 - 10t$ pounds. What is the resultant of these forces? According to Newton's Second Law of Motion, this force equals the product of the mass, $\dfrac{1500 - 10t}{32}$, and the acceleration v' of the rocket. Thus we have the initial value problem

$$v' = -32\,\frac{t + 50}{t - 150} \quad \text{and} \quad v = 0 \quad \text{when } t = 0.$$

What is the velocity of the rocket at burnout? How high is it then? How high does it finally go?

12. Let $r(x)$ be the distance between x and the nearest prime number. If $y' = r(x)/y$ and $y = 2$ when $x = 0$, find y when $x = 4$. (*Hint:* Sketch the curve $z = r(t)$ for t in the interval $[0, 4]$, and perform your integration graphically.)

61. INITIAL VALUE PROBLEMS WHOSE DIFFERENTIAL EQUATIONS ARE LINEAR

Suppose that we put our money in a bank that pays 5% interest compounded "instantaneously." How does it grow? This simple example is only one of many practical situations whose mathematical model is an initial value problem with a *linear* differential equation.

Assume that we start with $1000 and have y dollars in our account after x years. Our money grows at a rate that is proportional to the amount on hand, and the constant of proportionality is .05. We can express this statement and our initial condition in symbols as the initial value problem

(61-1) $y' = .05y \quad \text{and} \quad y = 1000 \quad \text{when } x = 0.$

If the bank harbors an embezzler who pilfers $100 per year from our account, the initial value problem that describes how much it contains becomes

(61-2) $y' = .05y - 100$ and $y = 1000$ when $x = 0$.

Let us set the stage for the general theory of linear differential equations by solving this initial value problem. When faced with a new problem, mathematicians instinctively try to reduce it to one they have solved before. For example, when we met separable differential equations, we made a change of variable which produced a situation in which Theorem 60-1 applied. We will do the same thing now.

We set

(61-3) $z = e^{-.05x}y$

so that

$$y = e^{.05x}z, \qquad y' = e^{.05x}z' + .05e^{.05x}z,$$

and

$$z = e^0 \cdot 1000 = 1000 \quad \text{when } x = 0.$$

When these quantities are substituted in Initial Value Problem 61-2, it becomes

$$e^{.05x}z' + .05e^{.05x}z = .05e^{.05x}z - 100$$

or

$$z' = -100e^{-.05x} \quad \text{and} \quad z = 1000 \quad \text{when } x = 0.$$

Now we have an initial value problem to which Theorem 60-1 applies, and we see that

$$z = 1000 - 100 \int_0^x e^{-.05t} \, dt = 1000 + 2000e^{-.05t} \Big|_0^x$$

$$= 2000e^{-.05x} - 1000$$

satisfies it. We use Equation 61-3 to replace z in terms of y,

$$y = 2000 - 1000e^{.05x},$$

and our problem is solved. This solution tells us, for example, that when $x = 10$ (after 10 years) our account contains

$$y = 2000 - 1000e^{.5} = 2000 - 1648.70 = 351.30,$$

which is not the figure we had in mind when we deposited our money.

An initial value problem with a general **linear differential equation** looks like this:

(61-4) $y' = p(x)y + q(x)$ and $y = y_0$ when $x = x_0$.

We assume that p and q are continuous in some interval that contains x_0 and that y_0 is a given number. In Initial Value Problem 61-2, $p(x) = .05$, $q(x) = -100$, $x_0 = 0$, and $y_0 = 1000$. Our first step in solving the problem is to calculate

$$P(x) = \int_{x_0}^{x} p(u) \, du.$$

(In our specific example, $P(x) = \int_{0}^{x} .05 \, du = .05x$.) Notice that $P'(x) = p(x)$ and $P(x_0) = 0$. Now we introduce a new variable by means of the equation

(61-5) $$z = e^{-P(x)}y.$$

It follows that

$$y = e^{P(x)}z, \qquad y' = e^{P(x)}z' + P'(x)e^{P(x)}z = e^{P(x)}z' + p(x)e^{P(x)}z,$$

and

$$z = e^{0}y_0 = y_0 \quad \text{when } x = x_0.$$

When these quantities are substituted in Initial Value Problem 61-4, the equation becomes

$$e^{P(x)}z' + p(x)e^{P(x)}z = p(x)e^{P(x)}z + q(x)$$

or

$$z' = e^{-P(x)}q(x) \quad \text{and} \quad z = y_0 \quad \text{when } x = x_0.$$

Now we can apply Theorem 60-1 to find that

$$z = y_0 + \int_{x_0}^{x} e^{-P(t)}q(t) \, dt.$$

When we use Equation 61-5 to replace z in terms of y, we complete the proof of the following theorem.

Theorem 61-1. *Initial Value Problem 61-4 is satisfied by*

(61-6) $$y = e^{P(x)}\left(y_0 + \int_{x_0}^{x} e^{-P(t)}q(t) \, dt \right),$$

$$\text{where } P(x) = \int_{x_0}^{x} p(u) \, du.$$

Equation 61-6 looks pretty complicated, and so do the steps we used to derive it. But applying this solution formula to a particular problem is not very difficult. No one is expected to memorize this formula. Either use it with the book open or derive it for each initial value problem you want to solve, starting with Equation 61-5.

Example 61-1. Solve the initial value problem

$$xy' = 2y + 8x^6 \quad \text{and} \quad y = 3 \quad \text{when } x = 1.$$

Solution. When we write this differential equation in the standard form illustrated in Equation 61-4, we see that

$$P(x) = 2/x \quad \text{and} \quad q(x) = 8x^5.$$

Hence, since $x_0 = 1$,

$$P(x) = \int_1^x (2/u)\ du = 2 \ln u \Big|_1^x = 2 \ln x - 2 \ln 1 = 2 \ln x = \ln x^2.$$

Therefore, $e^{P(x)} = e^{\ln x^2} = x^2$ (and so $e^{-P(t)} = 1/t^2$), and Equation 61-6 becomes

$$y = x^2 \left(3 + \int_1^x 8t^3\ dt \right) = 2x^6 + x^2.$$

As our original example in this section and a similar example in Section 14 have suggested, linear differential equations are associated with simple growth situations. (That is, our mathematical *model* is simple. If the real life situation doesn't correspond to the model, we may be talking nonsense!)

Example 61-2. In 1900 the rabbit population of East Rodentia was 1000 and by 1910 this number had doubled. What was the population in 1915? When had the population doubled again?

Solution. Let us suppose that there are r rabbits t years after 1900 and that *the rate of increase of the population is proportional to the population.* In symbols, this italicized statement is the differential equation

$$\frac{dr}{dt} = kr,$$

where k is some constant of proportionality. We are also told that $r = 1000$ when $t = 0$. So we have an initial value problem with a linear differential equation. Here $p(t) = k$, so $P(t) = \int_0^t k\ du = kt$. Since $q(t) = 0$, Equation 61-6 is especially simple and gives us

(61-7) $r = 1000e^{kt}.$

We still don't know k; we calculate that number from other data in our problem. Thus we were told that $r = 2000$ when $t = 10$, so $2000 = 1000e^{10k}$, and $e^{10k} = 2$. We could solve this equation for k, but we will solve it for e^k instead: $e^k = 2^{1/10}$. Since $e^{kt} = (e^k)^t$, we now see that Equation 61-7 becomes

(61-8) $r = 1000 \cdot 2^{t/10}.$

In 1915, $t = 15$, so there were

$$r = 1000 \cdot 2^{15/10} = 1000 \cdot 2^{1 \cdot 5} = 1000 \cdot 2\sqrt{2} = 2828$$

rabbits. We were also asked to find when the rabbit population reached 4000, so we must solve the equation

$$4000 = 1000 \cdot 2^{t/10} \quad \text{or} \quad 2^{t/10} = 4 = 2^2.$$

Hence $t/10 = 2$ and $t = 20$. The population reached 4000 by 1920. How does Equation 61-8 show that population doubles every 10 years?

PROBLEMS   61

1. (a) Solve Initial Value Problem 61-1 for the amount of money in an unembezzled account after x years.
    (b) How long does it take money to double at 5%?
    (c) At what point would our embezzled account be empty?

2. Solve the initial value problem.
    (a) $y' = -2y + 3$
        $y = 1$ when $x = 0$
    (b) $y' = 2y - 3$
        $y = 0$ when $x = 1$
    (c) $y' = y - e^x$
        $y = 1$ when $x = 0$
    (d) $y' = 5y + e^{6x}$
        $y = -2$ when $x = 0$
    (e) $y' = (\tan x)y$
        $y = 3$ when $x = -\pi$
    (f) $y' = (\cot x)y$
        $y = 2$ when $x = \frac{1}{2}\pi$

3. Solve the initial value problem.
    (a) $xy' = 3y$
        $y = 2$ when $x = 1$
    (b) $xy' = -3y$
        $y = 0$ when $x = 1$
    (c) $xy' = 5y + 4x^7 e^{x^2}$
        $y = 0$ when $x = 1$
    (d) $xy' = -5y + 12x \sin x^6$
        $y = -3$ when $x = 1$

4. Suppose $y = w(x)$ satisfies the initial value problem $y' = -2y + 8$ and $y = 100$ when $x = 0$. Approximately what is $w(100)$?

5. Solve the initial value problem $y' = \ln x^y$ and $y = 1$ when $x = 1$.

6. Let (x_0, y_0) and (x_1, y_1) be two given points. Solve the initial value problem $y' = (y - y_1)/(x - x_1)$ and $y = y_0$ when $x = x_0$, and discuss your result.

7. At each point of a certain curve, the slope is proportional to the product of the coordinates. If the graph contains the points $(0, 3)$ and $(1, 9)$, what is its equation? (*Hint:* Example 61-2 is very similar.)

8. If a certain commodity sells for p dollars, buyers will want to buy $20 - p$ units and sellers will want to sell $15 + p$ units. (Notice that demand decreases and supply increases when the price goes up.) Suppose that the rate of change of price with respect to time (measured in weeks) equals the excess of demand over supply $[(20 - p) - (15 + p)]$. Write this last sentence as a differential equation. If our commodity initially sells for A dollars, what is the price t weeks later? What would you say its equilibrium price is? Does the equilibrium price depend on the initial price?

9. The Dean of Women fills a 5-gallon container with fruit punch (no alcohol) for the freshman party. It is consumed at the rate of 2 quarts per minute, and the evil senior who is keeping the container full is pouring in 100 proof whiskey (50% alcohol). How long does it take for the punch to become 20% alcohol? If $A(t)$ is the number of quarts of alcohol in the punch t minutes after the party starts, what does $A(t)$ approximate if t is very large? (Answer this question in two ways: (a) Find a formula for $A(t)$ and look at it, and (b) use common sense.)

10. If $r(x)$ is the distance between x and the nearest prime number, and if

$$y' = 3y/x + x^3 r(x) \quad \text{and} \quad y = 3 \quad \text{when } x = 1,$$

what is y when $x = 4$?

If we can find an antiderivative $F(x)$ of $f(x)$, we can evaluate the integral $\int_a^b f(x)\,dx$ by subtraction; it is simply the number $F(b) - F(a)$. Of course, that's a big "if," and this chapter was devoted to a few of the standard techniques of finding antiderivatives. Basically we look them up in a table (Table V), but frequently we have to make various substitutions to reduce our integrand to a form that is tabulated.

Because we spend so much time on these techniques, you may get the impression that they are the only way to evaluate integrals or find antiderivatives. But they aren't. The integrand of $\int_0^1 e^{-x^3}\,dx$ does not appear in integral tables, and no amount of substitution will produce an equivalent integrand that does. Nevertheless, this integral is a perfectly definite number, one that we can easily find with a computer. In the old days students believed that this integral was a "bad" integral, whereas $\int_0^1 x^2 e^{-x^3}\,dx$ was a "good" one, since a simple substitution reduces it to a tabulated integral. So think of the techniques of this chapter as constituting some, *but by no means all*, of the tools one uses to evaluate integrals. A totally different, but perhaps even more useful, tool is the electronic computer.

Review Problems, Chapter Seven

1. Find.

(a) $\displaystyle\int x^2 e^{x^3}\,dx$

(b) $\displaystyle\int x^3 e^{x^2}\,dx$

(c) $\displaystyle\int \frac{x^3 - 8}{2x - 1}\,dx$

(d) $\displaystyle\int \frac{2x - 1}{x^3 - 8}\,dx$

(e) $\displaystyle\int x^2 \sqrt{x^3 + 1}\,dx$

(f) $\displaystyle\int x^3 \sqrt{x^2 + 1}\,dx$

(g) $\displaystyle\int x \cos^2 x^2\,dx$

(h) $\displaystyle\int x^2 \cos x\,dx$

2. To an "outsider," which of the integrals

$$I_1 = \int_0^1 e^{x^2}\, dx, \quad I_2 = \int_0^1 xe^{x^2}\, dx, \quad I_3 = \int_0^1 x^3 e^{x^2}\, dx$$

    looks easiest to evaluate? We can use the methods of this chapter to evaluate two of these integrals. Which are they, and what are their values? How would you evaluate the third?

3. Find $\displaystyle\int_0^1 xe^{tx^2}\, dt$ and $\displaystyle\int_0^1 xe^{tx^2}\, dx$.

4. Express $F(x)$ without using an integral sign.

    (a) $\displaystyle F(x) = \int_0^{\sin x} 4te^{t^2}\, dt$

    (b) $\displaystyle F(x) = \int_0^x \ln (\cos t)^{\sin t}\, dt$

    (c) $\displaystyle F(x) = \int_0^x x \sin xt\, dt$

    (d) $\displaystyle F(x) = \int_0^x t \sin xt\, dt$

    (e) $\displaystyle F(x) = \int_0^x \frac{\cos t}{x^2 + \sin^2 t}\, dt$

    (f) $\displaystyle F(x) = \int_0^x \frac{e^t}{1 + x^2 e^{2t}}\, dt$

5. Show that $\displaystyle\int_\pi^x |\sin t|\, dt = \cos x - 3$ if $x \in [-\pi, 0]$.

6. Show that $\displaystyle\int_0^x \frac{\cos t}{\cos^2 x + \sin^2 t}\, dt = \frac{x}{\cos x}$ if $x \in (-\tfrac{1}{2}\pi, \tfrac{1}{2}\pi)$.

7. Evaluate the integral $\displaystyle\int_0^1 \frac{dx}{\sqrt{x^2 + x + 1}}$ in two ways.

    (a) Complete the square in the radicand of the denominator and make a suitable trigonometric substitution.

    (b) Use the substitution $u = 2x + 1 + 2\sqrt{x^2 + x + 1}$.

8. Evaluate the following integrals.

    (a) $\displaystyle\int_0^{2\pi} |\sin x \cos x \cos 2x|\, dx$

    (b) $\displaystyle\int_0^{2\pi} |\sin x \cos x \sin 2x|\, dx$

    (c) $\displaystyle\int_0^{\pi/6} \cos x \sec 2x\, dx$

    (d) $\displaystyle\int_2^\pi \cos [\![x]\!]\, x\, dx$

9. Our voltmeter reads $A \sin \omega t$ volts t seconds after we throw the switch in a certain ac circuit.

    (a) What is ω if we are dealing with 60-cycle current?

    (b) What is the maximum voltage A if the root mean square voltage (the square root of the average value of the square of the voltage over one cycle) is 115?

10. Show that $\displaystyle\int_a^{a+2\pi} f(\cos x)\, dx = \int_0^{2\pi} f(\cos x)\, dx$ for every number a and continuous function f.

11. Show that t seconds after a man jumps out of a window 80 feet high he is $8[5 - t^2 + (t + \sqrt{5})|t - \sqrt{5}|]$ feet above the sidewalk.

12. Suppose that f is continuous in an interval that contains a point a and let $F(x) = \int_a^x f(t)\, dt + c$, where c is some number.

    (a) Show that if f is an odd function, then F is an even function.

    (b) Show that if f is even, then F is odd, provided that $c = \int_0^a f(t)\, dt$.

13. If f is continuous in an interval I, use integration by parts to show that for each pair of points a and b of I,

    $$\int_a^b \int_a^x f(t)\, dt\, dx = \int_a^b (b - x) f(x)\, dx.$$

    (*Hint:* If $u = \int_a^x f(t)\, dt$, then $du = f(x)\, dx$.)

14. Suppose that y satisfies the initial value problem

    $$\frac{d^2 y}{dx^2} = \left(\frac{dy}{dx}\right)^2 \quad \text{and} \quad y = r, \quad \frac{dy}{dx} = s \quad \text{when } x = a.$$

    (a) Show that $u = \dfrac{dy}{dx}$ satisfies the initial value problem $u' = u^2$ and $u = s$ when $x = a$.

    (b) Solve the initial value problem in part (a) to obtain $u = (s^{-1} + a - x)^{-1}$.

    (c) Use the result of part (b) to write an initial value problem with a differential equation of the first order that is satisfied by y, and solve it.

15. Suppose that f is continuous and that f^{-1} exists.

    (a) Sketch what the graph of f might look like, and by interpreting integrals geometrically, convince yourself that

    $$\int_a^b f^{-1}(x)\, dx = xf^{-1}(x)\Big|_a^b - \int_{f^{-1}(a)}^{f^{-1}(b)} f(x)\, dx.$$

    (b) Now apply the integration-by-parts procedure with $u = f^{-1}(x)$ and $dv = dx$, together with Equation 45-5, to show that

    $$\int_a^b f^{-1}(x)\, dx = xf^{-1}(x)\Big|_a^b - \int_a^b \frac{x\, dx}{f'[f^{-1}(x)]}.$$

    In the last integral, make the substitution $t = f^{-1}(x)$ to produce the formula of part (a). (In this part of the problem, we are assuming that f is differentiable.)

    (c) Show that if $\{x_0, x_1, \ldots, x_n\}$ is a partition of $[a, b]$, then

    $$\sum_{i=0}^{n-1} f^{-1}(x_i)(x_{i+1} - x_i) + \sum_{i=1}^{n} x_i [f^{-1}(x_i) - f^{-1}(x_{i-1})] = bf^{-1}(b) - af^{-1}(a).$$

    In the second sum, make the substitution $t_i = f^{-1}(x_i)$ and see if the result suggests the equation of part (a).

16. What is the largest interval in which we can apply Theorem 60-1 to the initial value problem $y' = x/|x|$ and $y = 2$ when $x = -1$? What happens if we try to apply the theorem beyond this interval?

17. It costs a manufacturer $c(u)$ dollars to produce u units of his product. He thinks of his *marginal cost of production* $m(u)$ as the "cost of producing one more unit"; that is, it is the number $c(u + 1) - c(u)$. Economists actually define marginal cost as $m(u) = c'(u)$, which comes out to about the same thing, since $c(u + 1) - c(u) \approx c'(u)$. For such reasons as overhead and startup costs, $c(0)$ is seldom 0. Find the cost function c in the following cases and think about whether or not our formulas make sense.

(a) $m(u) = 5$, $c(0) = 100$

(b) $m(u) = 50 - 2u - 2|u - 5|$, $c(0) = 400$

(c) $m(u) = 10(1 + e^{-2u})$, $c(0) = 200$

(d) $m(u) = 10(1 + e^{-u^2})$, $c(0) = 200$

(e) $m(u) = 5 - \text{Tan}^{-1} u$, $c(0) = 80$

(f) $m(u) = 10 - 6 \tanh 3u$, $c(0) = 50$

8 | Polar Coordinates. Vectors in the Plane

We now return to the subject of analytic geometry. This chapter has a twofold purpose. First of all, we will take a new approach toward the problem of representing points in the plane by means of pairs of numbers. Later it will turn out that some of the tools that we use in this new approach—vector methods—are also useful in the study of three-dimensional space. So a second objective of this chapter is to lay the groundwork for the study of analytic geometry in higher dimensions.

Some of the topics in this chapter, however, can be omitted without affecting one's understanding of later material. For example, the work on polar coordinates (or parts of it, such as Section 64) could be skipped. And we will consider rotation of axes in a more general setting in Chapter 12, so the treatment given here can be omitted.

62. POLAR COORDINATES

By introducing a pair of perpendicular lines and a unit of distance, we are able to assign to each point in the plane a pair of real numbers called the rectangular *cartesian coordinates* of the point. There are other ways to associate pairs of numbers with points, and now we are going to study one of the most important of them. Let P be a point in a plane in which we have a system of cartesian coordinates (Fig. 62-1), and let r be the length of segment OP. Suppose that the positive X-axis is the initial side and OP is part of the terminal side of an angle θ, as shown. We associate the quantities r and θ with the point P and say that P has **polar coordinates** (r, θ). If we write just a number for the angular coordinate θ—for example, $(r, \theta) = (2, \frac{1}{4}\pi)$— we are supposing that our angle is measured in radians. To indicate degree measure, we use the symbol °—for example, $(r, \theta) = (2, 45°)$. We refer to the point O (the origin in Fig. 62-1) as the **pole** of our polar coordinate system, and the initial side of the polar angle is called the **polar axis**. It is possible to introduce a polar coordinate system without reference to a cartesian system. We simply choose a point to be the pole of our system and any half-line emanating from the pole to be the polar axis.

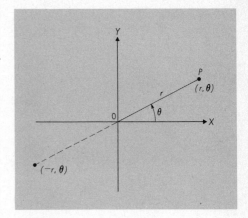

Figure 62-1

Not only do we allow angles greater than one revolution but negative angles as well, those angles formed by rotating the initial side in a clockwise direction. Therefore a given point has infinitely many pairs of polar coordinates. For example, the pairs $(2, 30°)$, $(2, 390°)$, and $(2, -330°)$ all represent the same point. We also occasionally find it convenient to use a negative number for the radial polar coordinate. In order to plot the point (r, θ) with $r > 0$, we proceed r units from the pole along the terminal side of the polar angle. To plot the point $(-r, \theta)$, we proceed r units from the pole along the extension of the terminal side through the pole (Fig. 62-1). Thus (r, θ) and $(-r, \theta + \pi)$ are polar coordinates of the same point. The coordinates $(0, \theta)$ represent the pole for every θ.

Example 62-1. Plot the points $(2, \frac{1}{2}\pi)$, $(3, -225°)$, $(-3, 225°)$, and $(0, 17°)$.

Solution. The listed points are shown in Fig. 62-2.

When a polar coordinate system is superimposed on a cartesian system, each point in the plane can be represented either by its cartesian coordinates (x, y) or by polar coordinates (r, θ) (Fig. 62-3). If $r > 0$, the basic Equations 6-2 concerning the values of trigonometric functions at angles tell us the fundamental relations between the cartesian and polar coordinates of a point:

(62-1) $$x = r \cos \theta \quad \text{and} \quad y = r \sin \theta.$$

You can verify that these equations are also valid if $r < 0$. It is easy to use Equations 62-1 to find the XY-coordinates of a point if we know a pair of polar coordinates (r, θ). Going the other way is a little more difficult.

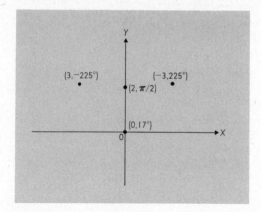

Figure 62-2

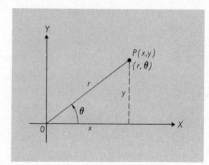

Figure 62-3

Example 62-2. Find a pair of polar coordinates of the point with cartesian coordinates $(-1, \sqrt{3})$.

Solution. Figure 62-4 shows the point $(-1, \sqrt{3})$; we have labeled the distance r and the angle θ that we are to determine. According to Equations 62-1, we must pick r and θ so that

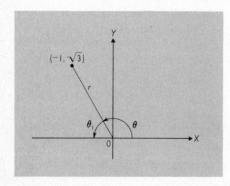

Figure 62-4

$$(62\text{-}2) \qquad -1 = r\cos\theta \quad \text{and} \quad \sqrt{3} = r\sin\theta.$$

If we square both of these equations and add, we see that

$$4 = r^2(\cos^2\theta + \sin^2\theta) = r^2.$$

Therefore $r = 2$ or $r = -2$. Let us take $r = 2$. Then Equations 62-2 tell us that we must choose θ so that $\cos\theta = -\frac{1}{2}$ and $\sin\theta = \frac{1}{2}\sqrt{3}$. From Fig. 62-4 we see that $\theta = \frac{2}{3}\pi$ is a suitable choice, so one pair of polar coordinates of our point is $(2, \frac{2}{3}\pi)$. Examples of other pairs are $(2, \frac{8}{3}\pi)$, $(2, -\frac{4}{3}\pi)$, and $(-2, -\frac{1}{3}\pi)$. You might check to see that each pair satisfies Equations 62-2.

As in the preceding example, it is often convenient to square both sides of Equations 62-1 and add, thereby obtaining a useful formula that relates the XY-coordinates of a point and its radial polar coordinate r:

$$(62\text{-}3) \qquad x^2 + y^2 = r^2.$$

The graph of an equation in x and y is the set of points whose XY-coordinates satisfy the equation. Similarly, the graph of an equation in the polar coordinates r and θ is the set of points that have polar coordinates that satisfy the equation.

For example, the graph of the equation $2x - 3y + 5 = 0$ is a line. If we replace x with $r \cos \theta$ and y with $r \sin \theta$, we obtain the same line in polar coordinates:

$$2r \cos \theta - 3r \sin \theta + 5 = 0.$$

The straightforward way to plot the graph of any equation is to begin with a table of values, plot the points in the table, and fill in the curve suggested by the plotted points.

Example 62-3. Sketch the curve

$$r = 2(1 - \cos \theta).$$

Solution. In this example we can lighten our labor by noting that since $\cos(-\theta) = \cos \theta$, the point $(r, -\theta)$ belongs to the graph if the point (r, θ) does. Therefore our graph is symmetric about the line lying along the polar axis. We can sketch the graph by plotting points whose angular coordinates lie between 0° and 180° and then reflect this portion about the line of symmetry. The table accompanying Fig. 62-5 lists the coordinates we used to sketch the curve. This heart-shaped curve is called a **cardioid**.

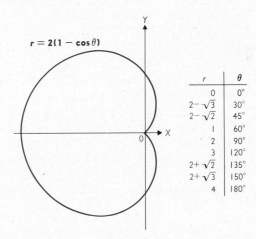

r	θ
0	0°
$2 - \sqrt{3}$	30°
$2 - \sqrt{2}$	45°
1	60°
2	90°
3	120°
$2 + \sqrt{2}$	135°
$2 + \sqrt{3}$	150°
4	180°

Figure 62-5

Frequently our knowledge of cartesian equations of curves helps us when dealing with polar equations.

Example 62-4. Discuss the curve

$$r = 4 \cos \theta - 2 \sin \theta.$$

Solution. We could plot the graph of this equation as we did in the preceding example, but instead we will transform the equation to one involving cartesian coordinates x and y. First, we multiply both sides by r:

$$r^2 = 4r \cos \theta - 2r \sin \theta.$$

Now we use Equations 62-1 and 62-3 to write our equation in cartesian coordinates:

$$x^2 + y^2 = 4x - 2y.$$

We then complete the square to obtain

$$(x^2 - 4x + 4) + (y^2 + 2y + 1) = 4 + 1,$$
$$(x - 2)^2 + (y + 1)^2 = 5,$$

and we find that our equation represents the circle whose center is the point $(2, -1)$ and whose radius is $\sqrt{5}$.

We have said that a point belongs to the graph of an equation in r and θ if it has a pair of polar coordinates that satisfy the equation. Since each point has infinitely many pairs of polar coordinates, it may be true that some of these pairs satisfy the equation while others don't. For example, the pair $(1, 0)$ satisfies the equation $r = e^{2\theta}$, so the point with polar coordinates $(1, 0)$ belongs to the graph of the equation. But the same point has polar coordinates $(-1, \pi)$ and $(1, 2\pi)$, and neither of these pairs satisfies the equation. Similarly, the graph of the equation $r = -6 \cos \theta$ contains the pole, since the coordinates $(0, \frac{1}{2}\pi)$ satisfy the equation, but other coordinates of the pole, such as $(0, \frac{1}{7}\pi)$, do not.

It is also true that different polar equations may have the same graph. For example, you can easily see that the graphs of the equations $r = 5$, $r = -5$, and $r^2 = 25$ are all the same circle. You can also verify that the graph of the equation $r = -2(1 + \cos \theta)$ is the cardioid of Example 62-3 (see Problem 62-14). We make these remarks to warn you that you must keep your eyes open when using polar coordinates. In the following example we will illustrate one type of pitfall to avoid.

Example 62-5.   Find the points of intersection of the cardioid $r = 2(1 - \cos \theta)$ and the circle $r = -6 \cos \theta$.

Solution. Figure 62-6 shows our two curves. We will attempt to find their points of intersection by finding simultaneous solutions of their equations. We equate the two expressions for r and get the trigonometric equation

$$2(1 - \cos \theta) = -6 \cos \theta,$$

from which we see that $\cos \theta = -\frac{1}{2}$. Two choices of θ that satisfy this last equation are $\theta = \frac{2}{3}\pi$ and $\theta = \frac{4}{3}\pi$. We see that $r = 3$ no matter which solution for θ is

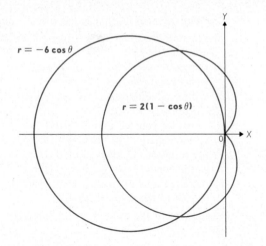

$r = -6 \cos \theta$

$r = 2(1 - \cos \theta)$

Figure 62-6

chosen, so we have found two points of intersection, $(3, \frac{2}{3}\pi)$ and $(3, \frac{4}{3}\pi)$. However, when we examine our graphs in Fig. 62-6, we see that there are actually three points of intersection. The pole is a point of intersection that we cannot find by solving the equations simultaneously. No pair of coordinates of the pole satisfies both equations. For example, the coordinates $(0, \frac{1}{2}\pi)$ satisfy the equation of the circle but not of the cardioid; and the coordinates $(0, 0)$ satisfy the equation of the cardioid but not of the circle. Nevertheless, the pole certainly does belong to both curves and thus is a point of intersection.

PROBLEMS 62

1. Find the cartesian coordinates of the points that have the following polar coordinates.

(a) $(8, 30°)$ (b) $(4, -420°)$ (c) $(-10, \frac{3}{4}\pi)$

(d) (π, π) (e) $(60, 60°)$ (f) $(-4, \frac{17}{6}\pi)$

2. Find a set of polar coordinates for each of the points with the following rectangular coordinates.

(a) $(-3, 0)$ (b) $(0, -3)$ (c) $(\sqrt{2}, -\sqrt{2})$

(d) (π, π) (e) $(4, -4\sqrt{3})$ (f) $(-4, 4\sqrt{3})$

3. Transform the following equations to cartesian coordinates and sketch their graphs.

(a) $r = 4 \cos \theta$ (b) $r = 2 \sin \theta$ (c) $r = 3 \sec \theta$

(d) $r = -2 \csc \theta$ (e) $\tan \theta = 2$ (f) $r \cos \theta = \tan \theta$

4. Sketch the graphs of the following equations.

(a) $r = 4 \cos 2\theta$ (b) $r = 4 \cos 4\theta$ (c) $r = 2 - \cos \theta$

(d) $r = 1 - 2 \cos \theta$ (e) $r = 1 - \cos \theta$ (f) $r^2 = 4 \cos \theta$

5. Convince yourself that the graph of an equation in r and θ is
   (a) symmetric with respect to the polar axis if $(r, -\theta)$ satisfies the equation whenever (r, θ) does.
   (b) symmetric with respect to the line $\theta = \frac{1}{2}\pi$ if $(r, \pi - \theta)$ satisfies the equation whenever (r, θ) does.
   (c) symmetric with respect to the pole if $(-r, \theta)$ or $(r, \theta + \pi)$ satisfies the equation whenever (r, θ) does.
   Convince yourself that the graphs of
   (d) all the equations of Problem 62-4 are symmetric with respect to the polar axis.
   (e) the equations of parts (a) and (b) of Problem 62-4 are symmetric with respect to the line $\theta = \frac{1}{2}\pi$.
   (f) the equations of parts (a), (b), and (f) of Problem 62-4 are symmetric with respect to the pole.

6. What point of the curve $r = 2^{\sin \theta}$ is closest to the pole? What point is farthest away?

7. Find the midpoint of the segment that terminates in the points $(3, \pi)$ and $(3, \frac{3}{2}\pi)$.

8. Find a polar equation of the line that contains the points with polar coordinates $(1, \pi)$ and $(2, \frac{1}{2}\pi)$. (*Hint*: You can write the cartesian coordinate equation first and then change to polar coordinates.)

9. Does it follow from Equations 62-1 that $r = \sqrt{x^2 + y^2}$ and $\theta = \text{Tan}^{-1}(y/x)$?

10. Find the points of intersection of the graphs of the following pairs of equations. Make a sketch, because this problem is somewhat tricky.
    (a) $r = 4 \cos 2\theta$, $r = 2$ (b) $r = 4 \cos 2\theta$, $r = 4 \cos \theta$
    (c) $r^2 = \sin^2 2\theta$, $\tan \theta = 1$ (d) $r = 1 - \cos \theta$, $r = \sin \frac{1}{2}\theta$
    (e) $r = 2(1 - \cos \theta)$, $r = -6 \cos \theta$ (f) $r = \cos \theta$, $r = \sin \theta$

11. Show that the distance between the points (r_1, θ_1) and (r_2, θ_2) is given by the formula
$$d = \sqrt{r_1^2 + r_2^2 - 2r_1 r_2 \cos(\theta_2 - \theta_1)}.$$

12. Sketch the curve $r\theta = 1$ (where θ is measured in radians). Convince yourself that if θ is close to 0, the Y-coordinate of a point (r, θ) of the graph is close to 1.

13. Sketch the graphs of the following equations.
    (a) $r = [\![\theta]\!]$ (b) $[\![r]\!] = \theta$ (c) $[\![r]\!] = [\![\theta]\!]$

14. Show that the equations $r = 2(1 - \cos \theta)$ and $r = -2(1 + \cos \theta)$ have the same graph.

15. Sketch the following sets.
    (a) $\{(r, \theta) \mid r > 0, 0 \le \theta \le \frac{1}{2}\pi\}$ (b) $\{(r, \theta) \mid r \in [1, 2], \theta \in [\frac{1}{2}\pi, \frac{3}{2}\pi]\}$
    (c) $\{(r, \theta) \mid r = 1, \sin \theta > 0\}$ (d) $\{(r, \theta) \mid r > 0, \sin \theta = 1\}$

63. LINES, CIRCLES, AND CONICS IN POLAR COORDINATES

In earlier sections we discussed the cartesian equations of certain common curves—lines, circles, and conics. Now we will consider the representation of these particular curves in polar coordinates. It helps tie our work with polar coordinates to our previous work with rectangular coordinates if we suppose that our polar coordinate system is superimposed on a cartesian system, as it was in the preceding section.

If α is a given angle, then it is clear that every point having polar coordinates (r, θ) such that

(63-1) $$\theta = \alpha$$

belongs to the line that contains the pole and makes an angle of α with the polar axis, and conversely. Thus the graph of Equation 63-1 is a line that contains the pole.

Now let us look at lines that do not contain the pole O (Fig. 63-1). A line is determined by a given point and a given direction. So we will start with a given point P and find the equation of the line that contains P and is perpendicular to the line segment joining P to the pole. Suppose that P has polar coordinates (p, β), where $p \neq 0$. Then the segment OP has slope $\tan \beta$, and the cartesian equation of any line perpendicular to OP has the form $(\cos \beta)x + (\sin \beta)y = c$ (since the slope of this line is $-1/\tan \beta$). Our line contains the point P whose cartesian coordinates are $(p \cos \beta, p \sin \beta)$. Hence $(\cos \beta)(p \cos \beta) + (\sin \beta)(p \sin \beta) = c$; that is, $p = c$. Therefore

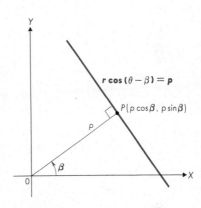

Figure 63-1

(63-2) $x \cos \beta + y \sin \beta = p$

is the cartesian equation of the line that contains the point P with polar coordinates (p, β) and is perpendicular to OP. By replacing x with $r \cos \theta$ and y with $r \sin \theta$ and simplifying, we obtain

(63-3) $r \cos (\theta - \beta) = p$

as the polar equation of our line. (What if $\beta = 0°$, $90°$, $180°$, or $270°$?)

Example 63-1.   Find the distance between the origin and the line $2x - 3y + 7 = 0$.

Solution.   We write

$$-2x + 3y = 7$$

and divide both sides of this equation by

$$\sqrt{(-2)^2 + 3^2} = \sqrt{13}$$

to obtain

$$-\frac{2}{\sqrt{13}} x + \frac{3}{\sqrt{13}} y = \frac{7}{\sqrt{13}}.$$

This last equation is in the form of Equation 63-2, with $p = 7/\sqrt{13}$, $\cos \beta = -2/\sqrt{13}$, and $\sin \beta = 3/\sqrt{13}$. Thus the distance between the origin and the line is $7/\sqrt{13}$, and one possible choice for β is (approximately) $\beta = 2.16$. This problem is illustrated in Fig. 63-2.

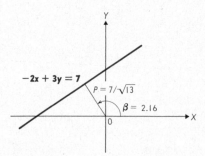

Figure 63-2

The cartesian equation of a circle with a radius of a and whose center is a point with polar coordinates (c, α) is

$$(x - c \cos \alpha)^2 + (y - c \sin \alpha)^2 = a^2.$$

If we replace x with $r \cos \theta$ and y with $r \sin \theta$ and simplify, we obtain the polar equation of our circle—namely,

$$(63\text{-}4) \qquad r^2 - 2rc \cos (\theta - \alpha) + c^2 = a^2.$$

If the center of the circle is the pole O, then $c = 0$ and Equation 63-4 reduces to $r^2 = a^2$, which is equivalent to the polar equation

$$(63\text{-}5) \qquad r = a.$$

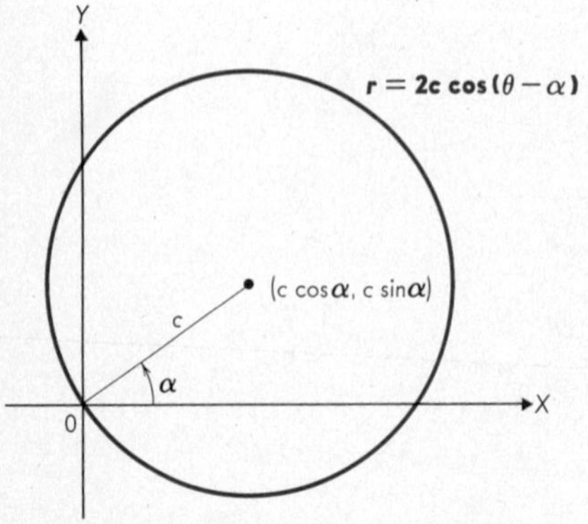

If the circle contains the origin (see Figure 63-3), then $c^2 = a^2$ and Equation 63-4 is equivalent to the equation

$$(63\text{-}6) \qquad r = 2c \cos (\theta - \alpha).$$

In particular, if we set $\alpha = 0$, we obtain the equation

$$(63\text{-}7) \qquad r = 2c \cos \theta$$

that represents a circle with a radius of $|c|$ and whose center is the point of the X-axis with cartesian coordinates $(c, 0)$. Similarly, if we set $\alpha = \frac{1}{2}\pi$, we get the

Figure 63-3

equation of a circle with a radius of $|c|$ and whose center is the point of the Y-axis with cartesian coordinates $(0, c)$:

(63-8)
$$r = 2c \sin \theta.$$

Polar coordinates are particularly well suited for representing conics. In Section 31 we found how a conic is determined by a point called a *focus*, a line called a *directrix*, and a positive number called the *eccentricity* of the conic. A point belongs to the conic if and only if the ratio of the distance between the point and the focus to the distance between the point and the directrix is the eccentricity. Suppose that we know the focus F, the corresponding directrix d, located p units from F, and the eccentricity e of a certain conic. Let us introduce polar coordinates so that the pole is the focus F and the directrix is perpendicular to the polar axis at the point with polar coordinates (p, π), where $p > 0$. We have sketched an arc of our conic in Fig. 63-4. If our conic is a hyperbola ($e > 1$), it will also have a branch lying to the left of d, but in the case of an ellipse ($e < 1$) or a parabola ($e = 1$), the entire conic lies to the right of the directrix. Suppose that P is a point of our conic and let (r, θ) be polar coordinates of P. The cartesian equation of the directrix is $x = -p$, so the distance between P and d is $|p + r \cos \theta|$. Thus the definition of a conic tells us that

$$\frac{|r|}{|p + r \cos \theta|} = e,$$

so that either

(63-9)
$$\frac{r}{p + r \cos \theta} = e \quad \text{or} \quad \frac{r}{p + r \cos \theta} = -e.$$

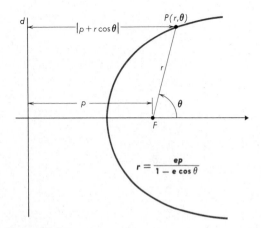

Figure 63-4

Now it is easy to show (we ask you to in Problem 63-13) that if the coordinates (r, θ) of a certain point satisfy either one of these equations, then the coordinates $(-r, \theta + \pi)$ *of the same point* satisfy the other equation. So we don't need both equations; either one will do. We will take the equation that is obtained by solving the first of Equations 63-9 for r to be the standard form of the polar equation of our conic:

$$(63\text{-}10) \qquad\qquad r = \frac{ep}{1 - e \cos \theta}.$$

By reasoning as we did above, you can find that if the focus is the pole, and if the directrix is to the right of the pole (see Fig. 63-5), then we can take the standard form of the equation of our conic to be

$$(63\text{-}11) \qquad\qquad r = \frac{ep}{1 + e \cos \theta}.$$

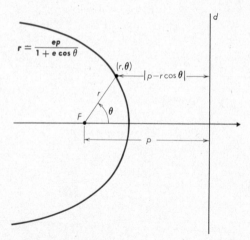

Figure 63-5

If the directrix is parallel to the polar axis, we take the standard polar equation of the conic to be

$$(63\text{-}12) \qquad\qquad r = \frac{ep}{1 + e \sin \theta}$$

if the directrix is above the focus, and

$$(68\text{-}13) \qquad\qquad r = \frac{ep}{1 - e \sin \theta}$$

if the directrix is below the focus (see Fig. 63-6).

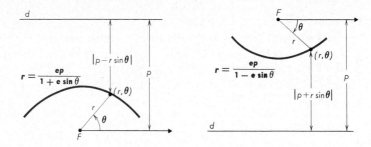

Figure 63-6

In each case, the focus is the pole of our coordinate system and the directrix that corresponds to this focus is p units away from it.

Example 63-2.   Write a polar equation of the parabola whose focus is the pole, whose axis lies along the polar axis, that opens to the right, and that contains the point $(2, \frac{1}{3}\pi)$.

Solution.   Since our parabola opens to the right, its directrix lies to the left of the focus (the pole), and hence Equation 63-10 is the equation we want. We are dealing with a parabola, so $e = 1$, and our equation is

$$r = \frac{p}{1 - \cos \theta}.$$

We determine p by setting $r = 2$ and $\theta = \frac{1}{3}\pi$, since the point $(2, \frac{1}{3}\pi)$ belongs to the parabola. Thus

$$2 = \frac{p}{1 - \cos \frac{1}{3}\pi} = \frac{p}{1 - \frac{1}{2}} = 2p,$$

and hence $p = 1$. Therefore the equation of our parabola is

$$r = \frac{1}{1 - \cos \theta}.$$

Example 63-3.   Discuss the curve

(63-14) $$r = \frac{16}{5 + 3 \sin \theta}.$$

Solution.   If we divide the numerator and the denominator of our fraction by 5, we can write Equation 63-14 in the form of Equation 63-12:

$$r = \frac{\frac{16}{5}}{1 + \frac{3}{5} \sin \theta} = \frac{\frac{3}{5} \cdot \frac{16}{3}}{1 + \frac{3}{5} \sin \theta}.$$

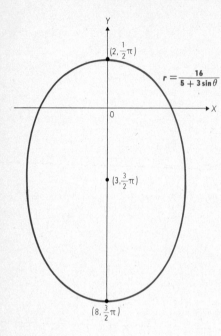

Figure 63-7

Here $e = \frac{3}{5}$ and $p = \frac{16}{3}$. Our curve is an ellipse, since $e < 1$, and its major diameter contains the pole and is perpendicular to the polar axis. We find the vertices of the ellipse by setting $\theta = \frac{1}{2}\pi$ and $\theta = \frac{3}{2}\pi$ in Equation 63-14. We obtain

$$r = \frac{16}{5 + 3 \cdot 1} = 2 \quad \text{when} \quad \theta = \tfrac{1}{2}\pi,$$

and

$$r = \frac{16}{5 + 3 \cdot (-1)} = 8 \quad \text{when} \quad \theta = \tfrac{3}{2}\pi.$$

Therefore the vertices are the points with polar coordinates $(2, \frac{1}{2}\pi)$ and $(8, \frac{3}{2}\pi)$. The major diameter is $8 + 2 = 10$ units long. In Section 31 we saw that the eccentricity of an ellipse is the ratio of the distance between its foci to the length of its major diameter. The major diameter of our present ellipse is $2b = 10$ units long, and if its foci are $2c$ units apart, we have the equation $e = 2c/10$; that is, $\frac{3}{5} = 2c/10$. Hence $c = 3$. Finally, if the minor diameter of our ellipse is $2a$ units long, then $a = \sqrt{b^2 - c^2} = \sqrt{25 - 9} = 4$. Our ellipse is shown in Fig. 63-7. You may readily verify that its cartesian equation is

$$\frac{x^2}{16} + \frac{(y + 3)^2}{25} = 1.$$

PROBLEMS 63

1. Find a polar equation of the line of which the given point is nearest the origin and sketch the situation.

   (a) $(6, 30°)$ (b) $(6, 150°)$ (c) $(8, -\tfrac{1}{4}\pi)$ (d) $(3, 27°)$

2. Find the distance between
   (a) the line $3x + 4y = 30$ and the origin.
   (b) the line $4x - 3y = 100$ and the origin.
   (c) the lines $3x + 4y = 30$ and $3x + 4y = -25$.
   (d) the line $3x + 4y = 30$ and the point $(2, 1)$.

3. Find a polar equation of the circle whose center has rectangular coordinates $(-1, 1)$ and
   (a) that contains the pole.
   (b) that has a radius of 3.
   (c) that contains the point with rectangular coordinates $(2, 3)$.
   (d) that contains the point with polar coordinates $(2\sqrt{2}, -\tfrac{1}{4}\pi)$.

4. Sketch the graphs of the following equations. In each case, label the vertices (or vertex) and the foci (or focus) of the conic.

(a) $r = \dfrac{16}{3 + 5 \sin \theta}$

(b) $r = \dfrac{16}{5 - 3 \cos \theta}$

(c) $r = \dfrac{4}{1 + \sin \theta}$

(d) $r = \dfrac{16}{5 + 3 \cos \theta}$

(e) $r \sin \theta = 1 - r$

(f) $3r \sin \theta = 5r - 16$

5. A conic has the origin of a cartesian coordinate system as a focus and a corresponding directrix whose equation is $x = -4$. Identify the conic and find a polar equation for it if it contains the following point (cartesian coordinates).

(a) $(5, 12)$ (b) $(5, 2\sqrt{14})$ (c) $(1, 2\sqrt{2})$ (d) $(0, -2)$

6. Show that the graph of the equation $r \sin (\theta - \alpha) = q \sin (\beta - \alpha)$ is the line that contains the point with polar coordinates (q, β) and that has a slope of $m = \tan \alpha$.

7. Describe the conic for which $p = 5{,}000{,}000$ and $e = 1/1{,}000{,}000$.

8. Find the minor diameter of the ellipse $r = ep/(1 - e \cos \theta)$.

9. Sketch the graph of the inequality $r > 3/(1 - \cos \theta)$.

10. Show that a latus rectum of the conic $r = ep/(1 - e \cos \theta)$ is ep units long.

11. Can you convince yourself that $\theta = \text{Cos}^{-1} 1/e$ is one of the asymptotes of the hyperbola $r = ep/(1 - e \cos \theta)$? Find the other asymptote.

12. Express Equation 63-10 in cartesian coordinates.

13. Show that if the coordinates (r, θ) of a certain point satisfy one of Equations 63-9, then the coordinates $(-r, \theta + \pi)$ of the same point satisfy the other equation.

14. Show that the standard form of the polar equation of a parabola can be written

$$r = \tfrac{1}{2} p \csc^2 \tfrac{1}{2}\theta.$$

15. A chord that contains a focus of a conic is divided into two segments by the focus. Prove that the sum of the reciprocals of the lengths of these two segments is the same no matter what chord is chosen.

16. Suppose that $e < 1$ in Equation 63-10 and let $2a$ denote the length of the major diameter and $2c$ the distance between the foci of the ellipse represented by that equation. Show that

$$a = \frac{ep}{1 - e^2} \quad \text{and} \quad c = \frac{e^2 p}{1 - e^2}.$$

64.   TANGENTS TO POLAR CURVES

After two sections devoted to the elementary notion of the graph of a polar equation, we are ready to apply some calculus to find out more about these graphs. In this section we discuss tangent lines, and in the next we will take up the area problem for polar curves.

Suppose that the tangent line to the curve $r = f(\theta)$ at the point (r, θ) makes an angle α with the polar axis. If we introduce XY-coordinates in the usual way, so that the positive X-axis is the polar axis, and transform our polar equation into an equation in x and y, then we know that

$$(64\text{-}1) \qquad\qquad \tan \alpha = \frac{dy}{dx}.$$

Since we wish to compute $\tan \alpha$ directly from the polar equation $r = f(\theta)$, we will express the derivative $\dfrac{dy}{dx}$ in terms of polar coordinates. We combine the Chain Rule Equation

$$\frac{dy}{dx} = \frac{dy}{d\theta}\frac{d\theta}{dx}$$

and the formula for the derivative of an inverse

$$\frac{d\theta}{dx} = 1 \bigg/ \frac{dx}{d\theta}$$

to obtain the equation

$$(64\text{-}2) \qquad\qquad \frac{dy}{dx} = \frac{dy}{d\theta} \bigg/ \frac{dx}{d\theta}.$$

In order to calculate $\dfrac{dy}{d\theta}$ at points of our curve $r = f(\theta)$, we replace r with $f(\theta)$ in the equation $y = r \sin \theta$ so that $y = f(\theta) \sin \theta$. Therefore,

$$\frac{dy}{d\theta} = f'(\theta) \sin \theta + f(\theta) \cos \theta.$$

Similarly, $x = r \cos \theta = f(\theta) \cos \theta$, and hence

$$\frac{dx}{d\theta} = f'(\theta) \cos \theta - f(\theta) \sin \theta.$$

When we substitute these values in Equation 64-1 to find $\dfrac{dy}{dx}$, Equation 64-1 becomes

$$(64\text{-}3) \qquad\qquad \tan \alpha = \frac{f'(\theta) \sin \theta + f(\theta) \cos \theta}{f'(\theta) \cos \theta - f(\theta) \sin \theta}.$$

Notice that the slope of the tangent line at a point of the polar curve $r = f(\theta)$ is *not* simply $\dfrac{dr}{d\theta}$.

Example 64-1.   Find the angle that the tangent to the cardioid $r = 2(1 - \cos \theta)$ at the point $(2, \tfrac{1}{2}\pi)$ makes with the polar axis.

Solution.   Here $f(\theta) = 2(1 - \cos \theta)$, so $f'(\theta) = 2 \sin \theta$.   Hence Equation 64-3 gives us

$$\tan \alpha = \frac{2 \sin \theta \sin \theta + 2(1 - \cos \theta) \cos \theta}{2 \sin \theta \cos \theta - 2(1 - \cos \theta) \sin \theta} = \frac{\cos \theta - \cos 2\theta}{\sin 2\theta - \sin \theta}.$$

Thus $\tan \alpha = -1$ at the point $(2, \tfrac{1}{2}\pi)$, and it follows that $\alpha = \tfrac{3}{4}\pi$.   You should check this result with Fig. 62-5.

If the pole belongs to the curve $r = f(\theta)$, it is particularly easy to find the tangent lines there. Suppose that the coordinates $(0, \theta_1)$ of the pole satisfy our equation; that is, suppose that $f(\theta_1) = 0$.   Furthermore, suppose that $f'(\theta_1) \neq 0$.   Then Equation 64-3 reduces to $\tan \alpha = \tan \theta_1$.   From this equation we can conclude that a tangent line at the point $(0, \theta_1)$ makes an angle θ_1 with the polar axis.   Therefore it has the equation $\theta = \theta_1$, and we find the tangents to the curve $r = f(\theta)$ at the pole by solving the equation $f(\theta) = 0$.

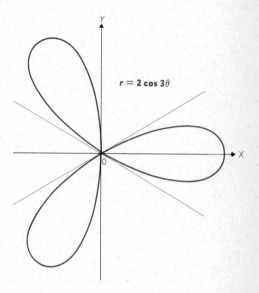

$r = 2 \cos 3\theta$

Example 64-2.   Find the tangents to the curve $r = 2 \cos 3\theta$ at the pole.

Solution.   Here $f(\theta) = 2 \cos 3\theta$, so we must solve the equation $2 \cos 3\theta = 0$.   Therefore $3\theta = \tfrac{1}{2}\pi + n\pi$; that is, $\theta = \tfrac{1}{6}\pi + \tfrac{1}{3}n\pi$.   There are three tangents at the pole, the lines $\theta = \tfrac{1}{6}\pi$, $\theta = \tfrac{1}{2}\pi$, and $\theta = \tfrac{5}{6}\pi$. Our curve, showing the tangent lines at the pole, appears in Fig. 64-1.

Figure 64-1

Equation 64-3 enables us to find the tangent of the angle that the tangent line to the curve $r = f(\theta)$ at the point (r, θ) makes with the polar axis.   In many cases we are interested in the angle ψ whose vertex is the point (r, θ), whose initial side is the radial line that contains this point, whose terminal side is the tangent line, and that is between 0 and π radians wide (Fig. 64-2).   The angles ψ, α, and θ are not independent of each other.   At each point of the graph there is an integer n such that ψ is related to α and θ by the equation

$$\psi = \alpha - \theta + n\pi.$$

For example, in the configuration on the left in Fig. 64-2, we have $\psi = \alpha - \theta$, and in the configuration on the right, $\psi = \alpha - \theta + \pi$.   But in any case,

$$\tan \psi = \tan \left[(\alpha - \theta) + n\pi \right] = \tan (\alpha - \theta).$$

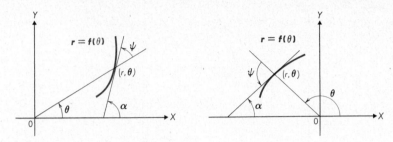

Figure 64-2

From trigonometry,

$$\tan(\alpha - \theta) = \frac{\tan \alpha - \tan \theta}{1 + \tan \alpha \tan \theta},$$

and if we replace $\tan \alpha$ in this quotient with the right-hand side of Equation 64-3, we will have an expression for $\tan(\alpha - \theta) = \tan \psi$ in terms of θ. In the problems we ask you to make the substitution and show that

(64-4)
$$\tan \psi = \frac{f(\theta)}{f'(\theta)} = r \left/ \frac{dr}{d\theta} \right. .$$

Example 64-3. Find the points of the curve $r = 2 \cos 3\theta$ at which the tangent line is perpendicular to the radial line.

Solution. The tangent line is perpendicular to the radial line at points where $\psi = \frac{1}{2}\pi$—that is, at points where $\tan \psi$ is not defined. From Equation 64-4 we see that these are the points for which $\dfrac{dr}{d\theta} = 0$. Since $\dfrac{dr}{d\theta} = -6 \sin 3\theta$, we must solve the equation

$$-6 \sin 3\theta = 0.$$

We find that $3\theta = n\pi$; that is, $\theta = \frac{1}{3}n\pi$. The graph in Fig. 64-1 tells us that we need only take the solutions $\theta = 0$, $\frac{1}{3}\pi$, and $\frac{2}{3}\pi$ to obtain the three points at which the tangent lines are perpendicular to the radial lines.

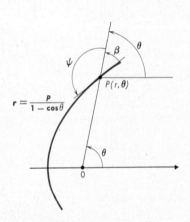

$$r = \frac{p}{1 - \cos\theta}$$

Figure 64-3

Example 64-4. Find the tangent of the angle that the tangent line to a parabola at a point P makes with the line that contains P and the focus.

Solution.   In Fig. 64-3 we have shown a parabola whose polar equation is (Equation 63-10 with $e = 1$)

(64-5) $$r = \frac{p}{1 - \cos \theta}.$$

We are looking for tan β. From Fig. 64-3 and Equation 64-4 we see that

$$\tan \beta = -\tan \psi = -r \bigg/ \frac{dr}{d\theta}.$$

From Equation 64-5 we have

$$\frac{dr}{d\theta} = \frac{d}{d\theta}\left(\frac{p}{1 - \cos \theta}\right) = \frac{-p \sin \theta}{(1 - \cos \theta)^2},$$

and so

$$\tan \beta = \frac{1 - \cos \theta}{\sin \theta}.$$

This equation gives us the answer to the question we originally asked, and from it we obtain some interesting information. There is a formula in trigonometry that says

$$\frac{1 - \cos \theta}{\sin \theta} = \tan \tfrac{1}{2}\theta.$$

Thus

$$\tan \beta = \tan \tfrac{1}{2}\theta.$$

This equation tells us that the horizontal line through P in Fig. 64-3 makes the same angle with the tangent line that the radial line through the focus and P does. Therefore we have another demonstration of the focusing property of the parabola that we discussed in Section 31.

PROBLEMS   64

1. Find ψ at the point $(4\pi, 2\pi)$ of the curve $r = 2\theta$.

2. Find the slope of the curve $r = 2 \cos 3\theta$ at the point with the given angular coordinate.
    (a) $\theta = 0$ (b) $\theta = \tfrac{2}{3}\pi$ (c) $\theta = \tfrac{4}{3}\pi$ (d) $\theta = -\tfrac{1}{4}\pi$

3. Find polar equations of the lines tangent at the origin to the following curves.
    (a) $r = 4 \cos 2\theta$ (b) $r = \sin 3\theta$
    (c) $r^2 = \cos 2\theta$ (d) $r = 1 - 3 \cos \theta$

4. Find the slope of the curve at the point with indicated angular coordinate.
    (a) $r = 1 - 2 \cos \theta,\ \theta = \tfrac{1}{2}\pi$ (b) $r = \sec^2 \theta,\ \theta = \tfrac{1}{3}\pi$
    (c) $r = 1 + 2\sqrt{\theta},\ \theta = \pi$ (d) $r = (1 + 2 \cos \theta)^{-1},\ \theta = 0$

5. Express $\tan \psi$ in terms of θ.
    (a) $r = e^{\sin \theta}$ (b) $r = \ln \theta$
    (c) $r = \sec \theta$ (d) $r^2 = \sin 2\theta$
    (e) $r^3 = 3 \sin 3\theta$ (f) $r = \ln |\sin \theta|$
    (g) $r = \exp (e^{\theta})$ (h) $r = \displaystyle\int_0^{\theta} \sin t^2 \, dt$

6. Sketch the curve $r = 2 \sin \theta$.
    (a) Find the equation of the tangent line at the pole.
    (b) Find ψ at an arbitrary point of the curve and interpret the result geometrically.

7. Show that the angular coordinate of a point of the curve $r = 2 \cos 3\theta$ at which the tangent line is parallel to the polar axis satisfies the equation
    $$3 \tan^4 \theta - 12 \tan^2 \theta + 1 = 0.$$

8. Find the points of the cardioid $r = 2(1 - \cos \theta)$ at which the tangent line is perpendicular to the polar axis.

9. Find the polar coordinates of the "highest" point of the cardioid $r = 2(1 - \cos \theta)$.

10. Sketch the curve $r = e^{\theta}$ and show that at any point the tangent line makes an angle of $45°$ with the radial line.

11. Are there any points of the spiral $r\theta = 1$ at which $\psi = \frac{1}{2}\pi$? Examine ψ for very large θ.

12. Show that $\tan \psi = \tan \frac{1}{2}\theta$ at points of the cardioid $r = 2(1 - \cos \theta)$.

13. Use the formula for $\tan (\psi_2 - \psi_1)$ to find the angle between the tangents at the points of intersection of the given curves.
    (a) $r = 2(1 - \cos \theta)$ (b) $r = \sin \theta$
        $r = -6 \cos \theta$ $r = \cos 2\theta$

14. Show that the curves $r = 3\theta$ and $r\theta = 3$ intersect at right angles.

15. What can you say about the angle between the radial line and the tangent line at a point of the curve $r = f(\theta)$ that is a maximum distance from the pole?

16. Derive Equation 64-4 by making the substitution suggested in the paragraph preceding that equation.

17. Show that if $r^n = f(\theta)$, then $\tan \psi = nf(\theta)/f'(\theta)$ and $\cot \psi = \dfrac{d \ln r}{d\theta}$.

18. Let P be a point of the ellipse $r = ep/(1 - e \cos \theta)$.
    (a) Show that
    $$\tan \psi = \frac{-p}{r \sin \theta} = \frac{-p}{y}.$$
    (b) Let F be the focus of the ellipse that is not the pole of the coordinate system chosen for part (a). Choose a new polar coordinate system whose pole is F so that if P has coordinates $(\bar{r}, \bar{\theta})$ in this new system, then
    $$\bar{r} = \frac{ep}{1 + e \cos \bar{\theta}}.$$
    Show that
    $$\tan \bar{\psi} = \frac{p}{y} = -\tan \psi,$$
    and conclude that
    $$\bar{\psi} = \pi - \psi.$$

(c) Draw a figure to help you deduce that the equation $\bar{\psi} = \pi - \psi$ means that the tangent line to the ellipse at a point P bisects the angle between the two lines that join P to the foci of the ellipse. (The focusing property of the ellipse can now be easily deduced.)

65. AREAS IN POLAR COORDINATES

Now we are going to find the area A of the region bounded by the polar curve $r = f(\theta)$ and radial lines $\theta = \alpha$ and $\theta = \beta$, as shown in Fig. 65-1. Let us suppose that $\alpha < \beta$ and that we use radian measure for our angles. We find areas of regions of the cartesian plane by "adding the areas of little rectangular regions" (Section 39). Here we "add areas of little triangular regions," like the shaded one in Fig. 65-1.

Actually, this region is a wedge $\Delta\theta$ radians wide cut from a circular disk of radius r. Since the central angle of the entire disk is 2π radians wide, the area of our wedge is $\dfrac{\Delta\theta}{2\pi}$ times the area of the disk (πr^2 square units); that is,

area of shaded region $= \tfrac{1}{2}r^2\,\Delta\theta.$

A sum of such terms, $\sum \tfrac{1}{2}r^2\,\Delta\theta$, approximates the area of the region we are measuring, and this sum suggests that the actual area A is given by an integral:

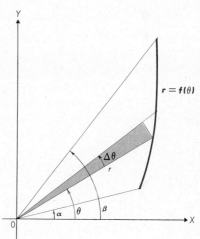

(65-1) $\quad A = \tfrac{1}{2}\displaystyle\int_\alpha^\beta r^2\,d\theta = \tfrac{1}{2}\displaystyle\int_\alpha^\beta f(\theta)^2\,d\theta.$

Figure 65-1

Example 65-1. Find the area of the region that is bounded by the cardioid $r = 2(1 - \cos\theta)$.

Solution. Because of the symmetry of the cardioid (Fig. 62-5), we may calculate the area of the part of the region for which $\theta \in [0, \pi]$ and multiply it by 2. Thus our desired area A is

$$A = 2 \cdot \tfrac{1}{2}\int_0^\pi r^2\,d\theta = 4\int_0^\pi (1 - \cos\theta)^2\,d\theta$$

$$= 4\int_0^\pi (1 - 2\cos\theta + \cos^2\theta)\,d\theta$$

$$= 6\pi.$$

The area represented by the integral in Equation 65-1 should be thought of as the area of the region that is "swept out" by the radial line segment joining the pole to the point (r, θ) as this point moves along the curve for $\theta \in [\alpha, \beta]$. Thus, for example, if we consider the circle $r = 5$ and let $\alpha = 0$, $\beta = 3\pi$, then the number

$$A = \tfrac{1}{2} \int_0^{3\pi} 25 \, d\theta = \tfrac{75}{2} \, \pi$$

is one and one-half times the area of the circle. You should make a sketch showing the region whose area you intend to find and carefully determine α and β so that this region is swept out once, and only once, by the radial line segment to the point (r, θ) for $\theta \in [\alpha, \beta]$. If you study the following example carefully, it will help you avoid a common error in determining α and β.

Example 65-2.   Find the area of the region that is cut out of the first quadrant by the curve $r = 2 \cos 3\theta$.

Solution.   We are interested in the shaded region shown in Fig. 65-2. The proper numbers α and β are *not* 0 and $\tfrac{1}{2}\pi$. You can easily convince yourself that the portion of the curve in the first quadrant is obtained by choosing θ from the interval $[0, \tfrac{1}{6}\pi]$. Thus the area we want is

$$A = \tfrac{1}{2} \int_0^{\pi/6} 4 \cos^2 3\theta = 2 \int_0^{\pi/6} \cos^2 3\theta \, d\theta = \tfrac{1}{6}\pi.$$

Equation 65-1 can be used to find areas of regions bounded by two polar curves. An example will show us how.

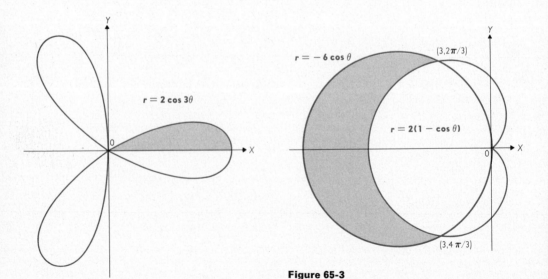

$r = 2 \cos 3\theta$

$r = -6 \cos \theta$

$(3, 2\pi/3)$

$r = 2(1 - \cos \theta)$

$(3, 4\pi/3)$

Figure 65-3

Figure 65-2

Example 65-3. Find the area of the region that is outside the cardioid $r = 2(1 - \cos \theta)$ and inside the circle $r = -6 \cos \theta$.

Solution. In Fig. 65-3 we have sketched the curves involved and have shaded the region whose area we seek. We found in Example 62-5 that our curves intersect in the points $(3, \frac{2}{3}\pi)$ and $(3, \frac{4}{3}\pi)$. So we see that the desired area is obtained by subtracting the area of the region swept out by the radial line of the cardioid from the area of the region swept out by the radial line of the circle as θ goes from $\frac{2}{3}\pi$ to $\frac{4}{3}\pi$. Making use of the symmetry of our region, we have

$$A = 2 \left[\frac{1}{2} \int_{2\pi/3}^{\pi} 36 \cos^2 \theta \; d\theta - \frac{1}{2} \int_{2\pi/3}^{\pi} 4(1 - \cos \theta)^2 \; d\theta \right]$$

$$= \int_{2\pi/3}^{\pi} [36 \cos^2 \theta - 4(1 - 2 \cos \theta + \cos^2 \theta)] \; d\theta$$

$$= 4 \int_{2\pi/3}^{\pi} (4 \cos 2\theta + 2 \cos \theta + 3) \; d\theta = 4\pi.$$

PROBLEMS 65

1. Find the area of the region that is bounded by one loop of the curve.
    (a) $r = \sin \theta$ (b) $r = \sin 2\theta$
    (c) $r^2 = \sin \theta$ (d) $r^2 = \sin 2\theta$
    (e) $r = \sin 8\theta$ (f) $r = \sin 9\theta$
    (g) $r = \cos 3\theta$ (h) $r^2 = \cos 3\theta$
    (i) $r = \sin^2 \theta$ (j) $r = \cos^2 \theta$

2. Find the area of the region bounded by the "first turn" of the given spiral $(0 \le \theta \le 2\pi)$ and the polar axis.
    (a) $r = \theta$ (b) $r = e^\theta$ (c) $r = \cosh \theta$

3. Find the area of the disk bounded by the circle $r = 6 \cos \theta + 8 \sin \theta$.

4. Find the area of the region enclosed by the polar curve $r = \sin \theta \cos^2 \theta$.

5. Evaluate the following integrals by interpreting them geometrically.

    (a) $\frac{1}{2} \int_0^{\pi/2} \cos^2 \theta \; d\theta$ (b) $\frac{1}{2} \int_0^{\pi/4} 4 \sec^2 \theta \; d\theta$ (c) $\frac{1}{2} \int_{\pi/4}^{3\pi/4} \csc^2 \theta \; d\theta$

6. Find the area of the region bounded by the inside loop of the curve $r = 1 - 2 \cos \theta$.

7. Find the area of the region bounded by the curve $r = 3 + 2 \cos \theta$.

8. How much material is needed to cover the face of a bow tie whose outline is the curve $r^2 = 8 \cos 2\theta$?

9. Find the area of the region outside the circle $r = -6 \cos \theta$ and inside the cardioid $r = 2(1 - \cos \theta)$ (see Fig. 65-3).

10. Show that the area of the region bounded by two radial lines and the spiral $r\theta = k$ is directly proportional to the difference of the lengths of the two radial segments.

11.   Find the area of the region that is enclosed by the curve $r^2 \cos^4 \theta - 5 \cos^2 \theta + 1 = 0$.

12.   Show that the area of the region enclosed by the *rose curve* $r = a \cos n\theta$ is one-half the area of the circular disk in which the rose is inscribed if n is an even positive integer and one-fourth the area of the disk if n is an odd positive integer.

13.   Interpret $\displaystyle\int_0^{2\pi} \frac{2\,d\theta}{(2 - \cos \theta)^2}$ as the area of a region enclosed by a certain ellipse and thereby find its value.

14.   Use polar coordinates to find the area of the right triangular region of the XY-plane whose vertices are the points $(0, 0)$, $(a, 0)$, and (a, b).

66.   ROTATION OF AXES

We now return to the problem of changing from one cartesian coordinate system to another. In Section 27 we found that we can often simplify the form of certain equations by translating the axes. For example, we saw that a proper translation puts the equation of a circle in the form $\bar{x}^2 + \bar{y}^2 = r^2$. At that time we only considered new coordinate systems whose axes were parallel to the original axes. Now we are ready to discuss more complicated transformations of axes.

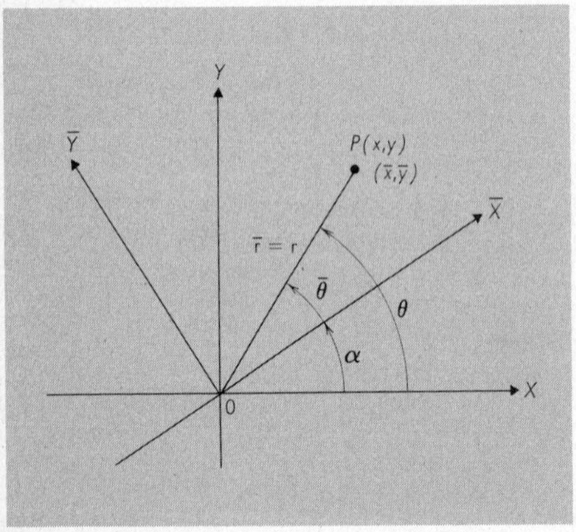

Figure 66-1

In Fig. 66-1 we show two cartesian coordinate systems—an XY-system and an $\overline{X}\overline{Y}$-system, with the same origin but such that the $\overline{X}$-axis makes an angle α with the X-axis. We say that the $\overline{X}\overline{Y}$-system is obtained from the XY-system by a **rotation** through α. Each point P will have two sets of coordinates, (x, y) and $(\bar{x}, \bar{y})$. We will now find the equations that relate these coordinates.

We introduce two polar coordinate systems, both systems having the origin O as the pole. In one system the positive X-axis is the polar axis, and in the other system the positive $\overline{X}$-axis is the polar axis. Then our point P will have polar coordinates (r, θ) and $(\bar{r}, \bar{\theta})$ in addition to its two pairs of cartesian coordinates. We see that

$$\bar{r} = r \quad \text{and} \quad \bar{\theta} = \theta - \alpha,$$

and therefore the equations

$$\bar{x} = \bar{r} \cos \bar{\theta} \quad \text{and} \quad \bar{y} = \bar{r} \sin \bar{\theta}$$

give us

(66-1) $\bar{x} = r \cos (\theta - \alpha) \quad \text{and} \quad \bar{y} = r \sin (\theta - \alpha).$

We can use the trigonometric identities

$$\cos (\theta - \alpha) = \cos \theta \cos \alpha + \sin \theta \sin \alpha$$

and

$$\sin (\theta - \alpha) = \sin \theta \cos \alpha - \cos \theta \sin \alpha$$

to write Equations 66-1 as

(66-2)
$$\bar{x} = r \cos \theta \cos \alpha + r \sin \theta \sin \alpha,$$
$$\bar{y} = r \sin \theta \cos \alpha - r \cos \theta \sin \alpha.$$

Now we replace $r \cos \theta$ with x and $r \sin \theta$ with y in Equations 66-2, and we obtain our transformation equations

(66-3)
$$\bar{x} = x \cos \alpha + y \sin \alpha,$$
$$\bar{y} = -x \sin \alpha + y \cos \alpha.$$

It is not hard to solve these equations for x and y and thus obtain the inverse transformation equations

(66-4)
$$x = \bar{x} \cos \alpha - \bar{y} \sin \alpha,$$
$$y = \bar{x} \sin \alpha + \bar{y} \cos \alpha.$$

Example 66-1.   Suppose that the $\bar{X}\bar{Y}$-axes are obtained by rotating the XY-axes through an angle of 45°. To what does the equation $xy = 1$ transform?

Solution.   Here $\alpha = 45°$, so Equations 66-4 become

$$x = \frac{\bar{x}}{\sqrt{2}} - \frac{\bar{y}}{\sqrt{2}}, \qquad y = \frac{\bar{x}}{\sqrt{2}} + \frac{\bar{y}}{\sqrt{2}}.$$

Hence

$$xy = \frac{(\bar{x} - \bar{y})(\bar{x} + \bar{y})}{2} = \frac{(\bar{x}^2 - \bar{y}^2)}{2}.$$

Thus the equation $xy = 1$ is transformed into the equation

$$\frac{\bar{x}^2}{2} - \frac{\bar{y}^2}{2} = 1.$$

This equation is the standard form for an equilateral hyperbola whose asymptotes bisect the quadrants in the $\bar{X}\bar{Y}$-coordinate system. Thus $xy = 1$ represents an equilateral hyperbola whose asymptotes are the X- and Y-axes (see Fig. 66-2).

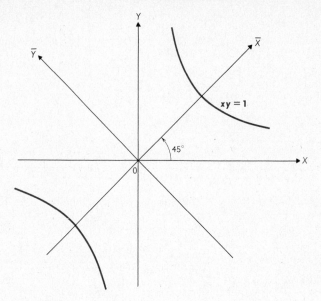

Figure 66-2

Example 66-2. Find the distance between the point $P(1, 4)$ and the line $2y - x = 2$.

Solution. Figure 66-3 shows the given point and line. The slope of the line is $\tan \alpha = \frac{1}{2}$, where α is the angle of inclination. Now let us introduce an $\bar{X}\bar{Y}$-coordinate system in such a way that the $\bar{X}$-axis is parallel to our given line. We can obtain such an $\bar{X}\bar{Y}$-system by a rotation of axes through α. Since $\tan \alpha = \frac{1}{2}$, we see that $\sin \alpha = 1/\sqrt{5}$ and $\cos \alpha = 2/\sqrt{5}$. Thus the Transformation Equations 66-3 and 66-4 become

$$\bar{x} = \frac{2x}{\sqrt{5}} + \frac{y}{\sqrt{5}}, \qquad \bar{y} = \frac{-x}{\sqrt{5}} + \frac{2y}{\sqrt{5}},$$

$$x = \frac{2\bar{x}}{\sqrt{5}} - \frac{\bar{y}}{\sqrt{5}}, \qquad y = \frac{\bar{x}}{\sqrt{5}} + \frac{2\bar{y}}{\sqrt{5}}.$$

When we substitute these expressions for x and y in the equation of our line, it becomes $\bar{y} = 2/\sqrt{5}$. Furthermore, these transformation equations tell us that the $\bar{Y}$-coordinate of P is $7/\sqrt{5}$. Since the line is parallel to the $\bar{X}$-axis, we find the distance d between it and P by subtraction:

$$d = \frac{7}{\sqrt{5}} - \frac{2}{\sqrt{5}} = \frac{5}{\sqrt{5}} = \sqrt{5}.$$

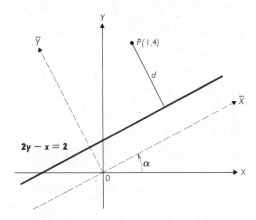

Figure 66-3

The **general rigid transformation** equations

(66-5)
$$\bar{x} = x \cos \alpha + y \sin \alpha - h,$$
$$\bar{y} = -x \sin \alpha + y \cos \alpha - k$$

are the transformation equations that result from a rotation through α *followed* by a translation.

Example 66-3. Sketch the two cartesian coordinate systems that are related by the equations

(66-6)
$$\bar{x} = \tfrac{3}{5}x + \tfrac{4}{5}y - 3,$$
$$\bar{y} = -\tfrac{4}{5}x + \tfrac{3}{5}y - 1.$$

Draw the XY-system in the "usual" position.

Solution. The points whose $\bar{X}\bar{Y}$-coordinates are $(0, 0)$ and $(1, 0)$ determine the $\bar{X}$-axis, and the points whose $\bar{X}\bar{Y}$-coordinates are $(0, 0)$ and $(0, 1)$ determine the $\bar{Y}$-axis. We will locate these points by finding their XY-coordinates from Equations 66-6 and then plotting them in the XY-coordinate system of Fig. 66-4. The XY-coordinates of the point whose $\bar{X}\bar{Y}$-coordinates are $(0, 0)$ satisfy the system of equations
$$0 = \tfrac{3}{5}x + \tfrac{4}{5}y - 3,$$
$$0 = -\tfrac{4}{5}x + \tfrac{3}{5}y - 1.$$

When we solve this system, we obtain $(x, y) = (1, 3)$. Similarly, we find that the point whose $\bar{X}\bar{Y}$-coordinates are $(1, 0)$ has XY-coordinates $(\tfrac{8}{5}, \tfrac{19}{5})$. The point whose

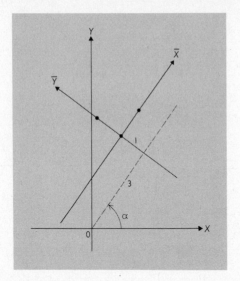

Figure 66-4

$\overline{X}\overline{Y}$-coordinates are $(0, 1)$ has XY-coordinates $(\frac{1}{5}, \frac{18}{5})$. Now we plot these points and draw in the $\overline{X}\overline{Y}$-axes. We see that the $\overline{X}\overline{Y}$-axes are obtained by first rotating the axes through an angle α, where $\cos \alpha = \frac{3}{5}$ and $\sin \alpha = \frac{4}{5}$, and then translating the origin 3 units in the "new X-direction" and 1 unit in the "new Y-direction."

PROBLEMS 66

1. Find the $\overline{X}\overline{Y}$-coordinates of the point $(-4, 6)$ under a rotation of axes through the given angle.

   (a) $30°$ (b) $135°$ (c) $240°$ (d) $-180°$

   (e) π (f) $-\frac{1}{2}\pi$ (g) 2π (h) $-\frac{1}{4}\pi$

2. Under a rotation of axes through an angle of $60°$, a point has the given $\overline{X}\overline{Y}$-coordinates. What are its XY-coordinates?

   (a) $(2, 0)$ (b) $(0, 2)$ (c) $(-2, 4)$ (d) $(-\sqrt{12}, 6)$

3. (a) Sketch the XY- and $\overline{X}\overline{Y}$-axes if they are related by the transformation equations

$$\bar{x} = .8x + .6y,$$

$$\bar{y} = -.6x + .8y.$$

   (b) Solve for x and y in terms of $\bar{x}$ and $\bar{y}$.
   (c) What angle does the $\overline{X}$-axis make with the X-axis?
   (d) What is the equation of the line $4x + 3y = 10$ in the $\overline{X}\overline{Y}$-system?
   (e) What is the equation of the circle $x^2 + y^2 = 16$ in the $\overline{X}\overline{Y}$-system? (You can answer this question with no computation!)

4. Find the distance between the line $3x - 4y = 10$ and
    (a) the origin.
    (b) the point $(-5, 15)$.
    (c) the line $3x - 4y + 15 = 0$.
    (d) the line $3x - 4y = 15$.
    (e) the center of the circle $x^2 + y^2 = 10x$.
    (f) the focus of the parabola $y^2 + 25 = 10y + 20x$.

5. (a) What are the transformation of axes equations for a rotation through $90°$?
    (b) What are the transformation of axes equations for a rotation through $180°$?
    (c) What are the transformation of axes equations for a rotation through $270°$?
    (d) Do the equations $\bar{x} = y$ and $\bar{y} = x$ represent a rotation of axes?

6. Suppose that the $\bar{X}\bar{Y}$-axes are obtained by a rotation through $30°$.
    (a) Express $\bar{x}$ and $\bar{y}$ in terms of x and y.
    (b) Express x and y in terms of $\bar{x}$ and $\bar{y}$.
    (c) Express the line $4x - 6y = 5$ in $\bar{X}\bar{Y}$-coordinates.
    (d) Express the circle $x^2 + y^2 - 2x = 2$ in $\bar{X}\bar{Y}$-coordinates. (*Hint:* You can find the center and radius very quickly.)

7. (a) Sketch the $\bar{X}\bar{Y}$- and XY-axes if they are related by the equation $\bar{x} = .9611x + .2764y$ and $\bar{y} = -.2764x + .9611y$.
    (b) What is the angle of rotation?
    (c) What is the slope of the $\bar{Y}$-axis in the XY-coordinate system?
    (d) What is the slope of the X-axis in the $\bar{X}\bar{Y}$-coordinate system?
    (e) What is the angle of inclination of the line $x + y = 3$ in the $\bar{X}\bar{Y}$-system?
    (f) What is the angle of inclination of the line $\bar{x} + \bar{y} = 3$ in the XY-system?

8. Explain why each system of transformation of coordinates equations does *not* represent a rotation.
    (a) $\bar{x} = x + y, \bar{y} = x - y$ (b) $\bar{x} = .1x + .2y, \bar{y} = -.2x + .1y$
    (c) $\bar{x} = y, \bar{y} = x$ (d) $\bar{x} = -.6x + .8y, \bar{y} = .8x + .6y$

9. Explain geometrically why moving the bars from the left- to the right-hand sides of Equations 66-3, and replacing α with $-\alpha$, produces Equations 66-4.

10. Solve Equations 66-5 for x and y.

11. Give a geometric description of the transformation that is determined by the equations $\bar{x} = hx$ and $\bar{y} = ky$, where h and k are given positive numbers.

12. Suppose that points P_1 and P_2 have coordinates (x_1, y_1), $(\bar{x}_1, \bar{y}_1)$ and (x_2, y_2), $(\bar{x}_2, \bar{y}_2)$ relative to two coordinate systems. It is geometrically obvious that if the coordinate transformation is a translation or a rotation, the distance formula should be the same in each coordinate system; that is,

$$\sqrt{(x_1 - x_2)^2 + (y_1 - y_2)^2} = \sqrt{(\bar{x}_1 - \bar{x}_2)^2 + (\bar{y}_1 - \bar{y}_2)^2}.$$

    Prove it.

67. THE GENERAL QUADRATIC EQUATION

For certain choices of the numbers A, B, C, D, E, and F, we already know that the graph of the quadratic equation

(67-1) $Ax^2 + Bxy + Cy^2 + Dx + Ey + F = 0$

is a circle, a conic, or a line. For example, if we choose $A = 1$, $B = 0$, $C = -1$, $D = 0$, $E = 0$, and $F = -1$, then Equation 67-1 becomes

$$x^2 - y^2 = 1,$$

which we recognize as the equation of an equilateral hyperbola. But other choices of A, B, C, D, E, and F lead to equations that we have not yet studied—for example, the equation

(67-2) $6x^2 + 24xy - y^2 - 12x + 26y + 11 = 0.$

In this section we shall see that the graph of a quadratic equation is always a familiar figure. In particular, it will turn out that the graph of Equation 67-2 is a hyperbola.

In order to find the graph of a quadratic equation, we will first make a transformation of axes that reduces the equation to one of the standard forms we have already studied. Thus in Example 66-1 we saw that a rotation of axes through $45°$ transforms the equation $xy = 1$ into the standard equation of an equilateral hyperbola

$$\frac{\bar{x}^2}{2} - \frac{\bar{y}^2}{2} = 1.$$

The question that naturally comes to mind is, "How did we decide to rotate through $45°$?" The following example shows how we select our angle of rotation.

Example 67-1. By a suitable rotation of axes, reduce the equation

(67-3) $8x^2 - 4xy + 5y^2 = 36$

to a standard form.

Solution. When we make the rotation given by Equations 66-4, our equation becomes

$8(\bar{x} \cos \alpha - \bar{y} \sin \alpha)^2 - 4(\bar{x} \cos \alpha - \bar{y} \sin \alpha) \cdot (\bar{x} \sin \alpha + \bar{y} \cos \alpha)$

$$+ 5(\bar{x} \sin \alpha + \bar{y} \cos \alpha)^2 = 36.$$

Now we expand and collect terms:

(67-4) $(8 \cos^2 \alpha - 4 \sin \alpha \cos \alpha + 5 \sin^2 \alpha)\bar{x}^2$

$$+ (4 \sin^2 \alpha - 6 \sin \alpha \cos \alpha - 4 \cos^2 \alpha)\bar{x}\bar{y}$$

$$+ (8 \sin^2 \alpha + 4 \sin \alpha \cos \alpha + 5 \cos^2 \alpha)\bar{y}^2 = 36.$$

This last equation has the form

(67-5) $\bar{A}\bar{x}^2 + \bar{B}\bar{x}\bar{y} + \bar{C}\bar{y}^2 = 36,$

where the numbers $\bar{A}$, $\bar{B}$, and $\bar{C}$ depend on α. If $\bar{B} = 0$, we would recognize the type of curve we are dealing with, so let us choose α so that $\bar{B} = 0$—that is, so

$$4 \sin^2 \alpha - 6 \sin \alpha \cos \alpha - 4 \cos^2 \alpha = 0.$$

Since $\cos^2 \alpha - \sin^2 \alpha = \cos 2\alpha$ and $\sin \alpha \cos \alpha = \frac{1}{2} \sin 2\alpha$, this equation can be written

$$-3 \sin 2\alpha - 4 \cos 2\alpha = 0,$$

from which we see that

$$\cot 2\alpha = -\tfrac{3}{4}.$$

We can select an angle 2α in the range $0° < 2\alpha < 180°$ that satisfies this equation, and thus $0° < \alpha < 90°$. To find the coefficients of $\bar{x}^2$ and $\bar{y}^2$, we find the numbers $\cos \alpha$ and $\sin \alpha$ from the trigonometric identities

$$\cos \alpha = \sqrt{\frac{1 + \cos 2\alpha}{2}} \quad \text{and} \quad \sin \alpha = \sqrt{\frac{1 - \cos 2\alpha}{2}}.$$

Since $\cot 2\alpha = -\tfrac{3}{4}$, it follows that the radial line containing the point $(-3, 4)$ makes an angle of 2α with the positive X-axis. Hence $\cos 2\alpha = -\tfrac{3}{5}$, and we have

$$\cos \alpha = \sqrt{\frac{1 - \tfrac{3}{5}}{2}} = \sqrt{\frac{1}{5}}$$

and

$$\sin \alpha = \sqrt{\frac{1 + \tfrac{3}{5}}{2}} = \sqrt{\frac{4}{5}}.$$

When we substitute these numbers in Equation 67-4, the equation becomes

$$4\bar{x}^2 + 9\bar{y}^2 = 36.$$

In other words,

$$\frac{\bar{x}^2}{9} + \frac{\bar{y}^2}{4} = 1.$$

So we see that the graph of Equation 67-3 is an ellipse whose major diameter is 6 units long and whose minor diameter is 4 units long. We have shown our ellipse, together with the rotated axes, in Fig. 67-1.

We can employ the same procedure to reduce any quadratic expression

$$Ax^2 + Bxy + Cy^2$$

in which $B \neq 0$ to the form

$$\bar{A}\bar{x}^2 + \bar{C}\bar{y}^2.$$

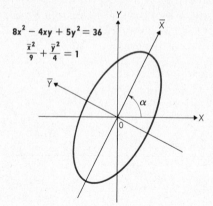

$8x^2 - 4xy + 5y^2 = 36$

$\dfrac{\bar{x}^2}{9} + \dfrac{\bar{y}^2}{4} = 1$

Figure 67-1

We simply replace x and y in terms of $\bar{x}$ and $\bar{y}$ according to the rotation Equations 66-4 and obtain the expression $A(\bar{x} \cos \alpha - \bar{y} \sin \alpha)^2 + B(\bar{x} \cos \alpha - \bar{y} \sin \alpha) \cdot (\bar{x} \sin \alpha + \bar{y} \cos \alpha) + C(\bar{x} \sin \alpha + \bar{y} \cos \alpha)^2$. When we multiply out and collect the coefficients of $\bar{x}^2$, $\bar{x}\bar{y}$, and $\bar{y}^2$, we find that

$$Ax^2 + Bxy + Cy^2 = \bar{A}\bar{x}^2 + \bar{B}\bar{x}\bar{y} + \bar{C}\bar{y}^2,$$

where

$$\bar{A} = A \cos^2 \alpha + B \sin \alpha \cos \alpha + C \sin^2 \alpha,$$

(67-6) $$\bar{B} = (C - A) \sin 2\alpha + B \cos 2\alpha,$$

$$\bar{C} = A \sin^2 \alpha - B \sin \alpha \cos \alpha + C \cos^2 \alpha.$$

We see from the second of Equations 67-6 that $\bar{B}$ will be 0 if we choose α so that

(67-7) $$\cot 2\alpha = \frac{A - C}{B}.$$

Clearly, we can always choose α so that $0° < 2\alpha < 180°$. We sum up our results as a theorem.

Theorem 67-1. *Every quadratic expression $Ax^2 + Bxy + Cy^2$ in which $B \neq 0$ can be reduced to the form $\bar{A}\bar{x}^2 + \bar{C}\bar{y}^2$ by rotating axes through an angle α, where $0° < \alpha < 90°$, and $\cot 2\alpha = (A - C)/B$.*

By rotating axes through the angle α, as described in Theorem 67-1, we can reduce any quadratic equation of the form of Equation 67-1 in which $B \neq 0$ to a quadratic equation

$$\bar{A}\bar{x}^2 + \bar{C}\bar{y}^2 + \bar{D}\bar{x} + \bar{E}\bar{y} + \bar{F} = 0.$$

If we don't recognize the graph of this equation, a suitable translation will bring it to a form that we do recognize. We will illustrate the entire procedure by an example. It will pay you to review our work on translation in Section 27.

Example 67-2. Describe the graph of Equation 67-2.

Solution. Here $A = 6$, $B = 24$, and $C = -1$. So we first rotate the axes through the acute angle α such that

$$\cot 2\alpha = \tfrac{7}{24}.$$

It follows that $\cos 2\alpha = \tfrac{7}{25}$, and hence

$$\cos \alpha = \sqrt{\frac{1 + \cos 2\alpha}{2}} = \frac{4}{5}$$

and

$$\sin \alpha = \sqrt{\frac{1 - \cos 2\alpha}{2}} = \frac{3}{5}.$$

Therefore our rotation equations are

$$x = \tfrac{4}{5}\bar{x} - \tfrac{3}{5}\bar{y},$$
$$y = \tfrac{3}{5}\bar{x} + \tfrac{4}{5}\bar{y}.$$

When we substitute these quantities in Equation 67-2, we obtain

(67-8) $15\bar{x}^2 - 10\bar{y}^2 + 6\bar{x} + 28\bar{y} + 11 = 0.$

Now we will translate the axes by means of the equations

$$\bar{x} = \bar{\bar{x}} + h \quad \text{and} \quad \bar{y} = \bar{\bar{y}} + k,$$

where h and k are to be determined so as to make our resulting equation "simpler" than Equation 67-8. In terms of $\bar{\bar{x}}$ and $\bar{\bar{y}}$, Equation 67-8 reads

(67-9) $15\bar{\bar{x}}^2 - 10\bar{\bar{y}}^2 + (30h + 6)\bar{\bar{x}} + (-20k + 28)\bar{\bar{y}} + 15h^2$
$$- 10k^2 + 6h + 28k + 11 = 0.$$

Now we select h and k so that $30h + 6 = 0$ and $-20k + 28 = 0$; that is, $h = -\tfrac{1}{5}$ and $k = \tfrac{7}{5}$. Then Equation 67-9 becomes

$$15\bar{\bar{x}}^2 - 10\bar{\bar{y}}^2 + 30 = 0.$$

Finally, we write this equation in the standard form

$$\frac{\bar{\bar{y}}^2}{3} - \frac{\bar{\bar{x}}^2}{2} = 1,$$

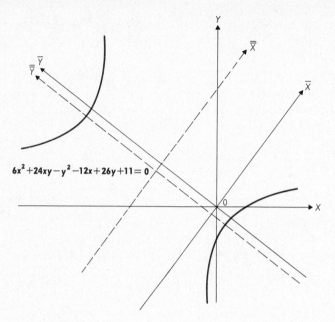

$$6x^2+24xy-y^2-12x+26y+11=0$$

Figure 67-2

and we see that our graph is a hyperbola. The "specifications" (diameter, length of latus rectum, focal length, and so on) of our hyperbola can be read from this equation. The graph of our equation is shown in Fig. 67-2, together with the three different coordinate axes involved.

In the preceding example we began with a quadratic equation

(67-10) $\qquad Ax^2 + Bxy + Cy^2 + Dx + Ey + F = 0$

and rotated axes to obtain an equation of the form

(67-11) $\qquad \bar{A}\bar{x}^2 + \bar{C}\bar{y}^2 + \bar{D}\bar{x} + \bar{E}\bar{y} + \bar{F} = 0.$

It is easy to verify, by using the proper translation, that if $\bar{A}$ and $\bar{C}$ have opposite signs ($\bar{A}\bar{C} < 0$), then the graph of Equation 67-11 is a hyperbola. If $\bar{A}$ and $\bar{C}$ have the same signs ($\bar{A}\bar{C} > 0$), our curve is an ellipse or a circle. And if one of the numbers $\bar{A}$ or $\bar{C}$ is 0 ($\bar{A}\bar{C} = 0$), we are dealing with a parabola. Here we are allowing for *degenerate* conics, such as the "hyperbola" $x^2 - 4y^2 = 0$ or the "ellipse" $\dfrac{x^2}{4} + \dfrac{y^2}{9} + 1 = 0$.

In Problem 67-10 we ask you to verify that

$$\bar{B}^2 - 4\bar{A}\bar{C} = B^2 - 4AC,$$

regardless of the angle of rotation. In particular, if we choose α so that $\bar{B} = 0$, then we find that

$$-4\bar{A}\bar{C} = B^2 - 4AC.$$

Therefore the sign of $\bar{A}\bar{C}$ is determined by the sign of $B^2 - 4AC$. And so from our remarks concerning the graph of Equation 67-11, we can conclude that *the graph of the Equation* 67-10 *is*

(1)   *a parabola if* $B^2 - 4AC = 0$,

(2)   *an ellipse if* $B^2 - 4AC < 0$,

(3)   *a hyperbola if* $B^2 - 4AC > 0$,

with the understanding that degenerate cases may occur.

The number $B^2 - 4AC$ is called the *discriminant* of the quadratic Equation 67-10. Thus the discriminant of Equation 67-2 is $(24)^2 - 4 \cdot 6 \cdot (-1) = 600$, and we found that the graph of this equation is a hyperbola. The discriminant of Equation 67-3 is $(-4)^2 - 4 \cdot 8 \cdot 5 = -144$, and we found that the graph of this equation is an ellipse.

PROBLEMS   67

1.   Reduce the following quadratic expressions to the form $\bar{A}\bar{x}^2 + \bar{C}\bar{y}^2$.

(a) $x^2 + 4xy + y^2$ (b) $9x^2 + 12xy + 4y^2$
(c) $4x^2 + 4xy + y^2$ (d) $6x^2 + 5xy - 6y^2$
(e) $19x^2 - 7xy - 5y^2$ (f) $2x^2 - 5xy - 10y^2$

2.   Rotate the axes so that the following equations take the form $\bar{A}\bar{x}^2 + \bar{C}\bar{y}^2 = \bar{F}$. Sketch the graphs of these equations. Find the algebraic sign of the discriminant of each equation and use the result to check the form of your graph.

(a) $11x^2 + 24xy + 4y^2 = 20$
(b) $25x^2 + 14xy + 25y^2 = 288$
(c) $3x^2 + 4xy = 4$
(d) $9x^2 - 24xy + 16y^2 - 400x - 300y = 0$

3.   Use translations and rotations to help you sketch the following curves. Find the algebraic sign of the discriminant of each equation and use the result to check the form of your graph.

(a) $5x^2 + 6xy + 5y^2 - 32x - 32y + 32 = 0$
(b) $x^2 - 4xy + 4y^2 + 5y - 9 = 0$
(c) $9x^2 - 24xy + 16y^2 - 56x - 92y + 688 = 0$
(d) $x^2 + 2xy + y^2 - 2x - 2y - 3 = 0$

4.   Use the discriminant to identify the "conic" $2x^2 + xy - y^2 + 6y - 8 = 0$. Write the equation in factored form $(x + y - 2)(2x - y + 4) = 0$ and describe its graph.

5. What are the graphs of the following equations (degeneracies allowed)?
   (a) $\pi x^2 + \pi y^2 + \sqrt{17}\, x - (\tan 5)y + \ln 7 = 0$
   (b) $(\cos 3)x^2 + 2(\sin 3)xy + (\cos 3)y^2 + 5x - 7y + \tan 5 = 0$
   (c) $(\cos 3)x^2 + 2(\cos 3)xy + (\sin 3)y^2 + 5x - 7y + \tan 5 = 0$
   (d) $(\ln 2)x^2 + 2(\ln 3)xy + (\ln 4)y^2 + 5x - 7y + \tan 5 = 0$

6. The following curves are degenerate conics. What does the discriminant say they are, and what are they really?
   (a) $3x^2 + 4y^2 + 5 = 0$ (b) $4x^2 - y^2 = 0$
   (c) $x^2 - 4 = 0$ (d) $x^2 - 2xy + y^2 + 3x - 3y + 2 = 0$

7. For what choices of p will the following curves be ellipses? Can we choose p so that the curve of part (a) is a parabola?
   (a) $px^2 + 6xy + 7y^2 = 8$ (b) $2x^2 - 4xy + py^2 = 9$
   (c) $2x^2 + pxy + 5y^2 = 7$ (d) $px^2 - 3xy - 4y^2 = 10$

8. Identify the conic that contains the given points.
   (a) $(-3, -2), (2, -2), (2, 1), (0, -5), (3, 0)$
   (b) $(1, 2), (1, 8), (-1, 6), (-1, 0), (-5, 0)$
   (c) $(3, 4), (-3, -2), (-3, 0), (-2, 0), (2, 4)$
   (d) $(0, 0), (4, 0), (1, 1), (3, 1), (0, -2)$

9. Describe the following polar curves.

   (a) $r = \dfrac{6}{2 - \cos(\theta - 45°)}$ (b) $r = \dfrac{6}{1 - 2\cos(\theta - 30°)}$

   (c) $r = \dfrac{6}{1 - \cos(\theta - 60°)}$ (d) $r = \dfrac{16}{3 + 5\sin(\theta + 45°)}$

   (e) $r = \dfrac{16}{5 - 3\cos(\theta + 30°)}$ (f) $r = \dfrac{4}{1 + \sin(\theta + 60°)}$

10. Use Equations 67-6 to show that $\bar{A} + \bar{C} = A + C$ and $\bar{B}^2 - 4\bar{A}\bar{C} = B^2 - 4AC$.

11. How must A, B, and C be related for the curve $Ax^2 + Bxy + Cy^2 + Dx + Ey + F = 0$ to be a circle (or a degenerate circle)?

12. Give a complete discussion of the curve $Ax^2 + Bxy + Cy^2 = 0$.

13. Explain why A and C must have the same sign if the graph of Equation 67-10 is to be an ellipse or a parabola. Must A and C have opposite signs if the graph is a hyperbola?

14. Show that if we pick α to satisfy Equation 67-7, then
$$\bar{A} = \tfrac{1}{2}[A + C + (B/|B|)\sqrt{(A - C)^2 + B^2}]$$
and
$$\bar{C} = \tfrac{1}{2}[A + C - (B/|B|)\sqrt{(A - C)^2 + B^2}].$$

68. VECTORS IN THE PLANE

Physicists and engineers talk and think a great deal in terms of *vectors*. These scientists describe a vector quantity as one that has both magnitude and direction, and so they use vectors to represent such things as velocity and force. Geometrically, a vector is represented as an arrow, the length of the arrow being the magnitude of the vector and the direction of the arrow being the assigned direction of the vector. For the present, we will think of vectors as arrows in the plane, but starting with Chapter 11 we will picture them in other ways, too. For example, some of our discussion of systems of equations will be expressed in terms of vectors. The geometric approach we take here gives us a valuable background for the more abstract way vectors are treated in advanced mathematics.

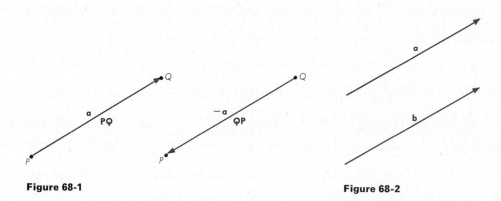

Figure 68-1   **Figure 68-2**

A **vector** is determined by a pair of points in the plane—the **initial point** and the **terminal point** (Fig. 68-1). We use boldface type to denote vectors. For instance, we have labeled the vector on the left in Fig. 68-1 as **a**, and we might also call it **PQ** if we want to emphasize that it is the vector from P to Q. We denote the magnitude of the vector **a** by the symbol $|\mathbf{a}|$. We often refer to this number as the **absolute value** of **a**. If **a** is the vector **PQ**, then the vector **QP** is denoted by $-\mathbf{a}$.

Two vectors are considered **equal** if and only if they have the same magnitude and the same direction. Thus for the vectors **a** and **b** in Fig. 68-2, we have $\mathbf{a} = \mathbf{b}$. Notice that we do not require that equal vectors coincide, but they must be parallel, have the same length, and point in the same direction.

The **sum a + b** of the vectors **a** and **b** shown on the left in Fig. 68-3 is obtained by making the initial point of **b** coincide with the terminal point of **a** and joining

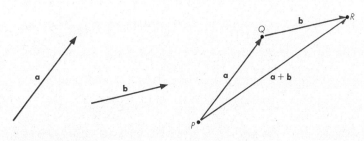

Figure 68-3

the initial point of **a** to the terminal point of **b**, as shown on the right. Thus if *P*, *Q*, and *R* are points such that **a** = **PQ** and **b** = **QR**, then **a** + **b** = **PR**; that is,

$$\mathbf{PQ} + \mathbf{QR} = \mathbf{PR}.$$

We get the sum **b** + **a** by making the initial point of **a** coincide with the terminal point of **b** and joining the initial point of **b** to the terminal point of **a**. Figure 68-4 shows how we construct **a** + **b** and **b** + **a**, and it is apparent that the **commutative law**

(68-1) $\mathbf{a} + \mathbf{b} = \mathbf{b} + \mathbf{a}$

holds for vector addition. Figure 68-4 suggests the name *parallelogram rule*, which is often used to describe the method of adding vectors.

If **a** = **PQ**, **b** = **QR**, and **c** = **RS** (see Fig. 68-5), then

$$(\mathbf{a} + \mathbf{b}) + \mathbf{c} = (\mathbf{PQ} + \mathbf{QR}) + \mathbf{RS} = \mathbf{PR} + \mathbf{RS} = \mathbf{PS}$$

and

$$\mathbf{a} + (\mathbf{b} + \mathbf{c}) = \mathbf{PQ} + (\mathbf{QR} + \mathbf{RS}) = \mathbf{PQ} + \mathbf{QS} = \mathbf{PS}.$$

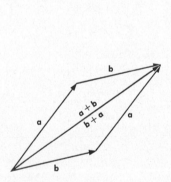

Figure 68-4

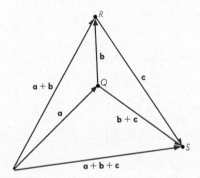

Figure 68-5

Thus the **associative law** of addition,

(68-2) $(\mathbf{a} + \mathbf{b}) + \mathbf{c} = \mathbf{a} + (\mathbf{b} + \mathbf{c})$,

is valid for vectors, and we write **a** + **b** + **c** for this sum.

We find it convenient to introduce the **zero vector 0**, which can be represented as **PP**, where *P* is any point. The magnitude of **0** is 0, and its direction is unspecified. If **a** is any vector, then

(68-3) $\mathbf{a} + \mathbf{0} = \mathbf{0} + \mathbf{a} = \mathbf{a}.$

Furthermore, our rules of vector addition tell us that

(68-4) $$\mathbf{a} + (-\mathbf{a}) = (-\mathbf{a}) + \mathbf{a} = \mathbf{0}.$$

Naturally, we define the difference $\mathbf{a} - \mathbf{b}$ by means of the equation

$$\mathbf{a} - \mathbf{b} = \mathbf{a} + (-\mathbf{b}).$$

Then it follows that

$$(\mathbf{a} - \mathbf{b}) + \mathbf{b} = \mathbf{a} + [(-\mathbf{b}) + \mathbf{b}] = \mathbf{a} + \mathbf{0} = \mathbf{a};$$

that is, the difference $\mathbf{a} - \mathbf{b}$ is the vector that must be added to $\mathbf{b}$ to obtain $\mathbf{a}$. If $\mathbf{a} = \mathbf{PQ}$ and $\mathbf{b} = \mathbf{PR}$, then

$$\mathbf{a} - \mathbf{b} = \mathbf{PQ} - \mathbf{PR} = \mathbf{PQ} + \mathbf{RP} = \mathbf{RP} + \mathbf{PQ} = \mathbf{RQ}.$$

Figure 68-6 shows the graphical relation among the vectors $\mathbf{a}$, $\mathbf{b}$, and $\mathbf{a} - \mathbf{b}$.

Some physical quantities, such as temperature and density, are not described by vectors but by numbers. Because he can measure these quantities on a number scale, the scientist calls them *scalar quantities* and the numbers that measure them, **scalars**. So, in the context of our vector analysis, the word *scalar* simply means *real number*. Notice that the absolute value $|\mathbf{a}|$ of the vector $\mathbf{a}$ is a scalar.

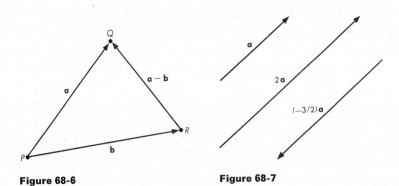

Figure 68-6   Figure 68-7

If r is a scalar (real number) and $\mathbf{a}$ is a vector, then we denote by $r\mathbf{a}$ or by $\mathbf{a}r$ *the vector that is $|r|$ times as long as* $\mathbf{a}$ *and in the direction of* $\mathbf{a}$ *if r is positive and in the opposite direction if r is negative* (see Fig. 68-7). When we think of vectors as geometric arrows, it is easy to convince ourselves that the following "natural" rules of arithmetic are valid:

$$1\mathbf{a} = \mathbf{a}$$
$$(r + s)\mathbf{a} = r\mathbf{a} + s\mathbf{a}$$
(68-5)
$$r(\mathbf{a} + \mathbf{b}) = r\mathbf{a} + r\mathbf{b}$$
$$(rs)\mathbf{a} = r(s\mathbf{a}).$$

Notice that our rules for operating with vectors, Equations 68-1 to 68-5, are like the usual rules of arithmetic; consequently, you should have no trouble in performing the various operations that we have introduced. In a formal development of the theory of vectors, we take these equations as the axioms that define a **vector space**.

The next example shows how we can use vectors to prove theorems of plane geometry.

Example 68-1. Show that the diagonals of a parallelogram bisect each other.

Solution. Figure 68-8 shows a parallelogram $OPQR$. We have labeled the midpoint of PR as M and the midpoint of OQ as N; we are to show that M and N are the same point. To prove that these points coincide, we will show that $\mathbf{ON} = \mathbf{OM}$. Since N is the midpoint of OQ, we see that $\mathbf{ON} = \frac{1}{2}\mathbf{OQ}$. But $\mathbf{OQ} = \mathbf{OR} + \mathbf{RQ}$, so we have

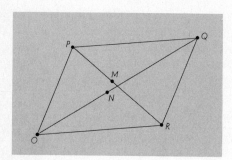

Figure 68-8

(68-6) $\mathbf{ON} = \frac{1}{2}\mathbf{OQ} = \frac{1}{2}(\mathbf{OR} + \mathbf{RQ})$.

On the other hand, $\mathbf{OM} = \mathbf{OP} + \mathbf{PM}$. Since M is the midpoint of PR, $\mathbf{PM} = \frac{1}{2}\mathbf{PR}$. But $\mathbf{PR} = \mathbf{OR} - \mathbf{OP}$, so $\mathbf{PM} = \frac{1}{2}(\mathbf{OR} - \mathbf{OP})$, and thus

(68-7) $\mathbf{OM} = \mathbf{OP} + \frac{1}{2}(\mathbf{OR} - \mathbf{OP})$
$= \frac{1}{2}(\mathbf{OR} + \mathbf{OP}) = \frac{1}{2}(\mathbf{OR} + \mathbf{RQ})$.

When we compare Equations 68-6 and 68-7, we see that $\mathbf{OM} = \mathbf{ON}$, and our assertion is proved.

Two vectors **a** and **b** are parallel if they have the same or opposite directions. For convenience, we say that the vector **0** is parallel to every vector. If **a** and **b** are parallel vectors, then one is a scalar multiple of the other. On the other hand, if **a** and **b** are not parallel, then one cannot be represented as a scalar multiple of the other, and it follows that the *single* vector equation

$$x\mathbf{a} + y\mathbf{b} = \mathbf{0}$$

is equivalent to the *two* scalar equations

$$x = 0 \quad \text{and} \quad y = 0.$$

In this case, the vectors **a** and **b** are said to be **linearly independent**.

Example 68-2. Show that two medians of a triangle intersect in the point of each median that is $\frac{2}{3}$ of the way from the vertex at which the median terminates.

Solution. A triangle OAB is determined by two non-parallel vectors **a** and **b**, as shown in Fig. 68-9. We will denote by M and N the midpoints of sides OA and OB and let P denote the intersection of the medians drawn to these midpoints. You can readily see that the medians terminating at M and N can be expressed as the vectors $\frac{1}{2}$**a** − **b** and $\frac{1}{2}$**b** − **a**. Now the vector **OP** can be considered either as the sum of **a** and a multiple $x(\frac{1}{2}$**b** − **a**) of the median terminating at N or as the sum of **b** and a multiple $y(\frac{1}{2}$**a** − **b**) of the median terminating at M. Therefore

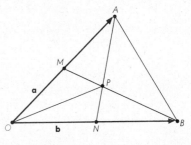

Figure 68-9

$$(68\text{-}8) \qquad \mathbf{a} + x(\tfrac{1}{2}\mathbf{b} - \mathbf{a}) = \mathbf{b} + y(\tfrac{1}{2}\mathbf{a} - \mathbf{b}),$$

and we are to show that $x = y = \frac{2}{3}$. Simple algebraic manipulation of Equation 68-8 gives us the vector equation

$$(1 - x - \tfrac{1}{2}y)\mathbf{a} + (\tfrac{1}{2}x + y - 1)\mathbf{b} = \mathbf{0}.$$

Since **a** and **b** are not parallel, their coefficients must both be zero, so

$$x + \tfrac{1}{2}y = 1 \quad \text{and} \quad \tfrac{1}{2}x + y = 1.$$

When we solve these equations, we find that $x = y = \frac{2}{3}$, as we were to prove.

PROBLEMS 68

1. Let **OP** = **OA** + **OB**. Find the coordinates of P if O is the origin and A and B are the given points.

   (a) (1, 0) and (0, 2)   (b) (1, 1) and (0, 3)
   (c) (−1, 0) and (2, 3)   (d) (−1, 2) and (3, −1)

2. Let P and Q be the points (−1, 2) and (2, 8).

   (a) Find |**PQ**|.   (b) Find R if **PR** = $\frac{1}{3}$**PQ**.
   (c) Find S if **PS** = 2**PQ**.   (d) Find T if **PT** = −**PQ**.

3. Suppose that P, Q, R, and S are the points (−1, 2), (2, 8), (0, 9), and (2, 13). Use slopes to show that the vectors **PQ** and **RS** are parallel and **PQ** and **QR** are perpendicular.

4. Express the given vector more simply.

   (a) 4**PQ** + 2**QP**   (b) 4**PQ** − 2**QP**
   (c) 2**PQ** + 2**QR** + **RP**   (d) 3**PQ** − 3**RQ**
   (e) −**PR** − **RQ**   (f) 6**PQ** + 4**RP** + 6**QR**

5. One diagonal of the parallelogram with sides **a** and **b** is **a** + **b** (see Fig. 68-4). What is the other diagonal?

6. Suppose that the vectors **a** and **b** are perpendicular. Express the following quantities in terms of the lengths $|\mathbf{a}|$ and $|\mathbf{b}|$.

   (a) $|\mathbf{a} + \mathbf{b}|$ (b) $|\mathbf{a} - \mathbf{b}|$

   (c) $|3\mathbf{a} + 4\mathbf{b}|$ (d) $\big||\mathbf{a}|^{-1}\mathbf{a}\big|$

   (e) $\big||\mathbf{a}|\mathbf{b} + |\mathbf{b}|\mathbf{a}\big|$ (f) $\big||\mathbf{a}|\mathbf{b} - |\mathbf{b}|\mathbf{a}\big|$

7. Suppose that the points P, Q, and R divide the circumference of the unit circle into three equal parts. Find $\mathbf{OP} + \mathbf{OQ} + \mathbf{OR}$.

8. Suppose that **a** and **b** are linearly independent vectors. Solve the following equations for t (if possible).

   (a) $(t^2 - 4)\mathbf{a} + (2t - 4)\mathbf{b} = 0$ (b) $\cos t\mathbf{a} + \sin t\mathbf{b} = 0$

9. Suppose that **a** and **b** are linearly independent vectors. Is there a number t such that the vector $(1 - t)\mathbf{a} + t\mathbf{b}$ is parallel to the vector $\mathbf{a} + \mathbf{b}$? $\mathbf{a} - \mathbf{b}$?

10. Interpret the following inequalities geometrically.

    (a) $|\mathbf{a} + \mathbf{b}| \le |\mathbf{a}| + |\mathbf{b}|$ (b) $|\mathbf{a} - \mathbf{b}| \ge |\mathbf{a}| - |\mathbf{b}|$

11. The initial point of a unit vector **n** (a "unit vector" is a vector that is 1 unit long) is the point $(1, 1)$ of the parabola $y = x^2$, and **n** is perpendicular to the tangent line to this parabola at the point $(1, 1)$. Find the coordinates of the possible terminal points of **n**.

12. Determine the conditions under which the statement is true.

    (a) $|r\mathbf{a}| = r|\mathbf{a}|$ (b) $|\mathbf{a} + \mathbf{b}| = |\mathbf{a}| + |\mathbf{b}|$

    (c) $|\mathbf{a} - \mathbf{b}| = |\mathbf{a}| - |\mathbf{b}|$ (d) $|r\mathbf{a} + s\mathbf{b}| = r|\mathbf{a}| + s|\mathbf{b}|$

13. Explain the difference in meaning of the expressions.

    (a) $|\mathbf{a}| > |\mathbf{b}|$ and $\mathbf{a} > \mathbf{b}$ (b) $|\mathbf{a}|^{-1}$ and $|\mathbf{a}^{-1}|$

    (c) $|\mathbf{a}|\,|\mathbf{b}|$ and $|\mathbf{ab}|$ (d) $|\mathbf{a}|/|\mathbf{b}|$ and $|\mathbf{a}/\mathbf{b}|$

69. BASIS VECTORS AND THE DOT PRODUCT

Now let us introduce a coordinate system into the plane. If we place a given vector **r** so that its initial point is the origin O of this coordinate system (Fig. 69-1), then its terminal point P will determine a pair of numbers (x, y). Conversely, a

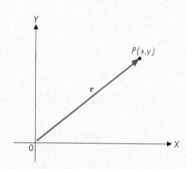

Figure 69-1

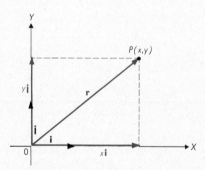

Figure 69-2

point P with coordinates (x, y) determines the vector $\mathbf{r} = \mathbf{OP}$. We will refer to a vector whose initial point is the origin as a **radius vector** or **position vector**. Thus we see that with reference to a particular coordinate system, a pair of numbers determines a position vector, and a position vector determines a pair of numbers.

If we let $\mathbf{i}$ and $\mathbf{j}$ be unit vectors in the directions of the positive X- and Y-axes, then (Fig. 69-2)

$$\text{(69-1)} \qquad\qquad \mathbf{r} = x\mathbf{i} + y\mathbf{j}.$$

The vectors $x\mathbf{i}$ and $y\mathbf{j}$ are the **projections** of $\mathbf{r}$ along $\mathbf{i}$ and $\mathbf{j}$. The unit vectors $\mathbf{i}$ and $\mathbf{j}$ form a **basis** for our system of vectors in the plane. Every vector can be written as a **linear combination** of $\mathbf{i}$ and $\mathbf{j}$, as in Equation 69-1. The coefficients of $\mathbf{i}$ and $\mathbf{j}$ are the **components** of the vector (relative to the given coordinate system).

If $\mathbf{a} = \mathbf{P}_1\mathbf{P}_2$, where P_1 is the point (x_1, y_1) and P_2 is the point (x_2, y_2), as shown in Fig. 69-3, we can determine the components of $\mathbf{a}$ very simply. We observe that $\mathbf{a} = \mathbf{OP}_2 - \mathbf{OP}_1 = (x_2\mathbf{i} + y_2\mathbf{j}) - (x_1\mathbf{i} + y_1\mathbf{j})$. Now we use the rules of arithmetic that we introduced in the last section, and we find that

$$\text{(69-2)} \quad \mathbf{a} = (x_2 - x_1)\mathbf{i} + (y_2 - y_1)\mathbf{j}.$$

We will denote the components of $\mathbf{a}$ by a_x and a_y, and so we see from Equation 69-2 that

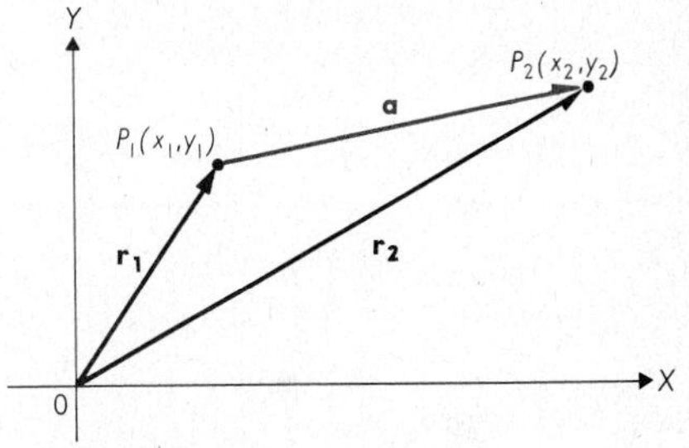

$$a_x = x_2 - x_1 \quad \text{and} \quad a_y = y_2 - y_1.$$

Figure 69-3

The distance between P_1 and P_2 is the magnitude of $\mathbf{a}$, so the distance formula allows us to express the magnitude of a vector in terms of its components:

$$\text{(69-3)} \qquad |\mathbf{a}| = \sqrt{(x_2 - x_1)^2 + (y_2 - y_1)^2} = \sqrt{a_x^2 + a_y^2}.$$

Since $\mathbf{i}$ and $\mathbf{j}$ are linearly independent, the vector equation $a_x\mathbf{i} + a_y\mathbf{j} = b_x\mathbf{i} + b_y\mathbf{j}$ is equivalent to the two scalar equations $a_x = b_x$ and $a_y = b_y$.

Example 69-1. What are the coordinates of the point Q that is $\frac{2}{3}$ of the way from the point $P(4, -2)$ to the point $R(7, 10)$?

Solution. Our problem is illustrated in Fig. 69-4; we must find the numbers x and y. We are told that $\mathbf{PQ} = \frac{2}{3}\mathbf{PR}$, and our first step is to express this vector equation in terms of components. According to Equation 69-2, $\mathbf{PQ} = (x - 4)\mathbf{i} + (y + 2)\mathbf{j}$ and $\mathbf{PR} = 3\mathbf{i} + 12\mathbf{j}$. Therefore our vector equation is

$$(x - 4)\mathbf{i} + (y + 2)\mathbf{j} = \tfrac{2}{3}(3\mathbf{i} + 12\mathbf{j}) = 2\mathbf{i} + 8\mathbf{j}.$$

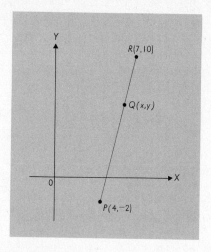

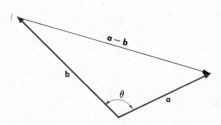

Figure 69-4   Figure 69-5

This vector equation is equivalent to the two scalar equations $x - 4 = 2$ and $y + 2 = 8$, from which we find that $x = 6$ and $y = 6$. Thus the desired point Q is $(6, 6)$.

We know how to associate with two given vectors a third *vector*, called their sum. Now we are going to associate a *number* with a pair of vectors. Two vectors **a** and **b** determine an angle θ, where $0 \le \theta \le \pi$, as shown in Fig. 69-5. We use the equation

(69-4) $$\mathbf{a} \cdot \mathbf{b} = |\mathbf{a}|\,|\mathbf{b}|\cos\theta$$

to define the **dot product** of **a** by **b**. Thus since the vectors **i** and **j** are each 1 unit long and meet at right angles, $\mathbf{i} \cdot \mathbf{j} = 1 \cdot 1 \cdot \cos\frac{1}{2}\pi = 1 \cdot 1 \cdot 0 = 0$. Similarly, $\mathbf{i} \cdot \mathbf{i} = 1 \cdot 1 \cdot \cos 0 = 1$. The dot product is a scalar, so it is sometimes called the **scalar product** of the two vectors.

From the defining Equation 69-4, we see that the dot product of a vector **a** with itself is given by the formula

$$\mathbf{a} \cdot \mathbf{a} = |\mathbf{a}|^2 \cos 0 = |\mathbf{a}|^2;$$

that is, the dot product of a vector with itself is the square of its magnitude. Some authors use the abbreviation $\mathbf{a}^2 = \mathbf{a} \cdot \mathbf{a}$. It also follows directly from the definition of the dot product that $\mathbf{a} \cdot \mathbf{b} = 0$ if and only if at least one of the following equations holds: $\mathbf{a} = \mathbf{0}, \mathbf{b} = \mathbf{0}$, or $\theta = \frac{1}{2}\pi$. Two vectors **a** and **b** are perpendicular, $\mathbf{a} \perp \mathbf{b}$, if $\theta = \frac{1}{2}\pi$; and if we agree that the vector **0** is perpendicular to every vector, we see that

$$\mathbf{a} \perp \mathbf{b} \quad \textit{if and only if} \quad \mathbf{a} \cdot \mathbf{b} = 0.$$

424

Our definition of the dot product is geometric; we have worded it in terms of the lengths of two line segments and the angle they determine. We also want an algebraic expression for the dot product of two vectors $\mathbf{a} = a_x\mathbf{i} + a_y\mathbf{j}$ and $\mathbf{b} = b_x\mathbf{i} + b_y\mathbf{j}$ in terms of their components. Therefore we apply the Law of Cosines to the triangle in Fig. 69-5 to obtain the equation

$$(69\text{-}5) \qquad |\mathbf{a} - \mathbf{b}|^2 = |\mathbf{a}|^2 + |\mathbf{b}|^2 - 2|\mathbf{a}|\,|\mathbf{b}|\cos\theta.$$

The numbers $|\mathbf{a} - \mathbf{b}|^2$, $|\mathbf{a}|^2$, and $|\mathbf{b}|^2$ are the squares of the side lengths of our triangle. If we use Equation 69-3 to express these lengths in terms of the components of the vectors involved, then $|\mathbf{a} - \mathbf{b}|^2 = (a_x - b_x)^2 + (a_y - b_y)^2$, $|\mathbf{a}|^2 = a_x^2 + a_y^2$, and $|\mathbf{b}|^2 = b_x^2 + b_y^2$. Furthermore, $|\mathbf{a}|\,|\mathbf{b}|\cos\theta = \mathbf{a}\cdot\mathbf{b}$, and when we make these substitutions in Equation 69-5 and simplify, we obtain our algebraic expression for the dot product:

$$(69\text{-}6) \qquad \mathbf{a}\cdot\mathbf{b} = a_xb_x + a_yb_y.$$

Example 69-2. What is the angle between the position vectors to the points $(5, 12)$ and $(3, 4)$?

Solution. Our two position vectors can be written as $\mathbf{r}_1 = 5\mathbf{i} + 12\mathbf{j}$ and $\mathbf{r}_2 = 3\mathbf{i} + 4\mathbf{j}$; let us suppose that they determine an angle θ. From Equation 69-4 we see that

$$\cos\theta = \frac{\mathbf{r}_1\cdot\mathbf{r}_2}{|\mathbf{r}_1|\,|\mathbf{r}_2|}.$$

Now we use Equation 69-6 to calculate $\mathbf{r}_1\cdot\mathbf{r}_2 = 15 + 48 = 63$, and we readily find that $|\mathbf{r}_1| = 13$ and $|\mathbf{r}_2| = 5$. Hence

$$\cos\theta = \tfrac{63}{65} = .97,$$

and so

$$\theta = \text{Arccos}\ .97 \approx .24\ \text{radian}.$$

Example 69-3. Verify the following case of the **Cauchy-Schwarz Inequality:** If $a_1, a_2, b_1,$ and b_2 are any four real numbers, then

$$(69\text{-}7) \qquad (a_1b_1 + a_2b_2)^2 \le (a_1^2 + a_2^2)(b_1^2 + b_2^2).$$

In words, we can read this inequality as "The square of the sum of the products is not greater than the product of the sum of the squares." Corresponding inequalities hold with sums of more than two terms.

Solution. If we set $\mathbf{a} = a_1\mathbf{i} + a_2\mathbf{j}$ and $\mathbf{b} = b_1\mathbf{i} + b_2\mathbf{j}$, then Inequality 69-7 can be written as

$$(\mathbf{a}\cdot\mathbf{b})^2 \le |\mathbf{a}|^2|\mathbf{b}|^2.$$

Since $\mathbf{a} \cdot \mathbf{b} = |\mathbf{a}|\,|\mathbf{b}|\cos\theta$, this last inequality is $|\mathbf{a}|^2|\mathbf{b}|^2\cos^2\theta \le |\mathbf{a}|^2|\mathbf{b}|^2$, which is obviously true because $\cos^2\theta \le 1$. Notice that the symbol $\le$ can be replaced by $=$ in case $\cos^2\theta = 1$; that is, when the vectors $\mathbf{a}$ and $\mathbf{b}$ are parallel. How can this condition be expressed in terms of the original numbers a_1, a_2, b_1, and b_2?

Finally, let us remark that we can use Equation 69-6 and elementary algebra to show that the dot product obeys the following fundamental laws of arithmetic:

$$\mathbf{a} \cdot \mathbf{b} = \mathbf{b} \cdot \mathbf{a} \qquad \text{(The Commutative Law)}$$

$$(69\text{-}8) \quad (\mathbf{a} + \mathbf{b}) \cdot \mathbf{c} = \mathbf{a} \cdot \mathbf{c} + \mathbf{b} \cdot \mathbf{c} \quad \text{(The Distributive Law)}$$

$$(r\mathbf{a}) \cdot \mathbf{b} = r(\mathbf{a} \cdot \mathbf{b}).$$

In these equations $\mathbf{a}$, $\mathbf{b}$, and $\mathbf{c}$ are any vectors and r is an arbitrary scalar.

Example 69-4. Express the components of the vector $\mathbf{a} = a_x\mathbf{i} + a_y\mathbf{j}$ in terms of the dot product.

Solution. We leave it to you to find the rule that justifies each step in the following calculation:

$$\mathbf{a} \cdot \mathbf{i} = (a_x\mathbf{i} + a_y\mathbf{j}) \cdot \mathbf{i} = (a_x\mathbf{i}) \cdot \mathbf{i} + (a_y\mathbf{j}) \cdot \mathbf{i}$$
$$= a_x(\mathbf{i} \cdot \mathbf{i}) + a_y(\mathbf{j} \cdot \mathbf{i}) = a_x.$$

A similar calculation shows that $a_y = \mathbf{a} \cdot \mathbf{j}$.

PROBLEMS   69

1. Write $\mathbf{a}$ in the form $\mathbf{a} = a_x\mathbf{i} + a_y\mathbf{j}$ and find $|\mathbf{a}|$ if $\mathbf{a}$ is the vector whose initial point is the first and whose terminal point is the second of the following points.

   (a) $(-1, 3)$, $(2, 7)$ \qquad\qquad (b) $(3, 5)$, $(-3, -3)$
   (c) $(0, 3)$, $(0, -2)$ \qquad\qquad (d) $(2, -1)$, $(-1, -2)$

2. Let $\mathbf{a} = 2\mathbf{i} - 3\mathbf{j}$ and $\mathbf{b} = 4\mathbf{i} + \mathbf{j}$. Find the components of the following vectors.

   (a) $2\mathbf{a}$ \qquad\qquad (b) $-3\mathbf{b}$ \qquad\qquad (c) $\mathbf{a} + \mathbf{b}$
   (d) $\mathbf{a} - \mathbf{b}$ \qquad\qquad (e) $2\mathbf{a} + 3\mathbf{b}$ \qquad\qquad (f) $\mathbf{a} - 2\mathbf{b}$

3. Use the vectors $\mathbf{a} = 3\mathbf{i} - 2\mathbf{j}$, $\mathbf{b} = 2\mathbf{i} + \mathbf{j}$, and $\mathbf{c} = \mathbf{i} - 2\mathbf{j}$, and the scalars $p = 3$ and $q = -2$ to illustrate the following laws of vector algebra.

   (a) $\mathbf{a} + \mathbf{b} = \mathbf{b} + \mathbf{a}$ \qquad\qquad (b) $(\mathbf{a} + \mathbf{b}) + \mathbf{c} = \mathbf{a} + (\mathbf{b} + \mathbf{c})$
   (c) $(\mathbf{a} + \mathbf{b}) \cdot \mathbf{c} = \mathbf{a} \cdot \mathbf{c} + \mathbf{b} \cdot \mathbf{c}$ \qquad (d) $p(q\mathbf{a}) = (pq)\mathbf{a}$
   (e) $(p + q)\mathbf{a} = p\mathbf{a} + q\mathbf{a}$ \qquad\qquad (f) $p(\mathbf{a} + \mathbf{b}) = p\mathbf{a} + p\mathbf{b}$
   (g) $\mathbf{a} \cdot \mathbf{b} = \mathbf{b} \cdot \mathbf{a}$ \qquad\qquad (h) $(p\mathbf{a}) \cdot \mathbf{b} = p(\mathbf{a} \cdot \mathbf{b})$

4. Let P be the point $(-5, -9)$ and Q be the point $(7, 7)$.

    (a) Find the point that is $\frac{3}{4}$ of the way from P to Q.
    (b) Find the point that is 5 units from P along the line from P to Q.
    (c) Find the point that is as far beyond Q as Q is from P.
    (d) Find a vector that has the same direction as **PQ** but that is 1 unit long.
    (e) Find the point R such that P is the midpoint of the segment QR.
    (f) If O (the origin), P, and Q are three vertices of a parallelogram, what is the fourth vertex?

5. Find the angles of the triangle whose vertices are the points $(-2, -1)$, $(2, 2)$, and $(-1, 6)$.

6. Find the equation that must be satisfied by the coordinates (x, y) of a point P such that **OP** is perpendicular to $2\mathbf{i} + 3\mathbf{j}$. Find a unit vector perpendicular to $2\mathbf{i} + 3\mathbf{j}$. Illustrate this problem with a sketch.

7. Let $\mathbf{r} = x\mathbf{i} + y\mathbf{j}$ and $\mathbf{a} = 3\mathbf{i} + 4\mathbf{j}$. Give a geometric description of the set of all pairs (x, y) such that **r** is a position vector that satisfies the given relation.

    (a) $\mathbf{r} \cdot \mathbf{a} = 12$ (b) $|\mathbf{r}| + |\mathbf{a}| = 12$
    (c) $|\mathbf{r} + \mathbf{a}| = 12$ (d) $\mathbf{r} \cdot \mathbf{a} = \frac{1}{2}|\mathbf{r}|\,|\mathbf{a}|$

8. Let P and R be the endpoints of a diameter of a circle and Q be a third point of the circle. Show that PQR is a right triangle in which Q is the vertex of the right angle.

9. Let $\mathbf{a} = 3\mathbf{i} + \mathbf{j}$. Let **b** be the unit vector that is perpendicular to **a** and that has a positive component along **i**. Let $\mathbf{c} = -\mathbf{i} + 2\mathbf{j}$. Find r and s such that $\mathbf{c} = r\mathbf{a} + s\mathbf{b}$.

10. (a) Show that $|\mathbf{a} + \mathbf{b}|^2 + |\mathbf{a} - \mathbf{b}|^2 = 2|\mathbf{a}|^2 + 2|\mathbf{b}|^2$.
    (b) Prove that the sum of the squares of the lengths of the diagonals of a parallelogram is equal to the sum of the squares of the lengths of its four sides.

11. Let **a** and **b** be unit position vectors that make angles α and β with **i**. Write **a** and **b** in component form. Compute $\mathbf{a} \cdot \mathbf{b}$ to obtain a formula for $\cos(\alpha - \beta)$.

12. Let $\mathbf{r}_1$ and $\mathbf{r}_2$ be unit position vectors that determine an angle θ, where $0 \le \theta \le \pi$. Show that $\sin \frac{1}{2}\theta = \frac{1}{2}|\mathbf{r}_2 - \mathbf{r}_1|$.

13. Show that a parallelogram is a rhombus (has sides of equal length) if and only if its diagonals intersect at right angles.

14. Suppose an $\overline{X}\overline{Y}$-coordinate system is obtained from an XY-system by rotation through an angle α. Show that the unit vectors $\overline{\mathbf{i}}$ and $\overline{\mathbf{j}}$ along the $\overline{X}$- and $\overline{Y}$-axes can be written as $\overline{\mathbf{i}} = \cos \alpha \mathbf{i} + \sin \alpha \mathbf{j}$ and $\overline{\mathbf{j}} = -\sin \alpha \mathbf{i} + \cos \alpha \mathbf{j}$. Use these equations to express the position vector $\bar{x}\overline{\mathbf{i}} + \bar{y}\overline{\mathbf{j}}$ of a point relative to the $\overline{X}\overline{Y}$-system in terms of **i** and **j**. Now equate this expression to the position vector $x\mathbf{i} + y\mathbf{j}$ of the same point relative to the XY-system, thereby obtaining Equations 66-4.

70. VECTOR-VALUED FUNCTIONS AND PARAMETRIC EQUATIONS

    In this section we will consider functions whose domains are sets of *numbers* and whose ranges are sets of *vectors*—that is, vector-valued functions. We will denote a vector-valued function by a boldfaced letter. An example is the function

f whose domain is the interval $[0, 2\pi]$ and whose rule of correspondence is the equation

$$\mathbf{f}(t) = \cos t\mathbf{i} + \sin t\mathbf{j}.$$

Here, corresponding to the *number* $t = 0$, we have the *vector* $\mathbf{f}(0) = \cos 0\mathbf{i} + \sin 0\mathbf{j} = \mathbf{i}$; corresponding to the *number* $t = \frac{1}{4}\pi$, we have the *vector*

$$\mathbf{f}(\tfrac{1}{4}\pi) = \tfrac{1}{2}\sqrt{2}\mathbf{i} + \tfrac{1}{2}\sqrt{2}\mathbf{j},$$

and so on.

We draw the graph of a vector-valued function **f** with domain D as follows. For each number $t \in D$, we plot the terminal point of the position vector $\mathbf{r} = \mathbf{f}(t)$ (Fig. 70-1). The graph of **f** is the set of all such terminal points:

$$\{(x, y) \mid x\mathbf{i} + y\mathbf{j} = \mathbf{f}(t),\ t \in D\}.$$

Example 70-1. Sketch the graph of the vector equation

$$\mathbf{r} = \cos t\mathbf{i} + \sin t\mathbf{j}, \qquad t \in [0, 2\pi].$$

Solution. We have tabulated some of our vectors and their terminal points and plotted these points in Fig. 70-2. If we plot enough points, they will form a good

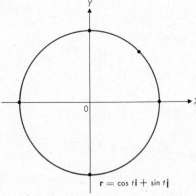

$$r = \cos t\mathbf{i} + \sin t\mathbf{j}$$

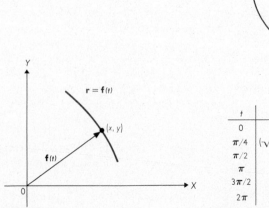

r = **f**(t)

(x, y)

f(t)

t	$\mathbf{r}$	(x,y)
0	$\mathbf{i}$	$(1,0)$
$\pi/4$	$(\sqrt{2}/2)\,\mathbf{i} + (\sqrt{2}/2)\,\mathbf{j}$	$(\sqrt{2}/2,\ \sqrt{2}/2)$
$\pi/2$	$\mathbf{j}$	$(0,1)$
π	$-\mathbf{i}$	$(-1,0)$
$3\pi/2$	$-\mathbf{j}$	$(0,-1)$
2π	$\mathbf{i}$	$(1,0)$

Figure 70-1

Figure 70-2

outline of the graph of our vector equation. But even without plotting points, we can see what the graph is. For each number t,

$$|\mathbf{r}| = \sqrt{\cos^2 t + \sin^2 t} = 1.$$

Thus each of our position vectors is 1 unit long, and hence its terminal point will be 1 unit from the origin. It is easy to see that since t can be chosen as any point of the interval $[0, 2\pi]$, the graph of our vector equation is the circle whose radius is 1 and whose center is the origin. If we had written $t \in [0, \pi]$ in place of $t \in [0, 2\pi]$, the graph would have formed only the upper semicircle in Fig. 70-2. On the other hand, if we had written $t \in [0, 4\pi]$ instead of $t \in [0, 2\pi]$, the graph would again be the entire circle. In this case, we might say that the curve is traced out twice.

If we write the vector equation

(70-1) $$\mathbf{r} = \mathbf{f}(t)$$

in terms of the components of the vectors $\mathbf{r} = x\mathbf{i} + y\mathbf{j}$ and $\mathbf{f}(t) = f_x(t)\mathbf{i} + f_y(t)\mathbf{j}$, we have

$$x\mathbf{i} + y\mathbf{j} = f_x(t)\mathbf{i} + f_y(t)\mathbf{j}.$$

Thus the vector Equation 70-1 is equivalent to the *two* scalar equations

(70-2) $$x = f_x(t) \quad \text{and} \quad y = f_y(t).$$

These equations assign a point (x, y) to each number t in the domain of $\mathbf{f}$. The set of these points is the graph of $\mathbf{f}$. Equations 70-2 are referred to as **parametric equations** of this graph, and t is called a **parameter**. Thus the equations

$$x = \cos t \quad \text{and} \quad y = \sin t, \qquad t \in [0, 2\pi],$$

are parametric equations of the circle whose "nonparametric" equation is $x^2 + y^2 = 1$.

We frequently change parametric equations to nonparametric form in order to use our knowledge of the graphs of equations in x and y.

Example 70-2. Suppose that a and b are two positive numbers. What is the graph of the equation $\mathbf{r} = a \cosh t\mathbf{i} + b \sinh t\mathbf{j}, t \in (-\infty, \infty)$?

Solution. Parametric equations of our graph are

$$x = a \cosh t \quad \text{and} \quad y = b \sinh t.$$

We must eliminate t from these equations so as to find a relation between x and y. A direct way to eliminate t would be to solve the second equation for t and substitute in the first. Thus $t = \sinh^{-1}(y/b)$, and so we have $x = a \cosh \sinh^{-1}(y/b)$. This form of the equation tells us no more about the graph than the original vector equation did. However, if we write the parametric equations as $x/a = \cosh t$ and $y/b = \sinh t$, and substitute in the identity $\cosh^2 t - \sinh^2 t = 1$, we see that

$$\frac{x^2}{a^2} - \frac{y^2}{b^2} = 1.$$

Thus the graph of our equation is a part of a hyperbola. Since $x/a = \cosh t > 0$, you can convince yourself that our graph consists of only one branch of the hyperbola.

Sometimes we wish to transform a nonparametric equation in x and y to a pair of parametric equations that have the same graph. The nonparametric equation

(70-3) $y = f(x)$,

for example, can always be replaced by the parametric equations

(70-4) $x = t$ and $y = f(t)$.

A given curve can be the graph of many different pairs of parametric equations. The choice of which pair to use is dictated by convenience more than any other factor.

Example 70-3.   Find a system of parametric equations whose graph is the curve $x^{2/3} + y^{2/3} = a^{2/3}$.

Solution.   We might set $x = t$ and then solve the equation $t^{2/3} + y^{2/3} = a^{2/3}$ for y to obtain the pair of parametric equations

$$x = t \quad \text{and} \quad y = (a^{2/3} - t^{2/3})^{3/2}.$$

These equations represent (for $t \in [-a, a]$) the portion of the graph of our given equation that lies above the X-axis. Because the given curve is symmetric about the X-axis, we can obtain the remainder of the graph by reflecting the upper half about the X-axis. But we can find a more "symmetric" pair of parametric equations. Suppose that we divide both sides of the given equation by $a^{2/3}$ to obtain the equation

$$\left(\frac{x^{1/3}}{a^{1/3}}\right)^2 + \left(\frac{y^{1/3}}{a^{1/3}}\right)^2 = 1.$$

Here we have a sum of two squares that equals 1, and this equation suggests that we set

$$\frac{x^{1/3}}{a^{1/3}} = \sin t \quad \text{and} \quad \frac{y^{1/3}}{a^{1/3}} = \cos t.$$

Hence we see that another pair of parametric equations of our curve is the pair

$$x = a \sin^3 t \quad \text{and} \quad y = a \cos^3 t.$$

This pair gives us (for $t \in [0, 2\pi]$) the entire curve.

Let us consider the problem of finding a vector equation of a line. A scalar equation of a line takes the form $y = mx + b$, where the number m determines the direction (slope) of the line and the number b determines a point (the Y-intercept) of the line. Now we will show that the similar-looking vector equation

(70-5) $\mathbf{r} = \mathbf{m}t + \mathbf{b}, \quad t \in (-\infty, \infty) \quad \text{and} \quad \mathbf{m} \neq \mathbf{0},$

is also the equation of a line. Not only does this equation resemble the scalar equation $y = mx + b$ in form, but the vector $\mathbf{m}$ determines the direction of the line and the position vector $\mathbf{b}$ determines a point of the line as well. Equation 70-5 states that the position vector $\mathbf{r}$ is obtained by adding the multiple $\mathbf{m}t$ of the vector $\mathbf{m}$ to the vector $\mathbf{b}$. Figure 70-3 shows how this statement implies that our curve is the line that contains the terminal point of $\mathbf{b}$ and is in the direction of $\mathbf{m}$. If $\mathbf{m} = m_x\mathbf{i} + m_y\mathbf{j}$ and $\mathbf{b} = b_x\mathbf{i} + b_y\mathbf{j}$, then parametric equations of the line are

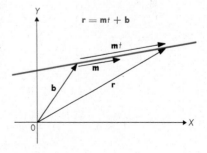

(70-6) $x = m_x t + b_x \quad \text{and} \quad y = m_y t + b_y,$

$$t \in (-\infty, \infty).$$

Figure 70-3

Now let us write vector and parametric equations of the line that contains two given points $P_1(x_1, y_1)$ and $P_2(x_2, y_2)$. In order to use Equation 70-5 we must find a vector $\mathbf{m}$ that gives the direction of the line and a vector $\mathbf{b}$, the position vector of a point of the line. Since P_1 is a point of the line, we may take the position vector $\mathbf{OP}_1$ as $\mathbf{b}$ (we could equally well have chosen $\mathbf{OP}_2$). The vector $\mathbf{P}_1\mathbf{P}_2$ determines the direction of our line, so we take $\mathbf{m} = \mathbf{P}_1\mathbf{P}_2$, and Equation 70-5 becomes

(70-7) $\mathbf{r} = \mathbf{P}_1\mathbf{P}_2 t + \mathbf{OP}_1, \quad t \in (-\infty, \infty).$

To obtain parametric equations of our line, we simply equate the components of the vectors in Equation 70-7:

$$x = (x_2 - x_1)t + x_1 \quad \text{and} \quad y = (y_2 - y_1)t + y_1.$$

We can write these parametric equations as

$$(70\text{-}8) \quad x = tx_2 + (1 - t)x_1 \quad \text{and} \quad y = ty_2 + (1 - t)y_1,$$

$$t \in (-\infty, \infty).$$

You can easily verify (refer to Equation 70-7) that if we restrict t to the interval $(0, 1)$, then Equations 70-8 are parametric equations of the line *segment* between the points (x_1, y_1) and (x_2, y_2).

Example 70-4. Find parametric equations of the line that contains the points $(-1, 2)$ and $(2, 3)$.

Solution. If we take as P_1 the point $(-1, 2)$ and as P_2 the point $(2, 3)$ and use Equations 70-8, we obtain the parametric equations

$$x = 2t + (1 - t)(-1) = 3t - 1 \quad \text{and} \quad y = 3t + (1 - t)2 = t + 2,$$

$$t \in (-\infty, \infty).$$

Frequently a curve that is defined in terms of a physical motion is most naturally described by means of parametric equations. We conclude this section with a famous example of this type.

Example 70-5. A circular hoop with a radius of a rolls along a line. Find parametric equations of the curve that is traced out by a given point of the hoop.

Solution. Let us suppose that our given point P originally is the origin of a cartesian coordinate system and that the hoop rolls along the X-axis. In Fig. 70-4 we have pictured the hoop after it has rolled through an angle of t radians. We must find an expression for the position vector **OP** in terms of t. From Fig. 70-4 we see that

$$\mathbf{OP} = \mathbf{OT} + \mathbf{TS} + \mathbf{SP}.$$

Because $|\mathbf{OT}| = \text{arc } TP = at$, it follows that $\mathbf{OT} = at\mathbf{i}$. Furthermore, simple trigonometry shows us that

$$\mathbf{TS} = (a - a \cos t)\mathbf{j} = a(1 - \cos t)\mathbf{j}$$

and

$$\mathbf{SP} = -\mathbf{PS} = -a \sin t\mathbf{i}.$$

Hence

$$\mathbf{OP} = at\mathbf{i} + a(1 - \cos t)\mathbf{j} - a \sin t\mathbf{i}$$

$$= a(t - \sin t)\mathbf{i} + a(1 - \cos t)\mathbf{j}.$$

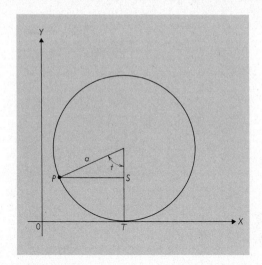

Figure 70-4

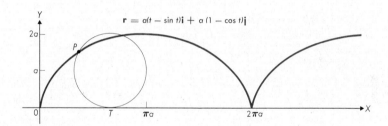

$$\mathbf{r} = a(t - \sin t)\mathbf{i} + a(1 - \cos t)\mathbf{j}$$

Figure 70-5

From this vector equation we obtain parametric equations for the coordinates (x, y) of the point P of our curve:

$$x = a(t - \sin t) \quad \text{and} \quad y = a(1 - \cos t), \qquad t \in [0, \infty].$$

This curve is called a **cycloid**; we have sketched part of it in Fig. 70-5.

PROBLEMS 70

1. Describe the graphs of the following vector equations
    (a) $\mathbf{r} = 3 \cos 2t\mathbf{i} + 3 \sin 2t\mathbf{j}$, $t \in [0, \tfrac{1}{2}\pi]$
    (b) $\mathbf{r} = t\mathbf{i} + t^2\mathbf{j}$, $t \in [-1, 1]$
    (c) $\mathbf{r} = \sin t\mathbf{i} + \sin t\mathbf{j}$, $t \in [0, \tfrac{1}{2}\pi]$

(d) $\mathbf{r} = t\mathbf{i} + 2t\mathbf{j},\ t \in [0, 1]$

(e) $\mathbf{r} = e^t\mathbf{i} + e^{-t}\mathbf{j},\ t \in (-\infty, \infty)$

(f) $\mathbf{r} = 5 \sin t\mathbf{i} + \cos t\mathbf{j},\ t \in [-\pi, \pi]$

2. Show that the following scalar and vector equations have the same graph.

(a) $2x - 3y + 4 = 0$ and $\mathbf{r} = (3t + 4)\mathbf{i} + (2t + 4)\mathbf{j},\ t \in (-\infty, \infty)$

(b) $y = e^x$ and $\mathbf{r} = \ln t\mathbf{i} + t\mathbf{j},\ t \in (0, \infty)$

(c) $y^2 = x^3$ and $\mathbf{r} = t^2\mathbf{i} + t^3\mathbf{j},\ t \in (-\infty, \infty)$

(d) $y = 2 \cosh x,\ \mathbf{r} = \ln t\mathbf{i} + \left(t + \dfrac{1}{t}\right)\mathbf{j},\ t \in (-\infty, \infty)$

3. Find a vector equation of

(a) the line containing the points $(2, 3)$ and $(-1, 4)$.

(b) the line containing the points $(0, 0)$ and $(-1, 2)$.

(c) the line $3x + 2y + 6 = 0$.

(d) the line $2x + 3y + 6 = 0$.

(e) the line of slope 3 that contains the point $(-1, 1)$.

(f) the line of slope 0 that contains the point $(1, 3)$.

4. Find the slope and Y-intercept of the given line.

(a) $\mathbf{r} = (3t - 1)\mathbf{i} + (1 - 5t)\mathbf{j}$ (b) $\mathbf{r} = (2t + 1)\mathbf{i} - (3t + 7)\mathbf{j}$

(c) $\mathbf{r} = (2t + 1)\mathbf{i} + (4t + 2)\mathbf{j}$ (d) $\mathbf{r} = (1 - 2t)\mathbf{i} + (3 - 4t)\mathbf{j}$

5. Find parametric equations for each of the following lines.

(a) $y = 3x + 7$ (b) $2x - 4y + 5 = 0$

(c) $x = 5$ (d) $y = -1$

6. Find two pairs of parametric equations corresponding to each of the following cartesian equations.

(a) $y^2 = x^3$ (b) $\dfrac{x^2}{4} + y^2 = 1$

(c) $y^2 + 1 = x^2$ (d) $y = 2x^2 - 1$

7. Explain why both vector equations have the same graph. [In each case, $t \in (-\infty, \infty)$.]

(a) $\mathbf{r} = \pi^t\mathbf{i} + \pi^{-t}\mathbf{j}$ and $\mathbf{r} = e^t\mathbf{i} + e^{-t}\mathbf{j}$

(b) $\mathbf{r} = \cos 2t\mathbf{i} + \sin 2t\mathbf{j}$ and $\mathbf{r} = \sin t\mathbf{i} + \cos t\mathbf{j}$

(c) $\mathbf{r} = (t^3 + 2t + 3)\mathbf{i} + (t^6 + 4t^4 + 6t^3 + 4t^2 + 12t + 9)\mathbf{j}$ and $\mathbf{r} = t\mathbf{i} + t^2\mathbf{j}$

(d) $\mathbf{r} = \cos^2 t\mathbf{i} + \sin^2 t\mathbf{j}$ and $\mathbf{r} = |\cos t|\mathbf{i} + (1 - |\cos t|)\mathbf{j}$

8. Sketch the graph of the equation in Example 70-3.

9. Show that the spiral $r\theta = 1$ is the graph of the vector equation

$$\mathbf{r} = \frac{\cos t}{t}\mathbf{i} + \frac{\sin t}{t}\mathbf{j}.$$

10. Sketch the graphs of the given system of parametric equations in polar coordinates.

(a) $r = 3,\ \theta = t,\ t \in [0, \pi]$

(b) $r = t,\ \theta = 3,\ t \in [0, \pi]$

(c) $r = 4 \cos t,\ \theta = \frac{1}{2}t,\ t \in [0, 4\pi]$

(d) $r = 10 \cos t,\ \theta = t + \frac{1}{4}\pi,\ t \in [0, 2\pi]$

11. (a) Convince yourself that the graph of the parametric equations

$$x = 3 \sin^2 t + 2 \cos^2 t, \quad y = 2 \sin^2 t + \cos^2 t, \qquad t \in [0, \tfrac{1}{2}\pi],$$

is a line segment and find its endpoints.
   (b) What if we replace $[0, \tfrac{1}{2}\pi]$ with $[0, \pi]$? With $[0, \tfrac{1}{4}\pi]$?

12. Under what conditions are $\mathbf{r} = \mathbf{m}t + \mathbf{b}$ and $\mathbf{r} = \mathbf{n}t + \mathbf{c}$ equations of the same line?

13. Find vectors $\mathbf{m}$ and $\mathbf{b}$ (expressed in terms of a, b, and c) so that the vector curve $\mathbf{r} = \mathbf{m}t + \mathbf{b}$ is the line $ax + by + c = 0$.

71. DERIVATIVES OF VECTORS. ARCLENGTH

Using the definition of the scalar derivative as a guide, we define the vector derivative $\mathbf{f}'(t)$ by the equation

(71-1)
$$\mathbf{f}'(t) = \lim_{u \to t} \frac{\mathbf{f}(u) - \mathbf{f}(t)}{u - t}.$$

Our difference quotient $\dfrac{\mathbf{f}(u) - \mathbf{f}(t)}{u - t}$ is the vector that results from multiplying the vector difference $\mathbf{f}(u) - \mathbf{f}(t)$ by the scalar $(u - t)^{-1}$, and Equation 71-1 says that $\mathbf{f}'(t)$ is the vector that is approximated by this quotient when u is close to t. One vector approximates another if the absolute value of their difference is small, and a precise definition of the vector Equation 71-1 is that it is equivalent to the scalar equation

(71-2)
$$\lim_{u \to t} \left| \frac{\mathbf{f}(u) - \mathbf{f}(t)}{u - t} - \mathbf{f}'(t) \right| = 0.$$

Notice that $\mathbf{f}'(t)$ is a *vector*, so it has magnitude and direction. Let us see what we can say about these quantities. We will first use a geometric argument to convince ourselves that the vector $\mathbf{f}'(t)$ is tangent to the graph of $\mathbf{f}$. We have sketched this graph in Fig. 71-1, which also shows the vectors $\mathbf{f}(t)$, $\mathbf{f}(u)$, and $\mathbf{f}(u) - \mathbf{f}(t)$. The vector $\mathbf{f}(u) - \mathbf{f}(t)$ forms a chord of the curve, and if u is close to t, the chord approximates the tangent to the graph. The difference quotient $\dfrac{\mathbf{f}(u) - \mathbf{f}(t)}{u - t}$ is a scalar multiple of $\mathbf{f}(u) - \mathbf{f}(t)$, so it lies along this chord. Therefore it is also a vector approximately tangent to the graph. Since the limit of the difference quotient as u approaches t is the vector $\mathbf{f}'(t)$, it thus appears that *the vector* $\mathbf{f}'(t)$ *is tangent to the curve* $\mathbf{r} = \mathbf{f}(t)$ *at the terminal point of* $\mathbf{f}(t)$.

The following theorem tells us that we can follow the usual rules of differentiation when calculating derivatives of vector-valued functions.

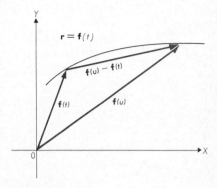

Figure 71-1

Theorem 71-1.   *If f_x and f_y are differentiable functions and $\mathbf{f}(t) = f_x(t)\mathbf{i} + f_y(t)\mathbf{j}$, then*

$$(71\text{-}3) \qquad\qquad \mathbf{f}'(t) = f_x'(t)\mathbf{i} + f_y'(t)\mathbf{j}.$$

PROOF

If we write the difference quotient that appears in Equation 71-2 in terms of its components and replace $\mathbf{f}'(t)$ with $f_x'(t)\mathbf{i} + f_y'(t)\mathbf{j}$, we must show that

$$\left| \left(\frac{f_x(u) - f_x(t)}{u - t} - f_x'(t) \right)\mathbf{i} + \left(\frac{f_y(u) - f_y(t)}{u - t} - f_y'(t) \right)\mathbf{j} \right|$$

approaches 0 as u approaches t. This number is not greater than the sum

$$\left| \frac{f_x(u) - f_x(t)}{u - t} - f_x'(t) \right| + \left| \frac{f_y(u) - f_y(t)}{u - t} - f_y'(t) \right|.$$

Since each term of this latter sum approaches 0, our theorem is proved.

We use our other notations for the derivative, too. Thus if $\mathbf{r} = \mathbf{f}(t)$, we write $\mathbf{r}'$ and $\dfrac{d\mathbf{r}}{dt}$ for the derivative $\mathbf{f}'(t)$. Also since, $\mathbf{r} = x\mathbf{i} + y\mathbf{j}$, where $x = f_x(t)$ and $y = f_y(t)$, we can write Equation 71-3 as

$$\mathbf{r}' = x'\mathbf{i} + y'\mathbf{j} \quad \text{or} \quad \frac{d\mathbf{r}}{dt} = \frac{dx}{dt}\mathbf{i} + \frac{dy}{dt}\mathbf{j},$$

and so on.

Example 71-1.   David whirls a stone in a counterclockwise direction around the circle $\mathbf{r} = \cos t\,\mathbf{i} + \sin t\,\mathbf{j}$. When $t = \tfrac{1}{6}\pi$, he releases the stone and it flies off on a tangent. Does it hit Goliath, who is standing at the point $(-1, \sqrt{3})$?

Solution.   We first find a vector that is tangent to the circle at the point that corresponds to $t = \tfrac{1}{6}\pi$. According to Equation 71-3, $\mathbf{r}' = -\sin t\,\mathbf{i} + \cos t\,\mathbf{j}$. Thus if $t = \tfrac{1}{6}\pi$, then

$$\mathbf{r} = \tfrac{1}{2}\sqrt{3}\,\mathbf{i} + \tfrac{1}{2}\mathbf{j} \quad \text{and} \quad \mathbf{r}' = -\tfrac{1}{2}\mathbf{i} + \tfrac{1}{2}\sqrt{3}\,\mathbf{j}.$$

Therefore the stone is at the point $(\tfrac{1}{2}\sqrt{3}, \tfrac{1}{2})$ of the circle when it is released, and it flies along a line parallel to the vector $-\tfrac{1}{2}\mathbf{i} + \tfrac{1}{2}\sqrt{3}\,\mathbf{j}$. We leave it to you to show that the vector joining the point $(\tfrac{1}{2}\sqrt{3}, \tfrac{1}{2})$ of the circle to the point in question,

$(-1, \sqrt{3})$, is not parallel to the tangent vector $-\frac{1}{2}\mathbf{i} + \frac{1}{2}\sqrt{3}\mathbf{j}$, so this time Goliath is spared. You should make a sketch illustrating this problem, and you might try to solve it graphically.

It is not hard to show, using Equation 71-3, that the usual Product Rule is valid when we compute the derivative of a scalar times a vector; that is,

$$(71\text{-}4) \qquad \frac{d}{dt}\left[a(t)\mathbf{f}(t)\right] = a(t)\mathbf{f}'(t) + a'(t)\mathbf{f}(t).$$

The Product Rule also applies to the dot product:

$$(71\text{-}5) \qquad \frac{d}{dt}\left[\mathbf{f}(t)\cdot\mathbf{g}(t)\right] = \mathbf{f}(t)\cdot\mathbf{g}'(t) + \mathbf{f}'(t)\cdot\mathbf{g}(t).$$

As a final rule of differentiation, let us write the Chain Rule as it applies to vector-valued functions. If $\mathbf{f}$ is differentiable and $t = g(s)$, then

$$(71\text{-}6) \qquad \frac{d\mathbf{f}(t)}{ds} = \frac{d\mathbf{f}(t)}{dt}\frac{dt}{ds}.$$

Example 71-2. Suppose that $|\mathbf{f}(t)| = c$, where c is a number that is independent of t in some interval. Show that $\mathbf{f}'(t)$ and $\mathbf{f}(t)$ are perpendicular vectors for each t in the interval.

Solution. Since $|\mathbf{f}(t)| = c$,

$$\mathbf{f}(t)\cdot\mathbf{f}(t) = c^2,$$

and now we use Equation 71-5 to find that

$$\mathbf{f}(t)\cdot\mathbf{f}'(t) + \mathbf{f}'(t)\cdot\mathbf{f}(t) = \frac{dc^2}{dt} = 0.$$

Therefore

$$\mathbf{f}(t)\cdot\mathbf{f}'(t) = 0,$$

which is equivalent to saying that $\mathbf{f}(t)$ and $\mathbf{f}'(t)$ are perpendicular.

The graph of a vector equation $\mathbf{r} = \mathbf{f}(t)$ for t in an interval $[a, b]$ in which $|\mathbf{f}'(t)| > 0$ is a curve in the plane. We will now find a formula for the length of such a curve, and in later sections we will discuss its "curvature" and the area of the surface that is generated when the graph is rotated about one of the coordinate axes. In these discussions we shall suppose that $\mathbf{f}''(t)$ exists for each t in $[a, b]$. The graph of such a function $\mathbf{f}$ is called a **smooth arc**.

We will treat the length of an arc in much the same way we treat the area of a plane region. Thus we will assume that we know what arclength is and what properties it possesses. On the basis of these assumptions, we will develop a formula for arclength, leaving the question of definition to courses in higher mathematics.

So let us suppose that the curve $\mathbf{r} = \mathbf{f}(t)$ for t in an interval $[a, b]$ is a smooth arc, as shown in Fig. 71-2, and see if we can find its length. If we had an arclength function R such that the number $R(t)$ is the length of the arc from the terminal point of $\mathbf{f}(a)$ to the terminal point of $\mathbf{f}(t)$, then the number $R(b)$ is the arclength we seek. We will find $R(b)$ indirectly—by finding an expression for $R'(t)$ and integrating it. Let u and t be numbers in $[a, b]$. If $u > t$, the terminal point of $\mathbf{f}(u)$ is "beyond" the terminal point of $\mathbf{f}(t)$, and hence $R(u) > R(t)$. If $u < t$, then $R(u) < R(t)$, and so the numbers $R(u) - R(t)$ and $u - t$ have the same sign. The arc that joins the terminal points of $\mathbf{f}(t)$ and $\mathbf{f}(u)$ is $|R(u) - R(t)|$ units long (Fig. 71-3).

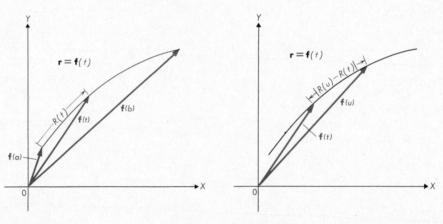

Figure 71-2   **Figure 71-3**

It seems natural to require that the ratio of the length of the arc joining two points of our curve to the length of the line segment or chord that joins these points should approach 1 as one point approaches the other. The chord that joins the terminal points of the vectors $\mathbf{f}(u)$ and $\mathbf{f}(t)$ is $|\mathbf{f}(u) - \mathbf{f}(t)|$ units long, so we will suppose that

$$\lim_{u \to t} \frac{|R(u) - R(t)|}{|\mathbf{f}(u) - \mathbf{f}(t)|} = 1.$$

From this basic assumption, we can determine what the derivative of arclength must be.

Because $R(u) - R(t)$ and $u - t$ have the same sign, the difference quotient $\dfrac{R(u) - R(t)}{u - t}$ is always positive, so we can write

$$\frac{R(u) - R(t)}{u - t} = \left| \frac{R(u) - R(t)}{u - t} \right| = \frac{|R(u) - R(t)|}{|\mathbf{f}(u) - \mathbf{f}(t)|} \cdot \frac{|\mathbf{f}(u) - \mathbf{f}(t)|}{|u - t|} .$$

Hence

$$R'(t) = \lim_{u \to t} \frac{R(u) - R(t)}{u - t} = \lim_{u \to t} \frac{|R(u) - R(t)|}{|\mathbf{f}(u) - \mathbf{f}(t)|} \cdot \lim_{u \to t} \left| \frac{\mathbf{f}(u) - \mathbf{f}(t)}{u - t} \right| = 1 \cdot |\mathbf{f}'(t)|.$$

Therefore the arclength function R must be such that

$$R'(t) = |\mathbf{f}'(t)|.$$

We find the length $R(b)$ of our arc simply by integrating both sides of this equation:

$$\int_a^b R'(t)\, dt = \int_a^b |\mathbf{f}'(t)|\, dt.$$

The Fundamental Theorem of Calculus tells us that $\displaystyle\int_a^b R'(t)\, dt = R(b) - R(a)$, and this expression simplifies to $R(b)$, since $R(a) = 0$ [it is the length of the "arc" that joins the terminal point of $\mathbf{f}(a)$ to the terminal point of $\mathbf{f}(a)$]. Thus if we write L for the arclength $R(b)$, we have developed the formula

(71-7) $$L = \int_a^b |\mathbf{f}'(t)|\, dt.$$

Example 71-3. Show that Equation 71-7 gives the correct length of the line segment $\mathbf{r} = \mathbf{m}t + \mathbf{b}$, $t \in [a, b]$.

Solution. The vectors $\mathbf{r}_a = \mathbf{m}a + \mathbf{b}$ and $\mathbf{r}_b = \mathbf{m}b + \mathbf{b}$ are position vectors of the endpoints of our segment, so its length is

$$|\mathbf{r}_b - \mathbf{r}_a| = |(\mathbf{m}b + \mathbf{b}) - (\mathbf{m}a + \mathbf{b})| = |\mathbf{m}b - \mathbf{m}a| = |\mathbf{m}|(b - a).$$

We are to show that Equation 71-7 also gives us this number. In our present example $\mathbf{f}(t) = \mathbf{m}t + \mathbf{b}$. Therefore $\mathbf{f}'(t) = \mathbf{m}$, and so Equation 71-7 tells us that $L = \displaystyle\int_a^b |\mathbf{m}|\, dt = |\mathbf{m}|(b - a)$, as it should be.

In our derivation of Equation 71-7 we ignored some subtle issues that deserve discussion. On the face of it, we should regard the expression we have just derived as a formula for *graph* length rather than *arc*length. We think of an arc as a set of points in the plane, such as the part of the parabola $y = x^2$ that joins the points $(0, 0)$ and $(1, 1)$, and this arc is the graph of any number of vector equations, $\mathbf{r} = t\mathbf{i} + t^2\mathbf{j}$, $t \in [0, 1]$, $\mathbf{r} = \sin t\mathbf{i} + \sin^2 t\mathbf{j}$, $t \in [0, \frac{1}{2}\pi]$, and so on. Since our formula for L is expressed in terms of a particular vector-valued function of which the given arc is the graph, we might expect to obtain different lengths for different functions. It is quite easy to show, however, that if a certain smooth arc is the graph of two different vector-valued functions, our formula gives the same value for its length when we use either function in the computation. We also get the same value if we change our coordinate system by a general rigid trans-formation. Consequently, it would be perfectly proper to regard our formula as the *definition* of the length of a smooth arc. Then we could use this definition to prove that the ratio of arclength to chord length approaches 1.

Whether you regard Equation 71-7 as a definition or as a theorem is not too important at this stage; all you have to know is that it is the formula for measuring the distance along an arc. Notice that this distance is a directed distance. If $b > a$, then L is positive, and if $b < a$, then L is negative. In order to obtain the number of units of arclength between two points, we can compute L from Equation 71-7 and take the absolute value of the result.

Since we can write $\mathbf{f}'(t) = \mathbf{r}' = x'\mathbf{i} + y'\mathbf{j}$, then $|\mathbf{f}'(t)| = \sqrt{x'^2 + y'^2}$, and our expression for arclength takes the form

$$(71\text{-}8) \qquad L = \int_a^b \sqrt{x'^2 + y'^2}\; dt.$$

Example 71-4. Show that the length of the curve $y = f(x)$ for $x \in [a, b]$ is given by the formula

$$(71\text{-}9) \qquad L = \int_a^b \sqrt{1 + f'(x)^2}\; dx.$$

Solution. Parametric equations of our given arc are

$$x = t \quad \text{and} \quad y = f(t), \qquad t \in [a, b].$$

Thus $x' = 1$ and $y' = f(t)$, and so Equation 71-9 follows immediately from Equation 71-8.

Example 71-5. How long is the arc of the *catenary* $y = \cosh x$ that joins the points $(0, 1)$ and $(1, \cosh 1)$?

Solution. We use Equation 71-9 with $a = 0$, $b = 1$, and $f(x) = \cosh x$. Then $f'(x) = \sinh x$, and $1 + f'(x)^2 = 1 + \sinh^2 x = \cosh^2 x$. Thus

$$L = \int_0^1 \cosh x\; dx = \sinh x \Big|_0^1 = \sinh 1 \approx 1.2.$$

If we write

(71-10) $s = R(t),$

then we have seen that

(71-11) $\dfrac{ds}{dt} = |\mathbf{f}'(t)|.$

For a smooth arc, $|\mathbf{f}'(t)| > 0$, and so our arclength function R is increasing and hence has an inverse. In other words, we can solve (theoretically) Equation 71-10 for t in terms of s and so express the radius vector $\mathbf{r} = \mathbf{f}(t)$ in terms of arclength s. The rule for differentiating an inverse says that $\dfrac{dt}{ds} = 1 \bigg/ \dfrac{ds}{dt}$, and therefore Equation 71-11 gives us $\dfrac{dt}{ds} = 1/|\mathbf{f}'(t)|$. Now we substitute in the Chain Rule Equation 71-6, and we have

$$\frac{d\mathbf{r}}{ds} = \frac{d\mathbf{f}(t)}{ds} = \frac{\mathbf{f}'(t)}{|\mathbf{f}'(t)|}.$$

The vector $\mathbf{f}'(t)$ is tangent to our graph, and when we divide this vector by its length, we obtain a *unit* tangent vector that we shall call $\mathbf{t}$. Thus

(71-12) $\mathbf{t} = \dfrac{d\mathbf{r}}{ds} = \dfrac{\mathbf{f}'(t)}{|\mathbf{f}'(t)|}.$

PROBLEMS 71

1. Find $\mathbf{f}'(1)$ if $\mathbf{f}(t) = f_x(t)\mathbf{i} + f_y(t)\mathbf{j}$ and
   (a) $f_x(t) = \ln t, f_y(t) = t.$ (b) $f_x(t) = e^t, f_y(t) = t^2 - 1.$
   (c) $f_x(t) = \cos^2 t, f_y(t) = \sin^2 t.$ (d) $f_x(t) = \cosh t, f_y(t) = \sinh(1 - t^2).$
2. Find a tangent vector to the curve $\mathbf{r} = e^t \sin t\mathbf{i} + e^{-t} \cos t\mathbf{j}$ at the point corresponding to $t = 0$. What is the slope of the tangent line to the curve at the point corresponding to $t = 1$?
3. Find unit vectors tangent to the curve at the point corresponding to the given parameter value.
   (a) $\mathbf{r} = t^3\mathbf{i} + t^4\mathbf{j}, t = 1$ (b) $\mathbf{r} = 3 \cos 2t\mathbf{i} + 3 \sin 2t\mathbf{j}, t = 5$
   (c) $\mathbf{r} = e^t\mathbf{i} + \ln t\mathbf{j}, t = 1$ (d) $\mathbf{r} = t \sin t\mathbf{i} + \cos t\mathbf{j}, t = \frac{1}{2}\pi$
4. Find a vector equation of the tangent line to the curve $\mathbf{r} = e^t\mathbf{i} + \ln t\mathbf{j}$ at the point at which $t = 1$.
5. The equation $\mathbf{r} = \cos t\mathbf{i} + \sin t\mathbf{j}$, with t in the interval $[0, 2\pi]$, and the equation $\mathbf{r} = \cos t^2\mathbf{i} + \sin t^2\mathbf{j}$, with t in the interval $[\sqrt{2\pi}, \sqrt{4\pi}]$, both represent the unit circle. Show that Equation 71-7 yields the circumference of this circle when we use either of these vector representations.

6. Find the length of one arch of the cycloid that we discussed in Example 70-5.

7. Let s denote the length of the arc of the parabola $y = x^2$ from the origin to the point where $x = c$. Express s in terms of c. Find the length of the arc from the origin to the point $(1, 1)$.

8. Find the total length of the curve with parametric equations $x = \cos^3 t, y = \sin^3 t$, $t \in [0, 2\pi]$.

9. Find the "circumference" of the *hypocycloid* $x^{2/3} + y^{2/3} = a^{2/3}$.

10. Find the length of the arc with parametric equations $x = t^3$, $y = 2t^2$, t in the interval $[0, 1]$.

11. Find the length of the part of the curve $y = x^{3/2}$ for x in the interval $[0, \frac{4}{3}]$.

12. Find the length of the curve $y = x^3/6 + 1/2x$ for x in the interval $[1, 2]$.

13. With the aid of Equation 71-3,

   (a) prove that Equation 71-4 is valid.   (b) prove that Equation 71-5 is valid.

14. Interpret the numbers $\int_a^b |\mathbf{f}'(t)| \, dt$ and $\left| \int_a^b \mathbf{f}'(t) \, dt \right|$ geometrically. Which is larger?

15. Show that Equation 71-10 gives the correct lengths of line segments.

16. In Section 6 we defined the trigonometric functions in terms of arclength along the unit circle, and in Section 18 we derived the basic equation $\lim_{t \to 0} (\sin t)/t = 1$ on the assumption that the limit of the ratio of arclength to chord length is 1. Thus we assumed that we knew what arclength along a circle is and what properties it possesses. On the basis of what we have said in this section, can you formulate an appropriate definition of arclength along the unit circle?

72. CURVATURE, VELOCITY, AND ACCELERATION

   In this section we look more closely at the geometry of plane curves and apply what we find to describe the path of a particle moving in the plane. As in the preceding section, we will work with a vector-valued function $\mathbf{f}$ such that the curve $\mathbf{r} = \mathbf{f}(t)$, $t \in [a, b]$, is a smooth arc.

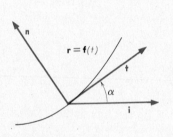

Figure 72-1

   Let $\mathbf{t} = \mathbf{r}'/|\mathbf{r}'|$ be a unit tangent vector to this curve. Suppose that the angle whose initial side contains the vector $\mathbf{i}$ and whose terminal side contains the vector $\mathbf{t}$ is α radians wide (Fig. 72-1). Because $\mathbf{t}$ is a unit vector, its components are $\cos \alpha$ and $\sin \alpha$, so

   (72-1) $\mathbf{t} = \cos \alpha \mathbf{i} + \sin \alpha \mathbf{j}.$

Therefore

   (72-2) $\dfrac{d\mathbf{t}}{ds} = (-\sin \alpha \mathbf{i} + \cos \alpha \mathbf{j}) \dfrac{d\alpha}{ds}.$

For short, let us write $\mathbf{u} = -\sin \alpha \mathbf{i} + \cos \alpha \mathbf{j}$. Then $\mathbf{t} \cdot \mathbf{u} = 0$, so $\mathbf{u}$ is a unit vector that is perpendicular to the tangent vector $\mathbf{t}$. Therefore $\dfrac{d\mathbf{t}}{ds}$ is perpendicular to $\mathbf{t}$, and $\left| \dfrac{d\mathbf{t}}{ds} \right| = \left| \dfrac{d\alpha}{ds} \right|$.

We will use the Greek letter κ to denote $\left| \dfrac{d\alpha}{ds} \right|$; that is,

$$(72\text{-}3) \qquad \kappa = \left| \frac{d\alpha}{ds} \right|.$$

Thus κ measures the absolute value of the rate of change of α with respect to s. Since it is the tangent vector $\mathbf{t}$ that makes the angle α with $\mathbf{i}$, we say that κ measures the absolute value of the "arc rate at which the tangent vector turns," and we call it the **curvature** of our curve at a point. Equation 72-2 tells us that the vector $\dfrac{d\mathbf{t}}{ds}$ can be written as the product of the curvature and a vector $\mathbf{n}$ that is perpendicular to the tangent vector:

$$(72\text{-}4) \qquad \frac{d\mathbf{t}}{ds} = \kappa \mathbf{n}.$$

This **unit normal** vector $\mathbf{n}$ is either the vector $\mathbf{u}$ or the vector $-\mathbf{u}$, depending on the sign of $\dfrac{d\alpha}{ds}$. You might try to convince yourself, by means of examples, that the vector $\dfrac{d\mathbf{t}}{ds}$ (and hence $\mathbf{n}$) is always directed, as shown in Fig. 72-1, into the convex region between an arc of the curve and its chord.

According to Equation 72-3, if we want to determine the curvature κ of a curve, we should express α in terms of s and then differentiate. That is easier said than done, so we will find a formula for κ that looks more complicated than Equation 72-3 but that is considerably easier to apply.

We first write the equation $\mathbf{t} = \mathbf{r}'/|\mathbf{r}'|$ as $\mathbf{r}' = |\mathbf{r}'|\mathbf{t}$. In the right-hand side of this equation we substitute the expression for $\mathbf{t}$ given by Equation 72-1, and we replace the left-hand side with $x'\mathbf{i} + y'\mathbf{j}$:

$$x'\mathbf{i} + y'\mathbf{j} = |\mathbf{r}'| (\cos \alpha \mathbf{i} + \sin \alpha \mathbf{j}).$$

This vector equation is equivalent to the two scalar equations

$$(72\text{-}5) \qquad x' = |\mathbf{r}'| \cos \alpha \quad \text{and} \quad y' = |\mathbf{r}'| \sin \alpha.$$

Now we differentiate both sides of these equations with respect to t:

$$x'' = |\mathbf{r}'|' \cos \alpha - |\mathbf{r}'| \sin \alpha \, \frac{d\alpha}{dt},$$

$$y'' = |\mathbf{r}'|' \sin \alpha + |\mathbf{r}'| \cos \alpha \, \frac{d\alpha}{dt}.$$

If we multiply the first of these equations by $\sin \alpha$, subtract the result from $\cos \alpha$ times the second, and use the identity $\cos^2 \alpha + \sin^2 \alpha = 1$, we obtain the equation

$$\cos \alpha y'' - \sin \alpha x'' = |\mathbf{r}'| \frac{d\alpha}{dt}.$$

Equations 72-5 allow us to replace $\cos \alpha$ and $\sin \alpha$ with $x'/|\mathbf{r}'|$ and $y'/|\mathbf{r}'|$, so we see that

$$\frac{x'y'' - y'x''}{|\mathbf{r}'|} = |\mathbf{r}'| \frac{d\alpha}{dt},$$

from which

$$\frac{d\alpha}{dt} = \frac{x'y'' - y'x''}{|\mathbf{r}'|^2}.$$

This equation gives us $\dfrac{d\alpha}{dt}$, rather than the number $\dfrac{d\alpha}{ds}$ that we want. However, according to the Chain Rule, these derivatives are related by the equation $\dfrac{d\alpha}{dt} = \dfrac{d\alpha}{ds} \dfrac{ds}{dt}$, and since (see Equation 71-11) $\dfrac{ds}{dt} = |\mathbf{r}'|$, we have $\dfrac{d\alpha}{dt} = |\mathbf{r}'| \dfrac{d\alpha}{ds}$. When we substitute this result into our expression for $\dfrac{d\alpha}{dt}$, we find that

$$\frac{d\alpha}{ds} = \frac{x'y'' - y'x''}{|\mathbf{r}'|^3}.$$

Now $|\mathbf{r}'| = \sqrt{x'^2 + y'^2}$, so our final equation for $\kappa = \left| \dfrac{d\alpha}{ds} \right|$ is

(72-6)
$$\kappa = \frac{|x'y'' - y'x''|}{(x'^2 + y'^2)^{3/2}}.$$

Example 72-1.   Find the curvature of the circle of radius a that has the parametric equations $x = a \cos t$ and $y = a \sin t$, $t \in [0, 2\pi]$.

Solution.   When we substitute in Equation 72-6, we find that

$$\kappa = \frac{|(-a \sin t)(-a \sin t) - (a \cos t)(-a \cos t)|}{[(-a \sin t)^2 + (a \cos t)^2]^{3/2}}$$

$$= \frac{a^2}{a^3} = \frac{1}{a}.$$

From the preceding example we see that the reciprocal of the curvature of a circle is the radius of the circle. In general, we define the number

$$\rho = \frac{1}{\kappa}$$

to be the **radius of curvature** of the graph of the equation $\mathbf{r} = \mathbf{f}(t)$ at the terminal point of $\mathbf{r}$. If you construct a circle whose radius is ρ and whose center is ρ units from the terminal point of $\mathbf{r}$ along the normal vector $\mathbf{n}$, then you obtain what is called the **circle of curvature** at the point. This circle is the circle that "best fits" the curve in the neighborhood of the given point.

Example 72-2.   Find the curvature of the graph of the equation $y = f(x)$.

Solution.   One pair of parametric equations for our curve is

$$x = t \quad \text{and} \quad y = f(t).$$

Then $x' = 1$ and $x'' = 0$, and Equation 72-6 becomes

(72-7) $$\kappa = \frac{|y''|}{(1 + y'^2)^{3/2}}.$$

Example 72-3.   Find the curvature of the parabola $y = x^2$ at the point $(1, 1)$. Show that this parabola is nearly a straight line at points far distant from the origin.

Solution.   Using Equation 72-7 with $y' = 2x$ and $y'' = 2$, we have

(72-8) $$\kappa = \frac{2}{(1 + 4x^2)^{3/2}}.$$

Thus the curvature of our parabola at $(1, 1)$ is

$$\kappa = 2/5^{3/2} \approx .18.$$

If a point is very far from the origin, the square of its X-coordinate is very large. And if x^2 is a large number, the number κ given by Equation 72-8 is nearly zero. Thus at points far from the origin, the curvature of the parabola is nearly zero; that is, the parabola is almost a straight line.

We often apply our theory of curves to describe the path of a particle moving in a plane. Usually in such cases, the parameter t measures time. Thus if the curve $\mathbf{r} = \mathbf{f}(t)$ is the path of some particle moving on a smooth arc in the plane, then t time units (seconds, for example) after we start to measure time, the particle is at the terminal point of $\mathbf{f}(t)$ (Fig. 72-2).

According to Equation 71-7, the distance the particle has moved in a given time interval $[a, t]$ is

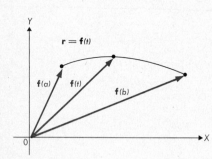

Figure 72-2

$$s = \int_a^t |\mathbf{f}'(u)|\ du.$$

The derivative of s with respect to t (that is, the number $\dfrac{ds}{dt}$) measures the rate of change of distance along our smooth arc with respect to time. It is natural to call this *number* $\dfrac{ds}{dt}$ the **speed** of our moving particle. The *vector* $\mathbf{f}'(t)$ is called the **velocity v** of the particle; thus

$$\mathbf{v} = \mathbf{f}'(t) = \frac{d\mathbf{r}}{dt}.$$

Furthermore, since $\dfrac{ds}{dt} = |\mathbf{f}'(t)|$, we see that

$$|\mathbf{v}| = |\mathbf{f}'(t)| = \frac{ds}{dt}.$$

In other words, the velocity of our particle is a *vector* quantity, and the length of the velocity vector is the *scalar* called the speed of the particle. We can also express the velocity vector $\mathbf{v}$ by the equation

(72-9) $$\mathbf{v} = \frac{d\mathbf{r}}{ds}\frac{ds}{dt} = \mathbf{t}\,\frac{ds}{dt},$$

where $\mathbf{t} = \dfrac{d\mathbf{r}}{ds}$ is our unit tangent vector.

The derivative of **v** with respect to t measures the rate of change of velocity; we call it the **acceleration a** of the moving particle:

$$\mathbf{a} = \frac{d\mathbf{v}}{dt} = \frac{d^2\mathbf{r}}{dt^2}.$$

Like velocity, acceleration is a *vector* quantity. The length of the velocity vector **v** is the speed of our moving particle, but the length of the acceleration vector is not, in general, the rate of change of the speed of the particle (why not?).

Example 72-4. In Chapter 2 Hiawatha learned that it is bad medicine to shoot an arrow straight up. So he launches his second arrow at an angle of 30° with the horizontal at a speed of 1000 feet per second. Describe its path.

Solution. In Fig. 72-3 we have introduced a co-ordinate system and have drawn in the initial velocity vector $\mathbf{v}_0$. We know that $|\mathbf{v}_0| = 1000$ and that $\mathbf{v}_0$ makes an angle of 30° with the positive X-axis, so

$$\mathbf{v}_0 = 1000(\cos 30°\mathbf{i} + \sin 30°\mathbf{j})$$

$$= 500\sqrt{3}\mathbf{i} + 500\mathbf{j}.$$

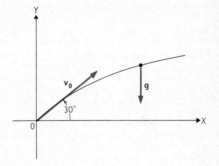

Figure 72-3

We will assume that the only force acting on the arrow in flight in the force of gravity and that this force is constant. Thus if we measure distance in feet and time in seconds, then the gravitational accelera-tion is $\mathbf{g} = -32\mathbf{j}$. Since acceleration is the derivative of velocity, $\dfrac{d\mathbf{v}}{dt} = \mathbf{g}$. In addition, $\mathbf{v} = \mathbf{v}_0$ when $t = 0$, so the velocity vector **v** satisfies the initial value problem

$$\mathbf{v}' = \mathbf{g} \quad \text{and} \quad \mathbf{v} = \mathbf{v}_0 \quad \text{when } t = 0.$$

Theorem 60-1 tells us that this problem is satisfied by

$$\mathbf{v} = \mathbf{v}_0 + \int_0^t \mathbf{g}\, ds = \mathbf{g}t + \mathbf{v}_0.$$

But **v** is the derivative of **r**, so $\mathbf{r}' = \mathbf{g}t + \mathbf{v}_0$. Furthermore, $\mathbf{r} = \mathbf{0}$ when $t = 0$, and thus **r** satisfies the initial value problem

$$\mathbf{r}' = \mathbf{g}t + \mathbf{v}_0 \quad \text{and} \quad \mathbf{r} = \mathbf{0} \quad \text{when } t = 0.$$

This problem is satisfied by

$$\mathbf{r} = \mathbf{0} + \int_0^t (g s + \mathbf{v}_0)\, ds$$

$$= \tfrac{1}{2} g t^2 + \mathbf{v}_0 t = 500\sqrt{3}\, t\mathbf{i} + (500t - 16t^2)\mathbf{j}.$$

We use this vector equation to find parametric equations of the arrow's path:

$$x = 500\sqrt{3}t \quad \text{and} \quad y = 500t - 16t^2.$$

To obtain a nonparametric equation of the path, we can solve the first of these equations for t in terms of x and substitute the result in the second:

$$y = \frac{x}{\sqrt{3}} - \frac{16x^2}{3(500)^2}.$$

You should recognize this curve as a parabola.

Let us denote the speed of a moving particle by v; that is, $v = \dfrac{ds}{dt}$. Then Equation 72-9 becomes

$$\mathbf{v} = v\mathbf{t},$$

and we can find the acceleration of the particle from this equation. Thus

$$\mathbf{a} = \frac{d}{dt}(v\mathbf{t}) = \frac{dv}{dt}\mathbf{t} + v\frac{d\mathbf{t}}{dt}.$$

We use the Chain Rule to write

$$\frac{d\mathbf{t}}{dt} = \frac{d\mathbf{t}}{ds}\frac{ds}{dt} = v\frac{d\mathbf{t}}{ds},$$

and hence our formula for the acceleration becomes

$$\mathbf{a} = \frac{dv}{dt}\mathbf{t} + v^2\frac{d\mathbf{t}}{ds}.$$

Finally, since $\dfrac{d\mathbf{t}}{ds} = \kappa\mathbf{n}$ (Equation 72-4), we see that

$$\mathbf{a} = \frac{dv}{dt}\mathbf{t} + \kappa v^2\mathbf{n}.$$

Thus the acceleration can be considered as the sum of the **tangential acceleration** $\dfrac{dv}{dt}\mathbf{t}$ and the **normal acceleration** $\kappa v^2\mathbf{n}$. The normal acceleration is also called the **centripetal acceleration**.

Example 72-5. A rocket plane makes a circular turn at a constant speed of 2400 miles per hour. If the radius of the turn is 10 miles, determine the magnitude of the centripetal acceleration. How many "G's" is the pilot subjected to?

Solution. The centripetal acceleration has a magnitude of κv^2. If we convert our units to feet and seconds and utilize the fact that the curvature of a circle is the reciprocal of its radius (see Example 72-1), then we find that the magnitude of the centripetal acceleration is

$$\kappa v^2 = \frac{1}{52{,}800} \left(\frac{2400 \cdot 5280}{60 \cdot 60} \right)^2 \approx 235 \text{ feet per second per second.}$$

Since 1 "G" is about 32 feet per second per second, we see that the pilot is subjected to slightly more than seven Gs as he makes his turn.

Example 72-6. Find differential equations satisfied by polar coordinates of a body moving in space if the only force acting on it is the gravitational attraction of the earth. (We will assume—it can be shown to be true—that the path of the body is a plane curve.)

Solution. In the plane of motion of the body, we set up a polar coordinate system whose pole is the center of the earth. We suppose that t hours after some initial instant the body is at the point (r, θ), where $r = f(t)$ and $\theta = g(t)$. Thus the body is at the terminal point of the position vector $\mathbf{r} = r \cos \theta \mathbf{i} + r \sin \theta \mathbf{j} = r\mathbf{u}$, where $\mathbf{u}$ is the unit vector $\cos \theta \mathbf{i} + \sin \theta \mathbf{j}$.

The differential equation we seek is a numerical expression of Newton's Second Law of Motion, "Force equals mass times acceleration." The acceleration of our body is $\mathbf{r}''$, and it is acted on by the force of gravity. This force, according to the Law of Gravitation, is directed along the radius vector and has a magnitude that is directly proportional to the mass m of the body and inversely proportional to the square of its distance from the center of the earth. Thus the force vector is $-kmr^{-2}\mathbf{u}$ for some number k, and the equation mass × acceleration = force becomes

(72-10) $$m\mathbf{r}'' = -kmr^{-2}\mathbf{u}.$$

Since $\mathbf{r} = r\mathbf{u}$, we have $\mathbf{r}' = r'\mathbf{u} + r\mathbf{u}'$. We see that $\mathbf{u}' = \theta'(-\sin \theta \mathbf{i} + \cos \theta \mathbf{j}) = \theta'\mathbf{v}$, where $\mathbf{v}$ is the unit vector $-\sin \theta \mathbf{i} + \cos \theta \mathbf{j}$, and therefore

$$\mathbf{r}' = r'\mathbf{u} + r\theta'\mathbf{v}.$$

Furthermore, $\mathbf{r}'' = r''\mathbf{u} + r'\mathbf{u}' + (r\theta')'\mathbf{v} + r\theta'\mathbf{v}'$, and since $\mathbf{v}' = -\theta'\mathbf{u}$, we obtain the equation

$$\mathbf{r}'' = r''\mathbf{u} + r'\theta'\mathbf{v} + (r\theta')'\mathbf{v} - r\theta'^2\mathbf{u}$$

$$= (r'' - r\theta'^2)\mathbf{u} + [r'\theta' + (r\theta')']\mathbf{v}.$$

Thus Equation 72-10 can be written as

$$(r'' - r\theta'^2)\mathbf{u} + [r'\theta' + (r\theta')']\mathbf{v} = -kr^{-2}\mathbf{u},$$

and this vector equation is equivalent to two scalar equations satisfied by the coordinates (r, θ) of our moving body:

(72-11) $$r'' = r\theta'^2 - kr^{-2}$$

and

(72-12) $$(r\theta')' + r'\theta' = 0.$$

Example 72-7. Use the equations we derived in the preceding example to discuss the motion of a body in space.

Solution. Equation 72-12 can be written as $r\theta'' + 2r'\theta' = 0$. Let us multiply the left-hand side of this equation by r to obtain the expression $r^2\theta'' + 2rr'\theta'$. This expression is the derivative $\dfrac{d}{dt}(r^2\theta')$, so Equation 72-12 is equivalent to the equation $\dfrac{d}{dt}(r^2\theta') = 0$. Thus $r^2\theta'$ is independent of time. The quantity $l = mr^2\theta'$ is the *angular momentum* of the body; we have just seen that it is constant in our problem. This fact has an interesting geometric interpretation. Suppose that when $t = a$, our body has polar coordinates (r_a, α), and when $t = b$, it has reached the point (r_b, β). Then the area of the region that is swept out by the radius vector is (Equation 65-1) the number $A = \frac{1}{2}\int_\alpha^\beta r^2\, d\theta = \frac{1}{2}\int_a^b r^2\theta'\, dt$. Since $r^2\theta' = l/m$, we therefore have

$$A = \frac{1}{2}\int_a^b \frac{l}{m}\, dt = \frac{l}{2m}(b - a).$$

This equation says that the area of the region that is swept out by the radius vector during any time interval is proportional to the length of that time interval. This statement is called **Kepler's First Law of Planetary Motion;** it is usually phrased by saying that "equal areas are swept out in equal times."

We have pointed out that $r^2\theta'$ is a number l/m. If the number l is 0, then $\theta' = 0$, and our body moves in a line that is an extension of a radius of the earth. The case in which $l \neq 0$ is more difficult, as well as more interesting. Let us show that then the path of our body is a conic. It is easy to show, using Equation 72-11 and the equation $r^2\theta' = l/m$, that

$$\frac{d}{dt}[(r^{-1} - km^2l^{-2})\cos\theta + mr'l^{-1}\sin\theta] = 0$$

and

$$\frac{d}{dt}[(r^{-1} - km^2l^{-2})\sin\theta - mr'l^{-1}\cos\theta] = 0.$$

Thus the expressions in brackets are numbers P and Q that are independent of t:

$$(r^{-1} - km^2 l^{-2}) \cos \theta + mr'l^{-1} \sin \theta = P$$

and

$$(r^{-1} - km^2 l^{-2}) \sin \theta - mr'l^{-1} \cos \theta = Q.$$

Now we multiply the first of these equations by $\cos \theta$ and the second by $\sin \theta$ and add, and we find that

$$r^{-1} - km^2 l^{-2} = P \cos \theta + Q \sin \theta.$$

When we solve this equation for r, we obtain

$$r = \frac{1}{km^2 l^{-2} + P \cos \theta + Q \sin \theta}.$$

We leave it to you to write this equation in the form

$$r = \frac{ep}{1 + e \cos (\theta - \phi)},$$

the polar equation of a conic whose focus is the pole and whose axis makes an angle ϕ with the polar axis.

Can you now discuss the problem of firing a satellite into orbit about the earth?

PROBLEMS   72

1. Find κ at the point corresponding to the given t.
    (a) $\mathbf{r} = \sin t\mathbf{i} + 2 \cos t\mathbf{j}, t = 0$ (b) $\mathbf{r} = (t^2 - 2t)\mathbf{i} + 3t\mathbf{j}, t = 1$
    (c) $\mathbf{r} = e^t\mathbf{i} + e^{-t}\mathbf{j}, t = 0$
    (d) $\mathbf{r} = 2(t - \sin t)\mathbf{i} + 2(1 - \cos t)\mathbf{j}, t = \frac{1}{3}\pi$
2. Find κ at the point indicated.
    (a) $y = x^2 - 2x + 3, (1, 2)$ (b) $y = e^{3x/4}, (0, 1)$
    (c) $y = \cos x, (\frac{1}{2}\pi, 0)$ (d) $y^2 = x + 3, (6, 3)$
3. At what points of the curve $y = x^3$ is the radius of curvature a minimum? At what points is the curvature a minimum?
4. Draw the parabola $y^2 = 6x$ and its circle of curvature at the point at which $y = 4$.
5. If the tangent to the curve $y = f(x)$ is nearly parallel to the X-axis, engineers regard the number $|y''|$ as an approximation of κ. Why?
6. Show that the curvature of a line is zero at every point. Is the converse true?
7. Let a and b be positive numbers with $a > b$. A point moves in an elliptical path $\mathbf{r} = a \cos t\mathbf{i} + b \sin t\mathbf{j}$. Sketch the path.
    (a) What is the velocity of the point when $t = 0$?
    (b) What is its maximum speed? Its minimum speed?
    (c) Find the acceleration vector $\mathbf{a}$. Describe its direction.

8.   A particle moves in a coordinate plane in such a way that t seconds after it starts, its position vector is $\mathbf{r} = (3t^2 - t + 1)\mathbf{i} + (2t - 3)\mathbf{j}$. Distances are measured in centimeters, the Y-axis points north, the X-axis points east, and the motion lasts 5 seconds.

(a) Where does the particle start?
(b) Where does it stop?
(c) Does it pass through the point (3, 1)?
(d) Does its path intersect the line $x = \pi^e$?
(e) Does its path intersect the line $y = 3\pi$?
(f) When does it reach the point (25, 3)?
(g) How fast is it going then?
(h) Which direction, N, S, E, or W, most nearly describes the direction of motion at that instant?
(i) What is the name of the path of the particle?
(j) In what direction does the particle start out?
(k) What is the acceleration vector for this motion?

9.   Find the magnitudes of the centripetal acceleration and tangential acceleration at the indicated point.

(a) $\mathbf{r} = 3t\mathbf{i} + 3 \ln t\mathbf{j}$, $t = 3$ (b) $\mathbf{r} = 2 \tan t\mathbf{i} + 2 \cot t\mathbf{j}$, $t = \frac{1}{4}\pi$

10. A bug sits on a hoop of radius 2 feet that rolls along a line at 1 revolution per second. Find its velocity and acceleration

(a) at the instant the bug is at the top of the hoop.
(b) at the instant the bug is at the bottom of the hoop (see Example 70-5).

11. Show that $|\mathbf{a}|^2 = \left(\dfrac{d}{dt}|\mathbf{v}|\right)^2 + \kappa^2|\mathbf{v}|^4$. Under what conditions is the length of the acceleration vector equal to the rate of change of speed?

12. Use the equation $r^2\theta' = l/m$ to eliminate θ from Equation 72-11 and show that the resulting equation is equivalent to

$$\frac{d}{dt}\left(\frac{mr'^2}{2}\right) = -\frac{d}{dt}\left(\frac{l^2}{2mr^2} - \frac{km}{r}\right).$$

This equation says that there is a number E such that

$$\frac{mr'^2}{2} + \frac{l^2}{2mr^2} - \frac{km}{r} = E.$$

73.   THE AREA OF A SURFACE OF REVOLUTION

If a smooth arc $\mathbf{r} = \mathbf{f}(t) = f_x(t)\mathbf{i} + f_y(t)\mathbf{j}$, $t \in [a, b]$, is rotated about the X-axis, it generates the surface of revolution shown in Fig. 73-1. We will now find the area of this surface. When discussing the area of plane figures in Section 33, we avoided a precise definition of what this concept means, and because defining the area of a two-dimensional figure in three-dimensional space is even harder, we will again omit the technical details. As before, we hope that your intuitive notion of what area should be will make our remarks seem reasonable.

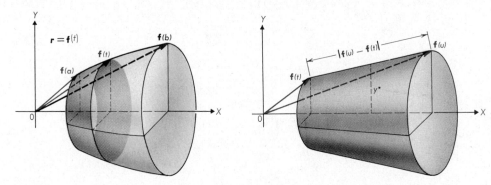

Figure 73-1   **Figure 73-2**

Let us therefore simply assume that there is a formula for the area of our surface of revolution and find it. Specifically, we will assume that there is a differentiable function S such that for each $t \in [a, b]$, the number $S(t)$ is the area of the surface that is obtained by rotating about the X-axis the arc that joins the terminal point of $\mathbf{f}(a)$ to the terminal point of $\mathbf{f}(t)$ (the colored region in Fig. 73-1). This figure is analogous to Fig. 71-2, which illustrates the arc of length $R(t)$ that generates our surface of revolution with an area of $S(t)$.

Figure 71-3 shows a short arc of length $|R(u) - R(t)|$. When rotated, it will generate a surface with an area of $|S(u) - S(t)|$ square units. This curved surface is approximated by the frustum of a cone (the slice shown in Fig. 73-2) formed by rotating the chord vector $\mathbf{f}(u) - \mathbf{f}(t)$, and so

(73-1) $|S(u) - S(t)| \approx$ *area of the frustum shown in Fig.* 73-2.

It is a simple matter of geometry (we sketch the steps in Problem 73-13) to show that the area of this frustum is 2π times the product of its slant height $|\mathbf{f}(u) - \mathbf{f}(t)|$ and its mean radius $|y^*|$; that is, *area* $= 2\pi|y^*|\,|\mathbf{f}(u) - \mathbf{f}(t)|$. Thus Approximation 73-1 can be written

$$|S(u) - S(t)| \approx 2\pi|y^*|\,|\mathbf{f}(u) - \mathbf{f}(t)|.$$

In the arclength problem we assumed that the ratio of arclength to chord length approaches 1 as u approaches t, and now we will make the meaning of our present approximation precise by assuming that the ratio of the area of the surface generated by the arc to the area of the surface generated by the chord approaches 1:

$$\lim_{u \to t} \frac{|S(u) - S(t)|}{2\pi|y^*|\,|\mathbf{f}(u) - \mathbf{f}(t)|} = 1.$$

Because $\lim_{u \to t} |y^*| = |y| = |f_y(t)|$, we can also express this assumption as

$$(73\text{-}2) \qquad \lim_{u \to t} \frac{|S(u) - S(t)|}{|f(u) - f(t)|} = 2\pi |y|.$$

Now we can find $S'(t)$. We first observe that the difference quotient $\dfrac{S(u) - S(t)}{u - t}$ is always positive, and then we write it in a form that enables us to make use of Equation 73-2:

$$\frac{S(u) - S(t)}{u - t} = \left| \frac{S(u) - S(t)}{u - t} \right| = \frac{|S(u) - S(t)|}{|f(u) - f(t)|} \cdot \left| \frac{f(u) - f(t)}{u - t} \right|.$$

Now we apply Equation 73-2, and we see that

$$S'(t) = \lim_{u \to t} \frac{S(u) - S(t)}{u - t} = 2\pi |y| \, |f'(t)|.$$

We have found a formula for $S'(t)$; what we want is the area A of the surface formed by rotating the arc joining the terminal points of $f(a)$ and $f(b)$ in Fig. 73-1. This number is $S(b)$, and we use the Fundamental Theorem of Calculus to find it. The Fundamental Theorem tells us that

$$S(b) - S(a) = \int_a^b S'(t) \, dt,$$

and since $S(b) = A$, $S(a) = 0$, and $S'(t) = 2\pi |y| \, |f'(t)|$,

$$(73\text{-}3) \qquad A = 2\pi \int_a^b |y| \, |f'(t)| \, dt.$$

Our derivation of Equation 73-3 rests on the unproved assumption that there is a differentiable area function S that satisfies Equation 73-2. Although this assumption may seem reasonable, it is by no means obvious. We will simply have to accept it without proof.

Since $f'(t) = r' = x'i + y'j$, we have $|f'(t)| = \sqrt{x'^2 + y'^2}$; as a result we often write Formula 73-3 as

$$(73\text{-}4) \qquad A = 2\pi \int_a^b |y| \sqrt{x'^2 + y'^2} \, dt.$$

In particular, if our arc is the graph of the equation $y = f(x)$, we can regard $x = t$ and $y = f(t)$ as parametric equations of the arc, and then Equation 73-4 becomes

$$(73\text{-}5) \qquad A = 2\pi \int_a^b |y| \sqrt{1 + y'^2} \, dx.$$

Example 73-1.   Use Equation 73-4 to find the surface area of a sphere whose radius is a.

Solution.   Our sphere is obtained by rotating about the X-axis the arc

$$x = a \cos t, \quad y = a \sin t, \quad t \in [0, \pi].$$

Here $x' = -a \sin t$ and $y' = a \cos t$. Substituting these numbers in Equation 73-4, we obtain

$$A = 2\pi \int_0^\pi a \sin t \sqrt{(-a \sin t)^2 + (a \cos t)^2}\, dt$$

$$= 2\pi a^2 \int_0^\pi \sin t\, dt = 4\pi a^2.$$

If we rotate an arc about the Y-axis instead of the X-axis, the area of the resulting surface can be found by interchanging x and y in the preceding formulas.

Example 73-2.   Find the area of the surface obtained by rotating the parabolic arc $x = t, y = t^2, t \in [0, 1]$, about the Y-axis.

Solution.   We use the formula

$$A = 2\pi \int_0^1 |x| \sqrt{x'^2 + y'^2}\, dt$$

$$= 2\pi \int_0^1 t\sqrt{1 + 4t^2}\, dt.$$

Now we change the variable of integration by the substitution $u = 1 + 4t^2$ to obtain

$$A = \frac{2\pi}{8} \int_1^5 \sqrt{u}\, du = \frac{\pi}{4}\frac{2}{3} u^{3/2}\Big|_1^5$$

$$\tfrac{1}{6}\pi(5^{3/2} - 1) \approx 5.33.$$

Example 73-3.   The upper half of the cardioid $r = 2(1 - \cos \theta)$ (Fig. 62-5) is rotated about the X-axis. Find the area of this surface of revolution.

Solution.   We recall that the cartesian coordinates (x, y) of a point are related to polar coordinates (r, θ) of the same point by the equations

$$x = r \cos \theta \quad \text{and} \quad y = r \sin \theta.$$

If we replace r with $2(1 - \cos \theta)$ in these equations, we obtain parametric equations for our cardioid;

$$x = 2(1 - \cos \theta) \cos \theta = 2(\cos \theta - \cos^2 \theta),$$

$$y = 2(1 - \cos \theta) \sin \theta = 2(\sin \theta - \cos \theta \sin \theta),$$

where θ is the parameter, and the parameter interval for the upper half is $[0, \pi]$. If we use primes to denote differentiation with respect to θ, we have

$$x' = 2(-\sin \theta + 2 \cos \theta \sin \theta) = 2(-\sin \theta + \sin 2\theta),$$

$$y' = 2(\cos \theta - \cos^2 \theta + \sin^2 \theta) = 2(\cos \theta - \cos 2\theta).$$

We then find that

$$x'^2 + y'^2 = 4(2 - 2 \sin \theta \sin 2\theta - 2 \cos \theta \cos 2\theta)$$

$$= 8[1 - (\cos 2\theta \cos \theta + \sin 2\theta \sin \theta)]$$

$$= 8(1 - \cos \theta).$$

Thus Equation 73-4 becomes

$$A = 2\pi \int_0^\pi 2(1 - \cos \theta) \sin \theta \sqrt{8(1 - \cos \theta)} \, d\theta$$

$$= 8\sqrt{2}\pi \int_0^\pi (1 - \cos \theta)^{3/2} \sin \theta \, d\theta.$$

To evaluate this integral, we change the variable of integration, letting

$$u = 1 - \cos \theta.$$

Then $du = \sin \theta \, d\theta$ and

$$A = 8\sqrt{2}\,\pi \int_0^2 u^{3/2} \, du = \tfrac{128}{5}\pi.$$

PROBLEMS 73

1. Find the area of the surface that results from rotating the curve about the X-axis.
    (a) $\mathbf{r} = (2t^2 + 1)\mathbf{i} + 3t\mathbf{j}, \, t \in [0, 1]$
    (b) $\mathbf{r} = (3e^t + 2)\mathbf{i} + (4e^t - 1)\mathbf{j}, \, t \in [0, \ln 3]$
2. Find the area of the surface that results from rotating the curve about the X-axis.

    (a) $y = 2\sqrt{x}, \, x \in [0, 8]$ (b) $y = \tfrac{1}{3}x^3, \, x \in [0, 2]$
    (c) $y = \cosh x, \, x \in [0, 1]$ (d) $y = \tfrac{1}{6}x^3 + \tfrac{1}{2}x^{-1}, \, x \in [1, 3]$
3. Find the area of the surface that results from rotating the curve about the X-axis.
    (a) $x = 3t^2 + 1, \, y = 2t, \, t \in [0, 1]$ (b) $x = t, \, y = \sin t, \, t \in [0, \pi]$

4. Find the area of the surface that is generated by rotating about the Y-axis the arc of the parabola $y = x^2$ that joins the point $(\sqrt{2}, 2)$ to the origin.

5. Find the area of the surface that is generated by rotating about the Y-axis the arc with parametric equations $x = t^2$ and $y = t^3$, $t \in [0, 2]$.

6. Find the area of the surface that is generated by rotating about the X-axis the arc with parametric equations $x = e^{-t} \cos t$ and $y = e^{-t} \sin t$ for $t \in [0, \frac{1}{2}\pi]$.

7. The given arc is rotated first about the X-axis and then about the Y-axis. Which surface of revolution has the larger area?

(a) $x = t$, $y = t^2$, $t \in [0, 1]$ (b) $x = \sin^4 t$, $y = \cos^4 t$, $t \in [\frac{1}{4}\pi, \frac{3}{4}\pi]$

(c) $x = \sin t$, $y = \tan t$, $t \in [0, \frac{1}{4}\pi]$ (d) $\mathbf{r} = \cosh t\,\mathbf{i} + e^t\mathbf{j}$, $t \in [0, 1]$

8. Show that Equation 73-5 gives the correct surface area of the cone generated by rotating about the X-axis the line segment that joins the origin to the point (h, r).

9. Find the area of the ellipsoidal surface that is generated by rotating about the X-axis the curve with parametric equations $x = a \cos t$ and $y = b \sin t$ for the given choices of a and b.

(a) $(a, b) = (5, 3)$ (b) $(a, b) = (3, 3)$ (c) $(a, b) = (3, 5)$

10. If the arc of the circle $x^2 + y^2 = a^2$ that lies above the interval $[x_1, x_1 + h]$ is rotated about the X-axis, a surface called a **spherical zone of altitude h** is generated. Show that the area of such a surface is $2\pi a h$.

11. Find the area of the surface that is generated by rotating about the X-axis the polar curve $r^2 = a^2 \cos 2\theta$.

12. (a) Interpret geometrically the number $\dfrac{1}{b-a} \displaystyle\int_a^b |y|\, dx$ [where $y = f(x)$, of course].

(b) Interpret geometrically the number that results when we multiply by $2\pi(b - a)$.

(c) Use Equation 73-5 to show that this latter number is usually less than the area of the surface we get when we rotate the curve $y = f(x)$ about the X-axis. When are the two numbers equal?

13. If a right circular cone is "slit and unrolled," it will form a sector of a circle, as shown in Fig. 73-3. The right-hand side of the figure shows a vertical cross section of the cone. We are interested in a formula for the area of a frustum of the cone—that is, the area A of the white region in our figure.

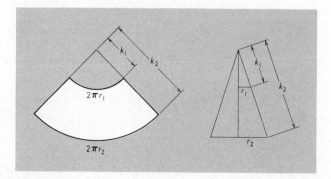

Figure 73-3

(a) Use the formula for the area of a circular sector (Appendix A) to show that $A = \pi(r_2 k_2 - r_1 k_1)$.

(b) From similar triangles, show that $r_1 k_2 = r_2 k_1$ and hence verify that $A = \pi(r_1 + r_2)(k_2 - k_1)$.

(c) The number $k_2 - k_1$ is the slant height s of our frustum, and if we write r for its mean radius $\frac{1}{2}(r_1 + r_2)$, we see that $A = 2\pi rs$.

Calculus goes hand in hand with geometry, as this chapter again emphasizes. Polar coordinates give us another natural way to locate points of the plane, and our introduction to vectors places a very powerful new tool into our hands. Not only do vectors describe plane curves, they also work equally well in three (and higher) dimensions. The material on arclength, curvature, surface area, and the like is not easy. However, it obviously deals with important geometric concepts, so try to make it seem as reasonable as you can.

Review Problems, Chapter Eight

1. Suppose that the points P and Q are symmetric with respect to the Y-axis, that P and R are symmetric with respect to the origin, and that P and S are symmetric with respect to the X-axis. If (r, θ) are polar coordinates of P, find polar coordinates of Q, R, and S.

2. Explain why the following polar and rectangular equations are equivalent.

   (a) $y = x$ and $\theta = \frac{1}{4}\pi$ (b) $x^2 + y^2 = 25$ and $r = 5$

   (c) $y = mx$ and $\tan \theta = m$ (d) $x + y = 3$ and $r \sin (\theta + \frac{1}{4}\pi) = \frac{3}{2}\sqrt{2}$

3. Show that the curve

   (a) $(r - 2)(r - 3) = 0$ consists of two concentric circles.

   (b) $\cos r = 0$ is a set of concentric circles.

   (c) $\sin 8\theta = 0$ is a set of radial lines.

   (d) $\cos r \sin 8\theta = 0$ is a "spider web."

4. Find a polar equation of the parabola whose focus is the origin and whose vertex is the point $(-2, \frac{1}{4}\pi)$.

5. At a maximum or minimum point of a curve $y = f(x)$ in cartesian coordinates, $\frac{dy}{dx} = 0$. At what points of a polar curve $r = f(\theta)$ is $\frac{dr}{d\theta} = 0$?

6. The area of the region bounded by the polar axis, a radial line making a positive acute angle of θ with the polar axis, and the curve $r = f(\theta)$ is given by the formula $A(\theta) = 2(e^{2\theta} - 1)$. Find the formula for $f(\theta)$.

7. If we rotate our cartesian axes through any angle θ, the expression $x^2 + y^2$ is transformed into $\bar{x}^2 + \bar{y}^2$. Through what angles of rotation of the axes is the expression $x^4 + y^4$ transformed into $\bar{x}^4 + \bar{y}^4$?

8. Find the asymptotes of the hyperbola $x^2 - 24xy - 6y^2 - 26x + 12y - 11 = 0$.

9. (a) What is the position vector to the point of the curve $\mathbf{r} = (t^3 + 1)\mathbf{i} + 2t^2\mathbf{j}$ that corresponds to $t = -1$?
   (b) What is a vector in a direction tangent to the curve at this point?
   (c) Find a vector equation of the tangent line at this point.
   (d) How long is the arc of our curve that corresponds to the interval $[-1, 0]$ of the T-axis?

10. Suppose that the curves $\mathbf{r} = \mathbf{f}(t)$ and $\mathbf{r} = \mathbf{g}(t)$ intersect in a certain point (x_0, y_0).
   (a) Explain why there must be numbers p and q such that $\mathbf{f}(p) = x_0\mathbf{i} + y_0\mathbf{j}$ and $\mathbf{g}(q) = x_0\mathbf{i} + y_0\mathbf{j}$.
   (b) Must $p = q$?
   (c) What does the equation $\mathbf{f}'(p) \cdot \mathbf{g}'(q) = 0$ mean geometrically?
   (d) What does the equation $3\mathbf{f}'(p) + 5\mathbf{g}'(q) = \mathbf{0}$ mean geometrically?

11. Why is it true that $\mathbf{f}'(t) \cdot \mathbf{j} = 0$ if $\mathbf{f}(t)$ is a maximum point of the curve $\mathbf{r} = \mathbf{f}(t)$?

12. Why isn't the "associative law of dot multiplication," $(\mathbf{a} \cdot \mathbf{b}) \cdot \mathbf{c} = \mathbf{a} \cdot (\mathbf{b} \cdot \mathbf{c})$, mentioned in the text?

13. (a) Show that $|\mathbf{a}|^2|\mathbf{b}|^2 - (\mathbf{a} \cdot \mathbf{b})^2 = |\mathbf{a}|^2|\mathbf{b}|^2 \sin^2 \theta$, where θ is the angle determined by $\mathbf{a}$ and $\mathbf{b}$.
   (b) Show that two vectors $\mathbf{a}$ and $\mathbf{b}$ are parallel if and only if $|\mathbf{a}|^2|\mathbf{b}|^2 = (\mathbf{a} \cdot \mathbf{b})^2$.

14. Let $\mathbf{a}$ and $\mathbf{b}$ be given position vectors and describe the line $\mathbf{r} = (\mathbf{a} - \mathbf{b})t + \mathbf{b}$.

15. Find a vector equation of the tangent line to the curve $\mathbf{r} = \mathbf{f}(t)$ at the point at which $t = a$.

16. Suppose that $\mathbf{a}$ and $\mathbf{b}$ are mutually perpendicular unit vectors and consider the arc $\mathbf{r} = t\mathbf{a} + t^2\mathbf{b}$ for $t \in [0, 1]$.
   (a) How long is this arc?
   (b) What is its curvature?
   (c) What is the area of the surface that is formed when this arc is rotated about a line that contains the origin and is parallel to the vector $\mathbf{b}$?

17. Suppose that the velocity vector $\mathbf{v}$ and position vector $\mathbf{r}$ of a particle moving in a coordinate plane are related by the equation $\mathbf{v} = 3\mathbf{r}$.
   (a) Discuss this equation geometrically.
   (b) Show that it implies that $\dfrac{d}{dt}(e^{-3t}\mathbf{r}) = \mathbf{0}$.
   (c) Show that this last equation means that $e^{-3t}\mathbf{r} = \mathbf{r}_0$, where $\mathbf{r}_0$ is the value of $\mathbf{r}$ when $t = 0$.
   (d) Now discuss the motion of our particle as time goes on.

18. Show that $\dfrac{d}{dt}|\mathbf{f}(t)| = \dfrac{\mathbf{f}(t) \cdot \mathbf{f}'(t)}{|\mathbf{f}(t)|}$. Explain why it follows that $\dfrac{d}{dt}|\mathbf{f}(t)| \leq \left|\dfrac{d}{dt}\mathbf{f}(t)\right|$.

9 l'Hospital's Rule and Improper Integrals

In one way or another, everything in calculus is based on the notion of a limit. The more we know about limits, the more we know about calculus. In this chapter we will learn a very useful rule for calculating limits; in addition, we will use limits to extend the idea of the integral. The techniques we learn here will be used in the next chapter, too.

74. l'HOSPITAL'S RULE

If f and g are functions with limits at a number a, then Theorem 11-3 tells us that

$$(74\text{-}1) \qquad \lim_{x \to a} \frac{f(x)}{g(x)} = \frac{\lim_{x \to a} f(x)}{\lim_{x \to a} g(x)},$$

provided, of course, that $\lim_{x \to a} g(x) \neq 0$. If $\lim_{x \to a} g(x) = 0$, the right-hand side of Equation 74-1 is meaningless although the left-hand side may be a perfectly definite number. The prime example of the breakdown of Equation 74-1 is given by the limit of the difference quotient. Thus if f is differentiable at a, then $\lim_{x\to a} \dfrac{f(x) - f(a)}{x - a} = f'(a)$, but we cannot find $f'(a)$ by dividing $\lim_{x\to a} [f(x) - f(a)]$ by $\lim_{x\to a} (x - a)$, for both limits are 0. Another simple example is furnished by the quotient

$$\frac{\sin^2 x}{1 - \cos x}.$$

As x approaches 0, both $\sin^2 x$ and $1 - \cos x$ approach 0, and hence we cannot use Equation 74-1 to find the limit of this quotient as x approaches 0. However, if we write the given quotient as

$$\frac{\sin^2 x}{1 - \cos x} = \frac{1 - \cos^2 x}{1 - \cos x} = \frac{(1 - \cos x)(1 + \cos x)}{1 - \cos x} = 1 + \cos x,$$

we see that

$$\lim_{x\to 0} \frac{\sin^2 x}{1 - \cos x} = \lim_{x\to 0} (1 + \cos x) = 1 + 1 = 2.$$

Equation 74-1 fails when $\lim_{x\to a} g(x) = 0$, and then the only possibility that $\lim_{x\to a} f(x)/g(x)$ exists is that $\lim_{x\to a} f(x) = 0$ also. In that event, **l'Hospital's Rule** tells us that we should look at the limit of the quotient $f'(x)/g'(x)$ in order to find the limit of the given quotient $f(x)/g(x)$.

Theorem   74-1.   *Let f and g be differentiable in a punctured neighborhood of a and suppose that $\lim_{x\to a} f(x) = 0$ and $\lim_{x\to a} g(x) = 0$. Suppose also that $g'(x) \neq 0$ in this neighborhood. Then*

(74-2)
$$\lim_{x\to a} \frac{f(x)}{g(x)} = \lim_{x\to a} \frac{f'(x)}{g'(x)}$$

if the limit on the right-hand side of this equation exists. The symbols $x \to a$ may be replaced by $x \uparrow a$ or by $x \downarrow a$ in the appropriate circumstances.

Figure 74-1 will help us see why l'Hospital's Rule works. It shows the arc $\mathbf{r} = f(t)\mathbf{i} + g(t)\mathbf{j}$ for t in an interval with endpoints a and a second point b of our given neighborhood of a. The hypotheses $\lim_{x\to a} f(x) = \lim_{x\to a} g(x) = 0$ tell us that $\lim_{t\to a} \mathbf{r} = \mathbf{0}$, as illustrated. It is clear from the figure that there must be a point of

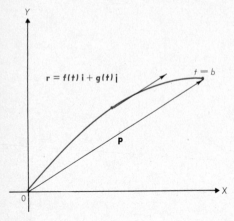

Figure 74-1

the arc (corresponding to some parameter value $t = m$) at which a tangent vector is parallel to the position vector

$$\mathbf{p} = f(b)\mathbf{i} + g(b)\mathbf{j}.$$

We can take as such a tangent vector the derivative

$$f'(m)\mathbf{i} + g'(m)\mathbf{j},$$

and to say that this vector is parallel to $\mathbf{p}$ is to say that there is a scalar r such that

$$f(b)\mathbf{i} + g(b)\mathbf{j} = rf'(m)\mathbf{i} + rg'(m)\mathbf{j}.$$

This single equation between vectors is equivalent to two equations between their components:

$$f(b) = rf'(m) \quad \text{and} \quad g(b) = rg'(m).$$

When we divide the second of these equations by the first, we find that

(74-3)
$$\frac{f(b)}{g(b)} = \frac{f'(m)}{g'(m)}.$$

Since the number m is "trapped" between a and b, it follows that

$$\lim_{b \to a} \frac{f(b)}{g(b)} = \lim_{m \to a} \frac{f'(m)}{g'(m)}$$

if the limit on the right-hand side of this equation exists. This equation is simply Equation 74-2 written in different letters.

Example 74-1. Use l'Hospital's Rule to find $\displaystyle\lim_{x \to 0} \frac{\sin^2 x}{1 - \cos x}$.

Solution. We notice that $\displaystyle\lim_{x \to 0} \sin^2 x = \lim_{x \to 0} (1 - \cos x) = 0$, so l'Hospital's Rule applies and

$$\lim_{x \to 0} \frac{\sin^2 x}{1 - \cos x} = \lim_{x \to 0} \frac{\dfrac{d}{dx} \sin^2 x}{\dfrac{d}{dx} (1 - \cos x)} = \lim_{x \to 0} \frac{2 \sin x \cos x}{\sin x}$$

$$= \lim_{x \to 0} 2 \cos x = 2,$$

as we found earlier.

Example 74-2. Find $\displaystyle\lim_{x \to \frac{1}{2}\pi} \frac{(2x - \pi)(\sin x - 1)}{\cos^2 x}$.

462

Solution.   When we apply l'Hospital's Rule, we get

$$\lim_{x \to \frac{1}{2}\pi} \frac{(2x - \pi)(\sin x - 1)}{\cos^2 x} = \lim_{x \to \frac{1}{2}\pi} \frac{(2x - \pi)\cos x + 2(\sin x - 1)}{-2\cos x \sin x}.$$

Again both the numerator and denominator have the limit 0 at $\frac{1}{2}\pi$, so we apply l'Hospital's Rule once more, and we have

$$\lim_{x \to \frac{1}{2}\pi} \frac{(2x - \pi)\cos x + 2(\sin x - 1)}{-2\cos x \sin x} = \lim_{x \to \frac{1}{2}\pi} \frac{4\cos x - (2x - \pi)\sin x}{2(\sin^2 x - \cos^2 x)} = \frac{0}{2} = 0.$$

We can often use l'Hospital's Rule even when the expression whose limit we seek is not presented to us as a quotient of two expressions with limit 0. For example, we might want the limit of the *difference* of two expressions that become large or the *product* of an expression that becomes small by one that becomes large. In such cases, we must rewrite the given expression as a quotient before we can apply l'Hospital's Rule.

Example 74-3.   Find $\lim\limits_{x \to 0} \left(\csc x - \dfrac{1}{x} \right).$

Solution.   If x is a number that is near 0, both $\csc x$ and $1/x$ have large absolute values. For example, when $x = .04$, then $\csc x = 25.01$ and $1/x = 25$. Their difference, however (as our numerical values suggest), is small, as we shall now see. We first write

$$\csc x - \frac{1}{x} = \frac{1}{\sin x} - \frac{1}{x} = \frac{x - \sin x}{x \sin x}.$$

Now we have a quotient, so we can apply l'Hospital's Rule (twice):

$$\lim_{x \to 0} \left(\csc x - \frac{1}{x} \right) = \lim_{x \to 0} \frac{x - \sin x}{x \sin x}$$

$$= \lim_{x \to 0} \frac{1 - \cos x}{\sin x + x \cos x}$$

$$= \lim_{x \to 0} \frac{\sin x}{2 \cos x - x \sin x}$$

$$= \frac{0}{2} = 0.$$

We are sometimes called on to evaluate limits of exponential expressions of the form $\lim\limits_{x \to a} f(x)^{g(x)}$, where the "natural" formula $[\lim\limits_{x \to a} f(x)]^{\lim\limits_{x \to a} g(x)}$ does not apply. For example, both $f(x)$ and $g(x)$ may approach 0, or $f(x)$ may approach 1 while $g(x)$ gets large. In such cases, we set $y = f(x)^{g(x)}$, and thus $\ln y = g(x) \ln f(x)$. Then we write $g(x) \ln f(x)$ as a quotient and apply l'Hospital's Rule to find

$$\lim_{x \to a} \ln y = L.$$

Since $y = e^{\ln y}$, and the exponential function is continuous,

$$\lim_{x \to a} y = e^{(\lim_{x \to a} \ln y)} = e^{L}.$$

Thus the limit of our original exponential expression is e^{L}.

Example 74-4. Find $\lim_{x \to \frac{1}{2}\pi} (1 + \cos x)^{\sec x}$.

Solution. If x is near $\frac{1}{2}\pi$, it is not obvious what number is approximated by the exponential $(1 + \cos x)^{\sec x}$. For example, if $x = 1.57$, then $1 + \cos x = 1.0008$ and $\sec x = 1256$. Are we to conclude that $(1.0008)^{1256}$ is near 1 because 1.0008 is near 1, or is it large because 1256 is large? Perhaps there is some middle ground. To find the required limit, we set

$$y = (1 + \cos x)^{\sec x},$$

and so

$$\ln y = \sec x \ln (1 + \cos x).$$

In order to use l'Hospital's Rule, we must now write this product as a quotient:

$$\ln y = \frac{\ln (1 + \cos x)}{\cos x}.$$

Therefore

$$\lim_{x \to \frac{1}{2}\pi} \ln y = \lim_{x \to \frac{1}{2}\pi} \frac{\ln (1 + \cos x)}{\cos x}$$

$$= \lim_{x \to \frac{1}{2}\pi} \frac{-\sin x/(1 + \cos x)}{-\sin x}$$

$$= \lim_{x \to \frac{1}{2}\pi} \frac{1}{(1 + \cos x)} = 1.$$

Thus we see that $\lim_{x \to \frac{1}{2}\pi} y = e^{1} = e$, and so we have shown that

$$\lim_{x \to \frac{1}{2}\pi} (1 + \cos x)^{\sec x} = e.$$

PROBLEMS 74

1.   For the following choices of f, g, a, and b, find m in Equation 74-3 and make a sketch like Fig. 74-1.

(a) $f(x) = 2 \cos x - 2$, $g(x) = \sin x$, $a = 0$, $b = \frac{1}{2}\pi$

(b) $f(x) = e^{4x} - 1$, $g(x) = e^{2x} - 1$, $a = 0$, $b = \ln 2$

(c) $f(x) = \sin x$, $g(x) = 1 - \cos x$, $a = 0$, $b = \frac{1}{2}\pi$

(d) $f(x) = \sin x$, $g(x) = 1 - \cos 2x$, $a = 0$, $b = \frac{1}{2}\pi$

2. Find the following limits.

(a) $\displaystyle \lim_{x \to 0} \frac{e^{3x} - \cos x}{\tan x}$

(b) $\displaystyle \lim_{x \to 0} \frac{x + \tan x}{\sin 3x}$

(c) $\displaystyle \lim_{x \to -1} \frac{x^2 - 3x - 4}{x^2 + 4x + 3}$

(d) $\displaystyle \lim_{x \to 3} \frac{x^3 + x^2 - 7x - 15}{x^3 - 5x^2 + 8x - 6}$

(e) $\displaystyle \lim_{x \to \pi} \frac{\ln x - \ln \pi}{\sin 2x}$

(f) $\displaystyle \lim_{x \to \pi} \frac{\ln x - \ln \pi}{\cos \frac{1}{2}x}$

(g) $\displaystyle \lim_{x \to 0} \frac{(1 + x)^3 - (1 - x)^3}{(1 + x)^5 - (1 - x)^5}$

(h) $\displaystyle \lim_{x \to 0} \frac{\sqrt[3]{1 + x} - \sqrt[3]{1 - x}}{\sqrt[5]{1 + x} - \sqrt[5]{1 - x}}$

3. Find the following limits.

(a) $\displaystyle \lim_{x \to 0} \frac{\tan x - x}{x - \sin x}$

(b) $\displaystyle \lim_{x \to \frac{1}{2}\pi} \frac{\ln \sin x}{(\pi - 2x)^2}$

(c) $\displaystyle \lim_{x \to 0} \frac{1 - \cos x}{\sin x \tan x}$

(d) $\displaystyle \lim_{x \to 0} \frac{\sinh x - \sin x}{\sin^3 x}$

(e) $\displaystyle \lim_{x \to \frac{1}{2}\pi} \frac{1 - \sin x}{1 + \cos 2x}$

(f) $\displaystyle \lim_{x \to 0} \frac{\sec x - 1}{x \sin x}$

4. Find the following limits.

(a) $\displaystyle \lim_{x \to 0} x \cot x$

(b) $\displaystyle \lim_{x \to \frac{1}{2}\pi} (x - \frac{1}{2}\pi) \tan x$

(c) $\displaystyle \lim_{x \to \frac{1}{2}\pi} (\sec x - \tan x)$

(d) $\displaystyle \lim_{x \to 0} (\csc x - \cot x)$

(e) $\displaystyle \lim_{x \to 0} [x^{-1} - (e^x - 1)^{-1}]$

(f) $\displaystyle \lim_{x \to 0} (x^{-2} - \csc^2 x)$

5. Find the following limits.

(a) $\displaystyle \lim_{x \to 0} (1 + x^{12})^{1/x}$

(b) $\displaystyle \lim_{x \to 0} (\sec x + \tan x)^{\csc x}$

(c) $\displaystyle \lim_{x \to 0} (1 + \sin x)^{1/x}$

(d) $\displaystyle \lim_{x \to \frac{1}{2}\pi} (\sin x)^{\tan x}$

6. Show that $\displaystyle \lim_{x \downarrow 0} (1 + x^r)^{1/x} = 1$ if $r > 1$. What is the limit if $r = 1$?

7. Find the following limits.

(a) $\displaystyle \lim_{x \to 0} \frac{\int_0^x \exp(-t^2)\, dt}{\int_0^x (\cos t^3 + \cos^3 t)\, dt}$

(b) $\displaystyle \lim_{x \to 0} \frac{1}{x^3} \int_0^x \sin t^2 \, dt$

8. Let $h(\theta)$ be the vertical distance from the polar axis to a point of the spiral whose polar coordinate equation is $r\sqrt{\theta} = 5$. Find $\displaystyle \lim_{\theta \downarrow 0} h(\theta)$.

9. At a point A of a circle with a radius of a (see Fig. 74-2), we draw a tangent line and a diameter. Next we pick a point P, different from A, of the circle. Then we choose a point Q of the tangent line so that the distance $\overline{AQ}$ equals the length of the arc AP. The line that contains P and Q intersects the diameter in a point R. Find the limiting position of R as P approaches A.

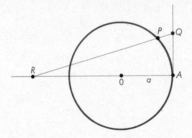

Figure 74-2

10. The current flowing t seconds after a switch is thrown in a series circuit containing a resistor of R ohms resistance and a coil of L henries inductance connected to a battery of E volts is given by the equation $I = (E/R)(1 - e^{-Rt/L})$. Find the limit of I as R approaches zero.

11. Find $\displaystyle\lim_{x \to a} \int_a^x \sin f(t)\, dt \Big/ \sin \int_a^x f(t)\, dt$ if f is continuous in some open interval that contains a.

12. Suppose that f and f' are continuous at 0 and that $f(0) = 0$ but $f'(0) \neq 0$. Find the following limits.

(a) $\displaystyle\lim_{x \to 0} \frac{f(\sin x)}{f(\tan x)}$ (b) $\displaystyle\lim_{x \to 0} \frac{\sin f(x)}{f(\tan x)}$ (c) $\displaystyle\lim_{x \to 0} \frac{f(\sin x)}{\tan f(x)}$

(d) $\displaystyle\lim_{x \to 0} \frac{f(\sin x)}{\sin f(x)}$ (e) $\displaystyle\lim_{x \to 0} \frac{f(\sin x)^2}{f(\tan x)}$ (f) $\displaystyle\lim_{x \to 0} \frac{f(\sin x^2)}{f(\tan x)}$

75. EXTENDED LIMITS

When we introduced the symbols ∞ and $-\infty$ in Chapter 1, we were careful to point out that *they do not stand for numbers.* Nevertheless, we find them useful in denoting intervals that do not have one or the other endpoint. Now let us carry that notation one step further.

A finite open interval (a, b) is a neighborhood of any one of its members. For example, $(3, 4)$ is a neighborhood of π. It is also geometrically natural to call (a, b) a **right neighborhood** of a and a **left neighborhood** of b. Analogously, let us say that an interval $(-\infty, c)$ is a right neighborhood of $-\infty$ and an interval (c, ∞) is a left neighborhood of ∞. This new terminology still does not make $-\infty$ and ∞ represent numbers, but we can use it to make worthwhile extensions of the limit concept.

If c and L are numbers and f is a function, the equation

$$(75\text{-}1) \qquad\qquad \lim_{x \uparrow c} f(x) = L$$

means that L is the number that is approximated by $f(x)$ when x is close to, but less than, c. When translated into mathematics, these words say that each neighborhood of L contains the image under f of a left neighborhood of c. But this statement also makes sense when c is the symbol ∞. Furthermore, if we agree that when the word "neighborhood" is applied to one of the symbols $-\infty$ or ∞ it automatically means right or left neighborhood, then L could be one of the symbols $-\infty$ or ∞. In this chapter and the next we will find it convenient to talk about such **extended limits**. Consequently, when you see Equation 71-1 or the similar equation $\lim\limits_{x\downarrow c} f(x) = L$, you should check the context to see if perhaps one or both the symbols $-\infty$ or ∞ is allowed.

Our definition of extended limits is too formal. We need a rough-and-ready version. Thus if L is a number and $c = \infty$, then Equation 75-1 says that L is the number that is approximated by $f(x)$ when x is large. In the same way, if $L = \infty$ and c is a number, the equation tells us that we can ensure that $f(x)$ is large by choosing x close to, but less than, c. It is more or less obvious that $\lim\limits_{x\uparrow\infty} f(x) = \lim\limits_{h\downarrow 0} f(1/h)$, a fact that is frequently useful when proving limit theorems. In many cases, the limit theorems of Chapter 1 apply directly to extended limits, and we will use these theorems without further comment.

Example 75-1.   Find   (a) $\lim\limits_{x\uparrow\infty} (x^2 + 2x - 1)/(3x^2 - 5)$,   (b) $\lim\limits_{x\downarrow 0} e^{1/x}$,   and   (c) $\lim\limits_{x\uparrow\infty} \ln x$.

Solution.   (a) Let us first divide the numerator and denominator of our given fraction by x^2:

$$\left(1 + \frac{2}{x} - \frac{1}{x^2}\right)\Big/\left(3 - \frac{5}{x^2}\right).$$

Then we observe that when x is very large, the numerator approximates 1 and the denominator approximates 3, so we conclude that

$$\lim\limits_{x\uparrow\infty} (x^2 + 2x - 1)/(3x^2 - 5) = \tfrac{1}{3}.$$

(b) When x is close to 0, but positive, $1/x$ is very large. So, then, is $e^{1/x}$, and hence we conclude that $\lim\limits_{x\downarrow 0} e^{1/x} = \infty$.

(c) We can ensure that $\ln x$ is large by taking x to be large, which is an informal way of saying that $\lim\limits_{x\uparrow\infty} \ln x = \infty$.

One reason we could find the limit of $\ln x$ at a glance is that the logarithm function is **nondecreasing** (that is, if $u < v$, then $\ln u \le \ln v$). It is relatively easy to give a complete description of the limits of such functions. Figure 75-1 shows the graph of a function f that is nondecreasing in an interval (a, b). The interval $[A, B]$ is the smallest closed interval that contains the image of (a, b) under f,

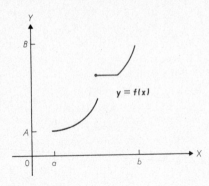

Figure 75-1

and it is apparent that $\lim_{x \uparrow b} f(x) = B$ and $\lim_{x \downarrow a} f(x) = A$. If you look at the graphs of specific nondecreasing functions—Arctan, $\sqrt{\ }$, $[\![\ ]\!]$, Tan—you will believe the following theorem. We will omit its proof.

Theorem 75-1. *Suppose that f is nondecreasing in an interval (a, b). According to the Completeness Property of the real numbers (Property 2-1), the image of (a, b) under f is contained in a smallest closed interval [A, B]. It is true that $\lim_{x \downarrow a} f(x) = A$ and $\lim_{x \uparrow b} f(x) = B$.*

Theorem 75-1 allows us to write an equation $\lim_{x \uparrow b} f(x) = B$ for a nondecreasing function f, but B may be either a number or the symbol ∞. For example, Arctan is a nondecreasing function, and $\lim_{x \uparrow \infty} \text{Arctan } x$ is the number $\frac{1}{2}\pi$. But $\lim_{x \uparrow \frac{1}{2}\pi} \text{Tan } x = \infty$. In general, either our function f is *bounded above* in the interval (a, b) [that is, there is a number M such that $f(x) \le M$ for each $x \in (a, b)$], or it isn't. If f is bounded above, then $f(x)$ can't get too large, and so $\lim_{x \uparrow b} f(x)$ is a number and not the symbol ∞. If f is unbounded above, then $\lim_{x \uparrow b} f(x) = \infty$. Because it is important, we state this obvious result as a theorem.

Theorem 75-2. *Suppose that f is nondecreasing in an interval (a, b). Then there is a number B such that $\lim_{x \uparrow b} f(x) = B$ if and only if f is bounded above in (a, b).*

Example 75-2. Apply the preceding theorem to $\lim_{x \uparrow \infty} (-x^3 e^{-x})$.

Solution. If we write $f(x) = -x^3 e^{-x}$, then $f'(x) = -3x^2 e^{-x} + x^3 e^{-x} = x^2 e^{-x}(x - 3)$. Hence $f'(x) > 0$ if $x > 3$, so f is nondecreasing in the interval $(3, \infty)$. Furthermore, it is clear that $f(x) \le 0$; that is, f is bounded above. Therefore Theorem 75-2 tells us that there is a number B such that $\lim_{x \uparrow \infty} (-x^3 e^{-x}) = B$. The theorem doesn't tell us what B is, and perhaps it is not obvious, since x^3 is large and e^{-x} is small when x is large. We will find B in Example 75-4.

One way to find limits "at infinity" is to use l'Hospital's Rule. In the problems at the end of the section we will give you a hint and ask you to show that Theorem 74-1 remains valid when the symbols $x \to a$ are replaced by $x \uparrow \infty$.

468

Example 75-3.   Find $\lim\limits_{x \uparrow \infty} x(\tfrac{1}{2}\pi - \text{Arctan } x)$.

Solution.   We apply l'Hospital's Rule, after first writing our given expression as a quotient and observing that both numerator and denominator approach 0:

$$\lim_{x \uparrow \infty} x(\tfrac{1}{2}\pi - \text{Arctan } x) = \lim_{x \uparrow \infty} (\tfrac{1}{2}\pi - \text{Arctan } x)/x^{-1}$$

$$= \lim_{x \uparrow \infty} \frac{d}{dx}(\tfrac{1}{2}\pi - \text{Arctan } x)\Big/ \frac{d}{dx}x^{-1}$$

$$= \lim_{x \uparrow \infty} [-(1 + x^2)^{-1}/(-x^{-2})] = \lim_{x \uparrow \infty} x^2/(1 + x^2) = 1.$$

l'Hospital's Rule also remains valid when the hypotheses $\lim\limits_{x \to a} f(x) = 0$ and $\lim\limits_{x \to a} g(x) = 0$ are replaced by the assumptions that $\lim\limits_{x \to a} f(x) = \infty$ (or $-\infty$) and $\lim\limits_{x \to a} g(x) = \infty$ (or $-\infty$). This theorem is considerably harder to prove than our original statement of l'Hospital's Rule. We will leave its proof to a course in advanced calculus and simply state it and show you how it works.

Theorem   75-3.   *Let f and g be differentiable in a punctured neighborhood of a and suppose that $\lim\limits_{x \to a} f(x) = \infty$ (or $-\infty$) and $\lim\limits_{x \to a} g(x) = \infty$ (or $-\infty$). Suppose also that $g'(x) \neq 0$ in this neighborhood. If*

$$\lim_{x \to a} \frac{f'(x)}{g'(x)} = L, \quad \text{then} \quad \lim_{x \to a} \frac{f(x)}{g(x)} = L.$$

The symbols $x \to a$ can be replaced by $x \uparrow a$ or $x \downarrow a$, and a or L (or both) may stand for $-\infty$ or ∞.

Example 75-4.   Find $\lim\limits_{x \uparrow \infty} x^m e^{-x}$, where m is a positive integer.

Solution.   We write $x^m e^{-x}$ as x^m/e^x. Since $\lim\limits_{x \uparrow \infty} x^m = \infty$ and $\lim\limits_{x \uparrow \infty} e^x = \infty$, Theorem 75-3 applies:

$$\lim_{x \uparrow \infty} x^m/e^x = \lim_{x \uparrow \infty} \frac{d}{dx}x^m \Big/ \frac{d}{dx}e^x = \lim_{x \uparrow \infty} mx^{m-1}/e^x.$$

If $m = 1$, this limit is clearly 0, but if $m > 1$, $\lim\limits_{x \uparrow \infty} mx^{m-1} = \infty$, and we must apply l'Hospital's Rule again:

$$\lim_{x \uparrow \infty} mx^{m-1}/e^x = \lim_{x \uparrow \infty} \frac{d}{dx}mx^{m-1} \Big/ \frac{d}{dx}e^x = \lim_{x \uparrow \infty} m(m-1)x^{m-2}/e^x.$$

If $m = 2$, this limit (and hence our original limit) is 0, but if $m > 2$,

$$\lim_{x \uparrow \infty} m(m - 1)x^{m-2} = \infty,$$

and we must apply l'Hospital's Rule once more. It should now be clear that to solve our problem, we must apply l'Hospital's Rule m times in all, and then we have

$$\lim_{x \uparrow \infty} x^m/e^x = \lim_{x \uparrow \infty} m!/e^x = 0.$$

Example 75-5. Find $f'(0)$ if $f(x) = e^{-1/x^2}$ for $x \neq 0$ and $f(0) = 0$.

Solution. By definition,

$$f'(0) = \lim_{h \to 0} \frac{f(h) - f(0)}{h} = \lim_{h \to 0} \frac{e^{-1/h^2} - 0}{h} = \lim_{h \to 0} \frac{e^{-1/h^2}}{h}.$$

Both the numerator and the denominator of this fraction approach 0 as h approaches 0, so we apply l'Hospital's Rule:

$$\lim_{h \to 0} \frac{e^{-1/h^2}}{h} = \lim_{h \to 0} \frac{(2/h^3)e^{-1/h^2}}{1} = \lim_{h \to 0} \frac{2e^{-1/h^2}}{h^3}.$$

Since this last limit is at least as hard to evaluate as the original, we must change our tactics. We set $x = 1/h$. Then

$$\lim_{h \to 0} \frac{e^{-1/h^2}}{h} = \lim_{|x| \uparrow \infty} xe^{-x^2} = \lim_{|x| \uparrow \infty} \frac{x}{e^{x^2}}.$$

Now we fare better when we apply l'Hospital's Rule:

$$\lim_{|x| \uparrow \infty} \frac{x}{e^{x^2}} = \lim_{|x| \uparrow \infty} \frac{1}{2xe^{x^2}} = 0.$$

Thus we have shown that $f'(0) = 0$.

PROBLEMS 75

1. Find the following limits "by inspection."

(a) $\displaystyle\lim_{x \uparrow \infty} \frac{3x^3 - 2x^2 + 1}{7x^3 + x - 3}$

(b) $\displaystyle\lim_{x \downarrow -\infty} \frac{3x^3 - 2x^2 + 1}{7x^3 + x - 3}$

(c) $\displaystyle\lim_{x \downarrow -\infty} \text{Arctan } 2x$

(d) $\displaystyle\lim_{x \uparrow \infty} 2 \text{ Arctan } x$

(e) $\lim_{x\uparrow\infty} \tanh x$

(f) $\lim_{x\downarrow-\infty} \tanh x$

(g) $\lim_{x\downarrow 0} \exp 1/x$

(h) $\lim_{x\uparrow 0} \exp 1/x$

(i) $\lim_{x\uparrow\infty} \text{Tan}^{-1} x^2$

(j) $\lim_{x\uparrow\infty} (\text{Tan}^{-1} x)^2$

2. Find the following limits.

(a) $\lim_{x\uparrow\infty} \dfrac{\ln(1492x + 1776)}{\ln(1776x + 1492)}$

(b) $\lim_{x\uparrow\infty} \dfrac{x - \sin x}{2x}$

(c) $\lim_{x\uparrow\infty} x^{234}e^{-x}$

(d) $\lim_{x\downarrow-\infty} x^{234}e^x$

(e) $\lim_{x\uparrow\infty} \dfrac{\sin e^{-x}}{\sin(1/x)}$

(f) $\lim_{x\uparrow\infty} e^{-x} \sin e^x$

3. Find the following limits.

(a) $\lim_{x\uparrow\infty} \dfrac{\ln x}{x}$

(b) $\lim_{x\uparrow\infty} \dfrac{\ln \sqrt{x}}{\sqrt{x}}$

(c) $\lim_{x\uparrow\infty} \dfrac{\ln \sqrt{x}}{x}$

(d) $\lim_{x\uparrow\infty} \dfrac{\ln x}{\sqrt{x}}$

4. Find the following limits.

(a) $\lim_{x\uparrow 0} xe^{1/x}$

(b) $\lim_{x\downarrow 0} xe^{1/x}$

(c) $\lim_{x\uparrow\infty} xe^{1/x}$

(d) $\lim_{x\downarrow-\infty} xe^{1/x}$

5. Find the following limits.

(a) $\lim_{x\downarrow 0} x^x$

(b) $\lim_{x\uparrow\infty} x^{1/x}$

(c) $\lim_{x\uparrow\infty} (\cos 2/x)^x$

(d) $\lim_{x\uparrow\infty} x^{\cos 2/x}$

6. Find the following limits.

(a) $\lim_{x\uparrow\infty} \dfrac{\displaystyle\int_0^x \exp t^2 \, dt}{\exp x^2}$

(b) $\lim_{x\uparrow\infty} \dfrac{\displaystyle\int_3^x e^t(4t^2 + 3t - 1) \, dt}{\displaystyle\int_5^x e^t(2t^2 + 5t + 6) \, dt}$

(c) $\lim_{x\uparrow\infty} \dfrac{\displaystyle\int_0^{[x]} t \, dt}{x^2}$

(d) $\lim_{x\uparrow\infty} \dfrac{\displaystyle\int_0^x [\![t]\!] \, dt}{x^2}$

7. Let $f(x) = (1 + 1/x)^x$ and $g(x) = (1 + 1/x)^{x+1}$. Show that f is increasing in $(1, \infty)$ and g is decreasing. Find $\lim_{x\uparrow\infty} f(x)$ and $\lim_{x\uparrow\infty} g(x)$.

8. Show that $\lim_{x\downarrow 0} \ln x = -\infty$. What is $\ln 1/10{,}000{,}000$?

9. Show that $\lim_{x\uparrow\infty} (1 + r^x)^{1/x} = 1$ if $0 \le r \le 1$. What is the limit if $r > 1$?

10. Find $\lim_{x\uparrow\infty} x^m e^{-x}$ if m is not necessarily a positive integer.

11. Prove l'Hospital's Rule with the symbols $x \to a$ replaced by $x \uparrow \infty$. [*Hint*: Set $h = 1/x$.]

12. Show that for each real number a, $\lim\limits_{x \uparrow \infty} [\![xa]\!](a/x) = a^2$.

13. (a) State the obvious analogs of Theorems 75-1 and 75-2 if f is nonincreasing in an interval (a, b).

(b) What can you say about $\lim\limits_{x \downarrow a} f(x)$ if f is nondecreasing in (a, b)? If f is nonincreasing?

76. IMPROPER INTEGRALS

An integral $\displaystyle\int_a^b f(x)\, dx$ is a *number*, the limit of approximating sums that we form according to the rules laid down in Chapter 5. If these sums do not have a limit, then f is not integrable on $[a, b]$, and we say that $\displaystyle\int_a^b f(x)\, dx$ "does not exist." Now let us look at an example of a different use of the integral sign.

If R is any positive number, then

$$\int_1^R \frac{1}{x^2}\, dx = -\frac{1}{x}\bigg|_1^R = 1 - \frac{1}{R}.$$

Thus since $\lim\limits_{R \uparrow \infty} (1 - 1/R) = 1$, we see that $\lim\limits_{R \uparrow \infty} \displaystyle\int_1^R x^{-2}\, dx = 1$, and we write this equation in the abbreviated form

$$\int_1^\infty \frac{1}{x^2}\, dx = 1.$$

Unlike "proper" integrals, this integral is not a limit of approximating sums. Nevertheless, such "improper integrals" are useful in mathematical applications. For example, suppose that the force between two unit charges x meters apart is x^{-2} newtons (the inverse square law). Then if the charges were initially 1 meter apart, our improper integral $\displaystyle\int_1^\infty x^{-2}\, dx$ represents the amount of work (in joules) necessary to move one of them "infinitely" far away from the other.

In general, suppose that a function f is integrable on the interval $[a, R]$ for every number $R > a$ and that $\lim\limits_{R \uparrow \infty} \displaystyle\int_a^R f(x)\, dx = L$. Then we write

(76-1)
$$\int_a^\infty f(x)\, dx = L.$$

If L is a number, we say that the **improper integral** $\displaystyle\int_a^\infty f(x)\,dx$ is **convergent**. If the limit does not exist or if L is one of the symbols $-\infty$ or ∞, the integral is **divergent**.

Example 76-1. Test the integrals $\displaystyle\int_0^\infty e^{-x}\,dx$ and $\displaystyle\int_{\pi/2}^\infty \cos x\,dx$ for convergence.

Solution. According to Equation 76-1, we are to find $\displaystyle\int_0^R e^{-x}\,dx$ and examine the limit of this expression. Thus

$$\int_0^R e^{-x}\,dx = 1 - e^{-R},$$

and so

$$\lim_{R\uparrow\infty}\int_0^R e^{-x}\,dx = \lim_{R\uparrow\infty}(1 - e^{-R}) = 1.$$

Therefore $\displaystyle\int_0^\infty e^{-x}\,dx = 1$; the integral converges.

On the other hand, $\displaystyle\int_{\pi/2}^R \cos x\,dx = \sin R - 1$. Since $\lim_{R\uparrow\infty}(\sin R - 1)$ does not exist, the improper integral $\displaystyle\int_{\pi/2}^R \cos x$ is divergent.

Example 76-2. For what choices of p is the integral $\displaystyle\int_1^\infty \frac{dx}{x^p}$ convergent?

Solution. We must consider three cases: $p > 1$, $p < 1$, and $p = 1$. If $p \neq 1$, then

$$\int_1^R \frac{dx}{x^p} = \int_1^R x^{-p}\,dx = \frac{x^{1-p}}{1-p}\Big|_1^R = \frac{R^{1-p} - 1}{1-p},$$

and so we must investigate

$$\lim_{R\uparrow\infty}\frac{R^{1-p} - 1}{1-p}.$$

In case $p > 1$, the exponent of R is negative, so $\lim_{R\uparrow\infty} R^{1-p} = 0$, and hence

$$\lim_{R\uparrow\infty}\frac{R^{1-p} - 1}{1-p} = \frac{-1}{1-p} = \frac{1}{p-1}.$$

If $p < 1$, the exponent of R is positive, and

$$\lim_{R\uparrow\infty}\frac{R^{1-p} - 1}{1-p} = \infty.$$

The case $p = 1$ must be handled separately. Here $\int_1^R x^{-1} \, dx = \ln x \Big|_1^R$, and hence $\lim\limits_{R\uparrow\infty} \int_1^R x^{-1} \, dx = \infty$. Summarizing,

$$\int_1^\infty \frac{dx}{x^p} \begin{cases} = \dfrac{1}{p-1} & \text{if } p > 1, \\ \text{diverges} & \text{if } p \le 1. \end{cases}$$

If f is integrable on each interval $[R, a]$ and if $\lim\limits_{R\downarrow-\infty} \int_R^a f(x) \, dx = L$, we write $\int_{-\infty}^a f(x) \, dx = L$. It is easy to see that $\int_{-\infty}^a f(x) \, dx = \int_{-a}^\infty f(-x) \, dx$. Also, let us observe that an improper integral $\int_a^\infty f(x) \, dx$ converges if and only if the integral $\int_b^\infty f(x) \, dx$ converges for each number $b \ge a$. Then

$$(76\text{-}2) \qquad \int_a^\infty f(x) \, dx = \int_a^b f(x) \, dx + \int_b^\infty f(x) \, dx.$$

For example,

$$\int_{-\infty}^0 e^x \, dx = \int_0^\infty e^{-x} \, dx = \int_0^1 e^{-x} \, dx + \int_1^\infty e^{-x} \, dx.$$

The integral $\int_a^\infty f(x) \, dx$ is improper because it does not have a finite interval of integration. Here is an example of another type of improper integral. If t is any number between 0 and 1, then

$$\int_0^t \frac{dx}{\sqrt{1-x^2}} = \text{Arcsin } x \Big|_0^t = \text{Arcsin } t.$$

Therefore

$$\lim_{t\uparrow 1} \int_0^t \frac{dx}{\sqrt{1-x^2}} = \lim_{t\uparrow 1} \text{Arcsin } t = \tfrac{1}{2}\pi.$$

Because of this limit relation, we will write $\int_0^1 \frac{dx}{\sqrt{1-x^2}} = \tfrac{1}{2}\pi$, but we must note that the "integral" $\int_0^1 \frac{dx}{\sqrt{1-x^2}}$ is not a proper integral as we defined the term in Chapter 5. The integrand $(1 - x^2)^{-1/2}$ is unbounded in our interval of integration, and unbounded functions are not integrable.

In general, suppose that f is integrable on every interval $[a, t]$, where $a < t < b$, but that f is unbounded in the interval $[a, b)$. If $\lim\limits_{t\uparrow b} \int_a^t f(x) \, dx = L$, we write $\int_a^b f(x)\, dx = L$, but we are dealing with an improper integral. Similarly, if f is integrable on every interval $[t, b]$, where $a < t < b$, but is unbounded in $(a, b]$, then $\int_a^b f(x)\, dx = L$ means $\lim_{t \downarrow a} \int_t^b f(x)\, dx = L$. If L stands for ∞ or $-\infty$, or if the limit does not exist, then the improper integral $\int_a^b f(x)\, dx$ is *divergent*.

Example 76-3. Test the integral $\int_0^1 \dfrac{dx}{x}$ for convergence.

Solution. The integrand $1/x$ is unbounded in the interval $(0, 1]$, and

$$\lim_{t \downarrow 0} \int_t^1 \frac{dx}{x} = \lim_{t \downarrow 0} (-\ln t) = \infty.$$

Thus the integral $\int_0^1 \dfrac{dx}{x}$ is divergent.

Example 76-4. For what choices of p is the integral $\int_0^1 \dfrac{dx}{x^p}$ convergent?

Solution. We have just seen that this integral is divergent for $p = 1$. Furthermore, it is a proper integral for $p \leq 0$, and its value is $1/(1 - p)$. Therefore we need only examine the case in which p is a positive number different from 1. In that case, the integrand is unbounded near 0, and so we look at the integral

$$\int_t^1 \frac{dx}{x^p} = \frac{x^{1-p}}{1 - p}\Big|_t^1 = \frac{1 - t^{1-p}}{1 - p}.$$

If $0 < p < 1$, then the exponent of t is positive, and so t^{1-p} approaches 0. Thus

$$\lim_{t \downarrow 0} \frac{1 - t^{1-p}}{1 - p} = \frac{1}{1 - p}.$$

But if $p > 1$, the exponent of t is negative, and so

$$\lim_{t \downarrow 0} \frac{1 - t^{1-p}}{1 - p} = \infty.$$

Hence we see that

$$\int_0^1 \frac{dx}{x^p} \begin{cases} = \dfrac{1}{1 - p} & \text{if } p < 1 \\ \text{diverges} & \text{if } p \geq 1. \end{cases}$$

If a function f is unbounded in a neighborhood of an interior point c of an interval $[a, b]$, we regard the improper integral $\int_a^b f(x)\, dx$ as the sum of two improper integrals, $\int_a^c f(x)\,dx + \int_c^b f(x)\,dx$. If both these integrals converge, we say that our original integral converges; otherwise it diverges. Thus

$$\int_0^\pi \sec^2 x\,dx = \int_0^{\frac12\pi} \sec^2 x\,dx + \int_{\frac12\pi}^\pi \sec^2 x\,dx.$$

Both of the improper integrals on the right-hand side of this equation diverge; for example,

$$\lim_{t\uparrow\frac12\pi} \int_0^t \sec^2 x\,dx = \lim_{t\uparrow\frac12\pi} \tan t = \infty.$$

Because both of the integrals into which the original integral was decomposed must converge for the original integral to converge, we see that $\int_0^\pi \sec^2 x\,dx$ is divergent. Notice that if we had ignored the fact that $\sec^2 x$ is not integrable on the interval $[0, \pi]$ and had simply tried to apply the Fundamental Theorem of Calculus, we would have obtained the incorrect result: $\int_0^\pi \sec^2 x\,dx$ is the number $\tan x\Big|_0^\pi = 0$.

PROBLEMS 76

1. Evaluate the improper integral or show that it diverges.

(a) $\displaystyle\int_4^\infty x^{-3/2}\,dx$
(b) $\displaystyle\int_4^\infty x^{-2/3}\,dx$
(c) $\displaystyle\int_5^\infty xe^{-x^2}\,dx$

(d) $\displaystyle\int_5^\infty xe^{-x}\,dx$
(e) $\displaystyle\int_5^\infty x\sin x^2\,dx$
(f) $\displaystyle\int_1^\infty x^{-2}\sin x^{-1}\,dx$

(g) $\displaystyle\int_1^\infty \frac{\ln x}{x}\,dx$
(h) $\displaystyle\int_1^\infty \frac{\ln x}{x^2}\,dx$

2. Evaluate the integral or show that it diverges.

(a) $\displaystyle\int_0^\infty (1 + x^2)^{-1}\,dx$
(b) $\displaystyle\int_0^\infty x(1 + x^2)^{-1}\,dx$

(c) $\displaystyle\int_0^\infty x^2(1 + x^2)^{-1}\,dx$
(d) $\displaystyle\int_0^\infty (1 + x^2)^{-2}\,dx$

3. Test for convergence.

(a) $\displaystyle\int_0^{\pi/2} \cot x\,dx$
(b) $\displaystyle\int_0^{\pi/2} \sec x\,dx$
(c) $\displaystyle\int_0^1 \frac{\cos\sqrt{x}}{\sqrt{x}}\,dx$

(d) $\displaystyle\int_0^1 \frac{\exp\sqrt{x}}{\sqrt{x}}\,dx$
(e) $\displaystyle\int_0^1 \ln x\,dx$
(f) $\displaystyle\int_0^1 \ln\sqrt{x}\,dx$

4. Evaluate the following improper integrals or show that they diverge.

(a) $\displaystyle\int_0^1 x^{-1/3}\, dx$

(b) $\displaystyle\int_0^1 x^{-2}\, dx$

(c) $\displaystyle\int_0^3 (9 - x^2)^{-1/2}\, dx$

(d) $\displaystyle\int_1^3 (x - 1)^{-3/2}\, dx$

(e) $\displaystyle\int_0^4 (16 - x^2)^{-1}\, dx$

(f) $\displaystyle\int_0^4 x(16 - x^2)^{-3/2}\, dx$

5. Evaluate the following improper integrals or show that they diverge. Notice that they are improper at both limits of integration.

(a) $\displaystyle\int_1^\infty (x \ln x)^{-1}\, dx$

(b) $\displaystyle\int_{-1}^1 (1 - x^2)^{-1/2}\, dx$

(c) $\displaystyle\int_{-\infty}^\infty (x^2 + 2x + 2)^{-1}\, dx$

(d) $\displaystyle\int_0^\infty x^{-1/2} e^{-x^{1/2}}\, dx$

6. Suppose that $r > 1$. Use the substitution $u = x^{1-r}$ to evaluate the integral $\displaystyle\int_1^\infty x^{-r} \ln x\, dx$. What if $r \le 1$?

7. Show that $\displaystyle\int_{-\infty}^\infty x(1 + x^2)^{-2}\, dx = 0$ but that $\displaystyle\int_{-\infty}^\infty x(1 + x^2)^{-1}\, dx$ diverges.

8. Solve for x: $\displaystyle\int_0^x \ln t\, dt = 0$.

9. (a) Explain why the integral $\displaystyle\int_{-1}^1 x^{-2}\, dx$ is improper. Does it converge?

(b) Is the integral $\displaystyle\int_0^{\pi/2} \frac{\sin \sqrt{t}}{\sqrt{t}}\, dt$ improper?

10. (a) If $\lim\limits_{R\uparrow\infty} f(R)e^{-R} = 0$, use integration by parts to derive the formula

$$\int_0^\infty f(x)e^{-x}\, dx = f(0) + \int_0^\infty f'(x)e^{-x}\, dx,$$

provided that the improper integrals converge.

(b) Use this formula to show that

$$\int_0^\infty e^{-x} \sin x\, dx = \int_0^\infty e^{-x} \cos x\, dx$$

and

$$\int_0^\infty e^{-x} \cos x\, dx = 1 - \int_0^\infty e^{-x} \sin x\, dx.$$

(c) Use part (b) to compute $\displaystyle\int_0^\infty e^{-x} \sin x\, dx$.

(d) Compute $\displaystyle\int_0^\infty x^n e^{-x}\, dx$ (n a positive integer).

11. The gravitational force of attraction between the earth and a body is inversely proportional to the square of the distance between the body and the center of the earth. Assume that the radius of the earth is 4000 miles and that the nose cone of a rocket weighs 1 ton on the surface of the earth. Find the work required to propel the nose cone out of the earth's gravitational field.

12. The force of repulsion between two positive charges q_1 and q_2 is given by the equation $F = kq_1q_2/d^2$, where d is the distance between the charges and k is a physical constant. The electric potential at a point is frequently defined as the work required to bring a unit charge from "infinity" to the point. Find a formula for the potential V at a point r units away from a charge of q units.

13. (a) Show that $\displaystyle\int_{-\infty}^{a} f(x)\, dx = \int_{-a}^{\infty} f(-x)\, dx$ (assuming that f is continuous).

    (b) Verify Equation 76-2.

14. Suppose that f is integrable (in the ordinary sense) on a finite interval $[a, b]$. Show that $\displaystyle\lim_{t\uparrow b}\int_{a}^{t} f(x)\, dx = \int_{a}^{b} f(x)\, dx$.

77. TESTING THE CONVERGENCE OF IMPROPER INTEGRALS

Equation 76-1 is very specific. It tells us to test the convergence of an improper integral by first integrating and then calculating the limit of the result. But what if we can't carry out one of these steps? For example, we can't check the convergence of the integral $\displaystyle\int_{0}^{\infty} e^{-x^2}\, dx$ by the methods of the last section, for no formula containing $\displaystyle\int e^{-x^2}\, dx$ appears in Table V.

We will now introduce some indirect techniques for testing the convergence of improper integrals. Rather than integrating and examining a limit, we convince ourselves by some means or other what the result of such an examination would be. For instance, in Example 77-2 we will show that $\displaystyle\int_{0}^{\infty} e^{-x^2}\, dx$ is, in fact, convergent. If we find that a given integral is divergent, our work is done. If it is convergent, we may be able to approximate its value or evaluate it by some of the elegant methods developed in courses in advanced calculus and complex variable theory. For brevity, we restrict our attention in this section to improper integrals of the form $\displaystyle\int_{a}^{\infty} f(x)\, dx$; it is easy to modify our technique to apply to other types.

Our integral will converge if $\lim_{R\uparrow\infty} F(R)$ is a number, where $F(R) = \displaystyle\int_{a}^{R} f(x)\, dx$. As we pointed out in Section 75, it is relatively easy to talk about limits of nondecreasing functions. So let us first consider integrals for which F is nondecreasing. We will suppose, therefore, that

$$f(x) \geq 0 \quad \text{for each } x \geq a.$$

Our situation is illustrated graphically in Fig. 77-1. The number $F(R)$ represents the area of the shaded region, and the larger R is, the larger this area is. Thus there are only two possibilites (Theorem 75-2); either

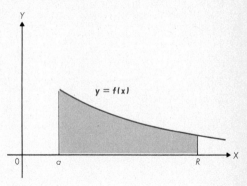

(1) $\lim\limits_{R\uparrow\infty} F(R) = \infty$, and the integral $\int_a^\infty f(x)\,dx$
    diverges, or

(2) $F(R)$ is bounded above, and the integral converges.

Example 77-1. Show that $\int_1^\infty \dfrac{dx}{x^2 + 2^x}$ converges.

Figure 77-1

Solution. In this example $F(R) = \int_1^R (x^2 + 2^x)^{-1}\,dx$, and since the integrand is not listed in Table V, we won't try to find a simpler formula for $F(R)$. Notice, however, that

$$(x^2 + 2^x)^{-1} \leq x^{-2} \quad \text{for each } x \geq 1,$$

and hence

$$F(R) \leq \int_1^R x^{-2}\,dx = 1 - \frac{1}{R} \leq 1.$$

Therefore F is nondecreasing and bounded, and so $\lim\limits_{R\uparrow\infty} F(R)$ is a number, and $\int_1^\infty (x^2 + 2^x)^{-1}\,dx$ converges. We haven't found the value of this improper integral, but we have shown that it *has* a value (in fact, that it is a number between 0 and 1).

Example 77-1 suggests the use of a straightforward convergence test, the **comparison test**.

Theorem 77-1. *Suppose that f and g are integrable on the interval $[a, R]$ for each $R > a$ and that $0 \leq g(x) \leq f(x)$ for each $x \geq a$. If the improper integral $\int_a^\infty f(x)\,dx$ converges, then $\int_a^\infty g(x)\,dx$ converges.*

PROOF

We simply observe that

$$\int_a^R g(x)\,dx \leq \int_a^R f(x)\,dx$$

for each $R > a$. If $\int_a^\infty f(x)\, dx$ converges, then $\int_a^R f(x)\, dx$ is bounded above. Therefore $\int_a^R g(x)\, dx$ is bounded above, and so $\int_a^\infty g(x)\, dx$ converges.

Example 77-2. Show that the improper integral $\int_0^\infty e^{-x^2}\, dx$ converges. Approximate its value.

Solution. As we pointed out in the last section, our given integral converges if and only if $\int_1^\infty e^{-x^2}\, dx$ is a number. We will compare this integral with the convergent (Example 76-1) integral $\int_1^\infty e^{-x}\, dx$. Since, for $x \geq 1$, $0 \leq e^{-x^2} \leq e^{-x}$, Theorem 77-1 tells us that convergence of the integral $\int_1^\infty e^{-x}\, dx$ implies convergence of $\int_1^\infty e^{-x^2}\, dx$.

In order to approximate the value of the given integral, we notice that e^{-x^2} is extremely small when x is even moderately large. Thus the integral $\int_0^\infty e^{-x^2}\, dx$ converges rapidly; that is, $\int_0^R e^{-x^2}\, dx$ approximates its limit $\int_0^\infty e^{-x^2}\, dx$ reasonably well even when R is fairly small. For example, according to Equation 76-2,

$$\int_0^\infty e^{-x^2}\, dx = \int_0^3 e^{-x^2}\, dx + \int_3^\infty e^{-x^2}\, dx.$$

If $x \geq 3$, then $e^{-x^2} \leq e^{-3x}$, and so

$$\int_3^\infty e^{-x^2}\, dx \leq \int_3^\infty e^{-3x}\, dx = \tfrac{1}{3}e^{-9} \approx .00004.$$

Thus you can be sure that the first four decimal places of $\int_0^3 e^{-x^2}\, dx$ and $\int_0^\infty e^{-x^2}\, dx$ differ by no more than one digit in the fourth decimal place. If you use Simpson's Rule with $n = 6$, compute with six-place tables, and round off your result to four places, you will find that the number .8862 is an approximation of $\int_0^3 e^{-x^2}\, dx$. It can be shown that $\int_0^\infty e^{-x^2}\, dx = \tfrac{1}{2}\sqrt{\pi}$, and this number is .8862 with four-place accuracy.

Theorem 77-1 can also be used to demonstrate divergence.

Example 77-3. Show that $\int_1^\infty \dfrac{dx}{\sqrt{x} - e^{-x}}$ is divergent.

Solution.   Observe that for each $x \geq 1$,

$$\frac{1}{\sqrt{x}} \leq \frac{1}{\sqrt{x} - e^{-x}}.$$

Therefore Theorem 77-1 says that *if our given integral were convergent*, then $\displaystyle\int_{1}^{\infty} \frac{dx}{\sqrt{x}}$
would converge. But we saw in Example 76-2 (here $p = \frac{1}{2} < 1$) that this latter integral is *divergent*. Hence our given integral must be divergent. Think about this example until you understand it.

Up to this point in the section we have been assuming that we are integrating a function f that does not take negative values. If $f(x)$ is positive for some choices of x and negative for others, the situation is more complicated (and more interesting). We don't have the simple alternative that either $F(R)$ is bounded above or it is not to tell us whether or not our integral converges. For example, $\displaystyle\int_{0}^{R} \cos x \, dx = \sin R$, and even though $\sin R$ is bounded, we see that it doesn't have a limit. Consequently, the improper integral $\displaystyle\int_{0}^{\infty} \cos x \, dx$ is divergent (test your understanding of this statement by making a sketch and using the area interpretation of the integral).

If the integrand changes sign, we apply the comparison test to the integral $\displaystyle\int_{a}^{\infty} |f(x)| \, dx$. Our next theorem tells us that if this integral converges, then the integral $\displaystyle\int_{a}^{\infty} f(x) \, dx$ converges.

Theorem   77-2.   *If f is integrable on the interval $[a, R]$ for each $R > a$ and if the integral $\displaystyle\int_{a}^{\infty} |f(x)| \, dx$ converges, then $\displaystyle\int_{a}^{\infty} f(x) \, dx$ converges.*

PROOF

For each number x,

$$0 \leq \tfrac{1}{2}[|f(x)| + f(x)] \leq |f(x)| \quad \text{and} \quad 0 \leq \tfrac{1}{2}[|f(x)| - f(x)] \leq |f(x)|.$$

Therefore since the integral $\displaystyle\int_{a}^{\infty} |f(x)| \, dx$ converges, our comparison test tells us that the integrals $\displaystyle\int_{a}^{\infty} \tfrac{1}{2}[|f(x)| + f(x)] \, dx$ and $\displaystyle\int_{a}^{\infty} \tfrac{1}{2}[|f(x)| - f(x)] \, dx$ converge.

For each number R, we can write

$$\int_a^R f(x)\,dx = \int_a^R \tfrac{1}{2}[|f(x)| + f(x)]\,dx - \int_a^R \tfrac{1}{2}[|f(x)| - f(x)]\,dx.$$

We have just shown that the limits of the two integrals on the right-hand side of this equation are numbers, so $\lim_{R\uparrow\infty}\int_a^R f(x)\,dx$ is a number; that is, the integral $\int_a^\infty f(x)\,dx$ converges.

Example 77-4. Test the integral $\int_1^\infty \dfrac{\sin x}{x^2}\,dx$ for convergence.

Solution. Since $0 \le \dfrac{|\sin x|}{x^2} \le \dfrac{1}{x^2}$, and since $\int_1^\infty \dfrac{dx}{x^2}$ converges, the comparison test tells us that $\int_1^\infty \dfrac{|\sin x|}{x^2}\,dx$ converges. Now we use Theorem 77-2 to see that the integral $\int_1^\infty \dfrac{\sin x}{x^2}\,dx$ converges.

Theorem 77-2 tells us that *if* $\int_a^\infty |f(x)|\,dx$ converges, then $\int_a^\infty f(x)\,dx$ converges. In this case, we say that the latter integral **converges absolutely**. It is possible that $\int_a^\infty f(x)\,dx$ converges but that $\int_a^\infty |f(x)|\,dx$ does not. We then say that our integral **converges conditionally**. For example, in Problem 77-10 we will help you show that $\int_0^\infty \dfrac{\sin x}{x}\,dx$ converges, but in Example 82-5, we will show that $\int_0^\infty \dfrac{|\sin x|}{x}\,dx$ diverges.

This situation is illustrated in Fig. 77-2. The curve $y = x^{-1}\sin x$ appears in color and the curve $y = x^{-1}|\sin x|$ is drawn in black. The number $F(R) = \int_0^R x^{-1}\sin x\,dx$ is the "net area" of the region bounded by the colored curve and the interval $[0, R]$ of the X-axis, areas of regions under the axis being considered negative and areas of regions above the axis being positive. Thus there is a "cancellation effect" as negative areas balance off positive ones. The result, in this case, is that $\lim_{R\uparrow\infty}\int_0^R x^{-1}\sin x\,dx$ is a number. For $\int_0^R x^{-1}|\sin x|\,dx$, the situation is different. Here the areas are all positive, and the larger R becomes, the

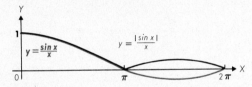

Figure 77-2

larger $\int_0^R x^{-1} |\sin x| \, dx$ gets. Although it is by no means obvious, we will see that

$$\lim_{R \uparrow \infty} \int_0^R x^{-1} |\sin x| \, dx = \infty;$$ the integral $\int_0^\infty x^{-1} \sin x \, dx$ is only conditionally

convergent. Many of the ideas and techniques of dealing with improper integrals will become clearer after you have studied sequences and infinite series in the next chapter.

PROBLEMS 77

1. Test the following integrals for convergence.

(a) $\int_2^\infty \dfrac{x^2}{x^7 + 3} \, dx$

(b) $\int_4^\infty \dfrac{x^3 - 4}{x^5 + 3x^2 + 5} \, dx$

(c) $\int_4^\infty \dfrac{\sqrt[5]{x^3 - 2x}}{\sqrt[3]{2x^5 + 4x + 1}} \, dx$

(d) $\int_4^\infty \sqrt{\dfrac{x^3 - 2x}{x^6 + 4x + 1}} \, dx$

(e) $\int_4^\infty \dfrac{x + 1}{x^2 - 3} \, dx$

(f) $\int_1^\infty \dfrac{\coth x}{\sqrt{x}} \, dx$

(g) $\int_2^\infty \dfrac{\ln x}{x^2} \, dx$

(h) $\int_2^\infty \dfrac{\ln x^2}{x} \, dx$

2. Do the following integrals converge absolutely?

(a) $\int_0^\infty e^{-x} \cos x \, dx$

(b) $\int_0^\infty e^{-3x} \sin (x + 3) \, dx$

(c) $\int_2^\infty x^{-4} \sin x^4 \, dx$

(d) $\int_2^\infty 4^{-x} \sin 4^x \, dx$

3. Test the following integrals for convergence, using Theorems 77-1 and 77-2 as necessary.

(a) $\int_0^\infty e^{-x} \sin x \, dx$

(b) $\int_0^\infty \sin e^{-x} \, dx$

(c) $\int_0^\infty e^{-\sin x - x} \, dx$

(d) $\int_0^\infty e^{-\sin x} \, dx$

4. Use a comparison test that is analogous to Theorem 77-1 to test the following integrals for convergence.

   (a) $\displaystyle\int_0^1 x^{-.3}e^{-x}\,dx$

   (b) $\displaystyle\int_0^1 x^{-1/2}\cos x\,dx$

   (c) $\displaystyle\int_2^3 \frac{dx}{x\sqrt{3-x}}$

   (d) $\displaystyle\int_5^7 \frac{dx}{(x^2+3)\sqrt[3]{7-x}\,\sqrt[5]{x-5}}$

   (e) $\displaystyle\int_0^1 x^{-1}e^x\,dx$

   (f) $\displaystyle\int_0^1 x^{-1}\cos x\,dx$

   (g) $\displaystyle\int_{-1}^1 x^{-4/3}\sin x\,dx$

   (h) $\displaystyle\int_{-2}^2 x^{-2}\sinh x\,dx$

5. Suppose that $f(x)$ and $g(x)$ are both positive for $x \geq a$ and that $\lim_{x\uparrow\infty} f(x)/g(x) = L$, where $L > 0$. Show that there is therefore a number c such that if $x \geq c$, then $\frac{1}{2}Lg(x) \leq f(x) \leq \frac{3}{2}Lg(x)$. From these inequalities, deduce that $\displaystyle\int_a^\infty f(x)\,dx$ converges if and only if $\displaystyle\int_a^\infty g(x)\,dx$ converges. Apply this result to test the following integrals for convergence. (Compare with Problem 77-1.)

   (a) $\displaystyle\int_2^\infty \frac{x^2}{x^7-3}\,dx$

   (b) $\displaystyle\int_4^\infty \frac{x^3+4}{x^5-3x^2-5}\,dx$

   (c) $\displaystyle\int_4^\infty \frac{\sqrt[5]{x^3+2x}}{\sqrt[3]{2x^5-4x-1}}\,dx$

   (d) $\displaystyle\int_4^\infty \sqrt{\frac{x^3+2x}{x^6-4x-1}}\,dx$

   (e) $\displaystyle\int_4^\infty \frac{x-1}{x^2+3}\,dx$

   (f) $\displaystyle\int_1^\infty \frac{\tanh x}{\sqrt{x}}\,dx$

6. Each of the following "infinite" regions has a finite "area." Find the area of the region that lies in the first quadrant and is bounded by the X-axis and
   (a) the Y-axis and the curve $y(x^2+4) = 8$.
   (b) the line $x = 3$ and the curve $y^2(x^2-1)^3 = 4x^2$.
   (c) the Y-axis and the curve $xy^2(x+4)^2 = 1$.
   (d) the line $x = 3$ and the curve $yx^2 = 1+4y$.

7. Let R be the infinite region bounded by the coordinate axes, the line $x = 1$, and the curve $y = 1/\sqrt{1-x}$. Express the area of R by means of an integral if you use (a) "vertical strips," and (b) "horizontal strips." Conclude from (a) and (b) that

$$\int_0^1 \frac{dx}{\sqrt{1-x}} = 1 + \int_1^\infty \frac{dy}{y^2}.$$

   Check this equality by actually computing both integrals.

8. Suppose that we rotate the region R of the preceding problem about the X-axis to obtain an "infinite" solid. Even though R has a finite area, show that the solid of revolution has an "infinite" volume.

9. As a contrast to the preceding problem, consider the infinite region R that lies in the first quadrant and is bounded by the Y-axis, the line $y = 1$, and the curve $y = 1/x$. Show that R has an "infinite" area but that the infinite solid obtained by rotating R about the Y-axis has a finite volume. Does the solid have a finite surface area?

10. Use integration by parts to show that

$$\int_{\pi/2}^{R} \frac{\sin x}{x}\, dx = -\frac{\cos R}{R} - \int_{\pi/2}^{R} \frac{\cos x}{x^2}\, dx.$$

From this equation infer that $\displaystyle\int_{0}^{\infty} \frac{\sin x}{x}\, dx$ converges.

11. From the inequalities $x - 1 < [\![x]\!] \le x$, deduce that $\displaystyle\int_{1}^{\infty} \frac{dx}{[\![x]\!]^p}$ converges if and only if $p > 1$. Why is it reasonable to express this integral as the "infinite sum"

$$1 + \frac{1}{2^p} + \frac{1}{3^p} + \frac{1}{4^p} + \cdots ?$$

This brief chapter has introduced you to some new uses of the symbol ∞, has presented a method for calculating limits, and has broadened the concept of the integral. In all these areas, particularly the last, we have only scratched the surface. The main things we expect you to retain are an ability to use l'Hospital's rule and an understanding that $\displaystyle\int_{a}^{\infty} f(x)\, dx$ means $\displaystyle\lim_{R \uparrow \infty} \int_{a}^{R} f(x)\, dx$.

Review Problems, Chapter Nine

1. Find the limit.
   (a) $\displaystyle\lim_{x \downarrow 0} (1 + rx)^{1/x}$
   (b) $\displaystyle\lim_{x \uparrow \infty} (1 + r/x)^x$
   (c) $\displaystyle\lim_{x \downarrow 0} (1 + x)^{r/x}$
   (d) $\displaystyle\lim_{x \uparrow \infty} (1 + 1/x)^{rx}$

2. Find the limit.
   (a) $\displaystyle\lim_{x \to 0} \frac{\ln (1 + x^2)}{[\ln (1 + x)]^2}$
   (b) $\displaystyle\lim_{x \to 0} \frac{\sin 3x^2}{\sin^2 3x}$
   (c) $\displaystyle\lim_{x \to 0} (1 - \cos x)^{\sin x}$
   (d) $\displaystyle\lim_{x \downarrow 0} (\sin x)^{1 - \cos x}$

3. When t is very large, the tangent vector to the curve $\mathbf{r} = t \ln t^4 \mathbf{i} + (t^2 + 2)\mathbf{j}$ is almost parallel to what unit vector?

4. Find $\displaystyle\lim_{x \downarrow 0} (1 + x^r)^{1/x}$ if $0 \le r < 1$.

5. Suppose that the functions f and g are continuous in the finite interval $[a, b]$ and that for each $x \in (a, b)$, $g(x) \neq 0$. Show that there is a point $m \in (a, b)$ such that

$$\frac{\int_a^b f(x)\, dx}{\int_a^b g(x)\, dx} = \frac{f(m)}{g(m)}.$$

6. Suppose that $\lim_{x \uparrow \infty} f(x) = A$ and $\lim_{x \uparrow \infty} g(x) = B$, where $A \neq 0$ and $B \neq 0$. Show that

$$\lim_{x \uparrow \infty} \frac{\int_a^x f(t)\, dt}{\int_a^x g(t)\, dt} = \frac{A}{B},$$

where a is any number such that f and g are continuous in the interval $[a, \infty)$.

7. Let $P(t)$ be t units from $(1, 0)$ along the unit circle and $Q(t)$ be t units to the left of $(1, 0)$ along the X-axis, where t is a small positive number. Find $\lim_{t \downarrow 0} b(t)$, where $b(t)$ is the Y-intercept of the line that contains $P(t)$ and $Q(t)$.

8. What happens when you try to apply l'Hospital's Rule to calculate $\lim_{x \uparrow \infty} e^x \ln (1 - e^{-x} \sin x)$?

9. Suppose that $f(x) = x^2 \sin (1/x)$ and $g(x) = x$. Show that $\lim_{x \to 0} f(x)/g(x)$ exists but that $\lim_{x \to 0} f'(x)/g'(x)$ does not. Does this fact contradict l'Hospital's Rule?

10. Test the given integral for convergence.

(a) $\displaystyle\int_1^\infty x^{-1} \sin x^{-1}\, dx$

(b) $\displaystyle\int_1^\infty x^{-1} \operatorname{Sin}^{-1} x^{-1}\, dx$

(c) $\displaystyle\int_1^\infty x^{-1} \cos x^{-1}\, dx$

(d) $\displaystyle\int_1^\infty x^{-1} \operatorname{Cos}^{-1} x^{-1}\, dx$

(e) $\displaystyle\int_1^\infty x^{-1} \tan x^{-1}\, dx$

(f) $\displaystyle\int_1^\infty x^{-1} \operatorname{Tan}^{-1} x^{-1}\, dx$

11. Do the following integrals converge?

(a) $\displaystyle\int_0^\infty e^{\sin x}\, dx$

(b) $\displaystyle\int_0^\infty \sin e^{\sin x}\, dx$

(c) $\displaystyle\int_0^\infty \sin e^x\, dx$

12. Test the following integrals for convergence.

(a) $\displaystyle\int_1^\infty \frac{\cos x}{1 + x^2}\, dx$

(b) $\displaystyle\int_1^\infty \frac{x}{1 + x^2}\, dx$

(c) $\displaystyle\int_1^\infty \frac{[\![x]\!]}{1 + x^2}\, dx$

(d) $\displaystyle\int_1^\infty \frac{x}{1 + [\![x]\!]^2}\, dx$

(e) $\displaystyle\int_1^\infty \sin \left(\frac{1}{1 + x^2}\right) dx$

(f) $\displaystyle\int_1^\infty \sin \left(\frac{x}{1 + x^2}\right) dx$

13. Convince yourself that $\displaystyle\int_1^\infty x^{-1} \sin x^{-1}\,dx = \int_0^1 x^{-1} \sin x\,dx.$

14. Show that the following integrals are convergent.

(a) $\displaystyle\int_0^{\pi/2} \sqrt{\cot x}\,dx$
(b) $\displaystyle\int_0^{\pi/2} \cot \sqrt{x}\,dx$

(c) $\displaystyle\int_0^{\pi/2} \sqrt{\tan x}\,dx$
(d) $\displaystyle\int_0^{\pi/2} \tan \sqrt{x}\,dx$

15. Evaluate the improper integral $\displaystyle\int_0^\infty [\![3e^{-x}]\!]\,dx.$

16. Show that $\displaystyle\int_0^\infty \frac{\sin ax}{x}\,dx = \frac{a}{|a|}\int_0^\infty \frac{\sin x}{x}\,dx$ if $a \neq 0.$

17. Show in two ways that $\displaystyle\int_0^1 \frac{dx}{1+x^2} = \int_1^\infty \frac{dx}{1+x^2}.$

(a) Evaluate the two integrals.

(b) Make the substitution $u = 1/x.$

18. (This problem shows that lim and $\displaystyle\int$ do not necessarily commute.)

(a) Show that $\displaystyle\int_0^1 \lim_{x\uparrow\infty} xe^{-xt}\,dt = \int_0^1 0\,dt = 0.$

(b) Show that $\displaystyle\lim_{x\uparrow\infty}\int_0^1 xe^{-xt}\,dt = \lim_{x\uparrow\infty}(1 - e^{-x}) = 1.$

19. Suppose that $x \in (0, \infty).$

(a) Show that $\displaystyle\frac{d}{dx}\int_0^\infty e^{-xt}\,dt = \frac{d}{dx}\left(\frac{1}{x}\right) = -x^{-2}.$

(b) Show that $\displaystyle\int_0^\infty \left(\frac{d}{dx}e^{-xt}\right)dt = \int_0^\infty (-te^{-xt})\,dt = -x^{-2}.$

10 Sequences and Series

You know that 3.14159 is only an approximation of the number π. But for most practical applications, this approximation is close enough, and if it isn't, we can replace it with a still closer approximation. The number π is the limit of the *sequence* of numbers 3, 3.1, 3.14, 3.141, 3.1415, ... Sometimes we express the terms of such a sequence as sums; for example, we might write 3, 3 + .1, 3 + .1 + .04, and so on. The more numbers we add together, the closer the sum approximates π. We are going to study such sequences in this chapter. We will find, for example, that another sequence whose limit is π is 4, $4 - \frac{4}{3}$, $4 - \frac{4}{3} + \frac{4}{5}$, $4 - \frac{4}{3} + \frac{4}{5} - \frac{4}{7}$, and so on. Thus we can get as accurate an approximation of π as we like by adding together sufficiently many terms in the "sum"

$$4 - \frac{4}{3} + \frac{4}{5} - \frac{4}{7} + \frac{4}{9} - \frac{4}{11} + \cdots.$$

Similar sums will give us the numbers e, ln 2, sin .7, and so on.

78. SEQUENCES AND THEIR LIMITS

A function f with domain D and range R is a set of pairs of the form $(x, f(x))$, where x is a member of D and $f(x)$ is the corresponding member of R. In most of our work D and R are sets of real numbers, but on occasion we find it convenient to use sets that consist of other mathematical objects, such as angles and vectors. In this chapter we will be working with functions whose domains are sets of *integers*; such functions are called **sequences**. The ranges of our sequences will be sets of real numbers.

Sequences have traditionally played a central role in mathematical analysis, and they have acquired a special notation of their own. If an integer n is in the domain of a function a, then the corresponding element in the range of a is denoted by $a(n)$. In sequence notation, we denote this element by a_n. Thus a sequence is a set of pairs of the form (n, a_n), where n is an integer. It is customary to abbreviate this notation and speak of "the sequence $\{a_n\}$." Usually we don't mention the domain of a sequence specifically. Most of our sequences will have as their domains the set of positive integers or perhaps the set of nonnegative integers. We can figure out the domain of a given sequence from the context. The numbers $a_1, a_2, a_3, \ldots$ are called the **terms** of the sequence $\{a_n\}$, and because they are paired with the integers $1, 2, 3, \ldots$, we speak of the *first term* a_1, the *second term* a_2, and so on.

Example 78-1. Find the second, third, and sixth terms of the sequence $\{(-1)^n \sin (\pi/n)\}$.

Solution. If we denote this sequence by $\{a_n\}$, where $a_n = (-1)^n \sin (\pi/n)$, then $a_2 = (-1)^2 \sin (\pi/2) = 1$, $a_3 = (-1)^3 \sin (\pi/3) = -\tfrac{1}{2}\sqrt{3}$, and $a_6 = \tfrac{1}{2}$.

The terms of a sequence need not be given by an elementary analytic formula. Any rule that assigns a number a_n to each integer n will define a sequence.

Example 78-2. Let $\{a_n\}$ be the sequence of digits in the decimal expression of $\sin \tfrac{1}{3}\pi$. Find the first four terms of the sequence.

Solution. We know that $\sin \tfrac{1}{3}\pi = \tfrac{1}{2}\sqrt{3} = .866025\ldots$. Thus $a_1 = 8$, $a_2 = 6$, $a_3 = 6$, and $a_4 = 0$. We have no simple formula to calculate a_n for each n, but the number a_{1066}, for example, is completely determined, and so are the other terms.

A sequence may also be given by specifying the first term and then stating a *recursion formula* that tells how to find each remaining term from the term that precedes it.

Example 78-3. Let $\{a_n\}$ be the sequence in which $a_1 = 1$, and $a_n = 3a_{n-1} - 1$ for $n > 1$. Find a_2, a_3, and a_4.

Solution. When we set $n = 2$ in the recursion formula $a_n = 3a_{n-1} - 1$, we see that $a_2 = 3a_1 - 1$. Since $a_1 = 1$, this equation tells us that $a_2 = 3 \cdot 1 - 1 = 2$. Now we set $n = 3$ in the recursion formula to find that $a_3 = 3a_2 - 1$. We have already found that $a_2 = 2$, so we see that $a_3 = 3 \cdot 2 - 1 = 5$. Similarly, $a_4 = 3a_3 - 1 = 14$. It is clear that we can find any term—for example, a_{1776}—simply by plodding along a step at a time.

Another common way to specify a sequence is to list the first few terms and let the reader use them to infer the general pattern. There is really no logical basis for saying that the fifth term of the sequence $\{1, \frac{1}{2}, \frac{1}{3}, \frac{1}{4}, \ldots\}$ is $\frac{1}{5}$, but that is the number that would come to most people's minds, so a sequence is often specified in this way.

As with other functions we study in calculus, we work with limits of sequences. If L is a number, the equation $\lim_{n \uparrow \infty} a_n = L$ says that L is approximated by a_n when n is large. If L is one of the symbols ∞ or $-\infty$, then the equation says that $|a_n|$ is large when n is large. The precise statement of what this limit equation means is similar to previous limit definitions.

Definition   78-1.   *The equation*

$$\lim_{n \uparrow \infty} a_n = L$$

*means that each neighborhood of L contains a **final segment** of $\{a_n\}$ (the set of all terms whose indices exceed a particular value). In general, the index N that determines this final segment depends on the given neighborhood of L; the narrower this neighborhood, the larger N is likely to have to be.*

We call L the **limit** of the sequence $\{a_n\}$. If L is a number, the sequence **converges**, whereas if L is one of the symbols ∞ or $-\infty$ or if the sequence does not have a limit, then $\{a_n\}$ **diverges**.

The simplest way (when it works) to find limits is by inspection. Then the formal Definition 78-1 can be used to *verify* that the limit found by inspection actually is the limit of the sequence.

Example 78-4.   What number seems to be the limit of the sequence of Example 78-1?

Solution.   Here $a_n = (-1)^n \sin (\pi/n)$. If n is a large number, π/n is practically 0 and hence $\sin (\pi/n)$ is practically 0. Therefore it appears that $\lim_{n \uparrow \infty} (-1)^n \sin (\pi/n) = 0$.

Frequently the terms of a sequence are the values at the positive integers of a function whose domain contains the set of positive real numbers. For example, the terms of the sequence $\{n(\frac{1}{2}\pi - \text{Arctan } n)\}$ are the numbers $f(1), f(2), \ldots$, where $f(x) = x(\frac{1}{2}\pi - \text{Arctan } x)$. In such a case, the limit of the sequence is

$\lim\limits_{x \uparrow \infty} f(x)$ if this latter limit exists. Thus in Example 75-3 we used l'Hospital's Rule to find that $\lim\limits_{x \uparrow \infty} x(\frac{1}{2}\pi - \text{Arctan } x) = 1$, and so the limit of the sequence $\{n(\frac{1}{2}\pi - \text{Arctan } n)\}$ is the number 1.

The next theorem is merely a rewording of some of our standard limit theorems in the language of sequences.

Theorem 78-1. *If $\{a_n\}$ and $\{b_n\}$ are convergent sequences, then*

(78-1)
$$\lim_{n \uparrow \infty} (a_n + b_n) = \lim_{n \uparrow \infty} a_n + \lim_{n \uparrow \infty} b_n,$$

(78-2)
$$\lim_{n \uparrow \infty} (a_n b_n) = \lim_{n \uparrow \infty} a_n \lim_{n \uparrow \infty} b_n,$$

and

(78-3)
$$\lim_{n \uparrow \infty} \frac{a_n}{b_n} = \frac{\lim\limits_{n \uparrow \infty} a_n}{\lim\limits_{n \uparrow \infty} b_n} \quad \text{(provided that } \lim_{n \uparrow \infty} b_n \neq 0\text{)}.$$

As a special case of Equation 78-2, we have

(78-4)
$$\lim_{n \uparrow \infty} ca_n = c \lim_{n \uparrow \infty} a_n,$$

where c may be any number.

If we can see at a glance that the limit of a given sequence $\{a_n\}$ is a number L, we can consider ourselves fortunate. In general, we would not expect to be able to tell whether or not a sequence converges, much less what its limit is. The sequence $\{(1 + 1/n)^n\}$ is an example. Is it obvious that this sequence converges, and if so, that its limit is the number e? If we set $h = 1/n$ in Equation 48-9, then we see that the sequence does converge to e, but this fact is not apparent from just looking at the sequence. We will now discuss some simple and useful criteria for determining whether or not sequences of certain types converge.

A sequence is **bounded** if its terms don't get too large or too small. More precisely, there is some finite interval $[A, B]$ that contains all terms of the sequence. Suppose $\{a_n\}$ is convergent; that is, suppose that there is a number L such that $\lim\limits_{n \uparrow \infty} a_n = L$. Then Definition 78-1 tells us that the neighborhood $(L - 1, L + 1)$ of L contains a final segment of $\{a_n\}$. This statement means that there is an integer N such that the number a_n belongs to the interval $(L - 1, L + 1)$ if its index n exceeds N. Now look at the set of numbers $\{a_1, a_2, \ldots, a_N, L - 1, L + 1\}$. It contains $N + 2$ members, among which are a smallest (call it A) and a largest (call it B). We claim that the interval $[A, B]$ contains *every* term of our sequence. For it clearly contains $a_1, \ldots, a_N$, and since it also contains the interval $(L - 1, L + 1)$, it contains the rest of the sequence as well. Therefore $\{a_n\}$ is bounded. We state this important result as a theorem.

Theorem 78-2. *If a sequence converges, then it is bounded.*

Example 78-5. Is the sequence $\{n \sin \frac{1}{2}n\pi\}$ convergent?

Solution. No. If n is an odd number, then $n \sin \frac{1}{2}n\pi$ is either n or $-n$. Consequently, no finite interval can hold all the terms of our sequence; it is unbounded. Hence it cannot converge. If it did, Theorem 78-2 would be violated.

Notice carefully what Theorem 78-2 *does not* say. It does not say that if a sequence is bounded, then it is convergent. For example, $\{(-1)^n\}$ is certainly a bounded sequence (all its terms belong to the interval $[-1, 1]$). But it is not convergent; $\lim_{n \uparrow \infty} (-1)^n$ does not exist.

A sequence $\{a_n\}$ is **nondecreasing** if $a_1 \le a_2 \le \cdots$; that is, if $a_n \le a_{n+1}$ for each positive integer n. If $a_1 \ge a_2 \ge \cdots$, then $\{a_n\}$ is **nonincreasing**. Together these two types of sequences form the class of **monotone** sequences. It is relatively easy to talk about limits of monotone sequences. The smallest closed interval that contains the terms of a nondecreasing sequence $\{a_n\}$ will have the form $[a_1, B]$, while the smallest closed interval that contains the terms of a nonincreasing sequence will be of the form $[A, a_1]$. By precisely the same type of argument that is used to prove Theorem 75-1, it follows that $\lim_{n \uparrow \infty} a_n = B$ in the first case and $\lim_{n \uparrow \infty} a_n = A$ in the second. The sequences converge if and only if A and B are numbers; that is, the sequences are bounded. Because we are dealing with *smallest* closed intervals, the lower bound A is called the **greatest lower bound** of its sequence and B is the **least upper bound** of its sequence.

Theorem. 78-3 *A monotone sequence converges if and only if it is bounded. The limit of a nondecreasing sequence is its least upper bound, and the limit of a nonincreasing sequence is its greatest lower bound.*

Example 78-6. Suppose that $a_1 = \frac{3}{4}$, and $a_n = [(4n^2 - 1)/4n^2]a_{n-1}$ for $n \ge 2$. Show that the sequence $\{a_n\}$ converges.

Solution. We have

$$a_1 = \tfrac{3}{4}, a_2 = \tfrac{3}{4} \cdot \tfrac{15}{16}, a_3 = \tfrac{3}{4} \cdot \tfrac{15}{16} \cdot \tfrac{35}{36}, \ldots.$$

In this sequence each term is a product of positive numbers; thus each term is positive. Therefore, the sequence is bounded below and its greatest lower bound is a number $A \geq 0$. We also notice that each term is obtained from the preceding term by multiplying it by a proper fraction. Consequently, each term is smaller than its predecessor, and so the sequence is nonincreasing. Since it is also bounded below by 0, Theorem 78-3 tells us that it converges to its greatest lower bound A. All we know about A is that it is a number that is not less than 0. A deeper analysis shows that $A = 2/\pi$. Thus as we calculate more and more terms of our sequence, we get closer and closer to $2/\pi$. It turns out that this sequence converges slowly. It takes many terms to get a good approximation to the limit.

We bring this section to a close by pointing out a self-evident, but useful, fact. Suppose that we construct a new sequence by lopping off an initial segment from a given sequence $\{a_n\}$. For example, if we delete the first five terms we have left a sequence that we denote as $\{a_{n+5}\}$. It is reasonably clear that the new sequence converges if and only if the original one does, and if the two series converge they have the same limit. For example, the first few terms of the sequence $\{n^3/n!\}$ are $1, 4, \frac{9}{2}, \frac{8}{3}, \frac{25}{24}, \ldots$. Therefore, the sequence is not monotone. However, if we delete the first two terms, we obtain a nonincreasing sequence of positive terms. Now we can apply Theorem 78-3 to see that the new sequence has a limit, and this limit is also the limit of the original sequence.

PROBLEMS 78

1. Find and simplify the fifth term of each sequence. Which of the sequences converge? Can you find the limits of the convergent sequences? Can you find the smallest closed interval that contains all the terms?

(a) $\{(-2)^n\}$

(b) $\{(-n)^2\}$

(c) $\{2^{-n}\}$

(d) $\{n^{-2}\}$

(e) $\{-2^n\}$

(f) $\{-n^2\}$

(g) $\{\cos \frac{1}{2} n\pi\}$

(h) $\{\cos 2\pi/n\}$

(i) $\{1 + (-1)^n\}$

(j) $\{1 + (-\frac{1}{2})^n\}$

(k) $\{(\sin n)/n\}$

(l) $\{n \ln (1 + 2/n)\}$

(m) $\left\{\dfrac{2n - 1}{n + 2}\right\}$

(n) $\left\{\dfrac{4n + 1}{2n^2 + 4n}\right\}$

(o) $\left\{\dfrac{n! \, 2^{n+3}}{(n + 3)! \, 2^n}\right\}$

(p) $\left\{\dfrac{2^n}{n!}\right\}$

(q) $\left\{\dfrac{1 \cdot 3 \cdot 5 \cdots (2n - 1)}{2 \cdot 4 \cdot 6 \cdots (2n)}\right\}$

(r) $\left\{\dfrac{(n!)^2}{(2n)!}\right\}$

(s) $\left\{\displaystyle\int_0^n e^{-x^2} \, dx\right\}$

(t) $\left\{\displaystyle\int_0^{2\pi} |\cos nx| \, dx\right\}$

2. Write the first four terms of the sequence $\{c_n\}$. Can you guess a formula for c_n?

   (a) $c_1 = 5, c_n = nc_{n-1}$ if $n > 1$

   (b) $c_1 = 2, c_n = c_{n-1}/n$ if $n > 1$

   (c) $c_1 = 3, c_n = \frac{1}{2}c_{n-1}$ if $n > 1$

   (d) $c_1 = 2, c_n = c_{n-1} + 5$ if $n > 1$

   (e) $c_1 = 2, c_n = c_{n-1}^2$ if $n > 1$

   (f) $c_1 = 2, c_n = c_{n-1}^{-1}$ if $n > 1$

   (g) $c_1 = 15, c_2 = 4, c_n = [\![\frac{1}{2}(c_{n-1} + c_{n-2})]\!]$ if $n > 2$

   (h) $c_1 = 5, c_2 = 4, c_n = \frac{1}{2}(c_{n-1} + c_{n-2})$ if $n > 2$

3. Find a simple formula for a_n that gives the following terms of $\{a_n\}$.

   (a) $1, 3, 7, 15, 31, \ldots$ $\qquad$ (b) $3, 6, 12, 24, 48, \ldots$

   (c) $0, 3, 8, 15, 24, \ldots$ $\qquad$ (d) $2, 2 \cdot 4, 2 \cdot 4 \cdot 6, 2 \cdot 4 \cdot 6 \cdot 8, \ldots$

   (e) $1, 0, -1, 0, 1, 0, -1, \ldots$ $\qquad$ (f) $1, -1, -1, 1, 1, -1, -1, 1, \ldots$

4. The first few terms of some sequences are given; use Theorem 78-3 to show that they converge.

   (a) $\dfrac{3}{1}, \dfrac{3^2}{2!}, \dfrac{3^3}{3!}, \dfrac{3^4}{4!}, \ldots$ $\qquad$ (b) $5, \sqrt{5}, \sqrt{\sqrt{5}}, \sqrt{\sqrt{\sqrt{5}}}, \ldots$

   (c) $1, \dfrac{2!}{2^2}, \dfrac{3!}{3^3}, \dfrac{4!}{4^4}, \ldots$ $\qquad$ (d) $\sin 1, \sin \sin 1, \sin \sin \sin 1, \ldots$

   (e) $\text{Tan}^{-1} 1, \text{Tan}^{-1} \text{Tan}^{-1} 1, \text{Tan}^{-1} \text{Tan}^{-1} \text{Tan}^{-1} 1, \ldots$

   (f) $\displaystyle\int_0^1 e^{x^2}\, dx, \int_0^1 \int_0^x e^{y^2}\, dy\, dx, \int_0^1 \int_0^x \int_0^y e^{z^2}\, dz\, dy\, dx, \ldots$

5. Find an index N beyond which all the terms of the given sequence are contained in the interval $(-.1, .1)$.

   (a) $\left\{\dfrac{\cos n}{n}\right\}$ $\qquad$ (b) $\{2^{-n}\}$

   (c) $\left\{\dfrac{n^2 + 3n - 2}{n^3}\right\}$ $\qquad$ (d) $\left\{\dfrac{n!}{n^n}\right\}$

6. (a) Find the formula for the nth term of the sequence of Example 78-3.

   (b) Show that a formula for the nth term of the sequence of Example 78-2 is
   $$a_n = [\![10^n \sin \tfrac{1}{3}\pi]\!] - 10[\![10^{n-1} \sin \tfrac{1}{3}\pi]\!].$$

7. Let a_n be the area of the regular n-sided polygon inscribed in the unit circle. Find a formula for a_n $(n \geq 3)$. What is $\lim\limits_{n \uparrow \infty} a_n$?

8. Let $a_1 = 1$ and $a_n = (1 - n^{-2})a_{n-1}$ for $n \geq 2$. Show that $\{a_n\}$ converges. Find a formula for a_n and show that $\lim\limits_{n \uparrow \infty} a_n = \frac{1}{2}$.

9. Use l'Hospital's Rule to show that the limit of the sequence $\{(1 + 1/n)^n\}$ is e.

10. Carefully explain why the sequence $\{(-1)^n\}$ is divergent.

11. (a) If the domain of the sequence $\{a_n\}$ is the set of positive integers and h is a positive integer, what is the "understood" domain of the sequences $\{a_{n+h}\}$ and $\{a_{n-h}\}$?

    (b) If $\lim\limits_{n \uparrow \infty} a_n = L$, show that $\lim\limits_{n \uparrow \infty} a_{n+h} = \lim\limits_{n \uparrow \infty} a_{n-h} = L$.

12. Suppose that r is a number in the interval $(0, 1)$ and let $a_n = r^n$. Use Theorem 78-3 to show that $\{a_n\}$ converges. Now take limits of both sides of the equation $a_n = ra_{n-1}$ (using the equation $\lim_{n \uparrow \infty} a_n = \lim_{n \uparrow \infty} a_{n-1}$ that we verified in the last problem) and show that $\lim_{n \uparrow \infty} a_n = 0$.

13. The domain of a sequence $\{a_n\}$ is the set of positive integers. Therefore if $f(x) = a_{[x]}$, the domain of f is the interval $[1, \infty)$. Sketch the graph of f for the following choices of $\{a_n\}$.

   (a) $\{(-1)^n\}$ (b) $\{(-2)^{-n}\}$ (c) $\{\sin n\}$ (d) $\{1 + (-1)^n\}$

79. INFINITE SERIES

From a given sequence $\{a_k\}$ we can construct a new sequence $\{S_n\}$ by means of the equation

(79-1) $$S_n = \sum_{k=1}^{n} a_k = a_1 + a_2 + \cdots + a_n.$$

This new sequence is the sequence of **sums** of $\{a_k\}$. Thus $S_1 = a_1$, $S_2 = a_1 + a_2 = S_1 + a_2$, $S_3 = a_1 + a_2 + a_3 = S_2 + a_3$, and so on. Clearly, for each $n > 1$,

(79-2) $$S_n = S_{n-1} + a_n.$$

Example 79-1.   Find the first five terms of the sequence of sums of $\{a_k\}$ if $a_k = 1/k(k + 1)$, and also find a formula for S_n in terms of n.

Solution.   We have

$$S_1 = a_1 = \frac{1}{1 \cdot 2} = \frac{1}{2},$$

$$S_2 = S_1 + a_2 = \frac{1}{2} + \frac{1}{2 \cdot 3} = \frac{2}{3},$$

$$S_3 = S_2 + a_3 = \frac{2}{3} + \frac{1}{3 \cdot 4} = \frac{3}{4},$$

$$S_4 = S_3 + a_4 = \frac{3}{4} + \frac{1}{4 \cdot 5} = \frac{4}{5},$$

$$S_5 = S_4 + a_5 = \frac{4}{5} + \frac{1}{5 \cdot 6} = \frac{5}{6}.$$

When we look at these terms, we are strongly tempted to conclude that $S_6 = \frac{6}{7}$, $S_7 = \frac{7}{8}$, and, in general, that $S_n = n/(n + 1)$. This conclusion is correct, for

$$a_k = \frac{1}{k(k + 1)} = \frac{1}{k} - \frac{1}{k + 1},$$

and therefore

$$S_n = a_1 + a_2 + a_3 + \cdots + a_n$$

$$= \left(1 - \frac{1}{2}\right) + \left(\frac{1}{2} - \frac{1}{3}\right) + \left(\frac{1}{3} - \frac{1}{4}\right) + \cdots + \left(\frac{1}{n} - \frac{1}{n + 1}\right).$$

When we remove the parentheses, this sum "collapses" to

$$S_n = 1 - \frac{1}{n + 1} = \frac{n}{n + 1}.$$

If the sums of the sequence $\{a_k\}$ converge to the number S, we write

(79-3)
$$S = \sum_{k=1}^{\infty} a_k = a_1 + a_2 + a_3 + \cdots.$$

Of course, this notation does not mean that we add together all the infinitely many terms of $\{a_k\}$. It is simply another way of saying that $\lim_{n \uparrow \infty} S_n = S$. We call the indicated sum

$$\sum_{k=1}^{\infty} a_k = a_1 + a_2 + a_3 + \cdots$$

an **infinite series**. The numbers $a_1, a_2, a_3, \ldots$ are its **terms**, and the sequence $\{S_n\}$, where $S_n = \sum_{k=1}^{n} a_k$, is the sequence of **partial sums** of the series. If $\{S_n\}$ converges, then the series **converges**; otherwise the series is **divergent**. The limit S of $\{S_n\}$ for a convergent series is the **sum** of the series. Notice that when we work with an infinite series $\sum_{k=1}^{\infty} a_k$, we are dealing with *two* sequences, the sequence $\{a_k\}$ of *terms* and the sequence $\{S_n\}$ of *partial sums*.

Example 79-2. Find the sum of the infinite series $\sum_{k=1}^{\infty} 1/k(k + 1)$.

Solution. The sequence of terms of this series is $\{1/k(k + 1)\}$, and in Example 79-1 we found that the corresponding sequence of partial sums is $\{n/(n + 1)\}$. Since $\lim_{n \uparrow \infty} n/(n + 1) = 1$, our given series converges, and its sum is 1:

(79-4)
$$\sum_{k=1}^{\infty} \frac{1}{k(k + 1)} = \frac{1}{1 \cdot 2} + \frac{1}{2 \cdot 3} + \frac{1}{3 \cdot 4} + \cdots = 1.$$

The straightforward way to test the convergence of a given infinite series is to find a formula for the nth partial sum S_n and calculate its limit. In general, this procedure is easier to describe than to carry out, but there is one important series that we can handle fairly simply. Suppose that x is some real number and consider the **geometric series**

$$(79\text{-}5) \qquad \sum_{k=0}^{\infty} x^k = 1 + x + x^2 + x^3 + \cdots.$$

The sequence of terms of this series is the *geometric progression* $\{x^k\}$. Our index of summation starts at 0 here, so we find it convenient to think of 1 as being the 0th term of the sequence $\{x^k\}$, x being the first term, and so on. Similarly, we will write $S_0 = 1$, $S_1 = 1 + x$, and, in general,

$$S_n = \sum_{k=0}^{n} x^k = 1 + x + x^2 + \cdots + x^n.$$

In this form it is not clear what $\lim_{n\uparrow\infty} S_n$ is (or even whether $\{S_n\}$ has a limit). However, you may remember that there is a simple formula for the sums of a geometric progression. To derive it, we write

$$S_n - xS_n = (1 + x + \cdots + x^n) - (x + x^2 + \cdots + x^{n+1}) = 1 - x^{n+1}.$$

Then we solve for S_n and find (if $x \neq 1$) that

$$S_n = \frac{1 - x^{n+1}}{1 - x} = \frac{1}{1 - x}(1 - x \cdot x^n).$$

This last equation tells us that

$$\lim_{n\uparrow\infty} S_n = \frac{1}{1 - x}(1 - x \lim_{n\uparrow\infty} x^n)$$

if $\lim_{n\uparrow\infty} x^n$ is a number. Whether or not it is depends on our original choice of x. If $|x| < 1$, then $\lim_{x\uparrow\infty} x^n = 0$ [for example, $\lim_{n\uparrow\infty} (\frac{1}{2})^n = 0$], and so $\lim_{n\uparrow\infty} S_n = 1/(1 - x)$. If $|x| > 1$ or if $x = -1$, then $\lim_{n\uparrow\infty} x^n$ is not a number (it may be ∞ or it may not exist), and hence $\{S_n\}$ is divergent. If $x = 1$, then $S_n = n + 1$, and so again the sequence $\{S_n\}$ is divergent. In summary, *the geometric series* $\sum_{k=0}^{\infty} x^k$ *converges if and only if* $|x| < 1$, *and then*

$$(79\text{-}6) \qquad \sum_{k=0}^{\infty} x^k = \frac{1}{1 - x} \qquad (|x| < 1).$$

Example 79-3.   Find the sum of the series $\sum_{k=0}^{\infty} (\frac{1}{10})^k$.

Solution. We are dealing with a geometric series, so we simply replace x with $\frac{1}{10}$ in Equation 79-6:

$$\sum_{k=0}^{\infty} (\tfrac{1}{10})^k = \frac{1}{1 - \frac{1}{10}} = \frac{10}{9}.$$

Example 79-4.   Half the money that is spent by the citizens of our town gets deposited in the local bank, which immediately lends it out to other citizens. A deposit of \$100 therefore generates total deposits of how much?

Solution. The original \$100 is lent out immediately. It is spent, and \$50 gets deposited back in the bank. This amount, in turn, is lent out, half comes back, and so on. Thus the original deposit generates total deposits of

$$100 + 50 + 25 + \cdots = 100 \sum_{k=0}^{\infty} (\tfrac{1}{2})^k = 200 \text{ dollars.}$$

Economists use sophisticated variations of this basic idea to develop many such "multipliers."

Although infinite series are not sums of numbers in the usual sense, they do enjoy some of the properties of ordinary sums. For example, if the sequences of partial sums of the series $\sum_{k=1}^{\infty} a_k$ and $\sum_{k=1}^{\infty} b_k$ are $\{A_n\}$ and $\{B_n\}$, and if c and d are any two real numbers, then the sequence of partial sums of the series $\sum_{k=1}^{\infty} (ca_k + db_k)$ is $\{cA_n + dB_n\}$. According to Theorem 78-1, if the sequences $\{A_n\}$ and $\{B_n\}$ converge, so does $\{cA_n + dB_n\}$, and its limit is the corresponding combination of their limits. In the language of series: *if $\sum_{k=1}^{\infty} a_k$ and $\sum_{k=1}^{\infty} b_k$ are convergent series and if c and d are any two real numbers, then the series $\sum_{k=1}^{\infty} (ca_k + db_k)$ converges, and*

$$(79\text{-}7) \qquad \sum_{k=1}^{\infty} (ca_k + db_k) = c \sum_{k=1}^{\infty} a_k + d \sum_{k=1}^{\infty} b_k.$$

Equation 79-7 suggests that the summation symbol $\sum_{k=1}^{\infty}$ and the integration symbol $\int_{1}^{\infty}$ behave in much the same manner. In fact, convergence or divergence of the infinite series $\sum_{k=1}^{\infty} a_k$ is equivalent to the convergence or divergence of the improper integral $\int_1^\infty a_{[x]}\, dx$. We won't dwell on this point, but we may note that the equation

(79-8)
$$\sum_{k=1}^\infty a_k = \sum_{k=1}^h a_k + \sum_{k=h+1}^\infty a_k$$

where h is any positive integer, corresponds to Equation 76-2.

Let us illustrate our remarks about sequences and infinite series by considering the decimal notation that we use to represent real numbers. An unending decimal expression for a real number a determines a sequence of which a is the limit. For example, the decimal 1.111... is the limit of the sequence $\{1, 1.1, 1.11, 1.111, \ldots\}$. We can regard the terms of this sequence as the partial sums of the infinite series

$$1 + .1 + .01 + .001 + .0001 + \cdots = \sum_{k=0}^\infty (\tfrac{1}{10})^k,$$

and then the limit of the sequence is the sum of the series. In Example 79-3 we found that the sum of this geometric series is $\tfrac{10}{9}$, and so 1.111... is the decimal expression for $\tfrac{10}{9}$.

Example 79-5.   Show that 2.999... = 3.

Solution.   The unending decimal 2.999... is the sum of the series

$$2 + .9 + .09 + .009 + \cdots = 2 + \tfrac{9}{10} + 9(\tfrac{1}{10})^2 + 9(\tfrac{1}{10})^3 + \cdots$$
$$= 2 + \tfrac{9}{10}[1 + \tfrac{1}{10} + (\tfrac{1}{10})^2 + \cdots].$$

The series in brackets is the geometric series whose sum we have found to be $\tfrac{10}{9}$. Thus the decimal 2.999... stands for the number

$$2 + \tfrac{9}{10} \cdot \tfrac{10}{9} = 2 + 1 = 3,$$

as was to be shown (compare this example with Problem 1-11).

To say that an infinite series $\sum_{k=1}^\infty a_k$ converges to a sum S is to say that its partial sums cluster about S. Therefore if n is large the consecutive sums S_{n-1} and S_n can't be far apart; that is, their difference must be small. But this difference is simply the nth term a_n of the series, and so we have established a simple but important theorem.

Theorem   79-1.   *If the series* $\sum_{k=1}^\infty a_k$ *converges, then* $\lim_{n\uparrow\infty} a_n = 0$.

Example 79-6.   Test the series

$$1 - \frac{3}{4} + \frac{4}{6} - \frac{5}{8} + \frac{6}{10} - \cdots + (-1)^{n+1} \left(\frac{n+1}{2n} \right) + \cdots$$

for convergence.

Solution.   The terms of this series are the terms of the sequence $\{(n+1)/2n\}$ alternately prefaced by plus and minus signs. Since $\lim_{n \uparrow \infty} (n+1)/2n = \frac{1}{2}$, you can see that a term with a large index is close to either $\frac{1}{2}$ or $-\frac{1}{2}$, so a_n does not approach 0. Thus Theorem 79-1 tells us our series cannot be convergent (otherwise a_n would approach 0). It must therefore be divergent.

What Theorem 79-1 says (and Example 79-6 illustrates) is that the condition $\lim_{n \uparrow \infty} a_n = 0$ is *necessary* in order that $\sum_{k=1}^{\infty} a_k$ be a convergent series. If this condition does not hold, the series must diverge. The condition, however, is not *sufficient* to guarantee convergence. Even if its terms approach 0, a series may still diverge. For example, consider the series $\sum_{k=1}^{\infty} \ln (1 + 1/k)$. For this series,

$$\lim_{n \uparrow \infty} a_n = \lim_{n \uparrow \infty} \ln (1 + 1/n) = \ln 1 = 0.$$

On the other hand,

$$S_n = \sum_{k=1}^{n} \ln (1 + 1/k) = \sum_{k=1}^{n} \left[\ln (k + 1) - \ln k \right]$$

$$= \left[\ln 2 - \ln 1 \right] + \left[\ln 3 - \ln 2 \right] + \cdots + \left[\ln (n + 1) - \ln n \right]$$

$$= \ln (n + 1).$$

Since $\lim_{n \uparrow \infty} S_n = \lim_{n \uparrow \infty} \ln (n + 1) = \infty$, not a number, the series $\sum_{k=1}^{\infty} \ln (1 + 1/k)$ is divergent, even though its terms approach 0.

PROBLEMS 79

1.   Find the fifth sum S_5 of each sequence. Can you find a formula for S_n?

   (a) $\{3\}$
   (b) $\{\ln k\}$
   (c) $\{(-1)^k\}$
   (d) $\{\cos k\pi\}$
   (e) $\{\sin (k + 1) - \sin k\}$
   (f) $\{(k + 1)! - k!\}$
   (g) $\{k\}$
   (h) $\{2k - 1\}$

2. From a given sequence $\{a_k\}$ we can find its sequence of sums. Conversely, show that we can find a sequence $\{a_k\}$ of which a given sequence $\{S_n\}$ is the sequence of sums. Find a formula for a_k for the given S_n. Does your formula give you a_1?

(a) n (b) $n/(n+1)$ (c) $1 + (-1)^n$

(d) $(-1)^n$ (e) 1 (f) $[\![\frac{1}{2}n]\!]$

3. (a) Show that

$$\sum_{k=1}^{\infty} \frac{1}{4k^2 - 1} = \frac{1}{2}\sum_{k=1}^{\infty}\left(\frac{1}{2k-1} - \frac{1}{2k+1}\right).$$

(b) By writing out a few partial sums, convince yourself that $S_n = n/(2n+1)$.

(c) What is the sum of our series?

4. Find the sum of each series. (Watch the lower limit of the index of summation.)

(a) $\sum_{k=0}^{\infty} (\frac{2}{3})^k$ (b) $\sum_{k=0}^{\infty} 4(\frac{1}{5})^k$

(c) $\sum_{k=0}^{\infty} 4(\frac{2}{3})^{2k}$ (d) $\sum_{k=0}^{\infty} [2(\frac{4}{5})^k + 3(\frac{1}{4})^k]$

(e) $\sum_{k=1}^{\infty} 3^{-k}$ (f) $\sum_{k=4}^{\infty} (-\frac{5}{8})^k$

(g) $\sum_{k=1}^{\infty} \sin^{2k} x$ (h) $\sum_{k=4}^{\infty} \cos^{2k} x$

5. Show that if h is any positive integer and x is a number in the interval $(-1, 1)$, then

$$\sum_{k=h}^{\infty} x^k = \frac{x^h}{1-x}.$$

6. Show that the following series diverge.

(a) $\sum_{k=1}^{\infty} \frac{k!}{100^k}$ (b) $\sum_{k=1}^{\infty} \frac{(-1)^k(k^2+k)}{4k^2 + 5k - 1}$

(c) $\sum_{k=1}^{\infty} \frac{\ln(k + e^k)}{k + \ln k}$ (d) $\sum_{k=1}^{\infty} \sin k$

7. Find the sum of the series

$$\frac{3}{4} + \frac{5}{36} + \frac{7}{144} + \cdots + \frac{2k+1}{k^2(k+1)^2} + \cdots.$$

[Notice that $a_k = 1/k^2 - 1/(k+1)^2$.]

8. (a) Find numbers A, B, and C so that

$$\sum_{k=1}^{\infty} \frac{1}{k(k+1)(k+2)} = \sum_{k=1}^{\infty}\left(\frac{A}{k} + \frac{B}{k+1} + \frac{C}{k+2}\right).$$

(b) Add a few terms of the resulting series to get an idea of what its partial sums look like; when you do, find the sum of the series.

9. Find an expression for the nth partial sum of the series

$$\ln 2 + \ln \frac{3}{4} + \ln \frac{8}{9} + \cdots + \ln \left(1 - \frac{1}{n^2} \right) + \cdots .$$

What is the sum of this series?

10. How can you tell, without computing S_n, that $\lim_{n \uparrow \infty} S_n$ is not a number for any of the series in Problem 79-1?

11. (a) Use the formula for the sum of the first k positive integers to simplify the terms of the series $\sum_{k=1}^{\infty} 1/(1 + 2 + \cdots + k)$.

(b) Now refer to Example 79-1 to find the formula for the partial sums of this series.

(c) What is the sum of the series?

12. (a) What is the sum of the infinite series $\sum_{k=1}^{\infty} \int_{k}^{k+1} e^{-x} dx$?

(b) Does the series $\sum_{k=1}^{\infty} \int_{k}^{k+1} e^{-x^2} dx$ converge?

13. If the series $\sum_{k=1}^{\infty} (a_k + b_k)$ converges, does it necessarily follow that the series $a_1 + b_1 + a_2 + b_2 + a_3 + b_3 + \cdots$ converges? (Examine the situation when $a_k = 1$ and $b_k = -1$ for all k.)

14. Show that if $\sum_{k=1}^{\infty} a_k$ converges, then $\sum_{k=1}^{\infty} \cos a_k$ diverges.

15. (a) Sketch the curve $y = a_{[\![x]\!]}$ if $a_k = 1/k$.

(b) Show that $S_n = \int_{1}^{n+1} a_{[\![x]\!]} dx$.

(c) Convince yourself that the series $\sum_{k=1}^{\infty} a_k$ converges if and only if the improper integral $\int_{1}^{\infty} a_{[\![x]\!]} dx$ converges.

80. **THE COMPARISON TEST**

When we examine an infinite series $\sum_{k=1}^{\infty} a_k$, three questions naturally arise:

(1) Does the series converge?

(2) If it does converge, what is its sum?

(3) For a given index n, how well does the partial sum S_n approximate the sum S?

For the geometric series $\sum_{k=0}^{\infty} x^k$, we can answer all these questions completely. We know that this series converges if $|x| < 1$; we know that if it converges, then its sum is $S = 1/(1 - x)$; and, finally, since the formula for the nth partial sum is

$S_n = (1 - x^{n+1})/(1 - x)$, we know that $S - S_n = x^{n+1}/(1 - x)$. For most series, however, we cannot answer all three questions so completely. Our attention will chiefly be devoted to finding the answer to the first question. For certain kinds of series, we will develop several tools that will help us answer the other two questions as well.

The convergence tests for infinite series that we develop in this section are completely parallel to the convergence tests for improper integrals that we have already met. In fact, because of the relationship between series and improper integrals that we pointed out earlier, these convergence tests for series are easy consequences of the convergence tests for integrals. But you will probably understand them better if we start from scratch and discuss them directly in terms of series. Because they are especially easy to work with, we will look first at series whose terms are nonnegative numbers.

Successive partial sums of a series $\sum_{k=1}^{\infty} a_k$ are formed by adding more and more terms. If these terms are not negative, the partial sums keep growing, and there are only two possibilities. Either they grow without limit or they don't. In the first case, our series diverges, and in the second case, it converges. Stating it more formally, we note that the sequence $\{S_n\}$ of partial sums is nondecreasing. Then Theorem 78-3 tells us that it converges if and only if it is bounded, and we have proved the basic theorem on series with nonnegative terms.

Theorem 80-1. *An infinite series of nonnegative terms converges if and only if its sequence of partial sums is a bounded sequence.*

Example 80-1. Test for convergence the series

(80-1) $\qquad 1 + \frac{1}{2} + \frac{1}{4} + \frac{1}{4} + \frac{1}{8} + \frac{1}{8} + \frac{1}{8} + \frac{1}{8} + \frac{1}{16} + \cdots,$

where there are 8 consecutive $\frac{1}{16}$'s, then 16 consecutive $\frac{1}{32}$'s, and so on.

Solution. We will show that the sequence $\{S_n\}$ of partial sums of this series is unbounded. Consider, for example, just those terms whose indices are powers of 2, such as S_1, S_2, S_4, and S_8. We have $S_1 = 1$, $S_2 = 1\frac{1}{2}$, $S_4 = 2$, $S_8 = 2\frac{1}{2}$, and it is not hard to see that, in general, $S_{2^r} = 1 + \frac{1}{2}r$. Since the sequence $\{S_n\}$ contains each of these numbers, for $r = 1, 2, 3, \ldots$, it is clear that $\{S_n\}$ is unbounded, and our given series diverges.

If each term of one series of nonnegative numbers is at least as large as the corresponding term of another such series, we say that the first series **dominates** the second. For example, the series $\sum_{k=1}^{\infty} 2^{-k}$ dominates $\sum_{k=1}^{\infty} 3^{-k}$. If the dominant series converges, its partial sums are bounded, so obviously the smaller partial sums of the dominated series are bounded and it too converges. This straight-forward observation is known as the **comparison test**, and it is important enough to state as a theorem.

Theorem 80-2. *Suppose that* $0 \leq a_k \leq b_k$ *for each positive integer* k. *If the series* $\sum_{k=1}^{\infty} b_k$ *converges, then the series* $\sum_{k=1}^{\infty} a_k$ *converges.*

The next two examples illustrate how we use the comparison test to demonstrate convergence or divergence of a given series. The logic behind the comparison test is perfectly simple, but beginners find it hard to apply because we have to choose our comparison series correctly, something that is not easy to do without a lot of experience.

Example 80-2. Show that the series

(80-2)
$$\sum_{k=1}^{\infty} \frac{1}{k^2} = 1 + \frac{1}{2^2} + \frac{1}{3^2} + \frac{1}{4^2} + \cdots$$

is convergent. (This series, like the harmonic series below, is frequently used as a comparison series, so you should remember this example.)

Solution. Series 80-2 converges if and only if the series

$$\frac{1}{2^2} + \frac{1}{3^2} + \frac{1}{4^2} + \cdots$$

that we obtain by dropping the first term converges. This latter series is dominated by

$$\frac{1}{1 \cdot 2} + \frac{1}{2 \cdot 3} + \frac{1}{3 \cdot 4} + \cdots,$$

which (see Example 79-2) converges and has 1 as its sum. Thus we see from Theorem 80-2 that Series 80-2 is convergent. Furthermore, since the sum of our comparison series is 1, it is clear that the sum of Series 80-2 is a number between 1 and 2. After this course you may study Fourier Series, and then you will learn that the sum of Series 80-2 is $\frac{1}{6}\pi^2$.

Example 80-3. Test for convergence the **harmonic series**

(80-3)
$$\sum_{k=1}^{\infty} \frac{1}{k} = 1 + \frac{1}{2} + \frac{1}{3} + \frac{1}{4} + \cdots.$$

Solution.   Let us write Series 80-1 on one line and Series 80-3 below it:

$$1 + \tfrac{1}{2} + \tfrac{1}{4} + \tfrac{1}{4} + \tfrac{1}{8} + \tfrac{1}{8} + \tfrac{1}{8} + \tfrac{1}{8} + \tfrac{1}{16} + \cdots,$$

$$1 + \tfrac{1}{2} + \tfrac{1}{3} + \tfrac{1}{4} + \tfrac{1}{5} + \tfrac{1}{6} + \tfrac{1}{7} + \tfrac{1}{8} + \tfrac{1}{9} + \cdots.$$

It is clear that the second series dominates the first. Therefore if it converges, then Theorem 80-2 tells us that the first series converges. But we found in Example 80-1 that the first series is divergent, and so the second series must diverge, too.

In Example 80-2 we used the obvious fact that the sum of a convergent series of nonnegative terms is larger than the sum of such a series with smaller terms. This observation frequently gives us an idea of how well partial sums approximate the sum of a series.

Example 80-4.   Estimate how well the sum of the first six terms approximates the sum of the series

$$\sum_{k=0}^{\infty} \frac{1}{k!} = 1 + \frac{1}{1!} + \frac{1}{2!} + \frac{1}{3!} + \cdots.$$

Solution.   The sum of our given series is the sum of the first six terms, plus the "remainder after six terms"

$$\sum_{k=6}^{\infty} \frac{1}{k!} = \frac{1}{6!} + \frac{1}{7!} + \frac{1}{8!} + \cdots.$$

This remainder series is dominated by

$$\frac{1}{6!} + \frac{1}{6!\,7} + \frac{1}{6!\,7^2} + \frac{1}{6!\,7^3} + \cdots = \frac{1}{6!}\left[1 + \left(\frac{1}{7}\right) + \left(\frac{1}{7}\right)^2 + \cdots \right].$$

But the series in brackets is a simple geometric series whose sum is $(1 - \tfrac{1}{7})^{-1} = \tfrac{7}{6}$, so the remainder series has a sum less than $\dfrac{1}{6!} \cdot \dfrac{7}{6} \approx .0016.$

PROBLEMS   80

1.   Use Series 80-2 or 79-4 (or a slight modification) as a comparison series to show that the following series converge.

(a) $\displaystyle\sum_{k=1}^{\infty} \frac{1}{k^2 + 1}$ (b) $\displaystyle\sum_{k=1}^{\infty} \frac{1}{(k + 1)^2}$ (c) $\displaystyle\sum_{k=1}^{\infty} \frac{1}{2k^2 - k}$

(d) $\displaystyle\sum_{k=5}^{\infty} \frac{k}{k^3 + 1}$ (e) $\displaystyle\sum_{k=1}^{\infty} \sin k^{-2}$ (f) $\displaystyle\sum_{k=1}^{\infty} \mathrm{Tan}^{-1} k^{-1}(k + 1)^{-1}$

2. Use a geometric series as a comparison series to show that the following series converge.

(a) $\displaystyle\sum_{k=0}^{\infty} \frac{3}{2^k + 1}$

(b) $\displaystyle\sum_{k=1}^{\infty} \frac{3}{2^k - 1}$

(c) $\displaystyle\sum_{k=0}^{\infty} \frac{3^k}{5^k + 7}$

(d) $\displaystyle\sum_{k=0}^{\infty} \frac{\sin^2 k}{2^k}$

(e) $\displaystyle\sum_{k=0}^{\infty} \operatorname{sech} k$

(f) $\displaystyle\sum_{k=1}^{\infty} \operatorname{csch} k$

(g) $\displaystyle\sum_{k=1}^{\infty} \operatorname{Sin}^{-1} 2^{-k}$

(h) $\displaystyle\sum_{k=1}^{\infty} \tan 2^{-k}$

3. Use the harmonic series (or a slight modification of it) to show that the following series are divergent.

(a) $\displaystyle\sum_{k=0}^{\infty} \frac{1}{2k + 1}$

(b) $\displaystyle\sum_{k=2}^{\infty} \frac{2}{3k - 4}$

(c) $\displaystyle\sum_{k=1}^{\infty} \frac{3}{k + 4}$

(d) $\displaystyle\sum_{k=1}^{\infty} \frac{5}{6k + 7}$

(e) $\displaystyle\sum_{k=1}^{\infty} \operatorname{Sin}^{-1} k^{-1}$

(f) $\displaystyle\sum_{k=2}^{\infty} \frac{1}{\ln k}$

4. Compare the series $\displaystyle\sum_{k=1}^{\infty} k^{-k}$ with Series 80-2 and with various geometric series to get an idea of its sum.

5. Test the following series for convergence.

(a) $\displaystyle\sum_{k=1}^{\infty} k^{-1} \sin k^{-1}$

(b) $\displaystyle\sum_{k=1}^{\infty} k^{-1} \operatorname{Sin}^{-1} k^{-1}$

(c) $\displaystyle\sum_{k=0}^{\infty} 3^k \sin^2 2^{-k}$

(d) $\displaystyle\sum_{k=0}^{\infty} 3^{-k} \sin^2 2^k$

6. In Example 80-2 we saw that when $p = 2$, the series $\displaystyle\sum_{k=1}^{\infty} \frac{1}{k^p}$ converges. Use this result to show that when $p > 2$ the series also converges. Show that the series diverges if $p < 1$.

7. Show that for $k \geq 4$, $3^k/k! \leq \frac{3}{2}(\frac{3}{4})^k$. Use this result to find a bound on the error that we incur when we use the first six terms of the series $\displaystyle\sum_{k=0}^{\infty} \frac{3^k}{k!}$ to approximate its sum.

8. Show that $\displaystyle\sum_{k=r}^{\infty} \frac{1}{k^2} < \frac{1}{r - 1}$ if r is an integer greater than 1.

9. We showed in Section 78 that $\displaystyle\lim_{k \uparrow \infty} k(\tfrac{1}{2}\pi - \operatorname{Arctan} k) = 1$.

(a) Show that this equation implies that $\tfrac{1}{2}\pi - \operatorname{Arctan} k \geq 1/2k$ if k is large enough and hence that $\displaystyle\sum_{k=1}^{\infty} (\tfrac{1}{2}\pi - \operatorname{Arctan} k)$ diverges.

(b) Show that $\displaystyle\sum_{k=1}^{\infty} (\tfrac{1}{2}\pi - \operatorname{Arctan} k^2)$ converges.

10. If $\{S_n\}$ is the sequence of partial sums of the harmonic series, we have seen that $\lim_{n \uparrow 0} S_n = \infty$. Thus if we add together sufficiently many terms of the harmonic series, our sum will exceed 1 million. From our work in this section, find how many terms would be "sufficiently many."

11. Show that for each positive integer n,

$$\int_1^n x^{-1} \, dx \le \sum_{k=1}^n \int_k^{k+1} x^{-1} \, dx \le \sum_{k=1}^n \int_k^{k+1} k^{-1} \, dx = \sum_{k=1}^n k^{-1}.$$

How do these inequalities tell us that the harmonic series diverges?

12. Write out enough cases to convince yourself that the terms of Series 80-1 are given by the formula $a_{k+1} = 2^{-\llbracket \log_2 2k \rrbracket}$ for $k = 1, 2, 3, \ldots$.

81. ALTERNATING SERIES

Many convergence tests, such as the comparison test of the last section, do not apply directly to a series $\sum_{k=1}^\infty a_k$ that has both positive and negative terms. However, the comparison test does apply to the series $\sum_{k=1}^\infty |a_k|$ that we get when we replace each term with its absolute value, and if this series converges, then our next theorem tells us that the original series converges. Of course, $\sum_{k=1}^\infty a_k$ and $\sum_{k=1}^\infty |a_k|$ do not have the same sum; our theorem is merely a test of convergence. If $\sum_{k=1}^\infty |a_k|$ converges, we say that $\sum_{k=1}^\infty a_k$ **converges absolutely**. The analogous test for improper integrals is Theorem 77-2.

Theorem 81-1. *If the series $\sum_{k=1}^\infty |a_k|$ converges, then $\sum_{k=1}^\infty a_k$ converges.*

PROOF

The key idea of the proof is to write our given series as a combination of two series to which the comparison test applies. These two series are

$$\sum_{k=1}^\infty \tfrac{1}{2}(|a_k| + a_k) \quad \text{and} \quad \sum_{k=1}^\infty \tfrac{1}{2}(|a_k| - a_k),$$

and we observe that they are series of nonnegative terms and are dominated by $\sum_{k=1}^\infty |a_k|$:

$$0 \le \tfrac{1}{2}(|a_k| + a_k) \le |a_k| \quad \text{and} \quad 0 \le \tfrac{1}{2}(|a_k| - a_k) \le |a_k| \qquad \text{for each index } k.$$

The comparison test therefore tells us that the dominated series converge. Now in Equation 79-7 we replace c with 1, d with -1, a_k with $\frac{1}{2}(|a_k| + a_k)$, and b_k with $\frac{1}{2}(|a_k| - a_k)$, and we see that

$$\sum_{k=1}^{\infty} \left[\tfrac{1}{2}(|a_k| + a_k) - \tfrac{1}{2}(|a_k| - a_k) \right] = \sum_{k=1}^{\infty} a_k$$

converges.

Example 81-1. Show that the series $\displaystyle\sum_{k=1}^{\infty} \frac{\sin k^2}{k^2}$ converges absolutely.

Solution. Since $|k^{-2} \sin k^2| \leq k^{-2}$, and since (Example 80-2) the series $\displaystyle\sum_{k=1}^{\infty} k^{-2}$ converges, the comparison test tells us that $\displaystyle\sum_{k=1}^{\infty} |k^{-2} \sin k^2|$ converges. Hence our given series converges absolutely.

The converse of Theorem 81-1 is *not* true; that is, just because a series $\displaystyle\sum_{k=1}^{\infty} a_k$ converges, it does not follow that the series of absolute values $\displaystyle\sum_{k=1}^{\infty} |a_k|$ converges. For example, let us change the sign of every other term in Series 80-1:

$$(81\text{-}1) \qquad 1 - \tfrac{1}{2} + \tfrac{1}{4} - \tfrac{1}{4} + \tfrac{1}{8} - \tfrac{1}{8} + \tfrac{1}{8} - \tfrac{1}{8} + \tfrac{1}{16} - \cdots.$$

In Example 80-1 we showed that Series 80-1 is divergent, so Series 81-1 is *not* absolutely convergent. It is convergent, however, as we will now see. We look at its sequence of partial sums and find that

$$S_1 = 1, \ S_2 = \tfrac{1}{2}, \ S_3 = \tfrac{3}{4}, \ S_4 = \tfrac{1}{2}, \ S_5 = \tfrac{5}{8}, \ S_6 = \tfrac{1}{2}, \ S_7 = \tfrac{5}{8}, \ S_8 = \tfrac{1}{2}, \ S_9 = \tfrac{9}{16},$$

and so on. We immediately observe that if n is an even index, then $S_n = \frac{1}{2}$. If n is an odd index, then $S_n = \frac{1}{2} + |a_{n+1}|$, where a_{n+1} is the $(n + 1)$st term of Series 81-1. Thus the inequality

$$(81\text{-}2) \qquad |S_n - \tfrac{1}{2}| \leq |a_{n+1}|$$

holds for every choice of n, odd or even. Since $\lim\limits_{n \uparrow \infty} a_n = 0$, we conclude that $\lim\limits_{n \uparrow \infty} S_n = \frac{1}{2}$. Therefore Series 81-1 converges, and its sum is $\frac{1}{2}$. A series that converges but that does not converge absolutely is said to be **conditionally convergent**; we have just shown that Series 81-1 is conditionally convergent.

We can't add up *all* the terms of a convergent infinite series; there are too many of them. After we have added n terms to find the partial sum S_n, we are left with the **remainder** $R_n = S - S_n$. This number is itself the sum of an infinite series, a "tail" of the original series. Since it represents the error we make by taking S_n as an approximation of the sum S, we naturally would like to know how big it is (as in Example 80-4). In the process of demonstrating the convergence of Series 81-1, we also found a bound on the remainder (Inequality 81-2).

Series 81-1 is typical of the class of infinite series that we are now going to study. An **alternating series** has the form

$$(81\text{-}3) \qquad \sum_{k=1}^{\infty} (-1)^{k+1} a_k = a_1 - a_2 + a_3 - a_4 + \cdots,$$

where each of the numbers $a_1, a_2, a_3, \ldots$ is positive. The most important theorem about such series applies when the a's shrink monotonically to zero.

Theorem 81-2. *Let $\{a_k\}$ be a nonincreasing sequence such that $\lim_{k \uparrow \infty} a_k = 0$. Then Series 81-3 converges to a sum S, and for each positive integer n, $|S - S_n| \leq a_{n+1}$.*

In place of a formal proof of this theorem, suppose that we simply try to explain the general idea. The first partial sum S_1 is the first term a_1, and we have plotted it on the number scale in Fig. 81-1. Then S_2 is calculated by subtracting a_2 from S_1, so S_2 appears to the left of S_1. To obtain S_3, we add a_3, thereby moving back to the right. But since a_3 *does not exceed* a_2, we don't go beyond S_1. Now we calculate S_4 by subtracting a_4, thus moving to the left. Again we don't move beyond S_2, since $a_4 \leq a_3$. As we keep calculating and plotting partial sums, we move first to the left and then to the right in an ever-narrowing range. Notice that the partial sums with even indices

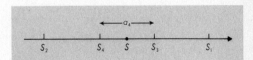

Figure 81-1

form a nondecreasing sequence, whereas the partial sums with odd indices form a nonincreasing sequence. Each sequence is bounded by the first term of the other, and so these monotone sequences converge. Furthermore, since $\lim_{k \uparrow \infty} a_k = 0$, the partial sums are getting closer together, and both sequences approach the same limit S. As Fig. 81-1 shows, the distance $|S - S_3|$ between S and S_3 is not greater than the distance a_4 between S_3 and S_4. In general, the distance between S and S_n is not greater than the distance a_{n+1} between S_n and S_{n+1}; in symbols, $|S - S_n| \leq a_{n+1}$. In other words, the remainder after n terms is no larger than the first neglected term.

Example 81-2. Show that the series

$$(81\text{-}4) \qquad\qquad 1 - \tfrac{1}{2} + \tfrac{1}{3} - \tfrac{1}{4} + \tfrac{1}{5} - \cdots$$

converges and approximate its sum with an error of less than .1.

Solution. The absolute values of the terms of our given alternating series form the decreasing sequence $\{1/k\}$ whose limit obviously is 0. Thus the hypotheses of Theorem 81-2 are satisfied and the series converges. According to the theorem, the remainder of such a convergent alternating series is less than the first unused term, so we make an error of less than .1 if we use the partial sum

$$1 - \tfrac{1}{2} + \tfrac{1}{3} - \tfrac{1}{4} + \tfrac{1}{5} - \tfrac{1}{6} + \tfrac{1}{7} - \tfrac{1}{8} + \tfrac{1}{9} = .746$$

to approximate the sum of the series. In Section 85 we will find that the sum is $\ln 2 = .693$; hence the actual remainder is $.693 - .746 = -.053$. This series converges "slowly"; we must use many terms to form a partial sum that gives a close approximation to the sum of the series.

In order to apply our alternating series test to a given series $\displaystyle\sum_{k=1}^{\infty} (-1)^{k+1}\, a_k$, we must show that each term of the sequence $\{a_k\}$ is not less than its successor; that is, we must show that $a_k \geq a_{k+1}$ for each index k. Sometimes it is more convenient to verify this inequality when it is written as

$$a_k - a_{k+1} \geq 0 \quad \text{or} \quad a_{k+1}/a_k \leq 1.$$

We can even use calculus, as in the following example.

Example 81-3. Test the series $\displaystyle\sum_{k=1}^{\infty} (-1)^{k+1}\, \frac{k}{1 + k^2}$ for convergence.

Solution. It is clear that $\displaystyle\lim_{k \uparrow \infty} k/(1 + k^2) = 0$. It only remains for us to show that the sequence $\{k/(1 + k^2)\}$ is nonincreasing. We observe that the terms of this sequence are the values at the positive integers of the function f, where $f(x) = x/(1 + x^2)$. We can readily calculate $f'(x) = (1 - x^2)/(1 + x^2)^2$, from which we see that $f'(x) < 0$ if $x > 1$. Hence f is decreasing in the interval $(1, \infty)$. It follows that $f(k + 1) < f(k)$ for each positive integer k; our sequence of terms is nonincreasing. Thus the hypotheses of Theorem 81-2 are satisfied and our series converges.

Although infinite series display many of the properties of ordinary sums, series and sums are not the same in all respects. Let us illustrate this statement by performing a few manipulations with Series 81-4. We have not yet shown that its sum is $\ln 2$, but we do know that the series converges, so let us denote its sum by S:

$$(81\text{-}5) \qquad S = 1 - \tfrac{1}{2} + \tfrac{1}{3} - \tfrac{1}{4} + \tfrac{1}{5} - \tfrac{1}{6} + \tfrac{1}{7} - \tfrac{1}{8} + \cdots.$$

Hence

$$\tfrac{1}{2}S = \qquad \tfrac{1}{2} \qquad - \tfrac{1}{4} \qquad + \tfrac{1}{6} \qquad - \tfrac{1}{8} + \cdots,$$

and so when we add these two series, we obtain

$$(81\text{-}6) \qquad \tfrac{3}{2}S = 1 + \tfrac{1}{3} - \tfrac{1}{2} + \tfrac{1}{5} + \tfrac{1}{7} - \tfrac{1}{4} + \cdots.$$

If you write out this addition for many terms, you will convince yourself that the series on the right-hand side of Equation 81-6 contains the same terms as the series on the right-hand side of Equation 81-5. Since the value of an ordinary sum is independent of the order in which its terms are written, we are naturally tempted to conclude that these series also have the same sum, and thus that $\tfrac{3}{2}S = S$. It would therefore follow that $S = 0$. But you can easily check (Theorem 81-2) that S is a number between $\tfrac{1}{2}$ and 1, so we must have gone wrong somewhere. We erred when we tried to conclude that just because the series in Equations 81-5 and 81-6 contain the same terms, they have the same sum. It can be shown that if two series have the same terms and if one of the series converges absolutely, then the other series also converges absolutely, and the two sums are the same. However, in the case of conditionally convergent series (such as the series in Equations 81-5 and 81-6), the sums of two series containing the same terms (but in different orders) need not be equal.

PROBLEMS 81

1. Test each series for convergence and absolute convergence.

    (a) $\displaystyle\sum_{k=1}^{\infty} \frac{(-1)^{k+1}}{k^2}$

    (b) $\displaystyle\sum_{k=1}^{\infty} \frac{(-1)^{k^2+1}}{k}$

    (c) $\displaystyle\sum_{k=1}^{\infty} \frac{(-1)^{k+1}}{2^k}$

    (d) $\displaystyle\sum_{k=1}^{\infty} \frac{(-1)^{2k+1}}{k}$

    (e) $\displaystyle\sum_{k=1}^{\infty} \int_0^{\pi} \sin k^2 x \, dx$

    (f) $\displaystyle\sum_{k=1}^{\infty} \int_0^{\pi} \sin^2 kx \, dx$

2. Test each series for absolute convergence.

    (a) $\displaystyle\sum_{k=1}^{\infty} \frac{\sin e^k}{e^k}$

    (b) $\displaystyle\sum_{k=1}^{\infty} \frac{\sin k^e}{k^e}$

    (c) $\displaystyle\sum_{k=1}^{\infty} \sin (-e)^{-k}$

    (d) $\displaystyle\sum_{k=1}^{\infty} \sin^k (-e)$

3. Test each series for absolute convergence and for convergence.

    (a) $\displaystyle\sum_{k=2}^{\infty} \frac{(-1)^k}{\ln k}$

    (b) $\displaystyle\sum_{k=1}^{\infty} \frac{(-1)^{k+1}}{\sqrt{k}}$

    (c) $\displaystyle\sum_{k=1}^{\infty} (-1)^{k+1} \ln \left(1 + \frac{1}{k} \right)$

    (d) $\displaystyle\sum_{k=1}^{\infty} \frac{(-1)^{k(k+1)/2}}{2k + 1}$

    (e) $\displaystyle\sum_{k=2}^{\infty} \left(\int_1^{1/k} u^{-1} \, du \right)^{-k}$

    (f) $\displaystyle\sum_{k=1}^{\infty} \int_0^{\pi/2} \cos kx \, dx$

4. Approximate the sum of the series with an error whose absolute value is no more than indicated.

    (a) $\dfrac{1}{3 \cdot 2!} - \dfrac{1}{5 \cdot 3!} + \dfrac{1}{7 \cdot 4!} - \dfrac{1}{9 \cdot 5!} + \cdots,\ |E| \leq .001$

    (b) $1 - \dfrac{1}{2!\ 2^2} + \dfrac{1}{4!\ 2^4} - \dfrac{1}{6!\ 2^6} + \cdots,\ |E| \leq .00005$

    (c) $1 - \dfrac{1}{3 \cdot 3^3} + \dfrac{1}{5 \cdot 3^5} - \dfrac{1}{7 \cdot 3^7} + \cdots,\ |E| \leq .00005$

5. For what choices of r does the series $\displaystyle\sum_{k=1}^{\infty} (-1)^{k+1} k^{-r}$ converge? Converge absolutely?

6. Theorem 81-2 states that the absolute value of the error we incur if we approximate the sum of the series $1 - \frac{2}{3} + (\frac{2}{3})^2 - (\frac{2}{3})^3 + \cdots$ by taking the sum of the first n terms is no larger than $(\frac{2}{3})^n$. What is the actual absolute value of the error?

7. Show that the geometric series $\displaystyle\sum_{k=0}^{\infty} x^k$ does not converge conditionally for any number x.

8. The following series satisfy only part of the hypotheses of Theorem 81-2; which part? Do they converge?

    (a) $\displaystyle\sum_{k=0}^{\infty} (-1)^{k+1} \dfrac{k}{3k - 1}$ (b) $\displaystyle\sum_{k=1}^{\infty} (-1)^{k+1} \dfrac{1 + \cos k\pi}{k}$

    (c) $\displaystyle\sum_{k=2}^{\infty} \dfrac{\cos k\pi}{\operatorname{Tan}^{-1} k\pi}$ (d) $\displaystyle\sum_{k=1}^{\infty} \int_{0}^{\pi} \sin kx\ dx$

9. The graph of a certain function f is the polygonal path that successively joins the points $(0, 0), (1, 1), (2, 0), (3, -\frac{1}{3}), (4, 0), (5, \frac{1}{5}), (6, 0), (7, -\frac{1}{7})$, and so on. Evaluate the integral $\displaystyle\int_{0}^{\infty} f(x)\ dx$ with an error of no more than $\frac{1}{10}$.

10. Does the series $\displaystyle\sum_{k=1}^{\infty} (-1)^{\llbracket \log k \rrbracket}/k$ converge? The series $\displaystyle\sum_{k=1}^{\infty} (-1)^{\llbracket \log k \rrbracket}/k^2$?

11. Suppose that $\{a_k\}$ is a nonincreasing sequence such that $\lim\limits_{k \uparrow \infty} a_k = 0$. Show that the series

    $$a_1 + a_2 - 2a_3 + a_4 + a_5 - 2a_6 + a_7 + a_8 - 2a_9 + \cdots$$

    converges.

12. Show that if f is continuous and nonincreasing in the interval $(1, \infty)$ and if $\lim\limits_{x \uparrow \infty} f(x) = 0$, then the integral $\displaystyle\int_{1}^{\infty} (-1)^{\llbracket x \rrbracket} f(x)\ dx$ converges.

13. Show that even though both of the series $\displaystyle\sum_{k=1}^{\infty} a_k$ and $\displaystyle\sum_{k=1}^{\infty} b_k$ converge, it may be possible that the series $\displaystyle\sum_{k=1}^{\infty} a_k b_k$ diverges. Show that if both series converge, at least one converging absolutely, then the series $\displaystyle\sum_{k=1}^{\infty} a_k b_k$ converges absolutely. [*Hint:* If $\displaystyle\sum_{k=1}^{\infty} a_k$ converges—conditionally or absolutely—then there is a number M such that $|a_k| \leq M$ for each index k.]

We have found throughout our study of calculus that drawing a sketch or two can often clear up a complicated-looking situation. Infinite series are no exception to this rule.

Each of the series $\sum_{k=1}^{\infty} \frac{1}{k}$, $\sum_{k=1}^{\infty} \frac{1}{2^k}$, and $\sum_{k=1}^{\infty} \frac{1}{k(k+1)}$ has the form $\sum_{k=1}^{\infty} f(k)$, where f is decreasing in the interval $(1, \infty)$ and $f(x) \geq 0$. In the first case, $f(x) = 1/x$; in the second, $f(x) = 2^{-x}$; and in the third, $f(x) = 1/x(x+1)$. Suppose that we have a general series of this type,

$$(82\text{-}1) \qquad \sum_{k=1}^{\infty} a_k = a_1 + a_2 + a_3 + \cdots,$$

where $a_k = f(k)$ for $k = 1, 2, \ldots$, with f being the function whose graph appears in Fig. 82-1. The terms of the series represent the lengths of the vertical segments that join the curve to the X-axis at the points $1, 2, \ldots$. By moving these segments one unit to the left, we "sweep out" the shaded rectangular regions, whose areas are therefore the terms of the series. Figure 82-1 also shows these rectangles shifted one unit to the right.

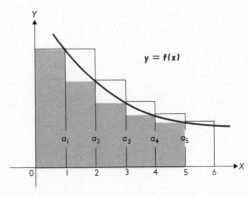

Figure 82-1

Since the area of each rectangle is given by a term of our series, we see that the fifth partial sum $S_5 = a_1 + a_2 + a_3 + a_4 + a_5$ is the area of the shaded step region. The last four steps of this region lie beneath our curve, and so it is clear that

$$S_5 \leq a_1 + \int_1^5 f(x)\, dx.$$

But when we shift the step region to the right, its upper boundary lies above the curve $y = f(x)$, and thus we have the inequality

$$\int_1^5 f(x) \le S_4 \le S_5.$$

These inequalities hold in general. That is, for each index n,

$$(82\text{-}2) \qquad \int_1^n f(x)\, dx \le S_n \le a_1 + \int_1^n f(x)\, dx.$$

The terms of Series 82-1 are nonnegative, so we know that it converges if and only if its sequence $\{S_n\}$ of partial sums is bounded. From Inequalities 82-2, it is clear that this sequence is bounded if and only if the integral $\int_1^\infty f(x)\, dx$ is convergent, and hence we have the **integral test** for the convergence of an infinite series.

Theorem   82-1.   *An infinite series whose terms are the values at the positive integers of a nonincreasing function f that does not take negative values in the interval $[1, \infty]$ converges if and only if the improper integral $\int_1^\infty f(x)\, dx$ converges.*

Example 82-1.   Apply the integral test to the series $\sum\limits_{k=1}^\infty \dfrac{1}{1 + k^2}$.

Solution. According to Theorem 82-1, the given series converges if and only if the improper integral $\int_1^\infty \dfrac{dx}{1 + x^2}$ converges. Therefore we investigate the limit of proper integrals of the form

$$\int_1^R \frac{dx}{1 + x^2} = \text{Tan}^{-1} x \bigg|_1^R = \text{Tan}^{-1} R - \tfrac{1}{4}\pi.$$

Since $\lim\limits_{R \uparrow \infty} \text{Tan}^{-1} R = \tfrac{1}{2}\pi$, we see that the integral $\int_1^\infty \dfrac{dx}{1 + x^2}$ converges, and so our given series converges.

The next example covers a number of cases (including the harmonic series) that we previously treated with special arguments.

Example 82-2.   For what choices of the exponent p is the *p*-series $\sum\limits_{k=1}^\infty \dfrac{1}{k^p}$ convergent?

Solution. The series obviously diverges if $p \leq 0$; if $p > 0$, the integral test says that the *p*-series will converge if and only if the improper integral $\int_1^\infty x^{-p} \, dx$ converges. We found in Example 76-2 that the improper integral converges if $p > 1$ and diverges if $p \leq 1$. Thus

$$\sum_{k=1}^\infty \frac{1}{k^p} \quad \begin{cases} \textbf{converges if } p > 1. \\ \textbf{diverges if } p \leq 1. \end{cases}$$

If Series 82-1 converges, we may take the limit of each term in Inequalities 82-2 and thereby gain an estimate of the sum S of the series:

$$(82\text{-}3) \qquad \int_1^\infty f(x) \, dx \leq S \leq a_1 + \int_1^\infty f(x) \, dx.$$

But it is easy to improve on this estimate. If m is a given positive integer, the remainder after m terms of Series 82-1 is the number $S - S_m = a_{m+1} + a_{m+2} + \cdots$. This number is the sum of what is left of Series 82-1 after an initial segment has been deleted. Thus the same argument that led to Inequalities 82-3 gives us the inequalities

$$\int_{m+1}^\infty f(x) \, dx \leq S - S_m \leq a_{m+1} + \int_{m+1}^\infty f(x) \, dx.$$

When we write these inequalities in the form

$$(82\text{-}4) \qquad 0 \leq S - \left(S_m + \int_{m+1}^\infty f(x) \, dx \right) \leq a_{m+1},$$

we see that the number $S_m + \int_{m+1}^\infty f(x) \, dx$ approximates the sum S of our series with an error no greater than a_{m+1}, the term of the series at which we stop adding and start integrating.

Example 82-3. How good is our approximation of the sum of the series $\sum_{k=3}^\infty \dfrac{1}{k^2 - 4}$ if we compute the sum of the first twelve terms and then integrate to estimate the remainder?

Solution. Since the index of summation starts at 3 in this example, the twelfth term is the one whose index is 14. Thus we choose $m = 14$, and Inequalities 82-4 tell us that the number

$$\tfrac{1}{5} + \tfrac{1}{12} + \cdots + \tfrac{1}{192} + \int_{15}^\infty \frac{dx}{x^2 - 4} \approx .4515 + .0671 = .5186$$

approximates the sum of our series with an error that lies between 0 and $1/(15^2 - 4) = \tfrac{1}{221}$.

Theorem 82-1 says that the sequence $\{S_n\}$ of partial sums of Series 82-1 and the sequence $\left\{\int_1^n f(x)\,dx\right\}$ of integrals of f either both converge or both diverge. Of course, if they both converge, then the sequence of differences $\{d_n\}$, where $d_n = S_n - \int_1^n f(x)\,dx$, converges. But the sequence of differences even converges when the series and the sequence of integrals diverge. We first observe from Inequalities 82-2 that $0 \le d_n$. Thus $\{d_n\}$ is bounded below; we will show that it is also nonincreasing, and then we will know that it converges. By definition,

$$d_n - d_{n+1} = S_n - \int_1^n f(x)\,dx - S_{n+1} + \int_1^{n+1} f(x)\,dx = \int_n^{n+1} f(x)\,dx - a_{n+1}.$$

Now for $x \le n + 1, f(x) \ge f(n + 1) = a_{n+1}$, and so

$$\int_n^{n+1} f(x)\,dx \ge \int_n^{n+1} a_{n+1}\,dx = a_{n+1}.$$

Therefore $d_n - d_{n+1} \ge 0$; that is, $\{d_n\}$ is a nonincreasing sequence.

Example 82-4. Discuss the sequence $\{1 + \frac{1}{2} + \cdots + n^{-1} - \ln n\}$.

Solution. We obtain this sequence by setting $f(x) = x^{-1}$ and calculating the differences between the partial sums of the harmonic series $1 + \frac{1}{2} + \frac{1}{3} + \cdots$ and the members of the sequence $\left\{\int_1^n x^{-1}\,dx\right\}$. Both the series and the improper integral $\int_1^\infty x^{-1}\,dx$ diverge, but we just saw that the sequence of differences converges, nevertheless. The limit of this sequence, the number

$$\lim_{n\uparrow\infty} (1 + \tfrac{1}{2} + \cdots + n^{-1} - \ln n),$$

is called **Euler's constant**. Its value is $.5772156\ldots$, and to three decimal places, $d_8 = .597$.

Not only can we use our knowledge of improper integrals to test the convergence of infinite series, but we can also use our knowledge of series to examine the convergence of integrals.

Example 82-5. Discuss the convergence of the improper integral $\int_0^\infty \dfrac{\sin x}{x}\,dx$.

Solution. In Problem 77-10 we asked you to use integration by parts to show that this integral converges; now we will use a different technique. If we break up the positive X-axis into the intervals $[0, \pi]$, $[\pi, 2\pi]$, $[2\pi, 3\pi]$, and so on, then our integral converges if and only if the following series converges:

$$(82\text{-}5) \qquad \int_0^\pi \frac{\sin x}{x}\, dx + \int_\pi^{2\pi} \frac{\sin x}{x}\, dx + \int_{2\pi}^{3\pi} \frac{\sin x}{x}\, dx + \cdots.$$

The kth term of this series is

$$a_k = \int_{(k-1)\pi}^{k\pi} \frac{\sin x}{x}\, dx,$$

which is the area of the shaded "hump" in Fig. 82-2 (if k is an odd index). The next term, a_{k+1}, is the negative of the area of the shaded "dip." Since this second area is smaller than the first, $|a_k| > |a_{k+1}|$, and we are dealing with an alternating series of decreasing terms. Now notice that if x is between $(k-1)\pi$ and $k\pi$, then

$$\frac{|\sin x|}{k\pi} \le \frac{|\sin x|}{x} \le \frac{|\sin x|}{(k-1)\pi},$$

and so

$$\int_{(k-1)\pi}^{k\pi} \frac{|\sin x|}{k\pi}\, dx \le \int_{(k-1)\pi}^{k\pi} \frac{|\sin x|}{x}\, dx \le \int_{(k-1)\pi}^{k\pi} \frac{|\sin x|}{(k-1)\pi}\, dx.$$

Therefore, since $\displaystyle\int_{(k-1)\pi}^{k\pi} |\sin x|\, dx = 2$,

$$(82\text{-}6) \qquad \frac{2}{k\pi} \le |a_k| \le \frac{2}{(k-1)\pi},$$

and hence $\displaystyle\lim_{k\uparrow\infty} a_k = 0$. So Theorem 81-2 tells us that Series 82-5 converges. But Inequalities 82-6 also say that if we replace each term of Series 82-5 with its absolute value, the resulting series dominates $2/\pi$ times the harmonic series. Since this latter series diverges, it follows that Series 82-5 (and hence our original improper integral) is only conditionally, not absolutely, convergent.

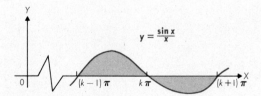

$$y = \frac{\sin x}{x}$$

Figure 82-2

PROBLEMS 82

1. Use the integral test to test the given series for convergence.

(a) $\displaystyle\sum_{k=1}^{\infty} e^{-k}$

(b) $\displaystyle\sum_{k=1}^{\infty} k^{-e}$

(c) $\displaystyle\sum_{k=1}^{\infty} 1/\sqrt{k} \; e^{\sqrt{k}}$

(d) $\displaystyle\sum_{k=1}^{\infty} \sqrt{k}/e^{\sqrt{k}}$

(e) $\displaystyle\sum_{k=1}^{\infty} k^{-2} \sin k^{-1}$

(f) $\displaystyle\sum_{k=1}^{\infty} k^{-2} \tan k^{-1}$

(g) $\displaystyle\sum_{k=2}^{\infty} \frac{\ln k}{k^2}$

(h) $\displaystyle\sum_{k=2}^{\infty} \frac{(\ln k)^2}{k}$

(i) $\displaystyle\sum_{k=3}^{\infty} \left(\frac{\ln k}{k}\right)^2$

(j) $\displaystyle\sum_{k=3}^{\infty} \frac{\ln k^2}{k}$

(k) $\displaystyle\sum_{k=1}^{\infty} \operatorname{sech}^2 k$

(l) $\displaystyle\sum_{k=1}^{\infty} \operatorname{csch}^2 k$

2. Use the integral test to check the convergence of the given series.

(a) $\displaystyle\sum_{k=1}^{\infty} (k + k^{-1})^{-1}$

(b) $\displaystyle\sum_{k=1}^{\infty} (1 + k^{-2})^{-1/2}$

(c) $\displaystyle\sum_{k=1}^{\infty} k^{-1} \sin k^{-1}$

(d) $\displaystyle\sum_{k=1}^{\infty} k^{-1} \operatorname{Sin}^{-1} k^{-1}$

3. (a) Look back in the book to see where we have previously tested the series $\displaystyle\sum_{k=1}^{\infty} (\tfrac{1}{2}\pi - \operatorname{Tan}^{-1} k)$ for convergence.

   (b) Now use the integral test (you might want to use integration by parts, as in Example 59-4, to find $\displaystyle\int_0^R \operatorname{Tan}^{-1} x \, dx$).

4. Show that the series $\displaystyle\sum_{k=2}^{\infty} \frac{1}{k(\ln k)^p}$ converges if $p > 1$ and diverges if $p \le 1$.

5. Use the integral test to find values of r for which the series $\displaystyle\sum_{k=1}^{\infty} k(1 + k^2)^r$ converges.

6. (a) Explain why the series $\displaystyle\sum_{k=1}^{\infty} (1 - \tanh k)$ converges if and only if $\displaystyle\lim_{R\uparrow\infty} (R - \ln \cosh R)$ is a number.

   (b) Evaluate this limit by showing that $R - \ln \cosh R = \ln 2/(1 - e^{-2R})$.

7. Estimate the sum of the series $\displaystyle\sum_{k=1}^{\infty} \frac{1}{1 + k^2}$ with an error of no more than .01.

8. Let $\zeta(p)$ be the sum of the p-series (Example 82-2) for $p > 1$.
   (a) Use Equation 82-3 to show that $1/(p - 1) \le \zeta(p) \le p/(p - 1)$.
   (b) Show that for each positive integer m,

$$\zeta(p) = \frac{1}{1^p} + \cdots + \frac{1}{m^p} + \frac{1}{(p - 1)(m + 1)^{p-1}} + E_m,$$

where

$$0 \le E_m \le \frac{1}{(m + 1)^p}.$$

9. Let r be a number between 0 and 1. Use the integral test to show that the geometric series $\sum_{k=0}^{\infty} r^k$ converges. Use Inequalities 82-3 to show that $1 - r \leq \ln 1/r \leq (1 - r)/r$.

10. Suppose that f is continuous and nondecreasing in the interval $(0, 1]$ and does not take negative values. Show that

$$\lim_{n \uparrow \infty} \left[f(1) + \cdots + f\left(\frac{1}{n}\right) - \int_{1/n}^{1} x^{-2} f(x) \, dx \right]$$

is a number.

11. Show that $\sum_{k=1}^{\infty} f'(k)$ converges if $f''(x) \leq 0$, $f'(x) \geq 0$, and there is a number M such that $f(x) \leq M$ for all x.

12. Let g be a function that is continuous and nonincreasing in an interval $[a, \infty]$ and is such that $\lim_{x \uparrow \infty} g(x) = 0$. Use the technique of Example 82-5 to show that the integral $\int_{a}^{\infty} g(x) \sin x \, dx$ converges and that this integral converges absolutely if and only if the integral $\int_{a}^{\infty} g(x) \, dx$ converges.

13. Find an example of a function that takes nonnegative values in the interval $[1, \infty]$ and is such that $\sum_{k=1}^{\infty} f(k)$ converges but $\int_{1}^{\infty} f(x) \, dx$ diverges. Can you find an example of another such function for which the integral converges but the series does not?

83. THE RATIO AND ROOT TESTS

In this section we will develop tests for divergence or absolute convergence, so in addition to a given series

(83-1) $$\sum_{k=1}^{\infty} a_k = a_1 + a_2 + a_3 + \cdots,$$

we consider the series of absolute values of its terms,

(83-2) $$\sum_{k=1}^{\infty} |a_k| = |a_1| + |a_2| + |a_3| + \cdots.$$

Because we will be dividing by the terms of our series, we will assume that none of them is 0. Therefore all the terms of Series 83-2 are positive.

Let us suppose that, after a certain index, the ratio of any term of Series 83-2 to the preceding term is less than or equal to a number q, which itself is less than 1. Thus we are assuming that there is some index K such that for each larger index k,

(83-3)
$$\frac{|a_{k+1}|}{|a_k|} \le q < 1.$$

For example, the ratio of successive terms of the series $\sum_{k=1}^{\infty} 3^k/k!$ is

$$\frac{a_{k+1}}{a_k} = \frac{3^{k+1}}{(k+1)!} \cdot \frac{k!}{3^k} = \frac{3}{k+1}.$$

If $k > 4$, this ratio does not exceed $\frac{1}{2}$, so the numbers $\frac{1}{2}$ and 4 can play the roles of q and K here.

Since $q = q^{k+1}/q^k$, the first of Inequalities 83-3 can be written as $|a_{k+1}|/|a_k| \le q^{k+1}/q^k$ or $|a_{k+1}|/q^{k+1} \le |a_k|/q^k$. Thus after the Kth term, the sequence $\{|a_k|/q^k\}$ of positive terms does not increase. In particular, no term can exceed $|a_{K+1}|/q^{K+1}$. For short, let us denote this number by p. So we are really assuming that there is a number p and an index K such that $|a_k|/q^k \le p$ for each $k \ge K+1$. These inequalities say that the series $\sum_{k=K+1}^{\infty} |a_k|$ is dominated by the series $\sum_{k=K+1}^{\infty} pq^k$. But this latter series is simply a geometric series, and it converges, since $q < 1$. The comparison test therefore tells us that Series 83-2 converges. Our example $\sum_{k=1}^{\infty} 3^k/k!$ is a series of the type we have been talking about, so it converges.

In using this *ratio test* to demonstrate that a series converges, we must show that the ratios of the absolute values of successive terms ultimately do not exceed a number q that itself is actually less than 1. The analogous test for divergence does not require us to be quite so careful. Suppose that there is an index K such that for each larger index k,

(83-4)
$$\frac{|a_{k+1}|}{|a_k|} \ge 1.$$

Then from some term onward the sequence of positive terms $\{|a_k|\}$ is nondecreasing and hence cannot have 0 as a limit. Therefore, according to Theorem 79-1, Series 83-1 diverges. We summarize our results as a theorem.

Theorem 83-1. *If after some term in the infinite series $\sum_{k=1}^{\infty} a_k$ we have $|a_{k+1}|/|a_k| \le q < 1$ for each index k, then the series is absolutely convergent. If after some term we have $|a_{k+1}|/|a_k| \ge 1$ for each index k, the series diverges.*

Example 83-1. Show that the series $\sum_{k=1}^{\infty} \frac{k}{2^k}$ converges.

Solution. Ratios of successive terms are given by the formula

$$\frac{k+1}{2^{k+1}} \cdot \frac{2^k}{k} = \frac{k+1}{k} \cdot \frac{1}{2}.$$

If $k > 1$, this ratio does not exceed $\frac{3}{4}$, and since $\frac{3}{4} < 1$, the convergence hypothesis of Theorem 83-1 is satisfied and the series converges.

Example 83-2. For what choices of x does the series

$$(83\text{-}5) \qquad \sum_{k=0}^{\infty} [2 + (-1)^k]x^k = 3 + x + 3x^2 + x^3 + 3x^4 + \cdots$$

converge, and for what choices does it diverge?

Solution. For each nonzero number x, the sequence of ratios of the absolute values of successive terms of our given series is

$$\{\tfrac{1}{3}|x|, \, 3|x|, \, \tfrac{1}{3}|x|, \, 3|x|, \dots \}.$$

No term of this sequence is greater than $3|x|$, so this number can serve as the number q of Theorem 83-1 if it is less than 1. Thus if $|x| < \frac{1}{3}$, our series converges. For example, the series

$$3 + \frac{1}{4} + \frac{3}{4^2} + \frac{1}{4^3} + \frac{3}{4^4} + \cdots$$

converges; it is the series we get when we replace x with $\frac{1}{4}$ in Series 83-5, and $\frac{1}{4} < \frac{1}{3}$.

   No term of our sequence of ratios is less than $\frac{1}{3}|x|$, and this number will be greater than or equal to 1 if $|x| \geq 3$. Therefore, according to the second part of Theorem 83-1, our series diverges if $|x| \geq 3$.

   In summary, our theorem tells us that the given series diverges if $x \leq -3$, converges if x is a number in the interval $(-\frac{1}{3}, \frac{1}{3})$, and diverges if $x \geq 3$. If $x \in (-3, -\frac{1}{3}] \cup [\frac{1}{3}, 3)$, Theorem 83-1 does not apply, and we must use other tests (see Example 83-4).

In order to apply Theorem 83-1, we must show that the appropriate one of Inequalities 83-3 or 83-4 holds for each index k that is larger than some particular index K. Thus we must really verify infinitely many inequalities. One method of showing that all these inequalities hold is to calculate the limit of the quotient $|a_{k+1}|/|a_k|$ if this limit exists. So let us suppose that

$$\lim_{k \uparrow \infty} \frac{|a_{k+1}|}{|a_k|} = L.$$

Thus if (p, q) is any neighborhood of L, there is an index K such that if $k > K$, then

$$\frac{|a_{k+1}|}{|a_k|} \in (p, q).$$

If L is a number that is less than 1, we can choose our neighborhood (p, q) so that q is also less than 1. Therefore Inequalities 83-3 hold, and Theorem 83-1 tells us that our series converges absolutely. If $L > 1$ or if L is the symbol ∞, we can take as (p, q) the interval $(1, \infty)$ and hence $|a_{k+1}|/|a_k| > 1$ for $k > K$. Thus Inequality 83-4 holds, so our series diverges.

If $L = 1$, we do not have enough information to tell whether our series converges or diverges. For example, look at the p-series $\sum_{k=1}^{\infty} k^{-p}$. Here the ratio $|a_{k+1}|/|a_k|$ is $k^p/(k + 1)^p = [k/(k + 1)]^p$, and its limit is 1, regardless of the choice of p. But we know that the p-series converges if $p > 1$ and diverges if $p \leq 1$, so we see that there are both convergent and divergent series for which $\lim_{k\uparrow\infty} |a_{k+1}|/|a_k| = 1$. Thus we have the following **ratio test**.

Theorem 83-2. *Suppose that* $\lim_{k\uparrow\infty} \dfrac{|a_{k+1}|}{|a_k|} = L$. *Then*

(i) *if $L < 1$, the series $\sum_{k=1}^{\infty} a_k$ converges absolutely,*

(ii) *if $L > 1$ or if L is the symbol ∞, the series $\sum_{k=1}^{\infty} a_k$ diverges, and*

(iii) *if $L = 1$, we need additional information to determine whether or not the series converges.*

Example 83-3. Test the series $\sum_{k=1}^{\infty} \dfrac{10^{2k}}{k!}$ for convergence.

Solution. Here

$$\frac{|a_{k+1}|}{|a_k|} = \frac{10^{2k+2}}{(k + 1)!} \cdot \frac{k!}{10^{2k}} = \frac{100}{k + 1},$$

so

$$\lim_{k\uparrow\infty} \frac{|a_{k+1}|}{|a_k|} = \lim_{k\uparrow\infty} \frac{100}{k + 1} = 0.$$

Certainly this limit, $L = 0$, is less than 1, so the ratio test tells us that the series converges.

The basic technique that we have used so far in this section to test a given series for convergence is to compare it with a geometric series. There are many other applications and refinements of this idea; let us mention one that we will use when we study power series in the next section.

With our given Series 83-1 we can associate the sequence

$$\{|a_k|^{1/k}\} = \{|a_1|, |a_2|^{1/2}, |a_3|^{1/3}, \ldots\}.$$

Suppose that there is a number q and an index K such that for each index k beyond K,

(83-6) $$|a_k|^{1/k} \leq q < 1.$$

These inequalities tell us two things: (1) that the series $\sum\limits_{k=K+1}^{\infty} |a_k|$ is dominated by the geometric series $\sum\limits_{k=K+1}^{\infty} q^k$, and (2) that this geometric series converges. It follows that Series 83-1 converges absolutely. On the other hand, suppose that for infinitely many indices,

(83-7) $$|a_k|^{1/k} \geq 1.$$

Then the sequence $\{a_k\}$ cannot have a limit of 0, and so Series 83-1 diverges.

To give us a criterion for determining which (if either) of Inequalities 83-6 or 83-7 holds, we now introduce the concept of the *upper limit* of a sequence.

Definition   83-1.   *We write* $\overline{\lim\limits_{k\uparrow\infty}} \, r_k = L$ *and call L the* **upper limit** *of a sequence* $\{r_k\}$ *if for every neighborhood* (p, q) *of L,*
  (i)   *there are infinitely many terms of the sequence in* (p,q), *and*
  (ii) *there is an index K such that* $r_k \in (-\infty, q)$ *for each* $k > K$.
  In other words, infinitely many terms of the sequence $\{r_{K+1}, r_{K+2}, \ldots\}$ *are greater than p, and none exceeds q.*

For example, 1 is the upper limit of the sequence $\{(-1)^k\}$. If we look at the neighborhood $(\frac{1}{2}, \frac{3}{2})$ of 1, we see that infinitely many terms of the sequence exceed $\frac{1}{2}$ and all terms are exceeded by $\frac{3}{2}$. These same remarks apply to any neighborhood of 1.

If the sequence $\{r_k\}$ actually has a limit L, then to every neighborhood (p, q) of L there corresponds an index K such that all terms of the sequence $\{r_{K+1}, r_{K+2}, \ldots\}$ belong to (p, q), and hence statements (i) and (ii) of Definition 83-1 are satisfied. Thus *if a sequence has a limit, then that limit is also its upper limit.* But a sequence can have an upper limit without having a limit, as the example $\{(-1)^k\}$ shows. With the aid of the Completeness Property 2-1, we can show that every sequence has an upper limit.

Now we can prove another convergence theorem, the **root test**.

Theorem   83-3.   *Suppose that* $\overline{\lim}_{k\uparrow\infty} |a_k|^{1/k} = L.$ *Then*

(i)   *if* $L < 1$, *the series* $\sum_{k=1}^{\infty} a_k$ *converges absolutely*,

(ii)   *if* $L > 1$ *or if* L *is the symbol* ∞, *the series* $\sum_{k=1}^{\infty} a_k$ *diverges, and*

(iii)   *if* $L = 1$, *we need additional information to determine whether or not the series converges.*

PROOF

The equation $\overline{\lim}_{k\uparrow\infty} |a_k|^{1/k} = L$ means that to each neighborhood (p, q) of L there corresponds an index K such that for each larger index k we have $|a_k|^{1/k} \in (-\infty, q)$, and there are infinitely many terms of the sequence such that $|a_k|^{1/k} > p$. If $L < 1$, we can choose $q < 1$, so Inequalities 83-6 hold and our series converges. If $L > 1$ or if L is the symbol ∞, we can choose $p = 1$; therefore Inequality 83-7 holds and our series diverges. We will leave to you the problem of finding examples of convergent and divergent series for which $L = 1$, thus showing that in this case we have insufficient information from which to conclude convergence or divergence.

Example 83-4.   Apply the root test to the series of Example 83-2.

Solution.   For convenience of notation, let us simply discard the first term of Series 83-5; thus for a given number x our sequence $\{|a_k|^{1/k}\}$ is

$$\{|x|, \ 3^{1/2}|x|, \ |x|, \ 3^{1/4}|x|, \ |x|, \ 3^{1/6}|x|, \ |x|, \dots\}.$$

Since $\lim_{n\uparrow\infty} 3^{1/2n} = 1$, we see that $\lim_{k\uparrow\infty} |a_k|^{1/k} = |x|$. Therefore $\overline{\lim}_{k\uparrow\infty} |a_k|^{1/k} = |x|$, so Theorem 83-3 tells us that our series converges if $|x| < 1$ and diverges if $|x| > 1$. This result, of course, does not conflict with the result of Example 83-2; it is simply more complete. Now we know that our series diverges if $x < -1$, converges if $x \in (-1, 1)$, and diverges if $x > 1$. Neither test tells us what happens if x is equal to 1 or -1. When you replace x in Series 83-5 with either 1 or -1, you can easily see that you have a divergent series. Thus the complete answer to the question raised in Example 83-2 is that Series 83-5 converges if $|x| < 1$ and diverges otherwise.

PROBLEMS   83

1.   Use the ratio test to see if the series converges.

(a) $\sum_{k=1}^{\infty} 2^{-k/2}$ (b) $\sum_{k=1}^{\infty} 2^{-2/k}$ (c) $\sum_{k=1}^{\infty} \frac{k^2}{2^k}$

(d) $\sum_{k=1}^{\infty} \frac{k^{\ln 2}}{(\ln 2)^k}$ (e) $\sum_{k=1}^{\infty} \frac{2^k}{k!}$ (f) $\sum_{k=1}^{\infty} \frac{(k!)^2}{(2k)!}$

(g) $\dfrac{1}{3} + \dfrac{1 \cdot 3}{3 \cdot 6} + \dfrac{1 \cdot 3 \cdot 5}{3 \cdot 6 \cdot 9} + \cdots$

(h) $1 + \dfrac{2!}{1 \cdot 3} + \dfrac{3!}{1 \cdot 3 \cdot 5} + \dfrac{4!}{1 \cdot 3 \cdot 5 \cdot 7} + \cdots$

$\underline{(2n)!}$

(i) $\displaystyle\sum_{k=1}^{\infty} \dfrac{\exp \displaystyle\int_0^k \sin x^2 \, dx}{k!}$

(j) $\displaystyle\sum_{k=1}^{\infty} \dfrac{\exp \displaystyle\int_0^k \cos e^x \, dx}{\exp k}$

2. Recall that $\lim_{k \uparrow \infty} (1 + k^{-1})^k = e$ and hence use the ratio test to check the convergence of the series $\displaystyle\sum_{k=1}^{\infty} k!/k^k$.

3. Apply the root test to the following series.

(a) $\displaystyle\sum_{k=1}^{\infty} 3^{-k/2}$

(b) $\displaystyle\sum_{k=1}^{\infty} (.3)^{-k/.2}$

(c) $\displaystyle\sum_{k=2}^{\infty} \dfrac{(\ln k)^{-k}}{k}$

(d) $\displaystyle\sum_{k=1}^{\infty} k^{-k}$

(e) $\displaystyle\sum_{k=2}^{\infty} (\mathrm{Tan}^{-1} k)^{-k}$

(f) $\displaystyle\sum_{k=1}^{\infty} (\mathrm{Cos}^{-1} k^{-1})^{-k}$

(g) $\displaystyle\sum_{k=1}^{\infty} [3 + (-1)^k]^{-k}$

(h) $\displaystyle\sum_{k=1}^{\infty} [.3 + (-.1)^k]^{-k}$

4. Show that neither the ratio test nor the root test helps to decide whether or not the given series converges, but that the integral test does.

(a) $\displaystyle\sum_{k=1}^{\infty} k^{-2}$

(b) $\displaystyle\sum_{k=1}^{\infty} e^{-\sqrt{k}}$

5. Is the ratio test applicable to the series

$$\frac{1}{2} + \frac{1 \cdot 3}{2 \cdot 4} + \frac{1 \cdot 3 \cdot 5}{2 \cdot 4 \cdot 6} + \cdots ?$$

Does the series converge?

6. Use the ratio test to determine x so that the given series converges.

(a) $\displaystyle\sum_{k=1}^{\infty} k x^k$

(b) $\displaystyle\sum_{k=1}^{\infty} \dfrac{k^2}{x^k}$

(c) $\displaystyle\sum_{k=1}^{\infty} \dfrac{e^{kx}}{k}$

(d) $\displaystyle\sum_{k=0}^{\infty} \sin^k x$

(e) $\displaystyle\sum_{k=0}^{\infty} \dfrac{(2x)^k}{(x^2 + 1)^k}$

(f) $\displaystyle\sum_{k=1}^{\infty} k^3 x^{3k}$

(g) $\displaystyle\sum_{k=1}^{\infty} (\ln x)^k$

(h) $\displaystyle\sum_{k=1}^{\infty} \exp \left(k \int_0^x |\sin t| \, dt \right)$

7. Suppose that the ratio test or the root test shows that the series $\sum\limits_{k=1}^{\infty} a_k$ converges. Show that then $\sum\limits_{k=1}^{\infty} ka_k$ also converges. Is $\sum\limits_{k=1}^{\infty} k^2 a_k$ convergent?

8. Find $\overline{\lim\limits_{k \uparrow \infty}} \, r_k$.

   (a) $r_k = \sin \tfrac{1}{2}k\pi$ (b) $r_k = \sin \tfrac{1}{3}k\pi$

   (c) $r_k = \sin^k \tfrac{1}{3}\pi$ (d) $r_k = \left(-1 - \dfrac{1}{k}\right)^k$

9. The ratios of successive terms of the harmonic series $\sum\limits_{k=0}^{\infty} 1/k$ form the sequence $\{k/(k + 1)\}$. Each term of this sequence is less than 1, yet the harmonic series diverges. Does this fact contradict Theorem 83-1?

10. Let $\sum\limits_{k=0}^{\infty} a_k$ be a convergent series of positive terms and suppose that $a_{k+1}/a_k \leq q < 1$ for all $k \geq m$. Show that $0 \leq R_m \leq a_{m+1}/(1 - q)$. Find an upper bound for the error you incur when you use the first five terms of the series $\sum\limits_{k=0}^{\infty} \dfrac{2^k}{k!}$ to approximate its sum.

11. Find a convergent series $\sum\limits_{k=1}^{\infty} a_k$ of positive terms and a divergent series $\sum\limits_{k=1}^{\infty} b_k$ of positive terms such that $\overline{\lim\limits_{k \uparrow \infty}} \, a_k^{1/k} = 1$ and $\overline{\lim\limits_{k \uparrow \infty}} \, b_k^{1/k} = 1$.

12. Make up several examples of series to which Theorem 83-1 applies but Theorem 83-2 does not. Find examples of series to which Theorem 83-3 applies but Theorem 83-2 does not. Can you find a series to which Theorem 83-2 applies but Theorem 83-3 does not?

13. (a) Show that if $\overline{\lim\limits_{k \uparrow \infty}} \, |a_{k+1}|/|a_k| < 1$, the series $\sum\limits_{k=0}^{\infty} a_k$ converges absolutely.

    (b) Show that if $\overline{\lim\limits_{k \uparrow \infty}} \, (-|a_{k+1}|/|a_k|) < -1$, the series $\sum\limits_{k=1}^{\infty} a_k$ diverges.

14. Give a definition for the *lower limit* of a sequence. Give an example of a sequence for which the lower limit exists but which does not have a limit. Prove that a sequence has a limit if and only if its upper limit and its lower limit are the same.

84. POWER SERIES

Suppose that we select a sequence $\{c_k\}$ of numbers and a number a. Then for each number x we can form the following **power series** in $(x - a)$:

$$(84\text{-}1) \qquad \sum_{k=0}^{\infty} c_k(x - a)^k = c_0 + c_1(x - a) + c_2(x - a)^2 + \cdots .$$

The terms of the sequence $\{c_k\}$ are the **coefficients** of the series. Thus in the first of the power series

(84-2) $$\sum_{k=0}^{\infty} \frac{x^k}{k!} \quad \text{and} \quad \sum_{k=1}^{\infty} \frac{(-1)^k}{k} (x - 3)^k,$$

$a = 0$, $c_0 = 1/0! = 1$, $c_1 = 1$, $c_2 = 1/2!$, and so on. In the second series, $a = 3$, $c_0 = 0$, $c_1 = -1$, $c_2 = \frac{1}{2}$, and so on. Power series play a large role in certain branches of mathematical analysis, and they are widely used in applied mathematics. We will obtain some idea of why they are important as we study the next few sections.

We can use any number x when we form a power series, and we naturally want to know what numbers yield convergent series. For example, if we replace x with 1 in the first of Series 84-2, we obtain the convergent series $\sum_{k=0}^{\infty} 1/k!$, whereas substituting 1 for x in the second series gives us the divergent series $\sum_{k=1}^{\infty} 2^k/k$. We observe that when we replace x with a in Series 84-1, we obtain the convergent series $c_0 + 0 + 0 + \cdots$, so every power series converges for at least this one choice of x. But suppose that $x \neq a$; does our series converge or diverge? An example will suggest an answer.

Example 84-1. For what choices of x does the second of Series 84-2 converge?

Solution. The easiest answer to this question is probably furnished by the ratio test. For the given series,

$$\frac{|a_{k+1}|}{|a_k|} = \frac{|x - 3|^{k+1}}{k + 1} \frac{k}{|x - 3|^k} = \frac{k}{k + 1} |x - 3|,$$

and the limit of this ratio (as $k \uparrow \infty$) is clearly $|x - 3|$. Hence our series converges if $|x - 3| < 1$ (that is, x belongs to the interval $(2, 4)$) and it diverges if $|x - 3| > 1$ (that is, x lies outside the interval $[2, 4]$). If $|x - 3| = 1$, we must use some other convergence test. Thus if $x = 2$, our series is the divergent harmonic series, while if $x = 4$, it is the convergent alternating series $\sum_{k=1}^{\infty} (-1)^k/k$. Finally, then, the convergence set of the second of Series 84-2 is the half-open interval $(2, 4]$.

The behavior of the specific series we just studied is typical of power series. In general, we expect Series 84-1 to converge in some **interval of convergence** $(a - r, a + r)$ whose center is a. The number r (half the width of the interval) is the **radius of convergence**. The series diverges outside the closed interval $[a - r, a + r]$, and some series converge at one or both endpoints and some don't. It is possible for a series to converge only for the choice $x = a$, but a series may converge for every choice of x (in which case, we say that $(-\infty, \infty)$ is the interval of convergence). The precise state of affairs is summed up in the next theorem, a proof of which appears at the end of the section.

Theorem 84-1.   *For a power series $\sum_{k=0}^{\infty} c_k(x - a)^k$, one of the following statements is true.*

  (i)   *The series converges only for the "trivial choice" $x = a$.*

  (ii) *The series converges absolutely for every number x.*

  (iii) *There is a positive number r such that the series converges absolutely for every number x in the open interval $(a - r, a + r)$ and diverges for every number x outside the closed interval $[a - r, a + r]$.*

Example 84-2.   Find the interval of convergence of the power series $\sum_{k=0}^{\infty} \dfrac{x^k}{k!}$.

Solution.   The ratios of the absolute values of successive terms of our series have the form

$$\frac{|x|^{k+1}}{(k + 1)!} \cdot \frac{k!}{|x|^k} = \frac{k!}{(k + 1)!} \frac{|x|^{k+1}}{|x|^k} = \frac{1}{k + 1} |x|.$$

No matter what x is, $\lim_{k\uparrow\infty} \dfrac{1}{k + 1} |x| = 0$, and since this number is less than 1, Theorem 83-2 tells us that our series converges. Thus the interval of convergence of the given series is $(-\infty, \infty)$.

The example we just worked has a byproduct that will serve us in Section 86, so let us list it here. We have just shown that if x is any number, then the series $\sum_{k=0}^{\infty} x^k/k!$ converges. Hence its sequence of terms has the limit 0; that is, *for each number x,*

$$(84\text{-}3) \qquad\qquad \lim_{k \uparrow \infty} \frac{x^k}{k!} = 0.$$

Example 84-3.   Discuss the convergence of the power series

$$\sum_{k=1}^{\infty} \frac{(-1)^k 2^k (x - 5)^{2k}}{k^2}.$$

Solution.   We first find the limit of the ratios of the absolute values of successive terms. Here $|a_k| = 2^k(x - 5)^{2k}/k^2$, so

$$\lim_{k\uparrow\infty} \frac{|a_{k+1}|}{|a_k|} = \lim_{k\uparrow\infty} \frac{2^{k+1}(x - 5)^{2k+2}}{(k + 1)^2} \cdot \frac{k^2}{2^k(x - 5)^{2k}}$$

$$= \lim_{k\uparrow\infty} 2(x - 5)^2 \left(\frac{k}{k + 1}\right)^2 = 2(x - 5)^2.$$

Our series will converge if the number $2(x - 5)^2$ is less than 1, and it will diverge if $2(x - 5)^2 > 1$. The inequality $2(x - 5)^2 < 1$ is equivalent to the inequality $|x - 5| < \frac{1}{2}\sqrt{2}$. Thus the radius of convergence of our series is $\frac{1}{2}\sqrt{2}$, and its interval of convergence is $(5 - \frac{1}{2}\sqrt{2}, 5 + \frac{1}{2}\sqrt{2})$. If we replace x with $5 - \frac{1}{2}\sqrt{2}$ or with $5 + \frac{1}{2}\sqrt{2}$, we get the convergent alternating series $\sum_{k=1}^{\infty} (-1)^k k^{-2}$. Therefore our given series converges for x in the closed interval $[5 - \frac{1}{2}\sqrt{2}, 5 + \frac{1}{2}\sqrt{2}]$, and it diverges for x outside this interval.

We define the sum (and similarly the difference) of two power series $\sum_{k=0}^{\infty} c_k(x - a)^k$ and $\sum_{k=0}^{\infty} d_k(x - a)^k$ by means of the equation

$$(84\text{-}4) \quad \sum_{k=0}^{\infty} c_k(x - a)^k + \sum_{k=0}^{\infty} d_k(x - a)^k = \sum_{k=0}^{\infty} (c_k + d_k)(x - a)^k.$$

Thus the sum of two power series is again a power series, and Equation 79-7 tells us that if x is any number for which the summand series converge, then the sum series converges. Therefore the radius of convergence of the sum series is at least as large as the smaller of the radii of convergence of the summand series.

The product of two power series is also a power series, and again we define its coefficients in the "natural way," although here the "natural way" is not quite as obvious as in the case of the sum. The term that contains $(x - a)^k$ in the product of the two series $\sum_{k=0}^{\infty} c_k(x - a)^k$ and $\sum_{k=0}^{\infty} d_k(x - a)^k$ is the sum of the products $c_0 \cdot d_k(x - a)^k$, $c_1(x - a) \cdot d_{k-1}(x - a)^{k-1}$, $c_2(x - a)^2 \cdot d_{k-2}(x - a)^{k-2}$, and so on; that is,

$$(c_0 d_k + c_1 d_{k-1} + c_2 d_{k-2} + \cdots + c_k d_0)(x - a)^k = \left(\sum_{r=0}^{k} c_r d_{k-r} \right)(x - a)^k.$$

Thus we define the product of two power series by means of the equation

$$(84\text{-}5) \quad \left(\sum_{k=0}^{\infty} c_k(x - a)^k \right)\left(\sum_{k=0}^{\infty} d_k(x - a)^k \right) = \sum_{k=0}^{\infty} \left(\sum_{r=0}^{k} c_r d_{k-r} \right)(x - a)^k.$$

We won't prove it, but it can be shown that the radius of convergence of the product series is at least as large as the smaller of the radii of convergence of the factor series.

Instead of using Equation 84-5, which is sometimes rather complicated to apply, we frequently find it easier to find an initial segment of a product series directly. To obtain the initial segment that ends with the term containing $(x - a)^k$, we multiply the initial segments of the factor series that end with the terms containing $(x - a)^k$ and then discard all terms that contain $(x - a)$ to a power larger than k.

Example 84-4. Discuss the product of the power series $\sum\limits_{k=0}^{\infty} x^k$ and $\sum\limits_{k=1}^{\infty} kx^{k-1}$.

Solution. As we just said, perhaps the easiest way to find a given initial segment of the product is by straightforward algebra rather than by relying on Formula 84-5. Thus suppose that we want to find the initial segment of the product series that ends with the term containing x^3. We multiply initial segments of the factor series:

$$
\begin{array}{r}
1 + x + x^2 + x^3 \\
1 + 2x + 3x^2 + 4x^3 \\
\hline
1 + 3x + 6x^2 + 10x^3 + 9x^4 + 7x^5 + 4x^6.
\end{array}
$$

Now we discard the terms that contain x to a power higher than 3, and we obtain the desired initial segment of the product:

$$ 1 + 3x + 6x^2 + 10x^3. $$

If we want to use Equation 84-5, our first task is to write the power series $\sum\limits_{k=1}^{\infty} kx^{k-1}$ in "standard form"—that is, as $\sum\limits_{k=0}^{\infty} d_k x^k$. You can readily check that $\sum\limits_{k=1}^{\infty} kx^{k-1} = \sum\limits_{k=0}^{\infty} (k + 1)x^k$. Now we apply Equation 84-5 with $c_k = 1$ for all k and $d_k = k + 1$, and we see that the coefficients of our product series are given by the formula

$$ \sum_{r=0}^{k} c_r d_{k-r} = \sum_{r=0}^{k} 1 \cdot (k - r + 1) = (k + 1) + k + (k - 1) + \cdots + 1. $$

Here we have the sum of the first $k + 1$ positive integers, an arithmetic progression whose sum, you may recall, is $\frac{1}{2}(k + 1)(k + 2)$. It therefore follows that

$$ \sum_{k=0}^{\infty} x^k \sum_{k=0}^{\infty} (k + 1)x^k = \sum_{k=0}^{\infty} \tfrac{1}{2}(k + 1)(k + 2)x^k. $$

You will observe that the initial segment that contains the first four terms of this series agrees with the result we obtained in the preceding paragraph.

We will not discuss quotients of power series except to say that here again we proceed "naturally." Once more, in order to find the initial segment of the quotient series that ends with the term containing $(x - a)^k$, we divide corresponding initial segments of the divisor and dividend series and discard all terms of the resulting quotient that contain $(x - a)$ to a power larger than k. There are some complications in the theory of division. For example, we require that the first coefficient of the divisor series be different from 0, and the radius of convergence of the quotient series may be less than the radius of convergence of either the divisor or the dividend.

Now we present our proof of Theorem 84-1. To test the convergence of the general Power Series 84-1 in case $x \neq a$, we can use the root test (Theorem 83-3).

Our first step is to calculate $L = \overline{\lim_{k \uparrow \infty}} |c_k(x - a)^k|^{1/k}$. Since $|c_k(x - a)^k|^{1/k} = |c_k|^{1/k} |x - a|$,

$$L = (\overline{\lim_{k \uparrow \infty}} |c_k|^{1/k}) |x - a|.$$

Now we consider the three possibilities corresponding to the three cases of Theorem 84-1, (i) $\overline{\lim_{k \uparrow \infty}} |c_k|^{1/k} = \infty$, (ii) $\overline{\lim_{k \uparrow \infty}} |c_k|^{1/k} = 0$, and (iii) $\overline{\lim_{k \uparrow \infty}} |c_k|^{1/k} = c$, where c is a positive number. In the first case (since we are supposing that $x \neq a$), $L = \infty$, so the series surely diverges. In case (ii), $L = 0$ regardless of x, and since $L < 1$, the root test says that Series 84-1 converges absolutely. Finally, in case (iii), $L = c|x - a|$. According to the root test, the series will converge absolutely if $L < 1$ (and hence $|x - a| < 1/c$) and will diverge if $L > 1$ ($|x - a| > 1/c$). So part (iii) of Theorem 84-1 is established with the number $1/c$ playing the role of r.

PROBLEMS 84

1. Find the radius of convergence, the interval of convergence, and the set of points for which the following power series converge.

(a) $\displaystyle\sum_{k=0}^{\infty} \frac{x^k}{2^k}$

(b) $\displaystyle\sum_{k=1}^{\infty} k(x - 2)^k$

(c) $\displaystyle\sum_{k=0}^{\infty} \frac{(-1)^k}{3^k} (x + 1)^k$

(d) $\displaystyle\sum_{k=1}^{\infty} \frac{k^5}{5^k} (x - 5)^k$

(e) $\displaystyle\sum_{k=1}^{\infty} \frac{(x - 4)^k}{\sqrt{3k}}$

(f) $\displaystyle\sum_{k=1}^{\infty} \frac{x^k}{k(k + 1)}$

(g) $\displaystyle\sum_{k=0}^{\infty} e^{\sin k} x^k$

(h) $\displaystyle\sum_{k=1}^{\infty} 2^{\ln k} x^k$

(i) $\displaystyle\sum_{k=0}^{\infty} \int_0^x t^k \, dt$

(j) $\displaystyle\sum_{k=0}^{\infty} \int_2^x k^2(t - 2)^k \, dt$

(k) $\displaystyle\sum_{k=0}^{\infty} \frac{d}{dx} (x - 2)^k$

(l) $\displaystyle\sum_{k=0}^{\infty} \frac{d^2}{dx^2} (x + 1)^k/k!$

2. Find the interval of convergence of the series.

(a) $\displaystyle\sum_{k=0}^{\infty} \frac{(x - 2)^k}{2^k\sqrt{k + 1}}$

(b) $\displaystyle\sum_{k=0}^{\infty} \frac{(x - 2)^{4k}}{2^k}$

(c) $\displaystyle\sum_{k=0}^{\infty} \frac{(-1)^k}{2^{2k+12}k! \, (k + 12)!} x^{2k+12}$

(d) $\displaystyle\sum_{k=1}^{\infty} \frac{(-1)^k 5 \cdot 6 \cdots (k + 4)}{k!} x^{2k}$

(e) $\displaystyle\sum_{k=1}^{\infty} \frac{k^k}{k!} x^k$

(f) $\displaystyle\sum_{k=1}^{\infty} [3^k + (-3)^k](x + 1)^{3k-2}$

(g) $\displaystyle\sum_{k=1}^{\infty} (\text{Tan}^{-1} k)(x - 2^k)$

(h) $\displaystyle\sum_{k=1}^{\infty} (\text{csch } k) x^k$

3. Find the set of points for which $\displaystyle\sum_{k=1}^{\infty} k^p x^k$ converges in the case that

(a) $p < -1$

(b) $p \in [-1, 0)$

(c) $p \geq 0$

4. If $|t| < 1$, we know that the sum of the geometric series $\sum\limits_{k=0}^{\infty} t^k$ is $1/(1 - t)$. Use this fact to find the sums of the following power series.

   (a) $\sum\limits_{k=0}^{\infty} (-x)^k$

   (b) $\sum\limits_{k=0}^{\infty} x^{2k}$

   (c) $\sum\limits_{k=0}^{\infty} 3(x + 1)^k$

   (d) $\sum\limits_{k=0}^{\infty} (\sin \tfrac{1}{2}k\pi)x^k$

5. Replace t in the equation $\dfrac{1}{1 - t} = \sum\limits_{k=0}^{\infty} t^k$ with x, $-x$, and x^2 to obtain power series whose sums are $1/(1 - x)$, $1/(1 + x)$, and $1/(1 - x^2)$. Show that the third series is the product of the other two.

6. (a) Divide the "power series" 1 by the "power series" $2 - x$ to obtain a power series whose sum is $1/(2 - x)$.

   (b) Multiply this series by a series in powers of x whose sum is $1/(1 - x)$ to obtain a series whose sum is $1/(1 - x)(2 - x)$.

   (c) Achieve the same result by writing $1/(1 - x)(2 - x)$ in the form $A/(1 - x) + B/(2 - x)$ and then adding two power series that have these numbers as sums.

7. Show that if $\lim\limits_{k\uparrow\infty} |c_{k+1}|/|c_k| = L > 0$, then the radius of convergence of the power series $\sum\limits_{k=0}^{\infty} c_k(x - a)^k$ is $1/L$. What if $L = 0$?

8. (a) Find two power series such that the radius of convergence of their sum is greater than the radius of convergence of either summand.

   (b) Find two power series such that the radius of convergence of their product is greater than the radius of convergence of either factor.

   (c) Find two power series such that the radius of convergence of the quotient of the first by the second is less than the radius of convergence of either.

9. Use Equation 84-5 to show that $\sum\limits_{k=0}^{\infty} \dfrac{x^k}{k!} \sum\limits_{k=0}^{\infty} \dfrac{(-x)^k}{k!} = 1$.

10. Suppose that $\sum\limits_{k=0}^{\infty} c_k(x - a)^k$ and $\sum\limits_{k=0}^{\infty} d_k(x - a)^k$ are two power series such that for each k, $|c_k| \le |d_k|$. Show that the radius of convergence of the first series is at least as large as the radius of convergence of the second.

11. Show that the series $\sum\limits_{k=1}^{\infty} kc_k(x - a)^{k-1}$ and $\sum\limits_{k=0}^{\infty} \dfrac{c_k}{k + 1}(x - a)^{k+1}$ have the same radius of convergence as the series $\sum\limits_{k=0}^{\infty} c_k(x - a)^k$. (You may use the fact that if $\lim\limits_{k\uparrow\infty} p_k = P$, where $P > 0$, then $\overline{\lim\limits_{k\uparrow\infty}} p_k q_k = P \overline{\lim\limits_{k\uparrow\infty}} q_k$.)

12. (a) By drawing a few pictures, convince yourself that if $|\sin k| < .1$, then $|\sin (k + 1)| > .1$.

   (b) Use this result to convince yourself that there are infinitely many positive integers such that $(.1)^{1/k} \le |\sin k|^{1/k} \le 1$.

   (c) From these inequalities, infer that $\overline{\lim\limits_{k\uparrow\infty}} |\sin k|^{1/k} = 1$ and hence find the radius of convergence of the power series $\sum\limits_{k=0}^{\infty} \sin kx^k$.

   (d) Does the series converge at either endpoint of the interval of convergence?

Let us agree not to bother with power series that converge at only one point. Then the interval of convergence of a power series

$$(85\text{-}1) \qquad \sum_{k=0}^{\infty} c_k(x - a)^k$$

is either the infinite interval $(-\infty, \infty)$ or a finite interval $(a - r, a + r)$. In the latter case, we may also have to adjoin one or both of the endpoints $a - r$ or $a + r$ to obtain the set of points for which the series is convergent. Let us denote by I the *interval of convergence* of our series and by D the *set of points for which the series converges*. These two sets may coincide, and, at worst, D can be obtained from I by adjoining two more points. So the difference between the two sets is small. Nevertheless, in the statements of the various theorems that follow it is necessary to distinguish between the sets D and I even though it looks as if we are splitting hairs.

Corresponding to each number $x \in D$, we obtain a number, the sum of Series 85-1. Thus our power series defines a function; we will call it f. The domain of f is the convergence set D, and the values of f are the sums of the series:

$$(85\text{-}2) \qquad f(x) = \sum_{k=0}^{\infty} c_k(x - a)^k.$$

Among the functions whose values can be expressed as sums of power series are most of the important functions of mathematics. For example, the trigonometric, logarithmic, and exponential functions belong to this class, as we shall see. As a result, this class of functions has been studied extensively, and we now state three important theorems that apply to such functions. Proofs are best left to a course in advanced calculus, where more mathematical machinery is available.

Our first theorem tells us that these functions are continuous.

Theorem 85-1. *If $f(x) = \displaystyle\sum_{k=0}^{\infty} c_k(x - a)^k$, then f is continuous at each point of the convergence set D of the infinite series. That is, if b is any point of D, then*

$$(85\text{-}3) \qquad \lim_{x \to b} f(x) = f(b) = \sum_{k=0}^{\infty} c_k(b - a)^k.$$

If b is an endpoint of the interval of convergence, this limit is a limit from the right or from the left.

We will give an application of this theorem after we have stated the theorems on differentiation and integration of series.

If $f(x)$ were an ordinary sum of terms of the form $c_k(x - a)^k$, we could find its derivative by differentiating each term in the sum and then adding; that is, it is immaterial whether we add and then differentiate or differentiate and then add. But $f(x)$ is *not* an ordinary sum; it is the "sum" of an infinite series. Clearly, then, it is not obvious that we can obtain derivatives or integrals of $f(x)$ by differentiating or integrating each term of the series and then calculating the sum of the resulting series. Such term-by-term differentiation or integration is not always possible for some kinds of infinite series. When we deal with a *power series*, however, we may differentiate or integrate "under the summation sign" according to the following theorems.

Theorem   85-2.   *If* $f(x) = \displaystyle\sum_{k=0}^{\infty} c_k(x - a)^k$, *then* f *is differentiable at every point* x *of the interval of convergence of the series, and*

$$(85\text{-}4) \qquad\qquad f'(x) = \sum_{k=1}^{\infty} kc_k(x - a)^{k-1}.$$

In Problem 84-11 we asked you to show that the interval of convergence of the series in Equation 85-4 is the interval of convergence of the original series.

Theorem   85-3.   *If* x *is any point of the convergence set of the infinite series of Equation 85-2, then*

$$(85\text{-}5) \qquad\qquad \int_a^x f(t)\, dt = \sum_{k=0}^{\infty} \frac{c_k(x - a)^{k+1}}{k + 1}.$$

In Problem 84-11 we asked you to show that the interval of convergence of the series in Equation 85-5 is the interval of convergence of the original series.

In order to give you an idea of how these theorems can be used, as well as to help fix them in your mind, we will apply them to some specific series. At the moment, the only power series whose sum we know is the geometric series;

$$(85\text{-}6) \qquad\qquad \frac{1}{1 - x} = \sum_{k=0}^{\infty} x^k,$$

where $D = I = (-1, 1)$. According to Equation 85-5, if x is any number in the interval $(-1, 1)$, then

$$\int_0^x \frac{dt}{1 - t} = \sum_{k=0}^{\infty} \frac{x^{k+1}}{k + 1} = x + \frac{x^2}{2} + \frac{x^3}{3} + \cdots.$$

But we know that

$$\int_0^x \frac{dt}{1-t} = -\ln(1-t)\Big|_0^x = -\ln(1-x),$$

so for each $x \in (-1, 1)$,

(85-7) $-\ln(1-x) = x + \dfrac{x^2}{2} + \dfrac{x^3}{3} + \dfrac{x^4}{4} + \cdots .$

Thus, for example, if we set $x = \frac{1}{2}$ and recall that $-\ln\frac{1}{2} = \ln 2$, we see that

$$\ln 2 = \tfrac{1}{2} + \tfrac{1}{8} + \tfrac{1}{24} + \tfrac{1}{64} + \cdots .$$

The sum of the first four terms of this series is .68, which is a close approximation of the value $\ln 2 = .69$. Of course, the more terms we add, the more nearly we will approximate $\ln 2$, and *we can approximate* $\ln 2$ *with any degree of accuracy we wish simply by adding together enough terms.* In practice, we would probably use a machine to help us with the computation.

We are only sure that Equation 85-7 is valid when $|x| < 1$, but it is not hard to develop a series that will give us the logarithm of any positive number. In Equation 85-7 we replace x with $-x$ and obtain the equation

(85-8) $-\ln(1+x) = -x + \dfrac{x^2}{2} - \dfrac{x^3}{3} + \dfrac{x^4}{4} + \cdots .$

Then we subtract the corresponding sides of Equations 85-7 and 85-8:

$$\ln(1+x) - \ln(1-x) = 2\left(x + \frac{x^3}{3} + \frac{x^5}{5} + \cdots\right).$$

Thus since $\ln(1+x) - \ln(1-x) = \ln(1+x)/(1-x)$,

(85-9) $\ln \dfrac{1+x}{1-x} = 2\left(x + \dfrac{x^3}{3} + \dfrac{x^5}{5} + \cdots\right).$

This series will give us the logarithm of any positive number N, for when we solve the equation $(1+x)/(1-x) = N$ for x, we find that $x = (N-1)/(N+1)$, and therefore $|x| < 1$. Thus we may replace x in Equation 85-9 with the number $(N-1)/(N+1)$ and obtain a convergent series whose sum is $\ln N$:

(85-10) $\ln N = 2\left[\left(\dfrac{N-1}{N+1}\right) + \dfrac{1}{3}\left(\dfrac{N-1}{N+1}\right)^3 + \dfrac{1}{5}\left(\dfrac{N-1}{N+1}\right)^5 + \cdots\right].$

We know that Equation 85-7 is valid for each x in the open interval $(-1, 1)$. But when we substitute the endpoints -1 and 1 in the series on the right-hand side of the equation, we find that it also converges when $x = -1$. In order to calculate the sum of the resulting series, we turn to Theorem 85-1. This theorem tells us that the sum at -1 is $\lim_{x \downarrow -1} [-\ln (1 - x)]$—that is, $-\ln 2$. Therefore $-\ln 2 = -1 + \frac{1}{2} - \frac{1}{3} + \frac{1}{4} - \cdots$, and so we have obtained the interesting equation

$$(85\text{-}11) \qquad \ln 2 = 1 - \tfrac{1}{2} + \tfrac{1}{3} - \tfrac{1}{4} + \tfrac{1}{5} - \cdots$$

that we mentioned in Example 81-2.

Finally, let us consider an application of Theorem 85-2.

Example 85-1. Find a power series in x whose sum is $1/(1 - x)^2$.

Solution. Equation 85-6 shows us a power series in x whose sum is $(1 - x)^{-1}$. According to Theorem 85-2, the derivative of this sum is the sum of the series of derivatives of the terms of our original geometric series. Thus since $\dfrac{d}{dx} (1 - x)^{-1} = (1 - x)^{-2}$,

$$\frac{1}{(1 - x)^2} = \sum_{k=1}^{\infty} kx^{k-1} = 1 + 2x + 3x^2 + 4x^3 + \cdots,$$

and this equation is valid if $|x| < 1$.

PROBLEMS   85

1. Suppose that $f(x) = \displaystyle\sum_{k=1}^{\infty} \frac{x^k}{k^2}$.

    (a) What is the domain of f?
    (b) Calculate $f(-\tfrac{1}{2})$ with an error of less than .01.
    (c) Show that f is an increasing function.
    (d) Find $\lim_{x \to 0} f(x)$.
    (e) Show that the range of f is a subset of the interval $(-1, 2)$.

2. Suppose that $f(x) = \displaystyle\sum_{k=0}^{\infty} \frac{x^{2k+1}}{(2k + 1)!}$.

    (a) Find the domain of f.
    (b) Find the range of f.
    (c) Show that f is an odd function.
    (d) Show that f is increasing.
    (e) Show that $y = f(x)$ satisfies the differential equation $y'' - y = 0$.
    (f) Compare $f'(x)$ and $\displaystyle\int_0^x f(t)\, dt$.

3. Use the rules for combining power series, together with Equation 85-6, to find the sum of the series $\displaystyle\sum_{k=0}^{\infty} [3 - (-1)^k]x^k$.

4. Use a power series to find an approximation to $\ln N$ and check your results with Table II.

(a) $N = .9$ (b) $N = 1.1$ (c) $N = 5$ (d) $N = .3$

5. Equation 85-11 gives us one infinite series whose sum is $\ln 2$, and we can obtain two others by replacing x with $\frac{1}{2}$ in Equation 85-7 and N with 2 in Equation 85-10. Which of these three series is "best"?

6. Let $g(x) = 1 - \frac{1}{2}x^3 + \frac{1}{4}x^6 - \frac{1}{8}x^9 + \cdots$.

(a) Find a two-place decimal approximation to $\displaystyle\int_{-1/2}^{1/2} g(x)\, dx$.

(b) Find a two-place decimal approximation to $g'(\frac{1}{2})$.

7. Suppose that $\displaystyle f(x) = \sum_{k=0}^{\infty} \frac{2^k}{k!} x^k$.

(a) Show that $f'(x) = 2f(x)$.

(b) Hence show that $\dfrac{d}{dx}\, [e^{-2x}f(x)] = 0$.

(c) Hence conclude that $e^{-2x}f(x) = f(0) = 1$ and find a simple formula for $f(x)$.

8. Use the equation developed in Example 85-1 to find the sum of the series $\displaystyle\sum_{k=1}^{\infty} \frac{k}{2^k}$.

Can you find the sum of the series $\displaystyle\sum_{k=0}^{\infty} \frac{1}{(k + 1)2^k}$?

9. In Example 85-1 we differentiated a series whose sum is $(1 - x)^{-1}$ to find a series whose sum is $(1 - x)^{-2}$. Show that we get the same result by squaring the original series.

10. Find a power series in t of which $1/(1 - t^2)$ is the sum. Now integrate from 0 to x and obtain Equation 85-9.

11. If we replace x with $-t^2$, Equation 85-6 becomes $\displaystyle\frac{1}{1 + t^2} = \sum_{k=0}^{\infty} (-1)^k t^{2k}$.

(a) Use this equation and Theorem 85-3 to show that

$$\text{Arctan } x = \sum_{k=0}^{\infty} \frac{(-1)^k x^{2k+1}}{2k + 1} \qquad \text{if } |x| < 1.$$

(b) Use the result of part (a) and Theorem 85-1 to show that

$$\tfrac{1}{4}\pi = 1 - \tfrac{1}{3} + \tfrac{1}{5} - \tfrac{1}{7} + \cdots.$$

(c) Use power series to show that $\displaystyle\int_0^x \text{Arctan } t\, dt = x \text{ Arctan } x - \tfrac{1}{2} \ln (1 + x^2)$.

12. What is the domain of f if $\displaystyle f(x) = \sum_{k=1}^{\infty} \frac{\sin k^2 x}{k^2}$? Can we find $f'(x)$ by term-by-term differentiation?

The equation

$$(86\text{-}1) \quad f(x) = \sum_{k=0}^{\infty} c_k(x-a)^k = c_0 + c_1(x-a) + c_2(x-a)^2 + \cdots$$

says that $f(x)$ is the sum of an infinite series. If we set $x = a$, however, the series is particularly simple. All the terms beyond c_0 are 0, and so $f(a) = c_0$. Since

$$f'(x) = c_1 + 2c_2(x-a) + 3c_3(x-a)^2 + \cdots,$$

we see that $f'(a) = c_1$. Similarly,

$$f''(x) = 2c_2 + 2 \cdot 3c_3(x-a) + 3 \cdot 4(x-a)^2 + \cdots,$$

from which it follows that $f''(a) = 2c_2$. If you continue these calculations, you will find that $f'''(a) = 2 \cdot 3c_3$, $f^{iv}(a) = 2 \cdot 3 \cdot 4c_4$, and, in general,

$$(86\text{-}2) \qquad f^{(k)}(a) = k!\, c_k, \qquad k = 0, 1, 2, \ldots.$$

We will agree that $f^{(0)}(x) = f(x)$ and $0! = 1$.

Example 86-1. Suppose that $\displaystyle\sum_{k=0}^{\infty} c_k(x-a)^k = 0$ for each x in some interval $(a - r, a + r)$. Show that each coefficient $c_k = 0$.

Solution. In this case, $f(x) = 0$, so each of the numbers $f^{(0)}(a)$, $f'(a)$, $f''(a)$, and so on, is 0. Therefore Equation 86-2 tells us that $c_k = 0$ for each index k.

Example 86-1 leads to a very simple but important theorem about power series. *One* equation between power series is equivalent to *infinitely many* equations between their coefficients.

Theorem 86-1. *If* $\displaystyle\sum_{k=0}^{\infty} a_k(x-a)^k = \sum_{k=0}^{\infty} b_k(x-a)^k$ *for each number x in some neighborhood of a, then $a_k = b_k$ for each index k.*

PROOF

To prove this theorem, we simply note that the equation

$$\sum_{k=0}^{\infty} a_k(x-a)^k = \sum_{k=0}^{\infty} b_k(x-a)^k$$

implies that

$$\sum_{k=0}^{\infty} (a_k - b_k)(x - a)^k = 0,$$

and we have just seen that if the sum of a power series in $(x - a)$ is 0 in a neighborhood of a, then all its coefficients must be 0. Thus $a_0 - b_0 = 0$, $a_1 - b_1 = 0$, and so on, as the theorem claims.

If we start with a power series $\sum_{k=0}^{\infty} c_k(x - a)^k$, Equation 86-1 defines a function f that has derivatives of all orders in the interval of convergence of the series. Now let us proceed in the opposite direction. Suppose that we start with a function f that has derivatives of all orders in a neighborhood of a and see if we can find a power series in $(x - a)$ of which $f(x)$ is the sum. For example, we might want to express e^x as the sum of a series in powers of x.

From the preceding material, it follows that the coefficients of such a power series (if there is one) must be the numbers given by Equation 86-2. That is, we must have $c_k = f^{(k)}(a)/k!$. So let us form the power series in $(x - a)$ that has these numbers as coefficients,

$$(86\text{-}3) \quad \sum_{k=0}^{\infty} \frac{f^{(k)}(a)}{k!} (x - a)^k = f(a) + f'(a)(x - a) + \frac{f''(a)}{2!} (x - a)^2 + \cdots,$$

and see whether or not it converges to $f(x)$ at each x of a neighborhood of a. Series 86-3 is called **Taylor's Series** for $f(x)$ at a. We can form Taylor's Series for any function that has derivatives of all orders at a, and it turns out that Taylor's Series converges to the value of the function for many of the familiar functions of mathematics.

Example 86-2. Find Taylor's Series for e^x about 0.

Solution. Here $f(x) = e^x$ and $a = 0$. Since $f(x)$ and all its derivatives equal e^x, we have $f(0) = f'(0) = f''(0) = \cdots = 1$. Thus in this case, $f^{(k)}(a) = 1$ for each index k, and so Taylor's Series for e^x about 0 is

$$\sum_{k=0}^{\infty} \frac{x^k}{k!} = 1 + x + \frac{x^2}{2!} + \frac{x^3}{3!} + \cdots.$$

The ratio test shows (Example 84-2) that this series converges for every number x.

By itself, the fact that Taylor's Series for e^x about 0 converges for each x is not a particularly useful piece of information. But the fact that its sum is actually e^x is important. Thus when we establish that

$$(86\text{-}4) \quad e^x = \sum_{k=0}^{\infty} \frac{x^k}{k!} = 1 + x + \frac{x^2}{2!} + \frac{x^3}{3!} + \cdots, \quad x \in (-\infty, \infty),$$

we will have an equation from which to calculate approximations to e^x by algebraic operations alone. Let us turn, then, to the task of verifying Equation 86-4.

To find out whether or not Taylor's Series for a given $f(x)$ converges to $f(x)$, we use Taylor's Formula. If x belongs to a neighborhood of a in which f has derivatives of all orders, and if n is any positive integer, Equation 21-5 states that

$$(86\text{-}5) \quad f(x) = f(a) + f'(a)(x - a) + \cdots + \frac{f^{(n)}(a)}{n!}(x - a)^n + R_n(x)$$

$$= S_n(x) + R_n(x).$$

In this equation $S_n(x)$ is the nth partial sum of Taylor's Series, and

$$(86\text{-}6) \qquad\qquad R_n(x) = \frac{f^{(n+1)}(m)}{(n + 1)!}(x - a)^{n+1},$$

where m is some number between a and x. Equation 86-5 tells us that $\lim_{n\uparrow\infty} S_n(x) = f(x)$ if and only if $\lim_{n\uparrow\infty} R_n(x) = 0$, and so we have proved the basic theorem of Taylor's series.

Theorem 86-2. *A necessary and sufficient condition that*

$$(86\text{-}7) \qquad f(x) = f(a) + f'(a)(x - a) + \frac{f''(a)}{2!}(x - a)^2 + \cdots$$

is that $\lim_{n\uparrow\infty} R_n(x) = 0.$

Example 86-3. Show that for each real number x, Taylor's Series for e^x about 0 converges to e^x.

Solution. In this example $R_n(x) = e^m x^{n+1}/(n + 1)!$, where m is a number such that $|m| \le |x|$. Therefore

$$|R_n(x)| \le \frac{e^{|x|}|x|^{n+1}}{(n + 1)!}.$$

Now we refer to Equation 84-3 to see that the right-hand side of this inequality approaches 0 as $n \uparrow \infty$. Thus $\lim_{n\uparrow\infty} R_n(x) = 0$, and we have verified Equation 86-4.

Example 86-4. Find Taylor's Series for $\cos x$ about a and show that its sum is $\cos x$ for each real number x.

Solution. If $f(x) = \cos x$, then $f'(x) = -\sin x$, $f''(x) = -\cos x$, $f'''(x) = \sin x$, and so on. Therefore $f(a) = \cos a$, $f'(a) = -\sin a$, $f''(a) = -\cos a$, $f'''(a) = \sin a$, and when these numbers are substituted in Equation 86-7, it becomes

$$(86\text{-}8) \quad \cos x = \cos a - (\sin a)(x - a) - \cos a\,\frac{(x - a)^2}{2!} + \sin a\,\frac{(x - a)^3}{3!} + \cdots.$$

To show that this equation is correct, we must still demonstrate that $\lim_{n \uparrow \infty} R_n(x) = 0$, where $R_n(x)$ is given by Equation 86-6. Depending on n, $f^{(n+1)}(m)$ will be one of the numbers $\cos m$, $\sin m$, $-\cos m$, or $-\sin m$. Whatever it is, $|f^{(n+1)}(m)| \leq 1$, and so $|R_n(x)| \leq |x - a|^{n+1}/(n + 1)!$. Now we let $|x - a|$ play the role of x in the equation $\lim_{k \uparrow \infty} x^k/k! = 0$, and we see that $\lim_{n \uparrow \infty} R_n(x) = 0$.

Example 86-5.   Use Equation 86-8 to calculate $\cos 3$.

Solution.   We will let $x = 3$ and choose a so that
 (i) we know the numbers $\cos a$ and $\sin a$ and
(ii) $|3 - a|$ is small and hence the terms of our series decrease rapidly.
These considerations suggest that we take $a = \pi$. Then $\cos a = -1$ and $\sin a = 0$, and Equation 86-8 becomes

$$\cos 3 = -1 + \frac{(\pi - 3)^2}{2!} - \frac{(\pi - 3)^4}{4!} + \frac{(\pi - 3)^6}{6!} - \cdots .$$

This series is an alternating series, so the error that we make by taking the sum of the first three terms as an approximation of $\cos 3$ is less than the fourth term. It is not hard to show that the fourth term is less than .0000001. Then we find that the sum of the first three terms gives us $\cos 3 = -.989992$ with six-place accuracy.
    If we set $a = 0$, Equation 86-8 reduces to

(86-9) $$\cos x = 1 - \frac{x^2}{2!} + \frac{x^4}{4!} - \frac{x^6}{6!} + \cdots .$$

We get $\cos 3$ by putting $x = 3$ in this series, but to be sure of an error of less than .0000001 (according to our rule for alternating series), we would have to add the first nine terms. Equation 86-9 would be more suitable for calculating a number like $\cos .1$.

PROBLEMS   86

1.   Use some of the series we worked with in the last section to find $f^{(9)}(0)$.

   (a) $f(x) = \dfrac{1}{(1 - x)^2}$ (b) $f(x) = \ln (1 + x)$

   (c) $f(x) = \text{Arctan } x$ (d) $f(x) = \text{Arctan } x^2$

2.   Find Taylor's Series for $x^3 - 3x - 2$ about a.

   (a) $a = 0$ (b) $a = 1$ (c) $a = -1$ (d) $a = 2$

3.   Find Taylor's Series for x^{-2} about 1. What is the radius of convergence of this series?

4.   Find Taylor's Series for $\sin x$ about 0. Show that this series converges to $\sin x$ for each $x \in (-\infty, \infty)$. Use this series to compute $\sin .1$ and $\sin 1$, and compare your results with tabulated values. If $f(x) = \sin 3x$, find $f^{(7)}(0)$.

5. Find Taylor's Series for e^x about an arbitrary a in two ways: (1) by using Equation 86-7 and (2) by replacing x with $x - a$ in Equation 86-4.

6. Find Taylor's Series for e^{-x} about the origin by replacing x with $-x$ in Equation 86-4 and use the resulting series to find Taylor's Series for sinh x and cosh x about 0. Now use Equation 86-7 to get the same result.

7. Let a be an arbitrary positive number and find Taylor's Series for ln x about a. What is the radius of convergence of this series? Can you show directly that for each $x \in (0, 2a)$, $\lim_{n \uparrow \infty} R_n(x) = 0$? In Equation 85-8, replace x with $(x - a)/a$ to show that the series we have just calculated does have ln x as its sum.

8. See how many terms you can find of Taylor's Series for tan x about the origin.

9. Show that for an arbitrary real number a, $\cos x = \sum_{k=0}^{\infty} \dfrac{\cos (a + \frac{1}{2}k\pi)}{k!} (x - a)^k$ and

$\sin x = \sum_{k=0}^{\infty} \dfrac{\sin (a + \frac{1}{2}k\pi)}{k!} (x - a)^k$ for each $x \in (-\infty, \infty)$.

10. Let x and y be any given numbers. Find Taylor's Series for $f(x) = \sin (x + y)$ about 0 to show that

$$\sin (x + y) = \sin y + (\cos y)x - (\sin y) \frac{x^2}{2!} - (\cos y) \frac{x^3}{3!} + \cdots .$$

Now group the terms that contain sin y and those that contain cos y, recognize series you have already found whose sums are cos x and sin x, and thus find the addition formula for sin $(x + y)$.

11. Prove the following theorem. If f has derivatives of all orders at each number in a neighborhood of a and if there is a number B such that for each positive integer n the values of $f^{(n)}$ are bounded by B^n throughout the neighborhood, then Taylor's Series for $f(x)$ converges to $f(x)$ for each x in the neighborhood.

12. Suppose that Taylor's Series for $f(x)$ and $g(x)$ about a converge to $f(x)$ and $g(x)$ in some neighborhood of a. Multiply these series, using Equation 84-5, to find Taylor's Series for the product $p(x) = f(x)g(x)$. From this result, show that for each positive integer k,

$$p^{(k)}(a) = \sum_{r=0}^{k} \frac{k!}{r! \, (k - r)!} f^{(r)}(a) g^{(k-r)}(a).$$

87. FINDING TAYLOR'S SERIES.
THE BINOMIAL SERIES

The straightforward way to find Taylor's Series for $f(x)$ about a is to calculate $f(a), f'(a), f''(a)$, and so on, and then substitute these numbers in Formula 86-3. Even if the resulting series converges, we must still investigate the remainder $R_n(x)$ to see if the sum of the series is $f(x)$. As we saw in some of the problems at the end of the last section, however, it is often easier to find a desired Taylor's Series by "operating on"—with calculus or algebra—some series we already know, such as the series for e^x, $(1 - x)^{-1}$, and cos x.

Example 87-1.   Find Taylor's Series for $\sin x$ about 0.

Solution.   Equation 86-9 gives us Taylor's Series for $\cos x$ about 0, and we may differentiate both sides of this equation and then multiply by -1 to see that

$$(87\text{-}1) \qquad\qquad \sin x = x - \frac{x^3}{3!} + \frac{x^5}{5!} - \frac{x^7}{7!} + \cdots.$$

Example 87-2.   Find Taylor's Series for $\tan x$ about 0.

Solution.   If you worked Problem 86-8, you found that calculation of higher derivatives of $\tan x$ becomes quite complicated. We can ease the difficulties somewhat by using the trigonometric identity $\sec^2 x = 1 + \tan^2 x$ to see that $f(x) = \tan x$ satisfies the differential equation $f'(x) = 1 + f(x)^2$. From this equation we successively find that $f''(x) = 2f(x)f'(x), f'''(x) = 2f(x)f''(x) + 2f'(x)^2$, and so on. Since $f(0) = \tan 0 = 0$, then $f'(0) = 1 + 0 = 1, f''(0) = 2f(0)f'(0) = 0, f'''(0) = 2f(0)f''(0) + 2f'(0)^2 = 2$, and we could continue step by step to find $f^{(4)}(0), f^{(5)}(0)$, and so on. Now we may substitute in Formula 86-3 and we find that the first few terms of Taylor's Series for $\tan x$ are

$$0 + x + 0 \cdot x^2 + \frac{2}{3!} x^3 = x + \frac{1}{3} x^3 + \cdots.$$

We can even do away with differentiation altogether if we regard $\tan x$ as $\sin x/\cos x$ and divide the series for $\sin x$ by the series for $\cos x$:

$$\tan x = \frac{x - \dfrac{x^3}{3!} + \dfrac{x^5}{5!} - \cdots}{1 - \dfrac{x^2}{2!} + \dfrac{x^4}{4!} - \cdots}.$$

When we perform this division, we find that

$$\tan x = x + \tfrac{1}{3} x^3 + \tfrac{2}{15} x^5 + \cdots.$$

In elementary algebra we learned the Binomial Theorem, which is a rule for writing a power of a sum as a sum of powers. We can use it to write $(1 + x)^p$ as a sum of powers of x. In our elementary work we had to assume that the exponent p is a positive integer, but even when p is not a positive integer, we can still formally apply the rules of the Binomial Theorem to the expression $(1 + x)^p$. Then the rules generate an infinite series rather than a finite sum. If we assume that the sum of this **binomial series** really is $(1 + x)^p$, we have the equation

$$(87\text{-}2) \quad (1 + x)^p = 1 + px + \frac{p(p-1)}{2!} x^2 + \cdots$$

$$+ \frac{p(p-1)\cdots(p-k+1)}{k!} x^k + \cdots.$$

So now we have two tasks: (1) We must find the interval of convergence of the binomial series and (2) we must show that its sum is $(1 + x)^p$. Our series has the form $\sum_{k=0}^{\infty} c_k x^k$, where

$$(87\text{-}3) \quad c_0 = 1 \quad \text{and} \quad c_{k+1} = \frac{p - k}{k + 1}\, c_k, \quad k = 0, 1, 2, \ldots .$$

Thus

$$\frac{|c_{k+1} x^{k+1}|}{|c_k x^k|} = \frac{|p - k|}{k + 1}\, |x|.$$

Since p is independent of k, the limit as $k \uparrow \infty$ of this ratio of consecutive terms is $|x|$. The ratio test therefore tells us that the series converges when $|x| < 1$ and diverges when $|x| > 1$; the interval of convergence of the binomial series is $(-1, 1)$.

We could treat the binomial series as Taylor's Series for $(1 + x)^p$ (which it is) and hence establish Equation 87-2 by showing that the remainder $R_n(x)$ approaches 0. The details of this procedure become rather complicated, however, and we will outline a simpler argument.

We have just seen that the binomial series converges in the interval $(-1, 1)$, so let us call its sum $f(x)$. We want to show that this number is $(1 + x)^p$, and our first step in this direction will be to show that $\dfrac{d}{dx}\left[f(x)(1 + x)^{-p}\right] = 0$. Since $\dfrac{d}{dx}\left[f(x)(1 + x)^{-p}\right] = f'(x)(1 + x)^{-p} - pf(x)(1 + x)^{-p-1}$, the condition that the derivative is 0 is

$$(1 + x)f'(x) = pf(x).$$

Now observe that

$$f'(x) = p + p(p - 1)x + \frac{p(p - 1)(p - 2)}{2!}\, x^2 + \cdots,$$

and so

$$xf'(x) = px + p(p - 1)x^2 + \frac{p(p - 1)(p - 2)}{2!}\, x^3 + \cdots .$$

When we add these two equations, we get

$$(1 + x)f'(x) = p + p^2 x + \frac{p^2(p - 1)}{2!}\, x^2 + \cdots = pf(x).$$

Now that we have shown that $\dfrac{d}{dx}\left[f(x)(1 + x)^{-p}\right] = 0$ for each $x \in (-1, 1)$, we are virtually done, for this equation says that the value of $f(x)(1 + x)^{-p}$ at any point x is the same as its value at 0; that is, $f(x)(1 + x)^{-p} = f(0)$. Since $f(0)$

is the first term of the binomial series, it is simply the number 1, and so Equation 87-2 is verified.

Example 87-3. Find an infinite series of which the integral $\int_0^{1/2} \dfrac{dx}{\sqrt{1-x^2}}$ is the sum.

Solution. To find a power series whose sum is the integrand, we simply replace p with $-\frac{1}{2}$ and x with $-x^2$ in Equation 87-2:

$$(87\text{-}4) \qquad \frac{1}{\sqrt{1-x^2}} = 1 + \tfrac{1}{2}x^2 + \tfrac{3}{8}x^4 + \tfrac{5}{16}x^6 + \cdots .$$

Then we integrate term by term:

$$\int_0^{1/2} \frac{dx}{\sqrt{1-x^2}} = \tfrac{1}{2} + \tfrac{1}{48} + \tfrac{3}{1280} + \cdots .$$

Can you show that our integral is actually $\frac{1}{6}\pi$, so that we have developed a way to calculate π to any degree of accuracy we choose?

PROBLEMS 87

1. Multiply or divide series you know to find a few terms of Taylor's Series about 0.
   (a) $e^x \sin x$ (b) $e^x \cos x$ (c) $\sec x$
   (d) e^{-x} (e) $\dfrac{e^x}{1+x}$ (f) $e^x \sec x$

2. Use Equation 87-1 to find Taylor's Series for $\sin(x-3)$ about 3. What is the series for $\sin(x-3)$ about 0?

3. Use Equation 86-8 to find Taylor's Series for $\sin x$ about a.

4. Use the identity $\sin^2 x = \frac{1}{2}(1 - \cos 2x)$ to find Taylor's Series for $\sin^2 x$ about 0. Differentiate the result to find Taylor's Series for $\sin 2x$ about 0. How else could you find this series?

5. Use the binomial formula to find a familiar series of powers of x whose sum is $1/(1-x)$.

6. Use the binomial series to compute the following numbers.
   (a) $\sqrt{1.02}$ (b) $\sqrt{3.92}$ (c) $\sqrt[3]{.98}$
   (d) $\sqrt[3]{30}$ (e) $10^{-2/3}$ (f) $\sqrt[4]{17}$

7. Use series and term-by-term integration to approximate the following integrals.
   (a) $\int_0^1 \cos\sqrt{x}\,dx$ (b) $\int_0^{.5} x \ln(1+x)\,dx$
   (c) $\int_0^{.4} x\sqrt{1-x^2}\,dx$ (d) $\int_0^{.1} (1+x^2)^{.9}\,dx$

8. Expand $(1 - t^2)^{-1/2}$ by the binomial formula and integrate from 0 to x term by term to obtain a series of powers of x whose sum is Arcsin x. Can you find a series whose sum is Arcos x? What is $\dfrac{d^5}{dx^5}$ Arcsin x when $x = 0$?

9. Find the area of the region $\{(x, y) \mid 1 \le x \le 2, y \ge 0, xy = \sin x\}$.

10. Find the first five terms of Taylor's Series about 0 for $f(x)$ if you know that $y = f(x)$ satisfies the differential equation $y'' + y = 0$ and the initial conditions $y = 1$ and $y' = -1$ when $x = 0$.

11. The **error function** erf is defined by the equation erf $x = \dfrac{2}{\sqrt{\pi}} \displaystyle\int_0^x e^{-t^2}\, dt$. Find Taylor's Series for erf x about 0.

12. Show that if $p \le -1$, the binomial series diverges when $x = 1$ and when $x = -1$.

13. In Example 75-5 we found that if $f(x) = e^{-1/x^2}$ when $x \ne 0$ and $f(0) = 0$, then $f'(0) = 0$. Can you show that $f''(0) = 0$? By continuing the argument, one can show that $f^{(k)}(0) = 0$ for each positive integer k. What is Taylor's Series for $f(x)$ about 0? Does it have $f(x)$ as its sum?

The equation $\sin x = x - \dfrac{x^3}{3!} + \dfrac{x^5}{5!} - \dfrac{x^7}{7!} + \cdots$ makes evaluating a trigonometric function an algebraic operation. The price for calculating a number like sin .3 by means of simple arithmetic operations seems to be high, however, because we are told to perform "infinitely many" of them. But we don't pay this price, for we only calculate approximations, and it takes just a few terms of our series to give a very good approximation of sin .3.

That idea is the main point you should remember about infinite series. However good an approximation to the sum of a convergent series you want, you can obtain it by adding up enough terms. How many terms is "enough" is an important and sometimes difficult question that we haven't had time to dwell on. It should be apparent that in processes that require large amounts of computation (as some series do) a high-speed computer is a valuable aid.

An idea of how important Taylor's Series are can be gained from the fact that the values of so many common functions—trigonometric, logarithmic, exponential—are sums of power series. Notice, too, that since it is easy to differentiate and integrate powers of x, we can easily perform these calculus operations on functions whose values are sums of power series. Although you probably should not memorize our series for $\sin x$, $\cos x$, e^x, and so on, you should at least know that they are available.

Review Problems, Chapter Ten

1. The first term of a certain sequence $\{a_n\}$ is 1 and each of the other terms is the square of one-half of its predecessor. Express a_n in terms of n. What is $\lim_{n \uparrow \infty} a_n$?

2. Find a formula for the nth partial sum S_n of the infinite series

$$\sum_{k=1}^{\infty} \ln \left[\frac{\text{Arctan } (k + 1)}{\text{Arctan } k} \right].$$

What is the sum of this series?

3. Test for convergence.

(a) $\displaystyle\sum_{k=1}^{\infty} e^{-k} \sin 2^k$ (b) $\displaystyle\sum_{k=1}^{\infty} e^k \sin 2^{-k}$ (c) $\displaystyle\sum_{k=1}^{\infty} e^k \sin^2 2^{-k}$

(d) $\displaystyle\sum_{k=1}^{\infty} 2^k \text{Sin}^{-1} e^{-k}$ (e) $\displaystyle\sum_{k=1}^{\infty} 2^{\ln 1/k}$ (f) $\displaystyle\sum_{k=1}^{\infty} \frac{1}{\ln 2^k}$

(g) $\displaystyle\sum_{k=1}^{\infty} \frac{\sin \frac{1}{3} k\pi}{k}$ (h) $\displaystyle\sum_{k=1}^{\infty} \frac{\tan \frac{1}{4} k\pi}{k}$

4. Find c_{1066} if $\sin 3x^2 = \displaystyle\sum_{k=0}^{\infty} c_k x^k$.

5. Suppose that $f(x) = 1 + c_1 x + c_2 x^2 + c_3 x^3 + \cdots$ and that $f'(x) = f(x)$. Show that for each positive integer k, $c_k = 1/k!$.

6. Suppose that $f(x) = \displaystyle\sum_{k=0}^{\infty} \frac{(x + 1)^{k+1}}{k + 1}$.

(a) What is the domain of f?

(b) Show that $\dfrac{d}{dx} [(x + 1)f(x)] = -1/x$ for $x \in (-2, 0)$.

(c) Now explain why $f(x) = -\ln |x|/(x + 1)$.

7. (a) Divide 1 by the series for $\cos x$ to find a couple of terms of Taylor's Series for $\sec x$ about 0.

(b) Now square to get an initial segment of Taylor's Series for $\sec^2 x$.

(c) Now integrate to get a few terms of Taylor's Series for $\tan x$.

8. (a) Show that $\displaystyle\sum_{k=1}^{\infty} k^{-rk} k!$ converges if $r \geq 1$ and diverges if $r < 1$.

(b) Show that $\displaystyle\sum_{k=1}^{\infty} (rk)^{-k} k!$ converges if $r > 1/e$.

9. Find the set of points for which the power series $\displaystyle\sum_{k=1}^{\infty} (\cos k)x^k$ converges.

10. Evaluate the following integrals.

(a) $\displaystyle\int_0^{\infty} e^{-[x]} dx$ (b) $\displaystyle\int_0^{\infty} (-1)^{[x]} e^{-x} dx$ (c) $\displaystyle\int_1^{\infty} \frac{e^x}{[x]!} dx$

11. Show that we can use the Binomial Theorem to expand $(a + b)^p$, where p is any real number, if $|b| < |a|$.

12. Convince yourself that the number of terminal zeros in $n!$ is $\displaystyle\sum_{k=1}^{\infty} [n5^{-k}]$. For example, $6! = 720$ has one terminal 0, and $1 = \displaystyle\sum_{k=1}^{\infty} [6 \cdot 5^{-k}]$.

13. Show that if a series $\sum_{k=1}^{\infty} a_k$ of positive terms converges, then $\sum_{k=1}^{\infty} a_k^2$ converges. Does it necessarily follow that $\sum_{k=1}^{\infty} \sqrt{a_k}$ converges? What if we delete the words "of positive terms" in the first sentence?

14. Show that if $\lim_{k\uparrow\infty} k a_k = L$, then the series $\sum_{k=1}^{\infty} a_k$ is divergent unless $L = 0$. Give an example of a divergent series $\sum_{k=1}^{\infty} a_k$ such that $\lim_{k\uparrow\infty} k a_k = 0$ and of a convergent series that satisfies the same condition.

15. Can you find a function f such the improper integral $\int_{1}^{\infty} f(x)\, dx$ diverges but the series $\sum_{k=1}^{\infty} \int_{k}^{k+1} f(x)\, dx$ converges? A function for which the integral converges but the series diverges? Replace $f(x)$ with $|f(x)|$ and answer these questions.

16. Replace x with ix (where $i^2 = -1$) in Equation 86-4 and write the result as $e^{ix} = A(x) + iB(x)$, where $A(x)$ and $B(x)$ are series in powers of x. Now refer to Equations 86-9 and 87-1 to find what the sums of these series are. The resulting equation $e^{ix} = \cos x + i \sin x$ is due to Euler, and we take it to be the *definition* of e^{ix}.

17. Use the definition of e^{ix} in the preceding problem to answer the following questions.

    (a) Show that $e^{\pi i} = -1$, $e^{2\pi i} = 1$, and $e^{-\pi i/2} = -i$.

    (b) Set $x = \ln u$, where u is a positive number, in Euler's equation and hence infer that
    $$u^i = \cos \ln u + i \sin \ln u.$$

    (c) Successively replace x in Euler's equation with u, v, and $u + v$. Then use the equation $e^{i(u+v)} = e^{iu}e^{iv}$ to "derive" addition formulas for $\cos (u + v)$ and $\sin (u + v)$. Discuss the legitimacy of this procedure.

    (d) Show that our definition of e^{ix} leads to the natural equation $\dfrac{d}{dx} e^{ix} = i e^{ix}$.

    (e) Replace x with $-x$ in Euler's equation and thus show that
    $$\cos x = \tfrac{1}{2}(e^{ix} + e^{-ix}) = \cosh ix.$$

    Find a similar expression for $\sin x$.

Matrices and Systems of Equations | **11**

Vector notation is a splendid language for expressing some of the ideas of plane geometry, and we will find it even more useful as we move into higher dimensions in the remainder of the book. In this chapter we lay some groundwork for that study. Our approach here is purely algebraic; we will save the geometry for Chapter 12. Much of one's work with vectors can be cast in terms of systems of linear equations, so even though a chapter on that subject may seem out of place in a book on calculus and analytic geometry, it isn't.

88. SYSTEMS OF LINEAR EQUATIONS. TRIANGULAR FORM

A **linear equation** in n unknowns $x_1, \ldots, x_n$ has the form

$$a_1 x_1 + a_2 x_2 + \cdots + a_n x_n = b,$$

where $a_1, \ldots, a_n$, and b are given numbers. For example,

$$3x - 4y = 5,$$

$$5s - 46t + 7u = 9,$$

and

$$43x_1 - 57x_2 + 4x_3 - x_4 = 29$$

are linear equations in two, three, and four unknowns. In this section we will introduce methods for solving a system of n linear equations in n unknowns. A **solution** of such a system consists of a set of n numbers that satisfy all the equations of the system. For example, you can readily verify by substitution that $x_1 = 1$, $x_2 = 0$, and $x_3 = 2$ satisfy each of the equations of the system

$$(88\text{-}1) \qquad \begin{aligned} 3x_1 - 4x_2 + x_3 &= 5 \\ 6x_1 - 8x_2 - 2x_3 &= 2 \\ x_1 + 3x_2 - x_3 &= -1. \end{aligned}$$

It is economical to let one equals sign do the work of three, and so we say that $(x_1, x_2, x_3) = (1, 0, 2)$ satisfies System 88-1. We denote by R^3 the set of ordered triples of real numbers, so our solution $(1, 0, 2)$ is a member of R^3.

Two systems of equations are said to be **equivalent** if every solution of one system is also a solution of the other. The systems

$$\begin{aligned} 7x + 3y &= 4 \\ x + y &= 0 \end{aligned}$$

and

$$\begin{aligned} x &= 1 \\ x + y &= 0, \end{aligned}$$

for example, are equivalent, since $(x, y) = (1, -1)$ satisfies each system and no other pair satisfies either.

When faced with a system of linear equations, we naturally try to replace it with an equivalent system that is easier to solve. Each of the following operations transforms a system of equations into an equivalent system.

Operation 88-1. Interchanging the position of two equations.

Operation 88-2. Replacing an equation with a nonzero multiple of itself. (A *multiple* of an equation is the equation that results when we multiply both sides by the same number.)

Operation 88-3. Adding to one equation a multiple of another. (*Adding* two equations means, of course, adding corresponding sides.)

In elementary algebra we solved two equations in two unknowns by first eliminating one of the unknowns. This method is simply a scheme for using Operations 88-1 to 88-3 to replace a system of equations with an equivalent system that is easier to solve.

Example 88-1. Solve the linear system

$$(88\text{-}2) \qquad \begin{aligned} 2x + 3y &= 4 \\ 3x - y &= -5. \end{aligned}$$

Solution. If we multiply the second equation by 3 and add it to the first, we obtain an equation in x alone, which we use as the new first equation of the equivalent system

(88-3)
$$11x \quad\;\; = -11$$
$$3x - y = -5.$$

Now we find $x = -1$ from the first equation and then replace x with -1 in the second to find $y = 2$. Since System 88-3 is equivalent to System 88-2, we conclude that $(x, y) = (-1, 2)$ satisfies our given system.

System 88-3 is in what is called *triangular form*. A system of n linear equations in n unknowns is said to be in **triangular form** if the first equation involves only x_1; the second equation involves x_2 and perhaps x_1, but not $x_3, x_4, \ldots,$ and x_n; the third equation involves x_3 and perhaps x_1 and x_2, but not $x_4, \ldots, x_n$; and so on. The following system, for example, is in triangular form:

(88-4)
$$12x \quad\qquad = 24$$
$$-9x + 2y \quad\;\; = -20$$
$$2x - \;y + z = 6.$$

From the first equation it is apparent that $x = 2$. Once we know that $x = 2$, the second equation tells us that $y = -1$, and then we can solve the third equation to find $z = 1$. Thus the solution of System 88-4 is $(x, y, z) = (2, -1, 1)$.

It is easy to solve a system of equations that is in triangular form, so our method of attack on a given system of linear equations will be to use Operations 88-1, 88-2, and 88-3 to try to replace it with an equivalent triangular system. "Most" systems can be reduced to triangular form in this way, and as we shall see in the next section, our efforts will not be wasted if we attempt the reduction on a system that is not equivalent to a system in triangular form. We can best describe the procedure by applying it to some typical examples.

Example 88-2. Solve the system of equations

$$\tfrac{1}{2}x - \tfrac{2}{3}y + z = \tfrac{8}{3}$$
$$2x - \;y + z = 6$$
$$3x + 2y \quad\;\; = 4.$$

Solution. (*Note.* It takes longer to describe the solution process than it does to carry it out. You are warned, however, against trying to take too many steps at once—it is all too easy to make little errors when solving linear systems.) To simplify things, we multiply the first equation by 6 and change the order of the equations, and we obtain the system

$$3x + 2y \quad\qquad = 4$$
$$3x - 4y + 6z = 16$$
$$2x - \;y + \;z = 6.$$

Now we multiply the third equation by 6 and subtract it from the second to obtain a system in which z is missing from the second equation:

(88-5)
$$
\begin{aligned}
3x + 2y &= 4 \\
-9x + 2y &= -20 \\
2x - y + z &= 6.
\end{aligned}
$$

Next we get an equivalent triangular system by subtracting the second of Equations 88-5 from the first:

$$
\begin{aligned}
12x &= 24 \\
-9x + 2y &= -20 \\
2x - y + z &= 6.
\end{aligned}
$$

Now we are back to System 88-4, which we have already solved: $(x, y, z) = (2, -1, 1)$. Notice that the last equation was used to eliminate one of the unknowns from all the other equations; then the next to the last equation was used to eliminate another unknown from the equation above it. Although this procedure stops after two steps in our example, it is clear that it is a direct method that can be used to reduce a system consisting of more linear equations in more unknowns to triangular form.

All the operations involved in solving a system of linear equations are the simplest possible—multiplication, division, addition, and subtraction. But many operations are involved. Because these systems have great practical importance, many schemes have been, and are still being, devised to solve them in the most efficient way possible. Some of these methods are especially suited to modern electronic computers. Each method has its advantages and its drawbacks. All involve large amounts of computation. As general methods go, the practice of reducing the system to triangular form is probably as efficient as any, and in the next section we introduce some notation that simplifies this method.

PROBLEMS 88

1. Find (x, y) by first finding an equivalent triangular system.

(a) $2x + 4y = 0$
$3x - 2y = -8$

(b) $x - 5y = 3$
$2x + y = -8$

(c) $5x - 4y = 1$
$6x - 3y = 3$

(d) $y - x = -5$
$2x - y = 7$

(e) $\frac{1}{3}x + y = -3$
$x - \frac{1}{4}y = 4$

(f) $\frac{1}{2}x - \frac{1}{3}y = -5$
$x - .1y = 24$

2. Solve the system of equations.

(a) $x + 2y = 5$
$x - 2y = -3$
$2x + 4y - 2z = 12$

(b) $x + 2y = 4$
$x - 2y = 0$
$2x + 4y - 2z = 6$

(c) $\quad x + 3y - 4z = -2$
$\quad\quad 2x - y + 2z = 6$
$\quad\quad 4x - 6y + z = 9$

(d) $\quad x + 3y - 4z = 2$
$\quad\quad 2x - y + 2z = 1$
$\quad\quad 4x - 6y + z = 9$

(e) $3x - 4y + 2w = 11$
$\quad 2x - y + w = 5$
$\quad x + 5y - z = -4$
$\quad\quad y - 3z + w = 1$

(f) $3x - 4y + 2w = -7$
$\quad 2x - y + w = -1$
$\quad x + 5y - z = 13$
$\quad\quad y - 3z + w = 7$

3. Solve the "homogeneous" systems.

(a) $\quad x + 3y - 4z = 0$
$\quad\quad 2x - y + 2z = 0$
$\quad\quad 4x - 6y + z = 0$

(b) $\quad x + 2y - 2z = 0$
$\quad\quad 2x - y + z = 0$
$\quad\quad 4x + y - 3z = 0$

4. Find (x, y) by first making a substitution that yields a linear system; for example, let $u = 1/x$ and $v = 1/y$ in part (a).

(a) $3/x + 4/y = -6$
$\quad 1/x - 5/y = 17$

(b) $2x^2 + 3y^2 = 5$
$\quad\quad x^2 + 2y^2 = 3$

(c) $5 \ln x - \ln y = 11$
$\quad \ln x + 3 \ln y = -1$

(d) $4 \sin x - \sin y = 2$
$\quad 2 \sin x + 3 \sin y = 10$

5. Express x and y in terms of a, b, c, d, u, and v if

$$ax + by = u$$

$$cx + dy = v.$$

6. Use a system of two equations in two unknowns to find what quantities of coffee worth 75¢ a pound and \$1.15 a pound are needed to produce a blend worth 91¢ a pound.

7. The sum of the digits of a two-digit integer is 13. The number formed by reversing the digits is 27 greater than the original number. Find the number.

8. Show that the three lines $x - 2y - 3 = 0$, $2x + y + 1 = 0$, and $3x + 4y + 5 = 0$ intersect in a point.

9. Let A, B, and C denote the degree measures of the angles of a triangle. One angle is $20°$ less than the sum of the other two, and $10°$ more than the positive difference of the other two. Find the widths of the angles.

10. (a) Show that both $(1, 2, -1)$ and $(6, 5, -8)$ are solutions of the system

$$2x - y + z = -1$$

$$x + 3y + 2z = 5$$

$$x - 4y - z = -6.$$

(b) In fact, show that $(1 + 5t, 2 + 3t, -1 - 7t)$ is a solution of this system for each number t.

(c) Show that if (x, y, z) and (u, v, w) are solutions of our system, then $(x - u, y - v, z - w)$ is a solution of the system we get when we replace each of the numbers on the right-hand sides of our equations with zero.

11. Solve for x and y.

(a) $3[\![x]\!] - 2[\![y]\!] = -7$
$\quad 2[\![x]\!] + 5[\![y]\!] = 8$

(b) $3|x| - 2|y| = -7$
$\quad 2|x| + 5|y| = 8$

(c)　$\ln \dfrac{x^3}{y^2} = -7$　　　　　　(d)　$3\dfrac{dx}{dt} - 2\dfrac{dy}{dt} = -7$

　　　$\ln x^2 y^5 = 8$　　　　　　　　　　$2\dfrac{dx}{dt} + 5\dfrac{dy}{dt} = 8$

(e)　$3e^x - 2e^y = -7$　　　　　　(f)　$3 \sin x + 2 \cos y = -7$
　　　$2e^x + 5e^y = 8$　　　　　　　　　$2 \sin x + 5 \cos y = 8$

12. To what extent can you use Operations 88-1, 88-2, and 88-3 to solve the system of inequalities

$$3x - 2y < -7$$
$$2x + 5y < 8?$$

13. Show that Operation 88-1 can be accomplished by several applications of Operations 88-2 and 88-3.

89. MATRICES. DETERMINATIVE SYSTEMS

Reducing a system of linear equations to triangular form requires us to write a number of equivalent systems. We can save a great deal of ink by using a notation in which the symbols for the unknowns need not be copied down each time we write out a new equivalent system.

To illustrate this notation, consider the system

$$
\begin{aligned}
\tfrac{1}{2}x - 3y + \ \ z &= -4 \\
x + 4y - 3z &= 2 \\
2x \ \ \ \ \ \ + \tfrac{1}{4}z &= -4.
\end{aligned}
$$

(89-1)

The coefficients of the unknowns can be exhibited as an array of numbers:

(89-2)
$$
\begin{bmatrix}
\tfrac{1}{2} & -3 & 1 \\
1 & 4 & -3 \\
2 & 0 & \tfrac{1}{4}
\end{bmatrix}.
$$

Such an array is called a **matrix**. In particular, if there are the same number of rows and columns in the array, it is a **square matrix**. The square Matrix 89-2 is the **coefficient matrix** of System 89-1. The numbers on the right-hand sides of the equations in System 89-1 can be displayed as a matrix with three rows and one column:

$$
\begin{bmatrix}
-4 \\
2 \\
-4
\end{bmatrix}.
$$

We combine this matrix with the coefficient matrix to get the **augmented matrix** of a system. Thus the matrix

$$\begin{bmatrix} \frac{1}{2} & -3 & 1 & -4 \\ 1 & 4 & -3 & 2 \\ 2 & 0 & \frac{1}{4} & -4 \end{bmatrix}$$

is the augmented matrix of System 89-1. (The vertical bar between the third and fourth columns is not really a part of matrix notation; we use it here to remind us that part of our matrix represents one side of a system of equations and part the other.)

It is easy to construct the augmented matrix of a given system of linear equations. We must arrange the system so that each column of the matrix represents the coefficients of the same unknown, of course, and we must decide which column corresponds to which unknown. But once we have disposed of these points, we have a unique matrix that represents the system. Conversely, it is easy to find the system of equations of a given augmented matrix. The augmented matrix and the system of equations say the same thing, but the matrix says it more concisely.

The operations that we use to obtain equivalent systems of equations from a given system can easily be expressed as operations to be performed on the rows of numbers in the augmented matrix of the system. These operations always produce a matrix of an equivalent system of equations and can be summarized as follows.

Operation 89-1. Interchanging two rows in a matrix.

Operation 89-2. Multiplying all the elements of some row by the same nonzero number. (We say that we multiply the row by the number.)

Operation 89-3. Adding to one row a multiple of another. (*Adding two rows* means, of course, adding corresponding elements.)

We will solve systems of linear equations by using these operations to produce a triangular coefficient matrix—that is, a matrix of a system of equations in triangular form.

Example 89-1. Use matrix operations to solve System 89-1.

Solution. To clear of fractions, we multiply the first row of our augmented matrix by 2 and the third row by 4:

$$\begin{bmatrix} 1 & -6 & 2 & -8 \\ 1 & 4 & -3 & 2 \\ 8 & 0 & 1 & -16 \end{bmatrix}.$$

Then we add three times the last row to the second and subtract twice the last row from the first:

$$\left[\begin{array}{ccc|c} -15 & -6 & 0 & 24 \\ 25 & 4 & 0 & -46 \\ 8 & 0 & 1 & -16 \end{array}\right].$$

Multiplying the first row by $-\frac{1}{3}$ (dividing by -3) and interchanging with the second set the stage for our final transformation:

$$\left[\begin{array}{ccc|c} 25 & 4 & 0 & -46 \\ 5 & 2 & 0 & -8 \\ 8 & 0 & 1 & -16 \end{array}\right].$$

We arrive at triangular form by subtracting twice the second row from the first:

$$\left[\begin{array}{ccc|c} 15 & 0 & 0 & -30 \\ 5 & 2 & 0 & -8 \\ 8 & 0 & 1 & -16 \end{array}\right].$$

This matrix is the matrix of the triangular system

$$15x \qquad\qquad = -30$$
$$5x + 2y \qquad = -8$$
$$8x \qquad + z = -16,$$

from which we see that $(x, y, z) = (-2, 1, 0)$.

Example 89-2. Solve the system of equations

$$2x - y + z - w = -1$$
$$x + 3y - 2z \qquad = -5$$
$$3x - 2y \qquad + 4w = 1$$
$$-x + y - 3z - w = -6.$$

Solution. The matrix of this system is

$$\left[\begin{array}{cccc|c} 2 & -1 & 1 & -1 & -1 \\ 1 & 3 & -2 & 0 & -5 \\ 3 & -2 & 0 & 4 & 1 \\ -1 & 1 & -3 & -1 & -6 \end{array}\right].$$

Our goal is to reduce the coefficient matrix to triangular form, and we will start by making the first three elements of the fourth column 0 (replacing the colored numbers with 0's). To do so, we multiply the last row by 4 and add the resulting row to the third; then we subtract the last row from the first to produce the matrix

$$\begin{bmatrix} 3 & -2 & 4 & 0 & \bigm| & 5 \\ 1 & 3 & -2 & 0 & \bigm| & -5 \\ -1 & 2 & -12 & 0 & \bigm| & -23 \\ -1 & 1 & -3 & -1 & \bigm| & -6 \end{bmatrix}.$$

The fourth column is now in satisfactory form, so we concentrate on the first three rows and work to make the first two (colored) elements in the third column 0. We multiply the second row by 6. Then we subtract the third row from this new second row. Now we multiply the first row by 3 and add the third row. These operations yield the matrix

$$\begin{bmatrix} 8 & -4 & 0 & 0 & \bigm| & -8 \\ 7 & 16 & 0 & 0 & \bigm| & -7 \\ -1 & 2 & -12 & 0 & \bigm| & -23 \\ -1 & 1 & -3 & -1 & \bigm| & -6 \end{bmatrix}.$$

Finally, we add the second row to four times the first to cancel the first (colored) element of the second column, and we obtain the matrix

$$\begin{bmatrix} 39 & 0 & 0 & 0 & \bigm| & -39 \\ 7 & 16 & 0 & 0 & \bigm| & -7 \\ -1 & 2 & -12 & 0 & \bigm| & -23 \\ -1 & 1 & -3 & -1 & \bigm| & -6 \end{bmatrix}.$$

This matrix is the matrix of the system of equations in triangular form

$$39x \qquad\qquad\qquad\quad = -39$$
$$7x + 16y \qquad\qquad\quad = -7$$
$$-x + 2y - 12z \qquad\quad = -23$$
$$-x + y - 3z - w = -6.$$

From this system we readily find the solution to be $(x, y, z, w) = (-1, 0, 2, 1)$.

Each system of linear equations that we have seen thus far has had exactly one solution. We will call such a system of linear equations a **determinative** system. Not all systems are determinative. A system may have no solution, in which case we say that the system is **inconsistent**. At the other extreme, a system may have infinitely many solutions, in which case the system is **dependent**. The method of solving systems of linear equations by trying to reduce them to triangular form is still applicable to inconsistent and dependent systems. In fact, this method will enable us to tell when we are dealing with an inconsistent or a dependent system. The following examples illustrate how we can detect these situations.

Example 89-3. Solve the system of equations

$$2x - y + z = 1$$

(89-3)
$$x + 2y - z = 3$$

$$x + 7y - 4z = 2.$$

Solution. In matrix notation, the reduction to triangular form appears as

$$\begin{bmatrix} 2 & -1 & 1 & | & 1 \\ 1 & 2 & -1 & | & 3 \\ 1 & 7 & -4 & | & 2 \end{bmatrix} \rightarrow \begin{bmatrix} 9 & 3 & 0 & | & 6 \\ 3 & 1 & 0 & | & 10 \\ 1 & 7 & -4 & | & 2 \end{bmatrix} \rightarrow \begin{bmatrix} 0 & 0 & 0 & | & -24 \\ 3 & 1 & 0 & | & 10 \\ 1 & 7 & -4 & | & 2 \end{bmatrix}.$$

To go from the second matrix to the third, we subtract three times the second row from the first. The purpose of this operation, of course, is to remove the 3 in the first row, but in the process we also remove the 9. The final matrix can be considered as the matrix of the system

$$0 \qquad\qquad = 24$$

$$3x + y \qquad = 10$$

$$x + 7y - 4z = 2.$$

Since this system contains the false statement $0 = 24$, some explanation is in order. Logically, we did nothing wrong. We are simply asserting that *if* there is a solution (x, y, z) of System 89-3, *then* $0 = 24$. But $0 \neq 24$, so we must conclude that *the system does not have a solution.* System 89-3 is inconsistent.

Example 89-4. Solve the system of equations

$$2x - y + z = 1$$

(89-4)
$$x + 2y - z = 3$$

$$x + 7y - 4z = 8.$$

Solution. (Notice that this system is very similar to the system in Example 89-3.) In this case, the matrix reduction is

$$\begin{bmatrix} 2 & -1 & 1 & | & 1 \\ 1 & 2 & -1 & | & 3 \\ 1 & 7 & -4 & | & 8 \end{bmatrix} \rightarrow \begin{bmatrix} 9 & 3 & 0 & | & 12 \\ 3 & 1 & 0 & | & 4 \\ 1 & 7 & -4 & | & 8 \end{bmatrix} \rightarrow \begin{bmatrix} 0 & 0 & 0 & | & 0 \\ 3 & 1 & 0 & | & 4 \\ 1 & 7 & -4 & | & 8 \end{bmatrix}.$$

The last matrix is the matrix of the system

$$0 \qquad\qquad = 0$$

(89-5)
$$3x + y \qquad = 4$$

$$x + 7y - 4z = 8.$$

Unlike Example 89-3 (which yielded $0 = 24$), the first equation here is of absolutely no help. There is nothing false about the statement that $0 = 0$, but it doesn't tell us anything that we didn't already know. There are many solutions of System 89-5. For instance, we could choose $x = 1$; then $y = 1$ and $z = 0$. Or if $x = 0$, then $y = 4$ and $z = 5$. Indeed, if t is any number, then a solution of System 89-5 (and hence of System 89-4) is $(x, y, z) = (t, 4 - 3t, 5 - 5t)$. We often write this triple as $(x, y, z) = t(1, -3, -5) + (0, 4, 5)$, where the operations of adding two triples and multiplying one by a number t are the obvious ones. System 89-4 is dependent. Observe that if we first choose $y = s$, we obtain our solution triple in the form

$$(x, y, z) = (\tfrac{1}{3}(4 - s), s, \tfrac{5}{3}(s - 1)) = s(-\tfrac{1}{3}, 1, \tfrac{5}{3}) + (\tfrac{4}{3}, 0, -\tfrac{5}{3}).$$

This expression for the solution seems to differ from our previous one. Can you show that it doesn't?

If a system consists of two linear equations in two unknowns, and if all the given numbers are real, then there is a simple geometric interpretation of what it means for the system to be determinative. Since a linear equation $ax + by = c$ represents a line in a plane, the *pair* of equations

$$ax + by = c$$
$$dx + ey = f$$

represents *two* lines. A solution of this system is a pair of real numbers, which we can interpret as the coordinates of a point that belongs to both lines. Thus *our system is determinative if and only if the lines represented by that system are distinct and nonparallel and hence intersect in one point.*

PROBLEMS   89

1. Use matrix notation to solve the systems of equations.

(a) $2x + 3y = 7$
$3x - y = 5$

(b) $2x + 3y = 1$
$3x - y = -4$

(c) $2x + 3y = 1$
$x + 2y = 1$

(d) $2x + 3y = -1$
$x + 2y = -1$

(e) $ 2y = 6$
$x - y = 1$

(f) $ 2y = 2$
$x - y = 3$

2. Use matrix notation to solve the systems of equations.

(a) $2x + 3y - z = -2$
$x - y + 2z = 4$
$x + 2y + z = 0$

(b) $2x + 3y - z = 0$
$x - y + 2z = 0$
$x + 2y + z = 0$

(c) $x + 4y - 2z = 0$
$-2x + y = 0$
$x - y + z = 0$

(d) $x + 4y - 2z = -1$
$-2x + y = 3$
$x - y + z = 0$

(e)
$$x + y = 2$$
$$y + z = -1$$
$$z + w = 0$$
$$x - y + z - w = 0$$

(f)
$$x + y = 3$$
$$y + z = 5$$
$$z + w = 7$$
$$x - y + z - w = -2$$

3. Use matrix notation to solve the "homogeneous" systems.

(a)
$$2x - y + z = 0$$
$$x + 3y + 2z = 0$$
$$x - 4y - z = 0$$

(b)
$$2x - y + z = 0$$
$$x + 3y + 2z = 0$$
$$3x + 2y + 3z = 0$$

4. The augmented matrix of a system of equations is

$$\begin{bmatrix} 24 & 0 & 0 & | & 0 \\ 1 & 2 & 0 & | & 2 \\ 3 & 4 & 1 & | & 5 \end{bmatrix}.$$

Is the system determinative?

5. Solve, if possible, the systems.

(a)
$$2x - 3y + z = -4$$
$$x - 4y - z = -3$$
$$x - 9y - 4z = -5$$

(b)
$$2x - 3y + z = 4$$
$$x - 4y - z = 3$$
$$x - 9y - 4z = 5$$

(c)
$$2x - 3y + z = 3$$
$$x - 4y - z = 4$$
$$x - 9y - 4z = 5$$

(d)
$$2x - 3y + z = 0$$
$$x - 4y - z = 0$$
$$x - 9y - 4z = 0$$

6. The sum of the digits of a three-digit number is 14 and the middle digit is the sum of the other two digits. If the last two digits are interchanged, the number obtained is 27 less than the original number. Find the number. Can you solve the problem if the number obtained when the last two digits are interchanged is 72 less than the original number?

7. Solve the system of equations

$$x \cos \alpha + y \sin \alpha = u$$

$$-x \sin \alpha + y \cos \alpha = v.$$

8. Three pipelines supply an oil reservoir. The reservoir can be filled by pipes A and B running for 10 hours, by pipes B and C running for 15 hours, or by pipes A and C running for 20 hours. How long does it take to fill the tank if (a) all three pipes run? (b) pipe A is used alone?

9. You are told that a bag of 30 coins contains nickels, dimes, and quarters amounting to \$3 and that there are twice as many nickels as there are dimes. Should you believe it?

10. Four high schools, South, East, North, and West, have a total enrollment of 1000 students. South reports that 10% of the students make A's, 25% B's, and 50% C's; East reports 15% A's, 25% B's, and 55% C's; North reports 25% A's, 15% B's, and 35% C's; West reports 15% A's, 20% B's, and 40% C's. Is this information consistent with the fact that of the total student enrollment in all four schools, 15% receive A's, 25% B's, and 50% C's?

11. (a) Let $P(x) = x^3 + ax^2 + bx + c$. If we know that $P(-1) = -1$, $P(1) = 5$, and $P(2) = 11$, find the values of a, b, and c.

    (b) Discuss the problem of finding the coefficients a, b, and c so that the graph of the polynomial equation $y = ax^2 + bx + c$ contains three given points (x_1, y_1), (x_2, y_2), and (x_3, y_3).

12. Find a simple function f such that $f(1) = 1, f'(1) = 2, f''(1) = 3$, and $f'''(1) = 4$.

13. For simplicity, we have confined our attention to systems of equations with the same number of unknowns as equations. Nothing we have done so far requires us to make this restriction. If we have a different number of equations and unknowns, our reduction technique leads to "echelon form," rather than triangular form, but there is no real conceptual difference. Solve the following systems of equations.

(a) $3x - 2y + z = 2$
    $x + 3y - 2z = 2$

(b) $3x + y = 4$
    $2x - 3y = -1$
    $x - 2y = -1$

(c) $3x + y = -1$
    $2x - 3y = 4$
    $x - 2y = -1$

(d) $x_1 + x_2 + x_3 + x_4 + x_5 + x_6 + x_7 + x_8 = 2$
    $x_1 - x_2 + x_3 - x_4 + x_5 - x_6 + x_7 - x_8 = 0$

90. VECTORS AND MATRICES

An element of R^3 is a triple of real numbers. For example, $(1, -3, \pi)$ and $(0, 0, -2)$ are elements of R^3. We will sometimes find it a notational convenience to write these triples as columns, $\begin{pmatrix} 1 \\ -3 \\ \pi \end{pmatrix}$ and $\begin{pmatrix} 0 \\ 0 \\ -2 \end{pmatrix}$, instead of rows. But we are still dealing with triples of numbers and hence with elements of R^3. This column notation merely helps us to perform more easily some of the rules of calculation that we will introduce later in the section.

At this point, R^3 is just a set. It has no arithmetic or other mathematical structure. But we have already made use of two arithmetic operations on triples that we will now introduce formally. Because we are thinking of elements of R^3 as individual entities here, we use a single boldface letter, rather than a triple, to represent an element. Thus we will write $\mathbf{a} = (a_1, a_2, a_3)$. Since we use the same notation for elements of R^2, R^4, and so on, you must look at the context in order to tell whether $\mathbf{a}$ stands for (a_1, a_2, a_3), (a_1, a_2), or whatever. The numbers a_1, a_2, and a_3 are the **components** of the element (a_1, a_2, a_3) of R^3.

We define **addition** of elements in R^3 in the natural way, by components. Thus if $\mathbf{a} = (a_1, a_2, a_3)$ and $\mathbf{b} = (b_1, b_2, b_3)$, then

$$\mathbf{a} + \mathbf{b} = (a_1 + b_1, a_2 + b_2, a_3 + b_3).$$

We will write $\mathbf{0} = (0, 0, 0)$ and $-\mathbf{a} = (-a_1, -a_2, -a_3)$. Then straightforward calculation shows us that

(90-1)
$$\mathbf{a} + \mathbf{0} = \mathbf{a},$$
$$\mathbf{a} + (-\mathbf{a}) = \mathbf{0},$$
$$(\mathbf{a} + \mathbf{b}) + \mathbf{c} = \mathbf{a} + (\mathbf{b} + \mathbf{c}),$$
$$\mathbf{a} + \mathbf{b} = \mathbf{b} + \mathbf{a}.$$

We also have a natural definition of what it means to **multiply** the element $\mathbf{a} = (a_1, a_2, a_3)$ of R^3 by a real number r:

$$r\mathbf{a} = (ra_1, ra_2, ra_3).$$

You can check to see that

$$1\mathbf{a} = \mathbf{a},$$

(90-2)
$$(r + s)\mathbf{a} = r\mathbf{a} + s\mathbf{a},$$

$$r(\mathbf{a} + \mathbf{b}) = r\mathbf{a} + r\mathbf{b},$$

$$(rs)\mathbf{a} = r(s\mathbf{a}).$$

In Section 68 we stated that a mathematical system in which Equations 90-1 and 90-2 are valid is called a vector space over the real numbers. So with the operations we have introduced, our *set* R^3 becomes a *vector space*; and when we have these operations in mind, we speak of the elements of R^3 as **vectors**. The real numbers r, s, and so on, are called **scalars**.

Although our definitions are written in terms of R^3, it is perfectly clear that the same sort of thing could be done to make a vector space out of R^2, R^4, R^{1492}, and so on. There is an obvious connection between the "algebraic" vector space R^2 and the "geometric" vector space of arrows in the plane that we introduced in Section 68, a connection that readily extends to three-dimensional space. We will discuss these points in the next chapter, confining this chapter to the purely algebraic aspects of R^2, R^3, and so on.

Example 90-1. Solve for $\mathbf{x}$ and $\mathbf{y}$:

$$3\mathbf{x} + 2\mathbf{y} = (3, 1)$$

$$\mathbf{x} - \mathbf{y} = (1, -3).$$

Solution. We will solve this system of equations in the usual way, adding twice the second equation to the first to produce an equivalent triangular system:

$$5\mathbf{x} \quad = (5, -5)$$

$$\mathbf{x} - \mathbf{y} = (1, -3).$$

The first of these equations tells us that $\mathbf{x} = (1, -1)$, and then we find from the second that $\mathbf{y} = (0, 2)$.

A capital letter, such as A, will denote a matrix. Because we will be working in two or three dimensions, we will usually consider only 2 by 2 or 3 by 3 matrices—that is, matrices of *order* 2 or *order* 3. The number in row i and column j of the matrix A will be denoted by a_{ij}; for example,

$$A = \begin{bmatrix} a_{11} & a_{12} & a_{13} \\ a_{21} & a_{22} & a_{23} \\ a_{31} & a_{32} & a_{33} \end{bmatrix}.$$

It is often helpful to view a matrix as a *pile* of rows or a *file* of columns. In other words,

$$A = \begin{bmatrix} \mathbf{r}_1 \\ \mathbf{r}_2 \\ \mathbf{r}_3 \end{bmatrix} = [\mathbf{c}_1, \mathbf{c}_2, \mathbf{c}_3],$$

where $\mathbf{r}_1 = (a_{11}, a_{12}, a_{13})$, $\mathbf{r}_2 = (a_{21}, a_{22}, a_{23})$, and $\mathbf{r}_3 = (a_{31}, a_{32}, a_{33})$ are the *row vectors* of A, and

$$\mathbf{c}_1 = \begin{pmatrix} a_{11} \\ a_{21} \\ a_{31} \end{pmatrix}, \qquad \mathbf{c}_2 = \begin{pmatrix} a_{12} \\ a_{22} \\ a_{32} \end{pmatrix}, \quad \text{and} \quad \mathbf{c}_3 = \begin{pmatrix} a_{13} \\ a_{23} \\ a_{33} \end{pmatrix}$$

are its *column vectors*.

We will now define the product of a vector by a matrix. To be explicit, we will suppose that we are dealing with a 3 by 3 matrix A and a vector $\mathbf{x} \in R^3$; the extension to other dimensions is self-evident. The product $A\mathbf{x}$ will be a vector (an element of R^3), so we must give the rule that assigns its three components. It is customary in this situation to write our vectors as columns, and we make the following definition of the **product** $A\mathbf{x}$:

$$(90\text{-}3) \quad \begin{bmatrix} a_{11} & a_{12} & a_{13} \\ a_{21} & a_{22} & a_{23} \\ a_{31} & a_{32} & a_{33} \end{bmatrix} \begin{pmatrix} x_1 \\ x_2 \\ x_3 \end{pmatrix} = \begin{pmatrix} a_{11}x_1 + a_{12}x_2 + a_{13}x_3 \\ a_{21}x_1 + a_{22}x_2 + a_{23}x_3 \\ a_{31}x_1 + a_{32}x_2 + a_{33}x_3 \end{pmatrix}.$$

Example 90-2. Find $A\mathbf{x}$ when $A = \begin{bmatrix} 2 & -1 \\ 3 & 2 \end{bmatrix}$ and $\mathbf{x} = (3, -2)$.

Solution.

$$\begin{bmatrix} 2 & -1 \\ 3 & 2 \end{bmatrix} \begin{pmatrix} 3 \\ -2 \end{pmatrix} = \begin{pmatrix} 2 \cdot 3 + (-1)(-2) \\ 3 \cdot 3 + 2(-2) \end{pmatrix} = \begin{pmatrix} 8 \\ 5 \end{pmatrix}.$$

The most important feature of this multiplication of a vector by a matrix is that the "general distributive law" is satisfied. Thus if $\mathbf{x}$ and $\mathbf{y}$ are any vectors and r and s are scalars, then

$$(90\text{-}4) \qquad A(r\mathbf{x} + s\mathbf{y}) = rA\mathbf{x} + sA\mathbf{y}.$$

So we can combine the vectors first and then multiply by A or multiply by A and then combine the resulting vectors. Observe the analogy with the similar behavior of the differential operator $\dfrac{d}{dx}$.

If we think of A as a file of columns, we can write Equation 90-3 as

$$(90\text{-}5) \qquad A\mathbf{x} = x_1\mathbf{c}_1 + x_2\mathbf{c}_2 + x_3\mathbf{c}_3,$$

a point of view that you might find useful later. We remark that A needn't be a square matrix. For example, the product of a vector in R^3 by a 2 by 3 matrix produces (by the obvious variation of Equation 90-3) a vector in R^2. Notice that we have not defined (and we will not define) the symbol $\mathbf{x}A$.

The concept of the product of a vector by a matrix provides us with an especially simple and useful notation for a system of linear equations. Thus suppose that we are given a matrix A and a vector $\mathbf{b} = (b_1, b_2, b_3)$, and we seek a vector $\mathbf{x}$ that satisfies the matrix-vector equation

$$A\mathbf{x} = \mathbf{b}.$$

This single equation between two vectors is equivalent to three equations between their components; that is (see Equation 90-3 for the components of the vector on the left-hand side), our matrix-vector equation is equivalent to the system of linear equations

$$a_{11}x_1 + a_{12}x_2 + a_{13}x_3 = b_1$$
$$a_{21}x_1 + a_{22}x_2 + a_{23}x_3 = b_2$$
$$a_{31}x_1 + a_{32}x_2 + a_{33}x_3 = b_3.$$

Clearly, we can use this scheme with matrices of arbitrary order n and vectors from R^n to represent a system of n linear equations in n unknowns as a single matrix-vector equation.

PROBLEMS 90

1. Suppose that $\mathbf{a} = (2, -1, 2)$, $\mathbf{b} = (4, 0, 3)$, and $\mathbf{c} = (2, 1, -1)$.
   (a) Verify that $3\mathbf{a} + 2\mathbf{a} = 5\mathbf{a}$.
   (b) Verify that $3(2\mathbf{a}) = (3 \cdot 2)\mathbf{a}$.
   (c) Verify that $3\mathbf{a} + 2\mathbf{b} = 2\mathbf{b} + 3\mathbf{a}$.
   (d) Verify that $3(\mathbf{a} + \mathbf{b}) = 3\mathbf{a} + 3\mathbf{b}$.
   (e) Verify that $(\mathbf{a} + \mathbf{b}) + \mathbf{c} = \mathbf{a} + (\mathbf{b} + \mathbf{c})$.
   (f) Verify that $(3\mathbf{a} + 2\mathbf{b}) - \mathbf{c} = 3\mathbf{a} + (2\mathbf{b} - \mathbf{c})$.
   (g) Find $\mathbf{a} - (\mathbf{b} - \mathbf{c})$.
   (h) Find $\mathbf{a} - \mathbf{b} - \mathbf{c}$.

2. Let $\mathbf{a}$, $\mathbf{b}$, and $\mathbf{c}$ be the vectors of the preceding problem.
   (a) Find x so that the third component of $x\mathbf{a} + \mathbf{b}$ is 0.
   (b) Find y so that the second component of $\mathbf{a} + y\mathbf{b}$ is 0.
   (c) Find x and y so that the first and second components of $x\mathbf{a} + y\mathbf{b}$ are 8 and -2.
   (d) Find x and y so that the first and second components of $x\mathbf{a} + y\mathbf{b}$ are both 0.
   (e) Show that there are no numbers x and y such that $x\mathbf{a} + y\mathbf{b} = \mathbf{c}$.
   (f) Show that if $x\mathbf{a} + y\mathbf{b} = \mathbf{0}$, then $x = y = 0$.

3. If we let $\mathbf{i} = (1, 0, 0)$, $\mathbf{j} = (0, 1, 0)$, and $\mathbf{k} = (0, 0, 1)$, explain why the vector $\mathbf{a}$ of the preceding problems can be written as $\mathbf{a} = 2\mathbf{i} - \mathbf{j} + 2\mathbf{k}$. How would you express $\mathbf{b}$ and $\mathbf{c}$ in terms of $\mathbf{i}$, $\mathbf{j}$, and $\mathbf{k}$? Find numbers x, y, and z so that $\mathbf{i} = x\mathbf{a} + y\mathbf{b} + z\mathbf{c}$.

4. Let A be the matrix whose row vectors are the vectors **a**, **b**, and **c** of the first problem and B be the matrix whose column vectors are **a**, **b**, and **c**. Find $A\textbf{a}$ and $B\textbf{b}$.

5. Let $A = \begin{bmatrix} 2 & 1 & 0 \\ 1 & -1 & 2 \\ 3 & 0 & 1 \end{bmatrix}$, $\textbf{a} = (-1, 2, 0)$, and $\textbf{b} = (1, -1, 1)$.

   (a) Verify that $A(3\textbf{a}) = 3A\textbf{a}$.
   (b) Verify that $A(3\textbf{a} + 2\textbf{b}) = 3A\textbf{a} + 2A\textbf{b}$.
   (c) Express $A\textbf{a}$ as a linear combination of the column vectors of A (Equation 90-5).
   (d) Express $A\textbf{b}$ as a linear combination of the column vectors of A.

6. Suppose that $A = \begin{bmatrix} -7 & 2 & 14 \\ -18 & 3 & 34 \\ -3 & 1 & 6 \end{bmatrix}$, $\textbf{u} = (2, 2, 1)$, $\textbf{v} = (1, -4, 1)$, and $\textbf{w} = (2, 1, 1)$. Show that $A\textbf{u}$, $A\textbf{v}$, and $A\textbf{w}$ are simply scalar multiples of **u**, **v**, and **w**.

7. Solve the equation $A\textbf{x} = \textbf{b}$ if $A = \begin{bmatrix} 2 & 1 & 2 \\ 2 & -4 & 1 \\ 1 & 1 & 1 \end{bmatrix}$.

   (a) $\textbf{b} = (5, -1, 3)$ (b) $\textbf{b} = (6, 5, 3)$
   (c) $\textbf{b} = (0, 0, 1)$ (d) $\textbf{b} = \textbf{0}$

8. In Problem 90-3 we introduced the vectors **i**, **j**, and **k**. Show that $A\textbf{i}$, $A\textbf{j}$, and $A\textbf{k}$ are the column vectors of A.

9. A set $\{\textbf{a}, \textbf{b}, \textbf{c}\}$ of vectors is said to be **linearly independent** if the only solution of the equation $x\textbf{a} + y\textbf{b} + z\textbf{c} = \textbf{0}$ is the triple $(x, y, z) = (0, 0, 0)$. Explain how this statement implies that no one of the vectors can be written as a linear combination of the other two (for example, we can't have $\textbf{c} = p\textbf{a} + q\textbf{b}$ for any numbers p and q).

10. The community judgment of the quality of our two local restaurants, Joe's and Joan's, is summed up in the matrix $T = \begin{bmatrix} .1 & .4 \\ .9 & .6 \end{bmatrix}$. One-tenth of Joe's customers return; the other .9 go to Joan's next time. Joan's customers are more loyal; .6 come back for their next meal, and only .4 defect to Joe's. Suppose that we plan to dine out every night next week. We can use the probability x that we eat at Joe's and the probability y that we eat at Joan's as the components of a probability vector (x, y). For example, if we choose our Sunday restaurant by a coin flip, the initial probability vector is $\textbf{p} = (.5, .5)$. The vector $T\textbf{p}$ is the probability vector for the location of Monday's dinner, the vector $T(T\textbf{p})$ gives us the probabilities for Tuesday's meal, and so on. What is the Monday probability vector? What is the probability that we will eat at Joe's on Tuesday? Show that if $\textbf{p} = (a, b)$, where $a + b = 1$, then the sum of the components of $T\textbf{p}$ is again 1.

91. THE ALGEBRA OF MATRICES

In the last section we introduced algebraic operations into our set R^3; now we will take up the algebra of square matrices. To be concrete, we frame our definitions in terms of 3 by 3 matrices, but they extend to matrices of other orders in an obvious manner.

Addition of matrices and **multiplication** by scalars follows the pattern set by elements of R^n. Thus if

$$A = \begin{bmatrix} a_{11} & a_{12} & a_{13} \\ a_{21} & a_{22} & a_{23} \\ a_{31} & a_{32} & a_{33} \end{bmatrix} \quad \text{and} \quad B = \begin{bmatrix} b_{11} & b_{12} & b_{13} \\ b_{21} & b_{22} & b_{23} \\ b_{31} & b_{32} & b_{33} \end{bmatrix}$$

are given matrices and r is a number, then

$$A + B = \begin{bmatrix} a_{11} + b_{11} & a_{12} + b_{12} & a_{13} + b_{13} \\ a_{21} + b_{21} & a_{22} + b_{22} & a_{23} + b_{23} \\ a_{31} + b_{31} & a_{32} + b_{32} & a_{33} + b_{33} \end{bmatrix}$$

and

$$rA = \begin{bmatrix} ra_{11} & ra_{12} & ra_{13} \\ ra_{21} & ra_{22} & ra_{23} \\ ra_{31} & ra_{32} & ra_{33} \end{bmatrix}.$$

The **zero matrix** is defined by the obvious equation

$$0 = \begin{bmatrix} 0 & 0 & 0 \\ 0 & 0 & 0 \\ 0 & 0 & 0 \end{bmatrix},$$

and $-A = (-1)A$. Then it is easy to see that Equations 90-1 and 90-2 continue to hold when matrices replace the elements of R^n.

Multiplication of matrices is more complicated. We define the **product** AB as the matrix whose column vectors are products of the column vectors of B by the matrix A. Thus if $B = [\mathbf{c}_1, \mathbf{c}_2, \mathbf{c}_3]$, we write

$$AB = [A\mathbf{c}_1, A\mathbf{c}_2, A\mathbf{c}_3].$$

We use Equation 90-3 to compute the components of these column vectors. For example, in the case of 2 by 2 matrices, our multiplication rule reads

$$(91\text{-}1) \quad \begin{bmatrix} a_{11} & a_{12} \\ a_{21} & a_{22} \end{bmatrix} \begin{bmatrix} b_{11} & b_{12} \\ b_{21} & b_{22} \end{bmatrix} = \begin{bmatrix} a_{11}b_{11} + a_{12}b_{21} & a_{11}b_{12} + a_{12}b_{22} \\ a_{21}b_{11} + a_{22}b_{21} & a_{21}b_{12} + a_{22}b_{22} \end{bmatrix}.$$

Example 91-1.   Compute both AB and BA if

$$A = \begin{bmatrix} 2 & -1 \\ -4 & 2 \end{bmatrix} \quad \text{and} \quad B = \begin{bmatrix} 1 & 5 \\ 2 & 10 \end{bmatrix}.$$

Solution.   According to Equation 91-1,

$$AB = \begin{bmatrix} 2 & -1 \\ -4 & 2 \end{bmatrix} \begin{bmatrix} 1 & 5 \\ 2 & 10 \end{bmatrix} = \begin{bmatrix} 2 \cdot 1 + (-1)2 & 2 \cdot 5 + (-1)10 \\ (-4)1 + 2 \cdot 2 & (-4)5 + 2 \cdot 10 \end{bmatrix} = \begin{bmatrix} 0 & 0 \\ 0 & 0 \end{bmatrix}$$

and

$$BA = \begin{bmatrix} 1 & 5 \\ 2 & 10 \end{bmatrix} \begin{bmatrix} 2 & -1 \\ -4 & 2 \end{bmatrix} = \begin{bmatrix} 1 \cdot 2 + 5(-4) & 1(-1) + 5 \cdot 2 \\ 2 \cdot 2 + 10(-4) & 2(-1) + 10 \cdot 2 \end{bmatrix}$$

$$= \begin{bmatrix} -18 & 9 \\ -36 & 18 \end{bmatrix}.$$

The preceding example shows that *it is not always true that AB and BA are equal.* Moreover, it is possible that $AB = 0$ without either A or B being the 0 matrix. Most of the other rules of the arithmetic of multiplication, however, are valid for matrices. For example,

$$A(BC) = (AB)C \quad \text{and} \quad A(B + C) = AB + AC.$$

Furthermore,

$$(AB)\mathbf{x} = A(B\mathbf{x}) \quad \text{and} \quad (A + B)\mathbf{x} = A\mathbf{x} + B\mathbf{x}.$$

These statements can be proved by writing the various expressions in terms of the components of the vectors and matrices that appear.

The matrix I that is defined by the equation

$$I = \begin{bmatrix} 1 & 0 & 0 \\ 0 & 1 & 0 \\ 0 & 0 & 1 \end{bmatrix}$$

acts like the number 1 of our real number system. Thus

$$IA = AI = A \quad \text{and} \quad I\mathbf{x} = \mathbf{x}.$$

We call I the **identity matrix**.

We have mentioned that matrix-vector notation allows us to express a system of linear equations as the single equation

(91-2) $$A\mathbf{x} = \mathbf{b}.$$

We are supposing that the coefficient matrix A and the vector $\mathbf{b}$ are given and that the unknowns of our system are the components of $\mathbf{x}$. The natural way to solve Equation 91-2 for $\mathbf{x}$ is to "divide" both sides by the matrix A—that is, to multiply by the *reciprocal*, or *inverse*, of A. We naturally call the matrix A^{-1} the **inverse** of a matrix A if

(91-3) $$AA^{-1} = A^{-1}A = I.$$

This definition tells us that $X = A^{-1}$ satisfies the *two* equations $AX = I$ and $XA = I$. In Problem 91-12 we ask you to show that these equations cannot have two different solutions; in other words, if a matrix has an inverse, it has only one. Therefore it is proper to speak of *the* inverse of a matrix rather than *an* inverse. Furthermore, later in this section we will see that we need only solve the ,single equation $AX = I$ to find the inverse of A; the equation $XA = I$ will automatically be satisfied. From the symmetry of the definition of the inverse, it is clear that A is the inverse of A^{-1}; that is, $(A^{-1})^{-1} = A$.

Example 91-2. Show that $\begin{bmatrix} 2 & 1 \\ 1 & 1 \end{bmatrix}^{-1} = \begin{bmatrix} 1 & -1 \\ -1 & 2 \end{bmatrix}$.

Solution. We need only verify that

$$\begin{bmatrix} 2 & 1 \\ 1 & 1 \end{bmatrix}\begin{bmatrix} 1 & -1 \\ -1 & 2 \end{bmatrix} = \begin{bmatrix} 1 & -1 \\ -1 & 2 \end{bmatrix}\begin{bmatrix} 2 & 1 \\ 1 & 1 \end{bmatrix} = \begin{bmatrix} 1 & 0 \\ 0 & 1 \end{bmatrix}.$$

Example 91-3. Show that the matrix $A = \begin{bmatrix} 1 & 1 \\ 1 & 1 \end{bmatrix}$ does not have an inverse.

Solution. Let us take an arbitrary matrix $X = \begin{bmatrix} a & b \\ c & d \end{bmatrix}$ and consider the product

$$AX = \begin{bmatrix} 1 & 1 \\ 1 & 1 \end{bmatrix}\begin{bmatrix} a & b \\ c & d \end{bmatrix} = \begin{bmatrix} a + c & b + d \\ a + c & b + d \end{bmatrix}.$$

It is clear that no choice of a, b, c, and d will make this product the identity matrix I. We simply cannot, for example, choose a and c so that $a + c = 1$ *and* $a + c = 0$. Therefore we cannot solve the equation $AX = I$, and so we cannot hope to satisfy Equations 91-3.

If the matrix A has an inverse, the solution of Equation 91-2 can be expressed in terms of the inverse matrix and the given vector **b**. We simply multiply both sides by A^{-1}:

$$A^{-1}A\mathbf{x} = A^{-1}\mathbf{b}$$

$$I\mathbf{x} = A^{-1}\mathbf{b}$$

$$\mathbf{x} = A^{-1}\mathbf{b}.$$

Example 91-4. Solve the system of equations

$$2x + y = 3$$

$$x + y = -4.$$

Solution. If we express this system in the form of the matrix-vector Equation 91-2, we will have

$$A = \begin{bmatrix} 2 & 1 \\ 1 & 1 \end{bmatrix}, \quad \mathbf{b} = \begin{pmatrix} 3 \\ -4 \end{pmatrix}, \quad \text{and} \quad \mathbf{x} = \begin{pmatrix} x \\ y \end{pmatrix}.$$

We are given A^{-1} in Example 91-2, and therefore

$$\mathbf{x} = A^{-1}\mathbf{b} = \begin{bmatrix} 1 & -1 \\ -1 & 2 \end{bmatrix} \begin{pmatrix} 3 \\ -4 \end{pmatrix} = \begin{pmatrix} 7 \\ -11 \end{pmatrix}.$$

You may verify this solution by substituting $(x, y) = (7, -11)$ in the original system of equations.

Let us put aside the question of the existence of the inverse of a given matrix for a moment and turn to the question of calculating it, assuming that there is one. We have seen that if a matrix A has an inverse A^{-1}, then $X = A^{-1}$ is a solution of the equation

(91-4) $AX = I.$

Suppose that we denote the column vectors of the matrix X by $\mathbf{x}_1, \mathbf{x}_2$, and $\mathbf{x}_3$. Then the column vectors of the matrix AX are $A\mathbf{x}_1, A\mathbf{x}_2$, and $A\mathbf{x}_3$, so Equation 91-4 is equivalent to the three matrix-vector equations

(91-5) $A\mathbf{x}_1 = \begin{pmatrix} 1 \\ 0 \\ 0 \end{pmatrix}, \qquad A\mathbf{x}_2 = \begin{pmatrix} 0 \\ 1 \\ 0 \end{pmatrix}, \qquad A\mathbf{x}_3 = \begin{pmatrix} 0 \\ 0 \\ 1 \end{pmatrix}.$

Each of these three matrix-vector equations is, in turn, equivalent to a system of three linear equations in three unknowns, and we can solve such systems by the methods we discussed in Section 89. In this way, we can find the solution $X = A^{-1}$ of Equation 91-4.

Let us use an example to show how we actually proceed. Suppose that A is the matrix

$$A = \begin{bmatrix} 2 & 1 & 2 \\ 2 & -4 & 1 \\ 1 & 1 & 1 \end{bmatrix}.$$

Then in order to solve Equations 91-5 by the methods of Section 89, we "manipulate" the augmented matrices

$$\begin{bmatrix} 2 & 1 & 2 & | & 1 \\ 2 & -4 & 1 & | & 0 \\ 1 & 1 & 1 & | & 0 \end{bmatrix}, \quad \begin{bmatrix} 2 & 1 & 2 & | & 0 \\ 2 & -4 & 1 & | & 1 \\ 1 & 1 & 1 & | & 0 \end{bmatrix}, \quad \text{and} \quad \begin{bmatrix} 2 & 1 & 2 & | & 0 \\ 2 & -4 & 1 & | & 0 \\ 1 & 1 & 1 & | & 1 \end{bmatrix}.$$

The operations we perform on these augmented matrices are determined by the square matrix on the left of the vertical line. Since this matrix is the same in each of the three cases, we won't treat them separately but will lump them together in the single augmented matrix

$$\begin{bmatrix} 2 & 1 & 2 & | & 1 & 0 & 0 \\ 2 & -4 & 1 & | & 0 & 1 & 0 \\ 1 & 1 & 1 & | & 0 & 0 & 1 \end{bmatrix}.$$

We treat this matrix just as we treated the augmented matrices of Section 89. By various row operations, we try to reduce the matrix on the left of the vertical line to triangular form. The matrix on the right will be changed in the process, but we ignore that fact:

$$\begin{bmatrix} 2 & 1 & 2 & | & 1 & 0 & 0 \\ 2 & -4 & 1 & | & 0 & 1 & 0 \\ 1 & 1 & 1 & | & 0 & 0 & 1 \end{bmatrix} \rightarrow \begin{bmatrix} 0 & -1 & 0 & | & 1 & 0 & -2 \\ 1 & -5 & 0 & | & 0 & 1 & -1 \\ 1 & 1 & 1 & | & 0 & 0 & 1 \end{bmatrix}$$

$$\rightarrow \begin{bmatrix} 0 & -5 & 0 & | & 5 & 0 & -10 \\ 1 & -5 & 0 & | & 0 & 1 & -1 \\ 1 & 1 & 1 & | & 0 & 0 & 1 \end{bmatrix}$$

$$\rightarrow \begin{bmatrix} -1 & 0 & 0 & | & 5 & -1 & -9 \\ 1 & -5 & 0 & | & 0 & 1 & -1 \\ 1 & 1 & 1 & | & 0 & 0 & 1 \end{bmatrix}.$$

This last matrix is the matrix of three systems of equations in triangular form, and we could write these equations, solve them, and thus obtain our desired inverse. It is more efficient, however (and we could have used this procedure in Section 89 also), to continue our row operations until the coefficient matrix is the identity. To accomplish this result, we apply row operations to make the elements below the main diagonal zero, and then we multiply as necessary to make the elements that remain in the diagonal 1:

$$\begin{bmatrix} -1 & 0 & 0 & | & 5 & -1 & -9 \\ 1 & -5 & 0 & | & 0 & 1 & -1 \\ 1 & 1 & 1 & | & 0 & 0 & 1 \end{bmatrix} \rightarrow \begin{bmatrix} -1 & 0 & 0 & | & 5 & -1 & -9 \\ 0 & -5 & 0 & | & 5 & 0 & -10 \\ 0 & 1 & 1 & | & 5 & -1 & -8 \end{bmatrix}$$

$$\rightarrow \begin{bmatrix} -1 & 0 & 0 & | & 5 & -1 & -9 \\ 0 & -1 & 0 & | & 1 & 0 & -2 \\ 0 & 1 & 1 & | & 5 & -1 & -8 \end{bmatrix}$$

$$\rightarrow \begin{bmatrix} -1 & 0 & 0 & | & 5 & -1 & -9 \\ 0 & -1 & 0 & | & 1 & 0 & -2 \\ 0 & 0 & 1 & | & 6 & -1 & -10 \end{bmatrix}$$

$$\rightarrow \begin{bmatrix} 1 & 0 & 0 & | & -5 & 1 & 9 \\ 0 & 1 & 0 & | & -1 & 0 & 2 \\ 0 & 0 & 1 & | & 6 & -1 & -10 \end{bmatrix}.$$

This final augmented matrix tells us that the first of Equations 91-5 is equivalent to the system

$$x = -5$$
$$y = -1$$
$$z = 6;$$

in other words, $\mathbf{x}_1 = (-5, -1, 6)$. Similarly, the other two columns of the solution X of Equation 91-4 are the other two columns of the matrix on the right of the vertical line. In short, this matrix is A^{-1}.

We can sum up our procedure for finding the inverse of a matrix A as follows. First, form the augmented matrix

$$[A \mid I].$$

Then apply operations 89-1, 89-2, and 89-3 until this augmented matrix takes the form

$$[I \mid A^{-1}].$$

As we indicated in our example, a systematic way to proceed is first to reduce the matrix on the left of the vertical line to triangular form. At that point, our augmented matrix will look like this:

$$(91\text{-}6) \qquad \begin{bmatrix} d_1 & 0 & 0 & * & * & * \\ * & d_2 & 0 & * & * & * \\ * & * & d_3 & * & * & * \end{bmatrix}.$$

Now, whether or not we can continue to reduce the matrix on the left to the identity depends on the diagonal numbers d_1, d_2, and d_3. If none of these numbers is zero, we can reduce the matrix on the left to the identity matrix, otherwise not. Thus the question of whether or not our method leads to a solution of the equation $AX = I$ becomes the question of whether or not the three numbers d_1, d_2, and d_3 are all different from zero.

Let us put this criterion into different words. If we were to attempt to solve the homogeneous (we use the word "homogeneous" when the vector on the right side of the equation is $\mathbf{0}$) matrix-vector equation

$$A\mathbf{x} = \mathbf{0}$$

by the methods we used in Section 89, we would set up the augmented matrix

$$[A \mid \mathbf{0}]$$

and reduce to triangular form

$$\begin{bmatrix} d_1 & 0 & 0 & 0 \\ * & d_2 & 0 & 0 \\ * & * & d_3 & 0 \end{bmatrix}.$$

If all the numbers d_1, d_2, and d_3 are different from zero, the equation $A\mathbf{x} = \mathbf{0}$ has only the solution $\mathbf{x} = \mathbf{0}$; otherwise it also has nonzero solutions. Since the numbers d_1, d_2, and d_3 are the same as those in the preceding paragraph, it follows that the matrix equation $AX = I$ has a solution if the homogeneous matrix-vector equation $A\mathbf{x} = \mathbf{0}$ has only the "trivial" solution $\mathbf{x} = \mathbf{0}$. Therefore we can state the following theorem.

Theorem 91-1. *If the homogeneous matrix-vector equation $A\mathbf{x} = \mathbf{0}$ has only the "trivial" solution $\mathbf{x} = \mathbf{0}$, then the matrix equation $AX = I$ has a solution.*

Now let us show that if X is a matrix such that $AX = I$, then $X = A^{-1}$; that is, the equation $XA = I$ holds, too. Suppose that $\mathbf{u}$ is a vector such that $X\mathbf{u} = \mathbf{0}$. Then $AX\mathbf{u} = \mathbf{0}$, and since $AX = I$, it follows that $\mathbf{u} = \mathbf{0}$. Therefore the homogeneous matrix-vector equation $X\mathbf{u} = \mathbf{0}$ has only the trivial solution, and so Theorem 91-1 (with X replacing A) tells us that there is a matrix U such that $XU = I$. But this equation implies that $AXU = A$, and since $AX = I$, it follows that $U = A$.

Notice that if A is invertible (has an inverse), then it is certainly true that the only solution of the equation $A\mathbf{x} = \mathbf{0}$ is the trivial solution. For if $A\mathbf{x} = \mathbf{0}$, then $A^{-1}A\mathbf{x} = \mathbf{x} = \mathbf{0}$. Together with Theorem 91-1 and the remarks that follow it, this statement gives us one of the basic theorems on systems of linear equations.

Theorem 91-2. *The matrix A is invertible if and only if the homogeneous matrix-vector equation $A\mathbf{x} = \mathbf{0}$ has only the trivial solution.*

For example, the matrix A of Example 91-3 is not invertible, and the equation $A\mathbf{x} = \mathbf{0}$ is satisfied by the nontrivial vector $\mathbf{x} = (1, -1)$. To say that the equation $A\mathbf{x} = \mathbf{0}$ has only the trivial solution is equivalent to saying that the equation $A\mathbf{x} = \mathbf{b}$ has a *unique* solution. For if $\mathbf{u}$ and $\mathbf{v}$ are solutions of this latter equation, then $A(\mathbf{u} - \mathbf{v}) = A\mathbf{u} - A\mathbf{v} = \mathbf{b} - \mathbf{b} = \mathbf{0}$; that is, $\mathbf{x} = \mathbf{u} - \mathbf{v}$ satisfies the homogeneous equation $A\mathbf{x} = \mathbf{0}$. Thus if $\mathbf{x}$ must be $\mathbf{0}$, then $\mathbf{u} = \mathbf{v}$, and conversely. Hence Theorem 91-2 says that *A is invertible if and only if the equation $A\mathbf{x} = \mathbf{b}$ has a unique solution for each vector $\mathbf{b}$.*

PROBLEMS 91

1. If $A = \begin{bmatrix} 0 & 1 & 0 \\ 0 & 0 & 1 \\ 1 & 0 & 0 \end{bmatrix}$, $B = \begin{bmatrix} 1 & 0 & 0 \\ 0 & 2 & 0 \\ 0 & 0 & 3 \end{bmatrix}$, and $C = \begin{bmatrix} 0 & 0 & 1 \\ 0 & 0 & 0 \\ 0 & 0 & 0 \end{bmatrix}$,

compute the following combinations.

(a) $AB - BA$ (b) $A(BC)$ (c) $(AB)C$ (d) $A(B + C)$
(e) $AB + AC$ (f) $(ABC)^2$ (g) $A^2B^2C^2$ (h) C^{1492}
(i) A^{1492} (j) B^{1492}

2. Verify that the members of the given pairs of matrices are inverses of each other.

(a) $\begin{bmatrix} 2 & 1 \\ 5 & 3 \end{bmatrix}, \begin{bmatrix} 3 & -1 \\ -5 & 2 \end{bmatrix}$

(b) $\begin{bmatrix} 1 & 0 & 0 \\ 0 & -2 & 0 \\ 0 & 0 & 3 \end{bmatrix}, \begin{bmatrix} 1 & 0 & 0 \\ 0 & -\frac{1}{2} & 0 \\ 0 & 0 & \frac{1}{3} \end{bmatrix}$

(c) $\begin{bmatrix} 0 & 0 & 1 \\ 0 & 1 & 0 \\ 1 & 0 & 0 \end{bmatrix}, \begin{bmatrix} 0 & 0 & 1 \\ 0 & 1 & 0 \\ 1 & 0 & 0 \end{bmatrix}$

(d) $\begin{bmatrix} 1 & 0 & 0 \\ 2 & 1 & 0 \\ 3 & 2 & 1 \end{bmatrix}, \begin{bmatrix} 1 & 0 & 0 \\ -2 & 1 & 0 \\ 1 & -2 & 1 \end{bmatrix}$

3. Find the inverse of each of the following matrices.

(a) $\begin{bmatrix} \frac{2}{3} & 0 & 0 \\ 0 & -1 & 0 \\ 0 & 0 & \frac{3}{5} \end{bmatrix}$

(b) $\begin{bmatrix} 0 & 0 & 1 \\ 0 & 2 & 0 \\ 3 & 0 & 0 \end{bmatrix}$

(c) $\begin{bmatrix} 2 & 2 & 1 \\ 1 & -4 & 1 \\ 2 & 1 & 1 \end{bmatrix}$

(d) $\begin{bmatrix} 1 & 2 & 2 \\ 1 & 1 & 2 \\ 1 & -4 & 1 \end{bmatrix}$

(e) $\begin{bmatrix} 1 & 2 & 0 & 0 \\ 4 & 9 & 0 & 0 \\ 0 & 0 & -3 & 1 \\ 0 & 0 & -8 & 3 \end{bmatrix}$

(f) $\begin{bmatrix} 13 & 4 & 0 & 0 \\ 3 & 1 & 0 & 0 \\ 0 & 0 & -2 & 3 \\ 0 & 0 & -1 & 2 \end{bmatrix}$

4. Use part (c) of the preceding problem to help you express x, y, and z in terms of u, v, and w if

$$2x + 2y + z = u$$
$$x - 4y + z = v$$
$$2x + y + z = w.$$

5. (a) Find a formula for A^{-1} if $A = \begin{bmatrix} a & b \\ c & d \end{bmatrix}$.

    (b) Use this formula to find the inverses of the matrices $\begin{bmatrix} 4 & 1 \\ 7 & 2 \end{bmatrix}$ and $\begin{bmatrix} 1 & 2 \\ 3 & 4 \end{bmatrix}$.

    (c) Under what conditions is $A^{-1} = A$?

    (d) Can you show that A has an inverse if and only if $ad \neq bc$?

6. Which of the following statements are true and which are false for arbitrary matrices A and B that have inverses? (To test whether a matrix X is the inverse of a matrix Y, we simply multiply them and see if we get the identity matrix I.)

    (a) $(A^{-1})^{-1} = A$

    (b) $(AB)^{-1} = A^{-1}B^{-1}$

    (c) $(AB)^{-1} = B^{-1}A^{-1}$

    (d) $(A + B)^{-1} = A^{-1} + B^{-1}$

    (e) $(3A)^{-1} = 3A^{-1}$

    (f) $(A^3)^{-1} = (A^{-1})^3$

7. Suppose that A is a matrix such that $A^4 = 0$. Show that $(I - A)^{-1} = I + A + A^2 + A^3$.

8. (a) A matrix of the form $\begin{bmatrix} a_{11} & a_{12} & a_{13} \\ 0 & a_{22} & a_{23} \\ 0 & 0 & a_{33} \end{bmatrix}$, with 0's below the "main diagonal," is

    an "upper triangular" matrix. Show that a product of two upper triangular matrices is an upper triangular matrix.

    (b) Find A^3 if $A = \begin{bmatrix} 0 & 1 & 2 \\ 0 & 0 & 3 \\ 0 & 0 & 0 \end{bmatrix}$.

9. Suppose that $S = \begin{bmatrix} 1 & 1 & 1 \\ 1 & 1 & 1 \\ 1 & 1 & 1 \end{bmatrix}$.

    (a) Find S^{10}.

    (b) Describe the most general matrix A such that $AS = SA$.

    (c) Let P be the matrix whose *column* vectors are $3^{-1/2}(1, 1, 1)$, $6^{-1/2}(2, -1, -1)$, and $2^{-1/2}(0, 1, -1)$, and let P^* be the matrix with these vectors as row vectors. Show that $P^* = P^{-1}$.

    (d) What is P^*SP?

10. Let $J = \begin{bmatrix} 0 & 1 \\ -1 & 0 \end{bmatrix}$.

    (a) Show that $J^2 = -I$.

    (b) Show that $X = 3I - 2J$ satisfies the equation $X^2 - 6X + 13I = 0$.

11. Make up an example (see Example 91-1) to show that the equation $AB = AC$ does not necessarily imply that $B = C$. What if A^{-1} exists?

12. Show that the equations $AX = XA = I$ have only one solution, if they have any.

13. Show that the set of all 2 by 2 matrices is a vector space. Find four specific matrices E_1, E_2, E_3, and E_4 such that any given 2 by 2 matrix can be expressed as a linear combination $a_1 E_1 + a_2 E_2 + a_3 E_3 + a_4 E_4$.

14. In our restaurant problem (Problem 90-10), show that the probability vector for meal 5 is $T^4(.5, .5)$. It is an interesting fact that $\lim_{n \uparrow \infty} T^n = \frac{1}{13} \begin{bmatrix} 4 & 4 \\ 9 & 9 \end{bmatrix}$. Show that $T \frac{1}{13}(4, 9) = \frac{1}{13}(4, 9)$. Show that if $\mathbf{x} = (a, b)$, where $a + b = 1$, then $\frac{1}{13} \begin{bmatrix} 4 & 4 \\ 9 & 9 \end{bmatrix} \mathbf{x} = \frac{1}{13}(4, 9)$. (These facts imply that people eat at Joe's about $\frac{1}{3}$ of the time.)

92. THE DETERMINANT OF A MATRIX

With each square matrix A we associate a number det A called the **determinant** of A. We will need to learn only a tiny part of the huge body of theory about determinants that has been built up over the years. And because computation of determinants is quite complicated, we will restrict ourselves to determinants of matrices of orders 2 and 3. Consequently, our work in this section will focus very narrowly on just those aspects of determinants that will be useful to us.

In addition to using the abbreviation det, it is also customary to denote determinants by means of vertical bars. Thus the defining equation

(92-1) $$\det \begin{bmatrix} a & b \\ c & d \end{bmatrix} = \begin{vmatrix} a & b \\ c & d \end{vmatrix} = ad - bc$$

tells us how to compute the determinant of a matrix of order 2. For example, the determinant of the matrix $\begin{bmatrix} 2 & 3 \\ 3 & 5 \end{bmatrix}$ is the number $2 \cdot 5 - 3 \cdot 3 = 1$.

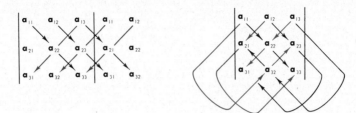

Figure 92-1

The determinant of a 3 by 3 matrix is defined by the equation

$$
(92\text{-}2) \quad \begin{vmatrix} a_{11} & a_{12} & a_{13} \\ a_{21} & a_{22} & a_{23} \\ a_{31} & a_{32} & a_{33} \end{vmatrix} = a_{11}a_{22}a_{33} + a_{12}a_{23}a_{31} + a_{13}a_{21}a_{32} - a_{13}a_{22}a_{31} \\ - a_{11}a_{23}a_{32} - a_{12}a_{21}a_{33}.
$$

This formula can be remembered by the devices illustrated in Fig. 92-1 (*which cannot be used to find determinants of matrices of higher orders*). In the scheme on the left-hand side of the figure, we copy down two extra columns of numbers (the first two columns of our given matrix) and from the resulting array obtain the terms in the sum in Equation 92-2 by multiplying as indicated by the arrows. To the terms that arise from multiplying along the colored arrows, we affix the positive sign, and to the terms that arise from multiplying along the black arrows, we affix the negative sign. On the right-hand side of the figure we have tried to show the same thing without copying the extra columns of numbers. It really isn't such a bad scheme, once you get used to it. For example, it is easy to see that

$$
\begin{vmatrix} 2 & 1 & -1 \\ 0 & 3 & 2 \\ -1 & 4 & 1 \end{vmatrix} = 6 - 2 + 0 - 3 - 16 - 0 = 6 - 21 = -15.
$$

Example 92-1. Write the formula for the determinant of a matrix of order 3 in terms of determinants of matrices of order 2.

Solution. Let us factor the numbers a_{11}, $-a_{12}$, and a_{13} from pairs of terms on the right-hand side of Equation 92-2:

$$
a_{11}(a_{22}a_{33} - a_{23}a_{32}) - a_{12}(a_{21}a_{33} - a_{23}a_{31}) + a_{13}(a_{21}a_{32} - a_{22}a_{31}).
$$

Now we use Equation 92-1 to write the expressions in parentheses as determinants of 2 by 2 matrices, and Equation 92-2 becomes

$$
(92\text{-}3) \quad \begin{vmatrix} a_{11} & a_{12} & a_{13} \\ a_{21} & a_{22} & a_{23} \\ a_{31} & a_{32} & a_{33} \end{vmatrix} = a_{11} \begin{vmatrix} a_{22} & a_{23} \\ a_{32} & a_{33} \end{vmatrix} - a_{12} \begin{vmatrix} a_{21} & a_{23} \\ a_{31} & a_{33} \end{vmatrix} + a_{13} \begin{vmatrix} a_{21} & a_{22} \\ a_{31} & a_{32} \end{vmatrix}.
$$

(Of course, we could have decomposed the right-hand side of Equation 92-2 in other ways, too, but we will use this particular decomposition later.)

In our work we are not so much concerned with the actual value of det A as we are with whether or not this number is 0. Suppose that we perform one of our elementary row operations 89-1, 89-2, or 89-3 on a matrix A and thus transform it to a matrix A'. For example, if we interchange the rows in the matrix

$$A = \begin{bmatrix} a & b \\ c & d \end{bmatrix},$$

we obtain

$$A' = \begin{bmatrix} c & d \\ a & b \end{bmatrix}.$$

Since the matrices A and A' are different, we expect their determinants to be different, and they are:

$$\det A = ad - bc \quad \text{and} \quad \det A' = bc - ad = -\det A.$$

However, because det A and det A' are negatives of each other, it follows that either *both* are 0 or *neither* is 0.

If A' is obtained from A by Operation 89-2, some row of A' is a nonzero multiple of the corresponding row of A; for example,

$$A' = \begin{bmatrix} a & b \\ kc & kd \end{bmatrix}.$$

Then det $A' = akd - bkc = k$ det A, and again the numbers det A and det A' are either both zero or both nonzero. Finally, Operation 89-3 might transform A to

$$A' = \begin{bmatrix} a + kc & b + kd \\ c & d \end{bmatrix}.$$

Here det $A' = (a + kc)d - (b + kd)c = ad - bc = $ det A. Of course, since det $A' = $ det A, it follows that one of these numbers is 0 if and only if the other is. Although we looked only at the two-dimensional case, the results are true for matrices in general.

Theorem   92-1.   *If A' is obtained from A by applying the elementary Operations 89-1, 89-2, or 89-3, then det $A' = 0$ if and only if det $A = 0$.*

To see why we are interested in whether det A is 0 or not (and also what role Theorem 92-1 plays), let us go back to our work in the last section. When we analyzed Matrix 91-6, we saw that the equation $AX = I$ has a solution if and only if none of the numbers d_1, d_2, d_3 that appear in the matrix

$$A' = \begin{bmatrix} d_1 & 0 & 0 \\ * & d_2 & 0 \\ * & * & d_3 \end{bmatrix}$$

is 0. So we want to know if the product $d_1 d_2 d_3$ is 0 or not. This product (Equation 92-2 tells us) is simply the number det A', and A' is obtained from A by elementary row operations. Hence, according to Theorem 92-1, $d_1 d_2 d_3 \neq 0$ if and only if det $A \neq 0$. But we saw in the last section that the inequality $d_1 d_2 d_3 \neq 0$ is precisely the condition that the homogeneous matrix-vector equation $A\mathbf{x} = \mathbf{0}$ has only the trivial solution. And this condition (according to Theorem 91-2) is equivalent to the statement that A is invertible. Hence we may state two more basic theorems about vectors and matrices (the second formulation will be useful to us in Section 103).

Theorem   92-2.   *The square matrix A has an inverse if and only if* det $A \neq 0$.

Theorem   92-3.   *The homogeneous matrix-vector equation $A\mathbf{x} = \mathbf{0}$ has a nontrivial solution if and only if* det $A = 0$.

Example 92-2.   For what choices of k does the matrix

$$A(k) = \begin{bmatrix} k - 1 & 1 \\ -2 & k - 4 \end{bmatrix}$$

fail to have an inverse?

Solution.   According to Theorem 92-2, we must choose k so that det $A(k) = 0$. This equation is

$$(k - 1)(k - 4) + 2 = k^2 - 5k + 6$$
$$= (k - 2)(k - 3)$$
$$= 0.$$

Hence $k = 2$ or $k = 3$. You might investigate $A(2)$ and $A(3)$ a little. For example, Theorem 92-3 tells us that the equations $A(2)\mathbf{x} = \mathbf{0}$ and $A(3)\mathbf{x} = \mathbf{0}$ have nontrivial solutions. Find some.

PROBLEMS   92

1. Evaluate the determinant.

(a) $\begin{vmatrix} 2 & 3 \\ 4 & 5 \end{vmatrix}$

(b) $\begin{vmatrix} 2 & 4 \\ -4 & -8 \end{vmatrix}$

(c) $\begin{vmatrix} a & 0 \\ b & c \end{vmatrix}$

(d) $\begin{vmatrix} a & b \\ 0 & c \end{vmatrix}$

(e) $\begin{vmatrix} \cos t & \sin t \\ -\sin t & \cos t \end{vmatrix}$

(f) $\begin{vmatrix} 1 & \tan x \\ \tan x & -1 \end{vmatrix}$

(g) $\begin{vmatrix} e^x & e^y \\ e^{-y} & e^{-x} \end{vmatrix}$

(h) $\begin{vmatrix} \ln x & \ln y \\ x & y \end{vmatrix}$

2. Evaluate the determinant.

(a) $\begin{vmatrix} 1 & 2 & -1 \\ 3 & 0 & 2 \\ 1 & 1 & 1 \end{vmatrix}$

(b) $\begin{vmatrix} 2 & 0 & 0 \\ 5 & 3 & 0 \\ -2 & 1 & 7 \end{vmatrix}$

(c) $\begin{vmatrix} 1 & 1 & 1 \\ 1 & 1 & 1 \\ 1 & 1 & 1 \end{vmatrix}$

(d) $\begin{vmatrix} 1 & 0 & 0 \\ 0 & 1 & 0 \\ 0 & 0 & 1 \end{vmatrix}$

(e) $\begin{vmatrix} 2 & 1 & 0 \\ -1 & 1 & 2 \\ 3 & 6 & -2 \end{vmatrix}$

(f) $\begin{vmatrix} 1 & 2 & 0 \\ 3 & 0 & 0 \\ -1 & 5 & 2 \end{vmatrix}$

3. For what choices of k does the matrix *fail* to have an inverse?

(a) $\begin{bmatrix} k & 2 \\ 12 & 3 \end{bmatrix}$

(b) $\begin{bmatrix} 2 & k \\ 3 & 12 \end{bmatrix}$

(c) $\begin{bmatrix} 2-k & 1 \\ 4 & 5-k \end{bmatrix}$

(d) $\begin{bmatrix} -2-k & 0 \\ 8 & 5-k \end{bmatrix}$

4. For what values of k does the system have nontrivial solutions?

(a) $kx + y = 0$
     $2x - y = 0$

(b) $kx + y = 0$
     $2kx - y = 0$

(c) $(6-k)x - 2y = 0$
     $-2x + (3-k)y = 0$

(d) $(6-k)x = 0$
     $-2x + (3-k)y = 0$

5. For what choices of k does the matrix *fail* to have an inverse?

(a) $\begin{bmatrix} k & 0 & 0 \\ 3 & 1 & 0 \\ 4 & 5 & 2 \end{bmatrix}$

(b) $\begin{bmatrix} 1 & 0 & 0 \\ k & 2 & 0 \\ 3 & 4 & 5 \end{bmatrix}$

(c) $\begin{bmatrix} -k & 0 & 0 \\ 1 & 2-k & 0 \\ 3 & 4 & 5-k \end{bmatrix}$

(d) $\begin{bmatrix} 0 & 0 & 1 \\ 0 & k & 3 \\ 2 & 4 & 5 \end{bmatrix}$

(e) $\begin{bmatrix} k & 1 & 1 \\ 1 & k & 1 \\ 1 & 1 & k \end{bmatrix}$

(f) $\begin{bmatrix} k & 1 & -2 \\ -2 & k+1 & 1 \\ -1 & 3 & k-6 \end{bmatrix}$

6. Compute the determinant

$$\begin{vmatrix} \begin{vmatrix} 1 & -2 \\ 5 & -3 \end{vmatrix} & \begin{vmatrix} 4 & 6 \\ -1 & 1 \end{vmatrix} \\ \begin{vmatrix} 7 & 8 \\ 1 & 1 \end{vmatrix} & \begin{vmatrix} 3 & -1 \\ 4 & -3 \end{vmatrix} \end{vmatrix}.$$

7. Let $A(t) = \begin{bmatrix} t & e^t \\ 1 & e^t \end{bmatrix}$.

(a) Find a number t_0 such that $A(t_0)$ *does not* have an inverse.
(b) Solve the equation $A(t_0)\mathbf{x} = \mathbf{0}$.
(c) If $t_1 \neq t_0$, solve the equation $A(t_1)\mathbf{x} = \mathbf{0}$.
(d) How do the results of parts (b) and (c) square with Theorem 92-3?

8. Show that

$$\frac{d}{dx} \begin{vmatrix} f(x) & g(x) \\ h(x) & k(x) \end{vmatrix} = \begin{vmatrix} f'(x) & g'(x) \\ h(x) & k(x) \end{vmatrix} + \begin{vmatrix} f(x) & g(x) \\ h'(x) & k'(x) \end{vmatrix}.$$

9. For what values of k does the system of equations

$$\begin{bmatrix} -2 & 2 & 2 \\ 6 & -2 & -6 \\ -3 & 2 & 3 \end{bmatrix} \mathbf{x} = k\mathbf{x}$$

have a solution other than the trivial solution $\mathbf{x} = \mathbf{0}$?

10. (a) Explain why

$$\begin{vmatrix} x_1 & y_1 & 1 \\ x_2 & y_2 & 1 \\ x & y & 1 \end{vmatrix} = 0$$

is an equation of the form $ax + by + c = 0$ and hence represents a line in the plane.
(b) Explain why (x_1, y_1) and (x_2, y_2) are points of this line.
(c) Use these ideas to find the line that contains the points $(1, 2)$ and $(3, 4)$.

11. For any two n by n matrices, it is true that $\det AB = \det A \det B$.

(a) Verify this equation for arbitrary 2 by 2 matrices.
(b) Use the equation to show that $\det A^{-1} = (\det A)^{-1}$.

12. (a) Show that for a 2 by 2 matrix A, $\det(-A) = \det A$, whereas if A is 3 by 3, $\det(-A) = -\det A$.
(b) Use this result and part (b) of the preceding problem to show that there is no 3 by 3 matrix A (of real numbers) such that $A^{-1} = -A$.
(c) Can you find a 2 by 2 matrix of real numbers such that $A^{-1} = -A$?

13. Use Theorem 92-3 and Equation 90-5 to show that $\det A = 0$ if and only if one of the columns of A is a linear combination of the others.

You should now know how to solve systems of linear equations and perform elementary algebraic operations with matrices and members of the vector spaces R^2 and R^3. These are the "practical" aspects of this chapter. Perhaps the most important "theoretical" results are contained in Theorems 91-2 and 92-3: The homogeneous matrix-vector equation $A\mathbf{x} = \mathbf{0}$ has a nontrivial solution if and only if the matrix A is not invertible, and this statement, in turn, is equivalent to the equation $\det A = 0$.

1. Find A^{1066}.

(a) $A = \begin{bmatrix} 0 & 1 \\ -1 & 0 \end{bmatrix}$

(b) $A = \begin{bmatrix} \frac{2}{5} & \frac{3}{5} \\ \frac{2}{5} & \frac{3}{5} \end{bmatrix}$

(c) $A = \begin{bmatrix} 0 & 0 \\ 1 & 0 \end{bmatrix}$

(d) $A = \begin{bmatrix} 1 & 1 \\ 1 & 1 \end{bmatrix}$

2. Find the inverse of the matrix.

(a) $\begin{bmatrix} 2 & 3 & -1 \\ 1 & -1 & 2 \\ 1 & 2 & 1 \end{bmatrix}$

(b) $\begin{bmatrix} -1 & 2 & 1 \\ 3 & -1 & 2 \\ 2 & 1 & 1 \end{bmatrix}$

(c) $\begin{bmatrix} 1 & 2 & -1 \\ 3 & 0 & 2 \\ 1 & 1 & 1 \end{bmatrix}$

(d) $\begin{bmatrix} 1 & 3 & 1 \\ 1 & 0 & 2 \\ 1 & 2 & -1 \end{bmatrix}$

3. Solve the system of equations

$$x_1 + x_2 = 10$$
$$x_2 + x_3 = 9$$
$$x_3 + x_4 = 8$$
$$x_4 + x_5 = 7$$
$$x_5 + x_6 = 6$$
$$x_6 + x_7 = 5$$
$$x_7 + x_8 = 4$$
$$x_8 + x_9 = 3$$
$$x_1 + x_9 = 2.$$

4. Let $S = \begin{bmatrix} \cos^2 x & \sin x \cos x \\ \sin x \cos x & \sin^2 x \end{bmatrix}$ and let S' be the matrix whose elements are the derivatives of the elements of S. Show that (a) $S^2 = S$ and (b) $SS'S = 0$.

5. We have only multiplied square matrices, but our rule for multiplication may also be applied to matrices with more rows than columns or more columns than rows. Compute the following products. Can you reverse the order of multiplication in these examples?

(a) $\begin{bmatrix} 1 & 2 & 3 \\ -1 & 1 & 2 \end{bmatrix} \begin{bmatrix} 1 & -1 \\ 0 & 2 \\ 3 & 1 \end{bmatrix}$

(b) $\begin{bmatrix} 1 & -1 \\ 0 & 2 \end{bmatrix} \begin{bmatrix} 1 & 2 & 3 \\ -1 & 1 & 2 \end{bmatrix}$

(c) $\begin{bmatrix} 1 & 2 & 3 \end{bmatrix} \begin{bmatrix} 4 \\ 5 \\ 6 \end{bmatrix}$

(d) $[3][1 \quad 2 \quad 3]$

6. Show that the following system is inconsistent unless $2r - s - 3t = 0$:

$$x + 3y = r$$
$$2x - 3z = s$$
$$2y + z = t.$$

7. (a) Show that if $A^2 = 0$, then $(I - A)^{-1} = I + A$.

   (b) Use the matrix $A = \begin{bmatrix} -4 & 16 \\ -1 & 4 \end{bmatrix}$ to illustrate this statement.

8. (a) If $A = \begin{bmatrix} a & b \\ c & d \end{bmatrix}$, show that $A^{-1} = \dfrac{1}{\det A} \begin{bmatrix} d & -b \\ -c & a \end{bmatrix}$.

   (b) Compute $P^{-1}AP$ if $P = \begin{bmatrix} 2 & 1 \\ 1 & 1 \end{bmatrix}$ and $A = \begin{bmatrix} 1 & 4 \\ -2 & 7 \end{bmatrix}$.

9. Let P and A be n by n matrices and k be a positive integer. Explain why $(P^{-1}AP)^k = P^{-1}A^kP$.

10. Suppose that you have a combination of ten coins consisting of nickels, dimes, and quarters whose total value is $1.25. How many of each coin do you have?

11. Show that the three lines $x - y + 1 = 0$, $2x + y - 2 = 0$, and $x + y - 3 = 0$ do not intersect in a point.

12. Suppose that four horses—A, B, C, and D—are entered in a race. If you buy a $1 win ticket on A, you can cash it for $6 if he wins, thus realizing a net gain of $5. You lose your dollar if A doesn't win. The odds are such that dollars bet on B, C, and D potentially yield net gains of $4, $3, and $2. How should you bet your money to guarantee that you win $12 no matter how the race comes out?

13. Suppose that you have a rectangular sheet of paper. Can you trim one edge of it to obtain a rectangle whose area and perimeter are half the area and perimeter of the original sheet? Can you do it by trimming two edges?

14. Show that $\begin{vmatrix} 1 & 1 & 1 \\ x_1 & x_2 & x_3 \\ x_1^2 & x_2^2 & x_3^2 \end{vmatrix} = (x_3 - x_1)(x_2 - x_1)(x_3 - x_2)$.

15. Show that if $\begin{vmatrix} a & b \\ c & d \end{vmatrix} = 0$ and $d \neq 0$, then there is a number r such that $(a, c) = r(b, d)$ and a number s such that $(a, b) = s(c, d)$.

16. Show that $Y = \begin{bmatrix} \cos t & \sin t \\ -\sin t & \cos t \end{bmatrix}$ satisfies the matrix differential equation $Y' = \begin{bmatrix} 0 & 1 \\ -1 & 0 \end{bmatrix} Y$.

17. Show that if $\mathbf{x} = (2e^{4t} + 3e^{-t}, 2e^{4t} - 2e^{-t})$, then $\mathbf{x}' = \begin{bmatrix} 1 & 3 \\ 2 & 2 \end{bmatrix} \mathbf{x}$.

18. Solve the equation $\begin{bmatrix} 2 & 1 & 3 \\ 1 & 0 & 1 \\ -1 & 2 & 1 \end{bmatrix} \mathbf{x} = \mathbf{0}$ and use the result (see Equation 90-5) to express the first column of the coefficient matrix as a combination of the other two.

19. Find matrices A and B for which the given equation is true and another pair for which it is false.

(a) $\det(A + B) = \det A + \det B$ (b) $(A + B)^2 = A^2 + 2AB + B^2$

20. It is not known if there exists an integer $n > 2$ and nonzero integers X, Y, and Z such that $X^n + Y^n = Z^n$. Show that for each n there are nonzero *matrices* (with integer components) that satisfy this equation.

21. If $ABC = I$, is it true that $CAB = BCA$?

22. Suppose that we are given two square matrices A and B of the same order and form the augmented matrix $[A \mid B]$. Furthermore, suppose that by applying our row operations, we can reduce this matrix to the form $[I \mid C]$. What is the matrix C?

23. Let A^* be the matrix whose rows are the columns (in the same order) of a 3 by 3 matrix A. How are the columns of A^* related to the rows of A? How are the numbers $\det A^*$ and $\det A$ related?

24. If A is a given matrix, we call $p(x) = \det(xI - A)$ the **characteristic polynomial** of A.

(a) Calculate $p(x)$ if $A = \begin{bmatrix} 2 & 0 & -3 \\ 0 & 1 & 0 \\ 1 & 2 & 0 \end{bmatrix}$. It will be of the form $p(x) = x^3 + p_2 x^2 + p_1 x + p_0$.

(b) Show that $A^3 + p_2 A^2 + p_1 A + p_0 I = 0$.

(c) Can you use this equation to find A^{-1}?

25. Suppose that each of the vectors $\mathbf{u}$ and $\mathbf{v}$ is a solution of the equation $A\mathbf{x} = \mathbf{b}$. Show that $\mathbf{u} - \mathbf{v}$ is a solution of the *homogeneous equation* $A\mathbf{x} = \mathbf{0}$. If $\mathbf{w}$ is a solution of the homogeneous equation, show that $t\mathbf{w}$ is also a solution of the homogeneous equation for any number t. Show that $\mathbf{x} = t\mathbf{w} + \mathbf{u}$ satisfies the equation $A\mathbf{x} = \mathbf{b}$.

Analytic Geometry in Three-Dimensional Space

12

Now we are going to put some of the vector theory we have learned to work. We start where vectors started, in three-dimensional space. Our first step will be to observe that the geometric vectors we introduced in Chapter 8 can easily be viewed in three-dimensions, too. When we bring coordinates into the picture, we can tie these geometric vectors to the algebraic vectors we have just been discussing. In this way, we will be able to extend the analytic geometry we talked about in Chapter 4 from the plane to space.

In Chapters 13 and 14 we will study the calculus of functions with domains and ranges that are subsets of R^2, R^3, and so on. The geometry we introduce in this chapter will also be useful to us then.

93. GEOMETRIC VECTORS

We discussed the vector space of arrows in the plane in Chapter 8, and it is no trouble to move up a dimension. Two points P and Q in three-dimensional space can be used as initial and terminal points of an arrow $\mathbf{PQ}$. We multiply such arrows by scalars (real numbers) and use tail-to-head addition just as in the plane.

The resulting collection of arrows forms a vector space, the members of which we will call **geometric vectors**. (To be precise, we shouldn't refer to a particular arrow as a vector; the vector is actually the *class* of all arrows equal to the given one. We won't be so careful with this distinction, however.)

Not surprisingly, geometric vectors are useful in solving problems in geometry.

Example 93-1. Let *ABCD* be a quadrilateral in space (notice that the points *A*, *B*, *C*, and *D* need not lie in the same plane) and denote by *P*, *Q*, *R*, and *S* the midpoints of the sides *AB*, *BC*, *CD*, and *AD*. Show that *PQRS* is a parallelogram.

Solution. Figure 93-1 shows the original quadrilateral, as well as the quadrilateral whose vertices are the midpoints *P*, *Q*, *R*, and *S*. To show that this latter figure is a parallelogram, we need only show that the sides *RS* and *QP* are parallel and of equal length, which will be true if **RS = QP**. We have **RS = DS − DR** = $\frac{1}{2}$**DA** − $\frac{1}{2}$**DC** = $\frac{1}{2}$(**DA** − **DC**) = $\frac{1}{2}$**CA**. Similarly, **QP = BP − BQ** = $\frac{1}{2}$**BA** − $\frac{1}{2}$**BC** = $\frac{1}{2}$(**BA** − **BC**) = $\frac{1}{2}$**CA**. Therefore **RS = QP**, as was to be shown.

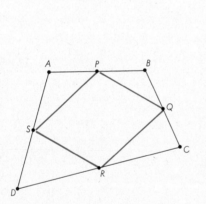

Figure 93-1

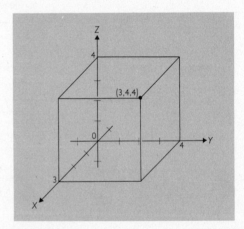

Figure 93-2

By introducing a coordinate system into the plane, we can turn certain geometric problems into algebraic ones and vice versa. The same situation holds in space, of course. Figure 93-2 shows a **right-handed** *XYZ*-coordinate system. A right-handed screw pointed along the *Z*-axis would advance if the positive *X*-axis were rotated 90° into the position of the positive *Y*-axis. We get a *left-handed* system by interchanging the labels on any two axes—for example, the *X*- and *Y*-axes. Most applied mathematicians use right-handed systems, and so shall we. The three planes determined by the various pairs of axes—the *XY*-plane, the

YZ-plane, and the *XZ*-plane—are called **coordinate planes**. Figure 93-2 illustrates how the point (3, 4, 4) is plotted in such a system, and you should plot a few points yourself. Of course, it is inherently difficult to draw three-dimensional figures on two-dimensional paper.

It is natural that we express vectors in *three*-dimensional space in terms of *three* basic vectors, the vectors **i**, **j**, and **k** that are one unit long and point in the direction of the coordinate axes (Fig. 93-3). Each vector **a** can be written as

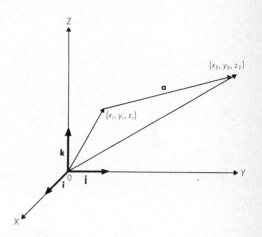

$$\mathbf{a} = a_x\mathbf{i} + a_y\mathbf{j} + a_z\mathbf{k},$$

where the numbers a_x, a_y, and a_z are the **components** of **a** with respect to the basic vectors. Clearly, if **a** is the vector whose initial point is (x_1, y_1, z_1) and whose terminal point is (x_2, y_2, z_2), then

$$\mathbf{a} = (x_2\mathbf{i} + y_2\mathbf{j} + z_2\mathbf{k}) - (x_1\mathbf{i} + y_1\mathbf{j} + z_1\mathbf{k}),$$

and so

(93-1) $\qquad a_x = x_2 - x_1, \qquad a_y = y_2 - y_1,$
$$a_z = z_2 - z_1.$$

Figure 93-3

Since a coordinate system allows us to associate with a geometric vector **a** a triple of numbers (a_x, a_y, a_z), we can think of our space of geometric vectors as a geometric image of R^3. In R^3 we work algebraically, and in ordinary space we work geometrically. But because these spaces are so closely related, we can shift back and forth between algebra and geometry as the need arises.

Example 93-2.   Find the coordinates of the midpoint of the segment *PQ*, where *P* is the point $(-1, 2, 5)$ and *Q* is the point $(3, 0, -1)$.

Solution.   Figure 93-4 shows the points *P* and *Q* and also the midpoint *M*, whose coordinates (x, y, z) we are to find. Let us write the vector equation $\mathbf{QM} = \frac{1}{2}\mathbf{QP}$ in terms of **i**, **j**, and **k**, using Equations 93-1 to calculate the components:

$$(x - 3)\mathbf{i} + y\mathbf{j} + (z + 1)\mathbf{k} = \tfrac{1}{2}(-4\mathbf{i} + 2\mathbf{j} + 6\mathbf{k}).$$

This single vector equation is equivalent to the three scalar equations

$$x - 3 = -2, \qquad y = 1, \quad \text{and} \quad z + 1 = 3,$$

from which we find that the coordinates of the midpoint *M* are (1, 1, 2).

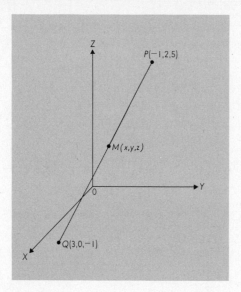

P(-1,2,5)

M(x,y,z)

Q(3,0,-1)

Figure 93-4

PROBLEMS 93

1. Let **a** and **b** be position vectors to the points $(2, -1, 2)$ and $(0, 3, 0)$. Sketch the following vectors as position vectors.

   (a) $\mathbf{a} + \mathbf{b}$ (b) $3\mathbf{a} + \mathbf{b}$ (c) $\mathbf{a} - \mathbf{b}$ (d) $3\mathbf{a} - \mathbf{b}$

2. Locate the points P and Q whose coordinates are $(-1, 2, -3)$ and $(11, 11, 3)$ in a right-handed coordinate system.

   (a) Express the vector **PQ** in terms of the vectors **i**, **j**, and **k**.
   (b) Express the vector $3\mathbf{OP} - 2\mathbf{OQ}$ in terms of the vectors **i**, **j**, and **k**.
   (c) Find the point that is one-third of the way from P to Q.
   (d) Find the point that is one-third of the way from Q to P.
   (e) Find the point R such that $\mathbf{PR} = 2\mathbf{PQ}$.
   (f) Find the point R such that $\mathbf{PQ} = 2\mathbf{PR}$.

3. Let $\mathbf{a} = 3\mathbf{i} - 2\mathbf{j} + \mathbf{k}$ and $\mathbf{b} = \mathbf{i} + 2\mathbf{j} - 3\mathbf{k}$. Find a scalar r such that

   (a) $\mathbf{a} + r\mathbf{b}$ is parallel to the XY-plane.
   (b) $r\mathbf{a} + \mathbf{b}$ is parallel to the YZ-plane.

4. Let P, Q, R, and S be the points $(1, 1, 1)$, $(2, 3, 0)$, $(3, 5, -2)$, and $(0, -1, 1)$. Show that the segments PQ and RS are parallel and find the ratio of their lengths.

5. In Example 68-2 we showed that the medians of a triangle intersect in the point of each median that is two-thirds of the way from the vertex in which the median terminates. Find the point of intersection of the medians of the triangle whose vertices are $(-1, 2, 3)$, $(3, 0, 1)$, and $(4, -2, 5)$.

6. Let $\mathbf{a} = 3\mathbf{i} - 2\mathbf{j} + \mathbf{k}$ and $\mathbf{b} = \mathbf{i} + 2\mathbf{j} - 3\mathbf{k}$. Find a scalar r such that

   (a) $\mathbf{a} + r\mathbf{b}$ is parallel to $3\mathbf{i} + 2\mathbf{j} - 4\mathbf{k}$.
   (b) $r\mathbf{a} + \mathbf{b}$ is parallel to $5\mathbf{i} - 2\mathbf{j}$.

7. The set $\{(x, y, z) \mid x > 0,\ y > 0,\ z > 0\}$ of points with positive coordinates is called the *first octant*. Sketch it and also picture the following sets.

(a) $\{(x, y, z) \mid x > 0, y < 0, z > 0\}$ (b) $\{(x, y, z) \mid x > 0\}$
(c) $\{(x, y, z) \mid x > 0, y < 0\}$ (d) $\{(x, y, z) \mid xyz \geq 0\}$
(e) $\{(x, y, z) \mid |xyz| = xyz\}$ (f) $\{(x, y, z) \mid xyz = 0\}$
(g) $\{(x, y, z) \mid |x| + |y| + |z| = 0\}$
(h) $\{(x, y, z) \mid |x| + |y| + |z| = x + y + z\}$

8. Suppose that P, Q, R, and S are points in space such that $\mathbf{OP} - \mathbf{OQ} = \mathbf{OS} - \mathbf{OR}$. Show that $PQRS$ is a parallelogram.

9. Two position vectors $\mathbf{a}$ and $\mathbf{b}$ are nonparallel sides of a parallelogram. Find the position vectors whose terminal points are the vertex opposite the origin and the center of the parallelogram.

10. Prove that the line segment joining the midpoints of two sides of a triangle is parallel to the third side and is half as long as the third side.

11. Do Problem 8-16 by vectors.

12. The origin and the terminal points of position vectors $\mathbf{a}$ and $\mathbf{b}$ determine a triangle in space. Find a similar triangle with twice the area.

13. Suppose that $\mathbf{a}, \mathbf{b}, \mathbf{c}$, and $\mathbf{d}$ are position vectors whose terminal points are the vertices of a parallelogram. Show that $\mathbf{d}$ must be one of the vectors $\mathbf{a} + \mathbf{b} - \mathbf{c}$, $\mathbf{a} + \mathbf{c} - \mathbf{b}$, or $\mathbf{b} + \mathbf{c} - \mathbf{a}$.

14. Suppose that $\mathbf{a} = a_x\mathbf{i} + a_y\mathbf{j} + a_z\mathbf{k}$, $\mathbf{b} = b_x\mathbf{i} + b_y\mathbf{j} + b_z\mathbf{k}$, and $\mathbf{c} = c_x\mathbf{i} + c_y\mathbf{j} + c_z\mathbf{k}$ are given vectors. Show that they generate three-dimensional space (that is, every vector in space can be expressed as a linear combination of $\mathbf{a}$, $\mathbf{b}$, and $\mathbf{c}$) if and only if

$$\begin{vmatrix} a_x & a_y & a_z \\ b_x & b_y & b_z \\ c_x & c_y & c_z \end{vmatrix} \neq 0.$$

94. DISTANCE AND THE DOT PRODUCT

A vector space consists of a set of objects that we can combine according to certain laws of *algebra* (Equations 90-1 and 90-2). But in our vector spaces of arrows it is natural to introduce the *geometric* concept of distance, too.

We will base our formula for distance in three-dimensional space on the idea of the length of a vector, so let us turn to Fig. 94-1 to see how long the vector $\mathbf{a} = a_x\mathbf{i} + a_y\mathbf{j} + a_z\mathbf{k}$ (shown as a position vector) is. A little study will convince you that the legs OQ and QR of right triangle OQR are $|a_x|$ and $|a_y|$ units long. Hence the square of the hypotenuse is $\overline{OR}^2 = a_x^2 + a_y^2$. This hypotenuse is one leg of right triangle ORP, and leg RP is $|a_z|$ units long. The hypotenuse of this latter triangle is just our vector $\mathbf{a}$, so if we denote its length by $|\mathbf{a}|$, we see that $|\mathbf{a}|^2 = a_x^2 + a_y^2 + a_z^2$. In other words, the vector $\mathbf{a} = a_x\mathbf{i} + a_y\mathbf{j} + a_z\mathbf{k}$ is

(94-1) $$|\mathbf{a}| = \sqrt{a_x^2 + a_y^2 + a_z^2}$$

units long.

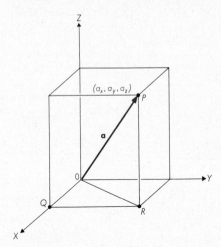

Figure 94-1

If the initial and terminal points of **a** are (x_1, y_1, z_1) and (x_2, y_2, z_2), then its components are given by Equations 93-1, and Equation 94-1 can be interpreted as saying: *the distance between two points (x_1, y_1, z_1) and (x_2, y_2, z_2) of three-dimensional space is given by the formula*

$$(94\text{-}2) \qquad d = \sqrt{(x_2 - x_1)^2 + (y_2 - y_1)^2 + (z_2 - z_1)^2}.$$

Of course, this distance formula is the obvious extension to three-space of Formula 3-1 for the distance between two points of the plane.

Example 94-1. What points of the XY-plane are less than 5 units distant from the point $(1, 2, 3)$?

Solution. A point that belongs to the XY-plane has coordinates $(x, y, 0)$. According to Equation 94-2 the distance between this point and the point $(1, 2, 3)$ is the number $\sqrt{(x - 1)^2 + (y - 2)^2 + (0 - 3)^2}$, and this number is to be less than 5; that is, $\sqrt{(x - 1)^2 + (y - 2)^2 + (0 - 3)^2} < 5$. When we write this inequality as $(x - 1)^2 + (y - 2)^2 < 16$, we see that our point must belong to the circular disk in the XY-plane whose center is $(1, 2)$ and whose radius is 4.

The **dot product** of two vectors in space is defined just as it is in the plane:

$$(94\text{-}3) \qquad\qquad \mathbf{a} \cdot \mathbf{b} = |\mathbf{a}|\,|\mathbf{b}|\cos\theta,$$

where θ is the angle (between $0°$ and $180°$) that the two vectors determine. As before, $\mathbf{a} \cdot \mathbf{a} = |\mathbf{a}|^2$, and $\mathbf{a} \cdot \mathbf{b} = 0$ if and only if $\mathbf{a}$ and $\mathbf{b}$ are perpendicular.

We apply the Law of Cosines as we did when we derived Equation 69-6, and we find that the dot product of the vector $\mathbf{a} = a_x\mathbf{i} + a_y\mathbf{j} + a_z\mathbf{k}$ by $\mathbf{b} = b_x\mathbf{i} + b_y\mathbf{j} + b_z\mathbf{k}$ is expressed in terms of components by the formula

(94-4) $$\mathbf{a} \cdot \mathbf{b} = a_xb_x + a_yb_y + a_zb_z.$$

Example 94-2. The vectors $\mathbf{a} = 2\mathbf{i} - \mathbf{j} + \mathbf{k}$ and $\mathbf{b} = \mathbf{i} + \mathbf{j} + 2\mathbf{k}$ determine an angle θ. Find it.

Solution. From Equation 94-3 we see that

$$\cos\theta = \frac{\mathbf{a} \cdot \mathbf{b}}{|\mathbf{a}|\,|\mathbf{b}|}.$$

Now we use Equation 94-4 to write $\mathbf{a} \cdot \mathbf{b} = 2 \cdot 1 + (-1) \cdot 1 + 1 \cdot 2 = 3$, and we readily see that $|\mathbf{a}| = |\mathbf{b}| = \sqrt{6}$. Hence $\cos\theta = 3/\sqrt{6}\,\sqrt{6} = \frac{1}{2}$, and so $\theta = 60°$.

We can use Equation 94-4 to establish the "natural" rules of arithmetic

$$\mathbf{a} \cdot \mathbf{b} = \mathbf{b} \cdot \mathbf{a}$$

$$(\mathbf{a} + \mathbf{b}) \cdot \mathbf{c} = \mathbf{a} \cdot \mathbf{c} + \mathbf{b} \cdot \mathbf{c}$$

$$(r\mathbf{a}) \cdot \mathbf{b} = r(\mathbf{a} \cdot \mathbf{b})$$

satisfied by arbitrary vectors $\mathbf{a}$, $\mathbf{b}$, and $\mathbf{c}$ and scalar r. Furthermore, Equation 94-4 suggests a definition of the dot product of two vectors in R^n for any n. If $\mathbf{a}$ and $\mathbf{b}$ belong to R^{1492}, for example, we define

$$\mathbf{a} \cdot \mathbf{b} = a_1b_1 + a_2b_2 + \cdots + a_{1492}b_{1492}.$$

Of course, it is a little difficult to interpret such a dot product geometrically, but that doesn't mean that it isn't a useful concept.

This definition of the dot product brings geometric terminology into R^n in a natural way. For example, we will say that the vector $\mathbf{v} = (2, 1, -2)$ of R^3 is $|\mathbf{v}| = \sqrt{2^2 + 1^2 + (-2)^2} = 3$ units long and that the vectors $\mathbf{a} = (3, -5)$ and $\mathbf{b} = (10, 6)$ of R^2 are **orthogonal** (another word for perpendicular), since $\mathbf{a} \cdot \mathbf{b} = 0$.

Example 94-3. Let $\mathbf{a}$ be an element of R^6 whose last component is not 0. How do you know that there is a real number x such that the vector

$$\mathbf{v} = (1, x, x^2, x^3, x^4, x^5)$$

is "perpendicular" to $\mathbf{a}$?

Solution. If $\mathbf{a} = (a_0, a_1, a_2, a_3, a_4, a_5)$, then $\mathbf{a} \cdot \mathbf{v}$ is a polynomial in x:

$$\mathbf{a} \cdot \mathbf{v} = a_0 + a_1x + a_2x^2 + a_3x^3 + a_4x^4 + a_5x^5.$$

We are assuming that a_5 is not 0, so this polynomial has degree 5. But every polynomial of odd degree (with real coefficients) has at least one real zero; thus there is a real number x such that $\mathbf{a} \cdot \mathbf{v} = 0$.

PROBLEMS   94

1. How far apart are the points?
    (a) $(-2, 1, 3)$, $(-1, 3, 1)$ (b) $(2, -1, -3)$, $(3, 1, -5)$
    (c) $(\ln 4, \ln 6, \ln 10)$, $(\ln 2, \ln 3, \ln 5)$
    (d) $(e^x, \sin t, -\cos t)$, $(e^{-x}, -\sin t, \cos t)$

2. Let $\mathbf{a} = \mathbf{i} + 2\mathbf{j} - 2\mathbf{k}$ and $\mathbf{b} = 3\mathbf{i} + \mathbf{j} + 2\mathbf{k}$ and find the following quantities.
    (a) $|\mathbf{a} + \mathbf{b}| - |\mathbf{a}| - |\mathbf{b}|$ (b) $|\mathbf{a} - \mathbf{b}| - |\mathbf{a}| + |\mathbf{b}|$
    (c) $|\mathbf{a}|\mathbf{b}$ (d) $(\mathbf{a} \cdot \mathbf{b})\mathbf{a}$
    (e) $\dfrac{1}{|\mathbf{a}|}\mathbf{a}$ (f) $\left|\dfrac{1}{|\mathbf{b}|}\mathbf{b}\right|$

3. Let P and Q be the points $(-1, 2, -3)$ and $(2, 8, -9)$. Find the point R that is one-third of the way from P to Q and use the distance formula to verify that $|\mathbf{PR}| = \frac{1}{3}|\mathbf{PQ}|$.

4. Find a number x such that the vectors $\mathbf{i} + 4\mathbf{j} + 3\mathbf{k}$ and $4\mathbf{i} + 2\mathbf{j} + x\mathbf{k}$ are perpendicular. Can you find a number x such that these vectors are parallel?

5. Show that each column vector of the matrix

$$M = \begin{bmatrix} \frac{2}{3} & \frac{1}{3} & -\frac{2}{3} \\ \frac{1}{3} & \frac{2}{3} & \frac{2}{3} \\ \frac{2}{3} & -\frac{2}{3} & \frac{1}{3} \end{bmatrix}$$

is one unit long and is orthogonal to each of the other two column vectors. Show that the same statement applies to the row vectors. By straightforward (but tedious) calculations, one can show that $|M\mathbf{x}| = |\mathbf{x}|$ for any vector $\mathbf{x} \in R^3$. Try it for $\mathbf{x} = (1, 1, 1)$, for example.

6. What is the angle determined by the vectors $2\mathbf{i} - \mathbf{j} + 2\mathbf{k}$ and $\mathbf{i} + 2\mathbf{j} - 2\mathbf{k}$?

7. Can you find a number x such that the vectors $2\mathbf{i} - \mathbf{j} + \mathbf{k}$ and $\mathbf{i} + 2\mathbf{j} + x\mathbf{k}$ determine an angle of $30°$?

8. Find a vector in the plane of the position vectors $\mathbf{a} = 2\mathbf{i} - \mathbf{j} + \mathbf{k}$ and $\mathbf{b} = \mathbf{i} + 2\mathbf{j} + 2\mathbf{k}$ that is perpendicular to $\mathbf{c} = \mathbf{i} - 2\mathbf{j} + 3\mathbf{k}$.

9. Suppose that $\mathbf{a} = 2\mathbf{i} - 2\mathbf{j} + \mathbf{k}$. Find a vector $\mathbf{x}$ such that $|\mathbf{x}| = 9$ and $|\mathbf{a} + \mathbf{x}|$ is a maximum. A minimum.

10. Let P, Q, and R be the points $(-1, 2, -3)$, $(3, 0, 1)$, and (x, y, z).
    (a) Express $\mathbf{PR} \cdot \mathbf{QR}$ in terms of x, y, and z.
    (b) Describe the set of points for which $\angle PRQ$ is a right angle.
    (c) Show that if $\angle PRQ$ is a right angle, then $|\mathbf{RM}|$ is independent of x, where M is the midpoint of PQ.

11. Show that the vector $(\mathbf{b} \cdot \mathbf{b})\mathbf{a} - (\mathbf{a} \cdot \mathbf{b})\mathbf{b}$ is perpendicular to $\mathbf{b}$.

12. If **a** and **b** determine θ, we say that the number $|\mathbf{a}| \cos \theta$ is the **component** of **a** along **b**.

   (a) Show that the component of **a** along **b** is $|\mathbf{b}|^{-1}(\mathbf{a} \cdot \mathbf{b})$.

   (b) Under what circumstances is the component of **b** along **a** equal to the component of **a** along **b**?

   (c) Find the component of $\mathbf{a} = 3\mathbf{i} - \mathbf{j} + 2\mathbf{k}$ along $\mathbf{b} = \mathbf{i} + 2\mathbf{j} - \mathbf{k}$.

13. The **projection** of a position vector **a** onto a position vector **b** is the vector that is obtained by dropping a perpendicular from the terminal point of **a** to the line determined by **b**. Show that it is the vector $|\mathbf{b}|^{-2}(\mathbf{a} \cdot \mathbf{b})\mathbf{b}$.

14. The inequality $|\mathbf{a} + \mathbf{b}| \le |\mathbf{a}| + |\mathbf{b}|$ is known as the **Triangle Inequality**.

   (a) Figure out a geometric explanation of where the name comes from and why it is valid in three-space.

   (b) Explain why it is equivalent to the inequality $(\mathbf{a} + \mathbf{b}) \cdot (\mathbf{a} + \mathbf{b}) \le (|\mathbf{a}| + |\mathbf{b}|)^2$ and hence to the inequality $(\mathbf{a} \cdot \mathbf{b})^2 \le |\mathbf{a}|^2 |\mathbf{b}|^2$.

   (c) This last inequality is the **Cauchy-Schwarz Inequality**. Use the argument of Example 69-3 to show that it is valid for vectors in three-space. (The Triangle Inequality and the Cauchy-Schwarz Inequality are valid in R^n for any n.)

95. THE CROSS PRODUCT

It is pretty hard to talk about the "orientation" of a mathematical figure without pointing to a picture and saying, "That's what we mean." For example, we have been drawing right-handed coordinate systems, and we hope that you know what we are talking about. You certainly wouldn't if we tried to describe the concept in words. Before introducing the *cross product* of two vectors, we need to explain what we mean when we say that three vectors form a "right-handed triple," and we will depend on pictures to make our point.

Roughly speaking, we say that three vectors **a**, **b**, and **c** (in that order) form a **right-handed triple** if they more nearly point in the directions of the basis vectors of a right-handed coordinate system than they point in the direction of the basis vectors of a left-handed system. The triple **a**, **b**, **c** shown in Fig. 95-1, for example, is right-handed. If we swing it around rigidly so that **a** lies along the positive X-axis and **b** lies in the XY-plane and points to the same side of the XZ-plane as the positive Y-axis does, then **c** will point to the same side of the XY-plane as the positive Z-axis does. Notice that if **a**, **b**, **c** is right-handed, then so are the triples **b**, **c**, **a** and **c**, **a**, **b**, whereas **b**, **a**, **c**, and **a**, **c**, **b**, and **c**, **b**, **a** are left-handed. Thus if we swing our vectors around so that **b** lies along the X-axis and **a** lies in the XY-plane and points to the same side of the XZ-plane that the positive Y-axis does, then **c** will point "more down than up"—that is, to the side of the XY-plane on which the negative Z-axis lies.

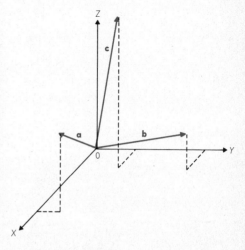

Figure 95-1

The dot product of two vectors is a scalar (a number). Now we are going to introduce a product of two vectors that is a *vector*. As far as this textbook is concerned, this product will be strictly a creature of three-dimensional space, so we will introduce it geometrically. The **cross** or **vector** product of **a** by **b** is the vector

(95-1)
$$\mathbf{a} \times \mathbf{b} = |\mathbf{a}|\,|\mathbf{b}|\,\sin\theta\,\mathbf{u},$$

where θ is the angle determined by **a** and **b**, and **u** is the unit vector that is perpendicular to the plane of **a** and **b** and is directed so that the triple **a**, **b**, **u** is right-handed. This statement does not determine **u** if **a** is parallel to **b** or if one of these vectors is **0**. But then we must have $\theta = 0°$, $\theta = 180°$, $|\mathbf{a}| = 0$, or $|\mathbf{b}| = 0$, and any one of these equations implies that $|\mathbf{a}|\,|\mathbf{b}|\,\sin\theta = 0$, so we can take $\mathbf{a} \times \mathbf{b} = \mathbf{0}$ in that case. The vector **u**, and hence $\mathbf{a} \times \mathbf{b}$, points in the direction a right-handed screw would advance if **a** were rotated into **b** through the angle θ.

You can use the definition of the cross product to check the following multiplication table for the unit basis vectors:

(95-2)

$$\mathbf{i} \times \mathbf{j} = \mathbf{k} \qquad \mathbf{j} \times \mathbf{k} = \mathbf{i} \qquad \mathbf{k} \times \mathbf{i} = \mathbf{j}$$
$$\mathbf{j} \times \mathbf{i} = -\mathbf{k} \qquad \mathbf{k} \times \mathbf{j} = -\mathbf{i} \qquad \mathbf{i} \times \mathbf{k} = -\mathbf{j}$$
$$\mathbf{i} \times \mathbf{i} = \mathbf{0} \qquad \mathbf{j} \times \mathbf{j} = \mathbf{0} \qquad \mathbf{k} \times \mathbf{k} = \mathbf{0}.$$

Example 95-1. Show that for any two vectors **a** and **b**,

(95-3)
$$|\mathbf{a}|^2|\mathbf{b}|^2 = (\mathbf{a} \cdot \mathbf{b})^2 + |\mathbf{a} \times \mathbf{b}|^2.$$

Solution. Since **u** is one unit long, Equation 95-1 tells us that $|\mathbf{a} \times \mathbf{b}|^2 = |\mathbf{a}|^2|\mathbf{b}|^2 \sin^2\theta$. Also, $(\mathbf{a} \cdot \mathbf{b})^2 = (|\mathbf{a}|\,|\mathbf{b}|\cos\theta)^2 = |\mathbf{a}|^2|\mathbf{b}|^2 \cos^2\theta$. When we add these two equations and take account of the trigonometric identity $\cos^2\theta + \sin^2\theta = 1$, we end up with Equation 95-3.

We have said that if **a** and **b** are parallel, then $\mathbf{a} \times \mathbf{b} = \mathbf{0}$. Conversely, if $\mathbf{a} \times \mathbf{b} = \mathbf{0}$, then either $\sin\theta = 0$ and the vectors are parallel, or one of the vectors is **0**. Let us agree that **0** is parallel to every vector. Then we can make the statement: *the vectors* **a** *and* **b** *are parallel if and only if* $\mathbf{a} \times \mathbf{b} = \mathbf{0}$. Even when the number $|\mathbf{a} \times \mathbf{b}|$ is not 0, we can give it a simple geometric interpretation. From Fig. 95-2 you can see that $|\mathbf{b}| \sin\theta$ is the altitude of the parallelogram that is determined by the vectors **a** and **b** and that $|\mathbf{a}|$ is its base length. Hence $|\mathbf{a}|\,|\mathbf{b}|\sin\theta$ is its area. Since $|\mathbf{a} \times \mathbf{b}| = |\mathbf{a}|\,|\mathbf{b}|\sin\theta$, we can therefore interpret the number $|\mathbf{a} \times \mathbf{b}|$ as the area of the parallelogram that is determined by **a** and **b**.

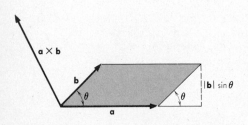

Figure 95-2

Although the usual rules of arithmetic work with the dot product, some of them *do not apply* to the cross product. For example, in Equations 95-2 we see that $\mathbf{i} \times \mathbf{j} = \mathbf{k}$, whereas $\mathbf{j} \times \mathbf{i} = -\mathbf{k}$. Since the cross products $\mathbf{i} \times \mathbf{j}$ and $\mathbf{j} \times \mathbf{i}$ are not the same, the commutative law is not valid for cross products. You can easily see that we have, in fact, the "anticommutative" rule

$$\mathbf{b} \times \mathbf{a} = -(\mathbf{a} \times \mathbf{b}).$$

The associative law of multiplication does not hold either; that is, the products $(\mathbf{a} \times \mathbf{b}) \times \mathbf{c}$ and $\mathbf{a} \times (\mathbf{b} \times \mathbf{c})$ are not necessarily the same vector. For example, a couple of short calculations will show you that $(\mathbf{i} \times \mathbf{i}) \times \mathbf{j} = \mathbf{0}$, but $\mathbf{i} \times (\mathbf{i} \times \mathbf{j}) = -\mathbf{j}$.

However, if $\mathbf{a}$, $\mathbf{b}$, and $\mathbf{c}$ are any vectors, and if r is a scalar, then

(95-4) $$\mathbf{a} \times (\mathbf{b} + \mathbf{c}) = \mathbf{a} \times \mathbf{b} + \mathbf{a} \times \mathbf{c}$$

and

(95-5) $$(r\mathbf{a}) \times \mathbf{b} = \mathbf{a} \times (r\mathbf{b}) = r(\mathbf{a} \times \mathbf{b}).$$

It is not hard to convince yourself, by a geometric argument based on the definition of the cross product, that the second set of equations is valid. It is harder to see why the first equation is true; we need some preliminary discussion first.

Figure 95-3 shows three vectors $\mathbf{a}$, $\mathbf{b}$, and $\mathbf{c}$ and the parallelepiped (squashed brick) that they determine. We will present a geometric argument to show that the number $\mathbf{a} \cdot (\mathbf{b} \times \mathbf{c})$, the **scalar triple product** of $\mathbf{a}$, $\mathbf{b}$, and $\mathbf{c}$, gives the volume of this brick. The definition of the dot product tells us that this number is $|\mathbf{a}| \, |\mathbf{b} \times \mathbf{c}| \cos \phi$, where the vectors $\mathbf{a}$ and $\mathbf{b} \times \mathbf{c}$ determine ϕ. The figure shows that $|\mathbf{a}| \cos \phi$ is the altitude of our parallelepiped. Since $|\mathbf{b} \times \mathbf{c}|$ is the area of the parallelogram that forms its base, the number

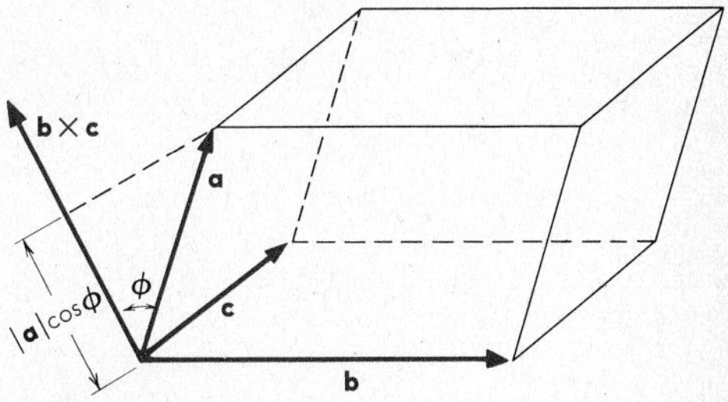

$$\mathbf{a} \cdot (\mathbf{b} \times \mathbf{c}) = |\mathbf{a}| \, |\mathbf{b} \times \mathbf{c}| \cos \phi$$

Figure 95-3

is the volume of the parallelepiped shown in Fig. 95-3. In that figure, $\mathbf{a}$, $\mathbf{b}$, $\mathbf{c}$ is a right-handed triple. If $\mathbf{a}$, $\mathbf{b}$, $\mathbf{c}$ had been left-handed, the scalar triple product $\mathbf{a} \cdot (\mathbf{b} \times \mathbf{c})$ would have been the negative of the volume. Soon we will have available (Equations 95-8 and 95-9) algebraic formulas for the scalar triple product, formulas that will give us an algebraic way to determine the orientation of a triple. If they yield a positive number as the scalar triple product, we are dealing with a right-handed triple; a negative number indicates a left-handed triple. The vectors $\mathbf{a}$, $\mathbf{b}$, and $\mathbf{c}$ are coplanar if and only if $\mathbf{a} \cdot (\mathbf{b} \times \mathbf{c}) = 0$.

Now let $\mathbf{a}$, $\mathbf{b}$, and $\mathbf{c}$ be any three vectors. We have said that the triples $\mathbf{a}$, $\mathbf{b}$,

c and **c**, **a**, **b** have the same orientation; either both are right handed or neither is. Thus both the scalar triple products $\mathbf{a} \cdot (\mathbf{b} \times \mathbf{c})$ and $\mathbf{c} \cdot (\mathbf{a} \times \mathbf{b})$ equal the volume of the parallelepiped determined by our three vectors, or both equal the negative of this volume. In any case, these products are equal. Since the dot product is commutative, $\mathbf{c} \cdot (\mathbf{a} \times \mathbf{b}) = (\mathbf{a} \times \mathbf{b}) \cdot \mathbf{c}$, and hence we see that *for any three vectors* **a**, **b**, *and* **c**,

$$\mathbf{a} \cdot (\mathbf{b} \times \mathbf{c}) = (\mathbf{a} \times \mathbf{b}) \cdot \mathbf{c}.$$

It is easy to remember this identity; it simply says that *one may interchange the dot and the cross in the scalar triple product.*

Now we are ready to return to the distributive law, Equation 95-4. Let **a**, **b**, and **c** be arbitrary vectors, and set $\mathbf{v} = \mathbf{a} \times (\mathbf{b} + \mathbf{c}) - \mathbf{a} \times \mathbf{b} - \mathbf{a} \times \mathbf{c}$. The distributive law says that $\mathbf{v} = \mathbf{0}$, so this is the equation we will verify. In fact, we will show that $|\mathbf{v}|^2 = 0$. We first write

$$|\mathbf{v}|^2 = \mathbf{v} \cdot \mathbf{v} = \mathbf{v} \cdot [\mathbf{a} \times (\mathbf{b} + \mathbf{c})] - \mathbf{v} \cdot (\mathbf{a} \times \mathbf{b}) - \mathbf{v} \cdot (\mathbf{a} \times \mathbf{c}).$$

Then we interchange the dot and cross in these scalar triple products:

$$\mathbf{v} \cdot \mathbf{v} = (\mathbf{v} \times \mathbf{a}) \cdot (\mathbf{b} + \mathbf{c}) - (\mathbf{v} \times \mathbf{a}) \cdot \mathbf{b} - (\mathbf{v} \times \mathbf{a}) \cdot \mathbf{c}.$$

Since the dot product obeys the distributive law, this last expression is 0. Therefore $\mathbf{v} = \mathbf{0}$, and the distributive law is verified for the cross product.

The general Equations 95-4 and 95-5 and the specific multiplication Formulas 95-2 enable us to express the cross product of $\mathbf{a} = a_x\mathbf{i} + a_y\mathbf{j} + a_z\mathbf{k}$ by $\mathbf{b} = b_x\mathbf{i} + b_y\mathbf{j} + b_z\mathbf{k}$ in terms of the components of these vectors. Thus when we use these equations to expand the product

$$\mathbf{a} \times \mathbf{b} = (a_x\mathbf{i} + a_y\mathbf{j} + a_z\mathbf{k}) \times (b_x\mathbf{i} + b_y\mathbf{j} + b_z\mathbf{k}),$$

we obtain the formula

$$(95\text{-}6) \quad \mathbf{a} \times \mathbf{b} = (a_yb_z - a_zb_y)\mathbf{i} - (a_xb_z - a_zb_x)\mathbf{j} + (a_xb_y - a_yb_x)\mathbf{k}$$

$$= \begin{vmatrix} a_y & a_z \\ b_y & b_z \end{vmatrix} \mathbf{i} - \begin{vmatrix} a_x & a_z \\ b_x & b_z \end{vmatrix} \mathbf{j} + \begin{vmatrix} a_x & a_y \\ b_x & b_y \end{vmatrix} \mathbf{k}.$$

By analogy with Equation 92-3, this last equation is sometimes symbolically expressed as

$$(95\text{-}7) \qquad\qquad \mathbf{a} \times \mathbf{b} = \begin{vmatrix} \mathbf{i} & \mathbf{j} & \mathbf{k} \\ a_x & a_y & a_z \\ b_x & b_y & b_z \end{vmatrix}.$$

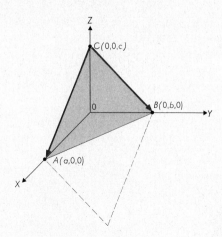

Figure 95-4

Example 95-2. Find the area of the triangle whose vertices are the points $A(a, 0, 0)$, $B(0, b, 0)$, and $C(0, 0, c)$.

Solution. The area of our triangle is one-half the area of the parallelogram with **CA** and **CB** as sides (Fig. 95-4), and the area of this parallelogram is $|\textbf{CA} \times \textbf{CB}|$. Now

$$\textbf{CA} = a\textbf{i} - c\textbf{k} \quad \text{and} \quad \textbf{CB} = b\textbf{j} - c\textbf{k},$$

and so

$$\textbf{CA} \times \textbf{CB} = \begin{vmatrix} 0 & -c \\ b & -c \end{vmatrix} \textbf{i} - \begin{vmatrix} a & -c \\ 0 & -c \end{vmatrix} \textbf{j} + \begin{vmatrix} a & 0 \\ 0 & b \end{vmatrix} \textbf{k}$$

$$= bc\textbf{i} + ac\textbf{j} + ab\textbf{k}.$$

Therefore $|\textbf{CA} \times \textbf{CB}|^2 = b^2c^2 + a^2c^2 + a^2b^2$, and hence the area of our triangle is

$$\tfrac{1}{2}|\textbf{CA} \times \textbf{CB}| = \tfrac{1}{2}\sqrt{b^2c^2 + a^2c^2 + a^2b^2}.$$

Now that we have a formula (Equation 95-6) for the cross product of two vectors in terms of their components, we can express the scalar triple product $\textbf{a} \cdot (\textbf{b} \times \textbf{c})$ in terms of the components of $\textbf{a}$, $\textbf{b}$, and $\textbf{c}$:

$$\textbf{a} \cdot (\textbf{b} \times \textbf{c}) = (a_x\textbf{i} + a_y\textbf{j} + a_z\textbf{k}) \cdot \left(\begin{vmatrix} b_y & b_z \\ c_y & c_z \end{vmatrix} \textbf{i} - \begin{vmatrix} b_x & b_z \\ c_x & c_z \end{vmatrix} \textbf{j} + \begin{vmatrix} b_x & b_y \\ c_x & c_y \end{vmatrix} \textbf{k} \right);$$

that is,

$$(95\text{-}8) \quad \textbf{a} \cdot (\textbf{b} \times \textbf{c}) = a_x \begin{vmatrix} b_y & b_z \\ c_y & c_z \end{vmatrix} - a_y \begin{vmatrix} b_x & b_z \\ c_x & c_z \end{vmatrix} + a_z \begin{vmatrix} b_x & b_y \\ c_x & c_y \end{vmatrix}.$$

Here we can legitimately use Equation 92-3, and so

$$(95\text{-}9) \qquad \mathbf{a} \cdot (\mathbf{b} \times \mathbf{c}) = \begin{vmatrix} a_x & a_y & a_z \\ b_x & b_y & b_z \\ c_x & c_y & c_z \end{vmatrix}.$$

Example 95-3. Show that the points (x_1, y_1), (x_2, y_2), and (x_3, y_3) are vertices of a triangle whose area is the absolute value of the number

$$(95\text{-}10) \qquad \tfrac{1}{2} \begin{vmatrix} 1 & 1 & 1 \\ x_1 & x_2 & x_3 \\ y_1 & y_2 & y_3 \end{vmatrix} = \tfrac{1}{2}[(x_2 y_3 - x_3 y_2) - (x_1 y_3 - x_3 y_1) + (x_1 y_2 - x_2 y_1)].$$

Solution. Let us consider the plane in which our given triangle lies to be the XY-plane of a three-dimensional space so that the vertices of the triangle are the points $A(x_1, y_1, 0)$, $B(x_2, y_2, 0)$, and $C(x_3, y_3, 0)$. The area of our triangle is (see Example 95-2) $\tfrac{1}{2}|\mathbf{CA} \times \mathbf{CB}|$. Since the vectors $\mathbf{CA}$ and $\mathbf{CB}$ lie in the XY-plane, the vector $\mathbf{CA} \times \mathbf{CB}$ is parallel to the Z-axis. Therefore the absolute value of the dot product of the vector $\mathbf{CA} \times \mathbf{CB}$ and the unit vector $\mathbf{k}$ is the product of the lengths $|\mathbf{CA} \times \mathbf{CB}|$ and $|\mathbf{k}| = 1$. Thus our desired area is simply the absolute value of the number $\tfrac{1}{2}\mathbf{k} \cdot (\mathbf{CA} \times \mathbf{CB})$. Now

$$\mathbf{CA} = (x_1 - x_3)\mathbf{i} + (y_1 - y_3)\mathbf{j} \quad \text{and} \quad \mathbf{CB} = (x_2 - x_3)\mathbf{i} + (y_2 - y_3)\mathbf{j},$$

so when we substitute in Equation 95-8, we get

$$\tfrac{1}{2}\mathbf{k} \cdot (\mathbf{CA} \times \mathbf{CB}) = \tfrac{1}{2} \cdot 1 \cdot \begin{vmatrix} x_1 - x_3 & y_1 - y_3 \\ x_2 - x_3 & y_2 - y_3 \end{vmatrix}.$$

We now expand this determinant and simplify, and we obtain the expression on the right-hand side of Equation 95-10.

PROBLEMS 95

1. If $\mathbf{a} = 2\mathbf{i} - \mathbf{j} + \mathbf{k}$, $\mathbf{b} = \mathbf{i} - 3\mathbf{j}$, and $\mathbf{c} = \mathbf{j}$, calculate the given quantity.
   (a) $\mathbf{a} \times \mathbf{c}$
   (b) $\mathbf{a} \times \mathbf{b}$
   (c) $(\mathbf{a} \times \mathbf{b}) \times \mathbf{c}$
   (d) $\mathbf{a} \times (\mathbf{b} \times \mathbf{c})$
   (e) $\mathbf{a} \times (\mathbf{b} + \mathbf{c})$
   (f) $\mathbf{a} \times \mathbf{b} + \mathbf{a} \times \mathbf{c}$
   (g) $\mathbf{a} \cdot (\mathbf{b} \times \mathbf{c})$
   (h) $(\mathbf{a} \times \mathbf{b}) \cdot \mathbf{c}$

2. The origin and the terminal points of the position vectors $\mathbf{i} - 2\mathbf{j}$ and $\mathbf{i} + 2\mathbf{j} + 2\mathbf{k}$ are the vertices of a triangle. What is its area?

3. Find a unit vector that is perpendicular to $\mathbf{a}$ and $\mathbf{b}$.
   (a) $\mathbf{a} = \mathbf{i} + \mathbf{j}$, $\mathbf{b} = \mathbf{i} - \mathbf{j}$
   (b) $\mathbf{a} = \mathbf{i} + \mathbf{j}$, $\mathbf{b} = \mathbf{i} - \mathbf{k}$
   (c) $\mathbf{a} = \mathbf{i} - 2\mathbf{j}$, $\mathbf{b} = \mathbf{i} + 2\mathbf{j} + 2\mathbf{k}$
   (d) $\mathbf{a} = 2\mathbf{i} + \mathbf{j} + 2\mathbf{k}$, $\mathbf{b} = \mathbf{i} - 2\mathbf{j} + 2\mathbf{k}$

4. Choose x and y so that $x\mathbf{i} + y\mathbf{j} + 3\mathbf{k}$ is parallel to $-2\mathbf{i} + 3\mathbf{j} - \mathbf{k}$.

5. What is the volume of the parallelepiped determined by $2\mathbf{i} + 2\mathbf{j} - \mathbf{k}$, $\mathbf{i} - 2\mathbf{j} + 2\mathbf{k}$, and $\mathbf{i} - \mathbf{j} - \mathbf{k}$?

6. The points $(1, 2, -3)$, $(2, 0, -1)$, and $(3, 1, -2)$ are vertices of a triangle in space. What is its area?

7. Let $\mathbf{a} = 2\mathbf{i} - \mathbf{j} + 3\mathbf{k}$ and $\mathbf{b} = \mathbf{i} - \mathbf{j} - \mathbf{k}$.

(a) Solve the equation $\mathbf{a} \times \mathbf{x} = \mathbf{b}$.   (b) Solve the equation $\mathbf{a} \cdot \mathbf{x} = |\mathbf{b}|$.

8. Find $\mathbf{a} \times \mathbf{b}$ if $\mathbf{a} = \cos \alpha \mathbf{i} + \cos \beta \mathbf{j} + \cos \gamma \mathbf{k}$, and $\mathbf{b} = \sin \alpha \mathbf{i} + \sin \beta \mathbf{j} + \sin \gamma \mathbf{k}$.

9. Show that

$$(\mathbf{a} \times \mathbf{b}) \cdot (\mathbf{a} \times \mathbf{b}) = \begin{vmatrix} \mathbf{a} \cdot \mathbf{a} & \mathbf{a} \cdot \mathbf{b} \\ \mathbf{a} \cdot \mathbf{b} & \mathbf{b} \cdot \mathbf{b} \end{vmatrix}.$$

10. Let $\mathbf{p}$, $\mathbf{q}$, and $\mathbf{r}$ be position vectors to points P, Q, and R. Show that the vector $(\mathbf{p} \times \mathbf{q}) + (\mathbf{q} \times \mathbf{r}) + (\mathbf{r} \times \mathbf{p})$ is perpendicular to the plane PQR.

11. By writing both sides of the equation in terms of the components of the vectors involved, one can show that for any three vectors $\mathbf{a}$, $\mathbf{b}$, and $\mathbf{c}$,

$$\mathbf{a} \times (\mathbf{b} \times \mathbf{c}) = (\mathbf{a} \cdot \mathbf{c})\mathbf{b} - (\mathbf{a} \cdot \mathbf{b})\mathbf{c}.$$

(a) Verify this equation for the vectors of Problem 95-1.

(b) Show that, in general, the vectors on both sides of the equation are perpendicular to $\mathbf{a}$.

(c) Use the equation to find a similar equation for $(\mathbf{a} \times \mathbf{b}) \times \mathbf{c}$.

12. Suppose that we write our vectors in the notation of R^3; that is, $\mathbf{a} = (a_1, a_2, a_3)$ and $\mathbf{b} = (b_1, b_2, b_3)$. Show that it is then natural to write

$$\mathbf{a} \times \mathbf{b} = \begin{bmatrix} 0 & -a_3 & a_2 \\ a_3 & 0 & -a_1 \\ -a_2 & a_1 & 0 \end{bmatrix} \mathbf{b}.$$

96.   PLANES

Graphs of equations in x and y, such as $3x + 2y = 5$ and $x^2 - 7y^2 = 3x$, are curves in the coordinate plane. Now we are going to see that graphs of equations in x, y, and z, such as $3x - 2y + 4z = 7$ and $x^2 + y^2 - z^2 = 16$, are surfaces in three-dimensional space. In this section we will take up the simplest surfaces—namely, planes.

In plane analytic geometry we determine a line by assigning its direction and specifying one of its points. Similarly, in solid analytic geometry we determine a plane by giving its orientation and specifying one of its points. Thus in Fig. 96-1 we have shown a point P_0 and a vector $\mathbf{n}$. These quantities determine the plane consisting of those points P for which the vector $\mathbf{P_0 P}$ is perpendicular to $\mathbf{n}$. The vector $\mathbf{n}$ is called a **normal** to the plane. For two vectors to be perpendicular, their dot product must be 0, so

$$(96\text{-}1) \qquad \mathbf{n} \cdot \mathbf{P_0 P} = 0.$$

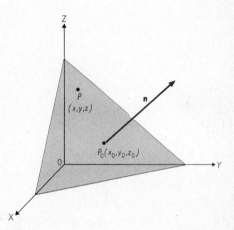

Figure 96-1

This equation is one form of the equation of the plane. Now suppose that the coordinates of P_0 are (x_0, y_0, z_0), the coordinates of P are (x, y, z), and $\mathbf{n} = a\mathbf{i} + b\mathbf{j} + c\mathbf{k}$. Then

$$\mathbf{P_0P} = (x - x_0)\mathbf{i} + (y - y_0)\mathbf{j} + (z - z_0)\mathbf{k},$$

and Equation 96-1 becomes

(96-2) $a(x - x_0) + b(y - y_0) + c(z - z_0) = 0.$

This form of the equation of our plane corresponds to the point-slope form of the equation of a line.

Example 96-1. Find the plane that contains the point (1, 2, 3) and has the vector $\mathbf{n} = 2\mathbf{i} - \mathbf{j} + 3\mathbf{k}$ as a normal.

Solution. Here we have $x_0 = 1$, $y_0 = 2$, $z_0 = 3$, $a = 2$, $b = -1$, and $c = 3$. Hence Equation 96-2 becomes $2(x - 1) - (y - 2) + 3(z - 3) = 0$, which simplifies to
$$2x - y + 3z - 9 = 0.$$

We can expand Equation 96-2 and write it as

(96-3) $ax + by + cz + d = 0,$

where $d = -ax_0 - by_0 - cz_0$. Notice the similarity between Equation 96-3 and the equation $ax + by + c = 0$ of a line in plane analytic geometry. We can easily determine the slope of such a line from the coefficients of x and y. Similarly, the coefficients of x, y, and z in Equation 96-3 are the components of a vector normal to the plane; that is, the vector $\mathbf{n} = a\mathbf{i} + b\mathbf{j} + c\mathbf{k}$ is normal to the plane $ax + by + cz + d = 0$.

Example 96-2. Discuss the plane $2x + y + 3z = 6$.

Solution. A plane is determined by three points, and we can find points of a surface by choosing two coordinates and solving for the third. Thus if we let $x = 0$ and $y = 0$, we find $z = 2$, and hence (0, 0, 2) is one of the points of our plane. In the same way, we find that two other points are (3, 0, 0) and (0, 6, 0). These points are the points in which the plane intersects the axes, and Fig. 96-2 shows how they determine the plane. The coefficients of x, y, and z give us a vector that is normal to the plane: $\mathbf{n} = 2\mathbf{i} + \mathbf{j} + 3\mathbf{k}$.

Example 96-3. Find the plane that contains the points $A(1, -1, 2)$, $B(3, 2, 1)$, and $C(2, 1, 3)$.

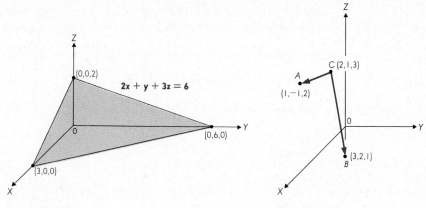

Figure 96-2

Figure 96-3

Solution. Figure 96-3 shows our three points, and we can use the coordinates of any one of them as the numbers x_0, y_0, and z_0 in Equation 96-2. The numbers a, b, and c are components of a normal to our plane—that is, a vector that is perpendicular to both of the vectors **CA** and **CB**. The cross product **CA** × **CB** is such a vector. Since **CA** $= -\mathbf{i} - 2\mathbf{j} - \mathbf{k}$ and **CB** $= \mathbf{i} + \mathbf{j} - 2\mathbf{k}$, we readily find (with the aid of Equation 95-6) that **CA** × **CB** $= 5\mathbf{i} - 3\mathbf{j} + \mathbf{k}$. We can therefore let $(a, b, c) = (5, -3, 1)$ and $(x_0, y_0, z_0) = (1, -1, 2)$, and Equation 96-2 gives us the equation of our plane:

$$5(x - 1) - 3(y + 1) + (z - 2) = 0.$$

We can rewrite this equation in the form of Equation 96-3 as

$$5x - 3y + z - 10 = 0.$$

Example 96-4. Find the distance between the point $(2, 3, 4)$ and the plane $2x + y + 2z = 2$.

Solution. Our point and plane are shown in Fig. 96-4. We sketched the plane by finding its intercepts $(1, 0, 0)$, $(0, 2, 0)$, and $(0, 0, 1)$. We know that the vector $\mathbf{n} = 2\mathbf{i} + \mathbf{j} + 2\mathbf{k}$ is perpendicular to this plane. Now let **v** be a vector from our given point to a point of the plane—for example, the point $(0, 0, 1)$. The distance we want is the projection of **v** onto the normal line. Figure 96-4 shows that this projection is $|\mathbf{v}| \cos \phi$, where ϕ is the acute angle between **v** and **n**. This number can be written as $|\mathbf{n}|^{-1}|\mathbf{v}| \, |\mathbf{n}| \cos \phi$, which we recognize as $|\mathbf{n}|^{-1}|\mathbf{v} \cdot \mathbf{n}|$. Since $\mathbf{v} = -2\mathbf{i} - 3\mathbf{j} - 3\mathbf{k}$, our desired distance is the number

$$(4 + 1 + 4)^{-1/2}|(-2\mathbf{i} - 3\mathbf{j} - 3\mathbf{k}) \cdot (2\mathbf{i} + \mathbf{j} + 2\mathbf{k})| = \tfrac{13}{3}.$$

At the end of Section 89 we pointed out that the problem of solving a system of two linear equations in x and y can be viewed geometrically as that of

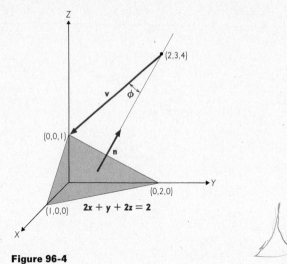

Figure 96-4

determining the intersection of two lines. Similarly, we can observe that solving the system

$$a_1 x + b_1 y + c_1 z = d_1$$

$$a_2 x + b_2 y + c_2 z = d_2$$

$$a_3 x + b_3 y + c_3 z = d_3$$

amounts to studying the intersection of three planes. One "expects" only one point of intersection, but it is possible that there is a whole line of intersection or that two of the planes are parallel and so on. You might think about the possibilities and relate your conclusions to what we learned about systems of equations in Chapter 11.

PROBLEMS 96

1. Sketch the plane and find a unit vector normal to it.
   (a) $2x + y - 2z = 6$ (b) $3x - 4y = 12$
   (c) $x + y + z = 0$ (d) $y - z - 4x - 7 = 0$

2. Find the plane for which the given vector is a normal vector and which contains the given point.
   (a) $\mathbf{n} = 2\mathbf{i} - 3\mathbf{j} + \mathbf{k}$, $(1, 2, 3)$ (b) $\mathbf{n} = 3\mathbf{i} - \mathbf{k}$, $(-1, -2, -3)$
   (c) $\mathbf{n} = \mathbf{i}$, $(0, 0, 0)$ (d) $\mathbf{n} = \mathbf{j} + \mathbf{k}$, $(0, 1, 0)$

3. Find the plane that
   (a) contains the points $(1, 0, -1)$, $(2, 3, 1)$, and $(-2, 1, 1)$.
   (b) contains the points $(2, 0, 0)$, $(0, 3, 0)$, and $(0, 0, 5)$.

600

(c) contains the points $(1, 2, 3)$ and $(-2, 1, 1)$ and that does not intersect the Y-axis.

(d) contains the points $(1, 2, 3)$ and $(-2, 1, 1)$ and is perpendicular to the plane $x + 2y - z = 3$.

(e) contains the point $(-1, 7, 9)$ and is parallel to the plane $3x + 4y - z + 6 = 0$.

(f) contains the point $(1, 0, 2)$ and is parallel to the plane $y = 3$.

4. (a) Find a vector that is parallel to the line of intersection of the planes $2x - y + 3z + 2 = 0$ and $x + 2y - z - 4 = 0$.

   (b) Find the plane that is perpendicular to this line of intersection at the point $(3, -1, -3)$.

5. Describe the set of planes that have the form $2x - 3y + 4z + d = 0$ for some number d.

6. Find the cosine of the angle between the (normals to the) planes $x - 2y + z + 4 = 0$ and $2x - 3y - z + 1 = 0$.

7. The set of all points equidistant from the two points $(2, 5, -1)$ and $(5, -2, -4)$ is a plane. Find it.

8. (a) How far apart are the parallel planes $2x - y + 2z = -3$ and $2x - y + 2z = 6$?

   (b) Find the plane halfway between them.

9. Find the coordinates of the point P of the plane $2x + y + 2z + 10 = 0$ such that the position vector $\mathbf{OP}$ is parallel to the vector $\mathbf{i} + \mathbf{j} + \mathbf{k}$.

10. Find the plane that is perpendicular to the plane $14x + 2y + 7 = 0$, that contains the origin, and whose normal makes a $45°$ angle with the Z-axis.

11. The plane $2x + 3y + 4z = 12$ "cuts off" a corner of the first octant. What is its volume?

12. Describe geometrically the problem of solving the system of equations

$$2x + y - z = 5$$
$$x + 3z = -2$$
$$4x + 2y + az = b$$

for the given choices of a and b.

(a) $a = -3, b = 10$ (b) $a = -2, b = 10$
(c) $a = -2, b = 5$ (d) $a = -3, b = 5$

13. Show that the distance between the plane $ax + by + cz + d = 0$ and the point (x_0, y_0, z_0) is the number

$$\frac{|ax_0 + by_0 + cz_0 + d|}{\sqrt{a^2 + b^2 + c^2}}.$$

14. Sketch the graphs of the following equations.

(a) $|x + y + z| = 1$ (b) $|x| + |y| + |z| = 1$
(c) $[x + y + z] = 1$ (d) $[x] + [y] + [z] = 1$

97. CYLINDERS. SURFACES OF REVOLUTION

The **cylinders** are among the simpler types of surfaces. We generate one by moving a line that is perpendicular to a given plane along a curve in that plane. For example, an ordinary right-circular cylinder (a beer can) is generated by moving a line that is perpendicular to the plane of a circle around the circle. A parabolic cylinder is generated by moving a line that is perpendicular to the plane of a parabola and so on. The lines are the **generators** of the cylinder.

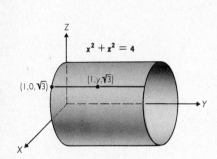

Figure 97-1

When the generators of a cylinder are parallel to one of the coordinate axes, the corresponding coordinate is missing from its equation. Thus the surface $x^2 + z^2 = 4$ is the right-circular cylinder whose generators are parallel to the Y-axis shown in Fig. 97-1. For example, for every number y, the point $(1, y, \sqrt{3})$ belongs to our surface, so the line

$$\{(1, y, \sqrt{3}) \mid y \in (-\infty, \infty)\}$$

(which is parallel to the Y-axis) is one of the generators. We obtain other generators by replacing 1 and $\sqrt{3}$ with various numbers whose squares add up to 4, such as 0 and 2, $\sqrt{2}$ and $-\sqrt{2}$. Our cylinder is generated by moving any one of these lines around the circle $x^2 + z^2 = 4$ in the XZ-plane, keeping the line parallel to the Y-axis during the motion. We say that the *surface* $x^2 + z^2 = 4$ in space is the cylinder whose *base curve* is $x^2 + z^2 = 4$ in the XZ-plane.

In order to plot a cylinder whose equation is missing a particular coordinate, we simply plot the base curve in the coordinate plane that is perpendicular to the axis of the missing coordinate and then "slide" it parallel to this axis.

Example 97-1. Sketch the surface $y^2 = x$.

Solution. Our equation does not involve z, so its graph is a cylinder whose generators are parallel to the Z-axis. We plot the base curve—the parabola $y^2 = x$ in the XY-plane (Fig. 97-2). Then we obtain the cylinder by sliding this parabola parallel to the Z-axis.

Another simple type of surface is formed by rotating a plane curve about a line. The equation of such a **surface of revolution** is easy to recognize if the axis of rotation is one of the coordinate axes. For example, the surface in Fig. 97-3 is obtained by rotating a plane curve about the Z-axis. Planes parallel to the XY-plane intersect this surface in circles. Suppose that such a plane is z units from the XY-plane and let $r(z)$ be the radius of the circle of intersection. If (x, y, z) is a point of our surface, we therefore see that

(97-1) $$x^2 + y^2 = r(z)^2.$$

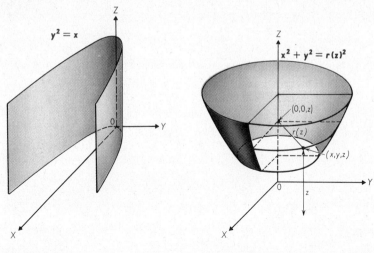

Figure 97-2   **Figure 97-3**

In other words, the equation of a surface of revolution about the Z-axis contains x and y in the combination $x^2 + y^2$.

If we set $y = 0$, Equation 97-1 reduces to $x^2 = r(z)^2$. This equation is thus the equation of the plane curve formed by the intersection of our surface with the XZ-plane. Similarly, the plane curve of intersection with the YZ-plane is obtained by setting $x = 0$; that is, $y^2 = r(z)^2$. It is therefore apparent that *if the curve $x = r(z)$ in the XZ-plane (or $y = r(z)$ in the YZ-plane) is rotated about the Z-axis, the resulting surface is the graph of Equation 97-1.*

Of course, for surfaces of revolution about other axes, we must make the obvious interchanges of letters. For example, if the curve $y = r(x)$ in the XY-plane is rotated about the X-axis, we square both sides and replace y^2 with $y^2 + z^2$ to obtain $y^2 + z^2 = r(x)^2$.

Example 97-2. Describe the surface $y = x^2 + z^2$.

Solution. We can think of the given equation as $x^2 + z^2 = r(y)^2$, with $r(y) = \sqrt{y}$, and so its graph is a surface of revolution about the Y-axis. We obtain the generating curve in the XY-plane simply by replacing z with 0 in our original equation, thereby obtaining the parabola $y = x^2$. Our surface, which may also be obtained by rotating the parabola $y = z^2$ about the Y-axis, is shown in Fig. 97-4.

Example 97-3. Describe the surface $z^2 = x^2 + y^2$.

Solution. The graph of this equation is a surface of revolution whose axis is the Z-axis. By setting $x = 0$, we obtain the equation of intersection of the surface with the YZ-plane, the curve $z^2 = y^2$. This "curve" consists of two lines in the YZ-plane—one bisecting the first quadrant and one bisecting the second quadrant. When we rotate these lines about the Z-axis, we obtain the cone shown in Fig. 97-5.

603

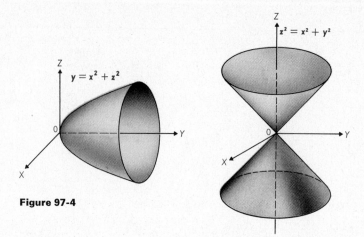

Figure 97-4

Figure 97-5

Example 97-4. Find the equation of the surface that is generated by rotating the circle $(y - 2)^2 + z^2 = 1$ about the Y-axis. What if we rotate it about the Z-axis?

Solution. In the first case, we simply replace z^2 with $x^2 + z^2$ to obtain the equation $(y - 2)^2 + x^2 + z^2 = 1$. The graph of this equation is the sphere shown in Fig. 97-6.

It is harder to find the equation of the surface of revolution about the Z-axis. According to the procedure outlined above, we should write the equation of the plane curve we are to rotate in the form $y^2 = r(z)^2$ and then replace y^2 with $x^2 + y^2$. When we solve the equation of our given circle for y, we get two solutions:

$$y = 2 + \sqrt{1 - z^2} \quad \text{and} \quad y = 2 - \sqrt{1 - z^2}.$$

Our surface of revolution is the union of the surfaces that are generated by rotating these curves about the Z-axis. The equations of these surfaces are

$$x^2 + y^2 = (2 + \sqrt{1 - z^2})^2 \quad \text{and} \quad x^2 + y^2 = (2 - \sqrt{1 - z^2})^2;$$

that is,

$$x^2 + y^2 + z^2 - 5 = 4\sqrt{1 - z^2} \quad \text{and} \quad x^2 + y^2 + z^2 - 5 = -4\sqrt{1 - z^2}.$$

A point belongs to the union of the graphs of these two equations if its coordinates satisfy one or the other of them, and this statement is equivalent to the statement that the coordinates satisfy the equation

$$(x^2 + y^2 + z^2 - 5)^2 = 16(1 - z^2).$$

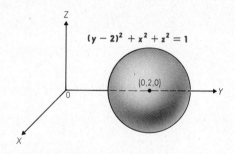

$(y - 2)^2 + x^2 + z^2 = 1$

$(0,2,0)$

Figure 97-6

We take this last equation to be the equation of the desired surface, and we leave it to you to sketch it. Such a doughnut-shaped surface is called a **torus**. These surfaces have certain interesting properties that make them play a more prominent role in mathematics than you might expect.

PROBLEMS 97

1. Sketch the cylinder.

    (a) $3y + 2z = 6$ (b) $yz = 6$ (c) $z = |x|$

    (d) $|z| = |x|$ (e) $z = 4 - y^2$ (f) $\dfrac{x^2}{4} + \dfrac{z^2}{9} = 1$

    (g) $z = \text{Sin}^{-1} y$ (h) $\text{Tan}\,(y + z) = 1$

2. Find the surfaces of revolution that are generated by rotating the base curves of the preceding problem about the coordinate axes of the planes in which they lie. (There will be two surfaces for each curve.)

3. Sketch the surface.

    (a) $4x^2 - y^2 - z^2 = 0$ (b) $4x^4 - y^2 - z^2 = 0$
    (c) $4x - y^2 - z^2 = 0$ (d) $4x^2 - y^2 - z^2 = 4$
    (e) $x^2 + z^2 = \sin^2 \pi y$ (f) $y^2 = \sin^2 \pi \sqrt{x^2 + z^2}$
    (g) $x^2 + z^2 = e^{2y}$ (h) $z^2 = \exp(-x^2 - y^2)$

4. Describe the surface $9(x - 1)^2 + 9(y + 1)^2 + 4(z - 3)^2 = 36$.

5. Find an equation of the torus (Example 97-4) that is obtained by rotating a circle of radius 3 about a line that is 5 units from its center.

6. (a) Show that the "cylinder" $(x + 1)^2 + (y - 2)^2 = 0$ is simply a line.
    (b) What is the set of points that are 5 units from this line?
    (c) What is the set of points that are no farther than 5 units from this line?

7. Describe the set of points that are equidistant from the Y-axis and the plane $x = 4$.

8. The cardioid in the XY-plane whose polar equation is $r = 2(1 - \cos \theta)$ is rotated about the X-axis. Find the equation of the resulting surface.

9. (a) Use the technique we discussed in Section 40 to find the volume of the region bounded by the surface of revolution $9x^2 + 9y^2 + 25z^2 = 225$.
    (b) Use the techniques we discussed in Section 73 to find the area of this surface.

10. Find the set of points, the sum of whose distances from $(0, 0, -3)$ and $(0, 0, 3)$ is 10.

11. The cylinders $x^2 + z^2 = 25$ and $y^2 + z^2 = 25$ intersect in a plane curve. Find it.

12. Show that if we rotate the curve $z = f(y)$ in the YZ-plane about the line $z = a$, then the equation of the resulting surface is $(z - a)^2 + x^2 = [f(y) - a]^2$.

13. Show that a surface that is generated by rotating a curve in the YZ-plane about the line $z = (\tan \alpha)y$ has an equation of the form

$$(y \sin \alpha - z \cos \alpha)^2 + x^2 = [r(y \cos \alpha + z \sin \alpha)]^2.$$

98. QUADRIC SURFACES

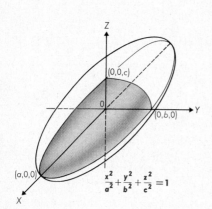

Figure 98-1

In our study of plane analytic geometry we considered the graphs of quadratic equations—that is, equations of the form $Ax^2 + Bxy + Cy^2 + Dx + Ey + F = 0$. We found that by suitably translating and rotating axes, we could reduce our study to equations in certain "standard" forms, and we are able to recognize the standard forms of the equations of ellipses, hyperbolas, and so on. Now we will briefly discuss the standard forms for analogous figures in space. These figures, the graphs of quadratic equations in x, y, and z, are called **quadric surfaces**. In Sections 103 and 104 we will consider the problem of rotating and translating the axes so as to put general quadratic equations in standard form.

The graph of an equation of the form

(98-1)
$$\frac{x^2}{a^2} + \frac{y^2}{b^2} + \frac{z^2}{c^2} = 1$$

is called an **ellipsoid**. We obtain the curves in which this figure intersects the coordinate planes by setting x, y, and z equal to 0, one at a time. These curves of intersection are ellipses. In fact, it is not hard to see that every plane that is parallel to one of the coordinate planes, and that intersects our ellipsoid, intersects it in an ellipse. Figure 98-1 shows the graph of Equation 98-1. You will notice that if $a = b = c$, the graph is just the sphere of radius a whose center is the origin.

The graph of Equation 98-1 intersects all three coordinate planes in ellipses. The surface

(98-2)
$$\frac{x^2}{a^2} + \frac{y^2}{b^2} - \frac{z^2}{c^2} = 1$$

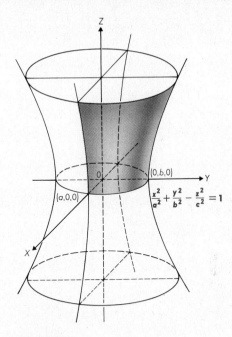

$$\frac{x^2}{a^2} + \frac{y^2}{b^2} - \frac{z^2}{c^2} = 1$$

Figure 98-2

intersects the XY-plane in an ellipse, but it intersects the other two coordinate planes in hyperbolas. We call this surface a **hyperboloid of one sheet** and it is shown in Fig. 98-2. If x and z (or y and z) are interchanged in Equation 98-2, we again have a hyperboloid of one sheet, but its axis is the X-axis (or the Y-axis).

The graph of the equation

(98-3)
$$\frac{x^2}{a^2} - \frac{y^2}{b^2} - \frac{z^2}{c^2} = 1$$

intersects the XY- and XZ-planes in hyperbolas, but it doesn't intersect the YZ-plane at all [since no point with coordinates $(0, y, z)$ can satisfy Equation 98-3]. However, a plane $x = k$, where $k^2 > a^2$, intersects the surface in an ellipse. For if (k, y, z) is a point of such a plane, then y and z satisfy the equation

$$\frac{k^2}{a^2} - \frac{y^2}{b^2} - \frac{z^2}{c^2} = 1,$$

which can be written in the standard form of the equation of an ellipse:

$$\frac{y^2}{b^2(k^2 - a^2)/a^2} + \frac{z^2}{c^2(k^2 - a^2)/a^2} = 1.$$

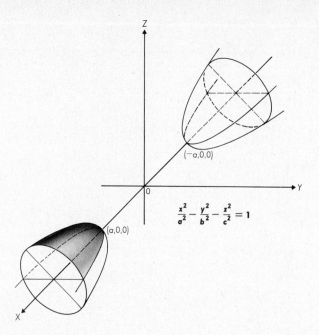

$$\frac{x^2}{a^2} - \frac{y^2}{b^2} - \frac{z^2}{c^2} = 1$$

Figure 98-3

The graph of Equation 98-3 is called a **hyperboloid of two sheets**, and Fig. 98-3 shows what such a surface looks like.

Corresponding to the parabola in two dimensions, we have the **elliptic paraboloid** in three dimensions. The graph of the equation

$$(98\text{-}4) \qquad\qquad z = \frac{x^2}{a^2} + \frac{y^2}{b^2}$$

is such a surface. We notice that the XZ-plane cuts the surface in the parabola $z = x^2/a^2$, and the YZ-plane cuts it in the parabola $z = y^2/b^2$. Each plane $z = k$, where $k > 0$, cuts the surface in an ellipse. In Fig. 98-4 we have drawn the graph of Equation 98-4.

If the elliptical cross sections of an ellipsoid, hyperboloid, or elliptical paraboloid are circles, the surface is merely a surface of revolution. (What relations among a, b, and c in the above equations guarantee that we have such a surface?) Our next surface does not resemble a surface of revolution in any way. The surface

$$(98\text{-}5) \qquad\qquad z = \frac{y^2}{b^2} - \frac{x^2}{a^2}$$

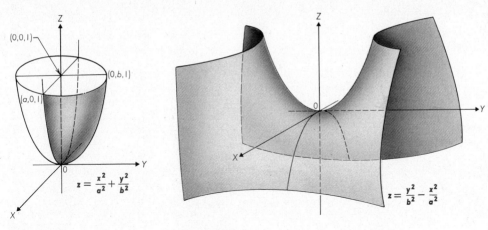

$$z = \frac{x^2}{a^2} + \frac{y^2}{b^2}$$

$$z = \frac{y^2}{b^2} - \frac{x^2}{a^2}$$

Figure 98-4   **Figure 98-5**

is a **hyperbolic paraboloid**. It intersects the YZ-plane in the parabola $z = y^2/b^2$ and the XZ-plane in the parabola $z = -x^2/a^2$. It intersects planes parallel to the XY-plane in hyperbolas, as shown in Fig. 98-5.

As a final quadratic surface, we consider the **elliptical cone**

(98-6)
$$\frac{x^2}{a^2} + \frac{y^2}{b^2} = \frac{z^2}{c^2}.$$

If $x = 0$, we see that either $z = cy/b$ or $z = -cy/b$. Thus the surface intersects the YZ-plane in the two lines represented by the last equations. Similarly, the surface intersects the XZ-plane in the lines $z = cx/a$ and $z = -cx/a$. Planes parallel to the XY-plane cut the surface in ellipses. Figure 98-6 shows the graph of Equation 98-6. If $a = b$, the surface is a right-circular cone.

Example 98-1. Discuss the surface

$$\frac{x^2}{25} + \frac{y^2}{16} - \frac{z|z|}{9} = 1.$$

Solution. If $z \le 0$, we can replace $|z|$ with $-z$, and so the part of our graph that lies below the XY-plane is the bottom half of the ellipsoid $x^2/25 + y^2/16 + z^2/9 = 1$ (Fig. 98-1). When $z > 0$, we replace $|z|$ with z, and then we have the upper half of the hyperboloid $x^2/25 + y^2/16 - z^2/9 = 1$ (Fig. 98-2). You might try to sketch part of our figure, say for $z \le 5$. It looks like a squashed badminton bird.

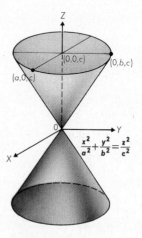

$$\frac{x^2}{a^2} + \frac{y^2}{b^2} = \frac{z^2}{c^2}$$

Figure 98-6

609

Example 98-2. The plane $z = 1$ intersects the hyperbolic paraboloid $8x = y^2 - 16z^2$ in a parabola. What is the focus of this parabola?

Solution. To obtain our parabola of intersection, we set $z = 1$ and we find that $8x = y^2 - 16$. The standard form of this parabola is $y^2 = 4 \cdot 2(x + 2)$, so we know that its focus is 2 units from the vertex. Since the vertex is the point $(-2, 0, 1)$, the X-coordinate of the focus is $-2 + 2 = 0$. The Y-coordinate is 0, and the Z-coordinate is, of course, 1. Therefore the point we were asked to find is $(0, 0, 1)$.

It was natural to introduce the equations of quadric surfaces by simply tacking on a "Z-term" to the equations of conics that we studied in Chapter 4. But soon we will find it convenient to write these equations in vector-matrix form, too. If $\mathbf{r} = (x, y, z)$ is a vector in R^3 and D is the matrix

$$D = \begin{bmatrix} \lambda_1 & 0 & 0 \\ 0 & \lambda_2 & 0 \\ 0 & 0 & \lambda_3 \end{bmatrix},$$

then $D\mathbf{r} = (\lambda_1 x, \lambda_2 y, \lambda_3 z)$. Now we dot this vector with $\mathbf{r} = (x, y, z)$ to obtain $(D\mathbf{r}) \cdot \mathbf{r}$. We customarily omit the parentheses when writing such a dot product:

$$D\mathbf{r} \cdot \mathbf{r} = \lambda_1 x^2 + \lambda_2 y^2 + \lambda_3 z^2.$$

Therefore Equations 98-1, 98-2, and 98-3 all have the form

(98-7) $$D\mathbf{r} \cdot \mathbf{r} = 1,$$

where $(\lambda_1, \lambda_2, \lambda_3) = (1/a^2, 1/b^2, 1/c^2)$ in Equation 98-1, $(\lambda_1, \lambda_2, \lambda_3) = (1/a^2, 1/b^2, -1/c^2)$ in Equation 98-2, and $(\lambda_1, \lambda_2, \lambda_3) = (1/a^2, -1/b^2, -1/c^2)$ in Equation 98-3. Equations 98-4 and 98-5 can be written as

$$D\mathbf{r} \cdot \mathbf{r} = \mathbf{r} \cdot \mathbf{k},$$

where $(\lambda_1, \lambda_2, \lambda_3) = (1/a^2, 1/b^2, 0)$ in the first case and $(\lambda_1, \lambda_2, \lambda_3) = (-1/a^2, 1/b^2, 0)$ in the second. Of course, we are thinking of $\mathbf{k}$ as the vector $(0, 0, 1)$. Equation 98-6 has the form

$$D\mathbf{r} \cdot \mathbf{r} = 0,$$

where $(\lambda_1, \lambda_2, \lambda_3) = (1/a^2, 1/b^2, -1/c^2)$.

Example 98-3. What does the graph of Equation 98-7 look like if

$$D = \begin{bmatrix} \frac{1}{4} & 0 & 0 \\ 0 & -1 & 0 \\ 0 & 0 & \frac{1}{9} \end{bmatrix} ?$$

Solution.   In component form, Equation 98-7 becomes

$$\frac{x^2}{4} - y^2 + \frac{z^2}{9} = 1,$$

which (see Equation 98-2) is a hyperboloid of one sheet whose "core" is the Y-axis.

PROBLEMS   98

1. Sketch the part of the surface that lies in the first octant.

(a) $\dfrac{x^2}{4} + y^2 + \dfrac{z^2}{9} = 1$

(b) $\dfrac{x^2}{4} + y^2 - \dfrac{z^2}{9} = 1$

(c) $\dfrac{x^2}{4} - y^2 - \dfrac{z^2}{9} = 1$

(d) $\dfrac{x^2}{4} + y^2 + \dfrac{z}{9} = 1$

2. Eight equations result when the $*$'s in the equation $*x^2 * 4y^2 * 9z^2 = 36$ are replaced by $+$ or $-$. Describe their graphs.

3. Use the results of the preceding problem to describe the following surfaces. (You will probably want to investigate each octant separately.)

(a) $x^2 + 4y^2 + 9z|z| = 36$

(b) $x|x| + 4y^2 + 9z^2 = 36$

(c) $x^2 + (|y| + y)^2 + 9z^2 = 36$

(d) $x^2 + 4y|y| + 9z|z| = 36$

4. Sketch the surface $4x = 4y^2 + z^2$ and use the result to describe the following surfaces.

(a) $-4x = 4y^2 + z^2$

(b) $4(x - 2) = 4y^2 + z^2$

(c) $4|x| = 4y^2 + z^2$

(d) $2(|x| + x) = 4y^2 + z^2$

5. For the given matrix D, describe the surface $D\mathbf{r} \cdot \mathbf{r} = 1$ [where $\mathbf{r} = (x, y, z)$].

(a) $D = \begin{bmatrix} \frac{1}{4} & 0 & 0 \\ 0 & 0 & 0 \\ 0 & 0 & 1 \end{bmatrix}$

(b) $D = \begin{bmatrix} \frac{1}{4} & 0 & 0 \\ 0 & -\frac{1}{9} & 0 \\ 0 & 0 & \frac{1}{25} \end{bmatrix}$

(c) $D = \begin{bmatrix} \frac{1}{9} & 0 & 0 \\ 0 & \frac{1}{9} & 0 \\ 0 & 0 & \frac{1}{4} \end{bmatrix}$

(d) $D = \begin{bmatrix} -1 & 0 & 0 \\ 0 & -\frac{1}{4} & 0 \\ 0 & 0 & 1 \end{bmatrix}$

6. When the hyperbolic paraboloid $z = x^2/4 - y^2/9$ is intersected by the following planes, we obtain familiar plane curves. Name these curves, find the coordinates of their vertices, and use the resulting information to sketch the hyperbolic paraboloid.

(a) the YZ-plane

(b) the XY-plane

(c) the plane parallel to the XZ-plane and 3 units to its right

(d) the plane parallel to the XY-plane and 1 unit below it

(e) the plane parallel to the YZ-plane and 4 units in front of it

7. Find numbers A, B, C, and D such that the quadric surface $Ax^2 + By^2 + Cz^2 + D = 0$ contains the given points. Name the surface.

(a) $(1, 1, -1)$, $(2, 1, 0)$, $(5, -5, 3)$

(b) $(2, -1, 1)$, $(-3, 0, 0)$, $(1, -1, -2)$

(c) $(1, 2, -1)$, $(0, 1, 0)$, $(3, 1, 2)$

(d) $(-1, 2, 2)$, $(2, -1, -2)$, $(0, 0, 3)$

8. Find the center and radius of the sphere.
   (a) $x^2 + y^2 + z^2 - 2x + 2y - 2 = 0$
   (b) $x^2 + y^2 + z^2 - 9x - 2y + 4z - 11 = 0$

9. Find the plane that is tangent to the given sphere at the given point.
   (a) $x^2 + y^2 + z^2 = 9$, $(1, 2, 2)$
   (b) $x^2 + y^2 + z^2 = 9$, $(2, -1, -2)$
   (c) $x^2 + y^2 + z^2 - 10x + 4y - 6z - 187 = 0$, $(-9, 3, 1)$
   (d) $x^2 + y^2 + z^2 - 2x + 4y + 8z - 11 = 0$, $(1, 2, 0)$

10. Discuss the graphs of the following equations. (First complete the square.)
    (a) $x^2 + 4y^2 + 9z^2 - 2x - 16y - 19 = 0$
    (b) $x^2 + 4y^2 - 9z^2 - 2x - 16y - 19 = 0$
    (c) $x^2 - 4y^2 + 9z^2 - 2x + 16y - 19 = 0$
    (d) $x^2 - 4y^2 - 9z^2 - 2x + 16y - 19 = 0$

11. Find the equation of the surface that contains the point (x, y, z) and describe it if
    (a) the square of the distance from the point to the Y-axis is $\frac{2}{5}$ the distance from the point to the XZ-plane.
    (b) the distance from the point to the origin is $\frac{3}{2}$ its distance from the Z-axis.

12. Let the point (x, y, z) be at a distance r_1 from the point $(0, 2, 0)$ and at a distance r_2 from the plane $y = -2$. Find the equation of the surface that contains (x, y, z) and describe it if r_1 and r_2 are related as follows.
    (a) $r_1 = r_2$ (b) $r_1 = 2r_2$ (c) $r_2 = 2r_1$

13. Prove that a point, the sum of whose distances from two given points is a given number, belongs to an ellipsoid of revolution.

14. Prove that a point, the difference of whose distances from two given points is a given number, belongs to a hyperboloid of revolution.

99. SPACE CURVES

We think of a curve as a one-dimensional set of points, the image of a line segment. So we may describe curves in space just as we described them in the plane, as graphs of vector equations. If $\mathbf{f}$ is a vector-valued function whose domain contains a segment $[a, b]$ of the T-axis, then the set of terminal points of position vectors of the form $\mathbf{f}(t)$ for $t \in [a, b]$ constitutes the image of $[a, b]$ under $\mathbf{f}$. This arc is the graph of the vector equation $\mathbf{r} = \mathbf{f}(t)$, where we are thinking of $\mathbf{r}$ as the position vector $x\mathbf{i} + y\mathbf{j} + z\mathbf{k}$. If $\mathbf{f}(t) = f_x(t)\mathbf{i} + f_y(t)\mathbf{j} + f_z(t)\mathbf{k}$, the vector equation $\mathbf{r} = \mathbf{f}(t)$ is equivalent to the set of parametric equations

$$x = f_x(t), \qquad y = f_y(t), \qquad z = f_z(t).$$

Choosing t in the base interval $[a, b]$ gives us the point $(x, y, z) = (f_x(t), f_y(t), f_z(t))$ of our curve in space.

Example 99-1. Describe the curve $\mathbf{r} = 3 \cos t\mathbf{i} + 4 \sin t\mathbf{j} + 4t\mathbf{k}$.

Solution. The vector **r** in Fig. 99-1 is a typical position vector to a point of our curve. The table of values lists four specific position vectors. When we plot their terminal points and join them, we obtain the illustrated **helix**. Parametric equations of the helix are

$$x = 3 \cos t, \qquad y = 4 \sin t, \qquad z = 4t.$$

From these equations we see that if (x, y, z) is a point of our helix, then

$$\frac{x^2}{9} + \frac{y^2}{16} = \cos^2 t + \sin^2 t = 1.$$

Therefore the helix lies in an elliptical cylinder whose generators are parallel to the Z-axis. It forms a "barber's pole stripe" in this cylinder. If we were to replace $4 \sin t$ with $3 \sin t$, our helix would lie in the right-circular cylinder $x^2 + y^2 = 9$, and it would then be a *circular helix*.

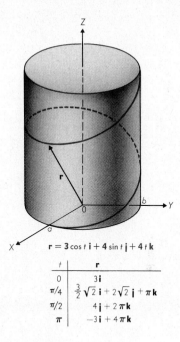

$$\mathbf{r} = 3 \cos t\,\mathbf{i} + 4 \sin t\,\mathbf{j} + 4t\,\mathbf{k}$$

t	$\mathbf{r}$
0	$3\mathbf{i}$
$\pi/4$	$\frac{3}{2}\sqrt{2}\,\mathbf{i} + 2\sqrt{2}\,\mathbf{j} + \pi\mathbf{k}$
$\pi/2$	$4\mathbf{j} + 2\pi\mathbf{k}$
π	$-3\mathbf{i} + 4\pi\mathbf{k}$

Figure 99-1

Differentiation of vector-valued functions that take values in three-dimensional space follows the pattern established for the two-dimensional case in Section 71. In particular,

$$\mathbf{f}'(t) = f_x'(t)\mathbf{i} + f_y'(t)\mathbf{j} + f_z'(t)\mathbf{k},$$

so that

(99-1) $$\frac{d}{dt}(3 \cos t\mathbf{i} + 4 \sin t\mathbf{j} + 4t\mathbf{k}) = -3 \sin t\mathbf{i} + 4 \cos t\mathbf{j} + 4\mathbf{k}.$$

The usual rules concerning the derivatives of sums and products, plus the Chain Rule, are valid. For example, we can differentiate the cross product and the scalar triple product in the "natural way":

$$\frac{d}{dt}(\mathbf{f}(t) \times \mathbf{g}(t)) = \mathbf{f}(t) \times \mathbf{g}'(t) + \mathbf{f}'(t) \times \mathbf{g}(t),$$

$$\frac{d}{dt}\left[\mathbf{f}(t) \cdot (\mathbf{g}(t) \times \mathbf{h}(t))\right]$$
$$= \mathbf{f}'(t) \cdot (\mathbf{g}(t) \times \mathbf{h}(t)) + \mathbf{f}(t) \cdot (\mathbf{g}'(t) \times \mathbf{h}(t)) + \mathbf{f}(t) \cdot (\mathbf{g}(t) \times \mathbf{h}'(t)).$$

Example 99-2. Show that

$$\frac{d}{dt}|\mathbf{f}(t)| = \frac{1}{|\mathbf{f}(t)|}\mathbf{f}(t) \cdot \mathbf{f}'(t).$$

Solution. Recall that $|\mathbf{f}(t)|^2 = \mathbf{f}(t) \cdot \mathbf{f}(t)$, and so

(99-2) $$\frac{d}{dt}|\mathbf{f}(t)|^2 = \frac{d}{dt}(\mathbf{f}(t) \cdot \mathbf{f}(t)).$$

We find the first of these derivatives by replacing u with $|\mathbf{f}(t)|$ in the differentiation formula $\dfrac{d}{dt}u^2 = 2u\dfrac{du}{dt}$:

$$\frac{d}{dt}|\mathbf{f}(t)|^2 = 2|\mathbf{f}(t)|\frac{d}{dt}|\mathbf{f}(t)|.$$

The Product Rule tells us that $\dfrac{d}{dt}(\mathbf{f}(t) \cdot \mathbf{f}(t)) = \mathbf{f}(t) \cdot \mathbf{f}'(t) + \mathbf{f}'(t) \cdot \mathbf{f}(t) = 2\mathbf{f}(t) \cdot \mathbf{f}'(t).$ Therefore, Equation 99-2 can be written as

$$2|\mathbf{f}(t)|\frac{d}{dt}|\mathbf{f}(t)| = 2\mathbf{f}(t) \cdot \mathbf{f}'(t),$$

which is essentially the equation we were to verify.

By the same geometric reasoning that we used in the plane situation, we conclude that *the vector $\mathbf{f}'(t)$ is tangent to the curve $\mathbf{r} = \mathbf{f}(t)$ at the terminal point of $\mathbf{f}(t)$.*

Example 99-3. A bug runs along the helix shown in Fig. 99-1, and at the point $P(\frac{3}{2}\sqrt{2}, 2\sqrt{2}, \pi)$ it flies off on a tangent. Does it land in an observer's eye at the point $E(-3\sqrt{2}, 8\sqrt{2}, 12 + \pi)$?

Solution. According to the table, the bug flies off at the point that corresponds to $t = \frac{1}{4}\pi$. At this point, a tangent vector is (Equation 99-1)

$$\mathbf{r}' = -3\sin\tfrac{1}{4}\pi\mathbf{i} + 4\cos\tfrac{1}{4}\pi\mathbf{j} + 4\mathbf{k} = -\tfrac{3}{2}\sqrt{2}\,\mathbf{i} + 2\sqrt{2}\,\mathbf{j} + 4\mathbf{k}.$$

We are asking whether or not this vector is parallel to the vector

$$\mathbf{PE} = -\tfrac{9}{2}\sqrt{2}\,\mathbf{i} + 6\sqrt{2}\,\mathbf{j} + 12\mathbf{k}.$$

But $\mathbf{PE}$ is obviously just $3\mathbf{r}'$, so these vectors are parallel, and the observer gets a bug in the eye.

As in the plane case, we will confine our study to *smooth arcs*, graphs of functions with continuous second derived functions and for which $|\mathbf{f}'(t)| > 0$ for each t in the basic interval. In Section 71 we explained why *the part of the smooth arc* $\mathbf{r} = \mathbf{f}(t)$ *that joins the terminal points of* $\mathbf{f}(a)$ *and* $\mathbf{f}(b)$ *is*

$$(99\text{-}3) \qquad L = \int_a^b |\mathbf{f}'(t)|\, dt = \int_a^b \sqrt{x'^2 + y'^2 + z'^2}\, dt$$

units long.

Example 99-4. The portion of the circular helix $\mathbf{r} = \cos t\mathbf{i} + \sin t\mathbf{j} + t\mathbf{k}$ that corresponds to the *T*-interval $[0, 2\pi]$ makes one complete circuit about the *Z*-axis. How long is this arc?

Solution. Here $\mathbf{f}'(t) = -\sin t\mathbf{i} + \cos t\mathbf{j} + \mathbf{k}$. Hence

$$|\mathbf{f}'(t)| = \sqrt{\sin^2 t + \cos^2 t + 1} = \sqrt{2},$$

and Equation 99-3 gives us

$$L = \int_0^{2\pi} \sqrt{2}\, dt = 2\pi\sqrt{2} \approx 8.9.$$

If we use our vector-valued function $\mathbf{f}$ to describe the position of a point moving in space, then the *number* $|\mathbf{f}'(t)|$ represents its **speed**, and the *vector* $\mathbf{v} = \mathbf{f}'(t)$ is its **velocity**. The vector $\mathbf{a} = \mathbf{f}''(t)$ is the point's **acceleration**.

Example 99-5. If distances are measured in feet and time in seconds, how fast did the bug in Example 99-3 hit the observer in the eye?

Solution. We will suppose that our bug ran along the helix picking up speed until it took off at the end of $\tfrac{1}{4}\pi$ seconds $(t = \tfrac{1}{4}\pi)$. Its speed at that time was

$$|\mathbf{r}'| = \sqrt{\tfrac{9}{4}\cdot 2 + 4\cdot 2 + 16} = \sqrt{\tfrac{57}{2}} \text{ feet per second.}$$

If the bug just coasted from then on, it hit the observer at this speed, about 5.3 feet per second.

PROBLEMS 99

1. Find a unit tangent vector at the point corresponding to $t = \tfrac{1}{2}\pi$. Also find the length of the arc corresponding to the *T*-interval $[0, \pi]$.

(a) $\mathbf{r} = (2t + 1)\mathbf{i} - t\mathbf{j} + (2t + 3)\mathbf{k}$

(b) $\mathbf{r} = (5 - 2t)\mathbf{i} - (2 + t)\mathbf{j} - (3 - 2t)\mathbf{k}$

(c) $\mathbf{r} = \cos 3t\mathbf{i} + \sin 3t\mathbf{j} + 4t\mathbf{k}$

(d) $\mathbf{r} = 3 \cos t\mathbf{i} + 3 \sin t\mathbf{j} + 4t\mathbf{k}$

(e) $\mathbf{r} = e^t \cos t\mathbf{i} + e^t \sin t\mathbf{j} + e^t\mathbf{k}$

(f) $\mathbf{r} = \cos 2t\mathbf{i} + \sin 2t\mathbf{j} + 4t\mathbf{k}$

(g) $\mathbf{r} = 3 \cos t\mathbf{i} + 3 \sin t\mathbf{j} + \cosh 3t\mathbf{k}$

(h) $\mathbf{r} = \cos \frac{1}{4}t\mathbf{i} + \sin \frac{1}{4}t\mathbf{j} + \ln (\cos \frac{1}{4}t)\mathbf{k}$

2. Find a unit vector that is tangent to the curve $\mathbf{r} = (t - 1)\mathbf{i} + (t^2 + t)\mathbf{j} + t^4\mathbf{k}$ at the point where it intersects the YZ-plane.

3. The graphs of the equations $y = \sin x$ and $z = \cos x$ in three-dimensional space are cylinders. Write a vector equation of the curve of intersection of these cylinders.

4. (a) At what point(s) of the curve $\mathbf{r} = (t^2 - 2t + 1)\mathbf{i} + (t + 3)\mathbf{j} - (3t + 4)\mathbf{k}$ is the tangent line perpendicular to the vector $\mathbf{a} = \mathbf{i} - \mathbf{j} + 3\mathbf{k}$?

   (b) At what point(s) is the tangent line parallel to $\mathbf{a}$?

5. The helix $\mathbf{r} = \cos t\mathbf{i} + \sin t\mathbf{j} + t\mathbf{k}$ intersects the curve $\mathbf{r} = (1 - 2t)\mathbf{i} - (t^2 - 2t)\mathbf{j} + (t^3 + t)\mathbf{k}$ in the point $(1, 0, 0)$. What is the angle between the tangents at this point?

6. How long is the arc $\mathbf{r} = t\mathbf{i} + \frac{1}{2}\sqrt{2}\, t^2\mathbf{j} + \frac{1}{3}t^3\mathbf{k}$ for t in the interval $[0, 2]$?

7. The acceleration vector of a certain moving point t seconds after it starts is $\mathbf{a} = 6t\mathbf{i} + \cos t\mathbf{j} + e^t\mathbf{k}$.

   (a) Express the velocity vector $\mathbf{v}$ in terms of t if you know that $\mathbf{v} = \mathbf{k}$ when $t = 0$.

   (b) Express the position vector $\mathbf{r}$ in terms of t if you know that $\mathbf{r} = \mathbf{i} - \mathbf{j} + 2\mathbf{k}$ when $t = 0$.

8. Let $\mathbf{c}$ be a given vector, independent of t; for example, $\mathbf{c} = 3\mathbf{i} + 4\mathbf{j} + \mathbf{k}$. Describe the relation between the curves $\mathbf{r} = \mathbf{f}(t)$ and $\mathbf{r} = \mathbf{f}(t) + \mathbf{c}$.

9. The position vector of a particle moving in space is $\mathbf{r} = \cos t\mathbf{i} + \sin t\mathbf{j} + t\mathbf{k}$, where we measure time in seconds and distance in feet. When $t = \frac{1}{4}\pi$, we release the particle and it flies off on a tangent. Where is it when $t = \frac{1}{2}\pi$?

10. Solve the initial value problem $\mathbf{x}' = 2\mathbf{x}$ and $\mathbf{x} = \mathbf{i} - \mathbf{j} + \mathbf{k}$ when $t = 0$. (*Hint*: Make the substitution $\mathbf{x} = e^{2t}\mathbf{z}$ and solve for $\mathbf{z}$.)

11. Suppose that $\mathbf{x} = \mathbf{u}(t)$ satisfies the vector differential equation $\mathbf{x}' = \mathbf{a}(t) \times \mathbf{x}$, where $\mathbf{a}$ is some given vector-valued function. Show that $\dfrac{d}{dt}\, |\mathbf{u}(t)|^2 = 0$—that is, $\mathbf{u}(t)$ has constant length.

12. Find.

   (a) $\dfrac{d}{dt}\, [\mathbf{f}(t) \times \mathbf{f}'(t)]$

   (b) $\dfrac{d}{dt}\, [\mathbf{f}(t) \cdot \mathbf{f}'(t)]$

   (c) $\dfrac{d}{dt}\, [\mathbf{f}(t) \cdot (\mathbf{f}'(t) \times \mathbf{f}''(t))]$

   (d) $\dfrac{d}{dt}\, [(\mathbf{f}(t) \times \mathbf{f}'(t)) \cdot \mathbf{f}'(t)]$

13. We say that the **projection** of the curve $\mathbf{r} = \mathbf{f}(t)$ onto the XY-plane is the curve $\mathbf{r} = f_x(t)\mathbf{i} + f_y(t)\mathbf{j}$.

   (a) What is the projection of the curve onto the YZ-plane?

   (b) Show that the projection of an arc onto a coordinate plane is no longer than the given arc.

   (c) Is a tangent line of a projection a projection of a tangent line?

14. The Theorem of the Mean says that at some point of a smooth plane arc, the tangent line is parallel to the chord that joins the endpoints. Show by means of an example that this theorem is *not true* in space.

As in the plane (Section 70), a line in space can be regarded as the graph of a vector equation

(100-1) $\mathbf{r} = \mathbf{m}t + \mathbf{b}$, $t \in (-\infty, \infty)$ and $\mathbf{m} \neq \mathbf{0}$.

Figure 100-1 illustrates how the position vector $\mathbf{r} = x\mathbf{i} + y\mathbf{j} + z\mathbf{k}$ is the sum of a given position vector $\mathbf{b}$ and a multiple $\mathbf{m}t$ of a "direction vector" $\mathbf{m}$.

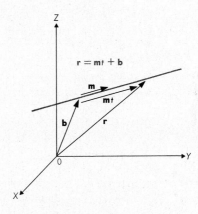

Figure 100-1

Example 100-1.   Find an equation of the line that contains the points $P(-1, 2, 3)$ and $Q(2, 1, -3)$.

Solution.   Our answer will be an equation $\mathbf{r} = \mathbf{m}t + \mathbf{b}$, so the problem amounts to finding the direction vector $\mathbf{m}$ and the position vector $\mathbf{b}$. Of course, $\mathbf{PQ}$ points along our line, so we can take $\mathbf{m} = \mathbf{PQ} = 3\mathbf{i} - \mathbf{j} - 6\mathbf{k}$. Since P belongs to the line, the position vector $\mathbf{OP} = -\mathbf{i} + 2\mathbf{j} + 3\mathbf{k}$ can serve as $\mathbf{b}$. Thus an equation of our line is

$$\mathbf{r} = (3\mathbf{i} - \mathbf{j} - 6\mathbf{k})t + (-\mathbf{i} + 2\mathbf{j} + 3\mathbf{k}) = (3t - 1)\mathbf{i} - (t - 2)\mathbf{j} - (6t - 3)\mathbf{k}.$$

We can choose $\mathbf{b}$ to be a position vector to *any* point of the line, so we could also have chosen $\mathbf{b} = \mathbf{OQ} = 2\mathbf{i} + \mathbf{j} - 3\mathbf{k}$. Then we would have ended up with

$$\mathbf{r} = (3t + 2)\mathbf{i} - (t - 1)\mathbf{j} - (6t + 3)\mathbf{k}.$$

Although our two vector equations are different, the graph of each is the line that contains the points P and Q, and hence either equation will serve as a solution of our problem.

Example 100-2. Find an equation of the line that contains the point $(2, -1, 3)$ and meets the plane $x + 2y - 4z + 12 = 0$ at right angles.

Solution. Our equation will have the form $\mathbf{r} = \mathbf{m}t + \mathbf{b}$, where $\mathbf{b}$ is the position vector to a point of the given line and $\mathbf{m}$ determines its direction. Since $(2, -1, 3)$ belongs to our line, we can take

$$\mathbf{b} = 2\mathbf{i} - \mathbf{j} + 3\mathbf{k}.$$

The direction of the line is that of a normal to the given plane, and we know that the components of a normal are the coefficients of x, y, and z. Therefore we can take $\mathbf{i} + 2\mathbf{j} - 4\mathbf{k}$ to be our direction vector $\mathbf{m}$, and an equation of our line is

$$\mathbf{r} = (\mathbf{i} + 2\mathbf{j} - 4\mathbf{k})t + (2\mathbf{i} - \mathbf{j} + 3\mathbf{k}) = (t + 2)\mathbf{i} + (2t - 1)\mathbf{j} - (4t - 3)\mathbf{k}.$$

Example 100-3. Under what conditions do two lines $\mathbf{r} = \mathbf{m}t + \mathbf{b}$ and $\mathbf{r} = \mathbf{n}t + \mathbf{c}$ intersect? When do these two equations represent the same line?

Solution. The condition that the two lines have a common point is that there be numbers u and v such that $\mathbf{m}u + \mathbf{b} = \mathbf{n}v + \mathbf{c}$; in other words,

(100-2) $$\mathbf{b} - \mathbf{c} = \mathbf{n}v - \mathbf{m}u.$$

Two intersecting lines coincide if and only if they have the same direction—that is, if $\mathbf{m}$ and $\mathbf{n}$ are parallel. Then there is a number k such that $\mathbf{n} = k\mathbf{m}$, and Equation 100-2 becomes $\mathbf{b} - \mathbf{c} = (kv - u)\mathbf{m}$, which tells us that $\mathbf{b} - \mathbf{c}$ is also a multiple of $\mathbf{m}$. Therefore our equations represent the same line if the vectors $\mathbf{b} - \mathbf{c}$, $\mathbf{m}$, and $\mathbf{n}$ are all parallel.

If our intersecting lines are not parallel, Equation 100-2 is equivalent to saying that $\mathbf{b} - \mathbf{c}$ belongs to the plane determined by $\mathbf{m}$ and $\mathbf{n}$, and this geometric statement can be expressed as the equation $(\mathbf{b} - \mathbf{c}) \cdot (\mathbf{m} \times \mathbf{n}) = 0$.

If $\mathbf{m} = l\mathbf{i} + m\mathbf{j} + n\mathbf{k}$ and $\mathbf{b} = a\mathbf{i} + b\mathbf{j} + c\mathbf{k}$, then the vector equation $\mathbf{r} = \mathbf{m}t + \mathbf{b}$ becomes

$$x\mathbf{i} + y\mathbf{j} + z\mathbf{k} = (lt + a)\mathbf{i} + (mt + b)\mathbf{j} + (nt + c)\mathbf{k}.$$

Thus

(100-3) $$x = lt + a, \quad y = mt + b, \quad z = nt + c, \quad t \in (-\infty, \infty),$$

are parametric equations of our line. In case the numbers l, m, and n are not 0, we can solve for t:

$$t = \frac{x - a}{l}, \quad t = \frac{y - b}{m}, \quad \text{and} \quad t = \frac{z - c}{n}.$$

You will frequently see these equations of the line that has the **direction numbers** *l*, *m*, and *n* and that contains the point (a, b, c) combined and written as

(100-4) $$\frac{x - a}{l} = \frac{y - b}{m} = \frac{z - c}{n}.$$

These equations can be interpreted as saying that our line is the line of intersection of any pair of the planes

$$\frac{x - a}{l} = \frac{y - b}{m}, \qquad \frac{x - a}{l} = \frac{z - c}{n}, \quad \text{and} \quad \frac{y - b}{m} = \frac{z - c}{n}.$$

Suppose that the direction vector **m** of our line is a unit vector that makes angles α, β, and γ with **i**, **j**, and **k** (Fig. 100-2). Then the components of **m** are the numbers

$$m_x = \mathbf{m} \cdot \mathbf{i} = \cos \alpha, \qquad m_y = \mathbf{m} \cdot \mathbf{j} = \cos \beta, \quad \text{and} \quad m_z = \mathbf{m} \cdot \mathbf{k} = \cos \gamma.$$

Since these numbers give its direction, they are called **direction cosines** of the line $\mathbf{r} = \mathbf{m}t + \mathbf{b}$. We are assuming here that **m** is a unit vector ($m_x^2 + m_y^2 + m_z^2 = 1$), and so the direction cosines of a line satisfy the equation

$$\cos^2 \alpha + \cos^2 \beta + \cos^2 \gamma = 1.$$

In addition to Equations 100-1, 100-3, and 100-4, a line in space can be specified as the intersection of two planes.

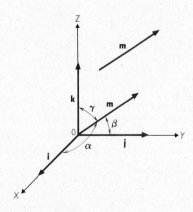

Figure 100-2

Example 100-4. Find direction cosines of the line of intersection of the planes $2x - 3y + z = 4$ and $x + 4y - 2z = 6$.

Solution. The vector $\mathbf{n}_1 = 2\mathbf{i} - 3\mathbf{j} + \mathbf{k}$ is normal to the plane $2x - 3y + z = 4$, and $\mathbf{n}_2 = \mathbf{i} + 4\mathbf{j} - 2\mathbf{k}$ is normal to $x + 4y - 2z = 6$. The vector $(\mathbf{n}_1 \times \mathbf{n}_2)/|\mathbf{n}_1 \times \mathbf{n}_2|$ is a unit vector along the line of intersection of the two planes (why?). We find that

$$\mathbf{n}_1 \times \mathbf{n}_2 = \begin{vmatrix} -3 & 1 \\ 4 & -2 \end{vmatrix} \mathbf{i} - \begin{vmatrix} 2 & 1 \\ 1 & -2 \end{vmatrix} \mathbf{j} + \begin{vmatrix} 2 & -3 \\ 1 & 4 \end{vmatrix} \mathbf{k} = 2\mathbf{i} + 5\mathbf{j} + 11\mathbf{k}.$$

Thus

$$|\mathbf{n}_1 \times \mathbf{n}_2| = \sqrt{150} = 5\sqrt{6},$$

and so

$$\frac{2}{5\sqrt{6}}\mathbf{i} + \frac{1}{\sqrt{6}}\mathbf{j} + \frac{11}{5\sqrt{6}}\mathbf{k}$$

is a unit vector along our line. The direction cosines of the line are simply the components of this vector, so to four decimal places

$$\cos \alpha = .1633, \quad \cos \beta = .4082, \quad \text{and} \quad \cos \gamma = .8981.$$

It follows that

$$\alpha \approx 81°, \quad \beta \approx 66°, \quad \text{and} \quad \gamma \approx 26°.$$

PROBLEMS 100

1. Find vector equations of the lines that contain the given pairs of points.
   (a) $(-1, 2, -4)$ and $(2, 0, 3)$ (b) $(1, 2, 3)$ and $(-1, -3, -2)$
   (c) $(0, 0, 0)$ and $(0, 1, 0)$ (d) $(0, 0, 1)$ and $(1, 0, 0)$
   (e) $(\ln 8, \ln 25, \ln 3)$ and $(\ln 4, \ln 5, \ln 6)$
   (f) $(\ln 6, \ln 12, \ln 8)$ and $(\ln 3, \ln 6, \ln 4)$

2. Find the line that contains the given point and is parallel to the given line.
   (a) $(2, -1, 3)$, $\mathbf{r} = (3t - 2)\mathbf{i} + 2t\mathbf{j} + (4 - t)\mathbf{k}$
   (b) $(1, 2, -1)$, $\mathbf{r} = (2 - 3t)\mathbf{i} - 2t\mathbf{j} + (t - 4)\mathbf{k}$
   (c) $(0, 0, 0)$, $\mathbf{r} = 3t\mathbf{i} + \mathbf{j} - (2t + 1)\mathbf{k}$
   (d) $(0, 1, 0)$, $\mathbf{r} = (3 - t)\mathbf{i} + (2t + 5)\mathbf{j} - 5\mathbf{k}$

3. For the points and lines of the preceding problem, find the plane that is perpendicular to the given line at the given point.

4. Find vector equations of the given lines.

   (a) $\dfrac{x + 2}{2} = \dfrac{y - 1}{1} = \dfrac{z - 2}{2}$ (b) $\dfrac{x - 2}{3} = \dfrac{y}{2} = \dfrac{z - 4}{1}$

   (c) $\dfrac{2 - x}{5} = \dfrac{y + 3}{2} = \dfrac{1 - 2z}{4}$ (d) $\dfrac{2x + 1}{3} = \dfrac{1 - y}{2} = 2z$

5. Find a vector equation of the line of intersection of the pair of planes.
    (a), $2x - 3y + z = 5$ and $x + 2z = 7$
    (b) $x - 2y + z = 3$ and $2x + y - z = 4$
    (c) $x = 0$ and $y = 2$
    (d) $y = 0$ and $x = 2$

6. Find the line that is perpendicular to the lines $\mathbf{r} = (3t - 1)\mathbf{i} + (2t + 2)\mathbf{j} + (1 - t)\mathbf{k}$ and $\mathbf{r} = (t - 3)\mathbf{i} + (8 - 3t)\mathbf{j} + (2t - 3)\mathbf{k}$ at their point of intersection.

7. Show that the vector equations $\mathbf{r} = (3t - 1)\mathbf{i} + (1 - 2t)\mathbf{j} + 4t\mathbf{k}$, $\mathbf{r} = (3t + 2)\mathbf{i} - (2t + 1)\mathbf{j} + (4t + 4)\mathbf{k}$, and $\mathbf{r} = (6t - 10)\mathbf{i} + (7 - 4t)\mathbf{j} + (8t - 12)\mathbf{k}$ all represent the same line.

8. Does the line that is tangent to the curve $\mathbf{r} = t\mathbf{i} + t^2\mathbf{j} + t^3\mathbf{k}$ at the point at which $t = 2$ contain the point $(1, 0, -4)$?

9. (a) Find the direction cosines of the lines
    $$\frac{x - 2}{-1} = \frac{y + 1}{3} = z \quad \text{and} \quad \frac{x + 1}{7} = \frac{y - 8}{-3} = z - 3.$$
    (b) Do these lines intersect?

10. How wide is the smaller angle between the normal to the plane $2x + y + z = 7$ and the line $x - 3 = \tfrac{1}{2}(y + 1) = 4 - z$?

11. Determine x so that the line intersects the line $\mathbf{r} = (3t - 2)\mathbf{i} + (t - 1)\mathbf{j} + 2t\mathbf{k}$.
    (a) $\mathbf{r} = (t - x)\mathbf{i} + 3t\mathbf{j} + (1 - t)\mathbf{k}$ (b) $\mathbf{r} = (t - 1)\mathbf{i} + 3t\mathbf{j} + (x - t)\mathbf{k}$
    (c) $\mathbf{r} = (xt - 1)\mathbf{i} + 3t\mathbf{j} + (1 - t)\mathbf{k}$ (d) $\mathbf{r} = (t - 1)\mathbf{i} + 3xt\mathbf{j} + (1 - t)\mathbf{k}$

12. The point $(-1, 2, 2)$ belongs to the ellipsoid $9x^2 + 4y^2 + z^2 = 29$.
    (a) Show that the planes $x = -1$ and $y = 2$ intersect this surface in the curves
    $$\mathbf{r} = -\mathbf{i} + t\mathbf{j} + 2\sqrt{5 - t^2}\,\mathbf{k} \quad \text{and} \quad \mathbf{r} = t\mathbf{i} + 2\mathbf{j} + \sqrt{13 - 9t^2}\,\mathbf{k}.$$
    (b) Find tangent vectors to these curves (and hence to our surface) at the point $(-1, 2, 2)$.
    (c) Use these vectors to find the tangent plane to the surface at this point.

13. Let $\mathbf{a}$ and $\mathbf{b}$ be given position vectors and describe the graph of the equation $\mathbf{r} = (1 - t)\mathbf{a} + t\mathbf{b}$, $t \in [0, 1]$.

14. For a given pair of vectors $\mathbf{m}$ and $\mathbf{b}$, and $t \in (-\infty, \infty)$, describe the graph of the equation.
    (a) $\mathbf{r} = \mathbf{m}|t| + \mathbf{b}$ (b) $\mathbf{r} = \mathbf{m}\sin t + \mathbf{b}$
    (c) $\mathbf{r} = \mathbf{m}e^t + \mathbf{b}$ (d) $\mathbf{r} = \mathbf{m}e^{-t^2} + \mathbf{b}$
    (e) $\mathbf{r} = \mathbf{m}[\![t]\!] + \mathbf{b}$ (f) $\mathbf{r} = \mathbf{m}(|t| + t) + \mathbf{b}$

15. The position vector of a particle in space t seconds after some initial instant is given by the equation $\mathbf{r} = \mathbf{f}(t)$. When $t = a$, the particle flies off on a tangent. Express the position vector $\mathbf{r}$ in terms of t for $t \geq a$.

101. CYLINDRICAL COORDINATES. SPHERICAL COORDINATES

Cartesian coordinates are not the only basis for associating numbers with points in space. In this section we discuss two other useful coordinate systems—cylindrical coordinates and spherical coordinates.

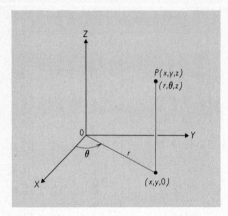

Figure 101-1

Let P be a point in space and suppose that its cartesian coordinates are (x, y, z). Let r and θ be polar coordinates of the point $(x, y, 0)$ in the XY-plane (Fig. 101-1). Then (r, θ, z) are **cylindrical coordinates** of P. From our previous study of polar coordinates, we know that $x = r \cos \theta$ and $y = r \sin \theta$, and therefore the cartesian coordinates of P are related to its cylindrical coordinates by the equations

$$(101\text{-}1) \quad x = r \cos \theta, \qquad y = r \sin \theta, \qquad z = z.$$

We obtain a **coordinate surface** in a given coordinate system by "fixing" one of the coordinates. For example, in a cartesian coordinate system the coordinate surface consisting of all points of the form $(x, 2, z)$ is a plane that is parallel to the XZ-plane and 2 units away from it. Every coordinate surface in a cartesian coordinate system is a plane, but there are two kinds of coordinate surfaces in a cylindrical coordinate system. A coordinate surface that consists of points of the form (r, θ, a) is a plane that is parallel to the XY-plane and is $|a|$ units away from it. A coordinate surface that consists of all points of the form (r, b, z) is also a plane, one that contains the Z-axis. But a coordinate surface made up of points of the form (c, θ, z) is a right-circular cylinder, which is what leads to the name *cylindrical coordinates.*

The graph of an equation in r, θ, and z is, of course, the set of points having cylindrical coordinates that satisfy the equation.

Example 101-1. Describe the surface $\theta = z$.

Solution. In order to plot a graph, we must find some points. If we let $\theta = 0$, then $z = 0$, and so any point $(r, 0, 0)$ belongs to our graph, no matter what r is. The set of all these points forms the X-axis. Similarly, if we let $\theta = \frac{1}{4}\pi$, we find that $z = \frac{1}{4}\pi$, and therefore all points of the set $\{(r, \frac{1}{4}\pi, \frac{1}{4}\pi) \mid r \in (-\infty, \infty)\}$ belong to our graph. This set is the line that is parallel to the XY-plane, is $\frac{1}{4}\pi$ units above it, and is directed so that it bisects the angle between the XZ-plane and the YZ-plane. For other choices of θ, we obtain similar lines. In fact, our surface is generated by a moving line that is kept parallel to the XY-plane while it is simultaneously moved along the Z-axis and twisted about it. A surface that is generated by a moving line is called a **ruled surface**, and our ruled surface here is a **helicoid**. (The cylinders we discussed in Section 97 are also ruled surfaces, of course.) In Fig. 101-2 we have shown part of our helicoid.

Just as a curve in space can be represented by three parametric equations for cartesian coordinates x, y, and z, so we can also represent the equation of a space

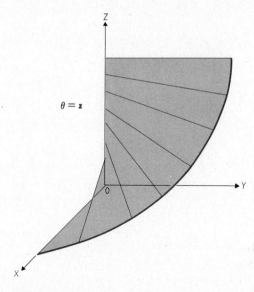

$\theta = z$

Figure 101-2

curve by giving three parametric equations for the cylindrical coordinates r, θ, and z. For example, the curve represented by the parametric equations

$$r = 1, \quad \theta = t, \quad z = t, \qquad t \in (-\infty, \infty),$$

is the circular helix of Example 99-4.

Equation 99-3 says that the length of an arc in space is given by the formula

$$(101\text{-}2) \qquad L = \int_a^b \sqrt{x'^2 + y'^2 + z'^2} \; dt.$$

In order to use this formula to find the length of the arc with parametric equations

$$r = f(t), \quad \theta = g(t), \quad z = h(t) \qquad \text{for } t \in [a, b],$$

we must express the derivatives x' and y' in terms of r' and θ'. So we differentiate both sides of the first two of Equations 101-1 to obtain

$$x' = r' \cos \theta - r\theta' \sin \theta \quad \text{and} \quad y' = r' \sin \theta + r\theta' \cos \theta.$$

Then you can easily check to see that

$$x'^2 + y'^2 + z'^2 = r'^2 + r^2\theta'^2 + z'^2,$$

and therefore Equation 101-2 becomes

$$(101\text{-}3) \qquad L = \int_a^b \sqrt{r'^2 + r^2\theta'^2 + z'^2} \; dt.$$

Example 101-2. How long is the arc with parametric equations $r = 2e^t$, $\theta = t$, $z = e^t$, $t \in [0, 1]$?

Solution. Here

$$r'^2 = 4e^{2t}, \qquad r^2 = 4e^{2t}, \qquad \theta'^2 = 1, \quad \text{and} \quad z'^2 = e^{2t}.$$

Therefore Formula 101-3 tells us that the length of our arc is

$$L = \int_0^1 \sqrt{4e^{2t} + 4e^{2t} + e^{2t}} \; dt = \int_0^1 3e^t \; dt = 3(e - 1).$$

Again let (x, y, z) be the cartesian coordinates of a point P and let r and θ (with $r > 0$) be polar coordinates of the point $(x, y, 0)$ in the XY-plane. Suppose that the position vector $\mathbf{OP}$ and $\mathbf{k}$ make an angle of ϕ, where $0 \le \phi \le \pi$, and let $|\mathbf{OP}| = \rho$ (Fig. 101-3). Then the numbers (ρ, θ, ϕ) are called **spherical coordinates** of P. We see that $z = \mathbf{OP} \cdot \mathbf{k} = \rho \cos \phi$. Furthermore, $\mathbf{k} \times \mathbf{OP} = \mathbf{k} \times (x\mathbf{i} + y\mathbf{j} + z\mathbf{k}) = x\mathbf{j} - y\mathbf{i}$, so

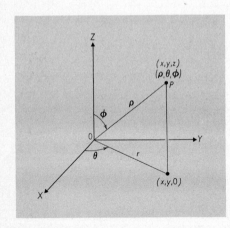

Figure 101-3

$$\rho \sin \phi = |\mathbf{k} \times \mathbf{OP}| = \sqrt{x^2 + y^2} = r.$$

When we substitute this result in the equations $x = r \cos \theta$ and $y = r \sin \theta$, we obtain the first two of the following relations between our spherical coordinates and the cartesian coordinates of the point P:

$$(101\text{-}4) \qquad x = \rho \cos \theta \sin \phi, \qquad y = \rho \sin \theta \sin \phi,$$
$$z = \rho \cos \phi.$$

In our spherical coordinate system the coordinate surface consisting of points of the form (ρ, a, ϕ) is a plane that contains the Z-axis. The coordinate surface made up of points of the form (ρ, θ, b) is a cone whose vertex is the origin (unless $b = 0$, $b = \pi$, or $b = \frac{1}{2}\pi$, in which case, the cone "degenerates"). The coordinate surface consisting of points of the form (c, θ, ϕ), where $c > 0$, is a sphere—hence the name *spherical coordinates*.

Of course, a space curve may be represented by parametric equations in spherical coordinates. For example, the parametric equations

$$(101\text{-}5) \qquad \rho = t, \quad \theta = t, \quad \phi = \tfrac{1}{4}\pi, \qquad t \in [0, \infty],$$

represent a curve that lies in the cone $\phi = \frac{1}{4}\pi$ and winds around this cone in much the same way that a helix winds around a cylinder (see Example 99-1).

To find a formula for the length of the arc with parametric equations

$$\rho = f(t), \quad \theta = g(t), \quad \phi = h(t), \quad t \in [a, b],$$

we must express x', y', and z' in terms of ρ', θ', and ϕ' and then substitute these expressions in Equation 101-2. When we differentiate both sides of each of Equations 101-4, we obtain

$$x' = \rho' \cos \theta \sin \phi - \rho\theta' \sin \theta \sin \phi + \rho\phi' \cos \theta \cos \phi,$$
$$y' = \rho' \sin \theta \sin \phi + \rho\theta' \cos \theta \sin \phi + \rho\phi' \sin \theta \cos \phi,$$
$$z' = \rho' \cos \phi - \rho\phi' \sin \phi.$$

It is now a simple matter of elementary trigonometry and algebra to show that

$$x'^2 + y'^2 + z'^2 = \rho'^2 + (\rho^2 \sin^2 \phi)\theta'^2 + \rho^2\phi'^2.$$

We substitute this result in Equation 101-2 to obtain the arclength formula

$$(101\text{-}6) \qquad L = \int_a^b \sqrt{\rho'^2 + (\rho^2 \sin^2 \phi)\theta'^2 + \rho^2\phi'^2} \ dt.$$

Example 101-3. Find the length of the first winding of the conical helix given by Equations 101-5; that is, find the length of the arc that corresponds to the parameter interval $[0, 2\pi]$.

Solution. Here

$$\phi' = 0, \qquad \rho' = \theta' = 1, \qquad \sin^2 \phi = \frac{1}{2}, \quad \text{and} \quad \rho = t.$$

Therefore Formula 101-6 becomes

$$L = \int_0^{2\pi} \sqrt{1 + \frac{1}{2}t^2} \ dt = \frac{1}{\sqrt{2}} \int_0^{2\pi} \sqrt{t^2 + 2} \ dt.$$

We look up this integral in Table V, and we find that

$$L = \pi\sqrt{2\pi^2 + 1} + \frac{1}{2}\sqrt{2} \ln (\pi\sqrt{2} + \sqrt{2\pi^2 + 1}) \approx 16.$$

PROBLEMS 101

1. Plot the points with the given cylindrical coordinates. Find their cartesian coordinates, find another set of cylindrical coordinates, and find a set of spherical coordinates.

   (a) $(2, 30°, -2)$ (b) $(0, 90°, 5)$ (c) $(-3, 45°, -3)$
   (d) $(-2, -30°, 2)$ (e) $(2, \frac{1}{3}\pi, 1)$ (f) $(1, \frac{1}{2}\pi, 3)$
   (g) (π, π, π) (h) $(-\pi, -\pi, -\pi)$

2. The following equations are written in terms of cylindrical coordinates. Describe their graphs. In some cases, you may want to write an equivalent equation in cartesian coordinates to make use of your previous knowledge of graphs.

   (a) $z = 2r$ (b) $z = 2r^2$ (c) $\dfrac{r^2}{4} + \dfrac{z^2}{9} = 1$

   (d) $\dfrac{r^2}{9} + \dfrac{z^2}{4} = 1$ (e) $\dfrac{r^2}{4} - \dfrac{z^2}{9} = 1$ (f) $\dfrac{z^2}{9} - \dfrac{r^2}{4} = 1$

   (g) $z = \sin r$ (h) $r = \sin z$

3. The following equations are written in terms of spherical coordinates. Describe their graphs.

   (a) $\rho = 3$ (b) $\phi = \tfrac{1}{2}\pi$ (c) $\rho \cos \phi = 2$
   (d) $\rho \sin \phi = 2$ (e) $\rho = 2 \cos \phi$ (f) $\rho = -2 \cos \phi$
   (g) $(3 \cos \theta - 2 \sin \theta) \sin \phi = 5 \cos \phi$
   (h) $\sin (\theta - \tfrac{1}{4}\pi) \sin \phi = \sqrt{2} \cos \phi$

4. Express the following equations in cylindrical and in spherical coordinates.

   (a) $x^2 + y^2 = z$ (b) $x^2 + y^2 = z^2$
   (c) $3x + 4y + 5z = 6$ (d) $y = z$
   (e) $x^2 + y^2 + z^2 - 4z = 0$ (f) $(x^2 + y^2 + z^2)^2 = 4(1 - z^2)$

5. The graphs of these equations in cylindrical or spherical coordinates are really quite easy to describe. Do it.

   (a) $\sin \pi r = 0$ (b) $\sin \pi \rho = 0$ (c) $\sin 4\theta = 0$
   (d) $\sin 4\phi = 0$ (e) $[\![r]\!] = 0$ (f) $[\![\rho]\!] = 0$

6. To get an idea of what the graph of the equation $\rho = 2 \sin \phi$ looks like, multiply both sides by ρ and observe that $\rho \sin \phi = r$. Now notice that we are dealing with a surface of revolution and describe it.

7. (a) What is the graph of the equation $(\sin \pi r)(\sin 4\theta)(\sin \pi z) = 0$ in cylindrical coordinates?

   (b) What is the graph of the equation $(\sin \pi \rho)(\sin 4\theta)(\sin 4\phi) = 0$ in spherical coordinates?

8. How long are the curves with the given parametric equations in cylindrical coordinates?

   (a) $r = 5 \cos t,\ \theta = \sec t,\ z = 5 \sin t,\ t \in [0, \tfrac{1}{6}\pi]$
   (b) $r = \cos t,\ \theta = \sqrt{2} \sec t,\ z = \sec t,\ t \in [0, \tfrac{1}{4}\pi]$

9. Find parametric equations in spherical coordinates that represent the curve of intersection of the surfaces $\rho = 2$ and $\rho \sin^2 \phi = 3 \cos \phi$. How long is this curve?

10. Find the length of the curve with parametric equations in spherical coordinates of $\rho = 1,\ \theta = k \ln (\sec t + \tan t),\ \phi = \tfrac{1}{2}\pi - t,\ t \in [0, \tfrac{1}{2}\pi)$. Describe this curve.

11. Depending on whether we consider the triple (t, t, t) as representing a point in cartesian, cylindrical, or spherical coordinates, the set $\{(t, t, t) \mid t \in [0, 1]\}$ can represent three different curves in space. Which is longest?

12. What can you say about the graph of an equation in cylindrical or spherical coordinates that does not contain θ?

13. Verify the equations used in the derivations of Equations 101-3 and 101-6.

(a) $x'^2 + y'^2 + z'^2 = r'^2 + r^2\theta'^2 + z'^2$ (cylindrical coordinates)

(b) $x'^2 + y'^2 + z'^2 = \rho'^2 + \rho^2\theta'^2 \sin^2 \phi + \rho^2 \phi'^2$ (spherical coordinates)

14. (a) Find the distance between two points with cylindrical coordinates (r_1, θ_1, z_1) and (r_2, θ_2, z_2).

(b) Find the distance between two points with spherical coordinates $(\rho_1, \theta_1, \phi_1)$ and $(\rho_2, \theta_2, \phi_2)$.

102. ROTATION OF AXES

In specifying a rotation of axes in the plane, we need only state the angle through which one axis is rotated; the other axis tags along behind. But in space there is much more freedom of movement, and rotating axes is inherently more complicated. Consequently, this section is a rather long one.

To help us explain rotation, we need the concept of the *transpose* of a matrix

$$A = \begin{bmatrix} a_{11} & a_{12} & a_{13} \\ a_{21} & a_{22} & a_{23} \\ a_{31} & a_{32} & a_{33} \end{bmatrix}.$$

The **transpose** of A is the matrix A^* whose rows are the columns of A (and whose columns are therefore the rows of A); that is,

$$A^* = \begin{bmatrix} a_{11} & a_{21} & a_{31} \\ a_{12} & a_{22} & a_{32} \\ a_{13} & a_{23} & a_{33} \end{bmatrix}.$$

Notice that $(A^*)^*$ is the matrix whose rows are the columns of A^*, which puts us back where we started: $(A^*)^* = A$. The definition of A^* is easy enough to state, and now we want to see what the transpose is good for.

The heart of the matter is the following theorem.

Theorem 102-1. *If A is a given matrix and $\mathbf{x}$ and $\mathbf{y}$ are vectors, then*

(102-1) $$A\mathbf{x} \cdot \mathbf{y} = \mathbf{x} \cdot A^*\mathbf{y}.$$

This theorem says that "when we transfer A across the dot, it becomes A^*," and its proof simply consists of replacing A and A^* with the matrices that are written out above and writing $\mathbf{x} = (x_1, x_2, x_3)$ and $\mathbf{y} = (y_1, y_2, y_3)$ in Equation 102-1. Rather than go through the mechanical details of the calculation, let us illustrate it with a two-dimensional example.

Example 102-1. Verify Equation 102-1 for $A = \begin{bmatrix} 1 & 2 \\ 3 & 4 \end{bmatrix}$, $\mathbf{x} = (1, 1)$, and $\mathbf{y} = (1, -1)$.

Solution. We first calculate $A\mathbf{x} = (3, 7)$ and then dot that vector with $\mathbf{y}$:

$$A\mathbf{x} \cdot \mathbf{y} = (3, 7) \cdot (1, -1) = 3 - 7 = -4.$$

To get A^*, we replace the rows of A with its columns:

$$A^* = \begin{bmatrix} 1 & 3 \\ 2 & 4 \end{bmatrix}.$$

Then we see that $A^*\mathbf{y} = (-2, -2)$, and we dot this vector with $\mathbf{x}$:

$$\mathbf{x} \cdot A^*\mathbf{y} = (1, 1) \cdot (-2, -2) = -4.$$

Now suppose that we think of a matrix A expressed in terms of its row vectors:

$$A = \begin{bmatrix} \mathbf{r}_1 \\ \mathbf{r}_2 \\ \mathbf{r}_3 \end{bmatrix}.$$

The columns of A^* are the rows of A, so in terms of its column vectors,

$$A^* = [\mathbf{r}_1, \mathbf{r}_2, \mathbf{r}_3].$$

Therefore the product AA^* can be written as $[A\mathbf{r}_1, A\mathbf{r}_2, A\mathbf{r}_3]$. Let us take a closer look at the column vector $A\mathbf{r}_1$. A careful study of the rule for multiplying a vector by a matrix (Equation 90-3) shows us that the first component of $A\mathbf{r}_1$ is the dot product $\mathbf{r}_1 \cdot \mathbf{r}_1$, the second component is $\mathbf{r}_2 \cdot \mathbf{r}_1$, and the third component is $\mathbf{r}_3 \cdot \mathbf{r}_1$. Thus $A\mathbf{r}_1 = \begin{pmatrix} \mathbf{r}_1 \cdot \mathbf{r}_1 \\ \mathbf{r}_2 \cdot \mathbf{r}_1 \\ \mathbf{r}_3 \cdot \mathbf{r}_1 \end{pmatrix}$, and there are similar expressions for the column vectors $A\mathbf{r}_2$ and $A\mathbf{r}_3$. Therefore if A is a 3 by 3 matrix, then

$$(102\text{-}2) \qquad AA^* = \begin{bmatrix} \mathbf{r}_1 \cdot \mathbf{r}_1 & \mathbf{r}_1 \cdot \mathbf{r}_2 & \mathbf{r}_1 \cdot \mathbf{r}_3 \\ \mathbf{r}_2 \cdot \mathbf{r}_1 & \mathbf{r}_2 \cdot \mathbf{r}_2 & \mathbf{r}_2 \cdot \mathbf{r}_3 \\ \mathbf{r}_3 \cdot \mathbf{r}_1 & \mathbf{r}_3 \cdot \mathbf{r}_2 & \mathbf{r}_3 \cdot \mathbf{r}_3 \end{bmatrix}.$$

A matrix U is said to be **orthogonal** if $U^{-1} = U^*$—that is, if $UU^* = U^*U = I$. Since, according to Equation 102-2, the elements of UU^* are dot products of row vectors of U, the equation $UU^* = I$ says that the dot product of each row vector with itself is 1 and with each other row vector is 0. Thus we see that *each row vector of an orthogonal matrix is a unit vector that is orthogonal to the other row vectors.*

Example 102-2. Show that the matrix

$$U = \begin{bmatrix} \cos\theta & \sin\theta \\ -\sin\theta & \cos\theta \end{bmatrix}$$

is orthogonal, regardless of the choice of θ.

Solution. We must show that $U^* = U^{-1}$—that is, that $UU^* = I$:

$$\begin{bmatrix} \cos\theta & \sin\theta \\ -\sin\theta & \cos\theta \end{bmatrix} \begin{bmatrix} \cos\theta & -\sin\theta \\ \sin\theta & \cos\theta \end{bmatrix} = \begin{bmatrix} 1 & 0 \\ 0 & 1 \end{bmatrix}.$$

Suppose that U is orthogonal—that is, U is the inverse of U^*. But since $U = (U^*)^*$, this statement says that the transpose of U^* is the inverse of U^*. In other words, if U is orthogonal, so is U^*. Therefore its row vectors are unit vectors and each is orthogonal to the others. These *row* vectors of U^* are the *column* vectors of U, so we see that the column vectors of an orthogonal matrix share with the row vectors the property of being unit vectors that are orthogonal in pairs. Test this statement by applying it to the matrix of Example 102-2.

Conversely, if each of the row vectors of a matrix A is one unit long and is orthogonal to the other row vectors, then Equation 102-2 shows that $AA^* = I$. Hence $A^* = A^{-1}$, and so A is orthogonal. The same conclusion follows, of course, if we consider column rather than row vectors.

To find out how to rotate in three dimensions, let us look at the transformation of coordinates equations that describe a rotation of axes in the plane:

$$\bar{x} = x\cos\theta + y\sin\theta$$

$$\bar{y} = -x\sin\theta + y\cos\theta.$$

We are supposing that (x, y) and $(\bar{x}, \bar{y})$ are coordinates of the same point relative to an XY-coordinate system and an $\bar{X}\bar{Y}$-system that results from rotating the XY-system through the angle θ. If we set $\mathbf{r} = (x, y)$ and $\bar{\mathbf{r}} = (\bar{x}, \bar{y})$, we can write these transformation equations in the compact form

(102-3) $$\bar{\mathbf{r}} = U\mathbf{r},$$

where U is the matrix in Example 102-2. Since we know that $U^{-1} = U^*$, it is easy to solve Equation 102-3 for $\mathbf{r}$. We simply multiply both sides by U^* and use the fact that $U^*U = I$:

(102-4) $$\mathbf{r} = U^*\bar{\mathbf{r}}.$$

We will show that the same sort of relations hold in space.

We start with a three-dimensional cartesian coordinate system, where $\mathbf{i}$, $\mathbf{j}$, and $\mathbf{k}$ are the customary unit vectors in the direction of the X-, Y-, and Z-axes. If P is a point in space, the position vector $\mathbf{OP}$ is expressed as

$$(102\text{-}5) \qquad \mathbf{OP} = x\mathbf{i} + y\mathbf{j} + z\mathbf{k},$$

where (x, y, z) are the coordinates of P with respect to the XYZ-coordinate system. In other words, our coordinate system associates the element $\mathbf{r} = (x, y, z)$ of R^3 with our point P in space.

Now let us introduce a new coordinate system, an $\overline{X}\,\overline{Y}\overline{Z}$-system with the same origin as the old and in which the unit basis vectors are $\bar{\mathbf{i}}$, $\bar{\mathbf{j}}$, and $\bar{\mathbf{k}}$. These vectors can be expressed as combinations of the original basis vectors $\mathbf{i}$, $\mathbf{j}$, and $\mathbf{k}$:

$$(102\text{-}6) \qquad \begin{aligned} \bar{\mathbf{i}} &= u_{11}\mathbf{i} + u_{12}\mathbf{j} + u_{13}\mathbf{k} \\ \bar{\mathbf{j}} &= u_{21}\mathbf{i} + u_{22}\mathbf{j} + u_{23}\mathbf{k} \\ \bar{\mathbf{k}} &= u_{31}\mathbf{i} + u_{32}\mathbf{j} + u_{33}\mathbf{k}. \end{aligned}$$

From their components we construct the matrix

$$U = \begin{bmatrix} u_{11} & u_{12} & u_{13} \\ u_{21} & u_{22} & u_{23} \\ u_{31} & u_{32} & u_{33} \end{bmatrix}.$$

Just as the statement that P has XYZ-coordinates (x, y, z) led to Equation 102-5, so does the statement that P has $\overline{X}\,\overline{Y}\overline{Z}$-coordinates $(\bar{x}, \bar{y}, \bar{z})$ lead to the equation

$$\mathbf{OP} = \bar{x}\bar{\mathbf{i}} + \bar{y}\bar{\mathbf{j}} + \bar{z}\bar{\mathbf{k}}.$$

Together with Equation 102-5, this equation tells us that

$$x\mathbf{i} + y\mathbf{j} + z\mathbf{k} = \bar{x}\bar{\mathbf{i}} + \bar{y}\bar{\mathbf{j}} + \bar{z}\bar{\mathbf{k}},$$

and when we replace $\bar{\mathbf{i}}$, $\bar{\mathbf{j}}$, and $\bar{\mathbf{k}}$ in terms of $\mathbf{i}$, $\mathbf{j}$, and $\mathbf{k}$ according to Equations 102-6, we obtain

$$x\mathbf{i} + y\mathbf{j} + z\mathbf{k} = \bar{x}(u_{11}\mathbf{i} + u_{12}\mathbf{j} + u_{13}\mathbf{k})$$
$$+ \bar{y}(u_{21}\mathbf{i} + u_{22}\mathbf{j} + u_{23}\mathbf{k}) + \bar{z}(u_{31}\mathbf{i} + u_{32}\mathbf{j} + u_{33}\mathbf{k}).$$

Now we collect terms and equate coefficients of $\mathbf{i}$, $\mathbf{j}$, and $\mathbf{k}$:

$$\begin{aligned} x &= u_{11}\bar{x} + u_{21}\bar{y} + u_{31}\bar{z} \\ y &= u_{12}\bar{x} + u_{22}\bar{y} + u_{32}\bar{z} \\ z &= u_{13}\bar{x} + u_{23}\bar{y} + u_{33}\bar{z}. \end{aligned}$$

This system is the component form of Equation 102-4.

In order that Equation 102-4 be equivalent to Equation 102-3, U must be an orthogonal matrix, and it will be if our transformation of axes is a rotation; that is, the old axis system can be moved "rigidly" into the new. In particular, the new basis vectors $\bar{\mathbf{i}}$, $\bar{\mathbf{j}}$, and $\bar{\mathbf{k}}$ must be unit vectors that meet at right angles. Hence we must have $\bar{\mathbf{i}} \cdot \bar{\mathbf{i}} = 1$, $\bar{\mathbf{i}} \cdot \bar{\mathbf{j}} = 0$, and so on. According to Equations 102-6,

$$\bar{\mathbf{i}} \cdot \bar{\mathbf{i}} = (u_{11}\mathbf{i} + u_{12}\mathbf{j} + u_{13}\mathbf{k}) \cdot (u_{11}\mathbf{i} + u_{12}\mathbf{j} + u_{13}\mathbf{k})$$

and

$$\bar{\mathbf{i}} \cdot \bar{\mathbf{j}} = (u_{11}\mathbf{i} + u_{12}\mathbf{j} + u_{13}\mathbf{k}) \cdot (u_{21}\mathbf{i} + u_{22}\mathbf{j} + u_{23}\mathbf{k}).$$

When we use the rules of vector algebra to expand these dot products, we find that

$$\bar{\mathbf{i}} \cdot \bar{\mathbf{i}} = u_{11}^2 + u_{12}^2 + u_{13}^2 \quad \text{and} \quad \bar{\mathbf{i}} \cdot \bar{\mathbf{j}} = u_{11}u_{21} + u_{12}u_{22} + u_{13}u_{23}.$$

Thus the equation $\bar{\mathbf{i}} \cdot \bar{\mathbf{i}} = 1$ says that the row vector (u_{11}, u_{12}, u_{13}) is a unit vector, and the equation $\bar{\mathbf{i}} \cdot \bar{\mathbf{j}} = 0$ says that this row vector is orthogonal to the row vector (u_{21}, u_{22}, u_{23}), and so on. We therefore see that U is an orthogonal matrix, and hence Equations 102-3 and 102-4 are equivalent. Both represent our rotation of axes.

In addition to being orthogonal, the matrix of a rotation of axes has one more important property. Its determinant is 1. This statement follows from the fact that $\bar{\mathbf{i}}$, $\bar{\mathbf{j}}$, $\bar{\mathbf{k}}$ is a right-handed triple that determines a 1 by 1 by 1 "block." The volume of this block is 1, and we saw in Section 95 that it is given by the scalar triple product $\bar{\mathbf{i}} \cdot (\bar{\mathbf{j}} \times \bar{\mathbf{k}})$. According to Equation 95-9, this product can be written in terms of the components of its factors as the determinant

$$\begin{vmatrix} u_{11} & u_{12} & u_{13} \\ u_{21} & u_{22} & u_{23} \\ u_{31} & u_{32} & u_{33} \end{vmatrix},$$

and hence det $U = 1$.

We have said that a rotation of axes determines an orthogonal matrix with determinant 1. The converse is also true; an orthogonal matrix with determinant 1 determines a rotation of axes. We won't give a detailed proof of this assertion (it's easy). The key idea is expressed in the next example.

Example 102-3. Show that an orthogonal transformation preserves dot products (that is, lengths and angles).

Solution. An orthogonal matrix U transforms position vectors $\mathbf{r}$ and $\mathbf{s}$ to $\bar{\mathbf{r}} = U\mathbf{r}$ and $\bar{\mathbf{s}} = U\mathbf{s}$. Then

$$\begin{aligned} \bar{\mathbf{r}} \cdot \bar{\mathbf{s}} &= U\mathbf{r} \cdot U\mathbf{s} = U^*U\mathbf{r} \cdot \mathbf{s} \quad &\text{(Theorem 102-1)} \\ &= I\mathbf{r} \cdot \mathbf{s} = \mathbf{r} \cdot \mathbf{s} \quad &(U^*U = I \text{ by hypothesis}). \end{aligned}$$

Thus $\bar{\mathbf{r}} \cdot \bar{\mathbf{s}} = \mathbf{r} \cdot \mathbf{s}$; the transformation determined by an orthogonal matrix leaves dot products unchanged.

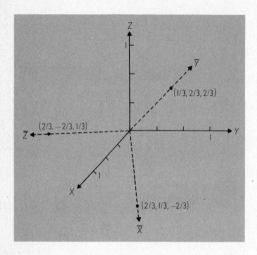

Figure 102-1

Example 102-4. Discuss the transformation determined by the matrix

$$U = \begin{bmatrix} \frac{2}{3} & \frac{1}{3} & -\frac{2}{3} \\ \frac{1}{3} & \frac{2}{3} & \frac{2}{3} \\ \frac{2}{3} & -\frac{2}{3} & \frac{1}{3} \end{bmatrix}.$$

Solution. You can easily verify that U is orthogonal and that det $U = 1$, so the transformation determined by this matrix is a rotation. To locate the $\bar{X}$-, $\bar{Y}$-, and $\bar{Z}$-axes relative to the X-, Y-, and Z-axes, we see from Equations 102-6 that the rows of U are the components of the basis vectors $\bar{\mathbf{i}}$, $\bar{\mathbf{j}}$, and $\bar{\mathbf{k}}$ relative to the XYZ-axes. In other words, the points $(\frac{2}{3}, \frac{1}{3}, -\frac{2}{3})$, $(\frac{1}{3}, \frac{2}{3}, \frac{2}{3})$, and $(\frac{2}{3}, -\frac{2}{3}, \frac{1}{3})$ are the terminal points of the vectors $\bar{\mathbf{i}}$, $\bar{\mathbf{j}}$, and $\bar{\mathbf{k}}$, and so they determine the $\bar{X}$-, $\bar{Y}$-, and $\bar{Z}$-axes, as shown in Fig. 102-1.

Example 102-5. Discuss the transformation determined by the orthogonal matrix

$$U = \begin{bmatrix} 0 & 0 & 1 \\ 0 & 1 & 0 \\ 1 & 0 & 0 \end{bmatrix}.$$

Solution. The terminal points of the vectors $\bar{\mathbf{i}}$, $\bar{\mathbf{j}}$, and $\bar{\mathbf{k}}$ are the points $(0, 0, 1)$, $(0, 1, 0)$, and $(1, 0, 0)$. Hence the $\bar{X}$-axis is the Z-axis, the $\bar{Y}$-axis is the Y-axis, and the $\bar{Z}$-axis is the X-axis. Notice that det $U = -1$ and that no rotation will bring the $\bar{X}\bar{Y}\bar{Z}$- and XYZ-axes into coincidence (Fig. 102-2).

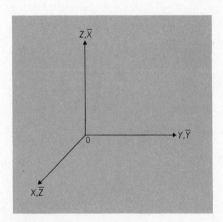

Figure 102-2

PROBLEMS 102

1. Verify the equation $A\mathbf{x} \cdot \mathbf{y} = \mathbf{x} \cdot A^*\mathbf{y}$ in the following cases.

(a) $A = \begin{bmatrix} 2 & -1 \\ -3 & 5 \end{bmatrix}$, $\mathbf{x} = (3, 2)$, $\mathbf{y} = (-1, 4)$

(b) $A = \begin{bmatrix} 2 & 1 \\ 1 & 0 \end{bmatrix}$, $\mathbf{x} = (1, 3)$, $\mathbf{y} = (-2, 2)$

(c) $A = \begin{bmatrix} 2 & -1 & 1 \\ 3 & 0 & 2 \\ 1 & 2 & -1 \end{bmatrix}$, $\mathbf{x} = (2, -1, 1)$, $\mathbf{y} = (1, 0, 2)$

(d) $A = \begin{bmatrix} 2 & -1 & 3 \\ -1 & 0 & 1 \\ 3 & 1 & 2 \end{bmatrix}$, $\mathbf{x} = (2, 1, 0)$, $\mathbf{y} = (0, 1, 2)$

2. The following equations are true for any matrices. Verify their validity for the choices

$$A = \begin{bmatrix} 2 & 1 \\ 1 & 1 \end{bmatrix} \quad \text{and} \quad B = \begin{bmatrix} 1 & 2 \\ -1 & 0 \end{bmatrix}.$$

(a) $(A^*)^* = A$ (b) $\det B^* = \det B$
(c) $(A + B)^* = A^* + B^*$ (d) $(5A + 3B)^* = 5A^* + 3B^*$
(e) $(A^{-1})^* = (A^*)^{-1}$ (f) $(AB)^* = B^*A^*$

3. Verify that the following matrices are orthogonal.

(a) $\begin{bmatrix} 0 & 1 & 0 \\ 1 & 0 & 0 \\ 0 & 0 & 1 \end{bmatrix}$ (b) $\begin{bmatrix} 1 & 0 & 0 \\ 0 & -1 & 0 \\ 0 & 0 & -1 \end{bmatrix}$

(c) $\begin{bmatrix} \cos\theta & 0 & \sin\theta \\ 0 & 1 & 0 \\ -\sin\theta & 0 & \cos\theta \end{bmatrix}$ (d) $\begin{bmatrix} \frac{2}{3} & \frac{1}{3} & -\frac{2}{3} \\ \frac{1}{3} & \frac{2}{3} & \frac{2}{3} \\ \frac{2}{3} & -\frac{2}{3} & \frac{1}{3} \end{bmatrix}$

(e) $\frac{1}{6}\begin{bmatrix} 2\sqrt{3} & 2\sqrt{6} & 0 \\ 2\sqrt{3} & -\sqrt{6} & 3\sqrt{2} \\ 2\sqrt{3} & -\sqrt{6} & -3\sqrt{2} \end{bmatrix}$ (f) $\frac{1}{5}\begin{bmatrix} 3 & 4 & 0 \\ 4 & -3 & 0 \\ 0 & 0 & 5 \end{bmatrix}$

4. Is there a number t such that $\begin{bmatrix} 1 & 0 \\ 4 & t \end{bmatrix}$ is an orthogonal matrix?

5. Determine numbers u, v, w, x, y, and z so that $\begin{bmatrix} u & 0 & 0 \\ v & w & 0 \\ x & y & z \end{bmatrix}$ is orthogonal.

6. Regardless of the numbers θ and ϕ, the matrices

$$A = \begin{bmatrix} \cos\theta & \sin\theta \\ -\sin\theta & \cos\theta \end{bmatrix} \quad \text{and} \quad B = \begin{bmatrix} \cos\phi & \sin\phi \\ -\sin\phi & \cos\phi \end{bmatrix}$$

are orthogonal. Show that AB is also orthogonal.

7. Let $\mathbf{r} = (x, y)$ and $\bar{\mathbf{r}} = (\bar{x}, \bar{y})$, the coordinates of a point relative to an XY- and an $\bar{X}\bar{Y}$-coordinate system, be related by the equation $\bar{\mathbf{r}} = U\mathbf{r}$. Draw the XY-axes in "standard position" and sketch the $\bar{X}\bar{Y}$-axes for the following choices of U.

(a) $\frac{1}{5}\begin{bmatrix} 3 & 4 \\ -4 & 3 \end{bmatrix}$ (b) $\begin{bmatrix} 0 & 1 \\ -1 & 0 \end{bmatrix}$ (c) $\begin{bmatrix} -1 & 0 \\ 0 & -1 \end{bmatrix}$

(d) $\begin{bmatrix} 1 & 0 \\ 0 & -1 \end{bmatrix}$ (e) $\begin{bmatrix} 0 & 1 \\ 1 & 0 \end{bmatrix}$ (f) $\frac{1}{2}\sqrt{2}\begin{bmatrix} 1 & 1 \\ -1 & 1 \end{bmatrix}$

8. The matrix $\begin{bmatrix} \frac{1}{2}\sqrt{3} & \frac{1}{2} \\ -\frac{1}{2} & \frac{1}{2}\sqrt{3} \end{bmatrix}$ represents a rotation of axes through an acute angle; find it.

9. Let the coordinates $\mathbf{r} = (x, y, z)$ and $\bar{\mathbf{r}} = (\bar{x}, \bar{y}, \bar{z})$ of a point relative to an XYZ- and an $\bar{X}\bar{Y}\bar{Z}$-coordinate system be related by the equation $\bar{\mathbf{r}} = U\mathbf{r}$. Draw the XYZ-axes in "standard position" and sketch the $\bar{X}\bar{Y}\bar{Z}$-axes for the following choices of U.

(a) $\begin{bmatrix} 0 & 1 & 0 \\ 1 & 0 & 0 \\ 0 & 0 & 1 \end{bmatrix}$ (b) $\begin{bmatrix} 1 & 0 & 0 \\ 0 & -1 & 0 \\ 0 & 0 & -1 \end{bmatrix}$

(c) $\begin{bmatrix} \frac{1}{3} & -\frac{2}{3} & \frac{2}{3} \\ \frac{2}{3} & \frac{2}{3} & \frac{1}{3} \\ -\frac{2}{3} & \frac{1}{3} & \frac{2}{3} \end{bmatrix}$ (d) $\begin{bmatrix} \frac{3}{5} & 0 & \frac{4}{5} \\ 0 & 1 & 0 \\ -\frac{4}{5} & 0 & \frac{3}{5} \end{bmatrix}$

(e) $\frac{1}{6}\begin{bmatrix} 2\sqrt{3} & 2\sqrt{6} & 0 \\ 2\sqrt{3} & -\sqrt{6} & 3\sqrt{2} \\ 2\sqrt{3} & -\sqrt{6} & -3\sqrt{2} \end{bmatrix}$ (f) $\frac{1}{6}\begin{bmatrix} 2\sqrt{3} & 2\sqrt{6} & 0 \\ 2\sqrt{3} & -\sqrt{6} & -3\sqrt{2} \\ 2\sqrt{3} & -\sqrt{6} & 3\sqrt{2} \end{bmatrix}$

10. What is the matrix of an orthogonal transformation such that the $\bar{X}$-axis is the Y-axis, and $\bar{Y}$-axis is the negative of the Z-axis, and the $\bar{Z}$-axis is the X-axis? Is this transformation a rotation?

11. Let the coordinates $\mathbf{r} = (x, y, z)$ and $\bar{\mathbf{r}} = (\bar{x}, \bar{y}, \bar{z})$ of a point relative to an XYZ- and an $\bar{X}\bar{Y}\bar{Z}$-coordinate system be related by the equation $\bar{\mathbf{r}} = U\mathbf{r}$, where

$$U = \begin{bmatrix} \frac{2}{3} & \frac{1}{3} & -\frac{2}{3} \\ \frac{1}{3} & \frac{2}{3} & \frac{2}{3} \\ \frac{2}{3} & -\frac{2}{3} & \frac{1}{3} \end{bmatrix}.$$

(a) Find the $\bar{X}\bar{Y}\bar{Z}$-coordinates of the point whose XYZ-coordinates are $(3, -6, 9)$.
(b) Find the XYZ-coordinates of the point whose $\bar{X}\bar{Y}\bar{Z}$-coordinates are $(3, -6, 9)$.
(c) The plane whose XYZ-coordinate equation is $6x - 9y + 3z + 1 = 0$ has a similar equation in terms of $\bar{x}$, $\bar{y}$, and $\bar{z}$. What is it?
(d) The plane whose $\bar{X}\bar{Y}\bar{Z}$-coordinate equation is $6\bar{x} - 9\bar{y} + 3\bar{z} + 1 = 0$ has a similar equation in terms of x, y, and z. What is it?
(e) What is the equation of the XY-plane in terms of $\bar{x}$, $\bar{y}$, and $\bar{z}$?
(f) Show that $\bar{x}^2 + \bar{y}^2 + \bar{z}^2 = x^2 + y^2 + z^2$.

12. It can be shown that every 3 by 3 rotation matrix U (an orthogonal matrix whose elements are real numbers and whose determinant is 1) leaves at least one vector fixed; that is, the equation $U\mathbf{x} = \mathbf{x}$ has a solution in addition to the "trivial" solution $\mathbf{x} = \mathbf{0}$. Solve this equation for the matrix U of the preceding problem.

13. The equation of a plane $ax + by + cz = d$ can be written as $\mathbf{n} \cdot \mathbf{r} = d$, where $\mathbf{n}$ is the normal vector (a, b, c) and $\mathbf{r}$ is the position vector (x, y, z). Under the transformation of coordinates equation $\bar{\mathbf{r}} = U\mathbf{r}$, where U is a given orthogonal matrix, the equation of this plane becomes $\bar{a}\bar{x} + \bar{b}\bar{y} + \bar{c}\bar{z} = d$. What is the new normal vector $(\bar{a}, \bar{b}, \bar{c})$ in the $\bar{X}\bar{Y}\bar{Z}$-coordinate system?

14. Explain why the elements of an orthogonal matrix U are the cosines of the angles between the X- and $\bar{X}$-, Y- and $\bar{Y}$, and Z- and $\bar{Z}$-axes.

15. Let $\begin{bmatrix} a & b \\ c & d \end{bmatrix}$ be an orthogonal matrix. Show that there is an angle θ such that $a = \cos\theta$ and $b = \sin\theta$. Then show that either $c = -\sin\theta$ and $d = \cos\theta$ or $c = \sin\theta$ and $d = -\cos\theta$. It follows that every 2 by 2 orthogonal matrix must either be a rotation matrix $\begin{bmatrix} \cos\theta & \sin\theta \\ -\sin\theta & \cos\theta \end{bmatrix}$ or a matrix $\begin{bmatrix} \cos\theta & \sin\theta \\ \sin\theta & -\cos\theta \end{bmatrix}$ that represents a rotation and a reflection.

15. What are the transformation of coordinates equations for a translation of axes in three dimensions? For a rotation followed by a translation? For a translation followed by a rotation?

103. CHARACTERISTIC VALUES AND DIAGONALIZATION OF A MATRIX

We saw in Section 98 that the standard form of the equations of certain quadric surfaces can be expressed as

$$(103\text{-}1) \qquad D\mathbf{r} \cdot \mathbf{r} = p,$$

where $\mathbf{r} = (x, y, z)$, p is a number, and D is a diagonal matrix:

$$(103\text{-}2) \qquad D = \begin{bmatrix} \lambda_1 & 0 & 0 \\ 0 & \lambda_2 & 0 \\ 0 & 0 & \lambda_3 \end{bmatrix}.$$

We get a more general quadratic equation if we replace D with a more general *symmetric matrix*:

$$S = \begin{bmatrix} a & d & e \\ d & b & f \\ e & f & c \end{bmatrix}.$$

The "symmetry" of S refers to the placement of the numbers d, e, and f relative to the numbers a, b, and c that form the *main diagonal* of S. Formally, S is **symmetric** if

$$S^* = S.$$

It is simply a matter of calculation to show that

$$\mathbf{Sr} \cdot \mathbf{r} = ax^2 + by^2 + cz^2 + 2dxy + 2exz + 2fyz,$$

so the equation

(103-3) $\mathbf{Sr} \cdot \mathbf{r} = p$

is quite a general quadratic equation. For example,

(103-4) $5x^2 + 2y^2 + 11z^2 + 20xy - 16xz + 4yz = 9$

is of this form, with

$$S = \begin{bmatrix} 5 & 10 & -8 \\ 10 & 2 & 2 \\ -8 & 2 & 11 \end{bmatrix}.$$

We will only sketch the broad outlines of our analysis of Equation 103-3 here, saving the details for the next section. We rotate the axes by means of the transformation

$$\mathbf{r} = U^*\mathbf{\bar{r}},$$

where U is a "suitably chosen" orthogonal matrix. Then Equation 103-3 becomes

$$\mathbf{Sr} \cdot \mathbf{r} = SU^*\mathbf{\bar{r}} \cdot U^*\mathbf{\bar{r}} = USU^*\mathbf{\bar{r}} \cdot \mathbf{\bar{r}} = p.$$

(To go from the second to the third member in this chain of equations, we transferred U^* across the dot, thereby turning it into U, in accordance with Theorem 102-1.) Our choice of U will be such that USU^* is a diagonal matrix, and hence we will have reduced Equation 103-3 to the standard form of Equation 103-1.

The matrix U we have just been talking about will be the transpose of another matrix V. Thus we are seeking an orthogonal matrix V such that

(103-5) $V^*SV = D.$

Of course, at this point, we have no guarantee that there is such a V. It is the purpose of this section to convince you that there is one and to show you how to find it. Because V is to be orthogonal, Equation 103-5 is equivalent to the equation $SV = VD$, or

$$S[\mathbf{v}_1, \mathbf{v}_2, \mathbf{v}_3] = [\mathbf{v}_1, \mathbf{v}_2, \mathbf{v}_3]D,$$

where we have written V in terms of its column vectors: $V = [\mathbf{v}_1, \mathbf{v}_2, \mathbf{v}_3]$. The definition of matrix multiplication tells us that the left-hand side of this equation is $[S\mathbf{v}_1, S\mathbf{v}_2, S\mathbf{v}_3]$, and an analysis of the right-hand side (taking into account the diagonal nature of D) shows that it is $[\lambda_1\mathbf{v}_1, \lambda_2\mathbf{v}_2, \lambda_3\mathbf{v}_3]$. Therefore each column of the matrix V is a solution of an equation of the form

$$(103\text{-}6) \qquad\qquad S\mathbf{v} = \lambda\mathbf{v},$$

where λ is a number.

If $\mathbf{v} = \mathbf{0}$, Equation 103-6 is valid for any number λ, but we want $\mathbf{v}$ to be a column of an orthogonal matrix, and hence it cannot be $\mathbf{0}$. So we want $\mathbf{v}$ to be a nonzero (nontrivial) solution of Equation 103-6. This equation has nontrivial solutions only for certain choices of λ, as we shall now see. We can write Equation 103-6 as $S\mathbf{v} = \lambda I\mathbf{v}$, or, equivalently,

$$(103\text{-}7) \qquad\qquad (S - \lambda I)\mathbf{v} = \mathbf{0}.$$

Now we apply the results of Section 92. Theorem 92-3 says that a homogeneous matrix-vector equation has a nontrivial solution if and only if the determinant of the coefficient matrix is 0. Therefore Equation 103-6 has a solution $\mathbf{v} \neq \mathbf{0}$ if and only if λ is a number that satisfies the **characteristic equation**

$$(103\text{-}8) \qquad\qquad \det (S - \lambda I) = 0.$$

Example 103-1. Solve Equation 103-6 if S is the matrix $\begin{bmatrix} 3 & 2 \\ 2 & 0 \end{bmatrix}$.

Solution. In this case, our characteristic Equation 103-8 is

$$\begin{vmatrix} 3 - \lambda & 2 \\ 2 & -\lambda \end{vmatrix} = (3 - \lambda)(-\lambda) - 4 = (\lambda - 4)(\lambda + 1) = 0.$$

Its solutions are the numbers $\lambda = 4$ and $\lambda = -1$. Corresponding to each of these numbers is a solution $\mathbf{v} = (x, y) \neq (0, 0)$ of Equation 103-6 (or the equivalent Equation 103-7). Let us choose $\lambda = 4$. Then the coefficient matrix of Equation 103-7 becomes

$$S - 4I = \begin{bmatrix} 3 - 4 & 2 \\ 2 & -4 \end{bmatrix} = \begin{bmatrix} -1 & 2 \\ 2 & -4 \end{bmatrix}.$$

Thus to find a vector $\mathbf{v}$ that corresponds to the number 4, we must find a solution of the homogeneous system

$$-x + 2y = 0$$
$$2x - 4y = 0.$$

One solution is the vector $\mathbf{v} = (2, 1)$, so for this vector

$$S\mathbf{v} = 4\mathbf{v}.$$

If we set $\lambda = -1$, we can take $\mathbf{v} = (1, -2)$, and then $S\mathbf{v} = (-1)\mathbf{v}$, as you can easily verify.

The preceding example illustrates the following facts about Equation 103-6. The characteristic equation is an algebraic equation whose degree is the order of the matrix S. Thus the matrix of our example was a 2 by 2 matrix, and λ satisfied a quadratic equation. Therefore the number of different possible choices for λ is, at most, the order of S. These numbers λ are called the **characteristic values** of S. Corresponding to each characteristic value, there is a **characteristic vector** **v**—that is, a vector $\mathbf{v} \neq \mathbf{0}$ that satisfies Equation 103-6. If **v** is a characteristic vector, so is any nonzero multiple of **v**—for example, 3**v**, 5**v**, and so on. Frequently we multiply a characteristic vector by the reciprocal of its length in order to get a *unit* characteristic vector. For instance, the vector $\mathbf{v} = (\frac{2}{5}\sqrt{5}, \frac{1}{5}\sqrt{5})$ is a characteristic vector of the matrix S of Example 103-1, and it is one unit long. The vector $\mathbf{v} = (-\frac{2}{5}\sqrt{5}, -\frac{1}{5}\sqrt{5})$ is another unit characteristic vector that corresponds to the characteristic value 4.

The theory of characteristic values and vectors is developed for general matrices, but we are going to use it only in connection with symmetric matrices, such as our matrix S of Example 103-1. The characteristic vectors $(2, 1)$ and $(1, -2)$ of this matrix are orthogonal, and the next theorem shows that this fact is no accident.

Theorem 103-1. *If **u** and **v** are characteristic vectors of a symmetric matrix S that correspond to different characteristic values λ and μ, then $\mathbf{u} \cdot \mathbf{v} = 0$.*

PROOF

We are supposing that

$$S\mathbf{u} = \lambda\mathbf{u} \quad \text{and} \quad S\mathbf{v} = \mu\mathbf{v}.$$

If we dot the first of these equations with **v** and the second with **u** and subtract, we obtain the equation

$$S\mathbf{u} \cdot \mathbf{v} - \mathbf{u} \cdot S\mathbf{v} = (\lambda - \mu)\mathbf{u} \cdot \mathbf{v}.$$

Because S is symmetric, $\mathbf{u} \cdot S\mathbf{v} = S^*\mathbf{u} \cdot \mathbf{v} = S\mathbf{u} \cdot \mathbf{v}$, and so the left-hand side of this equation is 0. Furthermore, we are assuming that $\lambda - \mu \neq 0$, and hence we see that $\mathbf{u} \cdot \mathbf{v} = 0$.

Theorem 103-1 states that characteristic vectors corresponding to different characteristic values *must* be orthogonal. A particular n by n symmetric matrix may not have n different characteristic values (for example, the 2 by 2 matrix $\begin{bmatrix} 3 & 0 \\ 0 & 3 \end{bmatrix}$ has only the single characteristic value 3). Then we have to *choose* orthogonal characteristic vectors (for example, $(1, 0)$ and $(0, 1)$). The important fact, which we are not going to take the time to prove, is that such a choice is always possible: *every n by n symmetric matrix has n mutually orthogonal characteristic vectors.*

We can multiply each of these characteristic vectors by the reciprocal of its length, thereby producing a set of mutually orthogonal unit vectors. Finally, we stack these unit vectors as columns of a matrix V, which will be an orthogonal matrix such that $V*SV$ is diagonal. For example, if we stack the unit characteristic vectors we found for the matrix of Example 103-1, we obtain the orthogonal matrix

$$V = \begin{bmatrix} \frac{2}{5}\sqrt{5} & \frac{1}{5}\sqrt{5} \\ \frac{1}{5}\sqrt{5} & -\frac{2}{5}\sqrt{5} \end{bmatrix}.$$

Then, as you can easily check,

(103-9)
$$V* \begin{bmatrix} 3 & 2 \\ 2 & 0 \end{bmatrix} V = \begin{bmatrix} 4 & 0 \\ 0 & -1 \end{bmatrix}.$$

Thus we have transformed S into a diagonal matrix whose diagonal elements are the characteristic values of S. Let us state these results as a theorem about 3 by 3 matrices.

Theorem 103-2. *If S is a symmetric matrix with characteristic values λ_1, λ_2, and λ_3, then there exists an orthogonal matrix V such that $V*SV = D$, where D is the diagonal matrix of Equation 103-2.*

Example 103-2. Diagonalize the symmetric matrix

$$S = \begin{bmatrix} 1 & 1 & 1 \\ 1 & 1 & 1 \\ 1 & 1 & 1 \end{bmatrix}.$$

Solution. Our first task is to find the characteristic values by solving the characteristic equation

$$\det(S - \lambda I) = \begin{vmatrix} 1-\lambda & 1 & 1 \\ 1 & 1-\lambda & 1 \\ 1 & 1 & 1-\lambda \end{vmatrix} = (1-\lambda)^3 - 3(1-\lambda) + 2 = 0.$$

Written out, this equation is $3\lambda^2 - \lambda^3 = 0$, so its solutions are 3, 0, and 0 (we take 0 to be a "double" characteristic value).

A characteristic vector corresponding to the characteristic value 3 must satisfy the equation $(S - 3I)\mathbf{v} = \mathbf{0}$, or

$$\begin{bmatrix} -2 & 1 & 1 \\ 1 & -2 & 1 \\ 1 & 1 & -2 \end{bmatrix} \begin{pmatrix} x \\ y \\ z \end{pmatrix} = \begin{pmatrix} 0 \\ 0 \\ 0 \end{pmatrix},$$

and you see immediately that $(1, 1, 1)$ is a solution. A *unit* solution is $\mathbf{v}_1 = (\frac{1}{3}\sqrt{3}, \frac{1}{3}\sqrt{3}, \frac{1}{3}\sqrt{3})$. A characteristic vector $\mathbf{v} = (x, y, z)$ corresponding to the characteristic value 0 must satisfy the equation

$$
\begin{bmatrix} 1 & 1 & 1 \\ 1 & 1 & 1 \\ 1 & 1 & 1 \end{bmatrix} \begin{pmatrix} x \\ y \\ z \end{pmatrix} = \begin{pmatrix} 0 \\ 0 \\ 0 \end{pmatrix}.
$$

If we reduce this equation to triangular form by our usual techniques, it becomes

$$
\begin{bmatrix} 0 & 0 & 0 \\ 0 & 0 & 0 \\ 1 & 1 & 1 \end{bmatrix} \begin{pmatrix} x \\ y \\ z \end{pmatrix} = \begin{pmatrix} 0 \\ 0 \\ 0 \end{pmatrix}.
$$

which is equivalent to the single scalar equation

$$
x + y + z = 0.
$$

There are many solutions of this equation: $(2, -1, -1)$, $(3, -1, -2)$, and so on. We will choose two that are orthogonal—for example, $(2, -1, -1)$ and $(0, 1, -1)$. Notice that these vectors (in accordance with Theorem 103-1) are orthogonal to $\mathbf{v}_1$. Then we multiply each by the reciprocal of its length:

$$
\mathbf{v}_2 = (\tfrac{1}{3}\sqrt{6}, -\tfrac{1}{6}\sqrt{6}, -\tfrac{1}{6}\sqrt{6}) \quad \text{and} \quad \mathbf{v}_3 = (0, \tfrac{1}{2}\sqrt{2}, -\tfrac{1}{2}\sqrt{2}).
$$

Finally, we "stack" $\mathbf{v}_1$, $\mathbf{v}_2$, and $\mathbf{v}_3$ to form the orthogonal matrix

$$
V = \begin{bmatrix} \tfrac{1}{3}\sqrt{3} & \tfrac{1}{3}\sqrt{6} & 0 \\ \tfrac{1}{3}\sqrt{3} & -\tfrac{1}{6}\sqrt{6} & \tfrac{1}{2}\sqrt{2} \\ \tfrac{1}{3}\sqrt{3} & -\tfrac{1}{6}\sqrt{6} & \tfrac{1}{2}\sqrt{2} \end{bmatrix}.
$$

Our theory says, and you may confirm, that

$$
V^*SV = \begin{bmatrix} 3 & 0 & 0 \\ 0 & 0 & 0 \\ 0 & 0 & 0 \end{bmatrix}.
$$

PROBLEMS   103

1. Let $S = \begin{bmatrix} 2 & 1 \\ 1 & 2 \end{bmatrix}$.

(a) What is the characteristic equation of S?
(b) What are the characteristic values of S?
(c) Find characteristic vectors that correspond to the characteristic values.
(d) Construct an orthogonal matrix V whose column vectors are characteristic vectors of S.
(e) Show that $V^*SV = D$, where D is a diagonal matrix.

2. Answer the questions of the first problem if $S = \begin{bmatrix} 1 & -2 \\ -2 & 1 \end{bmatrix}$.

3. Answer the questions of the first problem if $S = \begin{bmatrix} 7 & 3 \\ 3 & -1 \end{bmatrix}$.

4. For each of the given symmetric matrices S, we can find an orthogonal matrix V such that $V*SV = D$, a diagonal matrix. Find a candidate for D in each case without finding V.

(a) $\begin{bmatrix} 4 & 1 \\ 1 & 4 \end{bmatrix}$ (b) $\begin{bmatrix} 4 & 2 \\ 2 & 1 \end{bmatrix}$ (c) $\begin{bmatrix} 0 & 1 \\ 1 & 0 \end{bmatrix}$

(d) $\begin{bmatrix} 0 & 3 \\ 3 & 0 \end{bmatrix}$ (e) $\begin{bmatrix} 1 & 2 \\ 2 & 3 \end{bmatrix}$ (f) $\begin{bmatrix} 2 & 3 \\ 3 & -2 \end{bmatrix}$

5. Answer the questions of the first problem if $S = \begin{bmatrix} 1 & 1 & 1 \\ 1 & 2 & 0 \\ 1 & 0 & 2 \end{bmatrix}$.

6. Answer the questions of the first problem if $S = \begin{bmatrix} 2 & -2 & -2 \\ -2 & 1 & -1 \\ -2 & -1 & 1 \end{bmatrix}$.

7. For each of the given symmetric matrices S, we can find an orthogonal matrix V such that $V*SV = D$, a diagonal matrix. Find a candidate for D in each case without finding V.

(a) $\begin{bmatrix} 0 & 0 & 2 \\ 0 & 1 & 0 \\ 2 & 0 & 0 \end{bmatrix}$ (b) $\begin{bmatrix} 4 & 1 & 0 \\ 1 & 4 & 0 \\ 0 & 0 & 2 \end{bmatrix}$ (c) $\begin{bmatrix} 2 & 0 & 1 \\ 0 & 1 & 0 \\ 1 & 0 & 2 \end{bmatrix}$

(d) $\begin{bmatrix} 0 & 1 & 0 \\ 1 & 0 & 1 \\ 0 & 1 & 0 \end{bmatrix}$ (e) $\begin{bmatrix} 5 & -1 & -1 \\ -1 & 3 & 1 \\ -1 & 1 & 3 \end{bmatrix}$ (f) $\begin{bmatrix} -5 & 1 & 1 \\ 1 & -3 & -1 \\ 1 & -1 & -3 \end{bmatrix}$

8. If $\mathbf{v} = (3, 4)$ is one characteristic vector of a 2 by 2 symmetric matrix, what is another?

9. Find x so that 2 is a characteristic value of $\begin{bmatrix} 4 & x \\ x & 10 \end{bmatrix}$. What is the other characteristic value? What are corresponding characteristic vectors?

10. Find characteristic values and characteristic vectors for these nonsymmetric matrices. Symmetric matrices are much nicer to deal with.

(a) $\begin{bmatrix} 0 & 1 \\ -1 & 0 \end{bmatrix}$ (b) $\begin{bmatrix} 5 & 1 \\ 0 & 5 \end{bmatrix}$

(c) $\begin{bmatrix} 0 & 2 & 1 \\ -2 & 0 & 2 \\ -1 & -2 & 0 \end{bmatrix}$ (d) $\begin{bmatrix} 1 & 0 & 0 \\ 1 & 2 & 0 \\ 1 & 1 & 3 \end{bmatrix}$

11. Show that if λ is a characteristic value of a matrix A, then 3λ is a characteristic value of $3A$ and λ^2 is a characteristic value of A^2.

12. By writing out the product $(V*SV)^3 = V*SVV*SVV*SV$, you can see that it equals $V*S^3V$, and it is apparent that $(V*SV)^k = V*S^kV$ for any positive integer k. Use these ideas and Equation 103-9 to find $\begin{bmatrix} 3 & 2 \\ 2 & 0 \end{bmatrix}^5$ and check your result by direct calculation.

13. What can you conclude about a symmetric matrix if all its characteristic values are equal?

14. The characteristic values of the matrix $\begin{bmatrix} a & b \\ b & c \end{bmatrix}$ are solutions of a quadratic equation. Show that if a, b, and c are real numbers, then the solutions of this equation are real numbers. (It can be shown that the characteristic values of a real symmetric matrix of any size are real numbers.)

15. Is the sum of two symmetric matrices symmetric? A product? Can AB be symmetric and BA not be?

104. GRAPHS OF QUADRATIC EQUATIONS

In the last section we outlined our method of attack on a fairly general quadratic equation

$$(104\text{-}1) \qquad\qquad S\mathbf{r} \cdot \mathbf{r} = p,$$

where S is a symmetric 3 by 3 matrix and $\mathbf{r} = (x, y, z)$. We find an orthogonal matrix V such that $V^*SV = D$, a diagonal matrix. (By interchanging two columns if necessary, we can always arrange it so that det $V = 1$.) Then we set $U = V^*$, so that the transformation equation

$$(104\text{-}2) \qquad\qquad \mathbf{r} = U^*\bar{\mathbf{r}}$$

represents a rotation of axes. This substitution reduces Equation 104-1 to

$$(104\text{-}3) \qquad D\bar{\mathbf{r}} \cdot \bar{\mathbf{r}} = \lambda_1 \bar{x}^2 + \lambda_2 \bar{y}^2 + \lambda_3 \bar{z}^2 = p,$$

which is the standard form of a quadric surface.

To be specific, suppose that we apply these ideas to Equation 103-4:

$$5x^2 + 2y^2 + 11z^2 + 20xy - 16xz + 4yz = 9.$$

The matrix of this equation is written out in Section 103, and its characteristic equation is

$$\begin{vmatrix} 5 - \lambda & 10 & -8 \\ 10 & 2 - \lambda & 2 \\ -8 & 2 & 11 - \lambda \end{vmatrix} = 0.$$

When we expand the determinant, this equation becomes

$$\lambda^3 - 18\lambda^2 - 81\lambda + 1458 = 0.$$

Its solutions are $\lambda_1 = 18$, $\lambda_2 = 9$, and $\lambda_3 = -9$. Now we substitute these numbers and $p = 9$ in Equation 104-3 and simplify:

$$2\bar{x}^2 + \bar{y}^2 - \bar{z}^2 = 1.$$

This surface is a hyperboloid of one sheet in standard form relative to the $\overline{X}\,\overline{Y}\overline{Z}$-coordinate system. Thus the graph of Equation 103-4 is a hyperboloid. If we were merely interested in the form of the graph, we could stop now. However, if we actually want to sketch the graph, we must find how the $\overline{X}$-, $\overline{Y}$-, and $\overline{Z}$-axes are related to the X-, Y-, and Z-axes, and hence we need to know the transformation matrix U.

The columns of $V = U^*$ are unit characteristic vectors of S, so we must replace λ with 18, 9, and -9 in the equation

$$(S - \lambda I)\mathbf{v} = \mathbf{0}$$

and solve it. The first such equation is

$$\begin{bmatrix} -13 & 10 & -8 \\ 10 & -16 & 2 \\ -8 & 2 & -7 \end{bmatrix}\mathbf{v} = \mathbf{0}.$$

We leave it to you to show that $\mathbf{v}_1 = (2, 1, -2)$ is a solution, and when we divide this vector by its length, 3, we obtain the unit characteristic vector $\mathbf{v}_1 = (\frac{2}{3}, \frac{1}{3}, -\frac{2}{3})$ that corresponds to the characteristic value $\lambda_1 = 18$. Unit characteristic vectors that correspond to the characteristic values $\lambda_2 = 9$ and $\lambda_3 = -9$ are $\mathbf{v}_2 = (\frac{1}{3}, \frac{2}{3}, \frac{2}{3})$ and $\mathbf{v}_3 = (\frac{2}{3}, -\frac{2}{3}, \frac{1}{3})$. These vectors form the columns of our matrix V and therefore the rows of the transformation matrix $U = V^*$. Thus

$$U = \begin{bmatrix} \frac{2}{3} & \frac{1}{3} & -\frac{2}{3} \\ \frac{1}{3} & \frac{2}{3} & \frac{2}{3} \\ \frac{2}{3} & -\frac{2}{3} & \frac{1}{3} \end{bmatrix}.$$

This matrix is the transformation matrix we studied in Example 102-4, and the relation of the $\overline{X}$-, $\overline{Y}$-, and $\overline{Z}$-axes to the X-, Y-, and Z-axes is shown in Fig. 102-1. Our study of the graph of Equation 103-4 is therefore complete.

Example 104-1. Discuss the surface

$$2x^2 + 5y^2 + 3z^2 + 4xy = 3.$$

Solution. Associated with this equation is the symmetric matrix

$$S = \begin{bmatrix} 2 & 2 & 0 \\ 2 & 5 & 0 \\ 0 & 0 & 3 \end{bmatrix},$$

whose characteristic values are solutions of the equation

$$\begin{vmatrix} 2 - \lambda & 2 & 0 \\ 2 & 5 - \lambda & 0 \\ 0 & 0 & 3 - \lambda \end{vmatrix} = (3 - \lambda)[(2 - \lambda)(5 - \lambda) - 4]$$

$$= (3 - \lambda)(\lambda - 6)(\lambda - 1) = 0.$$

Hence $\lambda_1 = 6$, $\lambda_2 = 3$, and $\lambda_3 = 1$. There is a coordinate system, a rotation of the XYZ-system, in which our surface becomes $6\bar{x}^2 + 3\bar{y}^2 + \bar{z}^2 = 3$. This surface, we know, is an ellipsoid.

Let us sum up these results as a theorem.

Theorem   104-1. *The graph of the quadratic equation* (*Equation* 104-1)

$$ax^2 + by^2 + cz^2 + 2dxy + 2exz + 2fyz = p$$

*is a quadric surface (perhaps "degenerate"). The **principal axes** of this surface lie in the direction of the characteristic vectors of the coefficient matrix S, and the characteristic values of S determine the critical dimensions of the surface.*

The same sort of thing applies in two dimensions, of course.

Example 104-2.   What is the plane graph of the equation $x^2 - 4xy - 2y^2 = 3$?

Solution.   In this case, our matrix S is 2 by 2:

$$S = \begin{bmatrix} 1 & -2 \\ -2 & -2 \end{bmatrix}.$$

Its characteristic equation is

$$\begin{vmatrix} 1 - \lambda & -2 \\ -2 & -2 - \lambda \end{vmatrix} = \lambda^2 + \lambda - 6 = (\lambda + 3)(\lambda - 2) = 0,$$

and therefore $\lambda_1 = 2$ and $\lambda_2 = -3$. There is a rotation of axes under which our equation becomes

$$2\bar{x}^2 - 3\bar{y}^2 = 3,$$

which we recognize as a hyperbola.

In Section 67 we learned that the graph of the quadratic equation

(104-4) $$ax^2 + bxy + cy^2 + dx + ey + f = 0$$

is a conic, which one being determined by the *discriminant* $b^2 - 4ac$. To treat Equation 104-4 in accordance with the ideas we have been developing in this chapter, we associate with the quadratic form $ax^2 + bxy + cy^2$ the symmetric matrix

$$S = \begin{bmatrix} a & \tfrac{1}{2}b \\ \tfrac{1}{2}b & c \end{bmatrix}.$$

Its characteristic values λ_1 and λ_2 are solutions of the characteristic equation

$$(104\text{-}5) \quad \begin{vmatrix} a - \lambda & \frac{1}{2}b \\ \frac{1}{2}b & c - \lambda \end{vmatrix} = \lambda^2 - (a + c)\lambda + ac - \tfrac{1}{4}b^2 = 0.$$

After a suitable translation and rotation of axes, Equation 104-4 takes the form

$$\lambda_1 \bar{x}^2 + \lambda_2 \bar{y}^2 = p$$

(or $\bar{x}^2 = q\bar{y}$ if one of the characteristic values is 0). Thus except for "degenerate" cases, Equation 104-4 represents an ellipse if λ_1 and λ_2 have the same sign, a hyperbola if they have opposite signs, and a parabola if one of them is 0. Since the term $ac - \frac{1}{4}b^2$ in the quadratic Equation 104-5 is the product of the numbers λ_1 and λ_2, we see that λ_1 and λ_2 have the same sign if $4ac - b^2$ is positive, have opposite signs if $4ac - b^2$ is negative, and one of the characteristic values is 0 if $4ac - b^2$ is 0. Therefore (as we found earlier) the graph of Equation 104-4 is (taking degenerate cases into account)

(1)   an ellipse if $4ac - b^2 > 0$,

(2)   a hyperbola if $4ac - b^2 < 0$,

(3)   a parabola if $4ac - b^2 = 0$.

PROBLEMS   104

The first seven problems are closely related to the corresponding problems of the preceding section.

1.   Consider the quadratic equation $2x^2 + 2xy + 2y^2 = 1$.

(a) Under a suitable rotation of axes, the equation is reduced to what standard form?

(b) Find a rotation matrix (an orthogonal matrix of determinant 1) that produces this transformation.

(c) Through what angle are the axes rotated?

(d) Sketch the graph of our equation.

2.   Answer the questions of the first problem for the quadratic equation $x^2 - 4xy + y^2 = 1$.

3.   Answer the questions of the first problem for the equation $7x^2 + 6xy - y^2 = 8$.

4.   Express the quadratic equation in standard form (without finding the transformation equations).

(a) $4x^2 + 2xy + 4y^2 = 15$ 　　　　(b) $4x^2 + 4xy + y^2 = 20$

(c) $2xy = 1$ 　　　　(d) $2xy = 25$

(e) $x^2 + 4xy + 3y^2 = 9$ 　　　　(f) $2x^2 + 6xy - 2y^2 = \sqrt{13}$

5.   Answer parts (a), (b), and (d) of the first problem for the quadratic equation $x^2 + 2y^2 + 2z^2 + 2xy + 2xz = 1$.

6. Answer parts (a), (b), and (d) of the first problem for the quadratic equation $2x^2 + y^2 + z^2 - 4xy - 4xz - 2yz = 1$.

7. Express the quadratic equation in standard form (without finding the transformation equations).

(a) $x^2 + 4yz = 2$ (b) $2x^2 + 2y^2 + z^2 + xy = 1$
(c) $2x^2 + y^2 + 2z^2 + 2xz = 5$ (d) $xy + yz = 1$
(e) $5x^2 + 3y^2 + 3z^2 - 2xy - 2xz + 2yz = 6$
(f) $5x^2 + 3y^2 + 3z^2 - 2xy - 2xz + 2yz = -6$

8. Describe the graph of the equation $A\mathbf{r} \cdot \mathbf{r} = 1$, where $\mathbf{r} = (x, y)$, and A is the given matrix.

(a) $\begin{bmatrix} 1 & 4 \\ 0 & -2 \end{bmatrix}$ (b) $\begin{bmatrix} 1 & -4 \\ 0 & -2 \end{bmatrix}$ (c) $\begin{bmatrix} 0 & 8 \\ 0 & -6 \end{bmatrix}$ (d) $\begin{bmatrix} 4 & 1 \\ 9 & 4 \end{bmatrix}$

9. What is the graph of the equation $S\mathbf{r} \cdot \mathbf{r} + \mathbf{a} \cdot \mathbf{r} + 1 = 0$ if $S = \begin{bmatrix} 1 & 1 \\ 1 & 1 \end{bmatrix}$, $\mathbf{a} = (4\sqrt{2}, 8\sqrt{2})$, and $\mathbf{r} = (x, y)$?

10. Consider the vector-matrix initial value problem $\dfrac{d\mathbf{r}}{dt} = S\mathbf{r}$ and $\mathbf{r} = (1, 0)$ when $t = 0$, where $S = \begin{bmatrix} 0 & 12 \\ 12 & 7 \end{bmatrix}$ and $\mathbf{r} = (x, y)$.

(a) Show that the substitution $\bar{\mathbf{r}} = V^*\mathbf{r}$ (that is, $\mathbf{r} = V\bar{\mathbf{r}}$), where V is an orthogonal matrix such that $V^*SV = D$, a diagonal matrix, reduces our problem to $\dfrac{d\bar{\mathbf{r}}}{dt} = D\bar{\mathbf{r}}$ and $\bar{\mathbf{r}} = V^*(1, 0)$ when $t = 0$.

(b) Since this latter problem is equivalent to the *two* scalar problems

$$\frac{d\bar{x}}{dt} = \lambda_1\bar{x} \quad \text{and} \quad \bar{x} = \bar{x}_0 \quad \text{when } t = 0$$

and

$$\frac{d\bar{y}}{dt} = \lambda_2\bar{y} \quad \text{and} \quad \bar{y} = \bar{y}_0 \quad \text{when } t = 0,$$

we say that we have "uncoupled" the original problem. Find the specific numbers λ_1, λ_2, x_0, and y_0 and solve these problems by the techniques of Section 61.

(c) Now find the solution of our original initial value problem.

11. Suppose that the graph of Equation 104-1 is an ellipsoid. Express the lengths of its longest and shortest diameters in terms of characteristic values of S.

12. Show that if the matrix S has two equal characteristic values, then the graph of Equation 104-1 is a surface of revolution. What if S has three equal characteristic values?

13. What is the graph of Equation 104-1 if one of the characteristic values of S is 0? If two characteristic values are 0?

14. The numbers 18, 9, and -9 are the characteristic values of the matrix of Equation 103-4. We labeled these numbers λ_1, λ_2, and λ_3. What if we had labeled them λ_3, λ_2, and λ_1?

By now you should have a pretty good picture of vectors as arrows in the plane or in space. You can use them to find lines and planes and to discuss space curves. Notice, as we go along in the book, how often we return to familiar themes: a line is determined by a point and a direction; a derivative gives the direction of a tangent line; ellipses and ellipsoids, hyperbolas and hyperboloids are graphs of quadratic equations. Among the few really new things we took up in this chapter were the cross product and the use of matrices to generate rotations of axes.

Review Problems, Chapter Twelve

1. Suppose that the coordinates of a point P are $(3, 4, 5)$. What are the coordinates of the points Q, R, S, and T, where Q is symmetric to P with respect to the plane $z = 1$, R is symmetric to P with respect to the plane $x = -2$, S is symmetric to P with respect to the point $(1, 2, 3)$, and T is symmetric to P with respect to the plane $x + y = 0$?

2. Let $P(-1, -2, -3)$, $Q(-3, 2, 4)$, and $R(2, -1, 0)$ be three points in space. Find the point S such that $PQRS$ is the parallelogram in which Q and S are opposite vertices.

3. Suppose that $\mathbf{a}$ and $\mathbf{b}$ are nonparallel vectors. Find a number t such that the vectors $(1 - t)\mathbf{a} + t\mathbf{b}$ and $3\mathbf{a} + 4\mathbf{b}$ are parallel.

4. What is the equation of the plane that intersects the coordinate axes in the points $(a, 0, 0)$, $(0, b, 0)$, and $(0, 0, c)$?

5. Suppose that along the line $x + y = 1$ in the XY-plane we move a line that is parallel to the vector $\mathbf{i} + \mathbf{j} + \mathbf{k}$. What is the equation of the plane that the moving line generates?

6. At what point of the curve $\mathbf{r} = (3t + 1)\mathbf{i} + 2t^2\mathbf{j} + (6t - 1)\mathbf{k}$ is the tangent vector parallel to the vector $\mathbf{i} + 4\mathbf{j} + 2\mathbf{k}$?

7. Find a vector equation of the line that is tangent to the space curve $\mathbf{r} = \ln(t + 1)\mathbf{i} + e^{2t}\mathbf{j} + (t - 2)^2\mathbf{k}$ at the point at which $t = 0$. Find the plane that is perpendicular to the line at this point.

8. Let $P(a, \theta, \phi)$ be a point of the sphere $\rho = a$ and extend the segment OP until it intersects the cylinder $r = a$. What are the cylindrical coordinates of the point of intersection?

9. Sketch the surface.

    (a) $\dfrac{x^2}{25} + \dfrac{y^2}{16} + \dfrac{z|z|}{9} = 1$

    (b) $\dfrac{x^2}{25} + \dfrac{y^2}{16} + \dfrac{z|z|}{9} = 0$

    (c) $z|z| = x^2 + y^2$

    (d) $z + |z| = 8(x^2 + y^2)$

    (e) $\left[\!\left[\dfrac{x^2}{25} + \dfrac{y^2}{16} + \dfrac{z^2}{9} \right]\!\right] = 0$

    (f) $\ln\left(\dfrac{x^2}{25} + \dfrac{y^2}{16} + \dfrac{z^2}{9} \right) = 0$

10. Pick r, s, and t so that the planes

$$x + 3y = r, \quad 2x - 3z = s, \quad \text{and} \quad 2y + z = t$$

intersect. Can you choose r, s, and t so that they intersect in just one point?

11. Are the points $(-1, 2, 3)$ and $(2, -1, 1)$ on the same or opposite sides of the plane $3x - 2y + z = 5$?

12. Discuss the graph of the equation

$$5y^2 + 4z^2 + 4xy + 8xz + 12yz + 12y - 12z = 0.$$

13. Find the characteristic values and corresponding characteristic vectors for the matrix $S = \begin{bmatrix} 0 & 1 & 1 \\ 1 & 0 & 1 \\ 1 & 1 & 0 \end{bmatrix}$. Now find an orthogonal matrix V such that $V*SV$ is a diagonal matrix. Use this result to discuss the graph of the equation $xy + yz + zx - 2x - 5y - z + 1 = 0$.

14. The heart of a woolly mammoth occupies the region $4x^2 + 9y^2 + 36z^2 \le 36$, and Og hurls his spear along the path $\mathbf{r} = 3t\mathbf{i} - 6\sqrt{t}\,\mathbf{j} + (t + 2)\mathbf{k}$. Does Mrs. Og get her mammoth skin coat?

15. (a) Find the characteristic values and characteristic vectors for the matrix T of Problem 90-10.

    (b) Let P be the matrix whose columns are these characteristic vectors and calculate $P^{-1}TP$.

    (c) Conclude that $\lim_{n \uparrow \infty} (P^{-1}TP)^n = \begin{bmatrix} 1 & 0 \\ 0 & 0 \end{bmatrix} \left(\text{or } \begin{bmatrix} 0 & 0 \\ 0 & 1 \end{bmatrix} \right)$ and hence that

    $\lim_{n \uparrow \infty} T^n = P \begin{bmatrix} 1 & 0 \\ 0 & 0 \end{bmatrix} P^{-1} = \frac{1}{13} \begin{bmatrix} 4 & 4 \\ 9 & 9 \end{bmatrix}$ (Problem 91-14).

16. It is a fact that the determinant of a matrix is the product of its characteristic values and that the sum of its "main diagonal" elements (called the **trace** of the matrix) is the sum of its characteristic values. Show how these statements apply to the matrix

$$\begin{bmatrix} 0 & 2 & 1 \\ 2 & 0 & -1 \\ 1 & -1 & 0 \end{bmatrix}.$$

17. Convince yourself that the equation $(\mathbf{a} \times \mathbf{b}) \times (\mathbf{c} \times \mathbf{d}) = \mathbf{0}$ means that the vectors $\mathbf{a}$, $\mathbf{b}$, $\mathbf{c}$, and $\mathbf{d}$ are parallel to the same plane (or that $\mathbf{a} \parallel \mathbf{b}$ or $\mathbf{c} \parallel \mathbf{d}$).

18. Show that if a certain surface is a surface of revolution about *both* the X-axis and the Y-axis, then it must be a sphere whose center is the origin.

19. What is the distance between two parallel planes $ax + by + cz + d = 0$ and $ax + by + cz + e = 0$?

20. Show that if $\mathbf{r} = \mathbf{f}(t)$ and $\mathbf{r} = \mathbf{g}(t)$ satisfy the differential equation $\dfrac{d\mathbf{r}}{dt} = \mathbf{a}(t) \times \mathbf{r}$, where $\mathbf{a}$ is a given vector-valued function, then $\dfrac{d}{dt}\,[\mathbf{f}(t) \cdot \mathbf{g}(t)] = 0$.

21. Suppose that **a** and **b** are given vectors (with $\mathbf{a} \neq \mathbf{0}$) and consider the equation $\mathbf{a} \times \mathbf{x} = \mathbf{b}$. Show that this equation does not have a solution if $\mathbf{a} \cdot \mathbf{b} \neq 0$. If $\mathbf{a} \cdot \mathbf{b} = 0$, show that $\mathbf{x} = |\mathbf{a}|^2(\mathbf{b} \times \mathbf{a})$ satisfies the equation (use the formula of Problem 95-11). Explain why we can add a scalar multiple of **a** to a solution and obtain another solution. Can every solution be written as $r\mathbf{a} + |\mathbf{a}|^{-2}(\mathbf{b} \times \mathbf{a})$?

22. The equation of a plane, $ax + by + cz = d$, can be written as $\mathbf{n} \cdot \mathbf{r} = d$, where $\mathbf{n} = (a, b, c)$ and $\mathbf{r} = (x, y, z)$. Discuss the problem of choosing a rotation matrix U so that the transformation of coordinates equation $\mathbf{r} = U^*\bar{\mathbf{r}}$ reduces the equation of our plane to the simple form $e\bar{z} = d$. In other words, how do we rotate axes so that our plane is parallel to the $\bar{X}\bar{Y}$-plane?

23. Show that

$$\begin{vmatrix} x & y & z \\ x_1 & y_1 & z_1 \\ x_2 & y_2 & z_2 \end{vmatrix} + \begin{vmatrix} x & y & z \\ x_3 & y_3 & z_3 \\ x_1 & y_1 & z_1 \end{vmatrix} + \begin{vmatrix} x & y & z \\ x_2 & y_2 & z_2 \\ x_3 & y_3 & z_3 \end{vmatrix} = \begin{vmatrix} x_1 & y_1 & z_1 \\ x_2 & y_2 & z_2 \\ x_3 & y_3 & z_3 \end{vmatrix}$$

is the plane containing (x_1, y_1, z_1), (x_2, y_2, z_2), and (x_3, y_3, z_3).

24. (a) By writing each side of the equation in terms of the elements of the matrices that appear, show that $(AB)^* = B^*A^*$ for 2 by 2 matrices A and B.
    (b) Suppose that A and B are square matrices of the same order (not necessarily 2 by 2). Show that $\mathbf{x} \cdot (AB)^*\mathbf{y} = \mathbf{x} \cdot B^*A^*\mathbf{y}$ for arbitrary vectors **x** and **y**. Can you show that this identity implies that $(AB)^* = B^*A^*$?
    (c) Use the equation $(AB)^* = B^*A^*$ to show that the product of two orthogonal matrices is orthogonal.

25. Suppose that $A = \begin{bmatrix} a & b \\ c & d \end{bmatrix}$. What are the relations among the numbers a, b, c, and d if $AA^* = A^*A$?

13 | Partial Derivatives

The law of a falling body, $s = 16t^2$, defines a simple function. With each number t there is paired a number s. If the body falls for t seconds, it will travel s feet. Most physical situations cannot be described by such simple functions, however. In order to compute the pressure of a certain quantity of gas in a cylinder, for example, we must prescribe not one but two numbers, numbers that measure volume and temperature. So we must extend our notion of function to cover such cases and extend our knowledge of calculus to help us understand such functions.

105. FUNCTIONS ON R^n

By definition, a function is a set of pairs, no two of which have the same first member. If $f(x) = \sqrt{1 - x^2}$, for example, some of the pairs in f are $(1, 0)$, $(0, 1)$, $(-\frac{1}{2}, \frac{1}{2}\sqrt{3})$, and so on. The first members of the pairs in f constitute its *domain D* and the second members make up its *range R*. In most of the functions of elementary mathematics, D and R are sets of *numbers* (subsets of R^1). But in trigonometry we take D to be a set of angles, and earlier in this book we met vector-valued functions, functions whose ranges are sets of vectors.

Now we are going to look more systematically at functions whose domains and ranges are subsets of R^1, R^2, R^3, and so on. The topics we take up are extensions of ideas you have met before, but there are notational complications as we go from 1 to higher dimensions. As usual, we use boldface letters to denote elements of R^2 and R^3, as well as to denote functions whose values belong to one of these spaces. It is often convenient to assign as the name of a vector in R^n the first component written in boldface type. Thus we will write $\mathbf{x} = (x, y)$, $\mathbf{u} = (u, v, w)$, and so on. To aid in distinguishing between vectors and scalars, we continue our practice of writing vectors in roman, rather than italic, typeface. Thus we write $\mathbf{x}$, not x, for a vector.

Let us give an example of this notation in action. The equation $f(x, y, z) = 2x - 2y + z$ defines a function whose domain is R^3 and whose range is R^1. Thus $f(1, 2, 3) = 2 \cdot 1 - 2 \cdot 2 + 3 = 1$; that is, our function pairs the number 1 of R^1 with the triple $(1, 2, 3)$ of R^3. If we write $\mathbf{m} = (2, -2, 1)$ and $\mathbf{x} = (x, y, z)$, the defining equation for f becomes $f(\mathbf{x}) = \mathbf{m} \cdot \mathbf{x}$.

If the domain of f is a subset of R^n, we say that f is a **function on R^n**. Thus the function of the last paragraph is a function on R^3. The usual functions we have dealt with up to now (for example, the trigonometric functions) are functions on R^1. If the range of a function is a subset of R^1, we will say that the function is **scalar-valued**; if the range of $\mathbf{f}$ is a subset of R^n, where $n > 1$, then $\mathbf{f}$ is **vector-valued**.

In principle, there is nothing difficult about the idea of a vector-valued function on R^n, but in practice such a function may be rather complicated to describe and work with. Again let us look at an example. The equation

$$(105\text{-}1) \qquad \mathbf{g}(x, y, z) = (\ln z, x + y)$$

defines a vector-valued function $\mathbf{g}$ on R^3, one whose range is contained in R^2. Thus $\mathbf{g}(1, 1, 1) = (\ln 1, 1 + 1) = (0, 2)$. Since $\mathbf{g}(x, y, z)$ is a two-component vector [that is, $\mathbf{g}(x, y, z) = (g(x, y, z), h(x, y, z))$], our *single* vector Equation 105-1 is equivalent to *two* **component equations**

$$g(x, y, z) = \ln z \quad \text{and} \quad h(x, y, z) = x + y.$$

We live in a three-dimensional world and write on two-dimensional paper, so we run into certain technical problems when we try to graph functions on R^n if $n > 1$. The case of scalar-valued functions on R^2, however, is one that we can handle. Here we proceed just as we do with scalar-valued functions on R^1. The graph of a scalar-valued function f on R^2 is simply the graph of the equation $z = f(x, y)$, and we discussed such graphs in the last chapter. They are surfaces in space.

Example 105-1. Sketch the graph of f if $f(x, y) = 2\sqrt{4 - x^2 - 4y^2}$.

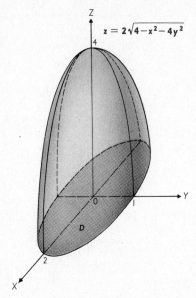

$z = 2\sqrt{4 - x^2 - 4y^2}$

Figure 105-1

Solution. The domain D of f is the subset of R^2 for which our defining formula makes sense; here it is the set of points that satisfy the inequality $4 - x^2 - 4y^2 \geq 0$. When we write this inequality as $\frac{1}{4}x^2 + y^2 \leq 1$, we see that D is the elliptical region shown in the XY-plane in Fig. 105-1. The graph of f is the surface $z = 2\sqrt{4 - x^2 - 4y^2}$, the upper half of the ellipsoid $\frac{1}{4}x^2 + y^2 + \frac{1}{16}z^2 = 1$, which we have sketched.

For a scalar-valued function on R^1, the equation $\lim_{x \to c} f(x) = L$ means (roughly) that L is the number that is approximated by $f(x)$ when x is close to c. And f is *continuous* at c if L can be replaced by $f(c)$; that is, the *limit* of f at c is the *value* of f at c. These same statements apply to functions on R^n.

Example 105-2. Explain why f is continuous at each $\mathbf{c} \in R^3$ if $f(\mathbf{x}) = \mathbf{m} \cdot \mathbf{x}$, where $\mathbf{m}$ is a given vector.

Solution. We must convince ourselves that $\lim_{x \to c} \mathbf{m} \cdot \mathbf{x} = \mathbf{m} \cdot \mathbf{c}$—that is, that $\mathbf{m} \cdot \mathbf{c}$ is the number that is approximated by $\mathbf{m} \cdot \mathbf{x}$ when $\mathbf{x}$ is close to $\mathbf{c}$. Since $\mathbf{m} \cdot \mathbf{x} - \mathbf{m} \cdot \mathbf{c} = \mathbf{m} \cdot (\mathbf{x} - \mathbf{c}) = |\mathbf{m}| \, |\mathbf{x} - \mathbf{c}| \cos \theta$ (where we are assuming that the vectors $\mathbf{m}$ and $\mathbf{x} - \mathbf{c}$ determine an angle θ), we see that $|\mathbf{m} \cdot \mathbf{x} - \mathbf{m} \cdot \mathbf{c}| \leq |\mathbf{m}| \, |\mathbf{x} - \mathbf{c}|$. This inequality clearly shows that $\mathbf{m} \cdot \mathbf{x}$ is close to $\mathbf{m} \cdot \mathbf{c}$ when $\mathbf{x}$ is close to $\mathbf{c}$, which is what we were trying to demonstrate.

A more precise definition of the limit for functions in higher dimensional space can be formulated along the lines we used for scalar-valued functions in Section 10. We first have to define what we mean by a neighborhood of a point in space, that is, an open interval that contains the point. We will agree that the inequality $\mathbf{a} < \mathbf{b}$ between vectors in R^n means inequality between corresponding components. For example, $(2, -1, 3) < (4, 0, 5)$. Then the **open interval** $(\mathbf{a}, \mathbf{b})$ is simply the set of points that satisfy the inequalities $\mathbf{a} < \mathbf{x} < \mathbf{b}$. Now we say, as before, that *the equation* $\lim_{x \to c} \mathbf{f}(\mathbf{x}) = \mathbf{L}$ *means that each neighborhood of* $\mathbf{L}$ *contains the image under* $\mathbf{f}$ *of a punctured neighborhood of* $\mathbf{c}$.

Proofs of limit theorems (limits of sums are sums of limits and the like) must all be based on some such formal definition of the limit. Details of these proofs do not belong in an introductory course, and we will simply use without proof certain natural extensions of the limit theorems we have already applied to functions on R^1.

PROBLEMS 105

1. In this problem f is defined by the statement, "To each point $(x, y, z) \in R^3$ let there correspond the number xy^2z^3."
   (a) What is the formula for $f(x, y, z)$?
   (b) Show that $f(x, y, z)^2 = f(x^2, y^2, z^2)$.
   (c) Is the equation $f(2x, 2y, 2z) = 2f(x, y, z)$ true for every point of R^3? Any point?
   (d) Compare the numbers $f(\mathbf{x})$ and $f(-\mathbf{x})$, where $\mathbf{x}$ is an arbitrary point of R^3.
   (e) Express the product $f(x, y, z)f(y, z, x)f(z, x, y)$ in terms of x, y, and z.
   (f) Let $\mathbf{x} = (3, 2, 1)$ and $\mathbf{j} = (0, 1, 0)$ and define the quotient function q on R^1 by means of the equation
   $$q(h) = \frac{f(\mathbf{x} + h\mathbf{j}) - f(\mathbf{x})}{h}.$$
   What is the domain of q? What is $\lim_{h \to 0} q(h)$?

2. Let $f(x, y) = x^y$.
   (a) Explain why the point $(-3, \frac{1}{2})$ is not in the domain of f.
   (b) Find a number y such that $(-3, y)$ is in the domain of f.
   (c) What is $\lim_{(x,y) \to (e, \ln 5)} f(x, y)$?
   (d) Is -3 in the range of f?
   (e) Show that $f(a, b + c) = f(a, b)f(a, c)$ and $f(ab, c) = f(a, c)f(b, c)$.
   (f) Is f continuous at $(-3, 3)$?

3. Sketch the graph of f and state its domain and range. At what points of its domain is f continuous?
   (a) $f(x, y) = 4x + 9y$ (b) $f(x, y) = 4x^2 + 9y^2$
   (c) $f(x, y) = \sin x$ (d) $f(x, y) = \text{Arcsin } y$
   (e) $f(\mathbf{x}) = |\mathbf{x}|^2$, $\mathbf{x} \in R^2$ (f) $f(\mathbf{x}) = (3\mathbf{i} - 4\mathbf{j}) \cdot \mathbf{x} + 12$, $\mathbf{x} \in R^2$

4. Each equation defines a function on R^3. What is the domain of f? Its range? At what points is f continuous?
   (a) $f(x, y, z) = \sqrt{x - y} \ln z$ (b) $f(x, y, z) = \sqrt{z} \ln (x - y)$
   (c) $f(x, y, z) = \sqrt{x + y + z}$ (d) $f(x, y, z) = \tan (x + y + z)$

5. Suppose that $f(x, y)$ is the slope (if any) of the line joining (x, y) and $(0, 0)$. What is the formula that defines f? Graph f. Is f continuous at each point of its domain?

6. Let $f(x, y) = (x - y)/|x - y|$ and sketch the graph of f.

7. The resistance of a piece of wire is directly proportional to its length and inversely proportional to the square of its radius. If a wire 10 centimeters long with a radius of 2 millimeters has a resistance of .1 ohm, what is the formula expressing the resistance in terms of the length and radius? In what units should the various quantities in your formula be expressed?

8. Let
   $$f(x, y, z) = \begin{vmatrix} 1 & 1 & 1 \\ x & y & z \\ x^2 & y^2 & z^2 \end{vmatrix}.$$
   Describe the set of points in space for which $f(x, y, z) = 0$.

9. Let **m** = 2**i** − 3**j** + 4**k**.

    (a) If $f(\mathbf{x}) = \mathbf{m} \cdot \mathbf{x}$, describe the set of points in R^3 such that $f(\mathbf{x}) = 7$.

    (b) If $\mathbf{f}(\mathbf{x}) = \mathbf{m} \times \mathbf{x}$, describe the set of points in R^3 such that $\mathbf{f}(\mathbf{x}) = 7\mathbf{i} - 2\mathbf{j} - 5\mathbf{k}$.

10. (a) Express the volume of a rectangular box as the value of a function on R^3.

    (b) If the box is made of 10 square feet of material, express its volume as the value of a function on R^2.

11. Let $M = \begin{bmatrix} 5 & 3 \\ 1 & 2 \end{bmatrix}$ and $f(\mathbf{x}) = \mathbf{x} \cdot M\mathbf{x}$.

    (a) What is the domain of f?

    (b) Write $f(\mathbf{x})$ in terms of x and y.

    (c) Discuss the graph of f.

12. Show that if $f(x, y) \geq f(y, x)$ for each point $(x, y) \in R^2$, then $f(x, y) = f(y, x)$. What can you conclude if you know that $f(x, y, z) \geq f(y, z, x)$ for each point $(x, y, z) \in R^3$?

13. Picture the interval (**a**, **b**) in the plane or in space.

    (a) **a** = (0, 0), **b** = (1, 2)   (b) **a** = (−1, 2), **b** = (2, 5)

    (c) **a** = (0, 0, 0), **b** = (1, 2, 3)   (d) **a** = (1, 2, 3), **b** = (4, 5, 6)

106. PARTIAL DERIVATIVES

Now let us introduce some of the ideas of calculus into our study of functions on R^n. From a given scalar-valued function f on R^1, we construct a derived function f' by means of the equation $f'(x) = \lim\limits_{h \to 0} \dfrac{f(x + h) - f(x)}{h}$. In a similar manner, we will now construct derived functions from a given function on R^n. Most of our discussion will refer to scalar-valued functions on R^2, but the concepts can be extended to functions on R^n for any n.

If f is a scalar-valued function on R^2, we define two **derived functions** f_1 and f_2:

$$(106\text{-}1) \qquad f_1(x, y) = \lim_{h \to 0} \frac{f(x + h, y) - f(x, y)}{h}$$

and

$$(106\text{-}2) \qquad f_2(x, y) = \lim_{h \to 0} \frac{f(x, y + h) - f(x, y)}{h}.$$

The numbers $f_1(x, y)$ and $f_2(x, y)$ are called the **(first) partial derivatives** of $f(x, y)$. Although the idea of partial derivatives is new to us, the process of computing them is just our usual process of differentiation, as the next example shows.

Example 106-1. Find $f_1(2, 3)$ if $f(x, y) = 2y^3 \sin x$.

Solution. According to Equation 106-1,

$$(106\text{-}3) \qquad f_1(2, 3) = \lim_{h \to 0} \frac{f(2 + h, 3) - f(2, 3)}{h}$$

$$= \lim_{h \to 0} \frac{2 \cdot 3^3 \sin (2 + h) - 2 \cdot 3^3 \sin 2}{h},$$

and we can use our knowledge of ordinary differentiation to find this limit. We will simply let $g(x) = f(x, 3) = 2 \cdot 3^3 \sin x$ and observe that Equation 106-3 then becomes

$$f_1(2, 3) = \lim_{h \to 0} \frac{g(2 + h) - g(2)}{h}.$$

Since this last limit is the number $g'(2)$, we have reduced the problem of finding the *partial* derivative $f_1(2, 3)$ to that of finding the *ordinary* derivative $g'(2)$. Of course, we know that

$$g'(x) = \frac{d}{dx} 2 \cdot 3^3 \sin x = 2 \cdot 3^3 \frac{d}{dx} \sin x = 54 \cos x$$

and therefore $g'(2) = 54 \cos 2$. Thus $f_1(2, 3) = 54 \cos 2$.

We found the derivative in our example by "fixing" the value of y in the formula $2y^3 \sin x$ at 3 and then applying the ordinary differential operator $\frac{d}{dx}$. In general, the definition of $f_1(x, y)$ tells us that when we find this derivative, we are to treat y as a "fixed" number and differentiate $f(x, y)$ with respect to x. For example, if $f(x, y) = 2y^3 \sin x$, then

$$f_1(x, y) = \frac{d}{dx} 2y^3 \sin x = 2y^3 \frac{d}{dx} \sin x = 2y^3 \cos x.$$

This equation is valid for any x and y. If we want $f_1(2, 3)$, we replace x with 2 and y with 3. Similarly, in order to find $f_2(x, y)$, we apply the differential operator $\frac{d}{dy}$, keeping x "fixed" when we do. In our present example, therefore, $f_2(x, y) = \frac{d}{dy} 2y^3 \sin x = 6y^2 \sin x$. What we are saying is that the derivative $f_1(x, y)$ is the result of differentiating $f(x, y)$ with respect to the *first* component of the element (x, y) of R^2. We obtain $f_2(x, y)$ by differentiating $f(x, y)$ with respect to the *second* component.

When dealing with partial derivatives, it is customary to replace the "straight" d in the symbol $\dfrac{d}{dx}$ with the "curved d" ∂ and thereby obtain the partial differential operator $\dfrac{\partial}{\partial x}$. This operator says, "Differentiate with respect to x, holding everything else fixed." If $w = f(x)$, we write $w' = \dfrac{dw}{dx} = f'(x)$. The corresponding notation in case $w = f(x, y)$ is

$$w_x = \frac{\partial w}{\partial x} = f_1(x, y) \quad \text{and} \quad w_y = \frac{\partial w}{\partial y} = f_2(x, y).$$

We might mention that some authors also write $f_x(x, y)$ and $f_y(x, y)$ where we have written $f_1(x, y)$ and $f_2(x, y)$. The symbol ∂ is a variant of the Greek letter delta.

The derivative

$$\frac{\partial^2}{\partial x^2} f(x, y) = \frac{\partial}{\partial x}\left[\frac{\partial}{\partial x} f(x, y) \right]$$

is a second partial derivative of $f(x, y)$, and we denote it by $f_{11}(x, y)$. (The symbol f_{11} can be read as "f sub one, one." If f were a function on R^n with $n > 11$, we might have to write $f_{1,1}$ to distinguish it from "f sub eleven," but we omit the comma when no confusion can arise.) The derivative

$$\frac{\partial^2}{\partial y\, \partial x} f(x, y) = \frac{\partial}{\partial y}\left[\frac{\partial}{\partial x} f(x, y) \right]$$

is a "mixed" second partial derivative that we denote by $f_{12}(x, y)$, and so on. Other symbols for second derivatives, if $w = f(x, y)$, are w_{xx}, w_{xy}, $\dfrac{\partial^2 w}{\partial y\, \partial x}$, and so on.

Of course, we also talk about derivatives of order higher than the second in similar notation.

Example 106-2. If $w = (x + y) \sin x$, find all its second derivatives.

Solution. The first derivatives of w are

$$\frac{\partial w}{\partial x} = \frac{\partial}{\partial x} [(x + y) \sin x] = \sin x \frac{\partial}{\partial x} (x + y) + (x + y) \frac{\partial}{\partial x} \sin x$$

$$= \sin x + (x + y) \cos x$$

and

$$\frac{\partial w}{\partial y} = \frac{\partial}{\partial y} [(x + y) \sin x] = \sin x \frac{\partial}{\partial y} (x + y) = \sin x.$$

Therefore

$$\frac{\partial^2 w}{\partial x^2} = \frac{\partial}{\partial x} [\sin x + (x + y) \cos x] = 2 \cos x - (x + y) \sin x,$$

$$\frac{\partial^2 w}{\partial y^2} = \frac{\partial}{\partial y} \sin x = 0,$$

$$\frac{\partial^2 w}{\partial x \, \partial y} = \frac{\partial}{\partial x} \sin x = \cos x,$$

$$\frac{\partial^2 w}{\partial y \, \partial x} = \frac{\partial}{\partial y} [\sin x + (x + y) \cos x] = \cos x.$$

You may have trouble remembering that $f_{12}(x, y)$ means $\dfrac{\partial}{\partial y}\left[\dfrac{\partial}{\partial x} f(x, y)\right]$ and not $\dfrac{\partial}{\partial x}\left[\dfrac{\partial}{\partial y} f(x, y)\right]$. In Example 106-2 we see that it doesn't matter; the two numbers are the same anyway. In fact, the two numbers $f_{12}(x, y)$ and $f_{21}(x, y)$ are the same for all the functions we meet in applications. The following theorem is proved in advanced calculus.

Theorem   106-1.   *If the derived functions f_{12} and f_{21} are continuous at a point (x, y), then $f_{12}(x, y) = f_{21}(x, y)$.*

Theorem 106-1 tells us that mixed partial derivatives of the second order are equal (under suitable conditions of continuity). It extends immediately to derivatives of higher order, too.

Example 106-3.   If $f(x, y) = e^{xy}$, show that $f_{112}(x, y) = f_{121}(x, y) = f_{211}(x, y)$.

Solution.   We have

$$f_1(x, y) = \frac{\partial}{\partial x} e^{xy} = y e^{xy} \quad \text{and} \quad f_2(x, y) = \frac{\partial}{\partial y} e^{xy} = x e^{xy}.$$

Hence

$$f_{11}(x, y) = \frac{\partial}{\partial x} y e^{xy} = y^2 e^{xy}$$

and

$$f_{12}(x, y) = \frac{\partial}{\partial y} y e^{xy} = (1 + xy) e^{xy} = f_{21}(x, y).$$

Thus

$$f_{112}(x, y) = \frac{\partial}{\partial y} y^2 e^{xy} = (2y + xy^2) e^{xy}$$

and

$$f_{121}(x, y) = f_{211}(x, y) = \frac{\partial}{\partial x} [(1 + xy) e^{xy}] = y e^{xy} + (1 + xy) y e^{xy} = (2y + xy^2) e^{xy}.$$

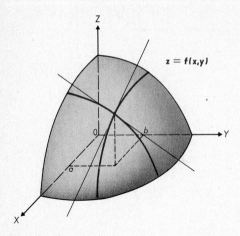

Figure 106-1

Recall that $\dfrac{dy}{dx}$ [or $f'(x)$] is interpreted geometrically as the slope of the tangent line to the curve $y = f(x)$ at the point (x, y). Partial derivatives have a similar geometric interpretation. If the graph of a scalar-valued function f on R^2 is a surface, the intersection of this surface with the plane $y = b$ is a curve, the graph of the equation $z = f(x, b)$ in the plane $y = b$. The number $f_1(a, b)$—which is a value of $\dfrac{\partial z}{\partial x}$—is the slope of this graph at the point where $x = a$ (Fig. 106-1). Similarly, the intersection of the graph of f with the plane $x = a$ is a curve $z = f(a, y)$. Since $\dfrac{\partial z}{\partial y} = f_2(a, y)$, we see that the number $f_2(a, b)$ is the slope of this curve at the point where $y = b$. Thus we may interpret the numbers $f_1(a, b)$ and $f_2(a, b)$ as the slopes of the profile curves obtained by cutting the graph of f with the planes $y = b$ and $x = a$.

One of the most used theorems in the calculus of functions on R^1 is the Chain Rule. You will recall that if f and g are scalar-valued differentiable functions on R^1, and if $u = g(x)$, then the Chain Rule Equation reads

$$\frac{d}{dx} f(u) = f'(u) \frac{du}{dx}.$$

If g is a function on R^n, where $n > 1$, differentiation with respect to x is performed in exactly the same way, but to remind us that we hold everything except x "fixed," we replace every straight d with a curved ∂:

$$(106\text{-}4) \qquad \frac{\partial}{\partial x} f(u) = f'(u) \frac{\partial u}{\partial x}.$$

For example, $\dfrac{\partial}{\partial x} \sin x^2 y = (\cos x^2 y) \dfrac{\partial}{\partial x} (x^2 y) = 2xy \cos x^2 y.$ Of course, we have a similar Chain Rule Equation for differentiation with respect to y:

$$(106\text{-}5) \qquad \frac{\partial}{\partial y} f(u) = f'(u) \frac{\partial u}{\partial y}.$$

Example 106-4. Show that if f is any differentiable function on R^1, then $w = f(3x + 4y)$ satisfies the *partial differential equation*

$$(106\text{-}6) \qquad\qquad 4w_x - 3w_y = 0.$$

Solution. Since $w = f(3x + 4y)$, then $w_x = \dfrac{\partial}{\partial x} f(3x + 4y)$ and $w_y = \dfrac{\partial}{\partial y} f(3x + y)$. We calculate these derivatives by replacing u with $3x + 4y$ in Equations 106-4 and 106-5:

$$w_x = f'(3x + 4y) \frac{\partial}{\partial x} (3x + 4y) = 3f'(3x + 4y)$$

and

$$w_y = f'(3x + 4y) \frac{\partial}{\partial y} (3x + 4y) = 4f'(3x + 4y).$$

Therefore

$$4w_x - 3w_y = 12f'(3x + 4y) - 12f'(3x + 4y) = 0.$$

What this example shows is that Equation 106-6 (like all partial differential equations) has many, many solutions. For example, $w = \sin(3x + 4y)$, $w = \ln(3x + 4y)$, and $w = \text{Arctan}(3x + 4y)^{1492}$ all satisfy it.

PROBLEMS 106

1. Find $f_1(3, -2)$.

(a) $f(x, y) = 3x^2 - 2y^3$

(b) $f(x, y) = 2x^3 - 3y^2$

(c) $f(x, y) = 3x^2y^3$

(d) $f(x, y) = 3x^3y^2$

(e) $f(x, y) = \dfrac{3x^2}{y^3}$

(f) $f(x, y) = \dfrac{3y^3}{x^2}$

(g) $f(x, y) = x^2e^y$

(h) $f(x, y) = e^xy^2$

(i) $f(x, y) = \text{Sin}^{-1} \dfrac{x}{y}$

(j) $f(x, y) = \text{Sin}^{-1} \dfrac{y}{x}$

(k) $f(x, y) = x^{-y}$

(l) $f(x, y) = (-y)^x$

(m) $f(x, y) = \ln x^y$

(n) $f(x, y) = (\ln x)^y$

(o) $f(x, y) = \displaystyle\int_y^x \cos \pi t^2 \, dt$

(p) $f(x, y) = \displaystyle\int_x^y \cos \pi t^2 \, dt$

2. Calculate $f_{12}(x, y)$ and $f_{21}(x, y)$ and show that they are equal for each function in the preceding problem.

3. Show that w satisfies *Laplace's* partial differential equation $w_{xx} + w_{yy} = 0$.

(a) $w = 2x + 3y - 1$ (b) $w = 3x + 2y + 5$ (c) $w = e^x \cos y$

(d) $w = e^y \sin x$ (e) $w = x^2 - y^2$ (f) $w = 2xy$

(g) $w = \ln(x^2 + y^2)$ (h) $w = \text{Tan}^{-1} \dfrac{y}{x}$ (i) $w = \cos x \cosh y$

(j) $w = \sin y \cosh x$

4. The surface of a mountain is the graph of the equation $z = 17 - 2x^2 - 3y^2$. Is the path down from the point $(1, 2, 3)$ steeper in the positive X-direction or in the positive Y-direction?

5. Let $w = \cos(3x - 4y + 5z)$ and form the vector $\mathbf{v} = \dfrac{\partial w}{\partial x}\mathbf{i} + \dfrac{\partial w}{\partial y}\mathbf{j} + \dfrac{\partial w}{\partial z}\mathbf{k}$. Show that $\mathbf{v}$ is perpendicular to the vector $8\mathbf{i} + \mathbf{j} - 4\mathbf{k}$.

6. Evaluate the determinant $\begin{vmatrix} u_x & v_x \\ u_y & v_y \end{vmatrix}$.

(a) $u = e^{x-y}, v = e^{y-x}$ (b) $u = e^x - e^y, v = e^y - e^x$

(c) $u = x \cos y, v = x \sin y$ (d) $u = x \cosh y, v = x \sinh y$

(e) $u = \displaystyle\int_x^y \cos t^2 \, dt, v = \int_x^y \sin t^2 \, dt$

(f) $u = \displaystyle\int_x^y \cosh^2 t \, dt, v = \int_x^y \sinh^2 t \, dt$

7. Determine from Fig. 106-1 whether the numbers $f_1(a, b)$ and $f_2(a, b)$ are positive or negative. What does the figure tell us about the signs of $f_{11}(a, b)$ and $f_{22}(a, b)$?

8. Let $z = \mathbf{x} \cdot S\mathbf{x}$, where S is a symmetric 2 by 2 matrix and $\mathbf{x} = (x, y)$, and show that

$$\begin{vmatrix} z_{xx} & z_{xy} \\ z_{yx} & z_{yy} \end{vmatrix} = 4 \det S.$$

9. The gas in a certain cylinder is at a pressure of P newtons per square meter when its volume is V cubic meters and its temperature is T degrees; the numbers P, T, and V are related by the equation $P = 10T/V$. How much does the pressure change if we change T from 100 to 101, keeping $V = 200$? Again let us start with $T = 100$ and $V = 200$ and this time hold T at 100. Approximately what change in volume will produce the same pressure change as before?

10. Is the equation $\dfrac{\partial}{\partial x} f(x^2, y^2) = f_1(x^2, y^2)$ true for *every* function f on R^2? For *any* function f on R^2?

11. If $w = h(x, y)$, we define $\Delta w = w_{xx} + w_{yy}$. Show that $\Delta r^{2m} = 4m^2 r^{2m-2}$, if $r^2 = x^2 + y^2$.

12. Let $w = f(u)$, where $u = g(x, y)$.

(a) Show that $\begin{vmatrix} w_x & u_x \\ w_y & u_y \end{vmatrix} = 0$.

(b) Show that $w_{xx} = f'(u)u_{xx} + f''(u)u_x^2$.
(c) Verify these statements if $w = u^2$ and $u = \sin xy$.

13. Suppose that $f(x, y) = xy/(x^2 + y^2)$ if $(x, y) \neq (0, 0)$, and $f(0, 0) = 0$. Use Equations 106-1 and 106-2 with $(x, y) = (0, 0)$ to find $f_1(0, 0)$ and $f_2(0, 0)$. What happens if you first find $\dfrac{\partial}{\partial x} f(x, y)$ and $\dfrac{\partial}{\partial y} f(x, y)$ and then replace (x, y) with $(0, 0)$?

From both practical and theoretical points of view, one of the most important theorems of differential calculus is the Chain Rule: *if* $u = g(x)$, *then*

(107-1)
$$\frac{d}{dx} f(u) = f'(u) \frac{du}{dx}.$$

In the last section we pointed out that if $u = g(x, y)$, we maintain the validity of Equation 107-1 simply by replacing the straight d of ordinary differentiation with the curved ∂ of partial differentiation. Things aren't so simple, however, if we have a vector-valued function **g** instead of the scalar-valued function g.

The difficulties lie mostly in the realms of theory and notation, however. In most specific cases, we can calculate a derivative by using what we already know. Thus to find $\frac{d}{dx} (u^2 + v^2)$, where $u = \sin x$ and $v = e^x$, we can merely substitute and then differentiate:

$$\frac{d}{dx} (u^2 + v^2) = \frac{d}{dx} (\sin^2 x + e^{2x}) = 2 \sin x \cos x + 2e^{2x} = \sin 2x + 2e^{2x}.$$

But often we need a formula for the derivative of a nonspecific $f(u, v)$, and we will now set about developing such a formula. In the process of finding it, we will need a "poor man's Theorem of the Mean for functions on R^2" (we will find a more elegant one later), so let us look into that matter first. This theorem is simply an application of the ordinary Theorem of the Mean for functions on R^1. You recall that if p is differentiable in an interval containing a and c, then there is a number m between a and c such that the tangent to the graph of p at the point $(m, p(m))$ is parallel to the chord joining the points $(a, p(a))$ and $(c, p(c))$. In symbols, this statement reads $\frac{p(c) - p(a)}{c - a} = p'(m)$; that is,

(107-2)
$$p(c) - p(a) = p'(m)(c - a).$$

Here is our generalization to functions on R^2.

Theorem 107-1. *If the partial derivatives $f_1(x, y)$ and $f_2(x, y)$ exist for each point of the right-angle path joining the points (a, b) and (c, d) shown in Fig. 107-1, then there is a number m between a and c and a number n between b and d such that*

(107-3)
$$f(c, d) - f(a, b) = f_1(m, d)(c - a) + f_2(a, n)(d - b).$$

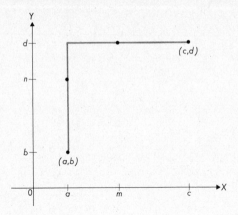

Figure 107-1

PROOF

We write

(107-4) $f(c, d) - f(a, b) = f(c, d) - f(a, d) + f(a, d) - f(a, b)$

and look at the differences $f(c, d) - f(a, d)$ and $f(a, d) - f(a, b)$ separately. If we let $p(x) = f(x, d)$, then $p'(x) = f_1(x, d)$, and so Equation 107-2 becomes

$$f(c, d) - f(a, d) = f_1(m, d)(c - a).$$

Exactly the same argument (with a slight shuffling of letters) shows that there is a number n between b and d such that

$$f(a, d) - f(a, b) = f_2(a, n)(d - b).$$

Now we substitute in Equation 107-4 and obtain Equation 107-3.

We introduced Theorem 107-1 as a tool for deriving a Chain Rule, but before using it for this purpose, we will show that it has other applications as well.

Example 107-1. Show that if $f_1(x, y) = 0$ and $f_2(x, y) = 0$ at each point $(x, y) \in R^2$, then f is a constant function on R^2.

Solution. We must show that the value of f at any point (a, b) is the same as its value at any other point (c, d). Since we are assuming that all first partial derivatives are 0, the numbers $f_1(m, d)$ and $f_2(a, n)$ in Equation 107-3 are 0. Therefore that equation reduces to $f(c, d) - f(a, b) = 0$; that is, $f(c, d) = f(a, b)$.

Now we return to our Chain Rule. We want to find a formula for $\frac{d}{dx} f(u, v)$, where $u = g(x)$ and $v = h(x)$, and we go back to first principles. By the very definition of the derivative,

$$\frac{d}{dx} f(u, v) = \frac{d}{dx} f(g(x), h(x)) = \lim_{z \to x} \frac{f(g(z), h(z)) - f(g(x), h(x))}{z - x}.$$

As it stands, we can't tell what the limit of this difference quotient is, so we rewrite it, using Equation 107-3 with c, d, a, and b replaced by $g(z)$, $h(z)$, $g(x)$, and $h(x)$:

$$(107\text{-}5) \quad \frac{f(g(z), h(z)) - f(g(x), h(x))}{z - x}$$

$$= f_1(m, h(z)) \frac{g(z) - g(x)}{z - x} + f_2(g(x), n) \frac{h(z) - h(x)}{z - x}.$$

All we know about the numbers m and n is that they lie between $g(x)$ and $g(z)$ and $h(x)$ and $h(z)$, but that is enough. For as $z \to x$, then $g(z) \to g(x)$ and $h(z) \to h(x)$, and hence $m \to g(x)$ and $n \to h(x)$. Thus (under the assumption that f_1 and f_2 are continuous and g and h are differentiable) the limit of the right-hand side of Equation 107-5 is $f_1(g(x), h(x))g'(x) + f_2(g(x), h(x))h'(x)$. Since $g(x) = u$, $h(x) = v$, $g'(x) = \frac{du}{dx}$, and $h'(x) = \frac{dv}{dx}$, we therefore see that

$$(107\text{-}6) \quad \frac{d}{dx} f(u, v) = f_1(u, v) \frac{du}{dx} + f_2(u, v) \frac{dv}{dx}.$$

Example 107-2. Use our newly developed Chain Rule Equation 107-6 to find the derivative we calculated in the second paragraph of this section.

Solution. In this example $f(u, v) = u^2 + v^2$, $u = \sin x$, and $v = e^x$. Hence $f_1(u, v) = 2u$, $f_2(u, v) = 2v$, $\frac{du}{dx} = \cos x$, and $\frac{dv}{dx} = e^x$. After substituting these quantities in Equation 107-6, we obtain

$$\frac{d}{dx} (u^2 + v^2) = 2u \cos x + 2ve^x = 2 \sin x \cos x + 2e^x e^x = \sin 2x + 2e^{2x},$$

as before.

One thing that makes the study of Chain Rules difficult for the beginner is that they come in a bewildering variety of notations. These notations aren't selected merely by whim, however; each is useful in particular circumstances. But it is still hard to keep track of all of them. For example, if we write $w = f(u, v)$, then Equation 107-6 takes the form

$$(107\text{-}7) \qquad \frac{dw}{dx} = \frac{\partial w}{\partial u}\frac{du}{dx} + \frac{\partial w}{\partial v}\frac{dv}{dx}.$$

We have been assuming that $u = g(x)$ and $v = h(x)$, where g and h are functions on R^1. If g and h are functions on R^2 [thus $u = g(x, y)$ and $v = h(x, y)$], then each straight d in Equation 107-7 becomes a curved ∂ to denote partial differentiation:

$$(107\text{-}8) \qquad \frac{\partial w}{\partial x} = \frac{\partial w}{\partial u}\frac{\partial u}{\partial x} + \frac{\partial w}{\partial v}\frac{\partial v}{\partial x}.$$

Example 107-3. Use a Chain Rule to compute $\dfrac{\partial}{\partial y}(x + y)^{xy}$.

Solution. Suppose that we think of $(x + y)^{xy}$ as u^v, where $u = x + y$ and $v = xy$. Then we obtain our desired derivative by replacing w with u^v and $\dfrac{\partial}{\partial x}$ with $\dfrac{\partial}{\partial y}$ in Equation 107-8:

$$\frac{\partial}{\partial y}u^v = \frac{\partial}{\partial u}u^v\frac{\partial u}{\partial y} + \frac{\partial}{\partial v}u^v\frac{\partial v}{\partial y}$$

$$= vu^{v-1}\frac{\partial}{\partial y}(x + y) + u^v(\ln u)\frac{\partial}{\partial y}(xy)$$

$$= xy(x + y)^{xy-1} + x(x + y)^{xy}\ln(x + y).$$

Can you figure out how to find this derivative without using a Chain Rule?

Our Chain Rule is closely linked with the process of implicit differentiation that we discussed in Section 26. Suppose that the curve $y = h(x)$ is part of the curve $f(x, y) = 0$. (For example, the semicircle $y = \sqrt{4 - x^2}$ is part of the circle $x^2 + y^2 - 4 = 0$.) Then we can think of x and y as playing the roles of u and v in Equation 107-6, and we have

$$\frac{d}{dx}f(x, y) = f_1(x, y)\frac{dx}{dx} + f_2(x, y)\frac{dy}{dx}.$$

Since $f(x, y) = 0$, then $\dfrac{d}{dx}f(x, y) = 0$. Of course, $\dfrac{dx}{dx} = 1$, so our equation reduces to $0 = f_1(x, y) + f_2(x, y)\dfrac{dy}{dx}$; that is,

$$(107\text{-}9) \qquad \frac{dy}{dx} = -\frac{f_1(x, y)}{f_2(x, y)}.$$

Thus if $f(x, y) = x^2 + y^2 - 4$, this equation tells us that

$$\frac{dy}{dx} = -\frac{2x}{2y} = -\frac{x}{y},$$

and we know that this number is indeed the slope of the circle $x^2 + y^2 = 4$ at the point (x, y).

Example 107-4. Suppose that u and v are functions on R^2 that satisfy the **Cauchy-Riemann** partial differential equations

$$u_1(x, y) = v_2(x, y) \quad \text{and} \quad u_2(x, y) = -v_1(x, y).$$

Show that the curves $u(x, y) = a$ and $v(x, y) = b$, where a and b are given numbers, meet at right angles.

Solution. First we let $u(x, y) - a$ play the role of $f(x, y)$ in Equation 107-9, and we find that the slope of the curve $u(x, y) = a$ is

$$m_1 = -u_1(x, y)/u_2(x, y).$$

Then we set $f(x, y) = v(x, y) - b$ in Equation 107-9 to find that the slope of the curve $v(x, y) = b$ is

$$m_2 = -v_1(x, y)/v_2(x, y).$$

In this last equation we replace $v_1(x, y)$ with $-u_2(x, y)$ and $v_2(x, y)$ with $u_1(x, y)$ in accordance with the Cauchy-Riemann equations:

$$m_2 = u_2(x, y)/u_1(x, y).$$

Therefore the product $m_1 m_2$ of the slopes of our graphs is -1; that is, the curves meet at right angles. It is this property of the graphs of solutions of the Cauchy-Riemann equations that makes them so useful in the study of electric fields, fluid flow, and similar topics.

PROBLEMS 107

1. Find $\dfrac{dw}{dx}$ in two ways. First, replace u and v in terms of x and differentiate. Then use the Chain Rule Equation 107-7.

(a) $w = uv$, $(u, v) = (x^3, x^{-2})$
(b) $w = u/v$, $(u, v) = (x^3, x^{-2})$
(c) $w = \sin(u + v)$, $(u, v) = (x, -x)$
(d) $w = \sin(u + v)$, $(u, v) = (x, x)$
(e) $w = e^v \ln u$, $(u, v) = (e^x, \ln x)$
(f) $w = \text{Sin } v \text{ Arcsin } u$, $(u, v) = (\text{Sin } x, \text{Arcsin } x)$
(g) $w = \begin{vmatrix} u & v \\ u^2 & v^2 \end{vmatrix}$, $(u, v) = (e^x, e^{-x})$

(h) $w = \begin{vmatrix} u & v \\ u^2 & v^2 \end{vmatrix}$, $(u, v) = (\tan x, \cot x)$

2. Suppose that $w = \dfrac{\sin u}{\cos v}$.

   (a) Find $\dfrac{dw}{dx}$ if $(u, v) = (x, x)$ by using Equation 107-7 and verify your answer by direct calculation.

   (b) Find $\dfrac{\partial w}{\partial x}$ if $(u, v) = (x + y, x - y)$ by using Equation 107-8 and verify your answer by direct calculation.

   (c) Find $\dfrac{\partial w}{\partial y}$ if $(u, v) = (x + y, x - y)$ by using Equation 107-8 and verify your answer by direct calculation.

3. We can write the equation $w = x^x$ as $w = u^v$, where $(u, v) = (x, x)$ and then use Equation 107-7 to find $\dfrac{dw}{dx}$. Do it. Do you remember how we found $\dfrac{dw}{dx}$ in an earlier chapter?

4. Use Equation 107-7 to find $\dfrac{d}{dx} \displaystyle\int_{x^2}^{x^3} \sin t^2 \, dt$.

5. Suppose that $f(u, v) = e^{uv}$ and that g and h are functions such that $g(2) = 3$, $g'(2) = 4$, $h(2) = 5$, and $h'(2) = 6$. Use Equation 107-6 to find $F'(2)$ if $F(x) = f(g(x), h(x))$.

6. If $u = g(x)$, $v = h(x)$, and $w = k(x)$, a slight extension of our arguments in this section leads to the Chain Rule Equation

$$\frac{d}{dx} f(u, v, w) = f_1(u, v, w) \frac{du}{dx} + f_2(u, v, w) \frac{dv}{dx} + f_3(u, v, w) \frac{dw}{dx}.$$

   Show that this formula gives the correct value of

$$\frac{d}{dx}\left(\frac{uw}{v}\right) \quad \text{if} \quad (u, v, w) = (x^4, x^{-3}, x^{-5}).$$

7. Use Equation 107-9 to find the slope of the curve at the point $(2, 3)$.

   (a) $4y = 3x^2$ (b) $9x = 2y^2$

   (c) $\dfrac{x^2}{8} + \dfrac{y^2}{18} = 1$ (d) $x^2 - \dfrac{y^2}{3} = 1$

   (e) $x^y = 8$ (f) $y^x = 9$

   (g) $\displaystyle\int_x^y e^{t^2} \, dt = \int_2^3 e^{t^2} \, dt$ (h) $\displaystyle\int_{3x}^{2y} e^{t^2} \, dt = 0$

8. Show that the tangent lines to the curve $f(x, y) = 0$ are horizontal at points where $f_1(x, y) = 0$ and $f_2(x, y) \neq 0$ and are vertical at points where $f_2(x, y) = 0$ and $f_1(x, y) \neq 0$. What can you say about tangent lines at points where $f_1(x, y) = f_2(x, y) \neq 0$?

9. It is the form of a Chain Rule, not the letters we use, that is important. In the following examples we use different letters from those found in Equation 107-6, but we are still talking about the same Chain Rule. Complete the statement of it.

   (a) If $x = g(u)$ and $y = h(u)$, then $\dfrac{d}{du} f(x, y) =$ _____.

   (b) If $p = r(t)$ and $q = s(t)$, then $\dfrac{d}{dt} f(p, q) =$ _____.

(c) If $g = u(x)$ and $h = v(x)$, then $\dfrac{d}{dx} f(g, h) =$ _____.

(d) If $x = u(g)$ and $f = y(g)$, then $\dfrac{d}{dg} h(x, f) =$ _____.

10. Show that if $f_{11}(x, y) = f_{12}(x, y) = f_{22}(x, y) = 0$ at every point $(x, y) \in R^2$, then there are numbers a, b, and c such that $f(x, y) = ax + by + c$.

11. (a) If we differentiate both sides of Equation 107-6 (assuming that $w = f(u, v)$) according to the usual rules of calculus, we obtain the equation

$$\frac{d^2w}{dx^2} = \left[\frac{d}{dx} f_1(u, v)\right] \frac{du}{dx} + f_1(u, v) \frac{d^2u}{dx^2} + \left[\frac{d}{dx} f_2(u, v)\right] \frac{dv}{dx} + f_2(u, v) \frac{d^2v}{dx^2}.$$

Now use Equation 107-6 to rewrite $\dfrac{d}{dx} f_1(u, v)$ and $\dfrac{d}{dx} f_2(u, v)$ and thus produce the equation

$$\frac{d^2w}{dx^2} = f_1(u, v) \frac{d^2u}{dx^2} + f_2(u, v) \frac{d^2v}{dx^2} + f_{11}(u, v) \left(\frac{du}{dx}\right)^2$$

$$+ 2f_{12}(u, v) \frac{du}{dx} \frac{dv}{dx} + f_{22}(u, v) \left(\frac{dv}{dx}\right)^2.$$

(b) Use this formula to find $\dfrac{d^2w}{dx^2}$ if $w = u^2 + v$, where $(u, v) = (e^x, \ln x)$. Check your answer by first expressing w in terms of x and then differentiating twice.

(c) Show that if $F(x) = f(mx + a, nx + b)$, then

$$F''(0) = f_{11}(a, b)m^2 + 2f_{12}(a, b)mn + f_{22}(a, b)n^2.$$

12. In Chapter 2 we learned the sum, product, and quotient rules of differentiation:

$$\frac{d}{dx} (u + v) = \frac{du}{dx} + \frac{dv}{dx}, \qquad \frac{d}{dx} (uv) = u \frac{dv}{dx} + v \frac{du}{dx},$$

and

$$\frac{d}{dx} \left(\frac{u}{v}\right) = v^{-2} \left(v \frac{du}{dx} - u \frac{dv}{dx}\right).$$

Explain how these rules follow from Equation 107-7.

108. THE GRADIENT AND DIRECTIONAL DERIVATIVES

If $x = g(t)$, the Chain Rule Equation says that

(108-1) $$\frac{d}{dt} f(x) = f'(x) \frac{dx}{dt}.$$

Although we used different letters, we discovered in the last two sections what this equation becomes in case (1) $x = g(t, u)$ and (2) $f(x)$ is replaced by $f(x, y)$; that is, f is a function on R^2. In case (1), we need only replace d with ∂, and in case (2), we have the equation [assuming that $x = g(t)$ and $y = h(t)$]

$$(108\text{-}2) \qquad \frac{d}{dt} f(x, y) = f_1(x, y) \frac{dx}{dt} + f_2(x, y) \frac{dy}{dt}.$$

As we said before, much of the work with the Chain Rule simply involves saying the same thing in different notation. While mere name-juggling may be an empty pastime, it is nonetheless true that we are likely to understand a concept better if we look at it from various angles. Consequently, in this section we will introduce some notation that makes Equation 108-2 look more like Equation 108-1. In the process, we will start to develop a geometric feeling for several of these ideas, which is always a good thing to do in calculus.

What we are really doing when we go from Equation 108-1 to Equation 108-2 is replacing the scalar x with the vector $\mathbf{x} = (x, y)$. If we made that replacement directly in Equation 108-1, we would have

$$\frac{d}{dt} f(\mathbf{x}) = f'(\mathbf{x}) \frac{d\mathbf{x}}{dt}.$$

Thus our problem is to find a substitute for $f'(\mathbf{x})$ so that the right-hand side of this equation says the same thing as the right-hand side of Equation 108-2. We naturally think of the derivative $\frac{d\mathbf{x}}{dt}$ as the vector $\left(\frac{dx}{dt}, \frac{dy}{dt} \right)$.

Both from a strictly logical point of view and because it generalizes to vector-valued functions, we should let the role of $f'(\mathbf{x})$ be played by the 1 by 2 *matrix* $[f_1(\mathbf{x}), f_2(\mathbf{x})]$. But because we prefer concepts we can visualize geometrically, we bend logic a little and use the vector parentheses ( ) instead of the matrix brackets [ ] and introduce the **gradient vector**

$$\nabla f(\mathbf{x}) = (f_1(\mathbf{x}), f_2(\mathbf{x})) = (f_1(x, y), f_2(x, y)).$$

Then the Chain Rule Equation 108-2 can be written as

$$(108\text{-}3) \qquad \frac{d}{dt} f(\mathbf{x}) = \nabla f(\mathbf{x}) \cdot \frac{d\mathbf{x}}{dt},$$

an equation that looks very much like Equation 108-1. If f is a function on R^3, then

$$\nabla f(\mathbf{x}) = (f_1(x, y, z), f_2(x, y, z), f_3(x, y, z)) = f_1(\mathbf{x})\mathbf{i} + f_2(\mathbf{x})\mathbf{j} + f_3(\mathbf{x})\mathbf{k},$$

and so on. The symbol ∇ is called "del" and some authors write **grad** f in place of our ∇f.

Example 108-1. Suppose that the temperature at each point (x, y, z) in space is the number $f(x, y, z)$ and suppose also that Santa Claus travels along the vector curve $\mathbf{x} = \mathbf{g}(t)$. Show that at the coldest point of his trip, Santa's path is perpendicular to the gradient of the temperature.

Solution.   The temperature at a point $\mathbf{x}$ of the path is the number $f(\mathbf{x})$, and at a minimum point the derivative of the temperature is 0. According to Equation 108-3, this statement is expressed by the equation

$$\nabla f(\mathbf{x}) \cdot \frac{d\mathbf{x}}{dt} = 0.$$

Since $\dfrac{d\mathbf{x}}{dt}$ is tangent to the curve $\mathbf{x} = \mathbf{g}(t)$, we see that the gradient vector is perpendicular to the path, as we were to show.

Example 108-2. Let u and v be functions on R^2 that satisfy the Cauchy-Riemann partial differential equations (Example 107-4). Show that at each point $(x, y) \in R^2$, $\nabla u(x, y) \cdot \nabla v(x, y) = 0$.

Solution.   By definition,

$$\nabla u(x, y) = (u_1(x, y), u_2(x, y)) \quad \text{and} \quad \nabla v(x, y) = (v_1(x, y), v_2(x, y)).$$

Hence
$$\nabla u(x, y) \cdot \nabla v(x, y) = u_1(x, y)v_1(x, y) + u_2(x, y)v_2(x, y).$$

The Cauchy-Riemann equations say that $v_1(x, y) = -u_2(x, y)$ and $v_2(x, y) = u_1(x, y)$, so this dot product is 0, as we were to show.

Let us suppose now that f is a function on R^3 and (x, y, z) is a point of its domain. Then the partial derivatives

$$\frac{\partial}{\partial x} f(x, y, z), \qquad \frac{\partial}{\partial y} f(x, y, z), \quad \text{and} \quad \frac{\partial}{\partial z} f(x, y, z)$$

measure the (instantaneous) rate of change of the functional values of f with respect to distance. These numbers are computed *at the point* (x, y, z) and they give us rates of change *in the directions* of the positive X-, Y-, and Z-axes. Now we want to calculate rates of change in other directions as well. Our first task is to decide what we mean by the rate of change of values of f at a point $\mathbf{x} = (x, y, z)$ and in a particular direction.

We will specify the direction by prescribing a unit vector $\mathbf{u}$. Then for each number s, the point $\mathbf{x} + s\mathbf{u}$ is s units away from $\mathbf{x}$ in the direction of $\mathbf{u}$. It follows that $\dfrac{d}{ds} f(\mathbf{x} + s\mathbf{u})$ gives us the rate of change of functional values of f with respect to distance in this direction. If we want the rate at the point **x**, we simply calculate this derivative and set $s = 0$. This rate is the **directional derivative** of $f(\mathbf{x})$ *at the point* **x** *and in the direction* of **u**, and we denote it by the symbol $D_{\mathbf{u}}f(\mathbf{x})$. Now according to Equation 108-3 (with s in place of t and $\mathbf{x} + s\mathbf{u}$ playing the role of **x**),

$$\frac{d}{ds}f(\mathbf{x} + s\mathbf{u}) = \nabla f(\mathbf{x} + s\mathbf{u}) \cdot \frac{d}{ds}(\mathbf{x} + s\mathbf{u}) = \nabla f(\mathbf{x} + s\mathbf{u}) \cdot \mathbf{u}$$

and hence when we set $s = 0$, we find that

(108-4) $$D_{\mathbf{u}}f(\mathbf{x}) = \nabla f(\mathbf{x}) \cdot \mathbf{u}.$$

Example 108-3. Find $D_{\mathbf{u}}f(1, 2)$ if $f(x, y) = x^2 + y^2$, and **u** makes an angle of 60° with the positive X-axis.

Solution. Figure 108-1 shows the point $(1, 2)$ and the vector **u**. Since **u** is 1 unit long, we see that $\mathbf{u} = (\cos 60°, \sin 60°) = (\frac{1}{2}, \frac{1}{2}\sqrt{3})$. Furthermore, $f_1(1, 2) = 2$ and $f_2(1, 2) = 4$, so $\nabla f(1, 2) = (2, 4)$. Therefore

$$D_{\mathbf{u}}f(1, 2) = (2, 4) \cdot (\tfrac{1}{2}, \tfrac{1}{2}\sqrt{3}) = 1 + 2\sqrt{3}.$$

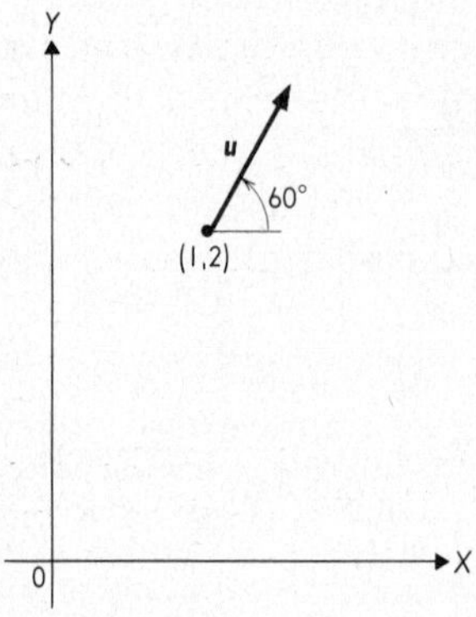

Example 108-4. Find $D_{\mathbf{u}}w$ at the point $(2, -1, 2)$ and in the direction from this point to the origin, if $w = 2x^2y + 3z$.

Solution. We must find the vectors ∇w and **u** and then compute their dot product. First we see that

$$\nabla w = \frac{\partial w}{\partial x}\mathbf{i} + \frac{\partial w}{\partial y}\mathbf{j} + \frac{\partial w}{\partial z}\mathbf{k} = 4xy\mathbf{i} + 2x^2\mathbf{j} + 3\mathbf{k}.$$

Therefore at the point $(2, -1, 2)$,

$$\nabla w = -8\mathbf{i} + 8\mathbf{j} + 3\mathbf{k}.$$

Figure 108-1

The vector **u** is 1 unit long and its direction is that of the vector $\mathbf{v} = -2\mathbf{i} + \mathbf{j} - 2\mathbf{k}$ whose initial point is $(2, -1, 2)$ and whose terminal point is the origin. Hence

$$\mathbf{u} = |\mathbf{v}|^{-1}\mathbf{v} = \tfrac{1}{3}(-2\mathbf{i} + \mathbf{j} - 2\mathbf{k}) = -\tfrac{2}{3}\mathbf{i} + \tfrac{1}{3}\mathbf{j} - \tfrac{2}{3}\mathbf{k}.$$

Thus we finally have

$$D_{\mathbf{u}}w = (-8\mathbf{i} + 8\mathbf{j} + 3\mathbf{k}) \cdot (-\tfrac{2}{3}\mathbf{i} + \tfrac{1}{3}\mathbf{j} - \tfrac{2}{3}\mathbf{k}) = \tfrac{16}{3} + \tfrac{8}{3} - \tfrac{6}{3} = 6.$$

The concept of the directional derivative is a generalization of the idea of partial derivatives. For example, if in Equation 108-4 we take as our unit directional vector **u** the basis vector **i**, then the resulting directional derivative $D_{\mathbf{i}}f(\mathbf{x})$ is simply the partial derivative $\dfrac{\partial}{\partial x}f(\mathbf{x})$, and so on.

When we calculate the directional derivatives of w at a point $\mathbf{x}$ by means of the formula $D_{\mathbf{u}}w = \nabla w \cdot \mathbf{u}$, we use the same vector ∇w for each direction vector $\mathbf{u}$. This formula yields different directional derivatives for different direction vectors, and it is natural to ask which direction gives us the *greatest* one. To answer this question, we write $\nabla w \cdot \mathbf{u} = |\nabla w| \, |\mathbf{u}| \cos \theta$ (where θ is the angle determined by ∇w and $\mathbf{u}$, of course). Since $|\mathbf{u}| = 1$, this formula for the directional derivative reduces to

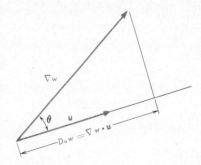

Figure 108-2

$$(108\text{-}5) \qquad D_{\mathbf{u}}w = |\nabla w| \cos \theta.$$

Figure 108-2 shows a geometric interpretation of the directional derivative $D_{\mathbf{u}}w$ in terms of the vectors ∇w and $\mathbf{u}$. Since $\cos \theta$ is a maximum when $\theta = 0$, the number $D_{\mathbf{u}}w$ is largest when $\mathbf{u}$ is selected as the unit vector in the direction of ∇w; that is, $\mathbf{u} = \dfrac{1}{|\nabla w|} \nabla w$. Then the maximum directional derivative is the number $(D_{\mathbf{u}}w)_{\max} = |\nabla w| \cos 0 = |\nabla w|$. Thus we see that *the maximum rate of change of w at a point occurs in the direction of the gradient vector calculated at that point, and the magnitude of the gradient of w is the maximum value of the rate of change of w with respect to distance.* Equation 108-5 also tells us that the rate of change of w with respect to distance is 0 in directions perpendicular to the gradient and that w decreases with distance most rapidly along the vector $-\nabla w$.

Example 108-5. The electric potential (voltage) at a point (x, y, z) of a certain region in space is given by the formula $V = x^2 - y^2 + 2xz$. If we put a unit positive charge at the point $(1, 2, 3)$, which way will it start to move?

Solution. It is a fact from electric field theory that the charge will start to move in the direction of the greatest potential drop—that is, in the direction in which V decreases most rapidly. Therefore the charge will move in the direction of $-\nabla V$. At the point $(1, 2, 3)$, $\nabla V = 8\mathbf{i} - 4\mathbf{j} + 2\mathbf{k}$, and hence the charge will start to move in the direction of the vector $-8\mathbf{i} + 4\mathbf{j} - 2\mathbf{k}$.

We close this section with the statement of the Theorem of the Mean for a scalar-valued function f on R^n. Suppose that $\mathbf{a}$ and $\mathbf{b}$ are two points of its domain and let $g(t) = f(\mathbf{a} + t(\mathbf{b} - \mathbf{a}))$. According to the ordinary Theorem of the Mean, there is a number $m \in (0, 1)$ such that

$$(108\text{-}6) \qquad\qquad g(1) - g(0) = g'(m).$$

Now we will express this equation in terms of our original function f. In the first place, $g(0) = f(\mathbf{a})$ and $g(1) = f(\mathbf{b})$, and so the left-hand side of Equation 108-6 is the number $f(\mathbf{b}) - f(\mathbf{a})$. According to the Chain Rule,

$$g'(t) = \frac{d}{dt} f(\mathbf{a} + t(\mathbf{b} - \mathbf{a})) = \nabla f(\mathbf{a} + t(\mathbf{b} - \mathbf{a})) \cdot (\mathbf{b} - \mathbf{a}),$$

so $g'(m) = \nabla f(\mathbf{a} + m(\mathbf{b} - \mathbf{a})) \cdot (\mathbf{b} - \mathbf{a})$. Since m is a number between 0 and 1, it follows that $\mathbf{a} + m(\mathbf{b} - \mathbf{a})$ is a point $\mathbf{m}$ of the line segment joining the points $\mathbf{a}$ and $\mathbf{b}$, so the right-hand side of Equation 108-6 is the number $\nabla f(\mathbf{m}) \cdot (\mathbf{b} - \mathbf{a})$. Thus we have shown that *there is a point* $\mathbf{m}$ *of the line segment between the points* $\mathbf{a}$ *and* $\mathbf{b}$ *such that*

$$(108\text{-}7) \qquad f(\mathbf{b}) - f(\mathbf{a}) = \nabla f(\mathbf{m}) \cdot (\mathbf{b} - \mathbf{a}).$$

We call this statement the **Theorem of the Mean** for scalar-valued functions on R^n, and you can see that it is a generalization of the Theorem of the Mean for functions on R^1.

PROBLEMS 108

1. Find $\nabla f(2, -3)$.
(a) $f(x, y) = x^3 y^2$
(b) $f(x, y) = x^3 + y^2$
(c) $f(x, y) = \sin \pi xy$
(d) $f(x, y) = \tan \frac{1}{2}\pi xy$
(e) $f(x, y) = |xy|$
(f) $f(x, y) = |x + y|$

2. Find the directional derivative of w at the given point and in the given direction.
(a) $w = e^{2x+3y}$, $(0, 0)$, direction of $\frac{3}{5}\mathbf{i} - \frac{4}{5}\mathbf{j}$
(b) $w = 2xy^2$, $(-1, 2)$, direction of $\mathbf{i} - 2\mathbf{j}$
(c) $w = e^x \sin (y + z)$, $(0, 0, 0)$, direction of the line segment from the origin to the point $(1, 2, -2)$
(d) $w = x^2 + y^2$, $(2, 1)$, direction of a vector that makes a positive angle of $60°$ with the positive X-axis
(e) $w = (y + x)/(y - x)$, $(1, 2)$, direction normal to the line $3x + 4y = 11$
(f) $w = \ln xyz^2$, $(1, 2, 1)$, direction of $\frac{1}{3}\mathbf{i} - \frac{2}{3}\mathbf{j} + \frac{2}{3}\mathbf{k}$
(g) $w = \sin \pi xyz$, $(\frac{1}{2}, \frac{1}{3}, 5)$, direction normal to the plane $x + 2y - 2z = 5$
(h) $w = xy$, $(3, 4)$, direction tangent to the circle $x^2 + y^2 = 25$

3. Suppose that you are standing at the point $(-1, 5, 8)$ on a hill whose equation is $z = 74 - x^2 - 7xy - 4y^2$. The Y-axis points north and the X-axis east, and distances are measured in meters.
(a) If you move to the south, are you ascending or descending? At what rate?
(b) If you move to the northwest, are you ascending or descending? At what rate?
(c) In what direction is the steepest downward path?
(d) In what directions is the path level?

4. Find a unit vector in the direction in which the maximum rate of change of w at the given point occurs if $w = \text{Tan}^{-1} xy + z$.
(a) $(0, 0, 0)$ (b) $(1, 0, 0)$ (c) $(0, 1, 0)$ (d) $(0, 0, 1)$

$(6xy + 2xz^2)\textbf{\i} + 6x^2\textbf{\j} + 6xz^2\textbf{\k}$

5. Suppose that $f(x, y, z) = 3x^2y + 2xz^3$ and calculate the following quantities.

   (a) $\nabla f(-1, -1, -1) \cdot \nabla f(1, 1, 1)$ (b) $\nabla f(-1, -1, -1) \times \nabla f(1, 1, 1)$

   (c) $\textbf{i} \cdot \nabla f(1, 2, 3)$ (d) $\textbf{i} \times \nabla f(1, 2, 3)$

   (e) $\textbf{i} \cdot \textbf{j} \times \nabla f(1, 1, 1)$ (f) $(\textbf{i} \times \textbf{j}) \times \nabla f(1, 1, 1)$

6. Suppose that the temperature at a point (x, y, z) in a region in space is given by the formula $w = e^{-x^2-y^2-z^2}$. Find the rate of change of temperature at the point $(1, 2, 3)$ and in the direction from that point to the origin.

7. Suppose that $\textbf{a}$ is a given vector of R^3 and let $f(\textbf{x}) = \textbf{a} \cdot \textbf{x}$ and $g(\textbf{x}) = |\textbf{x}|$. Show that $\nabla f(\textbf{x}) = \textbf{a}$ and $\nabla g(\textbf{x}) = |\textbf{x}|^{-1}\,\textbf{x}$.

8. (a) Use Equation 108-7 to show that if $\nabla f(\textbf{x}) = \textbf{0}$ for each $\textbf{x} \in R^n$, then f is a constant function.

   (b) Show that if $f(\textbf{a}) = f(\textbf{b})$, then there is a point $\textbf{m}$ of the line segment between $\textbf{a}$ and $\textbf{b}$ such that the vectors $\nabla f(\textbf{m})$ and $(\textbf{b} - \textbf{a})$ are perpendicular.

9. If $u = f(x, y)$ and $v = g(x, y)$, show that $\nabla(uv) = u\,\nabla v + v\,\nabla u$.

10. Suppose that f and g are given functions on R^n. Compare the directional derivative of $f(\textbf{x})$ in the direction of the gradient of g at $\textbf{x}$ with the directional derivative of $g(\textbf{x})$ in the direction of the gradient of f at $\textbf{x}$.

11. Suppose that $r = f(\textbf{x})$ and $s = g(\textbf{x})$, and that $\textbf{u}$ is a unit vector. Show that the following equations are correct.

   (a) $D_\textbf{u}(r + s) = D_\textbf{u}r + D_\textbf{u}s$ (b) $D_\textbf{u}rs = rD_\textbf{u}s + sD_\textbf{u}r$

   (c) $D_\textbf{u}\dfrac{r}{s} = \dfrac{sD_\textbf{u}r - rD_\textbf{u}s}{s^2}$

12. As we said in the text, if we define the gradient as a *matrix*, our Chain Rule generalizes to vector-valued functions. Thus if $\textbf{f}(\textbf{x}) = (f(\textbf{x}), g(\textbf{x}))$, where $\textbf{x} = (x, y)$,

$$\nabla \textbf{f}(\textbf{x}) = \begin{bmatrix} f_1(\textbf{x}) & f_2(\textbf{x}) \\ g_1(\textbf{x}) & g_2(\textbf{x}) \end{bmatrix}.$$

   (a) Show that the equation $\dfrac{d}{dt}\textbf{f}(\textbf{x}) = \nabla \textbf{f}(\textbf{x})\dfrac{d\textbf{x}}{dt}$ follows.

   (b) Try out this formula in case $\textbf{f}(\textbf{x}) = (x^2 + y^2, x + y)$ and $\textbf{x} = (\cos t, \sin t)$.

109. NORMALS TO CURVES AND SURFACES

    Suppose that c is a number in the range of a function f on R^2 and that the graph of the equation $f(x, y) = c$ is a curve in the plane. Such a curve is called a **level curve** of f. At all points of a given level curve, f has the same value. For example, if a geographical map is considered as a region in the XY-plane and if $f(x, y)$ is the altitude of the point (x, y) of the map, then the level curves of f are the contour lines of the map. If f measures temperature or voltage, the level curves are "isotherms" or "equipotential lines," and so on.

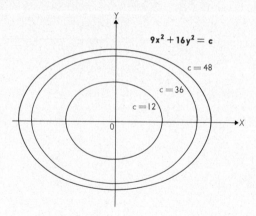

$9x^2 + 16y^2 = c$

$c = 48$

$c = 36$

$c = 12$

Figure 109-1

Example 109-1. Let $f(x, y) = 9x^2 + 16y^2$ and sketch the level curves that correspond to the functional values 12, 36, and 48.

Solution. The level curves are the ellipses $9x^2 + 16y^2 = c$—where c is replaced by 12, 36, and 48—shown in Fig. 109-1.

The situation in three dimensions is analogous to the two-dimensional case. If c is a number in the range of the function f on R^3 and the graph of the equation $f(x, y, z) = c$ is a surface in space, then this surface is a **level surface** of f. Every surface in space can be regarded as a level surface of some function on R^3.

Example 109-2. Find a function f for which the paraboloid of revolution $z + 4 = x^2 + y^2$ is a level surface.

Solution. Since our given equation is equivalent to $z - x^2 - y^2 + 4 = 0$, its graph is a level surface (corresponding to the choice $c = 0$) of the function defined by the equation $f(x, y, z) = z - x^2 - y^2 + 4$. Or we might let $g(x, y, z) = x^2 + y^2 - z$, and then the graph of our given equation is the level surface $g(x, y, z) = 4$. There is no end to the number of functions for which our surface is a level surface.

Let us consider a curve in a given level surface $f(x, y, z) = c$. A curve in space is specified by a vector equation $\mathbf{x} = \mathbf{g}(t)$. If we are thinking in terms of arrows, $\mathbf{x}$ is the position vector $x\mathbf{i} + y\mathbf{j} + z\mathbf{k}$. If we are thinking in terms of triples of numbers, $\mathbf{x}$ is simply the triple (x, y, z). In either case, to say that our curve lies in the given level surface is to say that $\mathbf{x}$ satisfies the equation $f(\mathbf{x}) = c$ for each t (see Fig. 109-2). Now we will differentiate both sides of this equation

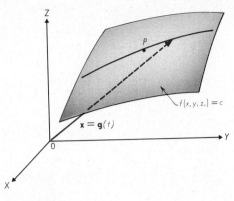

Figure 109-2

with respect to t. Equation 108-3 tells us that the derivative of the left-hand side is $\nabla f(\mathbf{x}) \cdot \dfrac{d\mathbf{x}}{dt}$, and the derivative of the right-hand side is 0:

$$(109\text{-}1) \qquad\qquad \nabla f(\mathbf{x}) \cdot \frac{d\mathbf{x}}{dt} = 0.$$

The vector $\dfrac{d\mathbf{x}}{dt}$ is tangent to the curve $\mathbf{x} = \mathbf{g}(t)$, and therefore we can interpret Equation 109-1 geometrically as saying: *The gradient of f at a given point P of our level surface is perpendicular to the tangent line to any curve in the surface that contains P.* Of course, there are infinitely many curves that lie in the surface and contain P, but Equation 109-1 says that their tangents at P all lie in the same plane, the plane that contains P and has the vector $\nabla f(\mathbf{x})$ as a normal. This plane is the **tangent plane** to our surface at P. A vector that is normal to this plane is **normal to the surface** at P. Thus *the vector $\nabla f(x, y, z)$ is normal to the surface $f(x, y, z) = c$.*

Example 109-3. Show that the position vector to a point (x, y, z) of the sphere $x^2 + y^2 + z^2 = 4$ is normal to the sphere.

Solution. If $f(x, y, z) = x^2 + y^2 + z^2$, our sphere is the level surface $f(x, y, z) = 4$. As we have just said, the vector $\nabla f(x, y, z)$ is normal to this surface. Since $\nabla f(x, y, z) = 2x\mathbf{i} + 2y\mathbf{j} + 2z\mathbf{k} = 2\mathbf{x}$, where $\mathbf{x}$ is the position vector to (x, y, z), we see that $\mathbf{x}$ is also normal to our sphere. Thus our definition of a normal vector to a surface leads to the natural result that a radius is normal to a sphere.

Example 109-4. The graph of a function f on R^2 is the surface $z = f(x, y)$. Find a vector normal to this surface.

Solution. Suppose that we let $F(x, y, z) = f(x, y) - z$. Then the graph of f is a level surface of F (its equation is $F(x, y, z) = 0$) and so $\nabla F(x, y, z)$ is normal to it. But

$$\nabla F(x, y, z) = \frac{\partial}{\partial x} F(x, y, z)\mathbf{i} + \frac{\partial}{\partial y} F(x, y, z)\mathbf{j} + \frac{\partial}{\partial z} F(x, y, z)\mathbf{k}.$$

Since

$$\frac{\partial}{\partial x} F(x, y, z) = f_1(x, y), \qquad \frac{\partial}{\partial y} F(x, y, z) = f_2(x, y),$$

and

$$\frac{\partial}{\partial z} F(x, y, z) = -1,$$

we therefore see that the vector $f_1(x, y)\mathbf{i} + f_2(x, y)\mathbf{j} - \mathbf{k}$ is normal to the surface $z = f(x, y)$.

Now we will find the equation of the tangent plane to a surface in space. Suppose that the point (x_0, y_0, z_0) belongs to the surface $f(x, y, z) = c$. In Section 96 we found that $a(x - x_0) + b(y - y_0) + c(z - z_0) = 0$ is the plane that has the vector $a\mathbf{i} + b\mathbf{j} + c\mathbf{k}$ as a normal and that contains the point (x_0, y_0, z_0). Our tangent plane has the gradient vector $f_1(x_0, y_0, z_0)\mathbf{i} + f_2(x_0, y_0, z_0)\mathbf{j} + f_3(x_0, y_0, z_0)\mathbf{k}$ as a normal, and so we see that

$$(109\text{-}2) \quad f_1(x_0, y_0, z_0)(x - x_0) + f_2(x_0, y_0, z_0)(y - y_0) \\ + f_3(x_0, y_0, z_0)(z - z_0) = 0$$

is the tangent plane to the surface $f(x, y, z) = c$ at the point (x_0, y_0, z_0).

Example 109-5. Find the plane that is tangent to the surface $x^2 y^3 z = 4 + x$ at the point $(1, 1, 5)$.

Solution. Before using Equation 109-2, we must write our given surface as a level surface—for example, as $x^2 y^3 z - x = 4$. Then we use Equation 109-2 with $f(x, y, z) = x^2 y^3 z - x$ and $(x_0, y_0, z_0) = (1, 1, 5)$. We calculate $f_1(x, y, z) = 2xy^3 z - 1$, and so $f_1(1, 1, 5) = 9$. Similarly, $f_2(1, 1, 5) = 15$ and $f_3(1, 1, 5) = 1$, and therefore the desired tangent plane has the equation $9(x - 1) + 15(y - 1) + (z - 5) = 0$, or, in simplified form,

$$9x + 15y + z = 29.$$

Our remarks concerning surfaces in space can be modified so as to apply to curves in a plane. For example, suppose that $f(x, y) = c$ is a level curve of a function f on R^2. Then the gradient vector $\nabla f(x, y)$ is perpendicular to this level curve at the point (x, y) (that is, the gradient vector is perpendicular to the tangent to the level curve at the point).

Example 109-6. Find the tangent line to the level curve $f(x, y) = c$ at the point (x_0, y_0).

Solution. The tangent line plays the same role in two dimensions that the tangent plane plays in space. In order to obtain the equation of the desired tangent, we simply drop the last term on the left-hand side of Equation 109-2:

$$f_1(x_0, y_0)(x - x_0) + f_2(x_0, y_0)(y - y_0) = 0.$$

Notice that the slope of this tangent line is $-f_1(x_0, y_0)/f_2(x_0, y_0)$, which agrees with Equation 107-9 for the slope of the curve $f(x, y) - c = 0$.

PROBLEMS 109

1. Describe the level curves of f. In particular, sketch the level curve that contains the point $(1, 1)$. Sketch the gradient vector at this point.

   (a) $f(x, y) = x^2 + y^2 - 1$ (b) $f(x, y) = 4x^2 - y^2 - 8x + 4y$
   (c) $f(x, y) = y - x^2$ (d) $f(x, y) = x^2 y$
   (e) $f(x, y) = e^y - e^x$ (f) $f(x, y) = e^{y-x}$
   (g) $f(x, y) = \text{Cos} \, \frac{1}{2}\pi xy$ (h) $f(x, y) = \text{Cos}^{-1} \frac{1}{2}xy$

2. Describe the level surface of f that contains the given point and find a normal vector to the level surface at that point.

   (a) $f(x, y, z) = 2x - 3y + 4z$, $(1, 1, 1)$
   (b) $f(x, y, z) = 2x + 4z$, $(1, 1, 1)$
   (c) $f(x, y, z) = 4x^2 + y^2 + z^2$, $(1, 1, 1)$
   (d) $f(x, y, z) = x^2 + y^2$, $(1, 2, 3)$
   (e) $f(x, y, z) = \ln(x^2 + y^2 - z)$, $(1, 1, 1)$
   (f) $f(x, y, z) = (x^2 + y^2 - z^2)^3$, $(1, 0, 1)$

3. Find the tangent plane at the given point.
   (a) $2x^2 - 3xy + 4y^2 = z$, $(-1, 1, 9)$
   (b) $x^2 y + y^2 z + z^2 x + 4 = 0$, $(2, -1, 0)$
   (c) $e^{xy} - 2 \sin z = 1$, $(0, 0, 0)$
   (d) $4 \, \text{Tan}^{-1} (y/z) = \pi \ln xyz$, $(1, 1, e)$
   (e) $x|x| + (y + |y|)^2 = |z|$, $(1, -1, 1)$
   (f) $x|x| + (y + |y|)^2 = |z|$, $(-2, 1, 0)$

4. At what point is the normal to the surface $z = 3x^2 - 2y^2$ parallel to the vector $2\mathbf{i} + 4\mathbf{j} + \frac{1}{3}\mathbf{k}$?

5. Find $\mathbf{m}$ and $\mathbf{b}$ such that the line $\mathbf{r} = \mathbf{m}t + \mathbf{b}$ is perpendicular to the surface $3x^2 - 4yz + xz^2 = 12$ at the point $(1, -2, 1)$.

6. At what points of the surface $x^2 + 4y^2 + 16z^2 - 2xy = 12$ are the tangent planes parallel to the XZ-plane?

7. Find the directional derivative of $w = 3x^2 yz + 2yz^2$ at the point $(1, 1, 1)$ and in a direction normal to the surface $x^2 - y + z^2 = 1$ at $(1, 1, 1)$.

8. Find a vector tangent to the curve of intersection of the surfaces $x^2 - 2xz + y^2 z = 1$ and $3xy + 2yz + 6 = 0$ at the point $(1, -2, 0)$. (*Hint:* This vector must be perpendicular to normals to both surfaces.)

9. What is the cosine of the acute angle between the tangent planes to the surfaces of the preceding problem at the given point?

10. A flat panel rests on a boulder whose equation is $9x^2 + 4y^2 + 36z^2 = 61$, touching it at the point $(1, 2, 1)$. Is the point $(5, 1, 1)$ in the shade of the panel?

11. Show that the line that is normal at the point (x_0, y_0, z_0) to the surface $f(x, y, z) = c$ has parametric equations

$$x = f_1(x_0, y_0, z_0)t + x_0, \qquad y = f_2(x_0, y_0, z_0)t + y_0, \qquad z = f_3(x_0, y_0, z_0)t + z_0.$$

12. (a) Let f and g be functions on R^2 that have the same level curves. Show that at each point $\mathbf{x}$,

$$\begin{vmatrix} f_1(\mathbf{x}) & f_2(\mathbf{x}) \\ g_1(\mathbf{x}) & g_2(\mathbf{x}) \end{vmatrix} = 0.$$

(b) Let f and g be functions on R^3 that have the same level surfaces. Show that at each point $\mathbf{x}$,

$$\begin{vmatrix} f_1(\mathbf{x}) & f_2(\mathbf{x}) \\ g_1(\mathbf{x}) & g_2(\mathbf{x}) \end{vmatrix} = \begin{vmatrix} f_2(\mathbf{x}) & f_3(\mathbf{x}) \\ g_2(\mathbf{x}) & g_3(\mathbf{x}) \end{vmatrix} = \begin{vmatrix} f_3(\mathbf{x}) & f_1(\mathbf{x}) \\ g_3(\mathbf{x}) & g_1(\mathbf{x}) \end{vmatrix} = 0.$$

13. Explain why $\nabla f(\mathbf{x}_0) \cdot (\mathbf{x} - \mathbf{x}_0) = 0$ is the equation of the tangent line to the curve $f(x, y) = c$ at the point $\mathbf{x}_0 = (x_0, y_0)$. Use this formula to find the tangent line to the ellipse $2x^2 + y^2 = 6$ at the point $(-1, 2)$.

110. MAXIMA AND MINIMA OF FUNCTIONS ON R^n

If a differentiable function f on R^1 has a maximum (or a minimum) value at an interior point a of its domain, then we know that

$$(110\text{-}1) \qquad\qquad\qquad f'(a) = 0.$$

Now we will find the analogous equation for functions on a general R^n.

We get a feeling for maximum and minimum problems by looking at the situation geometrically. From this point of view, Equation 110-1 simply states that the tangent line to the graph of f at the point $(a, f(a))$ is parallel to the X-axis. Now let us interpret the corresponding situation when f is a function on R^2. Figure 110-1 illustrates the case in which f takes a maximum value at the point

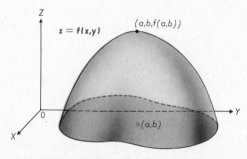

Figure 110-1

(a, b). Thus for each point (x, y) in some neighborhood of (a, b), we have $f(x, y) < f(a, b)$. At this maximum point (and the same remarks would apply to a minimum point) the tangent plane is parallel to the XY-plane. In other words, normals to the graph of f at the point $(a, b, f(a, b))$ are parallel to the Z-axis. In Example 109-4 we showed that the vector $f_1(a, b)\mathbf{i} + f_2(a, b)\mathbf{j} - \mathbf{k}$ is such a normal vector, and the condition that this vector be parallel to the Z-axis is that

$$(110\text{-}2) \qquad f_1(a, b) = 0 \quad \text{and} \quad f_2(a, b) = 0.$$

We can combine this pair of scalar equations into the vector equation

$$(110\text{-}3) \qquad \nabla f(a, b) = \mathbf{0}.$$

Equations 110-1 and 110-3 are completely analogous. *If* the given function has a maximum or a minimum value at the given point, then the equation holds.

We cannot draw pictures for functions on R^n if $n > 2$, but the results are the same, as the following informal argument suggests. Suppose that a function f takes a maximum value at a point $\mathbf{a}$. Then as we move away from $\mathbf{a}$, the values of f cannot increase; that is, the rate of change of the values of f with respect to distance cannot be positive. In other words, at the point $\mathbf{a}$ no directional derivative of $f(\mathbf{x})$ is positive. In particular, the maximum directional derivative, which we know to be $|\nabla f(\mathbf{a})|$, cannot be positive. But this number cannot be negative either, and so we must have

$$(110\text{-}4) \qquad \nabla f(\mathbf{a}) = \mathbf{0}.$$

Similar remarks lead to the same conclusion if $\mathbf{a}$ is a minimum point. Notice that the vector Equation 110-4 tells us that *all the first partial derivatives of f are zero at the point* $\mathbf{a}$. Thus if f is a function on R^n, Equation 110-4 is equivalent to n scalar equations.

You must keep clearly in mind that the condition we have mentioned is *necessary* in order that a differentiable function f should take a maximum or a minimum value at an interior point $\mathbf{a}$ of its domain. *If* the function takes a maximum or a minimum value, then its gradient is zero. But this condition is not *sufficient*; that is, the fact that the gradient of f is zero at a point does not mean that f must take a maximum or a minimum value there. In R^1, for example, if $f(x) = x^3$, then $f'(0) = 0$, but $f(0)$ is neither a maximum nor a minimum value of f. And similar remarks hold for functions on R^n if $n > 1$.

Example 110-1. Show that there are no maximum or minimum points of the hyperbolic paraboloid $z = y^2 - \frac{1}{4}x^2$. In other words, show that the equation $f(x, y) = y^2 - \frac{1}{4}x^2$ defines a function with no maximum or minimum values.

Solution. At a maximum or a minimum point, $\dfrac{\partial z}{\partial x}$ must equal 0 and so must $\dfrac{\partial z}{\partial y}$. Since $\dfrac{\partial z}{\partial x} = -\frac{1}{2}x$ and $\dfrac{\partial z}{\partial y} = 2y$, we see that $x = 0$ and $y = 0$ at an extreme point.

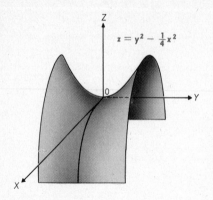

$$z = y^2 - \tfrac{1}{4}x^2$$

Figure 110-2

Hence $(0, 0, 0)$ is the only possible maximum or minimum point of our surface. But Fig. 110-2 makes it clear that $(0, 0, 0)$ is neither a maximum nor a minimum point, and, consequently, our surface has no such points.

At the end of the section we will say a few words about an analytic test (analogous to the Second Derivative Test that we apply to functions on R^1) for selecting maximum and minimum points from among the points at which the first partial derivatives of a function are 0. However, for many examples, we can use geometric or physical reasoning to decide when we have maxima or minima. It will be worthwhile for you to proceed by the following sequence of steps as you solve the "word problems" of this section.

(1) Decide what quantity is to be maximized (or minimized).

(2) Express this quantity as the value of a function on R^n; choose n as small as you can.

(3) Find the points at which the first partial derivatives are zero and determine whether or not the corresponding functional value is a maximum or a minimum.

(4) Reread the question and use the information from step (3) to answer it.

Example 110-2. Find the point of the plane $3x + 2y + z = 14$ that is nearest the origin.

Solution. The quantity that must be minimized is the distance d between the origin and a point (x, y, z) of the plane. According to the distance formula, $d = \sqrt{x^2 + y^2 + z^2}$. Since our point (x, y, z) belongs to the plane $z = 14 - 3x - 2y$, we can express d in terms of x and y only:

$$d = \sqrt{x^2 + y^2 + (14 - 3x - 2y)^2}.$$

The natural step now is to take partial derivatives, find out where they are 0, and so on. But it will simplify our differentiation if we recognize that d is a minimum where d^2 is a minimum, and so we will set

$$f(x, y) = x^2 + y^2 + (14 - 3x - 2y)^2$$

and find the point at which f takes its minimum value. We differentiate,

$$f_1(x, y) = 2x - 6(14 - 3x - 2y), \qquad f_2(x, y) = 2y - 4(14 - 3x - 2y),$$

680

equate these derivatives to 0, and simplify:

$$5x + 3y = 21$$
$$6x + 5y = 28.$$

Now we solve this system of linear equations and find that $(x, y) = (3, 2)$. Since the equations $f_1(x, y) = 0$ and $f_2(x, y) = 0$ have this single solution, it follows that *if f* has a minimum value, *then* that value must be the number $f(3, 2) = 14$. The geometry of the situation makes it clear that there *is* a point of the plane that is nearest the origin, and hence the point $(3, 2)$ *does* give a minimum value for f. Therefore the X- and Y-coordinates of the nearest point are $x = 3$ and $y = 2$. We obtain the Z-coordinate from the equation of the plane, $z = 14 - 3 \cdot 3 - 2 \cdot 2 = 1$, and so we have found that $(3, 2, 1)$ is the point of the given plane that is nearest the origin.

Example 110-3.   A rectangular box without a top is to be constructed from 900 square feet of material. What should its dimensions be in order that its volume be a maximum?

Solution.   We must maximize the volume V of the box. If the base of the box is x feet by y feet and it is h feet high, then

$$V = xyh.$$

Since we may use only 900 square feet of material, we can express one of the quantities x, y, or h in terms of the other two. For it requires $xy + 2xh + 2yh$ square feet of material to construct an x by y by h box without a top, and so

$$xy + 2xh + 2yh = 900.$$

From this equation we find that

(110-5)
$$h = \frac{900 - xy}{2(x + y)},$$

and hence

$$V = \frac{xy(900 - xy)}{2(x + y)}.$$

Now we must find the points at which $\dfrac{\partial V}{\partial x} = 0$ and $\dfrac{\partial V}{\partial y} = 0$. After differentiating and simplifying, we have

$$\frac{\partial V}{\partial x} = \frac{900y^2 - 2xy^3 - x^2y^2}{2(x + y)^2}$$

and

$$\frac{\partial V}{\partial y} = \frac{900x^2 - 2x^3y - x^2y^2}{2(x + y)^2}.$$

When we equate these derivatives to 0 and simplify, we arrive at the system of equations

$$900 - 2xy - x^2 = 0$$

$$900 - 2xy - y^2 = 0.$$

We see at once that $x^2 = y^2$ and therefore that $x = y$, since both these numbers must be positive. So we can replace y with x in the first equation and find that $x^2 = 300$. Thus $x = 10\sqrt{3}$ and $y = 10\sqrt{3}$. From Equation 110-5 we find that $h = 5\sqrt{3}$. As a result, the box with the largest volume has a square base measuring $10\sqrt{3}$ by $10\sqrt{3}$ feet, and it is $5\sqrt{3}$ feet high.

You recall from Chapter 3 that if $f'(a) = 0$ and $f''(a) > 0$, then $f(a)$ is a minimum value of f, whereas if $f'(a) = 0$ and $f''(a) < 0$, we are dealing with a maximum value. The first statement says that the graph of f is falling when x is to the left of a and rising when x is to the right of a, and the second statement says the opposite. The analogous Second Derivative Test in R^n sounds considerably more complicated although it says essentially the same thing. We possess the mathematical tools to tell you *what* this test is, but we lack the time to say exactly *why* it works.

Suppose that f is a function on R^3 and $\mathbf{a}$ is a point of its domain at which $\nabla f(\mathbf{a}) = \mathbf{0}$. Then we construct the matrix of second derivatives

$$H(\mathbf{a}) = \begin{bmatrix} f_{11}(\mathbf{a}) & f_{12}(\mathbf{a}) & f_{13}(\mathbf{a}) \\ f_{21}(\mathbf{a}) & f_{22}(\mathbf{a}) & f_{23}(\mathbf{a}) \\ f_{31}(\mathbf{a}) & f_{32}(\mathbf{a}) & f_{33}(\mathbf{a}) \end{bmatrix}.$$

Since $f_{21}(\mathbf{a}) = f_{12}(\mathbf{a})$, and so on, this matrix is symmetric. Hence its characteristic values are real numbers. If all the characteristic values are positive, then $f(\mathbf{a})$ is a minimum value of f; if all the characteristic values are negative, $f(\mathbf{a})$ is a maximum value. If some of the characteristic values are positive and some are negative, $f(\mathbf{a})$ is neither a maximum nor a minimum value. If the nonzero characteristic values have the same sign but some characteristic values are zero, then we need a more delicate test. This test applies to functions on R^2, R^4, and so on, but in those spaces $H(\mathbf{a})$ is a 2 by 2 matrix, a 4 by 4 matrix, and so on. Notice that this test reduces to the regular Second Derivative Test for functions on R^1, where we are dealing with the 1 by 1 matrix $[f''(a)]$.

Example 110-4. Apply this test to the functions of Examples 110-1 and 110-2.

Solution. In Example 110-1 $f_{11}(x, y) = -\frac{1}{2}$, $f_{12}(x, y) = f_{21}(x, y) = 0$, and $f_{22}(x, y) = 2$. Therefore

$$H(0, 0) = \begin{bmatrix} -\frac{1}{2} & 0 \\ 0 & 2 \end{bmatrix}.$$

Its characteristic values, $-\frac{1}{2}$ and 2, have opposite signs, so $f(0, 0)$ is not an extreme value.

In the case of Example 110-2, $f_{11}(x, y) = 20$, $f_{12}(x, y) = f_{21}(x, y) = 12$, and $f_{22}(x, y) = 10$. Therefore

$$H(3, 2) = \begin{bmatrix} 20 & 12 \\ 12 & 10 \end{bmatrix}.$$

The characteristic values of this matrix are the positive numbers 2 and 28, so $f(3, 2)$ is a minimum value of f.

You might try this test on some of the following problems.

PROBLEMS 110

1. Find the point of the plane $2x + y + z = 6$ that has positive coordinates and is such that
    (a) the product of the coordinates is a maximum.
    (b) the sum of the squares of the coordinates is a minimum.
2. Find the lowest point of the surface.
    (a) $z = 5x^2 - 6xy + 2y^2 - 8x - 32$
    (b) $9x^2 + 36y^2 + 4z^2 - 18x + 144y + 89 = 0$
3. Find the highest point of the surface.
    (a) $z = 6y - 2x - x^2 - y^2$
    (b) $9x^2 + 36y^2 + 4z^2 - 18x + 144y + 89 = 0$
4. Use calculus to find the distance between the point $(2, -1, 3)$ and the plane $2x - y + 2z = 5$. [Find the point of the plane that is nearest the point $(2, -1, 3)$.] Use a geometric argument to check your result.
5. Find the distance between the lines with parametric equations $x = t$, $y = 2t$, $z = t + 1$ and $x = s$, $y = s + 3$, $z = s$.
6. Find a point of the surface $z^2 = xy - 3x + 9$ that is closest to the origin.
7. Find the equation of the plane that contains the point $(1, 2, 1)$ and cuts off the least volume from the first octant.
8. Show that the product of the sines of the angles of a triangle is a maximum when the triangle is equilateral.
9. Suppose that we wish to make a rectangular box to hold 20 cubic feet of gold dust. The material used for the sides costs \$1 per square foot, the material used for the bottom costs \$2 per square foot, and the material used for the top costs \$3 per square foot. What are the dimensions of the cheapest box?
10. A long piece of tin 12 inches wide is to be made into a trough by bending up strips of equal width along the edges at equal angles with the horizontal. How wide should these strips be, and what angle must they make with the horizontal if the trough is to have a maximum carrying capacity?
11. Show that a rectangular box (with a top) made out of S square feet of material has a maximum volume if it is a cube.

12. Show that a rectangular box (without a top) made out of S square feet of material has a maximum volume if it has a square base and an altitude that is one-half the length of one side of the base.

13. One end of a house is to be built of glass in the shape of a rectangle surmounted by an isoceles triangle and is to have a given perimeter p. Find the slope of the roof if the house is constructed to admit a maximum amount of light.

14. Let f be the function whose domain is the unit sphere $x^2 + y^2 + z^2 = 1$ and whose rule of correspondence is $f(x, y, z) = x^2 y^2 z^2$. What are the maximum and minimum values of f?

15. Three points of the unit circle form the vertices of a triangle. Use calculus to determine how these points should be located so that we obtain a triangle with a maximum perimeter. Is there a triangle with minimum perimeter? What if we have more than three points?

Whatever difficulties there are in this chapter are mostly matters of notation. The basic ideas have been with us from the start. We calculate derivatives just as we have always done. But because we can differentiate with respect to various things now, our formulas get pretty complicated. These notations are summed in our basic Chain Rule Equation

$$\frac{\partial}{\partial x} f(\mathbf{u}) = \frac{\partial}{\partial u} f(\mathbf{u}) \frac{\partial u}{\partial x} + \frac{\partial}{\partial v} f(\mathbf{u}) \frac{\partial v}{\partial x}$$

$$= f_1(u, v) \frac{\partial u}{\partial x} + f_2(u, v) \frac{\partial v}{\partial x} = \nabla f(\mathbf{u}) \cdot \frac{\partial \mathbf{u}}{\partial x}.$$

Remember that the gradient vector $\nabla f(x)$ is normal to the level surface $f(\mathbf{x}) = c$ and that the directional derivative $D_{\mathbf{u}} f(\mathbf{x}) = \nabla f(\mathbf{x}) \cdot \mathbf{u}$ represents the rate of change of $f(\mathbf{x})$ with respect to distance at the point $\mathbf{x}$ and in the direction of $\mathbf{u}$. Since $\nabla f(\mathbf{x})$ plays the role of a vector-valued derivative, it is not surprising that maximum and minimum points are found by solving the equation $\nabla f(\mathbf{x}) = \mathbf{0}$.

Review Problems, Chapter Thirteen

1. If $f(x, y, z) = xy^2 z^3$, what is $f_{12}(1, 2, 3) - f_{21}(1, 2, 3) + f_3(2, 3, 1)$?

2. Find $\nabla f(\ln 3, 2)$ if $f(x, y)$ is the given expression.

   (a) $e^x + y^e$ (b) $e^x y^e$ (c) $e^y + x^e$
   (d) $e^y x^e$ (e) e^{x+y} (f) e^{xy}

3. The equation $f(x, y, z) = \text{Arcsin}\left(\frac{1}{6}x^2 + \frac{3}{2}y^2 + \frac{1}{24}z^2 - \frac{1}{2}\right)$ defines a scalar-valued function on R^3.

   (a) What is the domain of f?   (b) What is the range of f?
   (c) Sketch the level surface $f(x, y, z) = \frac{1}{6}\pi$.
   (d) Find the equation of the tangent plane to this level surface at the point $(1, \frac{1}{3}, -4)$.
   (e) At what rate do the functional values change as we start to move from the point $(1, \frac{1}{3}, -4)$ toward the point $(2, -\frac{5}{3}, -2)$?

4. Sketch the graph of f.
   (a) $f(x, y) = |x - y|$ (b) $f(x, y) = \frac{1}{2}(|x - y| + x + y)$

5. If $F(x, y) = \int_x^y \sqrt{t}\,\sin t\,dt$, find $F_1(0, \frac{1}{2}\pi)$, $F_2(0, \frac{1}{2}\pi)$, and explain why $F(y, x) = -F(x, y)$.

6. If f is a function on R^2 such that $f_1(0, 0) = 5$ and $f_2(0, 0) = 8$ and we write $F(x, y) = f(x - y, y - x)$, find the number $F_2(3, 3)$.

7. Suppose that $F(x) = f(3x, 4x)$, where $f_1(u, v) = u$ and $f_2(u, v) = 3v$. Find $F'(2)$, $F''(2)$, and $F'''(2)$.

8. Show that $\nabla|\mathbf{r}|^n = n|\mathbf{r}|^{n-2}\mathbf{r}$, where $\mathbf{r} = x\mathbf{i} + y\mathbf{j} + z\mathbf{k}$, and n is a given number.

9. Explain why $\nabla \cos w = -\sin w\,\nabla w$.

10. What is the relation between two functions f and g if for each $\mathbf{x}$, $\nabla f(\mathbf{x}) = \nabla g(\mathbf{x})$?

11. Find the maximum value of the quotient $\sqrt[4]{wxyz}/(w + x + y + z)$, where w, x, y, and z are positive numbers. Hence infer that the geometric mean of four positive numbers does not exceed their arithmetic mean:
$$\sqrt[4]{wxyz} \le \tfrac{1}{4}(w + x + y + z).$$
Is the same statement true when we are dealing with n positive numbers, $n \ne 4$?

12. Suppose that u and v satisfy the Cauchy-Riemann partial differential equations that we mentioned in Example 107-4. Show that they also satisfy Laplace's differential equation $w_{xx} + w_{yy} = 0$.

13. Suppose that $\mathbf{a}$ is a given vector and $\mathbf{x}$ is a position vector. Let $f(\mathbf{x}) = \mathbf{a} \cdot \mathbf{x}$ and evaluate $\lim_{h \to 0} \dfrac{f(\mathbf{x} + h\mathbf{u}) - f(\mathbf{x})}{h}$, where $\mathbf{u}$ is an arbitrary unit vector. Show that this limit is $D_\mathbf{u} f(\mathbf{x})$.

14. Suppose that $f(\mathbf{x}) = |\mathbf{x}|^2$ and that $\mathbf{b}$ and $\mathbf{a}$ are given vectors. What is the vector $\mathbf{m}$ that makes Equation 108-7 true?

15. If $f(x, y, z) = \ln x + z \ln y$ and $g(x, y, z) = \cos xy^z$, can you tell at a glance that $\nabla f(x, y, z) \times \nabla g(x, y, z) = \mathbf{0}$?

16. A hill has the equation $z = 2xy - x^2 - 2y^2 + 6x - 8y + 7$.
    (a) How high is it above sea level (the XY-plane)?
    (b) The base of our hill is an ellipse in the XY-plane. If the highest point of the hill is (p, q, r), show that $(p, q, 0)$ is the center of this ellipse.

17. A spherical stone of radius 1 rolls down the hill $z = 1000 - 50x^2 - 40y^2$, squashing a bug at the point $(2, -3, 440)$. Where is the center of the stone at the instant the bug is hit?

18. The temperature of a point (x, y, z) of Hell is $2x^2y^2 - z^3$ degrees. As a sinner descends the helical path $\mathbf{r} = \cos t\mathbf{i} + \sin t\mathbf{j} - t\mathbf{k}$ (t is the number of days after admission), how fast is his temperature changing π days after he starts his descent?

19. A groundhog is curled up in an elliptical ball $9x^2 + 3y^2 + 8z^2 = 111$, basking in the rays of the sun that are streaming in the direction $6\mathbf{i} - \mathbf{j} + 8\mathbf{k}$. What point of the groundhog do the rays strike at right angles?

20. An electron follows the path $\mathbf{r} = 3t^2\mathbf{i} - 2t\mathbf{j} + (t + 1)\mathbf{k}$ and hits the collecting plate $x^2 + 2y^2 + z^2 + 2xy - 3z = 3$ at the point $(3, -2, 2)$. What is the cosine of the angle of incidence of the electron (the angle the path makes with the surface of the plate)?

21. Suppose that $f(x, y, z) = x^2 + y^2 + z^2 + xy + xz + yz - 3x - y - 4z + 6$.

(a) Solve the equation $\nabla f(x, y, z) = \mathbf{0}$.

(b) If $\mathbf{u}$ is the solution vector of part (a), find the characteristic values of the matrix $H(\mathbf{u})$. (This matrix is defined in Section 110.)

(c) What is the minimum value of f?

22. Let $f(x, y) = 9x^3 - 18xy^2 + 8y^3 + 54x$.

(a) Solve the equation $\nabla f(\mathbf{x}) = \mathbf{0}$.

(b) For each solution of part (a), find the characteristic values of $H(\mathbf{x})$. What do these numbers tell you about maximum and minimum values of f?

23. Suppose that $f(x, y) = x^2 + 24xy + 19y^2$.

(a) Show that $\nabla f(0, 0) = \mathbf{0}$.

(b) Show that $f_{11}(0, 0) > 0$ and $f_{22}(0, 0) > 0$ and hence infer that $(0, 0, 0)$ is a minimum point of the cross-sectional curves $z = f(x, 0)$ and $z = f(0, y)$ in the XZ- and YZ-planes.

(c) Find the characteristic values and corresponding characteristic vectors of the matrix $H(0, 0)$ of second partial derivatives of $f(x, y)$.

(d) Explain why $z = f(t, 2t)$ can be considered to be the cross-sectional curve formed by cutting the surface $z = f(x, y)$ with the plane $y = 2x$. Is $(0, 0, 0)$ a maximum or a minimum point of this curve?

(e) Explain why $f(2t, -t)$ can be considered to be the cross-sectional curve formed by cutting the surface $z = f(x, y)$ with the plane $x = -2y$. Is $(0, 0, 0)$ a maximum or a minimum point of this curve?

(f) Is $(0, 0, 0)$ a maximum or a minimum point of our given surface?

24. (a) Suppose that

(i) $$f(tx, ty) = t^3 f(x, y)$$

for every point (x, y) and for every positive number t. Differentiate both sides with respect to t and then set $t = 1$ to obtain the equation

(ii) $$xf_1(x, y) + yf_2(x, y) = 3f(x, y).$$

(b) Conversely, suppose that f satisfies equation (ii). Show that it must also satisfy equation (i). (*Hint*: Set $g(t) = t^{-3}f(tx, ty)$, where x and y are "fixed," and find $g'(t)$. Conclude that $g'(t) = 0$ from equation (ii) with x and y replaced by tx and ty. It then follows that $g(t) = g(1)$ for each positive number t.)

(c) A function that satisfies equation (i) is said to be **homogeneous** of degree 3; for example, if $f(x, y) = x^3 \sin xy^{-1} + xy^2$, then f is homogeneous of degree 3. Show that the partial derivatives of $f(x, y)$ are homogeneous of degree 2.

25. Show that the differentiation formula we developed in Problem 107-11 can be written as

$$\frac{d^2w}{dx^2} = \nabla f(\mathbf{u}) \cdot \frac{d^2\mathbf{u}}{dx^2} + \frac{d\mathbf{u}}{dx} \cdot H(\mathbf{u}) \frac{d\mathbf{u}}{dx},$$

where $H(\mathbf{u})$ is the matrix of second partial derivatives that we mentioned in Section 110.

Multiple Integrals and Line Integrals | **14**

The two basic operations in the calculus of scalar-valued functions on R^1 are differentiation and integration. These same operations also play fundamental roles when we come to functions on R^n. We have just considered some aspects of the differential calculus of functions on R^n, and now we will take up integration of scalar- and vector-valued functions on R^2 and R^3.

111. DOUBLE INTEGRALS

Our introduction to multiple integrals will consist of answers to these questions:

(1)   What is an integral?

(2)   What functions are integrable?

(3)   How do we calculate integrals?

Since multiple integrals are analogous to the single integrals we met in Chapter 5, suppose that we start with a review of that situation.

The integral $\int_a^b f(x)\,dx$ of a function f over an interval $[a, b]$ is a *number*, the limit of approximating sums that are formed as follows. The interval of integration is first partitioned into a set of n subintervals. Then in each subinterval we select a point, the point in the ith subinterval being denoted by x_i^*. The value of f at this point x_i^* is multiplied by the length Δx_i of the subinterval to obtain the product $f(x_i^*)\,\Delta x_i$. We now add these numbers to obtain an *approximating sum*

$$s = \sum_{i=1}^{n} f(x_i^*)\,\Delta x_i.$$

Of course, we can get infinitely many such sums, but if our partition of the basic interval $[a, b]$ has a small norm (the length of the largest subinterval into which $[a, b]$ is partitioned), all approximating sums cluster about their limit $\int_a^b f(x)\,dx$.

Geometrically, our approximating sum s is the area of the region made up of the rectangular strips shown in Fig. 111-1, and the area of the region that is bounded by the graph of f, the X-axis, and the lines $x = a$ and $x = b$ is the limit of these sums—that is, the integral $\int_a^b f(x)\,dx$. It is all right for our function to have breaks in its graph (for example, f could be a step function). Unless f is unbounded in $[a, b]$ or is "very discontinuous," it will be integrable on $[a, b]$.

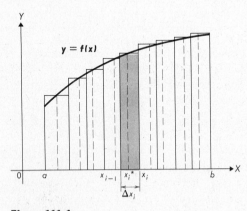

Figure 111-1

We define integrals of a function f on R^2 in a similar manner. In place of our one-dimensional interval $[a, b]$, we have a rectangular region $R = \{(x, y) \mid a \le x \le b, c \le y \le d\}$, as shown in Fig. 111-2. So we are going to define the integral of f over R. Our first step is to partition the region of integration R by drawing

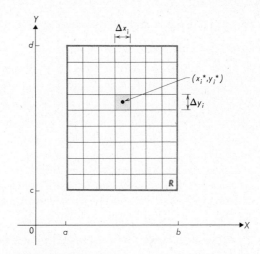

Figure 111-2

lines parallel to the coordinate axes, thereby forming a network of rectangular subregions that cover R. The **norm** of this partition is the length of the longest diagonal of any of these subregions. Hence if the norm of our partition is small, all the subregions are small, as in the one-dimensional case. Now we number the subregions of the partition from 1 to n. For each index i, we denote by Δx_i the width of the ith subregion and by Δy_i its height. Thus the product $\Delta x_i \, \Delta y_i$ is the area of the ith subregion, and we write

$$\Delta A_i = \Delta x_i \, \Delta y_i.$$

Now let us select a point (x_i^*, y_i^*) in the ith subregion, compute the number $f(x_i^*, y_i^*) \, \Delta A_i$, and then add all these products to obtain the sum

$$(111\text{-}1) \qquad s = \sum_{i=1}^{n} f(x_i^*, y_i^*) \, \Delta A_i.$$

We can compute many such sums, depending on how we partition the region R and choose the points (x_i^*, y_i^*). But if f is a "reasonable" function, all the sums that we can form when we use partitions with small norms will be close to one particular number, the limit of s as the norm of the partition of R approaches 0. This number is called the **double integral of f over R**, and we will denote it by one of the symbols

$$\iint\limits_{R} f(x, y) \, dA, \qquad \iint\limits_{R} f(x, y) \, dx \, dy, \qquad \text{or} \qquad \iint\limits_{R} f(x, y) \, dy \, dx.$$

Thus, in rough terms, *the double integral* $\iint\limits_R f(x, y)\, dA$ *is the number that is approximated by every sum s that we can form when we use a partition with a small norm.*

A precise definition of what the limit of s as the norm of our partition of R approaches 0 means follows the lines laid down in Chapter 5. Equation 111-1 defines a set-valued function whose limit from the right at 0 is the integral we are discussing. A proof of the existence of this limit and similar theoretical points can safely be left to a course in advanced calculus. The key fact is that the limit will exist (that is, f is integrable) if f is bounded in R and is not "too discontinuous." As with functions on R^1, step discontinuities of f are allowed (here the graph of f is a surface, and f will be integrable even though the surface is torn).

Example 111-1. Approximate the number $\iint\limits_R (8x^2 + 2y)\, dA$ if R is the square $\{(x, y) \mid 0 \le x \le 2, 0 \le y \le 2\}$.

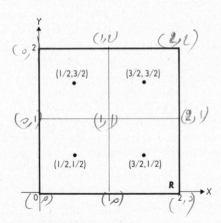

Figure 111-3

Solution. In Fig. 111-3 we show our region R partitioned into four subregions by the lines $x = 1$ and $y = 1$. Each subregion is a 1 by 1 square, and so $\Delta A_i = 1$ for each index i. In each square subregion we have selected the center as our point (x_i^*, y_i^*). Thus since $f(x, y) = 8x^2 + 2y$, a sum that approximates the desired integral is

$$s = f(\tfrac{1}{2}, \tfrac{1}{2}) \cdot 1 + f(\tfrac{1}{2}, \tfrac{3}{2}) \cdot 1 + f(\tfrac{3}{2}, \tfrac{3}{2}) \cdot 1 + f(\tfrac{3}{2}, \tfrac{1}{2}) \cdot 1$$

$$= 3 \cdot 1 + 5 \cdot 1 + 21 \cdot 1 + 19 \cdot 1 = 48.$$

In other words, $\iint\limits_R (8x^2 + 2y)\, dA \approx 48$. We will find in the next section (Example 112-1) that the integral is actually $\tfrac{152}{3} = 50\tfrac{2}{3}$.

As with single integrals, we can best get the feel of double integrals by interpreting them geometrically. Even when a single integral arises in a nongeometric problem, we know that we can interpret it in terms of the area of some plane region. Now we will see how we can interpret a double integral in terms of volume. Suppose that f is a function on R^2 and that $f(x, y) \geq 0$ for each point (x, y) of a rectangular region R of the XY-plane. Then the graph of the equation $z = f(x, y)$, for $(x, y) \in R$, is a surface that lies above the XY-plane, as shown in Fig. 111-4. In the figure we have shown one of the subregions into which R is partitioned when we form Approximating Sum 111-1. A typical term of this sum has the form

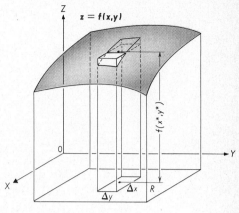

Figure 111-4

$$f(x^*, y^*) \, \Delta A = f(x^*, y^*) \, \Delta x \, \Delta y,$$

and this number is the volume of the rectangular block whose base measures Δx by Δy and whose altitude is $f(x^*, y^*)$. Approximating Sum 111-1 is a sum of such volumes, and it appears from the figure that this sum approximates the volume of the three-dimensional region that is bounded by R, planes that contain the boundaries of R, and the graph of f. Since our approximating sum also approximates the number $\iint_R f(x, y) \, dA$, it is natural to expect that the volume of the region under the graph of f is the integral. This expectation is correct, and hence we can use our ideas about volume to compute double integrals in exactly the same way that we use our ideas about area to compute single integrals.

Example 111-2. Evaluate the integral $\iint_R (\llbracket x \rrbracket + \llbracket y \rrbracket) \, dA$, where R is the square region shown in Fig. 111-3.

Solution. In Fig. 111-5 we have shown the part of the surface $z = \llbracket x \rrbracket + \llbracket y \rrbracket$ that lies above the region R. The three-dimensional region under this graph can be considered as two 1 by 1 by 1 blocks and one 1 by 1 by 2 block, so its volume is 4 cubic units. Hence we conclude that $\iint_R (\llbracket x \rrbracket + \llbracket y \rrbracket) \, dA = 4$.

Integrals are limits, and because limits of sums are sums of limits, it follows that integrals of sums are sums of integrals. You already know such general integral theorems for single integrals, and the analogous theorems are true for multiple integrals, too. Thus if f and g are integrable on a rectangular region R, and if m and n are any numbers, then

(111-2) $$\iint_R [mf(x, y) + ng(x, y)] \, dA = m \iint_R f(x, y) \, dA + n \iint_R g(x, y) \, dA.$$

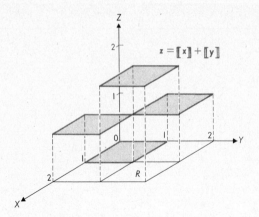

Figure 111-5

Furthermore, suppose that we split a rectangular region R into two nonoverlapping subregions R_1 and R_2; that is, suppose that $R = R_1 \cup R_2$, where $R_1 \cap R_2$ has zero area. Then

$$(111\text{-}3) \qquad \iint_R f(x, y)\, dA = \iint_{R_1} f(x, y)\, dA + \iint_{R_2} f(x, y)\, dA.$$

This last equation corresponds to the equation

$$\int_a^c f(x)\, dx = \int_a^b f(x)\, dx + \int_b^c f(x)\, dx$$

for single integrals.

PROBLEMS 111

1. Let R be the rectangular region $\{(x, y) \mid -1 \le x \le 1, 0 \le y \le 2\}$. The lines $y = 1$ and $x = 0$ partition this region into four square subregions, in each of which we take the center as our point (x^*, y^*). Based on this partition and choice of points within subregions, form the sums that approximate the given integral.

(a) $\displaystyle\iint_R (x + y)^2\, dA$ (b) $\displaystyle\iint_R (x^2 + y^2)\, dA$ (c) $\displaystyle\iint_R (|x| + |y|)\, dA$

(d) $\displaystyle\iint_R |x + y|\, dA$ (e) $\displaystyle\iint_R \cos^2 \pi xy\, dA$ (f) $\displaystyle\iint_R \sin^2 \pi xy\, dA$

2. In Example 111-1 we approximated $\displaystyle\iint_R (8x^2 + 2y)\, dA$ by choosing a particular partition of R and particular points (x^*, y^*) within the subregions of that partition.

What is the smallest approximating sum we could have calculated from that partition? The largest?

3. In each integral $\iint\limits_R f(x, y)\, dA$ of this problem, assume that

$$R = \{(x, y) \mid 0 \leq x \leq 2,\, 0 \leq y \leq 2\}.$$

Sketch the surface $z = f(x, y)$ for $(x, y) \in R$ and hence evaluate the integral geometrically.

(a) $\iint\limits_R x\, dA$ (b) $\iint\limits_R y\, dA$ (c) $\iint\limits_R (2 - x)\, dA$

(d) $\iint\limits_R (4 - y)\, dA$ (e) $\iint\limits_R |x - 1|\, dA$ (f) $\iint\limits_R |y - 1|\, dA$

(g) $\iint\limits_R e^{\pi}\, dA$ (h) $\iint\limits_R \ln \pi\, dA$ (i) $\iint\limits_R (x + y)\, dA$

(j) $\iint\limits_R (x + y + 2)\, dA$ (k) $\iint\limits_R [\![x]\!]\, dA$ (l) $\iint\limits_R [\![y]\!]\, dA$

(m) $\iint\limits_R [\![x]\!][\![y]\!]\, dA$ (n) $\iint\limits_R [\![x + y]\!]\, dA$

4. When you use the volume interpretation to evaluate the following double integrals, notice that the regions whose volumes you are to find are cylindrical solids generated by curves in the XZ-plane or the YZ-plane. You may use single integration to find the areas of the bases of the solids in these planes. In each case, R is the square region $\{(x, y) \mid -1 \leq x \leq 1,\, -1 \leq y \leq 1\}$.

(a) $\iint\limits_R x^2\, dA$ (b) $\iint\limits_R y^2\, dA$ (c) $\iint\limits_R \sin^2 \pi y\, dA$

(d) $\iint\limits_R \cos^2 \pi y\, dA$ (e) $\iint\limits_R \ln (x + 2)\, dA$ (f) $\iint\limits_R \ln (x^2 + 4x + 4)\, dA$

(g) $\iint\limits_R |4y^2 - 1|\, dA$ (h) $\iint\limits_R |4y - 1|\, dA$

5. Let $f(x, y) = \sqrt{4 - x^2 - y^2}$ if $x^2 + y^2 \leq 4$ and $f(x, y) = 0$ if $x^2 + y^2 > 4$.
    (a) Sketch the surface $z = f(x, y)$.
    (b) Find $\iint\limits_R f(x, y)\, dA$ if R is the rectangle $\{(x, y) \mid -5 \leq x \leq 3,\, -7 \leq y \leq 6\}$.
    (c) Find $\iint\limits_R f(x, y)\, dA$ if R is the rectangle $\{(x, y) \mid -5 \leq x \leq 0,\, -7 \leq y \leq 6\}$.

6. Let R be the rectangular region $\{(x, y) \mid a \le x \le b, c \le y \le d\}$.
   (a) Use the geometric interpretation of the integral to show that

   $$\iint_R (x - a) \, dA = \tfrac{1}{2}(b - a)^2(d - c).$$

   (b) Now write $x = (x - a) + a$ and use Equation 111-2 to show that $\iint_R x \, dA = \tfrac{1}{2}(b^2 - a^2)(d - c)$.

   (c) Show that $\iint_R y \, dA = \tfrac{1}{2}(d^2 - c^2)(b - a)$.

   (d) Show that for any three numbers p, q, and r,

   $$\iint_R (px + qy + r) \, dA = [\tfrac{1}{2}p(a + b) + \tfrac{1}{2}q(c + d) + r](b - a)(d - c).$$

   Interpret this result geometrically.

7. Suppose that $R = R_1 \cup R_2$, where $R_1 = \{(x, y) \mid -1 \le x \le 1, 0 \le y \le 1\}$ and $R_2 = \{(x, y) \mid 1 \le x \le 2, 0 \le y \le 1\}$. Use the formula of the preceding problem to verify that

   $$\iint_R (3x + 4y + 5) \, dA = \iint_{R_1} (3x + 4y + 5) \, dA + \iint_{R_2} (3x + 4y + 5) \, dA.$$

8. Same as Problem 111-3, but don't let the complicated formulas for $f(x, y)$ alarm you. The surfaces $z = f(x, y)$ are really quite simple.

   (a) $\displaystyle\iint_R \sqrt{2x - x^2} \, dA$ (b) $\displaystyle\iint_R (2 + \sqrt{2y - y^2}) \, dA$

   (c) $\displaystyle\iint_R \left(1 + \frac{x - y}{|x - y|}\right) dA$ (d) $\displaystyle\iint_R \text{Cos}^{-1}\left(\frac{x - y}{|x - y|}\right) dA$

   (e) $\displaystyle\iint_R [\![(x - 1)^2 + (y - 1)^2]\!] \, dA$ (f) $\displaystyle\iint_R ([\![x - 1]\!]^2 + [\![y - 1]\!]^2) \, dA$

9. Let $R = \{(x, y) \mid 0 \le x \le 1, 0 \le y \le 1\}$. From what we said in the text, you know that some of these integrals don't exist. Which ones?

   (a) $\displaystyle\iint_R \ln x \, dA$ (b) $\displaystyle\iint_R \tan(x + y) \, dA$

   (c) $\displaystyle\iint_R e^{[\![3x + 2y]\!]} \, dA$ (d) $\displaystyle\iint_R \sin\left(\frac{x}{y} + \frac{y}{x}\right) dA$

10. Explain why $\displaystyle\iint_R f(x, y) \, dA \ge 0$ if $f(x, y) \ge 0$ for each point $(x, y) \in R$. Use this result and Equation 111-2 to show that if $f(x, y) \le g(x, y)$ at each point of R, then

    $$\iint_R f(x, y) \, dA \le \iint_R g(x, y) \, dA.$$

We will now use our geometric interpretation of a double integral as a volume to develop a method of evaluating double integrals by using successive single integrals. We can thereby put to work the tricks for evaluating single integrals that we learned earlier. Our arguments in this section will be quite informal. Reference to pictures will take the place of mathematical proofs.

Let us suppose that we are given a function f that is integrable on a rectangular region $R = \{(x, y) \mid a \leq x \leq b, c \leq y \leq d\}$. Because we want to use our geometric interpretation of the double integral as a volume, we will assume that $f(x, y) \geq 0$ for each point (x, y) in R, but it is easy to extend our results to cover cases in which $f(x, y)$ is negative for some points of R. Figure 112-1 shows the graph of the equation $z = f(x, y)$ for $(x, y) \in R$. The number $\iint\limits_{R} f(x, y) \, dA$ that we seek is the volume of the solid region between this surface and R.

We will find this volume by thinking of our solid as a "union of narrow slices," a technique we practiced in Section 41. Thus suppose that we partition the interval $[a, b]$ of the X-axis and cut our solid with planes that contain the partition points and that are parallel to the YZ-plane. The figure shows the cross section that is cut by the plane containing a typical point x. Denote the area of this cross section by $C(x)$, multiply it by the distance Δx to the next partition point, and add the resulting numbers:

$$\sum C(x) \, \Delta x.$$

This sum approximates the volume of our solid, and a good approximation can be ensured by insisting that the norm of the partition of $[a, b]$ be small. In other words, the integral $\int_a^b C(x) \, dx$ is the volume of our solid. On the other hand,

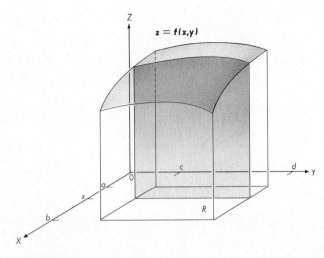

Figure 112-1

this volume is also the value of the double integral we are trying to calculate, and so

$$(112\text{-}1) \qquad \iint_R f(x, y)\, dA = \int_a^b C(x)\, dx.$$

Therefore we can evaluate the double integral of f over R by finding a single integral of $C(x)$.

So now we must find $C(x)$ for a given number x. Since $C(x)$ is the area of a plane region, we find it by integration. Figure 112-1 shows that the upper boundary of this plane region is the curve $z = f(x, y)$ for $y \in [c, d]$ (x is "fixed"), and so its area is $C(x) = \int_c^d f(x, y)\, dy$. When we substitute this result in Equation 112-1, we obtain the fundamental formula for evaluating double integrals:

$$(112\text{-}2) \qquad \iint_R f(x, y)\, dA = \int_a^b \left[\int_c^d f(x, y)\, dy \right] dx.$$

We usually omit the brackets in the **iterated integral** on the right-hand side of Equation 112-2 and write the equation as

$$(112\text{-}3) \qquad \iint_R f(x, y)\, dA = \int_a^b \int_c^d f(x, y)\, dy\, dx.$$

When we evaluate the "inner integral" in Equation 112-3, we must remember that y is the variable of integration, not x. In this integration x is considered fixed, just as it is fixed when we apply the differential operator $\dfrac{\partial}{\partial y}$ to find the partial derivative $\dfrac{\partial}{\partial y} f(x, y)$.

Example 112-1. Evaluate the integral $\iint_R (8x^2 + 2y)\, dA$, where R is the region $\{(x, y) \mid 0 \le x \le 2, 0 \le y \le 2\}$.

Solution. Here $a = 0$, $b = 2$, $c = 0$, and $d = 2$, so Equation 112-3 becomes

$$\iint_R (8x^2 + 2y)\, dA = \int_0^2 \int_0^2 (8x^2 + 2y)\, dy\, dx$$

$$= \int_0^2 \left[\int_0^2 (8x^2 + 2y)\, dy \right] dx$$

$$= \int_0^2 [8x^2 y + y^2]\Big|_0^2 dx$$

$$= \int_0^2 (16x^2 + 4)\, dx = 50\tfrac{2}{3}.$$

Example 112-2. Evaluate the integral $\displaystyle\iint\limits_{R} 2xye^{xy^2}\, dA$, where R is the 1 by 1 square whose vertices are $(0, 0)$, $(1, 0)$, $(1, 1)$, and $(0, 1)$.

Solution. If you draw a picture of R in the XY-plane, you will see that Equation 112-3 becomes

$$\iint\limits_{R} 2xye^{xy^2}\, dA = \int_0^1 \int_0^1 2xye^{xy^2}\, dy\, dx,$$

so our first task is to evaluate the inner integral

$$\int_0^1 2xye^{xy^2}\, dy,$$

keeping x fixed. We evaluate integrals by making a substitution that turns the integrand into one that has been tabulated. Thus, in this case, we let $u = xy^2$. Then $du = 2xy\, dy$, and $u = 0$ when $y = 0$ and $u = x$ when $y = 1$. Our inner integral therefore becomes

$$\int_0^x e^u\, du = e^u \Big|_0^x = e^x - 1.$$

Hence our double integral is

$$\iint\limits_{R} 2xye^{xy^2}\, dA = \int_0^1 (e^x - 1)\, dx = (e^x - x) \Big|_0^1 = e - 1 - 1 = e - 2.$$

Sometimes we have to use our General Integration Formulas 111-2 or 111-3 before applying Equation 112-3. In this way we can get rid of absolute value signs, greatest integer symbols, and the like. When calculating integrals, we always try to reduce a given integrand to one that is listed in Table V.

Example 112-3. Evaluate the iterated integral

$$I = \int_0^4 \int_{-1}^1 24 \left| y^3(x^2 + x - 6) \right|\, dy\, dx.$$

Solution. Let us think of this integral as the double integral $\displaystyle\iint\limits_{R} |f(x, y)|\, dA$, where R is the rectangular region in Fig. 112-2 and $f(x, y) = 24y^3(x^2 + x - 6) = 24y^3(x - 2)(x + 3)$. The plus and minus signs in the figure indicate the sign of the

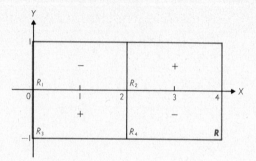

Figure 112-2

number $f(x, y)$ in the subregions R_1, R_2, R_3, and R_4. Hence we can use Equations 111-2 and 111-3 to write

$$I = \iint\limits_{R} f(x, y) \, dA$$

(112-4)
$$= -\iint\limits_{R_1} f(x, y) \, dA + \iint\limits_{R_2} f(x, y) \, dA$$

$$+ \iint\limits_{R_3} f(x, y) \, dA - \iint\limits_{R_4} f(x, y) \, dA.$$

We can use Equation 112-3 on each term in this sum—for example,

$$\iint\limits_{R_1} f(x, y) \, dA = \int_0^2 \int_0^1 24y^3(x^2 + x - 6) \, dy \, dx$$

$$= \int_0^2 6y^4(x^2 + x - 6) \Big|_0^1 \, dx$$

$$= \int_0^2 6(x^2 + x - 6) \, dx = -44.$$

When you carry out the calculation of the other integrals and substitute in Equation 112-4, you will find that $I = 240$.

The geometric reasoning by which we arrived at Equation 112-3 makes it obvious that our double integral is also equal to the iterated integral in which we first integrate with respect to x:

$$\iint\limits_{R} f(x, y) \, dA = \int_c^d \int_a^b f(x, y) \, dx \, dy.$$

The two iterated integrals may not be equally easy to calculate, however.

698

Example 112-4. We could have evaluated the double integral in Example 112-2 by writing it as the iterated integral $\int_0^1 \int_0^1 2xye^{xy^2}\, dx\, dy$. Try it.

Solution. Here our inner integral is $\int_0^1 2xye^{xy^2}\, dx$, in which we are to hold y fixed. If we set $u = xy^2$, then $du = y^2\, dx$, and this integral becomes

$$\int_0^{y^2} \frac{2u}{y}\, e^u\, \frac{1}{y^2}\, du = 2y^{-3} \int_0^{y^2} ue^u\, du.$$

From integral tables or by using the method of integration by parts, we find that $\int ue^u\, du = (u - 1)e^u$, and therefore our inner integral is

$$\int_0^1 2xye^{y^2}\, dx = 2y^{-3}[(y^2 - 1)e^{y^2} + 1].$$

So now we are to finish the evaluation of our double integral by finding

$$\int_0^1 2y^{-3}[(y^2 - 1)e^{y^2} + 1]\, dy,$$

which is easier said than done.

PROBLEMS   112

1. Evaluate the iterated integral.

(a) $\int_1^2 \int_0^1 (1 - y)x^2\, dy\, dx$

(b) $\int_1^2 \int_0^1 (1 - y)x^2\, dx\, dy$

(c) $\int_0^{\pi/2} \int_0^1 xy \cos x^2 y\, dx\, dy$

(d) $\int_0^{\pi/2} \int_0^1 xy \sin x^2 y\, dx\, dy$

(e) $\int_0^{\ln 3} \int_0^{\ln 3} e^{x+y}\, dx\, dy$

(f) $\int_0^{\ln 3} \int_0^{\ln 3} (e^x + e^y)\, dx\, dy$

2. Evaluate the iterated integral.

(a) $\int_0^{\pi/2} \int_0^{\pi/2} \sin x \sin y\, dx\, dy$

(b) $\int_0^{\pi/2} \int_0^{\pi/2} (\sin x + \sin y)\, dx\, dy$

(c) $\int_0^{\pi/2} \int_0^{\pi/2} \sin (x + y)\, dx\, dy$

(d) $\int_0^{\pi/2} \int_0^{\pi/2} \cos (x + y)\, dx\, dy$

(e) $\int_0^{\pi/2} \int_0^{\pi/2} \sin^2 (x + y)\, dx\, dy$

(f) $\int_0^{\pi/2} \int_0^{\pi/2} \cos^2 (x + y)\, dy\, dx$

3. Evaluate $\iint_R f(x, y)\, dA$ if $R = \{(x, y)\,|\,0 \le x \le 2, 0 \le y \le 1\}$ and

(a) $f(x, y) = xy^2$.

(b) $f(x, y) = x + y^2$.

(c) $f(x, y) = e^x(\cos y + \cos e^x)$.

(d) $f(x, y) = ye^x \cos ye^x$.

4. Evaluate the iterated integral.

(a) $\int_0^1 \int_0^1 ye^{xy}\, dx\, dy$

(b) $\int_0^1 \int_0^1 xye^y\, dy\, dx$

(c) $\int_0^1 \int_0^1 ye^{x+y}\, dx\, dy$

(d) $\int_0^1 \int_0^1 y(e^x + e^y)\, dy\, dx$

5. Evaluate the iterated integral.

(a) $\int_0^2 \int_0^1 2x\sqrt{y+x^2}\, dy\, dx$

(b) $\int_0^1 \int_0^2 2x\sqrt{y+x^2}\, dx\, dy$

(c) $\int_0^1 \int_0^2 e^x \cos(y+e^x)\, dx\, dy$

(d) $\int_0^2 \int_0^1 e^x \cos(y+e^x)\, dy\, dx$

6. Evaluate the given integral when $R = \{(x, y) \mid -1 \le x \le 1, -1 \le y \le 1\}$.

(a) $\iint_R (|x| + |y|)\, dA$

(b) $\iint_R |xy|\, dA$

(c) $\iint_R e^{|x|+|y|}\, dA$

(d) $\iint_R |e^{x+y}|\, dA$

7. Evaluate (if you can) the iterated integrals.

(a) $\int_0^1 \int_0^1 x^y\, dx\, dy$ and $\int_0^1 \int_0^1 x^y\, dy\, dx$

(b) $\int_0^1 \int_0^1 x \sin xy\, dy\, dx$ and $\int_0^1 \int_0^1 x \sin xy\, dx\, dy$

8. How far can you go in evaluating the integral $\int_0^{\pi/2} \int_0^{\pi/2} \sin xy\, dy\, dx$?

9. Explain why $\int_a^b \int_a^b e^{-(x^2+y^2)}\, dx\, dy = \left(\int_a^b e^{-x^2}\, dx \right)^2$.

10. If f is an *odd function* $[f(-u) = -f(u)]$, show that $\int_{-a}^a \int_{-b}^b f(x+y)\, dy\, dx = 0$.

11. The equation $F(x, y) = \int_0^x \int_0^y \sin(u^2 + v^2)\, du\, dv$ defines a function on R^2.

What is the domain of F? Is $F(1, 1)$ positive or negative? Is $F_1(1, 1)$ positive or negative? Calculate $F_{12}(x, y)$ and $F_{21}(x, y)$. Show that $F(y, x) = F(x, y)$.

12. If u is very large, $\text{Tan}^{-1} u \approx \frac{1}{2}\pi$. Can you use this fact to convince yourself that the integrals

$$\int_1^2 \int_0^{1000} \frac{1}{x^2+y^2}\, dx\, dy \quad \text{and} \quad \int_1^2 \int_0^{1000} \frac{1}{1+x^2y^2}\, dx\, dy$$

are both approximately equal to $\frac{1}{2}\pi \ln 2$?

In one dimension, we can only integrate over intervals, but all sorts of shapes are possible in the plane, the region R of Fig. 113-1, for example. What do we mean by the integral of f over R, and how do we calculate it? When faced with a new problem, mathematicians like to reduce it to one they have solved before, and that is what we will do now. We know how to integrate over rectangular regions, so we will reduce our present problem to such a case. We enclose R in a rectangular region $\bar{R}$ and extend f to a new function $\bar{f}$, where $\bar{f}(x, y) = f(x, y)$ if $(x, y) \in R$ and $\bar{f}(x, y) = 0$ if $(x, y) \notin R$. Since we know what the double integral of $\bar{f}$ over the rectangular region $\bar{R}$ means, we will simply *define* the integral of f over R by the equation

$$(113\text{-}1) \qquad \iint\limits_{R} f(x, y)\, dA = \iint\limits_{\bar{R}} \bar{f}(x, y)\, dA.$$

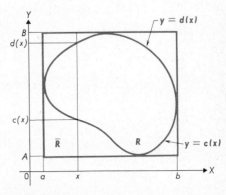

Figure 113-1

Because $\bar{f}$ takes the value 0 outside of R, we can choose as $\bar{R}$ *any* rectangular region that contains R, and Equation 113-1 will give us the same value of the integral of f over R.

Example 113-1. Evaluate the integral $\displaystyle\iint\limits_{R} \sqrt{x^2 + y^2}\, dA$, where R is the unit disk $x^2 + y^2 \leq 1$.

Solution. We can take $\bar{R}$ to be any rectangular region that contains the unit disk, so let us choose $\bar{R} = \{(x, y) \mid -1 \leq x \leq 1, -1 \leq y \leq 1\}$. The graph of $\bar{f}$ is shown in Fig. 113-2; it is the union of a cone and the part of the XY-plane that lies outside the unit circle. Our definition of an integral over R tells us that we are to evaluate $\displaystyle\iint\limits_{\bar{R}} \bar{f}(x, y)\, dA$. This integral is the volume of the region "between" the graph of $\bar{f}$ and the rectangular region $\bar{R}$. Since the graph of $\bar{f}$ coincides with the XY-plane except in R, we are actually to find the volume of the region between the cone and the disk R. When we calculate this volume, we find that

$$\iint\limits_{R} \sqrt{x^2 + y^2}\, dA = \tfrac{2}{3}\pi.$$

You will notice that unless $f(x, y) = 0$ at a boundary point of R, the function $\bar{f}$ will be discontinuous there, as it is in our example. Thus $\bar{f}$ is quite likely to be

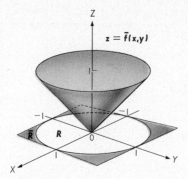

$z = \bar{f}(x,y)$

Figure 113-2

discontinuous on the boundary of R. But it takes "major" discontinuities to destroy the integrability of a function, and in all practical cases $\bar{f}$ is integrable. The general Integration Formulas 111-2 and 111-3 remain valid when R, R_1, and R_2 are nonrectangular regions.

   We reduce double integrals over rectangles to iterated integrals, and we can do the same for integrals over the region R shown in Fig. 113-1. According to Equation 113-1, the integral of a function f over R is the integral of its extension $\bar{f}$ over $\bar{R}$. Since $\bar{R}$ is rectangular, this latter integral can be expressed as an iterated integral:

$$(113\text{-}2) \qquad \iint\limits_{\bar{R}} \bar{f}(x, y)\, dA = \int_a^b \int_A^B \bar{f}(x, y)\, dy\, dx.$$

Our figure shows that the upper and lower boundaries of R are the curves $y = d(x)$ and $y = c(x)$. Thus if x is a given point of $[a, b]$, then $A \le c(x) \le d(x) \le B$. Now let us write the inner integral of the iterated integral in Equation 113-2 as

$$(113\text{-}3) \qquad \int_A^B \bar{f}(x, y)\, dy = \int_A^{c(x)} \bar{f}(x, y)\, dy$$

$$+ \int_{c(x)}^{d(x)} \bar{f}(x, y)\, dy + \int_{d(x)}^B \bar{f}(x, y)\, dy.$$

Our definition of $\bar{f}$ tells us that $\bar{f}(x, y) = 0$ if $A \le y < c(x)$ or $d(x) < y \le B$, and $\bar{f}(x, y) = f(x, y)$ if $c(x) \le y \le d(x)$. Therefore Equation 113-3 simplifies to

$$\int_A^B \bar{f}(x, y)\, dA = \int_{c(x)}^{d(x)} f(x, y)\, dy.$$

702

When this result is substituted in Equation 113-2 and then in Equation 113-1, we obtain an integration formula that expresses a double integral over our non-rectangular region R as an iterated integral:

(113-4)
$$\iint_R f(x, y) \, dA = \int_a^b \int_{c(x)}^{d(x)} f(x, y) \, dy \, dx.$$

Notice that this formula only applies if our region of integration is of the correct shape; specifically, we must have $R = \{(x, y) \mid a \le x \le b, \ c(x) \le y \le d(x)\}$. From this expression for R it is easy to read off the limits of integration on the iterated integral.

Example 113-2.   Evaluate $\displaystyle\iint_R x \, dA$, where R is the region that is bounded by the Y-axis and the semicircle $x = \sqrt{4 - y^2}$.

Solution.   The given region (Fig. 113-3) is the set

$$R = \{(x, y) \mid 0 \le x \le 2, \ -\sqrt{4 - x^2} \le y \le \sqrt{4 - x^2}\}.$$

Therefore Equation 113-4 says that

$$\iint_R x \, dA = \int_0^2 \int_{-\sqrt{4-x^2}}^{\sqrt{4-x^2}} x \, dy \, dx = \int_0^2 x \left[y \, \Big|_{-\sqrt{4-x^2}}^{\sqrt{4-x^2}} \right] dx$$

$$= \int_0^2 2x\sqrt{4 - x^2} \, dx = \tfrac{16}{3}.$$

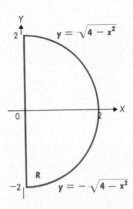

Figure 113-3

According to Equation 113-4, we can express a double integral as an iterated integral in which we first consider y as the variable of integration and then x. Of course, the roles of x and y can be interchanged (if R is of the right shape). Thus if $R = \{(x, y) \mid c \le y \le d, a(y) \le x \le b(y)\}$, then

$$(113\text{-}5) \qquad \iint\limits_R f(x, y)\, dA = \int_c^d \int_{a(y)}^{b(y)} f(x, y)\, dx\, dy.$$

If a given region R is not of the right shape for Equation 113-4 or 113-5 to apply, we may be able to write it as a union of such regions and use Equation 111-3 first.

Example 113-3.   Use Equation 113-5 to evaluate the integral in Example 113-2.

Solution.   The left boundary of our region is the line $x = 0$, and the right boundary is the semicircle $x = \sqrt{4 - y^2}$. Thus

$$R = \{(x, y) \mid -2 \le y \le 2, 0 \le x \le \sqrt{4 - y^2}\},$$

and Equation 113-5 becomes

$$\iint\limits_R x\, dA = \int_{-2}^2 \int_0^{\sqrt{4-y^2}} x\, dx\, dy = \int_{-2}^2 \tfrac{1}{2}x^2 \Big|_0^{\sqrt{4-y^2}} dy$$

$$= \tfrac{1}{2} \int_{-2}^2 (4 - y^2)\, dy = \tfrac{16}{3}.$$

The region of integration R determines the limits of integration on an iterated integral used to evaluate a double integral and, conversely, these limits determine R. Thus if $a \le b$ and $c(x) \le d(x)$ for each $x \in [a, b]$, then the iterated integral $\int_a^b \int_{c(x)}^{d(x)} f(x, y)\, dy\, dx$ is the double integral of f over the region

$$R = \{(x, y) \mid a \le x \le b, c(x) \le y \le d(x)\}.$$

Similarly, if $c \le d$ and $a(y) \le b(y)$ for each $y \in [c, d]$, then the iterated integral $\int_c^d \int_{a(y)}^{b(y)} f(x, y)\, dx\, dy$ is the double integral of f over the region

$$R = \{(x, y) \mid c \le y \le d, a(y) \le x \le b(y)\}.$$

Figure 113-4 shows how the boundary of the same region of integration looks from these points of view. It may be important to have both pictures in mind because you may want to shift from one iterated integral to the other, as the following example shows.

Example 113-4.   Evaluate the iterated integral $\int_0^1 \int_y^1 \tan x^2\, dx\, dy.$

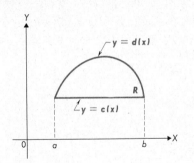
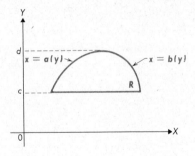

Figure 113-4

Solution. To evaluate the integral as written, our first task would be to find $\int_y^1 \tan x^2\, dx$. But none of the integration formulas in Table V applies, so we will change the order of integration in our given iterated integral. From its limits of integration we see that this integral equals the double integral of $\tan x^2$ over the region $R = \{(x, y) \mid 0 \leq y \leq 1,\, y \leq x \leq 1\}$. Figure 113-5 shows us that we can also write $R = \{(x, y) \mid 0 \leq x \leq 1,\, 0 \leq y \leq x\}$, and this expression tells how to write our double integral as an iterated integral in which we first integrate with respect to y:

$$\iint_R \tan x^2\, dA = \int_0^1 \int_0^x \tan x^2\, dy\, dx.$$

Therefore

$$\int_0^1 \int_y^1 \tan x^2\, dx\, dy = \int_0^1 \int_0^x \tan x^2\, dy\, dx$$

$$= \int_0^1 x \tan x^2\, dx = \tfrac{1}{2} \ln \sec x^2 \Big|_0^1$$

$$= \tfrac{1}{2} \ln \sec 1 \approx .308.$$

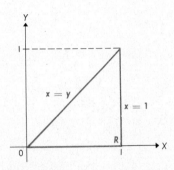

Figure 113-5

PROBLEMS 113

1. Evaluate the iterated integral.

(a) $\displaystyle\int_0^4 \int_0^x x \, dy \, dx$
(b) $\displaystyle\int_0^4 \int_0^y x \, dx \, dy$

(c) $\displaystyle\int_0^1 \int_0^{x^2} \sin \pi x^3 \, dy \, dx$
(d) $\displaystyle\int_0^1 \int_0^{x^2} \cos \pi x^3 \, dy \, dx$

(e) $\displaystyle\int_0^1 \int_{2x}^{3x} e^{x+y} \, dy \, dx$
(f) $\displaystyle\int_1^2 \int_0^{\sqrt{x}} 2y \ln x^2 \, dy \, dx$

2. Evaluate the iterated integral.

(a) $\displaystyle\int_0^2 \int_0^{\sqrt{4-x^2}} x \, dy \, dx$
(b) $\displaystyle\int_0^2 \int_0^{\sqrt{4-y^2}} x \, dx \, dy$

(c) $\displaystyle\int_0^{\pi/2} \int_0^{\sin y} \frac{dx \, dy}{\sqrt{1 - x^2}}$
(d) $\displaystyle\int_0^{\pi/4} \int_0^{\tan y} \frac{dx \, dy}{1 + x^2}$

3. Evaluate the integral $\displaystyle\iint_R xy \, dA$ if R is the given region.

(a) $\{(x, y) \mid 0 \le x \le 2, 0 \le y \le x\}$
(b) $\{(x, y) \mid 0 \le x \le 2, x \le y \le 2\}$
(c) $\{(x, y) \mid x \ge 0, y \ge 0, x^2 + y^2 \le 1\}$
(d) $\{(x, y) \mid 0 \le x \le 2, x^2 \le y \le 4\}$
(e) $\{(x, y) \mid |x| + |y| \le 1\}$
(f) $\{(x, y) \mid x^2 + |y| \le 4\}$

4. Sketch the region of integration R if the double integral $\displaystyle\iint_R f(x, y) \, dA$ can be expressed as the given iterated integral.

(a) $\displaystyle\int_{-1}^2 \int_3^4 f(x, y) \, dy \, dx$
(b) $\displaystyle\int_{-1}^2 \int_3^4 f(x, y) \, dx \, dy$

(c) $\displaystyle\int_0^\pi \int_0^{\sin x} f(x, y) \, dy \, dx$
(d) $\displaystyle\int_0^{2\pi} \int_0^{|\sin x|} f(x, y) \, dy \, dx$

(e) $\displaystyle\int_{-1}^1 \int_{x^2-1}^{1-x^2} f(x, y) \, dy \, dx$
(f) $\displaystyle\int_{-2}^2 \int_{|y|-2}^{2-|y|} f(x, y) \, dx \, dy$

5. Evaluate $\displaystyle\iint_R f(x, y) \, dA$ for the given $f(x, y)$ and R.

(a) $f(x, y) = 4 - x$, $R = \{(x, y) \mid 2 \le x \le 3, 0 \le y \le 1/x\}$.
(b) $f(x, y) = 1/(4 - x)$, R is the region of part (a).
(c) $f(x, y) = 2x$, R is the region bounded by the parabola $4y = x^2$ and the line $x - 2y + 4 = 0$.

(d) $f(x, y) = 2y$, R is the region of part (c).

(e) $f(x, y) = 2x^2e^{xy}$, R is the region bounded by the curves $y = x$, $y = 1/x$ and the line $x = 2$.

(f) $f(x, y) = 2/y$, R is the region of part (e).

6. Evaluate the following integrals by changing the order of integration.

(a) $\displaystyle\int_0^2 \int_{2y}^4 e^{x^2} \, dx \, dy$
    (b) $\displaystyle\int_0^1 \int_y^1 \sin x^2 \, dx \, dy$

(c) $\displaystyle\int_0^1 \int_0^{\text{Arccos }x} e^{\sin y} \, dy \, dx$
    (d) $\displaystyle\int_{-1}^0 \int_0^{\ln(2+x)} \sin(e^y - 2) \, dy \, dx$

7. Evaluate the integral $\displaystyle\int_{-1}^0 \int_{\text{Arccos }x}^{2\pi} e^y \, dy \, dx$.

8. Evaluate the integral $\displaystyle\int_1^2 \int_0^1 |xy - 1| \, dy \, dx$.

9. Show that $\displaystyle\iint_R f(x, y) \, dA = 0$ if $f(-x, -y) = -f(x, y)$ and R is the disk $x^2 + y^2 \le 1$.

10. Invert the order of integration in the iterated integral

$$\int_a^b \int_0^x f(x, y) \, dy \, dx, \qquad (0 < a < b).$$

11. Use the geometric interpretation of the double integral to evaluate $\displaystyle\iint_R f(x, y) \, dA$ if R is the unit disk $x^2 + y^2 \le 1$ and $f(x, y)$ is the given expression.

(a) 1
    (b) $\sqrt{1 - x^2 - y^2}$

(c) $ax + by + c$
    (d) $[\![5(x^2 + y^2)]\!]$

12. Show that both

$$\int_1^2 \int_0^y \frac{1}{x^2 + y^2} \, dx \, dy \quad \text{and} \quad \int_1^2 \int_0^{1/y} \frac{1}{1 + x^2y^2} \, dx \, dy$$

equal $\frac{1}{4}\pi \ln 2$. If these iterated integrals are viewed as double integrals over regions R_1 and R_2, describe R_1 and R_2.

114. FINDING VOLUMES AND AREAS BY DOUBLE INTEGRATION

Double integrals are widely used in science and engineering to calculate such things as work and force. Because of space limitations, we won't discuss the background material necessary for an understanding of these physical applications. Instead we will restrict our applications of multiple integrals to geometric problems. This restriction is simply for convenience, and you should realize that double integrals are not merely devices for finding volumes, areas, and the like, even though that is the only use we will make of them. These relatively minor applications will, however, give you experience in setting up and evaluating double integrals, and this experience will come in handy in case you are faced with a double integral in a real-life situation.

If $f(x, y) \geq 0$ at each point (x, y) of a region R of the XY-plane, then the region, the cylinder based on the boundary of R, and the surface $z = f(x, y)$ bound a region in space. We will speak of this three-dimensional region as the "region under the graph of f and above the plane region R," and we have said that its volume is the number

(114-1)
$$V = \int\!\!\int_R f(x, y) \, dA.$$

Example 114-1. Find the volume of the region that is bounded by the XY-plane, the plane $x + y + z = 2$, and the parabolic cylinder $y = x^2$.

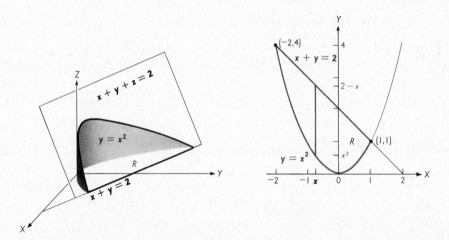

Figure 114-1

Solution. On the left-hand side of Fig. 114-1 we show the three-dimensional region whose volume we are to find. The parabolic cylinder intersects the XY-plane in the parabola $y = x^2$, and the plane $x + y + z = 2$ intersects the XY-plane in the line $x + y = 2$. These curves bound the plane region R that is shown on the right-hand side of the figure. The solid whose volume we seek is the region under the plane $x + y + z = 2$ (that is, $z = 2 - x - y$) and above R, so Equation 114-1 tells us that we are looking for the number

$$V = \int\!\!\int_R (2 - x - y) \, dA.$$

This integral is equivalent to an iterated integral in which we first integrate with respect to y and whose limits of integration may be found as follows. First project R onto the interval $[-2, 1]$ of the X-axis, thereby obtaining the outer limits of integration. Now fix a number $x \in [-2, 1]$ and draw the segment in which a perpendicular to the X-axis at x intersects R. This segment projects onto the interval $[x^2, 2 - x]$ of the Y-axis, thus giving us our inner limits of integration:

$$V = \int_{-2}^{1} \int_{x^2}^{2-x} (2 - x - y)\, dy\, dx = \int_{-2}^{1} (2y - xy - \tfrac{1}{2}y^2)\Big|_{x^2}^{2-x} dx$$

$$= \int_{-2}^{1} (2 - 2x - \tfrac{3}{2}x^2 + x^3 + \tfrac{1}{2}x^4)\, dx = \tfrac{81}{20}.$$

Example 114-2. What if we replace our double integral with the "reverse" iterated integral?

Solution. We have to work harder. We project R onto the interval $[0, 4]$ of the Y-axis, thereby obtaining the outer limits of integration. Now we fix a number $y \in [0, 4]$ and draw the segment in which a perpendicular to the Y-axis at y intersects R. This segment projects onto an interval $[a(y), b(y)]$ of the X-axis, so $a(y)$ and $b(y)$ are our inner limits of integration. It is easy to see that $a(y) = -\sqrt{y}$ for each $y \in [0, 4]$, but the formula for $b(y)$ depends on the value of y. Thus $b(y) = \sqrt{y}$ if $y \in [0, 1]$, and $b(y) = 2 - y$ if $y \in [1, 4]$. Taking these different formulas into account, we write our integral over the Y-interval $[0, 4]$ as the sum of integrals over $[0, 1]$ and $[1, 4]$, and Equation 114-1 becomes

$$V = \int_{0}^{1} \int_{a(y)}^{b(y)} (2 - x - y)\, dx\, dy + \int_{1}^{4} \int_{a(y)}^{b(y)} (2 - x - y)\, dx\, dy$$

$$\text{(114-2)} \qquad = \int_{0}^{1} \int_{-\sqrt{y}}^{\sqrt{y}} (2 - x - y)\, dx\, dy + \int_{1}^{4} \int_{-\sqrt{y}}^{2-y} (2 - x - y)\, dx\, dy$$

$$= \tfrac{131}{60} + \tfrac{28}{15} = \tfrac{81}{20}.$$

Our definition of the double integral of a function f over a plane region R can be stated informally as follows. First we cover R with a rectangular mesh as shown in Fig. 114-2. We call such a mesh a *covering* of R. In each subregion of the covering we choose a point; the figure shows a typical subregion of dimensions Δx by Δy, in which we have chosen the point (x^*, y^*). At this point, our function f has the value $f(x^*, y^*)$. (For subregions that contain boundary points of R, it is possible that our chosen point is not in R; then we use 0 in place of $f(x^*, y^*)$.) We multiply this number by the area $\Delta A = \Delta x\, \Delta y$ of the subregion and add up all these products:

$$\sum f(x^*, y^*)\, \Delta A = \sum f(x^*, y^*)\, \Delta x\, \Delta y.$$

The double integral is the limit of such approximating sums.

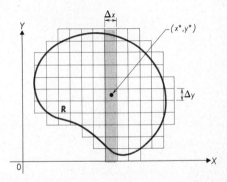

Figure 114-2

Although approximating sums and double integrals are not the same thing, we frequently find it convenient to think of the process of evaluating the double integral $\iint\limits_{R} f(x, y)\, dA$ as a matter of addition. We can consider an iterated integral as a device by which this addition is performed systematically. Thus we can view the iterated integral in which the first integration is with respect to y as a method of adding terms that correspond to rectangular subregions that form a strip parallel to the Y-axis, such as the one shown in Fig. 114-2. Then our second integration (with respect to x) adds up the contributions that correspond to these vertical strips. If we set up our iterated integral so that we first integrate with respect to x, we reverse the procedure—we first add terms that correspond to a horizontal strip of rectangular subregions and then we add up the contributions of these strips. Thus we evaluate a double integral by using an iterated integral to "sweep out" the region R, first in one direction and then in the other.

If f is the constant function with value 1, then $f(x^*, y^*) = 1$ for each point (x^*, y^*), and our approximating sum reduces to $\sum \Delta A$. It is apparent that the limit of this sum is the area A of R, and since this limit is also the double integral $\iint\limits_{R} dA$, we have the formula

$$(114\text{-}3) \qquad\qquad A = \iint\limits_{R} dA.$$

Example 114-3. Find the area of the plane region that lies in the first quadrant and is bounded by the X-axis, the circle $x^2 + y^2 = 18$, and the parabola $y^2 = 3x$.

Solution. In Fig. 114-3 we have shown two drawings of our region R covered with a rectangular mesh. The area we seek is $A = \iint\limits_{R} dA$. If we evaluate this double integral by means of an iterated integral, we must decide whether to use

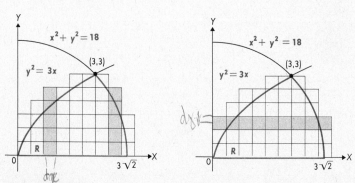

Figure 114-3

x or *y* first as the variable of integration. If we first integrate with respect to *y*, we are first finding the area of a vertical strip (see the picture on the left in Fig. 114-3). The disadvantage of using such vertical strips lies in the fact that two different equations describe the curve that forms the upper boundary of our region. If we first integrate with respect to *x*, we are first finding the area of a horizontal strip, as in the diagram on the right in Fig. 114-3. In this case, the right-boundary curve is $x = \sqrt{18 - y^2}$, and the left-boundary curve is $x = \frac{1}{3}y^2$. Thus we describe *R* as $\{(x, y) \mid 0 \le y \le 3, \frac{1}{3}y^2 \le x \le \sqrt{18 - y^2}\}$, and our iterated integral in this case (which appears to be the simpler one) is

$$A = \int_0^3 \int_{y^2/3}^{\sqrt{18-y^2}} dx\, dy = \int_0^3 (\sqrt{18 - y^2} - \tfrac{1}{3}y^2)\, dy$$

$$= \left[\frac{y\sqrt{18 - y^2}}{2} + 9 \, \mathrm{Sin}^{-1}\left(\frac{y}{3\sqrt{2}}\right) - \frac{y^3}{9} \right]\Big|_0^3$$

$$= \frac{6 + 9\pi}{4}.$$

PROBLEMS   114

1. Use double integration to find the area of the given region.
    (a) The region bounded by the curves $y^2 = x + 1$ and $x + y = 1$.
    (b) The first quadrant region bounded by the curves $x^2 = 4 - 2y$, $x = 0$, $y = 0$.
    (c) The region bounded by the curves $xy = 4$ and $x + y = 5$.
    (d) The region bounded by the curves $y^2 = x^3$ and $y = x$.
    (e) The region bounded by the curves $y = 2x - x^2$ and $y = 2x^3 - x^2$.
    (f) The region bounded by the curves $y^2 = 4x$ and $y^2 = 5 - x$.

2. Use double integration to find the area of the triangular region whose vertices are the points (0, 0,) (6, 4), and (2, 8).

3. Use double integration to find the volume of the solid tetrahedron whose vertices are the points (0, 0, 0), (*a*, 0, 0), (0, *b*, 0), and (0, 0, *c*).

4. Use double integration to find the volume of the region in the first octant and bounded by the given surfaces.
    (a) The planes $z = x$, $x = 0$, $x = 1$, $y = 0$, $y = 1$, $z = 0$.
    (b) The planes $x = 0$, $y = 0$, $z = 0$, $z = y$, and the cylinder $x^2 + y^2 = 4$.
    (c) The planes $x = 0$, $y = 0$, $z = 0$, $x + z = 4$, and the cylinder $y = 4 - x^2$.
    (d) The planes $z = 0$, $z = x + 4$, and the cylinder $|x| + |y| = 1$.

5. Use double integration to find the volume of one of the wedges bounded by the cylinder $x^2 + y^2 = a^2$, the *XY*-plane, and the plane $z = mx$.

6. Find the volume of the region bounded by the cylinder $x^2 + z = 1$ and by the planes $x + y = 1$, $y = 0$, and $z = 0$.

7. Find the volume of the wedge that is bounded by the cylinder $y^2 = x$, the *XY*-plane, and the plane $x + z = 1$.

8. The *XY*-plane and the surface $y^2 = 16 - 4z$ cut the cylinder $x^2 + y^2 = 4x$. Find the volume of the region bounded by these surfaces.

9. A haystack is bounded by the XY-plane and the paraboloid $x^2 + y^2 + z = 1$. How much hay is in the stack?

10. Use double integration to show that the volume of the region bounded by the ellipsoid $x^2/a^2 + y^2/b^2 + z^2/c^2 = 1$ is $\frac{4}{3}\pi abc$.

11. What does the surface $z = \sin(|x| + |y|)$ for (x, y) in the "diamond" $|x| + |y| \le \pi$ look like? What is the volume of the region bounded by this piece of surface and the XY-plane?

12. If $f(x, y) = y\sin x - y^2$, then $f(\frac{1}{2}\pi, \frac{1}{2}) = \frac{1}{4}$. Since $\frac{1}{4}$ is a positive number, there is a region containing the point $(\frac{1}{2}\pi, \frac{1}{2})$ and in which $f(x, y) > 0$. Let R be the largest such region and find $\displaystyle\iint_R f(x, y)\, dA$.

115. MOMENTS AND MOMENTS OF INERTIA OF PLANE REGIONS

Let us consider the rectangular region shown in Fig. 115-1. Suppose that the coordinates of its geometric center (the intersection of the diagonals of the bounding rectangle) are (x, y) and that its area is A square units. Then the numbers

$$(115\text{-}1) \qquad M_X = yA \quad \text{and} \quad M_Y = xA$$

are called the **moments** of the region about the X- and Y-axes, respectively. Notice that we have defined the moment of a rectangular region—that is, of a geometric object. However, if we think of our region as a plate made of some material of uniform density, then its mass will be proportional to its area A. So if we suppose that gravity acts in a direction perpendicular to the XY-plane, we can find the moments of the plate in the usual mechanical sense of force times lever arm by multiplying the right-hand side of Equations 115-1 by a factor of proportionality.

In order to define the moment of a general region R (Fig. 115-2), we first

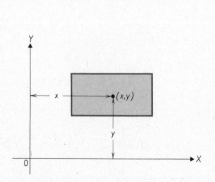

Figure 115-1

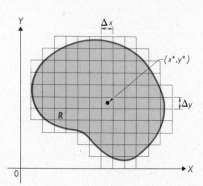

Figure 115-2

cover it with a rectangular covering. In a typical subregion of this covering (with an area of $\Delta A = \Delta x\, \Delta y$), we select the center (x^*, y^*), and thus the moment of this subregion about the X-axis is $y^*\, \Delta A$. The sum of these moments, $\sum y^*\, \Delta A$, is the moment of the covering of R. Now we take the limit of these sums as the norm of our collection of subregions approaches 0 to be the moment of R itself. Since this number is the integral of y over R, we have

$$(115\text{-}2) \qquad\qquad M_X = \iint\limits_R y\, dA.$$

Similarly, the moment of R about the Y-axis is defined to be the number

$$(115\text{-}3) \qquad\qquad M_Y = \iint\limits_R x\, dA.$$

Example 115-1.   Find the moments M_X and M_Y of the region that is bounded by the semicircle $y = \sqrt{a^2 - x^2}$ and the X-axis.

Solution.   Our region R is shown in Fig. 115-3. According to Equation 115-2,

$$M_X = \iint\limits_R y\, dA = \int_{-a}^{a} \int_{0}^{\sqrt{a^2-x^2}} y\, dy\, dx = \int_{-a}^{a} \frac{a^2 - x^2}{2}\, dx = \frac{2a^3}{3}.$$

We use Equation 115-3 to find M_Y:

$$M_Y = \iint\limits_R x\, dA = \int_{0}^{a} \int_{-\sqrt{a^2-y^2}}^{\sqrt{a^2-y^2}} x\, dx\, dy = \int_{0}^{a} 0\, dy = 0.$$

You will notice that for ease of calculation, we used different orders of integration when we replaced the double integrals for M_X and M_Y with iterated integrals.

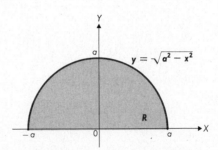

Figure 115-3

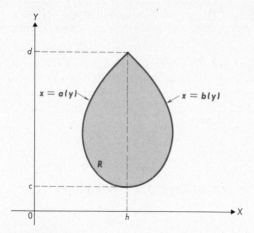

Figure 115-4

Our concept of moment gives us a "net moment" of a region about an axis. If the axis is vertical, for example, the moment of the part of the region to the right of the axis is positive and the moment of the part to the left is negative. In Example 115-1 we saw how these moments canceled each other when we calculated the moment about the Y-axis and found that $M_Y = 0$.

If a region is symmetric about a line that is parallel to one of the coordinate axes, it is easy to find its moment about that axis. Thus suppose that we have a region R, such as the one shown in Fig. 115-4, that is symmetric about the line $x = h$. This symmetry means that the average of the X-coordinates of the two boundary points with the same Y-coordinate is always h. So if the equations $x = a(y)$ and $x = b(y)$ represent the left and right boundaries of R, we have

$$\frac{a(y) + b(y)}{2} = h$$

for each y. Therefore

$$M_Y = \int_c^d \int_{a(y)}^{b(y)} x \, dx \, dy = \int_c^d \left(\frac{b(y)^2 - a(y)^2}{2}\right) dy$$

$$= \int_c^d \left(\frac{b(y) + a(y)}{2}\right)(b(y) - a(y)) \, dy = h \int_c^d (b(y) - a(y)) \, dy.$$

Now observe that the area A of R is given by the equation

$$A = \int_c^d \int_{a(y)}^{b(y)} dx \, dy = \int_c^d (b(y) - a(y)) \, dy,$$

and so $M_Y = hA$. Similarly, if our region were symmetric with respect to a line $y = k$, we would have $M_X = kA$.

714

In mechanics we define the **center of gravity** of a body as being "the point at which we can imagine the mass of the body to be concentrated." By this statement we mean that if we replace a given body with a "point mass" at the center of gravity, then the moment of this new system about an axis will be the same as the moment of the original body about that axis. The concept of the centroid of a plane region is similar. The **centroid** of a region R that has an area of A is the point $(\bar{x}, \bar{y})$ such that the "moments" $\bar{y}A$ and $\bar{x}A$ equal the moments of R about the X- and Y-axes. Thus $\bar{x}$ and $\bar{y}$ are numbers such that

$$(115\text{-}4) \qquad \bar{x}A = M_Y \quad \text{and} \quad \bar{y}A = M_X.$$

With the aid of Equations 115-2 and 115-3, we can rewrite these equations for $\bar{x}$ and $\bar{y}$ as

$$(115\text{-}5) \qquad \bar{x} = \frac{\iint\limits_R x\, dA}{A} \quad \text{and} \quad \bar{y} = \frac{\iint\limits_R y\, dA}{A}.$$

Example 115-2. Find the centroid of the region shown in Fig. 115-3.

Solution. In Example 115-1 we found that $M_X = \frac{2}{3}a^3$ and $M_Y = 0$. The area of our semicircular region whose radius is a is $A = \frac{1}{2}\pi a^2$. Therefore, according to Equations 115-4,

$$\bar{x} = \frac{0}{A} = 0 \quad \text{and} \quad \bar{y} = \frac{2a^3/3}{\pi a^2/2} = \frac{4a}{3\pi}.$$

Thus the centroid of R is the point $(0, 4a/3\pi)$. If we make a cardboard model of R, it will balance on a pin placed at this point.

If a region R with an area of A square units is symmetric with respect to the line $x = h$, we have seen that $M_Y = hA$, and hence (since M_Y is also $\bar{x}A$), $\bar{x}A = hA$. Thus $\bar{x} = h$; that is, the centroid is in the line of symmetry, as illustrated by the preceding example. Similarly, if a region is symmetric with respect to the line $y = k$, then $\bar{y} = k$. You should take advantage of symmetry when calculating moments and centroids.

Example 115-3. Find the centroid of the region that is bounded by the X-axis and one arch of the sine curve.

Solution. Since the region (shown in Fig. 115-5) is symmetric about the line $x = \frac{1}{2}\pi$, we know immediately that $\bar{x} = \frac{1}{2}\pi$. Furthermore,

$$A = \int_0^\pi \sin x\, dx = 2,$$

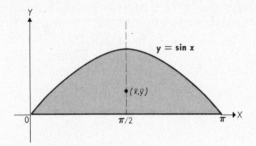

Figure 115-5

and

$$M_X = \int_0^\pi \int_0^{\sin x} y \, dy \, dx = \tfrac{1}{2} \int_0^\pi \sin^2 x \, dx = \tfrac{1}{4}\pi.$$

Thus

$$\bar{y} = \frac{M_X}{A} = \tfrac{1}{8}\pi,$$

so our centroid is the point $(\tfrac{1}{2}\pi, \tfrac{1}{8}\pi)$.

We can define moments of a region R about lines other than the coordinate axes. For example, the moment about the line L whose equation is $x = h$ is

$$M_L = \iint_R (x - h) \, dA.$$

Since

$$\iint_R (x - h) \, dA = \iint_R x \, dA - h \iint_R dA = M_Y - hA,$$

we have the equation

(115-6) $$M_L = M_Y - hA.$$

Similarly, the moment M_L of a region R about the line L whose equation is $y = k$ is

(115-7) $$M_L = M_X - kA.$$

Example 115-4. Find the centroid of the region R that consists of a circular disk D whose radius is 1 and a 2 by 2 square region S that touches it, as shown in Fig. 115-6.

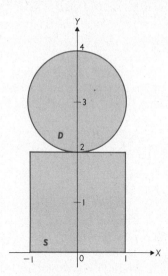

Figure 115-6

Solution. Our figure is symmetric about the *Y*-axis, so the *X*-coordinate of its centroid is 0. By definition, $M_X = \iint\limits_R y\,dA$. Since our given region is the union of the disk *D* and the square region *S*, we may use Equation 111-3 to write

$$M_X = \iint\limits_R y\,dA = \iint\limits_D y\,dA + \iint\limits_S y\,dA.$$

The two integrals on the right-hand side of this equation yield the moments M_{DX} and M_{SX} of the disk and the square region about the *X*-axis, so $M_X = M_{DX} + M_{SX}$. We find these moments by multiplying the areas of *D* and *S* by the distances of their lines of symmetry from the *X*-axis: $M_{DX} = 3\pi$, and $M_{SX} = 1 \cdot 4 = 4$. Therefore $M_x = 3\pi + 4$. The area of *R* is $A = \pi + 4$, so

$$\bar{y} = \frac{M_X}{A} = \frac{3\pi + 4}{\pi + 4} \approx 1.88.$$

The coordinates of the centroid of our region are approximately (0, 1.88).

Example 115-5. Find the moment of the region *R* of the preceding example about the line *L* whose equation is $y = 1$.

Solution. In the preceding example we found that $M_X = 3\pi + 4$ and $A = \pi + 4$. Then according to Equation 115-7 with $k = 1$,

$$M_L = (3\pi + 4) - (\pi + 4) = 2\pi.$$

Another idea that we meet in mechanics is the concept of the **moment of inertia** of a body. The moment of inertia about the X-axis of the rectangular region shown in Fig. 115-1 is the number $y^2 A$, its moment of inertia about the Y-axis is $x^2 A$, and its moment of inertia about the origin is $(x^2 + y^2)A$. In general, if I_X, I_Y, and I_O are the moments of inertia of the region shown in Fig. 115-2 about the X-axis, the Y-axis, and the origin, then, by definition,

$$(115\text{-}8) \quad I_X = \iint_R y^2 \, dA, \qquad I_Y = \iint_R x^2 \, dA, \qquad I_O = \iint_R (x^2 + y^2) \, dA.$$

Notice that $I_O = I_X + I_Y$.

If our region R were a plate made of material of uniform density, then its moments of inertia (as the term is used in mechanics) are obtained by multiplying our numbers I_X, I_Y, and I_O by the density. The moment of inertia of a body plays a role in rotational motion that is analogous to the role played by the mass of the body in linear motion.

Example 115-6. Find the moment of inertia of a right-triangular region about the vertex of the $90°$ angle.

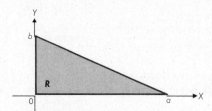

Figure 115-7

Solution. Suppose that the legs of our triangle are a and b units long and that we place it in a coordinate system as in Fig. 115-7. We must compute I_O, which is given by the formula

$$I_O = \iint_R (x^2 + y^2) \, dA.$$

Since the hypotenuse of our triangle is part of the line $x/a + y/b = 1$,

$$I_O = \iint_R (x^2 + y^2) \, dA = \int_0^b \int_0^{a - ay/b} (x^2 + y^2) \, dx \, dy = \frac{ab}{12}(a^2 + b^2).$$

Example 115-7. Find I_X and I_Y for the region R shown in Fig. 115-3.

Solution. According to Equations 115-8,

$$I_X = \iint_R y^2 \, dA = \int_{-a}^{a} \int_{0}^{\sqrt{a^2 - x^2}} y^2 \, dy \, dx$$

$$= \tfrac{1}{3} \int_{-a}^{a} (a^2 - x^2)^{3/2} \, dx$$

$$= \tfrac{1}{3} \int_{-a}^{a} a^2 \sqrt{a^2 - x^2} \, dx - \tfrac{1}{3} \int_{-a}^{a} x^2 \sqrt{a^2 - x^2} \, dx$$

$$= \tfrac{1}{8}\pi a^4 \qquad \text{(see Table V)}.$$

Similarly,

$$I_Y = \iint_R x^2 \, dA = \int_{0}^{a} \int_{-\sqrt{a^2 - y^2}}^{\sqrt{a^2 - y^2}} x^2 \, dx \, dy$$

$$= \tfrac{2}{3} \int_{0}^{a} (a^2 - y^2)^{3/2} \, dy = \tfrac{1}{8}\pi a^4.$$

PROBLEMS 115

1. Find the centroid of the given plane region.

   (a) The first quadrant quarter of the disk $x^2 + y^2 \leq a^2$.
   (b) The upper half of the elliptical region $4x^2 + 9y^2 \leq 36$.
   (c) The first quadrant region bounded by the coordinate axes and the parabola $y^2 = 4 - x$.
   (d) The region bounded by the parabola $x^2 = 4y$ and the line $y = 1$.
   (e) The triangular region whose vertices are the points $(0, 0)$, (a, b), and $(c, 0)$, where $b > 0$ and $c > 0$.
   (f) The region bounded by the curves $\sqrt{x} + \sqrt{y} = \sqrt{a}$, $x = 0$, and $y = 0$.
   (g) The region bounded by the parabolas $x^2 = y$ and $y^2 = x$.
   (h) The graph of the equation $[\![x]\!]^2 + [\![y]\!]^2 = 1$.

2. Find the centroid of the region $R = \{(x, y) \mid 0 \leq x \leq \tfrac{1}{2}\pi, 0 \leq y \leq \sin x\}$.

3. Show that the centroid of a triangular region is the intersection of the medians of the triangle that forms its boundary (see part (e) of Problem 115-1).

4. Suppose that the point $(\bar{x}, \bar{y})$ is the centroid of a certain region R that has an area of A square units. Show that the moment of R about the line $x = h$ is $(\bar{x} - h)A$, and the moment of R about the line $y = k$ is $(\bar{y} - k)A$. Show that if R is symmetric about the line $x = h$ or the line $y = k$, then the moment of R about the line is 0.

5. Use the result of the preceding problem to find the moment of the region R about the line L.

   (a) $R : x^2 + y^2 \leq a^2$, $L : x = 5a$.
   (b) $R : b^2 x^2 + a^2 y^2 \leq a^2 b^2$, $L :$ A latus rectum of this elliptical region.
   (c) $R :$ The region bounded by the parabolas $x^2 = 4y$ and $x^2 = 5 - y$, $L :$ A vertical line containing the right-hand intersection point of the given curves.

6. Find I_X, I_Y, and I_O for the given region.
   (a) The rectangular region $\{(x, y) \mid 0 \leq x \leq a, 0 \leq y \leq b\}$.
   (b) The region that is bounded by the parabola $y^2 = 8x$ and the line $x = 2$.
   (c) The diamond-shaped region $|x| + |y| \leq 1$.
   (d) The graph of the equation $[\![x]\!] + [\![y]\!] = 1$ for $x \geq 0$ and $y \geq 0$.

7. The moment of inertia of a region R about the line L whose equation is $x = h$ is defined to be

$$I_L = \iint\limits_R (x - h)^2 \, dA.$$

   (a) Show that $I_L = I_Y - 2hM_Y + h^2 A = I_Y + h(h - 2\bar{x})A$, where A is the area of R.
   (b) What can you conclude from the equation in part (a) if the centroid of R is a point of the Y-axis?
   (c) What can you conclude from the equation in part (a) if the centroid of R is a point of L?
   (d) If K is the line $x = k$, show that

$$I_K = I_L - 2M_Y(k - h) + (k^2 - h^2)A = I_L + (k + h - 2\bar{x})(k - h)A.$$

8. (a) Find I_Y for the elliptical region $b^2 x^2 + a^2 y^2 \leq a^2 b^2$.
   (b) Use the result you found in part (a) to find I_X and I_0.
   (c) Find I_Y for the disk $x^2 + y^2 \leq a^2$. (Compare your answer with our results in Example 115-7.)
   (d) Suppose that $a < b$ and use the formulas of Problem 115-7 to find the moment of inertia about a latus rectum of the elliptical region $b^2 x^2 + a^2 y^2 \leq a^2 b^2$.
   (e) Suppose that $a < b$ and use the formulas of Problem 115-7 to find the moment of inertia about a directrix of the elliptical region $b^2 x^2 + a^2 y^2 \leq a^2 b^2$.

9. Let $R = \{(x, y) \mid x^2 + y^2 - 2|x| - 2|y| + 1 \leq 0\}$. Refer to Problems 115-7 and 115-8 for the necessary formulas to find I_0 and I_L, where L is the line $x = -1$.

10. The moment of a region about a line through its centroid is zero. Can you find a line such that the moment of inertia of the region about the line is zero?

116. DOUBLE INTEGRALS IN POLAR COORDINATES

The double integral of a function f over a plane region R is the limit of sums of the form

(116-1) $$s = \sum f(x^*, y^*) \, \Delta A.$$

Here the number ΔA is the area of a rectangular subregion of a covering of R. It doesn't matter at what point (x^*, y^*) in this subregion we evaluate f; if the maximum dimension of the subregions is small, then the Sum 116-1 is close to the number $\iint\limits_R f(x, y) \, dA$. Even if the subregions of the covering are not rectangular (if they are of reasonably regular shape), it can be shown that the limit of s will

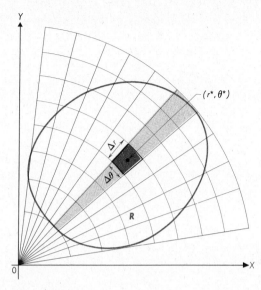

Figure 116-1

still be the integral of f over R. In this section we are going to investigate the non-rectangular covering we get when we use polar coordinates.

Suppose that we introduce a polar coordinate system into the coordinate plane, taking as the pole the origin of the cartesian coordinate system and as the polar axis the positive X-axis. We form a "polar covering" of a region R by drawing a network of radial lines and concentric circles, as shown in Fig. 116-1. The norm of this covering is the longest of the "diagonals" of the "curved rectangles" that make up the covering. We assert that if our covering has a small norm, then when we evaluate f at a point in each subregion, multiply this value of f by the area of the subregion, and add up the resulting numbers, the sum will approximate the integral of f over R.

Let us see what one such approximating sum looks like. We shall suppose that the two radial lines that form the sides of our typical subregion make an angle of $\Delta\theta$ and that the concentric circles that form its ends are Δr units apart. Now let (r^*, θ^*) be polar coordinates of the point of our subregion that is equidistant from its ends and from its sides. The XY-coordinates of this point are $(r^* \cos \theta^*, r^* \sin \theta^*)$, so the value of f there is $f(r^* \cos \theta^*, r^* \sin \theta^*)$. Our typical subregion can be considered to be the difference of two circular "wedges," both with a central angle of $\Delta\theta$, but one having a radius of $r^* - \frac{1}{2} \Delta r$ and the other a radius of $r^* + \frac{1}{2} \Delta r$. In Appendix A we point out that the area of a circular wedge is one-half the product of the square of its radius and the width in radians of its central angle. Therefore the area of our typical subregion is

$$\Delta A = \tfrac{1}{2}(r^* + \tfrac{1}{2} \Delta r)^2 \, \Delta\theta - \tfrac{1}{2}(r^* - \tfrac{1}{2} \Delta r)^2 \, \Delta\theta.$$

After a little simplification, this formula reduces to

(116-2) $\Delta A = r^* \, \Delta r \, \Delta \theta.$

Thus there is an approximating sum that is expressed in terms of polar coordinates as

(116-3) $s = \sum f(r^* \cos \theta^*, r^* \sin \theta^*) r^* \, \Delta r \, \Delta \theta.$

The limit of such sums is the double integral of f over R, and their form suggests that we write

(116-4) $$\iint\limits_{R} f(x, y) \, dA = \iint\limits_{R} f(r \cos \theta, r \sin \theta) r \, dr \, d\theta.$$

It is convenient to think that we obtain the right-hand side of this equation from the left-hand side by replacing x with $r \cos \theta$, y with $r \sin \theta$, and dA with $r \, dr \, d\theta$ (see Equation 116-2).

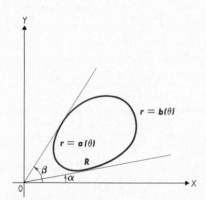

Figure 116-2

In order to evaluate our double integral, we use iterated integrals. Let us first assume that our region R lies between, and is tangent to, two radial lines $\theta = \alpha$ and $\theta = \beta$, as shown in Fig. 116-2. The remainder of the boundary of R consists of the two curves $r = a(\theta)$ and $r = b(\theta)$, $\theta \in [\alpha, \beta]$. Thus $R = \{(r, \theta) \mid \alpha \leq \theta \leq \beta, a(\theta) \leq r \leq b(\theta)\}$. Informally, we think of our double integral as a sum, each term of which corresponds to one of the subregions of the covering of R. We may first add the terms that correspond to subregions forming a radial wedge, like the shaded one in Fig. 116-1, and then add the contributions of the wedges.

From this point of view, it seems reasonable that our double integral should be given by an iterated integral:

$$(116\text{-}5) \qquad \iint\limits_{R} f(r\cos\theta, r\sin\theta)r\,dr\,d\theta = \int_{\alpha}^{\beta} \int_{a(\theta)}^{b(\theta)} f(r\cos\theta, r\sin\theta)r\,dr\,d\theta.$$

Now let us suppose that our region R lies between, and is tangent to, circles of radii a and b whose center is the origin, and that the remainder of the boundary of R consists of the curves $\theta = \alpha(r)$ and $\theta = \beta(r)$, $r \in [a, b]$, as shown in Fig. 116-3. Thus $R = \{(r, \theta) \mid a \le r \le b, \alpha(r) \le \theta \le \beta(r)\}$. Now we integrate with respect to θ first, and our double integral is given by another iterated integral:

$$(116\text{-}6) \qquad \iint\limits_{R} f(r\cos\theta, r\sin\theta)r\,dr\,d\theta = \int_{a}^{b} \int_{\alpha(r)}^{\beta(r)} f(r\cos\theta, r\sin\theta)r\,d\theta\,dr.$$

Example 116-1. Find the moment of inertia about the origin of the region bounded by two concentric circles of radii a and b, where $a < b$.

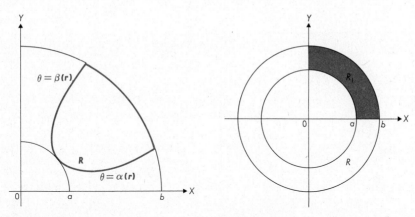

Figure 116-3   **Figure 116-4**

Solution. The given region is shown in Fig. 116-4. From Equations 116-4 and 116-5 we find that the required moment of inertia is

$$I_O = \iint\limits_{R} (x^2 + y^2)\,dA = \iint\limits_{R} r^2 r\,dr\,d\theta$$

$$= \int_{0}^{2\pi} \int_{a}^{b} r^3\,dr\,d\theta = \tfrac{1}{2}\pi(b^4 - a^4).$$

Example 116-2. Find the centroid of the part of the "washer" described in the last example that lies in the first quadrant.

Solution. Our region (we will label it R_1 in this example) is the shaded quarter of the region R of Fig. 116-4. If we denote the coordinates of the required centroid by $(\bar{x}, \bar{y})$, symmetry considerations tell us that $\bar{x} = \bar{y}$. So we need only compute $\bar{y}$. According to Equation 115-5,

(116-7)
$$\bar{y} = \frac{1}{A} \iint\limits_{R_1} y \, dA.$$

Now we use Equations 116-4 and 116-5 to see that

$$\iint\limits_{R_1} y \, dA = \iint\limits_{R_1} r \sin \theta \, r \, dr \, d\theta = \int_0^{\pi/2} \int_a^b r^2 \sin \theta \, dr \, d\theta$$

$$= \int_0^{\pi/2} \tfrac{1}{3}(b^3 - a^3) \sin \theta \, d\theta = \tfrac{1}{3}(b^3 - a^3).$$

It is a matter of simple geometry to show that the area A of the region R_1 is $\frac{1}{4}\pi(b^2 - a^2)$. Hence Equation 116-7 tells us that

$$\bar{y} = \frac{4(b^3 - a^3)}{3\pi(b^2 - a^2)}.$$

PROBLEMS   116

1. Find the area of the region that is bounded by the given polar curve.
    (a) $r = 2 \cos \theta$ (b) $r = \cos 2\theta$ (c) $r = 2 \cos 3\theta$
    (d) $r^2 = \cos 2\theta$ (e) $r = 2 + \cos \theta$ (f) $r = \cos^2 \theta$
    (g) $r = |\cos \theta|$ (h) $|r| = \cos \theta$ (i) $r = [\![\cos \theta]\!]$

2. Describe the region of integration R if the double integral $\iint\limits_R f(x, y) \, dA$ equals the given iterated integral.

    (a) $\displaystyle\int_1^2 \int_3^4 f(r \cos \theta, r \sin \theta) r \, dr \, d\theta$ (b) $\displaystyle\int_1^2 \int_3^4 f(r \cos \theta, r \sin \theta) r \, d\theta \, dr$

    (c) $\displaystyle\int_0^{\pi/2} \int_0^{\sin^2 \theta} f(r \cos \theta, r \sin \theta) r \, dr \, d\theta$

    (d) $\displaystyle\int_0^{2\pi} \int_0^{2+\sin \theta} f(r \cos \theta, r \sin \theta) r \, dr \, d\theta$

    (e) $\displaystyle\int_0^{\pi} \int_0^{[\![\theta]\!]} f(r \cos \theta, r \sin \theta) r \, dr \, d\theta$

    (f) $\displaystyle\int_0^{\pi} \int_0^{[\![r]\!]} f(r \cos \theta, r \sin \theta) r \, d\theta \, dr$

3. Use polar coordinates to evaluate the integral $\displaystyle\iint_R \sqrt{x^2 + y^2}\,dA$ over the given region R.

(a) $R = \{(x, y) \mid x^2 + y^2 \le a^2\}$ (b) $R = \{(x, y) \mid 9 \le x^2 + y^2 \le 16\}$

(c) $R = \{(r, \theta) \mid 0 \le \theta \le 2\pi, 0 \le r \le 1 - \cos\theta\}$

(d) $R = \{(r, \theta) \mid 0 \le \theta \le 2\pi, 0 \le r \le 1 + \cos\theta\}$

(e) $R = \{(r, \theta) \mid 0 \le \theta \le \pi, 0 \le r \le \sin\theta\}$

(f) $R = \{(r, \theta) \mid 0 \le \theta \le \tfrac{1}{2}\pi, \sqrt{3} \le r \le 2\sin 2\theta\}$

4. Replace the given iterated integral with an iterated integral in polar coordinates and evaluate.

(a) $\displaystyle\int_0^2 \int_0^{\sqrt{4-x^2}} e^{-x^2-y^2}\,dy\,dx$

(b) $\displaystyle\int_0^1 \int_y^{\sqrt{2-y^2}} x^4\,dx\,dy$

(c) $\displaystyle\int_0^3 \int_{-\sqrt{18-y^2}}^{-y} \sin(x^2 + y^2)\,dx\,dy$

(d) $\displaystyle\int_0^3 \int_{x/\sqrt{3}}^{\sqrt{12-x^2}} (1 + x^2 + y^2)^{-1}\,dy\,dx$

5. Use polar coordinates to calculate the volume of the region that is cut from the cylindrical region $\{(x, y, z) \mid x^2 + y^2 \le a^2\}$ by the given surfaces.

(a) The sphere $x^2 + y^2 + z^2 = 4a^2$.

(b) The half-planes $z = mx$ and $z = 0$, with $x \ge 0$. (See Problem 114-5.)

6. Let W be a wedge cut from a circular disk of radius a by an angle whose vertex is the center of the disk and that cuts an arc of length s from the boundary of the disk.

(a) What is the moment of inertia of W about its vertex?

(b) How far from the vertex is the centroid of W?

(c) What is the moment of inertia about one edge of W?

7. Use polar coordinates to evaluate the double integral that gives the volume of a right-circular cone of altitude h and base radius a.

8. Find the centroid of the given plane region.

(a) The interior of the cardioid $r = a(1 - \cos\theta)$.

(b) The region in the upper half-plane that is bounded by the curve $r^2 = a\cos\theta$.

(c) The interior of one loop of the curve $r = 4\sin 2\theta$.

(d) The region bounded by the rays $\theta = 0$ and $\theta = \tfrac{1}{4}\pi$, and the curve $r = \cos 2\theta$.

9. Examine the formula for the coordinates of the centroid of the region R of Example 116-2.

(a) Where is the centroid if $a = 0$?

(b) What is $\lim_{a \uparrow b} \bar{x}$?

(c) Can we choose a so that the centroid of R is a point of the inner boundary of R? Outside R?

10.   Look up the discussion of the polar coordinate area formula $A = \frac{1}{2} \int_\alpha^\beta r^2 \, d\theta$ in an earlier section of the book. Now show how to derive it by replacing the double integral in the equation $A = \iint\limits_R dA$ with an iterated integral.

117. TRIPLE INTEGRALS

The concept of the triple integral of a function on R^3 over a solid region in three-dimensional space is a straightforward generalization of the idea of the double integral of a function on R^2 over a plane region. To start with, we consider a function f whose domain contains a rectangular block R, as shown in Fig. 117-1.

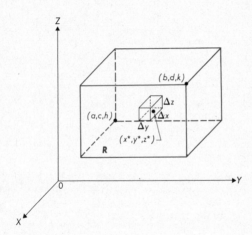

Figure 117-1

As a first step in the definition of the integral of f over R, we use a set of planes parallel to the coordinate planes to partition the block R into smaller blocks that we will call *subblocks* of R. The *norm* of this partition is the longest of the diagonals of these subblocks. We have labeled the dimensions of the typical subblock shown in Fig. 117-1 as Δx, Δy, and Δz, so its volume is $\Delta V = \Delta x \, \Delta y \, \Delta z$. In each subblock of the partition, we pick a point (x^*, y^*, z^*) and consider sums of the form

(117-1) $s = \sum f(x^*, y^*, z^*) \, \Delta V,$

where the sum has a term for each subblock of the partition of R. Then if the limit of s as the norm of our partition goes to 0 exists, we say that f is **integrable** over R, and the limit is called the **triple integral of f over R**. We denote this integral by one of the symbols

$$\iiint_R f(x, y, z)\, dV \quad \text{or} \quad \iiint_R f(x, y, z)\, dx\, dy\, dz.$$

The geometric argument that we used to convince ourselves that a double integral can be evaluated as an iterated integral can't be used for triple integrals, but the result is still true. Thus for the block $R = \{(x, y, z) \mid a \le x \le b, c \le y \le d, h \le z \le k\}$, the counterpart of Equation 112-3 is

$$(117\text{-}2) \qquad \iiint_R f(x, y, z)\, dV = \int_a^b \int_c^d \int_h^k f(x, y, z)\, dz\, dy\, dx.$$

Example 117-1. Evaluate the integral $\displaystyle\iiint_R xy \sin yz\, dV$, if R is the cubical block $\{(x, y, z) \mid 0 \le x \le \pi, 0 \le y \le \pi, 0 \le z \le \pi\}$.

Solution. In this case (if you are in doubt, make a sketch), the numbers a, c, and h in Equation 117-2 are all 0, and the numbers b, d, and k are all π. Thus

$$
\begin{aligned}
\iiint_R xy \sin yz\, dV &= \int_0^\pi \int_0^\pi \int_0^\pi xy \sin yz\, dz\, dy\, dx \\
&= \int_0^\pi \int_0^\pi -x \cos yz \Big|_0^\pi \, dy\, dx \\
&= \int_0^\pi \int_0^\pi x(1 - \cos \pi y)\, dy\, dx \\
&= \int_0^\pi x \left(y - \frac{1}{\pi} \sin \pi y \right) \Big|_0^\pi \, dx \\
&= \int_0^\pi x \left(\pi - \frac{\sin \pi^2}{\pi} \right) dx = \frac{\pi^3 - \pi \sin \pi^2}{2}.
\end{aligned}
$$

To extend the idea of a triple integral from a block-shaped region to a region of a different shape, such as the potato R in Fig. 117-2, we use the same technique that we employed in the two-dimensional case. We put R inside a block $\bar{R}$, "extend" our given function f to a new function $\bar{f}$ that has the values of f at points of R and takes the value 0 outside R, and then integrate $\bar{f}$ over the block $\bar{R}$. We won't go through the details of this extension; we will simply tell how the resulting triple integral can be expressed as an iterated integral. Suppose, as shown in the figure, that our given region R projects onto a region R_{XY} of the XY-plane and

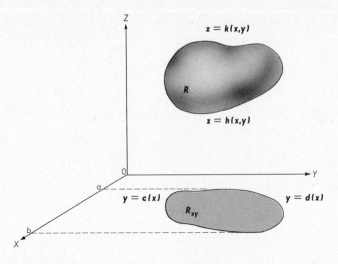

Figure 117-2

is bounded above and below by the surfaces $z = k(x, y)$ and $z = h(x, y)$. In other words, suppose that

$$R = \{(x, y, z) \mid (x, y) \in R_{XY}, h(x, y) \leq z \leq k(x, y)\}.$$

Then our triple integral is given by the iterated integral

$$(117\text{-}3) \quad \iiint\limits_{R} f(x, y, z)\, dV = \iint\limits_{R_{XY}} \left[\int_{h(x,y)}^{k(x,y)} f(x, y, z)\, dz \right] dA.$$

This equation says that the triple integral on the left is the double integral over R_{XY} whose integrand is

$$\int_{h(x,y)}^{k(x,y)} f(x, y, z)\, dz.$$

Now we use the ideas we learned in Section 113 to write this double integral over R_{XY} as an iterated integral. Thus if $R_{XY} = \{(x, y) \mid a \leq x \leq b, c(x) \leq y \leq d(x)\}$, then

$$(117\text{-}4) \quad \iiint\limits_{R} f(x, y, z)\, dV = \int_{a}^{b} \int_{c(x)}^{d(x)} \int_{h(x,y)}^{k(x,y)} f(x, y, z)\, dz\, dy\, dx.$$

When our region is of suitable shape, we can integrate in some other order, using limits that are appropriate to the order of integration.

728

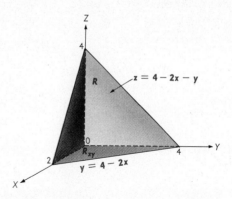

$$z = 4 - 2x - y$$

$$y = 4 - 2x$$

Figure 117-3

Example 117-2. Evaluate the integral $\iiint\limits_{R} e^{x+y+z} \, dV$, where R is the region that is bounded by the plane $2x + y + z = 4$ and the coordinate planes.

Solution. The region R is shown in Fig. 117-3. When we project R onto the XY-plane, we obtain the triangular region R_{XY} that is bounded by the X- and Y-axes and the line $2x + y = 4$. Thus

$$R = \{(x, y, z) \,|\, (x, y) \in R_{XY}, \, 0 \leq z \leq 4 - 2x - y\},$$

so Equation 117-3 tells us that

(117-5) $$\iiint\limits_{R} e^{x+y+z} \, dV = \iint\limits_{R_{XY}} \left(\int_0^{4-2x-y} e^{x+y+z} \, dz \right) dA.$$

Since $R_{XY} = \{(x, y) \,|\, 0 \leq x \leq 2, \, 0 \leq y \leq 4 - 2x\}$, we can express the double integral in Equation 117-5 as an iterated integral:

$$\iiint\limits_{R} e^{x+y+z} \, dV = \int_0^2 \int_0^{4-2x} \int_0^{4-2x-y} e^{x+y+z} \, dz \, dy \, dx.$$

We leave the evaluation of this integral to you; you should obtain the number $e^4 + 4e^2 - 1$.

We think of a triple integral $\iiint\limits_{R} f(x, y, z) \, dV$ as the limit of sums of the form $\sum f(x^*, y^*, z^*) \, \Delta V$, each term in the sum corresponding to one of a set of rectangular blocks that covers R. In particular, then, the integral $\iiint\limits_{R} dV$ is the limit of sums of the form $\sum \Delta V$, and so this integral appears to be the volume of R; that is,

$$V = \iiint_R dV.$$

We define the **centroid** of R to be the point $(\bar{x}, \bar{y}, \bar{z})$, where

$$\bar{x}V = \iiint_R x \, dV, \qquad \bar{y}V = \iiint_R y \, dV, \qquad \bar{z}V = \iiint_R z \, dV.$$

The **moments of inertia** of a three-dimensional region about the coordinate axes are given by the equations

$$I_X = \iiint_R (y^2 + z^2) \, dV, \qquad I_Y = \iiint_R (x^2 + z^2) \, dV,$$

$$I_Z = \iiint_R (x^2 + y^2) \, dV.$$

Example 117-3. Let R be the "wedge" that is bounded by the cylinder $y = x^2$, the XY-plane, and the plane $y + z = 1$, as shown in Fig. 117-4. Find the moment of inertia of R about the Z-axis.

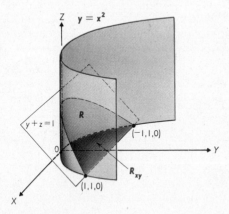

Figure 117-4

Solution. We will replace the triple integral $I_Z = \iiint_R (x^2 + y^2) \, dV$ with an iterated integral in which we first integrate with respect to z and then integrate the result over R_{XY}, the projection of R onto the XY-plane. Thus we write

$$R = \{(x, y, z) \mid (x, y) \in R_{XY}, 0 \le z \le 1 - y\},$$

so the limits of integration on z are 0 and $1 - y$. Now R_{XY} is the region in the XY-plane that lies above the parabola $y = x^2$ and below the line $y = 1$. In other words,

$$R_{XY} = \{(x, y) \mid -1 \le x \le 1, x^2 \le y \le 1\},$$

and therefore

$$I_Z = \iint_{R_{XY}} \left[\int_0^{1-y} (x^2 + y^2) \, dz \right] dA = \int_{-1}^1 \int_{x^2}^1 \int_0^{1-y} (x^2 + y^2) \, dz \, dy \, dx$$

$$= \int_{-1}^1 \int_{x^2}^1 (x^2 + y^2 - x^2 y - y^3) \, dy \, dx$$

$$= \int_{-1}^1 \left(\frac{x^8}{4} + \frac{x^6}{6} - x^4 + \frac{x^2}{2} + \frac{1}{12} \right) dx = \tfrac{64}{315}.$$

PROBLEMS   117

1. Evaluate the iterated integral.

(a) $\displaystyle \int_0^3 \int_0^{\pi/2} \int_0^1 y \cos zy \, dz \, dy \, dx$

(b) $\displaystyle \int_0^3 \int_0^{\pi/2} \int_0^1 y \sin zy \, dz \, dy \, dx$

(c) $\displaystyle \int_0^1 \int_0^1 \int_0^1 e^{x+y+z} \, dz \, dy \, dx$

(d) $\displaystyle \int_0^{\pi/3} \int_0^{\pi/3} \int_0^{\pi/3} \cos (x + y + z) \, dz \, dy \, dx$

(e) $\displaystyle \int_0^2 \int_0^{\sqrt{4-x^2}} \int_0^{2-x} y \, dz \, dy \, dx$

(f) $\displaystyle \int_0^2 \int_0^{\sqrt{4-x^2}} \int_0^{4y^2} y \, dz \, dy \, dx$

(g) $\displaystyle \int_{-\ln 2}^{\ln 2} \int_0^{\sqrt{x}} \int_0^{x+y^2} 2ye^z \, dz \, dy \, dx$

(h) $\displaystyle \int_0^4 \int_0^{\sqrt{x}} \int_0^{x+y^2} 2yz \, dz \, dy \, dx$

(i) $\displaystyle \int_0^3 \int_{-\pi/3}^{\pi/6} \int_{-1}^2 y \sin (y|z|) \, dz \, dy \, dx$

(j) $\displaystyle \int_0^3 \int_{-\pi/3}^{\pi/6} \int_{-1}^2 y \sin (\llbracket x \rrbracket z y) \, dz \, dy \, dx$

2.   Evaluate $\iiint\limits_{R} xy \, dV$ if R is

(a) the rectangular block whose faces lie in the coordinate planes and the planes $x = 2$, $y = 3$, and $z = 4$.

(b) the rectangular block whose faces lie in the planes $x = -1$, $x = 2$, $|y| = 3$, $z = 0$, and $z = 4$.

(c) the prism bounded by the coordinate planes, the plane $z = 2$, and the plane $x + y = 1$.

(d) the prism bounded by the coordinate planes, the plane $x = 2$, and the plane $y + z = 1$.

(e) the first octant region cut from the cylindrical region $x^2 + y^2 \le 1$ by the coordinate planes and the plane $z = x$.

(f) the first octant region cut from the cylindrical region $x^2 + y^2 \le 1$ by the coordinate planes and the plane $z = 2$.

3.   Describe the region R if the triple integral $\iiint\limits_{R} f(x, y, z) \, dV$ can be expressed as the given iterated integral.

(a) $\int_{0}^{4} \int_{0}^{3} \int_{0}^{2} f(x, y, z) \, dz \, dy \, dx$ (b) $\int_{-1}^{3} \int_{2}^{4} \int_{0}^{1} f(x, y, z) \, dz \, dy \, dx$

(c) $\int_{0}^{3} \int_{0}^{2} \int_{0}^{x} f(x, y, z) \, dz \, dy \, dx$ (d) $\int_{0}^{2} \int_{0}^{x} \int_{0}^{y} f(x, y, z) \, dz \, dy \, dx$

(e) $\int_{-1}^{1} \int_{0}^{1-x^2} \int_{0}^{y} f(x, y, z) \, dz \, dy \, dx$ (f) $\int_{-1}^{1} \int_{0}^{\sqrt{1-x^2}} \int_{0}^{y} f(x, y, z) \, dz \, dy \, dx$

4.   Find the moment of inertia about the edge of length b of a brick whose dimensions are a units by b units by c units.

5.   Find the centroid of the tetrahedron that is bounded by the plane $x/a + y/b + z/c = 1$ and the coordinate planes.

6.   Find the centroid of the region bounded by the planes $y + z = 1$, $z = 0$, and the cylinder $x^2 = 4y$.

7.   Find I_Z for the tetrahedron that is bounded by the coordinate planes and the plane $x + y + z = 1$.

8.   Use a geometric argument to evaluate $\iiint\limits_{R} [\![\sqrt{x^2 + y^2 + z^2}]\!] \, dV$ if R is

(a) the cube whose faces lie in the planes $x^2 = 1$, $y^2 = 1$, and $z^2 = 1$.

(b) the spherical ball $x^2 + y^2 + z^2 \le 4$.

9.   The moment of inertia of a region R about the line formed by the intersection of the planes $x = a$ and $y = b$ is defined to be $I_L = \iiint\limits_{R} [(x - a)^2 + (y - b)^2] \, dV$.

Show that $I_L = I_Z + (a^2 + b^2 - 2a\bar{x} - 2b\bar{y})V$. What is I_L if the centroid of R is a point of the Z-axis? If the centroid of R is a point of L?

10.   (a) Show that the moment of inertia of an ellipsoid whose diameters are $2a$, $2b$, and $2c$ about the diameter whose length is $2a$ is $\frac{1}{5}(b^2 + c^2)V$.

(b) Show that the moment of inertia about its axis of a right-circular cylinder whose altitude is h and whose base radius is a is $\frac{1}{2}a^2 V$.

In Section 116 we considered double integrals in polar coordinates; now we will move up a dimension and discuss triple integrals in cylindrical and spherical coordinates. We think of the triple integral of a function f on R^3 over a three-dimensional region R as the limit of sums of the form

$$(118\text{-}1) \qquad s = \sum f(x^*, y^*, z^*)\, \Delta V.$$

The number ΔV is the volume of a typical subblock of a covering of R. These subblocks are formed by slicing R with coordinate surfaces, which are planes parallel to the coordinate planes when we are working in rectangular coordinates. In other coordinate systems, the coordinate surfaces are not all planes; and so when we slice R, the resulting subregions are not rectangular blocks. Therefore we can't calculate the volume ΔV of a typical subregion simply by multiplying the lengths of its edges. We must develop (depending on the coordinate system) an appropriate formula for ΔV. The key to integration in cylindrical or spherical coordinates is this proper formula.

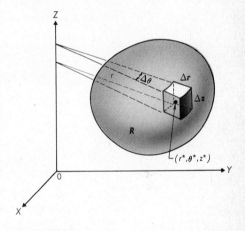

Figure 118-1

Let us introduce a cylindrical coordinate system into space as we did in Section 101. Now we slice our region R with coordinate surfaces—cylinders whose common axis is the Z-axis, planes that intersect along the Z-axis, and planes parallel to the XY-plane. A typical subregion that results is shown in Fig. 118-1. Its dimensions are indicated in the figure. Planes parallel to the XY-plane cut this subregion in figures that are congruent to its base, so its volume is simply

$$\Delta V = (\text{area of the base})\, \Delta z.$$

The base of this subregion looks like one of the polar subregions that are illustrated in Fig. 116-1. We know that the area of such a subregion is $r^* \, \Delta r \, \Delta \theta$, where r^* is the radial coordinate of a suitably chosen point (r^*, θ^*, z^*) inside it. Then $\Delta V = r^* \, \Delta r \, \Delta \theta \, \Delta z$, and one of our approximating sums is

$$\sum f(r^* \cos \theta^*, r^* \sin \theta^*, z^*) r^* \, \Delta r \, \Delta \theta \, \Delta z.$$

The limit of such sums is the triple integral of f over R, and their form suggests that we write

$$(118\text{-}2) \qquad \iiint\limits_{R} f(x, y, z)\, dV = \iiint\limits_{R} f(r \cos \theta, r \sin \theta, z) r \, dr \, d\theta \, dz.$$

733

It is convenient to think that we obtain the right-hand side of this equation from the left-hand side by replacing x with $r \cos \theta$, y with $r \sin \theta$, z with z, and dV with $r \, dr \, d\theta \, dz$.

We evaluate such integrals as iterated integrals, as the following examples show.

Example 118-1. Evaluate the integral $\iiint\limits_{R} x^2 \, dV$, where R is the region that is bounded by cylinders whose radii are 3 and 4 and the planes $z = 0$ and $z = 3$, as shown in Fig. 118-2.

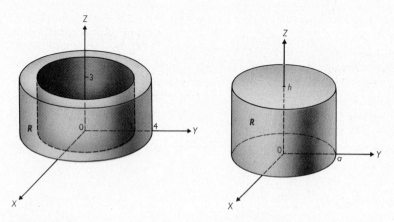

Figure 118-2   **Figure 118-3**

Solution. In this case, the limits of integration on r are 3 and 4, the limits on θ are 0 and 2π, and the limits on z are 0 and 3. Thus

$$\iiint\limits_{R} x^2 \, dV = \iiint\limits_{R} r^2 \cos^2 \theta \, r \, dr \, d\theta \, dz$$

$$= \int_{0}^{3} \int_{0}^{2\pi} \int_{3}^{4} r^3 \cos^2 \theta \, dr \, d\theta \, dz = \tfrac{525}{4}\pi.$$

Example 118-2. Use cylindrical coordinates to find the moment of inertia about its axis of a solid right-circular cylinder whose altitude is h and whose base radius is a.

Solution. We introduce a cylindrical coordinate system (Fig. 118-3) so that our solid cylinder is the region R that is bounded by the planes $z = h$ and $z = 0$, plus the cylindrical surface $r = a$. Then

$$I_z = \iiint\limits_{R} (x^2 + y^2) \, dV = \iiint\limits_{R} r^2 \, dV.$$

Now we express this integral as an iterated integral in cylindrical coordinates:

$$I_z = \int_0^a \int_0^{2\pi} \int_0^h r^3 \, dz \, d\theta \, dr = h \int_0^a \int_0^{2\pi} r^3 \, d\theta \, dr$$

$$= 2\pi h \int_0^a r^3 \, dr = \tfrac{1}{2}\pi h a^4 = \tfrac{1}{2} a^2 V.$$

If you use cartesian coordinates to find this moment of inertia (part (b) of Problem 117-10), you must work a lot harder.

The role of spherical coordinates in the study of triple integrals is similar to the role of cylindrical coordinates. We slice a given three-dimensional region R with spherical coordinate surfaces—concentric spheres whose center is the origin, planes containing the Z-axis, and cones whose axis is the Z-axis and whose vertex is the origin. One of the subregions that results is shown in Fig. 118-4. Suppose that the coordinates of the labeled point P are (ρ, θ, ϕ). The three edges of the subregion that meet at P have lengths $\Delta\rho$, $\rho\,\Delta\phi$, and $\rho \sin\phi\,\Delta\theta$, where the meaning of the symbols $\Delta\rho$, $\Delta\theta$, and $\Delta\phi$ is indicated on the figure. If our region were rectangular, its volume would be the product of the lengths of these edges—that is, the number

$$\rho^2 \sin\phi \, \Delta\rho \, \Delta\theta \, \Delta\phi.$$

Actually, this number is not quite equal to ΔV, but it can be shown that there is a point $(\rho^*, \theta^*, \phi^*)$ inside our region for which

$$\Delta V = \rho^{*2} \sin\phi^* \, \Delta\rho \, \Delta\theta \, \Delta\phi.$$

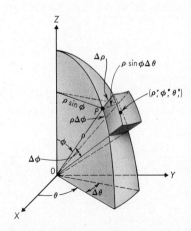

Figure 118-4

Furthermore, we know that

$$x^* = \rho^* \sin \phi^* \cos \theta^*,$$

$$y^* = \rho^* \sin \phi^* \sin \theta^*, \quad \text{and} \quad z^* = \rho^* \cos \phi^*$$

are the cartesian coordinates of our point $(\rho^*, \theta^*, \phi^*)$. Therefore one of the sums (Equation 118-1) that approximate the integral of f over R takes the form

$$\sum f(\rho^* \sin \phi^* \cos \theta^*, \rho^* \sin \phi^* \sin \theta^*, \rho^* \cos \phi^*)\rho^{*2} \sin \phi^* \, \Delta\rho \, \Delta\theta \, \Delta\phi.$$

This sum suggests that we write

$$(118\text{-}3) \quad \iiint_R f(x, y, z) \, dV$$

$$= \iiint_R f(\rho \sin \phi \cos \theta, \rho \sin \phi \sin \theta, \rho \cos \phi)\rho^2 \sin \phi \, d\rho \, d\theta \, d\phi.$$

It is convenient to think that we obtain the right-hand side of this equation from the left-hand side by replacing x with $\rho \sin \phi \cos \theta$, y with $\rho \sin \phi \sin \theta$, z with $\rho \cos \phi$, and dV with $\rho^2 \sin \phi \, d\rho \, d\theta \, d\phi$.

The following examples show how we evaluate such integrals by replacing them with iterated integrals.

Example 118-3.   Find the volume of the "ice cream cone" that is cut from a sphere whose radius is 6 by a cone with a half-angle of 30°, as shown in Fig. 118-5.

Solution.   The volume of R is the integral $V = \iiint_R dV$. From the figure we see

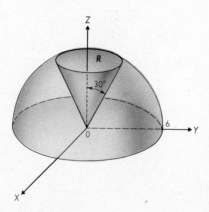

Figure 118-5

that the limits on the variable of integration ρ are 0 and 6, the limits on θ are 0 and 2π, and the limits on ϕ are 0 and $\frac{1}{6}\pi$. Thus

$$V = \int_0^{\pi/6} \int_0^{2\pi} \int_0^6 \rho^2 \sin \phi \, d\rho \, d\theta \, d\phi = 72\pi[2 - \sqrt{3}\,].$$

Example 118-4. Find the moment of inertia about a diameter of a spherical ball of radius a.

Solution. Let the ball be represented in spherical coordinates by the inequality $\rho \le a$ and let us compute I_z. Then

$$I_z = \iiint_R (x^2 + y^2) \, dV = \iiint_R \rho^2 \sin^2 \phi \, dV$$

$$= \int_0^\pi \int_0^{2\pi} \int_0^a \rho^4 \sin^3 \phi \, d\rho \, d\theta \, d\phi = \tfrac{1}{5}a^5 \int_0^\pi \int_0^{2\pi} \sin^3 \phi \, d\theta \, d\phi$$

$$= \tfrac{2}{5}\pi a^5 \int_0^\pi \sin^3 \phi \, d\phi = \tfrac{8}{15}\pi a^5.$$

To appreciate the simplicity of this calculation, try to use cartesian coordinates to find this moment of inertia.

PROBLEMS   118

1. Find the formula for the volume of a spherical ball S of radius a by evaluating the triple integral $\iiint_S dV$ as an iterated integral in rectangular, cylindrical, and spherical coordinates.

2. Use cylindrical coordinates to evaluate $\iiint_R x^4 yz^2 \, dV$, where R is the first octant region bounded by the plane $z = 2$ and the cylinder $x^2 + y^2 = 9$. Check your answer by using rectangular coordinates. Can you do the problem in spherical coordinates?

3. Use spherical coordinates to evaluate the integral $\iiint_R xyz \, dV$, where R is the first octant portion of the solid spherical ball $x^2 + y^2 + z^2 \le 9$. Check your answer by using cylindrical coordinates and rectangular coordinates.

4. Find the moment of inertia of a circular cylindrical shell whose length is L, inside radius is a, and outside radius is b

    (a) about the longitudinal axis.
    (b) about a line containing its center and perpendicular to its longitudinal axis. Find approximations to these moments of inertia if our shell is very thin, with a mean radius of c and thickness $2h$, by setting $b = c + h$ and $a = c - h$ and supposing that h is so small that h^2 can be neglected.

5. Find the moment of inertia about a diameter of a spherical shell bounded by two concentric spheres of radii a and b, where $a < b$. Use your result to check the solution of Example 118-4. Show that there is a number c between a and b such that this moment of inertia is $\frac{8}{3}\pi c^4 t$, where t is the thickness of the shell.

6. Let R be the region bounded by the XY-plane, the sphere $x^2 + y^2 + z^2 = 4$, and the cylinder $x^2 + y^2 = 1$. Find I_Z. What is the centroid of R?

7. (a) Sketch the region $R = \{(r, \theta, z) \mid 0 \le r \le 2, 0 \le \theta \le \pi, 0 \le z \le r\}$.
    (b) Find the volume of R.
    (c) Find the centroid of R.
    (d) Find the moment of inertia of R about the line $\sin \theta = 0, z = 0$.

8. (a) Sketch the region $R = \{(\rho, \theta, \phi) \mid 0 \le \rho \le 2, 0 \le \theta \le \frac{1}{6}\pi, 0 \le \phi \le \pi\}$.
    (b) Find the volume of R.
    (c) Find the centroid of R.
    (d) Find the moment of inertia about the line $\sin \phi = 0$.

9. Use cylindrical coordinates to find the moment of inertia about the axis of a solid right-circular cone whose base radius is a and whose altitude is h. How far above the base is the centroid of the cone?

10. The region $R = \{(x, y, z) \mid a^2 < x^2 + y^2 + z^2 < b^2, z \ge 0\}$ forms an inverted bowl lying on the XY-plane. Find the Z-coordinate $\bar{z}$ of the centroid of the bowl. What is the Z-coordinate of the centroid of a solid hemisphere? What is $\lim\limits_{a \uparrow b} \bar{z}$? Show that the Z-coordinate of the centroid of our bowl always lies between these two numbers.

119.   LINE INTEGRALS

So far in this chapter we have extended the concept of the integral from scalar-valued functions on R^1 to scalar-valued functions on R^2 and R^3. Now we will generalize the notion of the integral in yet another direction, to integrals of vector-valued functions. Although much of our discussion applies to functions that take values in a general R^n, we limit ourselves to R^2 for simplicity. Even so, some of the details and proofs are too long or too difficult (or both) to include in an elementary course in calculus, so we often resort to informal arguments to make our points. If you keep your eyes on our figures, and draw some of your own, you can understand what we are talking about here. The difficulties arise when we try to express in words and mathematical symbols things that seem obvious in the pictures.

As one of our notational conventions, we often designate the vector (x, y) of R^2 by writing its first component in boldface roman type, $\mathbf{x} = (x, y)$. When we picture (x, y) as a point of the coordinate plane, the vector $\mathbf{x}$ can be regarded as the position vector to the point. We can also introduce the basis vectors $\mathbf{i}$ and $\mathbf{j}$ and write $\mathbf{x} = x\mathbf{i} + y\mathbf{j}$. In the formulas and calculations that follow, we will use whatever notation seems to express our ideas in the simplest and clearest way.

The semicircle shown in Fig. 119-1 is the graph of the vector equation $\mathbf{x} = 5 \cos t\mathbf{i} + 5 \sin t\mathbf{j}$, $t \in [-\frac{1}{2}\pi, \frac{1}{2}\pi]$. We say that this equation "maps" the interval $[-\frac{1}{2}\pi, \frac{1}{2}\pi]$ onto the semicircle. Since the interval is part of a number scale, it is directed, the positive direction being from the point $-\frac{1}{2}\pi$ toward the point $\frac{1}{2}\pi$. Our equation transfers this direction to the semicircle; for example, the "first" point of the semicircle is $(0, -5)$ and the "last" point is $(0, 5)$ in this mapping. We will denote this directed semicircle by the boldface italic letter $\boldsymbol{C}$. Our semicircle can also be thought of as the image of the interval $[-5, 5]$ under the mapping that is determined by the parametric equations $x = \sqrt{25 - u^2}$ and $y = u$, $u \in [-5, 5]$. This mapping gives the semicircle the same direction that our previous one did, so we have different representa-

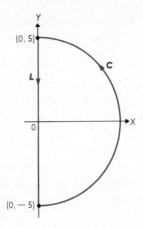

Figure 119-1

tions of the same directed semicircle. The parametric equations $x = \sqrt{25 - v^2}$ and $y = -v$, $v \in [-5, 5]$, also map the interval $[-5, 5]$ onto our semicircle, but now its direction is reversed. It will be important for us to distinguish this new directed semicircle from $\boldsymbol{C}$; we will call it $-\boldsymbol{C}$.

In general, a **directed arc** $\boldsymbol{C}$ is the graph of a vector equation $\mathbf{x} = \mathbf{r}(t)$, $t \in [a, b]$, where $\mathbf{r}$ is a differentiable vector-valued function such that $|\mathbf{r}'(t)| > 0$ for each $t \in [a, b]$. The direction of $\boldsymbol{C}$ is inherited from the parameter interval $[a, b]$.

Suppose that we are given such a directed arc $\boldsymbol{C}$ and a vector-valued function $\mathbf{f}$ on R^2 whose domain contains the points of $\boldsymbol{C}$. Then we define the **line integral** of $\mathbf{f}$ along $\boldsymbol{C}$ to be the number

$$(119\text{-}1) \qquad \int_{C} \mathbf{f}(\mathbf{x}) \cdot d\mathbf{x} = \int_{a}^{b} \mathbf{f}(\mathbf{r}(t)) \cdot \mathbf{r}'(t) \, dt.$$

Example 119-1. Find the line integral of $\mathbf{f}$ along the directed arc $\boldsymbol{C}$ of Fig. 119-1 if $\mathbf{f}(\mathbf{x}) = x^2\mathbf{i} + y^2\mathbf{j}$.

Solution. If we think of $\boldsymbol{C}$ as the graph of the vector equation

$$\mathbf{x} = \mathbf{r}(t) = 5 \cos t\mathbf{i} + 5 \sin t\mathbf{j} \quad \text{for} \quad t \in [-\tfrac{1}{2}\pi, \tfrac{1}{2}\pi],$$

then $x = 5 \cos t$ and $y = 5 \sin t$, so

$$\mathbf{f}(\mathbf{r}(t)) = 25 \cos^2 t\mathbf{i} + 25 \sin^2 t\mathbf{j},$$

and hence Equation 119-1 tells us that

$$\int_C \mathbf{f(x)} \cdot d\mathbf{x} = \int_{-\pi/2}^{\pi/2} (25 \cos^2 t\mathbf{i} + 25 \sin^2 t\mathbf{j}) \cdot (-5 \sin t\mathbf{i} + 5 \cos t\mathbf{j}) \, dt$$

$$= \int_{-\pi/2}^{\pi/2} (-125 \cos^2 t \sin t + 125 \sin^2 t \cos t) \, dt$$

$$= 125(\tfrac{1}{3} \cos^3 t + \tfrac{1}{3} \sin^3 t)\Big|_{-\pi/2}^{\pi/2} = \tfrac{250}{3}.$$

Line integrals can be written in many different notations. Let us suppose that $\mathbf{f(x)} = f(x, y)\mathbf{i} + g(x, y)\mathbf{j}$ and that the vector equation $\mathbf{x} = \mathbf{r}(t)$ that represents C can be expressed as the parametric scalar equations $x = r(t)$ and $y = s(t)$. Then $\mathbf{r}'(t) = r'(t)\mathbf{i} + s'(t)\mathbf{j}$, and so the expanded form of Equation 119-1 is

$$(119\text{-}2) \qquad \int_C \mathbf{f(x)} \cdot d\mathbf{x} = \int_a^b \left[f(r(t), s(t))r'(t) + g(r(t), s(t))s'(t) \right] dt.$$

Since $\mathbf{x} = x\mathbf{i} + y\mathbf{j}$, it is natural to replace $d\mathbf{x}$ with $dx\mathbf{i} + dy\mathbf{j}$. Then the dot product $\mathbf{f(x)} \cdot d\mathbf{x}$ becomes $f(x, y) \, dx + g(x, y) \, dy$, and we write

$$(119\text{-}3) \qquad \int_C \mathbf{f(x)} \cdot d\mathbf{x} = \int_C f(x, y) \, dx + g(x, y) \, dy.$$

Of course, when we compute the right-hand side of this equation, we must replace x with $r(t)$, y with $s(t)$, dx with $r'(t) \, dt$, dy with $s'(t) \, dt$, and integrate over the parameter interval $[a, b]$.

Example 119-2. For the function $\mathbf{f}$ and directed arc C of Example 119-1, evaluate $\int_C \mathbf{f(x)} \cdot d\mathbf{x}$ in case C is represented by the parametric equations $x = \sqrt{25 - u^2}$ and $y = u$ for $u \in [-5, 5]$.

Solution. From these equations we see that we should replace $f(x, y) = x^2$ with $25 - u^2$, dx with $-u(25 - u^2)^{-1/2} \, du$, $g(x, y) = y^2$ with u^2, and dy with du in Equation 119-3:

$$\int_C \mathbf{f(x)} \cdot d\mathbf{x} = \int_C x^2 \, dx + y^2 \, dy$$

$$= \int_{-5}^5 [-u(25 - u^2)^{1/2} + u^2] \, du$$

$$= [\tfrac{1}{3}(25 - u^2)^{3/2} + \tfrac{1}{3}u^3]\Big|_{-5}^5 = \tfrac{250}{3}.$$

Equation 119-1 expresses the line integral of a function **f** along a directed arc C in terms of values of a vector-valued function **r** of which C is the graph. Consequently, you might think that we obtain different values for the line integral, depending on which representation of C we use. That this is not so is illustrated in our examples. In Examples 119-1 and 119-2 we used two different representations of C to calculate the same line integral, and we obtained the same result each time. We will omit the proof and simply state that the line integral of a function **f** along a directed arc C depends only on **f** and C, not on any particular choice of a function **r** of which C is the graph. It can be shown that reversing the direction of an arc changes the sign of line integrals along the arc; that is,

$$(119\text{-}4) \qquad \int_{-C} \mathbf{f}(\mathbf{x}) \cdot d\mathbf{x} = -\int_{C} \mathbf{f}(\mathbf{x}) \cdot d\mathbf{x}.$$

It follows directly from our definition (Equation 119-1) that line integrals enjoy many of the properties of ordinary integrals. For example, integrals of sums are sums of integrals, and so forth. Furthermore, we may join a number of directed arcs to form a **directed path**. The line integral of a vector-valued function along the path is then defined as the sum of its line integrals along the arcs. For example, the line integral of **f** around a (directed) square is the sum of its integrals along the sides (properly directed) of the square.

Example 119-3. Find the line integral of the vector-valued function of Example 119-1 along the closed directed path P (shown in Fig. 119-1) that is formed by joining to the directed semicircle C the directed line segment L from the point $(0, 5)$ to the point $(0, -5)$.

Solution. We compute the desired number $\int_{P} \mathbf{f}(\mathbf{x}) \cdot d\mathbf{x}$ by adding to the line integral of **f** along C (which we know is $\frac{250}{3}$) the number $\int_{L} \mathbf{f}(\mathbf{x}) \cdot d\mathbf{x} = \int_{L} x^2 \, dx + y^2 \, dy$. Parametric equations of L are $x = 0$ and $y = -t$, $t \in [-5, 5]$, so

$$\int_{L} x^2 \, dx + y^2 \, dy = -\int_{-5}^{5} t^2 \, dt = -\tfrac{1}{3}t^3 \Big|_{-5}^{5} = -\tfrac{250}{3}.$$

Usually we don't go through this whole substitution routine. We simply notice that $x = 0$ on L and y varies from 5 to -5. We therefore write

$$\int_{L} x^2 \, dx + y^2 \, dy = \int_{5}^{-5} y^2 \, dy = \tfrac{1}{3}y^3 \Big|_{5}^{-5} = -\tfrac{250}{3}.$$

Now we add this number to the integral along the semicircle and see that

$$\int_{P} \mathbf{f}(\mathbf{x}) \cdot d\mathbf{x} = \tfrac{250}{3} - \tfrac{250}{3} = 0.$$

In Section 108 we saw how the gradient of a scalar-valued function f on R^2 plays the role of a derivative. Here is another illustration of that role. Suppose that $\mathbf{a}$ and $\mathbf{b}$ are two points of the plane. Let $\mathbf{x} = \mathbf{r}(t)$ for $t \in [t_1, t_2]$ be a vector equation of an arc from $\mathbf{a}$ to $\mathbf{b}$; that is, $\mathbf{r}(t_1) = \mathbf{a}$ and $\mathbf{r}(t_2) = \mathbf{b}$. The Chain Rule (Equation 108-3) tells us that

$$\frac{d}{dt} f(\mathbf{r}(t)) = \nabla f(\mathbf{r}(t)) \cdot \mathbf{r}'(t).$$

Therefore, according to the Fundamental Theorem of Calculus,

$$(119\text{-}5) \qquad \int_{t_1}^{t_2} \nabla f(\mathbf{r}(t)) \cdot \mathbf{r}'(t) \, dt = f(\mathbf{r}(t)) \Big|_{t_1}^{t_2}$$

$$= f(\mathbf{r}(t_2)) - f(\mathbf{r}(t_1)) = f(\mathbf{b}) - f(\mathbf{a}).$$

The left-hand side of this equation is the line integral of ∇f along a certain path joining $\mathbf{a}$ and $\mathbf{b}$. But the right-hand side depends only on the endpoints, not on the path itself. Hence we need not mention the path when we talk about line integrals of a gradient, and we write Equation 119-5 as

$$(119\text{-}6) \qquad \int_{\mathbf{a}}^{\mathbf{b}} \nabla f(\mathbf{x}) \cdot d\mathbf{x} = f(\mathbf{b}) - f(\mathbf{a}).$$

Because it resembles the ordinary Fundamental Theorem,

$$\int_{a}^{b} f'(x) \, dx = f(b) - f(a),$$

we sometimes refer to Equation 119-6 as the *Fundamental Theorem of Calculus for line integrals.*

Line integrals are frequently used to calculate work. In Section 42 we defined work to be the integral of force, an idea that is easy to understand if the motion is linear and the force acts along the line of motion (a body being lifted straight up against the force of gravity, for example). Now we will drop this restriction. Suppose that at each point (x, y) of the coordinate plane, a force vector $\mathbf{f}(x, y)$ is defined. For example, if a positive unit electrical charge is placed at the origin, it will exert (according to the Inverse Square Law) a force

$$(119\text{-}7) \qquad \mathbf{f}(x, y) = x(x^2 + y^2)^{-3/2}\mathbf{i} + y(x^2 + y^2)^{-3/2}\mathbf{j}$$

on a unit charge at the point (x, y). If this latter charge moves along a directed path P, then the equation

$$W = \int_{P} \mathbf{f}(\mathbf{x}) \cdot d\mathbf{x}$$

gives the work done as P is traversed.

Example 119-4. In our electrical example, how much work is required to move a positive unit charge along the straight path from the point $(-1, 1)$ to the point $(0, 1)$?

Solution. Parametric equations of our path are $x = r(t) = t$ and $y = s(t) = 1$, $t \in [-1, 0]$. After substituting these values in Equation 119-2 (we read the components of the force vector from Equation 119-7), we find that

$$W = \int_{-1}^{0} t(t^2 + 1)^{-3/2} \, dt = \tfrac{1}{2}\sqrt{2} - 1 \approx -.3 \text{ unit.}$$

The negative answer indicates that the net work has been done "against the force."

Frequently we wish to integrate a function f around a *closed* path—that is, one that "begins" and "ends" at the same point, such as the path in Example 119-3. Of course, we must distinguish between the two directions around the path, and we will use the symbol

$$\oint_{P} \mathbf{f(x)} \cdot d\mathbf{x}$$

to denote the line integral of $\mathbf{f}$ around a closed path P in a counterclockwise direction.

Example 119-5. Show that if the force vector is the gradient of a scalar-valued function $\mathbf{f}$, then the work done in traversing a closed directed arc P is zero.

Solution. Let $\mathbf{a}$ be any point of P. To find the work done in traversing P, starting at $\mathbf{a}$ and ending at $\mathbf{a}$, we must integrate the force vector $\nabla f(\mathbf{x})$ around the path. According to Equation 119-6, the result will simply be the number $f(\mathbf{a}) - f(\mathbf{a}) = 0$, as we were to show.

PROBLEMS   119

1. Let $\mathbf{f(x)} = y\mathbf{i} + 3\mathbf{j}$ and find $\int_{P} \mathbf{f(x)} \cdot d\mathbf{x}$, where P is the given directed path.

(a) The portion of the X-axis from -1 to 5.
(b) The portion of the Y-axis from -1 to 5.
(c) The line segment from the origin to the point $(2, 3)$.
(d) The line segment from $(-1, 2)$ to $(2, 3)$.
(e) The arc of the curve $y = e^x$ from the point $(0, 1)$ to the point $(\ln 2, 2)$.
(f) The arc of the vector curve $\mathbf{x} = \sin t\mathbf{i} + \cos t\mathbf{j}$ for t in the interval $[0, \tfrac{1}{4}\pi]$.

2. Let $f(x, y) = 3x^2y\mathbf{i} + (2x - y)\mathbf{j}$. Find $\int_P f(\mathbf{x}) \cdot d\mathbf{x}$, where P is the directed path from $(0, 0)$ to $(1, 1)$ described as follows.
   (a) The line segment.
   (b) The parabolic arc $y = x^2$, $x \in [0, 1]$.
   (c) The graph of the vector equation $\mathbf{x} = \sin t\mathbf{i} + (2t/\pi)\mathbf{j}$, $t \in [0, \frac{1}{2}\pi]$.
   (d) The graph of the parametric equations $x = t^2$ and $y = t^3$, $t \in [0, 1]$.
   (e) The union of the segment from $(0, 0)$ to $(1, 0)$ and the segment from $(1, 0)$ to $(1, 1)$.
   (f) The union of the segment from $(0, 0)$ to $(0, 1)$ and the segment from $(1, 0)$ to $(1, 1)$.

3. Find $\int_P 3x^2y \, dx + (2x - y) \, dy$ if P is the shortest path from $(0, 0)$ to $(2, 2)$ that also contains the points $(1, 0)$, $(1, 1)$, and $(2, 1)$.

4. Find $\int_C y^2 \, dx + x^2 \, dy$ along each of the following directed arcs.
   (a) The arc of the ellipse $x^2/a^2 + y^2/b^2 = 1$ from the point $(a, 0)$ to the point $(0, b)$.
   (b) The line segment from the point $(a, 0)$ to the point $(0, b)$.

5. Let $\mathbf{m}$ be a given vector that is independent of x; for example, $\mathbf{m} = 3\mathbf{i} + 4\mathbf{j}$.
   (a) Show that $\mathbf{m} = \nabla(\mathbf{m} \cdot \mathbf{x})$.
   (b) Show that $\int_C \mathbf{m} \cdot d\mathbf{x} = \mathbf{m} \cdot (\mathbf{b} - \mathbf{a})$ if C is any path from a given point $\mathbf{a}$ to a given point $\mathbf{b}$.

6. Let $f(\mathbf{x}) = \mathbf{x} \cdot \mathbf{x}$ and let P be the directed path consisting of the union of the directed line segments from $(0, 0)$ to $(1, 0)$ and from $(1, 0)$ to $(1, 1)$. Use Equation 119-3 to compute $\int_P \nabla f(\mathbf{x}) \cdot d\mathbf{x}$, and use Equation 119-6 to check your result.

7. Let P be the closed path consisting of line segments joining the points $(0, 0)$, $(1, 0)$, $(1, 1)$, $(0, 1)$, $(0, 0)$ in that order. Compute the work done as P is traversed, if the force function is defined by the equation $f(\mathbf{x}) = (x^2 - y^2)\mathbf{i} + 2xy\mathbf{j}$. Can you find a function g such that $\mathbf{f} = \nabla g$? Do the problem if the defining equation is $f(\mathbf{x}) = (x^2 + y^2)\mathbf{i} + 2xy\mathbf{j}$.

8. For each number a of the interval $[0, 1]$ we construct the path $P(a)$ from the point $(0, 0)$ to the point $(1, 1)$ by proceeding from $(0, 0)$ along the X-axis to $(a, 0)$, then parallel to the Y-axis to $(a, 1)$, and then parallel to the X-axis to $(1, 1)$. Then we calculate the number $I(a) = \int_{P(a)} x^2(y - 1) \, dx + xy^8 \, dy$. For what number a is $I(a)$ a maximum? What is the minimum value of $I(a)$?

9. Show that no work is done if we move through a force field along a path that is perpendicular to the lines of force—for instance, if we move along circular arcs whose center is the origin in our electrical example.

10. If we view an interval of the X-axis as a small portion of the surface of the earth and the Y-axis as pointing up, then the gravitational force on an object of mass m is expressed by the vector $\mathbf{G} = -mg\mathbf{j}$, where g is a number that depends on the units we use. Show that the work done by gravity as the object is moved from a point (a, b) to a point (c, d) along any path is $mg(d - b)$.

11. Show that $\displaystyle\int_C \mathbf{f}(\mathbf{x}) \cdot d\mathbf{x}$ is the length of C if $\mathbf{f}(\mathbf{x})$ is the unit tangent vector in the positive direction at each point $\mathbf{x}$ of C.

12. Let $\mathbf{f}$ be a vector-valued function on R^2 and let C be a directed arc in the domain of $\mathbf{f}$. Suppose that there is a number M such that for each point $(x, y) \in C$, $|\mathbf{f}(x, y)| \leq M$, and let L be the length of C. Show that $\left| \displaystyle\int_C \mathbf{f}(\mathbf{x}) \cdot d\mathbf{x} \right| \leq ML$.

120. GREEN'S THEOREM, THE FUNDAMENTAL THEOREM OF CALCULUS FOR DOUBLE INTEGRALS

In symbols, the Fundamental Theorem of Calculus says that

$$(120\text{-}1) \qquad \int_a^b f'(x)\, dx = f(x) \Big|_a^b = f(b) - f(a).$$

Thus the integral of f' over $[a, b]$ is expressed in terms of the values of f at the "boundary points" a and b of the interval. We are now going to derive an analogous formula for double integrals.

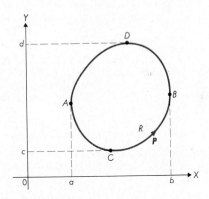

Figure 120-1

We will suppose that we are given a vector-valued function $\mathbf{f}$ (thus $\mathbf{f}(x, y) = f(x, y)\mathbf{i} + g(x, y)\mathbf{j}$) whose domain contains the region R shown in Fig. 120-1. This region is bounded by the closed, directed path $\mathbf{P}$, which we will view as the union of the directed arcs ACB and BDA, where ACB is the curve $y = r(x)$ and ADB is the curve $y = s(x)$ for $x \in [a, b]$. Then the double integral of f_2 over R (remember, $f_2(x, y) = \dfrac{\partial}{\partial y} f(x, y)$) can be expressed as an iterated integral:

$$\iint_R f_2(x, y)\, dA = \int_a^b \int_{r(x)}^{s(x)} f_2(x, y)\, dy\, dx.$$

The inner integral is simply the integral of a derivative, so we use the ordinary Fundamental Theorem (Equation 120-1 with some letters changed) to see that

$$\int_{r(x)}^{s(x)} f_2(x, y)\, dy = f(x, y)\Big|_{y=r(x)}^{y=s(x)} = f(x, s(x)) - f(x, r(x)).$$

Therefore

$$(120\text{-}2)\qquad \iint_R f_2(x, y)\, dx = \int_a^b f(x, s(x))\, dx - \int_a^b f(x, r(x))\, dx.$$

But we can think of these last two integrals as line integrals along the upper and lower boundaries of R:

$$\int_a^b f(x, s(x))\, dx = \int_{ADB} f(x, y)\, dx = -\int_{BDA} f(x, y)\, dx$$

and

$$\int_a^b f(x, r(x))\, dx = \int_{ACB} f(x, y)\, dx.$$

Since P is the union of the arcs ACB and BDA, we can use these last formulas to write Equation 120-2 as

$$(120\text{-}3)\qquad \iint_R f_2(x, y)\, dA = -\oint_P f(x, y)\, dx.$$

Now let us view our path P as the union of the arcs DAC and CBD, where CAD is the curve $x = p(y)$ and CBD is the curve $x = q(y)$ for $y \in [c, d]$. We will write the double integral of g_1 (where $g_1(x, y) = \dfrac{\partial}{\partial x} g(x, y)$) over R as an iterated integral:

$$\iint_R g_1(x, y)\, dA = \int_c^d \int_{p(y)}^{q(y)} g_1(x, y)\, dx\, dy = \int_c^d g(x, y)\Big|_{x=p(y)}^{x=q(y)} dy$$

$$= \int_c^d [g(q(y), y) - g(p(y), y)]\, dy.$$

This last integral can be thought of as the line integral $\oint_P g(x, y)\, dy$, and so

$$(120\text{-}4)\qquad \iint_R g_1(x, y)\, dA = \oint_P g(x, y)\, dy.$$

We combine Equations 120-3 and 120-4 to obtain the conclusion of **Green's Theorem** in the plane:

$$(120\text{-}5) \quad \iint_R [g_1(x, y) - f_2(x, y)]\, dA = \oint_P f(x, y)\, dx + g(x, y)\, dy$$

$$= \oint_P \mathbf{f(x)} \cdot d\mathbf{x}.$$

Like the ordinary Fundamental Theorem of Calculus, it expresses an integral of a combination of the derived functions g_1 and f_2 over a region in terms of values of f and g on the boundary path P. In developing Green's Theorem, we assumed that P had an especially simple shape. By "patching together" regions of the type we have considered, the range of validity of Equation 120-5 can be greatly extended. In fact, Green's Theorem is true for quite general regions.

Example 120-1. Explain why, if we had known Green's Theorem at that time, we could have anticipated that 0 would be the answer to our problem in Example 119-3.

Solution. In that example $f(x, y) = x^2$ and $g(x, y) = y^2$. Therefore $g_1(x, y) = \dfrac{\partial}{\partial x} y^2 = 0$ and $f_2(x, y) = \dfrac{\partial}{\partial y} x^2 = 0$ at each point (x, y) of the region that is bounded by our path of integration. Hence the left-hand side of Equation 120-5 is 0, so the right-hand side must be 0, too.

Example 120-2. Explain why $\frac{1}{2} \oint_P x\, dy - y\, dx$ is the area of the region R that is bounded by the closed path P.

Solution. Simply let $f(x, y) = -\frac{1}{2}y$ and $g(x, y) = \frac{1}{2}x$ in Equation 120-5. Since

$$g_1(x, y) = \frac{\partial}{\partial x}\, \tfrac{1}{2}x = \tfrac{1}{2} \quad \text{and} \quad f_2(x, y) = \frac{\partial}{\partial y}\, (-\tfrac{1}{2}y) = -\tfrac{1}{2},$$

our equation becomes

$$\iint_R [\tfrac{1}{2} - (-\tfrac{1}{2})]\, dA = \tfrac{1}{2} \oint_P x\, dy - y\, dx.$$

The left-hand side is the number $\displaystyle\iint_R dA$, the area of R.

A line integral $\int_P \mathbf{f(x)} \cdot d\mathbf{x}$ over a path P obviously depends on the given function $\mathbf{f}$ and on the path P. (We earlier stated that it doesn't depend on any particular *equations* used to describe P.) But there are many instances in nature in which the path doesn't matter either. Only the endpoints count. For example, if you lift a heavy weight from the floor to a tabletop, there is no way to lessen the amount of work by trying to maneuver the weight through some devious path or other. You might as well lift it straight up, because it takes the same amount of work in any case. Green's Theorem can tell us when a given line integral is *independent of path*.

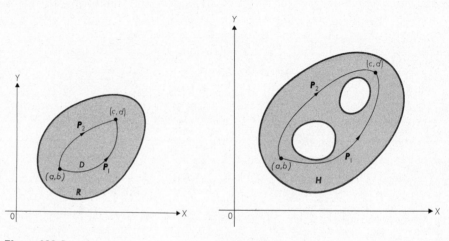

Figure 120-2   **Figure 120-3**

Suppose that (a, b) and (c, d) are two points in the plane, joined by paths P_1 and P_2 (Fig. 120-2). Then the equation

$$\int_{P_1} \mathbf{f(x)} \cdot d\mathbf{x} = \int_{P_2} \mathbf{f(x)} \cdot d\mathbf{x}$$

is equivalent to the equation $\oint_P \mathbf{f(x)} \cdot d\mathbf{x} = 0$, where the path of integration P is the closed path from (a, b) along P_1 to (c, d) and then back to (a, b) along $-P_2$. So the statement that line integrals are independent of path is equivalent to saying that line integrals around closed paths are 0. Let us assume that any such closed path P bounds a region D that lies entirely within the basic region R in which we are working. (For a region with holes in it, such as the region H shown in Fig. 120-3, we can find closed paths that lie in H but that bound a set that is not in H. So we will rule out such regions and restrict our attention to regions that don't contain holes. Such regions are said to be **simply connected**.) Then according to Green's Theorem,

$$\oint_P \mathbf{f}(\mathbf{x}) \cdot d\mathbf{x} = \iint_D [g_1(x, y) - f_2(x, y)] \, dA.$$

This double integral is 0 if for each point $(x, y) \in D$, $g_1(x, y) = f_2(x, y)$. Since D is a subset of our basic region R, this equation will surely hold throughout D if it holds throughout R. Thus we can state the following theorem.

Theorem 120-1. *If for each point (x, y) of a simply connected region R*

(120-6) $f_2(x, y) = g_1(x, y),$

then line integrals of **f** *are independent of path in R.*

Example 120-3. Evaluate the line integral $\displaystyle\int_P 2xy^3 \, dx + 3x^2y^2 \, dy$, where P is a directed path from the point $(-1, 2)$ to $(2, 0)$.

Solution. In this example $f(x, y) = 2xy^3$ and $g(x, y) = 3x^2y^2$, and so

$$f_2(x, y) = \frac{\partial}{\partial y}(2xy^3) = 6xy^2 \quad \text{and} \quad g_1(x, y) = \frac{\partial}{\partial x}(3x^2y^2) = 6xy^2.$$

Condition 120-6 is therefore satisfied in the whole XY-plane, and line integrals are independent of path. Hence we can take P as any path we like from $(-1, 2)$ to $(2, 0)$, and it may seem simplest to use the direct line

$$x = 3t - 1 \quad \text{and} \quad y = 2 - 2t, \qquad t \in [0, 1].$$

We make these replacements and also substitute $3dt$ and $-2dt$ for dx and dy, and our integral becomes

$$\int_0^1 [2(3t - 1)(2 - 2t)^3 \, 3 + 3(3t - 1)^2(2 - 2t)^2(-2)] \, dt.$$

You can check to see that this number works out to be -8.

If a line integral is independent of path, it is often easiest to evaluate it by "angle-paths" like those shown in Fig. 120-4. Thus suppose that **f** is a given function for which line integrals are independent of path and that (a, b) and

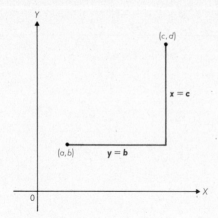

Figure 120-4

(c, d) are two points of the plane. Then we can denote the line integral of **f** from (a, b) to (c, d) as $\int_{(a,b)}^{(c,d)} \mathbf{f(x)} \cdot d\mathbf{x}$ and choose any path we like to evaluate this integral. For example, we might proceed from (a, b) to (c, b) along the line $y = b$ and then straight up along the line $x = c$ to (c, d). Or we could first integrate vertically along the line $x = a$ to (a, d) and then horizontally to (c, d) along the line $y = d$. After carrying out the necessary calculations, we find the formulas

$$(120\text{-}7) \qquad \int_{(a,b)}^{(c,d)} \mathbf{f(x)} \cdot d\mathbf{x} = \int_a^c f(x,\, b)\, dx + \int_b^d g(c,\, y)\, dy$$

$$= \int_b^d g(a,\, y)\, dy + \int_a^c f(x,\, d)\, dx.$$

Example 120-4. Use Equation 120-7 to evaluate the integral of Example 120-3.

Solution. We may simply replace $f(x, y)$ with $2xy^3$, $g(x, y)$ with $3x^2y^2$, (a, b) with $(-1, 2)$, and (c, d) with $(2, 0)$ in the first of Equations 120-7:

$$\int_{(-1,2)}^{(2,0)} 2xy^3\, dx + 3x^2y^2\, dy = \int_{-1}^2 2x \cdot 2^3\, dx + \int_2^0 3 \cdot 2^2 y^2\, dy$$

$$= 8x^2 \Big|_{-1}^2 + 4y^3 \Big|_2^0 = 32 - 8 - 32 = -8.$$

The calculations are even simpler if we use the second of Equations 120-7.

PROBLEMS   120

1. Show that the given integral is independent of path and use Equation 120-7 to evaluate it.

(a) $\displaystyle\int_{(-1,2)}^{(2,0)} 3y^2\, dx + (6xy - 2)\, dy$

(b) $\displaystyle\int_{(-1,2)}^{(2,0)} (6xy - 2)\, dx + 3x^2\, dy$

(c) $\displaystyle\int_{(0,0)}^{(\ln 2,\ln 3)} e^{x+2y}\, dx + 2e^{x+2y}\, dy$

(d) $\displaystyle\int_{(0,0)}^{(\ln 1/2,\ln 3)} 2e^{2x+y}\, dx + e^{2x+y}\, dy$

(e) $\displaystyle\int_{(0,0)}^{(-2,\pi/6)} 2xy \cos x^2 y\, dx + x^2 \cos x^2 y\, dy$

(f) $\displaystyle\int_{(0,0)}^{(\pi/6,\pi/3)} 2 \sin (2x + y)\, dx + \sin (2x + y)\, dy$

2. Show that the given integral is independent of path and evaluate it by integrating along the segment from the origin to the upper limit of integration. Would it be easier to use Equation 120-7?

(a) $\displaystyle\int_{(0,0)}^{(-1,2)} 3y^2 dx + (6xy - 2)\, dy$

(b) $\displaystyle\int_{(0,0)}^{(2,0)} (6xy - 2)\, dx + 3x^2\, dy$

(c) $\displaystyle\int_{(0,0)}^{(\sqrt{\pi/2},\sqrt{\pi/2})} y \cos xy\, dx + x \cos xy\, dy$

(d) $\displaystyle\int_{(0,0)}^{(\sqrt{\pi/2},\sqrt{\pi/2})} y \sec^2 xy\, dx + x \sec^2 xy\, dy$

3. Use Green's Theorem to evaluate the line integral of $\mathbf{f(x)} = 3y\mathbf{i} - 4x\mathbf{j}$ around the given path in the counterclockwise direction.
   (a) The circle $x^2 + y^2 = 1$.
   (b) The rectangle with vertices $(-1, 2)$, $(3, 2)$, $(3, 5)$, $(-1, 5)$.
   (c) The triangle with vertices $(-1, 2)$, $(3, 2)$, $(5, 5)$.
   (d) The triangle with vertices $(-1, 2)$, $(2, 3)$, $(5, 5)$.

4. If f and g are continuous functions on R^1 and (a, b) and (c, d) are any points of the plane, show that

$$\int_{(a,b)}^{(c,d)} f(x)\, dx + g(y)\, dy = \int_a^c f(t)\, dt + \int_b^d g(t)\, dt.$$

5. In each case, find $h(x, y)$ so that the line integral is independent of path. Is there more than one solution?

(a) $\int_a^b h(x, y)\, dx + 3x^2y\, dy$ (b) $\int_a^b 3x^2y\, dx + h(x, y)\, dy$

6. Suppose that a unit positive charge is placed at the origin (see Example 119-4). Use Green's Theorem to deduce that no net work is done if another unit positive charge is moved around the square whose vertices are $(1, -1)$, $(1, 1)$, $(3, 1)$, and $(3, -1)$. What if the square is replaced by any closed path that doesn't encircle the origin?

7. Let R^* be the coordinate plane with the origin removed. Let

$$f(x, y) = \frac{y}{x^2 + y^2} \quad \text{and} \quad g(x, y) = \frac{-x}{x^2 + y^2},$$

and let $\mathbf{f}(\mathbf{x}) = f(x, y)\mathbf{i} + g(x, y)\mathbf{j}$.

(a) Show that $f_2(x, y) = g_1(x, y)$ at each point of R^*.

(b) Compute the line integrals of $\mathbf{f}$ from the point $(1, 0)$ to the point $(-1, 0)$ along the upper half and along the lower half of the unit circle. Are line integrals of $\mathbf{f}$ independent of path in R^*? In view of part (a), does your answer violate Theorem 120-1?

8. Let $(\bar{x}, \bar{y})$ be the centroid of a region R that has an area of A square units and is bounded by a closed directed path P. Show that

$$\bar{x} = \frac{1}{2A} \oint_P x^2\, dy \quad \text{and} \quad \bar{y} = \frac{1}{A} \oint_P xy\, dy.$$

9. Show that two moments of inertia of a region bounded by a directed path P are

$$I_X = \oint_P xy^2\, dy \quad \text{and} \quad I_O = \oint_P xy^2\, dy - yx^2\, dx.$$

10. In Section 119 (Equation 119-6) we found that a line integral of a gradient is independent of path. Show that this statement also follows from Theorem 120-1.

121. FUNCTIONS DEFINED BY LINE INTEGRALS

The two basic processes of calculus—differentiation and integration—transform one function into another. We may start with a given function f and calculate the derived function f', or we can construct a function F by integration:

(121-1) $$F(x) = \int_a^x f(u)\, du.$$

For example, in Chapter 6 the logarithmic function was defined by the equation

$$\ln x = \int_1^x \frac{1}{u}\, du.$$

Because x is the upper limit of integration in Equation 121-1, we can't use it as the variable of integration as well. We use u instead. Thus we can think of the domain of F (an interval in which f is integrable) as an interval I of the U-axis, as shown in Fig. 121-1. To evaluate F at a point x of this interval, we simply integrate f from a to x.

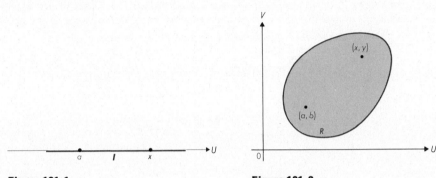

Figure 121-1   **Figure 121-2**

Now we are going to do the same thing for functions on R^2. In this case, we start with a vector-valued function $\mathbf{f}$ defined in a region R of the UV-plane like the one shown in Fig. 121-2. Thus there are scalar-valued functions f and g such that $\mathbf{f(u)} = f(u, v)\mathbf{i} + g(u, v)\mathbf{j}$ for each point $\mathbf{u} = (u, v) \in R$. Then if $\mathbf{a}$ is a point of R, we use the equation

$$(121\text{-}2) \qquad F(\mathbf{x}) = \int_{\mathbf{a}}^{\mathbf{x}} \mathbf{f(u)} \cdot d\mathbf{u}$$

to define a function F. Although this equation looks exactly like Equation 121-1, there is one new feature. Both equations tell us to integrate from one point to another. In the first case, there is only one path to follow, but there are many ways to get from the point $\mathbf{a} \in R^2$ to the point $\mathbf{x}$. So when we use Equation 121-2 to define a function F on R^2, *we will always assume that integrals of our given function $\mathbf{f}$ are independent of path.* We can use Theorem 120-1 to check this independence in any particular case. Furthermore, as in the usual applications of Equation 121-1, we will assume that our given function $\mathbf{f}$ is continuous in R.

Example 121-1.   Find a simpler formula for $F(\mathbf{x})$ if

$$(121\text{-}3) \qquad F(\mathbf{x}) = \int_{\mathbf{0}}^{\mathbf{x}} 15u^2v^2 \, du + 10u^3v \, dv.$$

Solution. We are dealing with a line integral of a function **f**, where

$$\mathbf{f(x)} = 15x^2y^2\mathbf{i} + 10x^3y\mathbf{j}.$$

Since

$$\frac{\partial}{\partial y}(15x^2y^2) = \frac{\partial}{\partial x}(10x^3y) = 30x^2y,$$

integrals of **f** are independent of path and Equation 121-3 does indeed define a function. In order to find a simpler formula for $F(\mathbf{x})$, we replace (a, b) with $(0, 0)$, (c, d) with (x, y) and (x, y) with (u, v) in the first of Equations 120-7:

$$F(\mathbf{x}) = \int_0^x 15u^2 \cdot 0 \, du + \int_0^y 10x^3v \, dv = 5x^3v^2\Big|_0^y = 5x^3y^2.$$

The most important attribute of the function defined by Equation 121-1 is that its derivative is easy to calculate:

$$F'(x) = f(x).$$

The function defined by Equation 121-2 has an analogous feature.

Theorem 121-1. *If*

(121-4) $$F(x, y) = \int_{(a,b)}^{(x,y)} f(u, v) \, du + g(u, v) \, dv,$$

then

$$F_1(x, y) = f(x, y) \qquad and \qquad F_2(x, y) = g(x, y).$$

PROOF

By definition, $F_1(x, y) = \dfrac{\partial}{\partial x} F(x, y)$, and in order to calculate this derivative, we hold y "fixed" and differentiate $F(x, y)$ with respect to x. Before we perform this operation, however, we will rewrite $F(x, y)$. In calculating this number, we must integrate from the point (a, b) along an arbitrary path to the point (x, y). Let us choose, as shown in Fig. 121-3, an intermediate point (c, y) with the same second coordinate as (x, y) and such that the horizontal line joining it to (x, y) lies entirely in our basic region R. Then

$$F(x, y) = \int_{(a,b)}^{(c,y)} f(u, v) \, du + g(u, v) \, dv + \int_{(c,y)}^{(x,y)} f(u, v) \, du + g(u, v) \, dv$$

$$= \int_{(a,b)}^{(c,y)} f(u, v) \, du + g(u, v) \, dv + \int_c^x f(u, y) \, du.$$

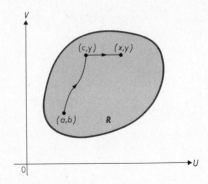

Figure 121-3

(We used Equation 120-7, with a slight change of lettering, to get the last integral.) Since $\int_{(a,b)}^{(c,y)} f(u, v)\, du + g(u, v)\, dv$ is independent of x, its partial derivative with respect to x is 0, and so

$$\frac{\partial}{\partial x} F(x,\, y) = \frac{\partial}{\partial x} \int_{c}^{x} f(u,\, y)\, du.$$

But now we use Theorem 43-2 (remember that y is fixed here) to see that this last derivative is simply $f(x, y)$. The proof that $F_2(x, y) = g(x, y)$ is similar.

Example 121-2. Illustrate Theorem 121-1 with the function of Example 121-1.

Solution. In Example 121-1 we generated $5x^3y^2$ by integrating the vector $15x^2y^2\mathbf{i} + 10x^3y\mathbf{j}$. It should not be surprising, then, that

$$\frac{\partial}{\partial x} 5x^3y^2 = 15x^2y^2 \quad \text{and} \quad \frac{\partial}{\partial y} 5x^3y^2 = 10x^3y,$$

which is what Theorem 121-1 says should be true.

In exactly the same way that Theorem 43-2 can be regarded as the "converse" of the Fundamental Theorem of Calculus (Theorem 37-1), Theorem 121-1 can be interpreted as the "converse" of the Fundamental Theorem of Calculus for line integrals, Equation 119-6. That equation says that *if* a vector is a gradient, then integrals of it are independent of path. Now we are saying that if integrals of $\mathbf{f}$ are independent of path, then there is a function F such that

$$\nabla F(\mathbf{x}) = \mathbf{f}(\mathbf{x}).$$

Thus if $\mathbf{f}(x, y) = f(x, y)\mathbf{i} + g(x, y)\mathbf{j}$ we let $F(x, y)$ be defined by Equation 121-4 and we see that

$$\nabla F(x, y) = F_1(x, y)\mathbf{i} + F_2(x, y)\mathbf{j} \qquad \text{(definition of } \nabla F(x, y))$$
$$= f(x, y)\mathbf{i} + g(x, y)\mathbf{j} \qquad \text{(Theorem 121-1)}.$$

In physics, we call a force **conservative** if the work that is required to move from one point to another in the force field does not depend on the path traversed (the force of gravity is a simple example). In other words, according to the preceding remarks, a conservative force vector is the gradient of a scalar-valued function. Such a function is called a **potential function**, and we can use a line integral to find one.

Example 121-3.   Suppose that $\mathbf{f}(x, y) = (xy \cos xy + \sin xy)\mathbf{i} + x^2 \cos xy\mathbf{j}$ is a force vector. Show that it is conservative and find a potential function that generates it.

Solution.   To show that $\mathbf{f}$ is conservative, we need to verify that its integrals are independent of path, and Theorem 120-1 tells us that they will be if

$$\frac{\partial}{\partial y}(xy \cos xy + \sin xy) = \frac{\partial}{\partial x}(x^2 \cos xy).$$

After performing the indicated differentiation, we find that both derivatives are $2x \cos xy - x^2 y \sin xy$, and so our force *is* conservative. It will thus be the gradient of the potential function defined by the line integral

$$F(x, y) = \int_{(0,0)}^{(x,y)} (uv \cos uv + \sin uv)\, du + u^2 \cos uv\, dv.$$

As usual, we use Equation 120-7 to evaluate this integral:

$$F(x, y) = \int_0^x 0\, du + \int_0^y x^2 \cos xv\, dv = x \sin xv \Big|_0^y = x \sin xy.$$

Example 121-4.   At each point (x, y) of a certain curve $F(x, y) = 0$, the vector $(2xye^{x^2y} - 2y)\mathbf{i} + (x^2 e^{x^2y} - 2x)\mathbf{j}$ is normal to the curve. Furthermore, the curve contains the point $(-1, 2)$. What is a possible formula for $F(x, y)$?

Solution.   Recall that in Section 109 we showed that the vector $\nabla F(x, y)$ is normal to the level surface $F(x, y) = 0$. Therefore our problem will be solved if we can find a function of which our given vector is the gradient. According to Theorem 121-1, such a function will be defined by the line integral

$$F(x, y) = \int_{(-1,2)}^{(x,y)} (2uve^{u^2v} - 2v)\, du + (u^2 e^{u^2v} - 2u)\, dv.$$

We chose the lower limit so that the point $(-1, 2)$ belongs to our curve $F(x, y) = 0$. We find a more explicit formula for $F(x, y)$ by using Equation 120-7 to evaluate this integral:

$$F(x, y) = \int_{-1}^{x} (4ue^{2u^2} - 4)\, du + \int_{2}^{y} (x^2 e^{x^2 v} - 2x)\, dv$$

$$= (e^{2u^2} - 4u)\Big|_{-1}^{x} + (e^{x^2 v} - 2xv)\Big|_{2}^{y}$$

$$= e^{2x^2} - 4x - e^2 - 4 + e^{x^2 y} - 2xy - e^{2x^2} + 4x$$

$$= e^{x^2 y} - 2xy - e^2 - 4.$$

PROBLEMS   121

1. Express $F(x, y)$ without using the integral sign.

(a) $F(x, y) = \displaystyle\int_{(1,0)}^{(x,y)} (1 + uv)e^{uv}\, du + u^2 e^{uv}\, dv$

(b) $F(x, y) = \displaystyle\int_{(1,1)}^{(x,y)} \ln v\, du + \frac{u}{v}\, dv$

(c) $F(x, y) = \displaystyle\int_{(1,1)}^{(x,y)} (3v^2 + 2)\, du + 6uv\, dv$

(d) $F(x, y) = \displaystyle\int_{(2,\pi)}^{(x,y)} -2uv^2 \sin u^2 v\, du + (\cos u^2 v - u^2 v \sin u^2 v)\, dv$

2. Solve for w if the equation has a solution.
(a) $\nabla w = 8xy^3 \mathbf{i} + 12x^2 y^2 \mathbf{j}$ (b) $\nabla w = 8x^3 y \mathbf{i} + 12x^2 y^2 \mathbf{j}$
(c) $\nabla w = \sin (x^2 + y)(2x\mathbf{i} + \mathbf{j})$ (d) $\nabla w = \sin 2xy(y\mathbf{i} + x\mathbf{j})$.

3. Find a potential function that generates the given force vector *if the force is conservative.*
(a) $\mathbf{f}(x, y) = e^{xy^2}[(1 + xy^2)\mathbf{i} + 2x^2 y\mathbf{j}]$
(b) $\mathbf{f}(x, y) = e^{xy^2}[y^3 \mathbf{i} + (1 + 2xy^2)\mathbf{j}]$
(c) $\mathbf{f}(x, y) = 2(x^2 + y^2)^{-1}(x\mathbf{i} + y\mathbf{j})$
(d) $\mathbf{f}(x, y) = 2(x^2 + y^2)^{-1}(y\mathbf{i} + x\mathbf{j})$

4. At a point (x, y) of the curve $F(x, y) = 0$, the given vector is normal to the curve. Furthermore, the curve contains the given point. Find a possible formula for $F(x, y)$.
(a) $y^2 \mathbf{i} + 2xy\mathbf{j}$, $(-1, 2)$ (b) $2xy\mathbf{i} + x^2 \mathbf{j}$, $(-1, 2)$
(c) $\sin xy\ (y\mathbf{i} + x\mathbf{j})$, $(\sqrt{\pi}, \sqrt{\pi})$ (d) $\cos xy\ (y\mathbf{i} + x\mathbf{j})$, $(\sqrt{\pi}, \sqrt{\pi})$

5. Suppose that we define

$$F(x, y) = \int_{P(x,y)} 8u^2 v\, du + 3uv\, dv,$$

where $P(x, y)$ is the line segment from $(0, 0)$ to (x, y). Find a simpler formula for $F(x, y)$ and calculate $F_1(x, y)$ and $F_2(x, y)$.

6. If $F(x, y) = \displaystyle\int_{(0,0)}^{(x^2,x+y)} 4u^3v^2 \, du + 2u^4v \, dv$, find $F_1(-1, 2)$ and $F_2(3, 0)$.

7. If $F(x, y) = \displaystyle\int_{(0,0)}^{(x,y)} ve^{u^2v^2} \, du + ue^{u^2v^2} \, dv$, show that $F(y, x) = F(x, y)$. Can you find a simpler expression for $F(x, y)$?

8. At what point (x, y) is $\displaystyle\int_{(3,4)}^{(x,y)} (5u + 7v - 9) \, du + (7u + 10v - 13) \, dv$ least?

9. Suppose that $\displaystyle\int_a^x \mathbf{f}(\mathbf{u}) \cdot d\mathbf{u} \leq \int_a^b \mathbf{f}(\mathbf{u}) \cdot d\mathbf{u}$ for every point $\mathbf{x}$ in some neighborhood of $\mathbf{b}$. How do you know that $\mathbf{f}(\mathbf{b}) = \mathbf{0}$?

10. (a) If $F(\mathbf{x}) = \displaystyle\int_a^x \mathbf{f}(\mathbf{u}) \cdot d\mathbf{u}$ and $G(\mathbf{x}) = \int_x^a \mathbf{f}(\mathbf{u}) \cdot d\mathbf{u}$, how are $F(\mathbf{x})$ and $G(\mathbf{x})$ related?

    (b) If $H(\mathbf{x}) = \displaystyle\int_b^x \mathbf{f}(\mathbf{u}) \cdot d\mathbf{u}$, how are $\nabla F(\mathbf{x})$ and $\nabla H(\mathbf{x})$ related?

    (c) If $K(\mathbf{x}) = \displaystyle\int_a^{-x} \mathbf{f}(\mathbf{u}) \cdot d\mathbf{u}$, how are $\nabla F(\mathbf{x})$ and $\nabla K(\mathbf{x})$ related?

122. LINE INTEGRALS AND DIFFERENTIAL EQUATIONS

In Chapter 7 we stated that a differential equation of the first order has the general form

(122-1) $$y' = f(x, y).$$

Whole courses are devoted to such equations, and we are not about to cover their entire theory in a few sections of a calculus book. Yet, one aspect of the study of differential equations is so intimately tied up with line integrals that this is really the best place to discuss it.

A solution of a differential equation is a function; and whenever we have a function, we naturally think of its graph. We will find it convenient to view solution curves of the Differential Equation 122-1 as being embedded in level curves of a function on R^2.

Example 122-1. Verify that $y = \sqrt{c - x^2}$ satisfies the differential equation

(122-2) $$y' = -\frac{x}{y}$$

for an arbitrary number c, and show that its graph is part of a level curve of a function on R^2.

Solution. We needn't remember that we learned how to solve such *separable* equations in Section 60. We are given an alleged solution, and we merely have to verify that it is one. We therefore replace y with $\sqrt{c - x^2}$ in Equation 122-2 and observe that equality continues to hold:

$$\frac{d}{dx} \sqrt{c - x^2} = \frac{-2x}{2\sqrt{c - x^2}} = \frac{-x}{\sqrt{c - x^2}}.$$

The graph of this solution function is part of many different level curves, but the one that comes to mind first is

$$x^2 + y^2 = c.$$

We can use line integrals to find functions whose level curves contain the graphs of solutions of Equation 122-1. Suppose that

$$(122\text{-}3) \qquad F(x, y) = \int_{(a,b)}^{(x,y)} p(u, v)\, du + q(u, v)\, dv,$$

where our line integral does not depend on the path from the point (a, b) to the point (x, y). According to Equation 107-9, the slope of the level curve $F(x, y) = c$ is given by the equation $y' = -F_1(x, y)/F_2(x, y)$. Theorem 121-1 tells us that

$$F_1(x, y) = p(x, y) \quad \text{and} \quad F_2(x, y) = q(x, y),$$

so $y' = -p(x, y)/q(x, y)$. Therefore, if $p(x, y)$ and $q(x, y)$ are chosen so that $-p(x, y)/q(x, y) = f(x, y)$, then arcs of level curves of the function defined by Equation 122-3 are solution curves of Differential Equation 122-1. Let us put these remarks together as a theorem.

Theorem 122-1. *If*

$$(122\text{-}4) \quad f(x, y) = -\frac{p(x, y)}{q(x, y)} \qquad \text{where} \quad p_2(x, y) = -q_1(x, y),$$

then arcs of the level curves

$$\int_{(a,b)}^{(x,y)} p(u, v)\, du + q(u, v)\, dv = c$$

are solution curves of the differential equation $y' = f(x, y)$. Here (a, b) is any conveniently chosen point. If we want the solution curve that contains a given initial point (x_0, y_0), we can replace (a, b) with (x_0, y_0) and c with 0. (The second of Conditions 122-4, of course, simply guarantees that the line integral is independent of path.)

Example 122-2. Use Theorem 122-1 to solve Equation 122-2.

Solution. Here we can take $p(x, y) = x$ and $q(x, y) = y$. Then

$$p_2(x, y) = \frac{\partial}{\partial y} x = 0 \quad \text{and} \quad q_1(x, y) = \frac{\partial}{\partial x} y = 0.$$

Thus solution curves are arcs of the level curves

$$\int_{(0,0)}^{(x,y)} u \, du + v \, dv = c \quad \text{or} \quad x^2 + y^2 = 2c \qquad \text{(Equation 120-7)}.$$

Example 122-3. Solve the differential equation $6x + y^3 + 3xy^2 y' = 0$.

Solution. We can write our given differential equation as

$$y' = -\frac{6x + y^3}{3xy^2},$$

which suggests that we try $p(x, y) = 6x + y^3$ and $q(x, y) = 3xy^2$. Since

$$p_2(x, y) = \frac{\partial}{\partial y} (6x + y^3) = 3y^2 \quad \text{and} \quad q_1(x, y) = \frac{\partial}{\partial x} (3xy^2) = 3y^2,$$

we see that the hypotheses of Theorem 122-1 are fulfilled and solution curves are therefore arcs of the level curves

$$\int_{(0,0)}^{(x,y)} (6u + v^3) \, du + 3uv^2 \, dv = c.$$

We use Equation 120-7 to evaluate this line integral:

$$c = \int_0^x 6u \, du + \int_0^y 3xv^2 \, dv = 3u^2 \Big|_0^x + xv^3 \Big|_0^y = 3x^2 + xy^3.$$

Hence $y = \left(\dfrac{c - 3x^2}{x} \right)^{1/3}$, where c is any number, satisfies our given differential equation.

We can write a given $f(x, y)$ in the form $-p(x, y)/q(x, y)$ for infinitely many choices of $p(x, y)$ and $q(x, y)$. When we solved Equation 122-2, we naturally chose $p(x, y) = x$ and $q(x, y) = y$, but, of course, we could have written $p(x, y) = y^{-1}$ and $q(x, y) = x^{-1}$, and so on. It is one of the basic theorems of differential equations that there is *always* a choice for which the second of Equations 122-4 holds, and, consequently, we can use Theorem 122-1 to solve Equation 122-1. This statement is important theoretically, but there is no practical way of finding a "proper" $p(x, y)$ and $q(x, y)$ in any given case. Experienced differential equation solvers are more apt to hit on a correct choice than beginners are.

Example 122-4. Solve the initial value problem

$$y' = \frac{3x - 2y}{x} \quad \text{and} \quad y = 3 \quad \text{when } x = -1.$$

Solution. We are to choose $p(x, y)$ and $q(x, y)$ so that

$$-\frac{p(x, y)}{q(x, y)} = \frac{3x - 2y}{x},$$

and the natural choice seems to be

$$p(x, y) = 2y - 3x \quad \text{and} \quad q(x, y) = x.$$

But then $p_2(x, y) = 2$ and $q_1(x, y) = 1$, so the second of Conditions 122-4 is violated. It turns out that we should choose

$$p(x, y) = 2xy - 3x^2 \quad \text{and} \quad q(x, y) = x^2.$$

We still get the same quotient, but now $p_2(x, y) = q_1(x, y) = 2x$, and we can apply Theorem 122-1. It says that the solution of our initial value problem is obtained by solving the equation

$$\int_{(-1,3)}^{(x,y)} (2uv - 3u^2)\, du + u^2\, dv = 0$$

for y. According to Equation 120-7, this line integral is

$$\int_{-1}^{x} (6u - 3u^2)\, du + \int_{3}^{y} x^2\, dv = 3u^2 - u^3 \Big|_{-1}^{x} + x^2 v \Big|_{3}^{y}$$

$$= 3x^2 - x^3 - 3 - 1 + x^2 y - 3x^2$$

$$= x^2 y - x^3 - 4.$$

Therefore, our initial value problem is satisfied by $y = (x^3 + 4)/x^2$.

In Section 60 we said that a differential equation is "separated" when it is written in the form

$$y' = \frac{f(x)}{g(y)};$$

all the x's are in the numerator and all the y's are in the denominator. Theorem 60-2 tells us how to solve such equations, and we can use the techniques of this section to give another proof of this theorem. We simply write $p(x, y) = f(x)$

and $q(x, y) = -g(y)$, and so $p_2(x, y) = q_1(x, y) = 0$. Thus solution curves that contain the initial point (x_0, y_0) are arcs of the curve

$$\int_{(x_0,y_0)}^{(x,y)} f(u) \, du - g(v) \, dv = 0.$$

According to Equation 120-7, this equation can be written as

(122-5) $$\int_{y_0}^{y} g(v) \, dv = \int_{x_0}^{x} f(u) \, du,$$

which is just Equation 60-9.

Example 122-5. Solve the initial value problem $y' = e^{x+y}$ and $y = 3$ when $x = 0$.

Solution. Here we can take $f(x) = e^x$ and $g(y) = e^{-y}$, so we must solve the equation

$$\int_{3}^{y} e^{-v} \, dv = \int_{0}^{x} e^{u} \, du \quad \text{or} \quad -e^{-v} \Big|_{3}^{y} = e^{u} \Big|_{0}^{x}$$

for y. This last equation simplifies to $-e^{-y} + e^{-3} = e^x - 1$, and when we solve for y, we find that $y = -\ln (e^{-3} + 1 - e^{x})$.

PROBLEMS 122

1. Use Equation 122-5 to solve the initial value problem.
   (a) $y' = x/y$ and $y = 2$ when $x = 0$
   (b) $y' = (1 + y)/(1 + x)$ and $y = 2$ when $x = 0$
   (c) $y' = (1 + y^2)/(1 + x^2)$ and $y = 2$ when $x = 0$
   (d) $y' = (1 + y)^2/(1 + x)^2$ and $y = 2$ when $x = 0$
   (e) $y' = \cos x \csc y$ and $y = \frac{1}{2}\pi$ when $x = 0$
   (f) $y' = 2x \cos^2 y$ and $y = \pi$ when $x = 0$

2. Use Theorem 122-1 to solve the initial value problem.
   (a) $2xy^3 + 3x^2y^2y' = 0$ and $y = 2$ when $x = -1$
   (b) $3x^2y^2 + 2x^3yy' = 0$ and $y = -1$ when $x = 2$
   (c) $e^x y + 1 + e^x y' = 0$ and $y = 0$ when $x = 1$
   (d) $2x + e^y + xe^y y' = 0$ and $y = 0$ when $x = 1$
   (e) $\text{Tan}^{-1} y - 3x^2 + \dfrac{x}{1 + y^2} y' = 0$ and $y = 1$ when $x = \frac{1}{2}\sqrt{\pi}$
   (f) $\ln y - \cos x + \dfrac{x}{y} y' = 0$ and $y = 1$ when $x = \pi$

3. Solve the initial value problem $y' = 6x^2 e^{4x^3 - y^2}/y$ and $y = 2$ when $x = 1$.

4. Solve the initial value problem $y' = 3x^2 \sec y$ and $y = \pi$ when $x = 0$.

5. Find the formula for $w(x)$ if $w'(x) = \exp [x + w(x)]$ and $w(\ln 3) = \ln \frac{1}{2}$.

6. Solve the differential equation $y' = y/x$ by using Theorem 122-1 with the choices $p(x) = -y/(x^2 + y^2)$ and $q(x, y) = x/(x^2 + y^2)$. Now use Equation 122-5.

7. At each point of a certain curve in the first quadrant, the slope is the geometric mean of the coordinates. If the curve contains the point (1, 4), what is its equation?

8. At each point (x, y) of an arc connecting the points (1, 3) and (2, 12), the slope is directly proportional to y and inversely proportional to x. Explain why this statement implies that there is a number k such that $y' = ky/x$. Now find the equation of the arc (you will have to determine k in the process).

9. If $y' = -(2x + y)/(x + 2y)$ and $y = 1$ when $x = 1$, what is y when $x = 0$?

10. Suppose that we start with a colony of 1000 bacteria on an agar plate and that t minutes later there are y organisms. It seems reasonable to assume that the "birth rate" of the bacteria is proportional to the number present, so there is a number r such that our colony tends to increase at the rate of ry bacteria per minute. On the other hand, members of the colony are constantly dying, and the death rate increases rapidly with the number of bacteria present (overcrowding). Let us suppose that the death rate is proportional to the square of the population. Hence there is a number s such that bacteria are dying at the rate of sy^2 per minute. Thus the net increase in population is expressed by the differential equation $y' = ry - sy^2$. Show that

$$y = \frac{1000r}{1000s + (r - 1000s)e^{-rt}} .$$

How would you go about finding r and s experimentally? What is the limiting population of our agar plate?

11. In Section 107 we met the Cauchy-Riemann partial differential equations $f_1(x, y) = g_2(x, y)$ and $f_2(x, y) = -g_1(x, y)$. If $f(x, y)$ and $g(x, y)$ satisfy these equations, show that

$$F(x, y) = \int_{(a,b)}^{(x,y)} f(u, v) \, du - g(u, v) \, dv$$

and

$$G(x, y) = \int_{(c,d)}^{(x,y)} g(u, v) \, du + f(u, v) \, dv$$

also satisfy them.

12. Suppose that $w = f(x, y)$ satisfies Laplace's partial differential equation $w_{xx} + w_{yy} = 0$. Let

$$g(x, y) = \int_{(a,b)}^{(x,y)} f_{-2}(u, v) \, du + f_1(u, v) \, dv$$

and show that $f(x, y)$ and $g(x, y)$ satisfy the Cauchy-Riemann equations.

As with differentiation, you will observe that our treatment of integration of functions on R^n is simply an extension of what you already know about integration of functions on R^1. In fact, when it comes to evaluating integrals, we reduced the problem to that of calculating integrals of functions on R^1. This technique is fine for those integrals that can be evaluated "by hand" (that is, by looking up relevant antiderivatives in a table), but you should understand that a multiple integral is a number, a limit of sums, and it can be evaluated by numerical methods, too. We have omitted a discussion of such methods (extensions of Simpson's Parabolic Rule, and so on), for a computer is necessary to apply them.

Furthermore, we have omitted a discussion of integrals over surfaces (just as line integrals are integrals along curves). Such integrals, together with the ones we have studied, play a large role in various sciences. So this chapter was simply an introduction to what higher-level integrals are and to some techniques for working with them. You should know that a multiple integral is a limit of sums and that it can be evaluated by iterated integration. Even though you will forget the exact formulas for integration in polar coordinates, you now know where to look them up. Finally, you have been introduced to line integrals and have seen some indications of their many uses.

Review Problems, Chapter Fourteen

1. Evaluate the given iterated integral.

   (a) $\displaystyle\int_{1}^{2}\int_{0}^{\ln x} e^{y}\, dy\, dx$

   (b) $\displaystyle\int_{1}^{2}\int_{0}^{\ln x} e^{x+y}\, dy\, dx$

   (c) $\displaystyle\int_{-1}^{2}\int_{0}^{e^{x}} 2ye^{x}\, dy\, dx$

   (d) $\displaystyle\int_{-1}^{2}\int_{0}^{e^{x}} e^{x+y}\, dy\, dx$

   (e) $\displaystyle\int_{-1}^{2}\int_{0}^{|x|} 2e^{x^{2}}\, dy\, dx$

   (f) $\displaystyle\int_{-1}^{2}\int_{0}^{|x|} 2e^{x+y}\, dy\, dx$

2. Let

$$I_{1} = \iint_{R_{1}} \exp(x^{2} + y^{2})\, dA, \qquad I_{2} = \iint_{R_{2}} \exp(x^{2} + y^{2})\, dA,$$

   and

$$I_{3} = \iint_{R_{3}} \exp(x^{2} + y^{2})\, dA,$$

   where $R_{1} = \{(x, y) \mid x^{2} + y^{2} \le 1\}$, $R_{2} = \{(x, y) \mid x^{4} + y^{4} \le 1\}$, and $R_{3} = \{(x, y) \mid |x| + |y| \le 1\}$. Rank the numbers I_{1}, I_{2}, and I_{3} according to size.

3. The lines $\theta = \frac{1}{3}\pi$, $\theta = \frac{2}{3}\pi$, and the circle $r = 1$ partition the polar region $R = \{(r, \theta) \mid 0 \le r \le 2, 0 \le \theta \le \pi\}$ into six subregions. Choose as (r^{*}, θ^{*}) the midpoint of the "center line" of each subregion and form a sum that approximates $\displaystyle\iint_{R} r \sin \theta\, dA$. What is the actual value of this integral?

4. Without integrating, solve the equation $\displaystyle\int_{\pi/2}^{3\pi/2}\int_{\cos x}^{0} (x - h)\, dy\, dx = 0$ for h.

5. Write the integral $\iint\limits_{R} (|x| + |y|)\, dA$, where R is the disk $x^2 + y^2 \leq 1$, as four times an integral in which the absolute value signs are missing, and evaluate it.

6. Express the given iterated integral as an integral in which we first integrate with respect to x.

    (a) $\displaystyle\int_{0}^{\pi/2} \int_{0}^{\sin x} f(x, y)\, dy\, dx$ (b) $\displaystyle\int_{0}^{\pi/2} \int_{0}^{\cos x} f(x, y)\, dy\, dx$

    (c) $\displaystyle\int_{0}^{\pi} \int_{0}^{\sin x} f(x, y)\, dy\, dx$ (d) $\displaystyle\int_{0}^{\pi} \int_{-1}^{\cos x} f(x, y)\, dy\, dx$

7. Find in two ways the moment about the Y-axis of the region that is bounded by the curve $y = \sin x$ and the interval $[0, 2\pi]$.

8. Evaluate the iterated integral $\displaystyle\int_{0}^{\pi/4} \int_{0}^{4\sec\theta} r\, dr\, d\theta$ geometrically by interpreting it as the area of a certain plane region.

9. Is the number $\displaystyle\int_{\pi/4}^{5\pi/4} \int_{\sin x}^{\cos x} \exp(xy)\, dy\, dx$ positive or negative?

10. (a) Explain why

    $$\int_{0}^{\pi} \int_{0}^{\sin x} f(x, y)\, dy\, dx = \int_{0}^{\pi} \int_{0}^{x} f(x, \sin y) \cos y\, dy\, dx.$$

    (b) Verify this equation in case $f(x, y) = x + y$.

11. If we rotate one arch of the sine curve $y = \sin x$ about the X-axis, we obtain a surface that bounds a spindle-shaped region. What is the moment of inertia of this region about the axis of rotation?

12. Show that the volume V of the region under the plane $z = ax + by + c$ and above a plane region R is given by the equation $V = aM_Y + bM_X + cA$, where M_X and M_Y are the moments of R about the X- and Y-axes and A is the area of R.

13. (a) What is $\dfrac{d^2}{dx^2} \displaystyle\int_{0}^{x} \int_{0}^{u} f(v)\, dv\, du$?

    (b) By changing the order of integration, show that

    $$\int_{0}^{x} \int_{0}^{u} f(v)\, dv\, du = \int_{0}^{x} (x - v)f(v)\, dv = \int_{0}^{x} vf(x - v)\, dv.$$

14. Evaluate the integral $\displaystyle\int_{0}^{\pi/2} \int_{0}^{\pi/2} |\sin(x - y)|\, dx\, dy$.

15. Let R be the diamond-shaped region $|x| + |y| \leq 1$ and use geometric reasoning to evaluate the given integral.

    (a) $\displaystyle\iint\limits_{R} [\![x]\!]\, dA$ (b) $\displaystyle\iint\limits_{R} [\![x + y]\!]\, dA$

    (c) $\displaystyle\iint\limits_{R} ([\![x]\!] + [\![y]\!])\, dA$ (d) $\displaystyle\iint\limits_{R} [\![x]\!][\![y]\!]\, dA$

16. Show that

$$\int_0^2 \int_0^{[x+1]} f(x, y) \, dy \, dx = \int_0^2 \int_{[y]}^2 f(x, y) \, dx \, dy$$

$$= \int_0^1 \int_0^1 f(x, y) \, dy \, dx + \int_1^2 \int_0^2 f(x, y) \, dy \, dx.$$

17. Evaluate the integral $\displaystyle\int_0^1 \int_0^1 \int_0^{|x-y|} \sin(x - y + z) \, dz \, dy \, dx.$

18. Evaluate the integral $\displaystyle\iiint_R \sqrt{x^2 + y^2} \, dV$, where R is the cube

$$\{(x, y, z) \mid 0 \le x \le a, 0 \le y \le a, 0 \le z \le a\}.$$

19. The iterated integral $\displaystyle\int_0^a \int_0^x \int_0^y f(z) \, dz \, dy \, dx$ equals a triple integral $\displaystyle\iiint_R f(z) \, dV$

over a region R in three-dimensional space.
(a) Sketch R.
(b) By changing the order of integration, show that our integral equals

$$\tfrac{1}{2} \int_0^a (a - z)^2 f(z) \, dz.$$

20. Use a geometric argument to show that for any positive number R,

$$\int_0^{R/\sqrt{2}} \int_0^{R/\sqrt{2}} e^{-x^2-y^2} \, dx \, dy < \int_0^{\pi/2} \int_0^R e^{-r^2} r \, dr \, d\theta$$

$$< \int_0^R \int_0^R e^{-x^2-y^2} \, dx \, dy.$$

From these inequalities and the equation $\displaystyle\int_0^a \int_0^a e^{-x^2-y^2} \, dx \, dy = \left(\int_0^a e^{-u^2} \, du\right)^2$

that we developed in Problem 112-9, show that

$$\int_0^{R/\sqrt{2}} e^{-u^2} \, du < \tfrac{1}{2}\sqrt{\pi}\sqrt{1 - e^{-R^2}} < \int_0^R e^{-u^2} \, du.$$

Now deduce that $\displaystyle\int_0^\infty e^{-u^2} \, du = \tfrac{1}{2}\sqrt{\pi}.$

21. Line integrals in three-dimensional space are defined just as they are in the plane. So use your knowledge of line integrals in two dimensions to evaluate the line integral $\displaystyle\int_P 2xy^3z \, dx + 3x^2y^2z \, dy + x^2y^3 \, dz$, where P is the direct linear path from the origin to the point $(1, 2, 3)$.

22. (a) How should $f(x, y)$ and $g(x, y)$ be related so that the line integral $\displaystyle\oint_P f(x, y) \, dx +$
    $g(x, y) \, dy$ around a closed path P gives the area of the region bounded by P?
    (b) For example, if $f(x, y) = x^2 + y^2$, what is a suitable $g(x, y)$?

23. Use the area formula $A = \frac{1}{2} \oint_P x\, dy - y\, dx$ to compute the area of the region $\{(r, \theta) \mid \alpha \le \theta \le \beta, 0 \le r \le f(\theta)\}$ and hence obtain the formula $A = \frac{1}{2} \int_\alpha^\beta f(\theta)^2 \, d\theta.$

24. (a) Show that $\int_a^b (px + qy)\, dx + (rx + sy)\, dy$ is independent of path if the matrix

$M = \begin{bmatrix} p & q \\ r & s \end{bmatrix}$ is symmetric.

(b) What is $\nabla(\mathbf{x} \cdot M\mathbf{x})$, where $\mathbf{x} = (x, y)$?

(c) Therefore conclude that the integral of part (a) equals $\frac{1}{2}\mathbf{b} \cdot M\mathbf{b} - \frac{1}{2}\mathbf{a} \cdot M\mathbf{a}$.

25. (a) As it produces more and more cars, the Fraud Motor Company has to reduce their price in order to sell them. Thus suppose that the xth car wholesales at $3000 - 10x$ dollars. (For example, the first car sells for \$2,990, but the tenth costs only \$2,900.) Can you convince yourself that if the company sells 100 cars, it takes in about $\int_0^{100} (3000 - 10x)\, dx$ dollars?

(b) Criminal Motors has a more complicated problem. It manufactures two kinds of cars, Shovrolets and Sadillacs. The latter are more expensive, so three Shovies are built for each Saddy that comes off the production line. The selling prices of the xth Shovy and yth Saddy are $3000 - 10x - 5y - 2xy$ and $5000 - 10x - 50y - 5xy$ dollars. Can you convince yourself that if 180 Shovies are sold, CM takes in about

$$\int_P (3000 - 10x - 5y - 2xy)\, dx + (5000 - 10x - 50y - 5xy)\, dy$$

dollars, where P is the arc of the curve $3y = x$ that joints the points $(0, 0)$ and $(180, 60)$?

Appendix

A. TRIGONOMETRIC FORMULAS

The following trigonometric identities are valid for any numbers t, u, and v for which the functions involved are defined.

(A-1)
$$\cos^2 t + \sin^2 t = 1$$
$$\tan^2 t + 1 = \sec^2 t$$
$$\cot^2 t + 1 = \csc^2 t$$

(A-2)
$$\sin(u \pm v) = \sin u \cos v \pm \cos u \sin v$$
$$\cos(u \pm v) = \cos u \cos v \mp \sin u \sin v$$
$$\tan(u \pm v) = \frac{\tan u \pm \tan v}{1 \mp \tan u \tan v}$$

(A-3)
$$\sin(-t) = -\sin t$$
$$\cos(-t) = \cos t$$
$$\tan(-t) = -\tan t$$

$$\sin 2t = 2 \sin t \cos t$$

(A-4) $$\cos 2t = \cos^2 t - \sin^2 t = 1 - 2 \sin^2 t = 2 \cos^2 t - 1$$

$$\tan 2t = \frac{2 \tan t}{1 - \tan^2 t}$$

(A-5)
$$\sin^2 t = \tfrac{1}{2}(1 - \cos 2t)$$
$$\cos^2 t = \tfrac{1}{2}(1 + \cos 2t)$$

The length s of the arc of a circle of radius r that is intercepted by a central angle that is t radians wide is given by the formula

$$s = rt.$$

The area of the sector that this angle cuts from the interior of this circle is given by the formula

$$K = \tfrac{1}{2}rs = \tfrac{1}{2}r^2t.$$

If an angle is A degrees and t radians wide, then

$$\frac{A}{180} = \frac{t}{\pi}.$$

The following brief table of values of the trigonometric functions is often useful; you should know it by heart.

t	$\cos t$	$\sin t$	$\tan t$
0	1	0	0
$\frac{1}{6}\pi$	$\frac{1}{2}\sqrt{3}$	$\frac{1}{2}$	$\frac{1}{3}\sqrt{3}$
$\frac{1}{4}\pi$	$\frac{1}{2}\sqrt{2}$	$\frac{1}{2}\sqrt{2}$	1
$\frac{1}{3}\pi$	$\frac{1}{2}$	$\frac{1}{2}\sqrt{3}$	$\sqrt{3}$
$\frac{1}{2}\pi$	0	1	—
$\frac{2}{3}\pi$	$-\frac{1}{2}$	$\frac{1}{2}\sqrt{3}$	$-\sqrt{3}$
$\frac{3}{4}\pi$	$-\frac{1}{2}\sqrt{2}$	$\frac{1}{2}\sqrt{2}$	-1
$\frac{5}{6}\pi$	$-\frac{1}{2}\sqrt{3}$	$\frac{1}{2}$	$-\frac{1}{3}\sqrt{3}$
π	-1	0	0

B. TABLES

TABLE
I

Trigonometric Functions

t	sin t	cos t	tan t	cot t	sec t	csc t
0.0	.0000	1.0000	.0000	—	1.000	—
0.1	.0998	.9950	.1003	9.967	1.005	10.02
0.2	.1987	.9801	.2027	4.933	1.020	5.033
0.3	.2955	.9553	.3093	3.233	1.047	3.384
0.4	.3894	.9211	.4228	2.365	1.086	2.568
0.5	.4794	.8776	.5463	1.830	1.139	2.086
0.6	.5646	.8253	.6841	1.462	1.212	1.771
0.7	.6442	.7648	.8423	1.187	1.307	1.552
0.8	.7174	.6967	1.030	.9712	1.435	1.394
0.9	.7833	.6216	1.260	.7936	1.609	1.277
1.0	.8415	.5403	1.557	.6421	1.851	1.188
1.1	.8912	.4536	1.965	.5090	2.205	1.122
1.2	.9320	.3624	2.572	.3888	2.760	1.073
1.3	.9636	.2675	3.602	.2776	3.738	1.038
1.4	.9854	.1700	5.798	.1725	5.883	1.015
1.5	.9975	.0707	14.101	.0709	14.137	1.003
1.6	.9996	−.0292	−34.233	−.0292	−34.25	1.000

Natural Logarithms **TABLE II**

x	ln x	x	ln x	x	ln x	x	ln x
0.0	–	3.0	1.0986	6.0	1.7918	9.0	2.1972
0.1	− 2.303	3.1	1.1314	6.1	1.8083	9.1	2.2083
0.2	− 1.609	3.2	1.1632	6.2	1.8245	9.2	2.2192
0.3	− 1.204	3.3	1.1939	6.3	1.8406	9.3	2.2300
0.4	− .916	3.4	1.2238	6.4	1.8563	9.4	2.2407
0.5	− .693	3.5	1.2528	6.5	1.8718	9.5	2.2513
0.6	− .511	3.6	1.2809	6.6	1.8871	9.6	2.2618
0.7	− .357	3.7	1.3083	6.7	1.9021	9.7	2.2721
0.8	− .223	3.8	1.3350	6.8	1.9169	9.8	2.2824
0.9	− .105	3.9	1.3610	6.9	1.9315	9.9	2.2925
1.0	.0000	4.0	1.3863	7.0	1.9459	10.0	2.3026
1.1	.0953	4.1	1.4110	7.1	1.9601		
1.2	.1823	4.2	1.4351	7.2	1.9741		
1.3	.2624	4.3	1.4586	7.3	1.9879		
1.4	.3365	4.4	1.4816	7.4	2.0015		
1.5	.4055	4.5	1.5041	7.5	2.0149		
1.6	.4700	4.6	1.5261	7.6	2.0282		
1.7	.5306	4.7	1.5476	7.7	2.0412		
1.8	.5878	4.8	1.5686	7.8	2.0541		
1.9	.6419	4.9	1.5892	7.9	2.0669		
2.0	.6931	5.0	1.6094	8.0	2.0794		
2.1	.7419	5.1	1.6292	8.1	2.0919		
2.2	.7885	5.2	1.6487	8.2	2.1041		
2.3	.8329	5.3	1.6677	8.3	2.1163		
2.4	.8755	5.4	1.6864	8.4	2.1282		
2.5	.9163	5.5	1.7047	8.5	2.1401		
2.6	.9555	5.6	1.7228	8.6	2.1518		
2.7	.9933	5.7	1.7405	8.7	2.1633		
2.8	1.0296	5.8	1.7579	8.8	2.1748		
2.9	1.0647	5.9	1.7750	8.9	2.1861		

TABLE III *Exponential and Hyperbolic Functions*

x	e^x	e^{-x}	sinh x	cosh x	tanh x
0.0	1.0000	1.0000	.0000	1.0000	.0000
0.1	1.1052	.9048	.1002	1.0050	.0997
0.2	1.2214	.8187	.2013	1.0201	.1974
0.3	1.3499	.7408	.3045	1.0453	.2913
0.4	1.4918	.6703	.4108	1.0811	.3799
0.5	1.6487	.6065	.5211	1.1276	.4621
0.6	1.8221	.5488	.6367	1.1855	.5370
0.7	2.0138	.4966	.7586	1.2552	.6044
0.8	2.2255	.4493	.8881	1.3374	.6640
0.9	2.4596	.4066	1.0265	1.4331	.7163
1.0	2.7183	.3679	1.1752	1.5431	.7616
1.1	3.0042	.3329	1.3356	1.6685	.8005
1.2	3.3201	.3012	1.5095	1.8107	.8337
1.3	3.6693	.2725	1.6984	1.9709	.8617
1.4	4.0552	.2466	1.9043	2.1509	.8854
1.5	4.4817	.2231	2.1293	2.3524	.9051
1.6	4.9530	.2019	2.3756	2.5775	.9217
1.7	5.4739	.1827	2.6456	2.8283	.9354
1.8	6.0496	.1653	2.9422	3.1075	.9468
1.9	6.6859	.1496	3.2682	3.4177	.9562
2.0	7.3891	.1353	3.6269	3.7622	.9640
2.1	8.1662	.1225	4.0219	4.1443	.9705
2.2	9.0250	.1108	4.4571	4.5679	.9757
2.3	9.9742	.1003	4.9370	5.0372	.9801
2.4	11.023	.0907	5.4662	5.5569	.9837
2.5	12.182	.0821	6.0502	6.1323	.9866
2.6	13.464	.0743	6.6947	6.7690	.9890
2.7	14.880	.0672	7.4063	7.4735	.9910
2.8	16.445	.0608	8.1919	8.2527	.9926
2.9	18.174	.0550	9.0596	9.1146	.9940
3.0	20.086	.0498	10.018	10.068	.9951

Differentiation Formulas   **TABLE IV**

IV-1 $\dfrac{d}{dx} u^r = ru^{r-1} \dfrac{du}{dx}$

IV-15 $\dfrac{d}{dx} \operatorname{Arctan} u = (1 + u^2)^{-1} \dfrac{du}{dx}$

IV-2 $\dfrac{d}{dx} |u| = u|u|^{-1} \dfrac{du}{dx}$

IV-16 $\dfrac{d}{dx} \cosh u = \sinh u \dfrac{du}{dx}$

IV-3 $\dfrac{d}{dx} \sin u = \cos u \dfrac{du}{dx}$

IV-17 $\dfrac{d}{dx} \sinh u = \cosh u \dfrac{du}{dx}$

IV-4 $\dfrac{d}{dx} \cos u = -\sin u \dfrac{du}{dx}$

IV-18 $\dfrac{d}{dx} \tanh u = \operatorname{sech}^2 u \dfrac{du}{dx}$

IV-5 $\dfrac{d}{dx} \tan u = \sec^2 u \dfrac{du}{dx}$

IV-19 $\dfrac{d}{dx} \coth u = -\operatorname{csch}^2 u \dfrac{du}{dx}$

IV-6 $\dfrac{d}{dx} \cot u = -\csc^2 u \dfrac{du}{dx}$

IV-20 $\dfrac{d}{dx} \operatorname{sech} u = -\operatorname{sech} u \tanh u \dfrac{du}{dx}$

IV-7 $\dfrac{d}{dx} \sec u = \sec u \tan u \dfrac{du}{dx}$

IV-21 $\dfrac{d}{dx} \operatorname{csch} u = -\operatorname{csch} u \coth u \dfrac{du}{dx}$

IV-8 $\dfrac{d}{dx} \csc u = -\csc u \cot u \dfrac{du}{dx}$

IV-22 $\dfrac{d}{dx} \sinh^{-1} u = \dfrac{1}{\sqrt{u^2 + 1}} \dfrac{du}{dx}$

IV-9 $\dfrac{d}{dx} e^u = e^u \dfrac{du}{dx}$

IV-23 $\dfrac{d}{dx} \operatorname{Cosh}^{-1} u = \dfrac{1}{\sqrt{u^2 - 1}} \dfrac{du}{dx}$

IV-10 $\dfrac{d}{dx} \ln |u| = u^{-1} \dfrac{du}{dx}$

IV-24 $\dfrac{d}{dx} \tanh^{-1} u = \dfrac{1}{1 - u^2} \dfrac{du}{dx}$

IV-11 $\dfrac{d}{dx} a^u = a^u \ln a \dfrac{du}{dx}$

IV-25 $\dfrac{d}{dx} \coth^{-1} u = \dfrac{-1}{u^2 - 1} \dfrac{du}{dx}$

IV-12 $\dfrac{d}{dx} \log_b |u| = u^{-1} \log_b e \dfrac{du}{dx}$

IV-26 $\dfrac{d}{dx} \operatorname{Sech}^{-1} u = \dfrac{-1}{u\sqrt{1 - u^2}} \dfrac{du}{dx}$

IV-13 $\dfrac{d}{dx} \operatorname{Arcsin} u = \dfrac{1}{\sqrt{1 - u^2}} \dfrac{du}{dx}$

IV-27 $\dfrac{d}{dx} \operatorname{csch}^{-1} u = \dfrac{-1}{u\sqrt{1 + u^2}} \dfrac{du}{dx}$

IV-14 $\dfrac{d}{dx} \operatorname{Arccos} u = -\dfrac{1}{\sqrt{1 - u^2}} \dfrac{du}{dx}$

IV-28 $\dfrac{d}{dx} \displaystyle\int_c^u f(t)\, dt = f(u) \dfrac{du}{dx}$

(if *f* continuous)

TABLE V *Basic Integration Formulas*

(*Note*: The number a is assumed to be positive in the following 25 basic formulas.)

V-1 $\quad \displaystyle\int u^r \, du = \frac{u^{r+1}}{r+1} \qquad (r \neq -1)$

V-10 $\quad \displaystyle\int a^u \, du = \frac{a^u}{\ln a}$

V-2 $\quad \displaystyle\int \cos u \, du = \sin u$

V-11 $\quad \displaystyle\int \ln |u| \, du = u(\ln |u| - 1)$

V-3 $\quad \displaystyle\int \sin u \, du = -\cos u$

V-12 $\quad \displaystyle\int \frac{du}{\sqrt{a^2 - u^2}} = \text{Arcsin}\left(\frac{u}{a}\right)$

V-4 $\quad \displaystyle\int \sec^2 u \, du = \tan u$

V-13 $\quad \displaystyle\int \frac{a \, du}{a^2 + u^2} = \frac{1}{a}\text{Arctan}\left(\frac{u}{a}\right)$

V-5 $\quad \displaystyle\int \csc^2 u \, du = -\cot u$

V-14 $\quad \displaystyle\int \sinh u \, du = \cosh u$

V-6 $\quad \displaystyle\int \sec u \tan u \, du = \sec u$

V-15 $\quad \displaystyle\int \cosh u \, du = \sinh u$

V-7 $\quad \displaystyle\int \csc u \cot u \, du = -\csc u$

V-16 $\quad \displaystyle\int \text{sech}^2 u \, du = \tanh u$

V-8 $\quad \displaystyle\int e^u \, du = e^u$

V-17 $\quad \displaystyle\int \text{csch}^2 u \, du = -\coth u$

V-9 $\quad \displaystyle\int \frac{du}{u} = \ln |u|$

V-18 $\quad \displaystyle\int \text{sech}\, u \tanh u \, du = -\text{sech}\, u$

V-19 $\quad \displaystyle\int \text{csch}\, u \coth u \, du = -\text{csch}\, u$

V-20 $\quad \displaystyle\int \frac{du}{\sqrt{u^2 + a^2}} = \sinh^{-1}\left(\frac{u}{a}\right) = \ln\left(\frac{u + \sqrt{u^2 + a^2}}{a}\right)$

V-21 $\quad \displaystyle\int \frac{du}{\sqrt{u^2 - a^2}} = \text{Cosh}^{-1}\left(\frac{u}{a}\right) = \ln\left(\frac{u + \sqrt{u^2 - a^2}}{a}\right)$

V-22 $\quad \displaystyle\int \frac{du}{a^2 - u^2} = \frac{1}{a}\tanh^{-1}\left(\frac{u}{a}\right) = \frac{1}{2a}\ln\left(\frac{a + u}{a - u}\right), \qquad (u^2 < a^2)$

V-23 $\quad \displaystyle\int \frac{du}{u^2 - a^2} = -\frac{1}{a}\coth^{-1}\left(\frac{u}{a}\right) = -\frac{1}{2a}\ln\left(\frac{u + a}{u - a}\right), \qquad (u^2 > a^2)$

V-24 $\quad \displaystyle\int \frac{du}{u\sqrt{a^2 - u^2}} = -\frac{1}{a}\text{Sech}^{-1}\left(\frac{u}{a}\right) = -\frac{1}{a}\ln\left(\frac{a + \sqrt{a^2 - u^2}}{u}\right),$
$$0 < u < a$$

V-25 $\quad \displaystyle\int \frac{du}{u\sqrt{a^2 + u^2}} = -\frac{1}{a}\text{csch}^{-1}\left(\frac{|u|}{a}\right) = -\frac{1}{a}\ln\left(\frac{a + \sqrt{a^2 + u^2}}{|u|}\right)$

Additional Integration Formulas

V-26 $\displaystyle\int \frac{u\,du}{au+b} = \frac{u}{a} - \frac{b}{a^2}\ln|au+b|$

V-27 $\displaystyle\int \frac{u\,du}{(au+b)^2} = \frac{b}{a^2(au+b)} + \frac{1}{a^2}\ln|au+b|$

V-28 $\displaystyle\int \frac{du}{u(au+b)} = \frac{1}{b}\ln\left|\frac{u}{au+b}\right|$

V-29 $\displaystyle\int \frac{du}{u(au+b)^2} = \frac{1}{b(au+b)} + \frac{1}{b^2}\ln\left|\frac{u}{au+b}\right|$

V-30 $\displaystyle\int \sqrt{a^2-u^2}\,du = \frac{u}{2}\sqrt{a^2-u^2} + \frac{a^2}{2}\,\text{Arcsin}\,\frac{u}{a}$

V-31 $\displaystyle\int \sqrt{u^2 \pm a^2}\,du = \frac{u}{2}\sqrt{u^2 \pm a^2} \pm \frac{a^2}{2}\ln|u + \sqrt{u^2 \pm a^2}|$

V-32 $\displaystyle\int u^2\sqrt{a^2-u^2}\,du = -\tfrac{1}{4}u(a^2-u^2)^{3/2} + \frac{a^2 u}{8}\sqrt{a^2-u^2} + \frac{a^4}{8}\,\text{Arcsin}\,\frac{u}{a}$

V-33 $\displaystyle\int \tan u\,du = \ln|\sec u|$

V-34 $\displaystyle\int \cot u\,du = \ln|\sin u|$

V-35 $\displaystyle\int \sec u\,du = \ln|\sec u + \tan u|$

V-36 $\displaystyle\int \csc u\,du = \ln|\csc u - \cot u|$

V-37 $\displaystyle\int \sec^3 u\,du = \tfrac{1}{2}(\tan u \sec u + \ln|\sec u + \tan u|)$

V-38 $\displaystyle\int e^{au}\sin bu\,du = \frac{e^{au}(a\sin bu - b\cos bu)}{a^2+b^2}$

V-39 $\displaystyle\int e^{au}\cos bu\,du = \frac{e^{au}(b\sin bu + a\cos bu)}{a^2+b^2}$

V-40 $\displaystyle\int u^n \ln u\,du = u^{n+1}\left[\frac{\ln u}{n+1} - \frac{1}{(n+1)^2}\right]$

V-41 $\displaystyle\int \sin^m u \cos^n u\,du = -\frac{\sin^{m-1} u \cos^{n+1} u}{m+n} + \frac{m-1}{m+n}\int \sin^{m-2} u \cos^n u\,du$

$\displaystyle\qquad\qquad\qquad = \frac{\sin^{m+1} u \cos^{n-1} u}{m+n} + \frac{n-1}{m+n}\int \sin^m u \cos^{n-2} u\,du$

V-42 $\displaystyle\int \tan^n u\,du = \frac{\tan^{n-1} u}{n-1} - \int \tan^{n-2} u\,du$

V-43 $\displaystyle\int u^n e^{au}\,du = \frac{u^n e^{au}}{a} - \frac{n}{a}\int u^{n-1} e^{au}\,du$

V-44 $\displaystyle\int u^n \sin au\,du = \frac{-u^n}{a}\cos au + \frac{n}{a}\int u^{n-1}\cos au\,du$

Additional Integration Formulas (continued)

V-45 $\displaystyle \int u^n \cos au \, du = \frac{u^n}{a} \sin au - \frac{n}{a} \int u^{n-1} \sin au \, du$

V-46 $\displaystyle \int \frac{du}{au^2 + bu + c} = \frac{2}{\sqrt{4ac - b^2}} \text{Tan}^{-1} \left(\frac{2au + b}{\sqrt{4ac - b^2}} \right), \qquad (b^2 < 4ac)$

V-47 $\displaystyle \int \frac{u \, du}{au^2 + bu + c} = \frac{1}{2a} \ln |au^2 + bu + c|$

$$- \frac{b}{a\sqrt{4ac - b^2}} \text{Tan}^{-1} \left(\frac{2au + b}{\sqrt{4ac - b^2}} \right), \qquad (b^2 < 4ac)$$

V-48 $\displaystyle \int \frac{du}{(au^2 + bu + c)^r} = \frac{2au + b}{(r - 1)(4ac - b^2)(au^2 + bu + c)^{r-1}}$

$$+ \frac{2(2r - 3)a}{(r - 1)(4ac - b^2)} \int \frac{du}{(au^2 + bu + c)^{r-1}}$$

V-49 $\displaystyle \int \frac{u \, du}{(au^2 + bu + c)^r} = \frac{-(2c + bu)}{(r - 1)(4ac - b^2)(au^2 + bu + c)^{r-1}}$

$$- \frac{b(2r - 3)}{(r - 1)(4ac - b^2)} \int \frac{du}{(au^2 + bu + c)^{r-1}}$$

V-50 $\displaystyle \int f^{-1}(u) \, du = uf^{-1}(u) - \int_c^{f^{-1}(u)} f(t) \, dt, \qquad (c \text{ arbitrary})$

Answers to Selected Problems

PROBLEMS 1, p. 4

1. (a) $\frac{7}{3}$; (c) -2. 2. $-1, 0, 1, 2$, no, no. 3. (a) $x < 3$; (c) $x < 2$; (e) $1 < x < 4$.
4. (a) $x = 4$ or $x = -4$; (c) $x \geq 0$. 5. (a) $-4 \leq x \leq 4$; (c) true for any x.
6. (a) and (c). 7. $|a| < |b|$. 8. (a) $x > y + 3$; (c) $|x - 3| < \frac{1}{2}$. 11. $b = \frac{142}{999}$,
$c = 3$. 13. (a) $x > \frac{1}{4}$; (c) $\frac{1}{3} < x < 1$; (e) $-3 < x < 3$.

PROBLEMS 2, p. 9

2. (a) $(-1, 7)$; (c) $(-\infty, 7]$; (e) $(-5, 8)$. 3. (a) $(-\infty, 4)$; (c) $(2, 7)$.
4. (a) $(-\infty, -1) \cup (1, \infty)$; (c) $(-\infty, 0) \cup (0, \infty)$. 5. (a) $(-\frac{3}{2}, \infty)$;
(c) $(-\infty, 0) \cup (5, \infty)$; (e) $[-3, 2]$; (g) $(0, 1) \cup (2, \infty)$. 6. $p = 3, q = -4$.
9. $a = -\infty, b = \infty$. 10. (a) $[-1, 2]$; (c) $[-\infty, \infty]$; (e) $[0, \infty]$. 11. No.
12. (a) $[-3, 6]$; (b) no, suppose $r < 0$; (c) $[4, 6]$; (d) $(-\infty, \infty)$; (f) no,
consider $A = B = [1, 1]$, for example.

PROBLEMS 3, p. 15

1. (a) $\sqrt{29}$; (c) 5; (e) 1; (g) $\sqrt{2}\log 2$. 2. (a) $2|t|$; (c) $|\sec t|$. 3. (a) II; (c) IV.
4. $\sqrt{2}$. 6. 8. 7. $(-7, 2)$ and $(-7, 6)$, $(1, 2)$ and $(1, 6)$, $(-5, 4)$ and $(-1, 4)$.
8. (a) $(2, 7)$; (c) $(\frac{1}{2}(x_1 + x_2), \frac{1}{2}(y_1 + y_2))$. 10. $(4, 0)$. 11. $x + y = 1$.
12. (a) Not collinear. 13. $(-\infty, -2) \cup (6, \infty)$. 14. $(-2, 2)$, $(4, 6)$, $(0, -4)$.
15. (a) 1; (c) 4.

PROBLEMS 4, p. 20

2. (a) $(x + 1)^2 + (y - 3)^2 = 9$; (c) $|y| = 7$; (e) $xy = 0$. 3. (a) The right half-plane;
(c) the coordinate axes; (e) the disk of radius 2 with center the origin.
4. (a) $(x - 1)^2 + (y + 2)^2 > 9$; (c) $xy > 0$; (e) $4 < x^2 + y^2 < 9$.
7. (a) $(x + 1)^2 + (y + 3)^2 = 4$; (c) $3x^2 + 3y^2 - 4x + 14y + 15 = 0$; (e) $x = \pi$.
8. (a) $\{(0, 0), (1, 1)\}$; (c) $\{(k\pi, 0)\}$; (e) $\{(1, 1), (-\frac{1}{2}, \frac{1}{4})\}$.

PROBLEMS 5, p. 26

1. (a) 2; (c) 37; (e) $\frac{17}{16}$; (g) 4; (i) $(\log 2)^2 + 1$; (k) $\sec^2 1$. 2. (a) 13; (c) -6;
(e) -202; (g) $x + 13$; (i) 81. 3. (a) $(0, \infty)$; (c) $(-\infty, 1)$; (e) $(-\infty, -1) \cup (1, \infty)$;
(g) $(-\infty, 0) \cup [1, \infty]$. 4. (a) Domain $= \{1, 2, \ldots\}$, range $= \{2, 3, \ldots\}$,
$f(n) = n + 1$; (c) domain $= (0, \infty)$, range $= [2, \infty]$, $h(x) = x + 1/x$;
(e) domain $= \{n \mid n$ a prime number$\}$, range $= \{0, 6\}$,
$q(x) = \frac{3}{20}(x - 3)(10 - 3x)(x - 7 - |x - 7|)$. 5. $p = q$.
6. (e) is true for every constant function. 8. Domain $= \{1, 2, \ldots, 15\}$,
range $= \{0, 1, 2\}$, $f(1) + f(4) - f(5) = 1$. 9. Domain $= [-3, 0]$,
range $= [0, 6]$, -8. 10. No. 11. $-2[\![-x]\!]$. 12. $2x^2 + 100/x$. 13. $22x^2$.
14. $c = .1\pi r^2(1 + .1\sqrt{1 + 81r^{-6}})$.

PROBLEMS 6, p. 31

1. (a) .84; (c) .07. 2. (a) F; (c) T; (e) F. 3. (a) $\frac{1}{2}\sqrt{2}$; (c) 1; (e) $\frac{1}{2}\sqrt{3}$.
4. (b), (c), (d) are true. 5. (a), (c), (f). 6. (a) $\sqrt{2}$; (c) $\sqrt{2}$; (e) 1. 8. (a) 1;
(c) $-\cos 4t$; (e) $-\cot 2t$; (g) 1. 10. (a) $\{\frac{1}{4}\pi, \frac{5}{4}\pi\}$; (c) $\{\frac{1}{4}\pi, \frac{3}{4}\pi, \frac{5}{4}\pi, \frac{7}{4}\pi\}$.
11. $\sin(-t) = -\sin t$, $\tan(-t) = -\tan t$. 13. $\frac{5}{2}$ radians, $450/\pi$ degrees.
14. $\frac{42}{5}\pi$ inches.

PROBLEMS 7, p. 36

1. (a) $3x - 2$; (c) $\frac{1}{3}\pi(1 - x)$; (e) $x + \log 5$. 3. $\{(-1, 1)\}$. 4. (a) $f(x) = mx$;
(c) $f(x) = m(x - 3) + 2$; (e) $f(x) = x + b$; (g) $f(x) = b$, where $b \geq 0$ or $f(x) = mx$,
where $m \geq 0$. 5. (a) $(0, -5)$; (b) $(\frac{5}{3}, 0)$; (c) $\frac{5}{3}\sqrt{10}$. 6. $f(x) = 3x + 2$.
7. (a) m miles per hour, b miles; (c) m books per student, b books. 8. 80 calories.
9. (b) $f(x) = x$ or $f(x) = -x + b$.

PROBLEMS 8, p. 41

1. (a) $m = -\frac{1}{2}, b = -\frac{1}{4}$; (c) $m = 2, b = \frac{7}{2}$. 2. (a) $y = 4x - 1$;
(c) $7y = 22x + 7\pi - 22$. 3. (a) $y = x + 1$; (c) $y = 1$. 4. (a) Not collinear.
5. $3x + 5y = 7, 5x - 3y = -11$. 6. No. 7. $x + \sqrt{3}y = 4$. 8. (a) $-\frac{1}{2}$; (b) 12;
(c) 0; (d) $-10, -5$. 9. (b) $\frac{1}{7}x + \frac{1}{11}y = 1$. 10. $(3, 1)$ and $(-27, 41)$. 11. $-1/\sqrt{15}$.
12. (a) $x = 1$; (b) $y = -x/\sqrt{3} + \sqrt{3}$. 14. 5. 16. $P_3 = P_1$.

PROBLEMS 9, p. 45

1. (a) Continuous everywhere; (b) $\lim_{x \uparrow 0} f(x) = -1$; (c) continuous everywhere;
(d) discontinuous at the integers, $\lim_{x \downarrow 3} f(x) = 3$; (e) continuous everywhere;
(f) discontinuous at $\frac{1}{2}\pi, \frac{3}{2}\pi$, and so on. 2. (a) -5; (c) $\frac{1}{2}\sqrt{3}$; (e) 4; (g) 3; (i) 6.
3. (a) 1; (b) 0; (c) 1; (d) 0; (e) 0; (f) 0. 4. (a) Continuous; (c) continuous;
(e) the limit is 0, but the value is 1. 5. (a) 1; (b) -1; (c) 0; (d) 0; (e) 0; (f) 1.
6. (a) 9; (c) no limit. 7. (a) 1; (c) 3. 8. (a) $\{-\frac{3}{2}, -1, 0, \frac{1}{2}, 1\}$; (b) $\lim_{x \to \frac{1}{2}} h(x) = 0$,
$h(\frac{1}{2}) = 1$; (c) no limit, $h(0) = 0$; (d) 1; (e) 1. 9. (b) $f(x) = |x - [\![x + .5]\!]|$.
10. Discontinuous.

PROBLEMS 10, p. 49

1. (a) $[-1, 1]$; (c) $[0, 1]$; (e) $[0, 1]$; (g) $[3, 3]$.
3. (a) Continuous in $(0, 1)$ but not in $[0, 1]$; (b) $f((0, 1)) = \{0\}, f([0, 1]) = \{0, 1\}$.
5. (c) 1. 6. 1. 7. (b) and (d) are punctured neighborhoods,
(e) and (f) are left neighborhoods, (g) is a right neighborhood, and (h) is the interval
$(2, 3]$, which is not (by our definition) a neighborhood. 12. (a) $(-1, 1)_1$;
(b) $(-.1, .1)_1$; (c)$(-.1, .1)_1$; (d) $(-.1, .1)_1$; (e) $(-10, 10)_1$; (f) same as (b);
(g) $(-.1, .1)_1$.

PROBLEMS 11, p. 54

1. (a) -2; (c) 2; (e) 6.   2. (a) $\frac{1}{2}$; (c) 3; (e) 0.   3. (a) 0; (b) 1; (c) 1; (d) sin 1;
(e) tan 3; (f) limit does not exist.   4. 0, $-\sin 1$.   5. (b) $f(x) = x + 1, g(x) = 2x$.
6. (a) $f(g(x)) = x$ if $x \geq 0$, $g(f(x)) = |x|$; (b) $f(g(x)) = g(f(x)) = x^2/x$;
(c) $g(f(x)) = f(g(x)) \cos x$; (d) $f(g(x)) = g(f(x)) = g(x)$; (e) $f(g(x)) = 0$,
$g(f(x)) = 2xf(x)$; (f) $g(f(x)) = 0$.   7. (a) 6; (b) 3; (c) 32; (d) 8; (e) 5; (f) 3;
(g) 405; (h) 81.   9. True in (a) and (c), $\lim_{x \to 0} f(x)$ does not exist in (b) and (d).
10. (b) No; (d) yes.

REVIEW PROBLEMS, CHAPTER 1, p. 56

1. Yes, no, no, yes.   4. (b) and (d).   8. (0, 0), (1, 2), and so on.   9. (a) $\{k\pi\}$;
(b) $\varnothing$; (c) $\{0 \leq x < 1\}$; (d) $\varnothing$.   14. (5, 2).   15. (c) Yes.   16. The tooth misses Babe.
17. (a) 0; (b) -1; (c) -2; (d) -1.   18. (a) $\{z \neq 1\}$; (c) 2, 3.   19. (b) Yes.
20. False only if x is an odd integer.

PROBLEMS 12, p. 64

1. (a) 0; (c) -1; (e) 0.   2. (a) $y = x$; (c) $y = 2x$.   3. (a) $\frac{1}{2}$; (c) 2.
5. (a) $-x/\sqrt{9 - x^2}$.   6. $f'(x) = 0$.   7. $\{k\pi\}$.   8. If $x \in (0, 1)$, $g'(x) = -x/\sqrt{1 - x^2}$,
if $x \in (-1, 0)$, $g'(x) = x/\sqrt{1 - x^2}$.   9. $f'(x) = (x - 2)/\sqrt{21 + 4x - x^2}$.
10. $\{0, 1, 2, 3\}$.   11. (a) $((2k - \frac{1}{2})\pi, (2k + \frac{1}{2})\pi)$.
13. (a) 0, 9, -6; (b) $\{x$ not an integer$\}$; (c) $3[\![x]\!]$.   14. $-1, -1, 1$.
15. $-1/[\![x]\!]([\![x]\!] + 1)$.

PROBLEMS 13, p. 71

1. (a) $f'(x) = 2, y = 2x - 3$; (c) $f'(x) = 2x, y = 4x - 8$; (e) $f'(x) = 3x^2$,
$y = 12x - 16$.   2. (a) 12; (c) 12; (e) 1.   3. $f'(x) = m$.   4. (a) $1/\sqrt{x}$;
(c) $1/2\sqrt{x} + 2$.   6. $f'(0) = 1$.   7. (a) $f'(2) = 3, y = 3x - 4$; (c) $f'(2) = 0, y = 0$;
(e) $f'(2) = 1, y = x$.   9. $f'(0)$ does not exist.   10. $g'(x) = f'(x) + 2x$.
11. (a) .43, .043, .0043, 4.3; (b) $\frac{100}{43}$; (c) $\frac{1}{43}10^{x+2}$.   13. $f'(c) \leq g'(c)$.

PROBLEMS 14, p. 76

1. (a) $2x$; (c) $5 - 3x^{-2}$; (e) $2x + 4$; (g) $1 - x^{-2}$. 2. (a) $2x$ and 1;
(c) $1/2\sqrt{x}$ and 1. 3. (a) 3; (c) $18r$; (e) $2/\sqrt{t}$. 4. (a) 1; (b) -2; (c) -3; (d) 0.

5. (a) $2, 2$; (c) $3, 5$. 6. $\dfrac{dy}{dx}\dfrac{dx}{dy} = 1$. 7. $\dfrac{dA}{dr} = 2\pi r$. 8. $-10,800\pi$ inches per minute.

9. (a) $-3, 0, \frac{3}{4}$; (b) to the left, stopped, to the right. 10. (a) Positive; (c) negative.
12. (a) Approximately 1, $.88$, $.54$; (b) 1.5^+; (c) toward; (d) when $t = 0$.

PROBLEMS 15, p. 81

1. (a) $7x^6 + 3x^2$; (c) $\frac{3}{5}t^{-4/5}$; (e) $6u^5$; (g) $3x^2 + \frac{1}{3}x^{-2/3}$; (i) $18x - 6$; (k) $\frac{1}{8}x^{-7/8}$.
2. (a) $x - 3y + 2 = 0, 3x + y = 4$; (c) $y = nx + 1 - n, x + ny = n + 1$.

3. (a) 0 or $\frac{2}{3}$; (c) $2^{-4/3}$. 4. $\frac{5}{16}\sqrt{41}$. 5. $\frac{1}{3}$. 6. (a), (b), (c) all no. 12. $\dfrac{dy}{dx}\dfrac{dx}{dy} = 1$.

15. (b) $|ra^r + 1/ra^{r-2}|$.

PROBLEMS 16, p. 88

1. (a) $14x^{5/2}$; (c) $14x^{13}$; (e) $\frac{7}{2}(x + 4)^{5/2}$. 2. (a) $3x + y = -6$; (c) $y = 2x + 2$;
(e) $6y = x + 17$. 3. (a) $32t$; (c) $512t$; (e) 0. 4. (a) $(2x + 9)^{-1/2}$;
(c) $-18(27x + 8)^{-5/3}$; (e) $3x(3x^2 - 2)^{-1/2}$; (g) $3x(3x^2 + 2)^{-1/2}$. 5. (a) $\frac{1}{3}$;
(c) -9; (e) $-\frac{4}{3}$. 8. $4\pi r^2 \dfrac{dr}{dt}$. 9. $2\pi g(t)g'(t)$. 10. (a) and (c) are correct. 11. $2xv$.

13. $x = 0$ and $x = \frac{1}{2}\pi$ are two solutions. 17. $\dfrac{d}{dx}f(3x + 4) = 3f'(3x + 4)$,

$\dfrac{d}{dx}f(x + 4) = f'(x + 4)$.

PROBLEMS 17, p. 92

1. (a) $2/\sqrt{4x - 9}$; (c) $1/2\sqrt{x + 3} - 1/2\sqrt{x}$; (e) $(x + 3)(x + 2)^2(5x + 13)$;
(g) $3(x + 2)/2\sqrt{x + 3}$. 2. (a) $\frac{3}{8}$; (c) 0. 3. (a) $(4, 1)$; (c) $(\frac{1}{16}, \frac{7}{2})$. 4. $(0, 0)$.

5. (a) $\dfrac{dV}{dr} = 4\pi r^2$; (b) $\dfrac{dV}{dr} = 2\pi rh$; (c) $\dfrac{dV}{dr} = \frac{2}{3}\pi rh$. 7. (a) $2|x|$;

(c) $3(x - 1)^3(x + 2)^4(x - 3)^5(5x^2 - 6x - 7)$. 8. (a) -5; (c) $-\frac{2}{5}$; (e) $\frac{24}{5}$.
9. (a) $t \in [0, 1)$; (b) $t > 1$; (c) $t = 1$; (d) $\frac{1}{2}$. 11. (b) 0; (d) 0.

PROBLEMS 18,   p. 99

1. (a) $(\cos x)/2\sqrt{\sin x} - (\cos \sqrt{x})/2\sqrt{x}$; (c) $2 \sec 2x \tan 2x - 2 \sec x \tan x$;
(e) $-2 \sin 2x + \cos 2 \sin x$.   2. (a) $(x \cos x - \sin x)x^{-2}$; (c) $6 \sin 3x \cos 3x$;
(e) $-2x \sin (2 \cos^2 x^2) \sin 2x^2$.   3. (a) $y = x, y = -x$; (c) $y = 1, x = \frac{1}{2}\pi$.
4. (a) 3; (c) $-6 \csc 1$; (e) $-6 \sqrt{\cos 1}/\sin 1$.   5. (a) $(\frac{2}{3}\pi, \frac{1}{2}\sqrt{3})$.
9. $v = Ab \cos (bt + c)$.   12. $4\pi\sqrt{3}$ units per minute.
13. $0°$ at $(\frac{1}{2}\pi, 0)$ and $90°$ at $(\frac{3}{2}\pi, 0)$.

PROBLEMS 19,   p. 106

2. (a) 1; (c) $1 - \frac{1}{3}\sqrt{3}$ and $1 + \frac{1}{3}\sqrt{3}$; (e) no.   3. (a) $\frac{1}{2}\pi$; (c) $\frac{1}{2}$; (e) $\frac{5}{3}$.
4. (a) Increasing in $[-\infty, 1]$ and $[3, \infty]$; (c) increasing in $(-1, 5]$;
(e) increasing in $[-1, 1]$.   5. (a) $[0, \frac{1}{4}\pi]$ and $[\frac{5}{4}\pi, 2\pi]$; (c) $[0, \frac{1}{4}\pi]$, $[\frac{1}{2}\pi, \frac{3}{4}\pi]$,
$[\pi, \frac{5}{4}\pi]$, $[\frac{3}{2}\pi, \frac{7}{4}\pi]$.   6. $g(x) = f(x) + C$, where (a) $C = -\frac{1}{2}$; (b) $C = -4$;
(c) $C = -1$; (d) $C = 1$; (e) $C = -1$; (f) $C = -\sin 3$.   7. (a) Yes; (b) no;
(c) no; (d) yes.   8. (a) $f(x) = 3/10^x$; (b) $f(x) \geq 3/10^x$.   13. (a) No.

PROBLEMS 20,   p. 109

1. (a) $4(1 + x)^{-3}$; (c) $-2 \sin x$; (e) $2x^{-3} \sec^2 1/x + 2x^{-4} \sec^2 1/x \tan 1/x$.
2. (a) 14; (c) $-\sqrt{3} \pi^2/162$; (e) 2.   3. (b) $(a, b) = (5, -2)$.   4. k an integer.
6. $k = 1$ or -1.   11. (b) There is a number C such that $y' = (C - x)^{-1}$.
We don't yet have the background to find y.

PROBLEMS 21,   p. 115

1. (a) 5.1; (c) $\frac{81}{40}$; (e) $\frac{191}{768}$; (g) $\frac{1}{180}\pi \approx .017$.   2. .7033.   3. $A_1 = 2\pi a(b - a)$,
$A = \pi(b^2 - a^2)$.   4. (a) $V \approx 2\pi hrt$; (b) $V \approx 4\pi r^2 t$.   7. (a) $\cos b \approx 1 - \frac{1}{2}b^2$;
(b) too small.   8. $10\frac{79}{320}$.   9. $(x - 1)^4 + 4(x - 1)^3 + 9(x - 1)^2 + 10(x - 1) + 6$.
11. (a) $\sin b^2 \approx b^2$.   12. .00035.   13. $m = \frac{9}{16}$ or $-\frac{9}{16}$.

REVIEW PROBLEMS, CHAPTER 2,   p. 117

1. (a) $3 \cos x \sin^2 x$; (b) $3x^2 \cos x^3$; (c) $-\cos x$; (d) $\cos^3 x$; (e) $3 \cos x$;
(f) $3 \cos 3x$.   2. (a) $\{0\} \cup \{x \geq 5\}$; (b) $\{x \leq -5\}$; (c) equal for all x;
(d) $(-\frac{1}{8}\pi, \frac{3}{8}\pi)$, and so on.   3. (a) -7; (b) -3; (c) -1; (d) 0; (e) -2; (f) $-\frac{1}{9}$;
(g) $(-1, 3)$; (h) $(-2, 1)$; (i) 1; (j) $f'(0)f'(1) \approx .8$; (k) 1.   4. (b).   5. $(1.38)4^x$.

6. No.   9. (a) $\frac{1}{4}$; (b) 2; (c) 0; (d) 4 cos 4; (e) 0; (f) 0.   10. $P'(r)$.   12. 10, 20.
13. (a) $<$; (b) $>$; (c) $<$, $>$, ≈ 0; (d) sad.   14. (a) $x = 16 \sin 6\pi t$; (b) $48\pi\sqrt{3}$;
(c) $288\pi^2$.   15. $(\pi, 2\pi)$, and so on.   18. $f(x) = g(x) + c_0 + c_1 x + \cdots + c_{n-1}x^{n-1}$.
19. $f'(\pi) = -1, f'(1776) = 1, f''(x) = 0$, numbers of the form k^2 and $k^2 + k + \frac{1}{2}$ are
not in the domain of f' or f''.   20. No.

PROBLEMS 22,   p. 128

1. (a) $y = -\frac{5}{2}, -2, -\frac{5}{2}, \frac{5}{2}, 2, \frac{5}{2}, y' = \frac{3}{4}, 0, -3, -3, 0, \frac{3}{4}, y'' = -\frac{1}{4}, -2, -16, 16,$
2, $\frac{1}{4}$; (b) concave up when $x \in (0, \infty)$; (c) maximum point $(-1, -2)$,
minimum point $(1, 2)$.   2. (a) Maximum point $(0, 1)$, minimum point $(\frac{2}{3}, \frac{23}{27})$;
(b) no maximum or minimum points; (c) minimum point $(1, 0)$;
(d) no maximum or minimum points.   3. (a) $(\frac{1}{3}, \infty)$; (b) $(-\infty, -\sqrt[4]{3}) \cup (\sqrt[4]{3}, \infty)$;
(c) nowhere; (d) $(-\infty, -1)$.   4. (a) maximum value $f(-2) = 33$,
minimum value $f(\frac{1}{2}) = \frac{7}{4}$; (b) maximum value $f(-1) = -6$,
minimum value $f(1) = -10$.   5. (a) Maximum point $(0, 1)$,
concave down in $(-\frac{1}{2}\sqrt{2}, \frac{1}{2}\sqrt{2})$; (c) minimum point $(1, 3)$, concave down in $(-\sqrt[3]{2}, 0)$;
(e) minimum point $(1, 0)$, concave down in $(0, \frac{2}{3})$.   6. (a) Minimum at 0,
maximum at $1/\sqrt[3]{2}$; (c) minimum at $\frac{3}{2}\pi + 2k\pi$, maximum at $\frac{1}{2}\pi + 2k\pi$;
(e) minimum at $\frac{5}{6}\pi + 2k\pi$ maximum at $\frac{1}{6}\pi + 2k\pi$.
9. (a) Minimum points $(-1, 0)$ and $(1, 0)$, maximum point $(0, 1)$;
(c) minimum point $(\frac{1}{2}, -\frac{1}{3})$; (e) minimum points $((k + \frac{1}{2})\pi, 0)$,
maximum points $(k\pi, 1)$; (g) minimum points $(0, 0)$ and $(\frac{1}{2}\pi, 0)$,
X-coordinate of maximum point satisfies equation $x = \cot x$.   10. (a) $(0, 1)$;
(c) $\{(k\pi, 0)\}$; (e) $(0, 0)$.   11. (a) $[2, \frac{10}{3}]$.   12. $[-\sqrt{5}, \sqrt{5}]$.

PROBLEMS 23,   p. 133

3. 333.   4. (a) 30, \$7.50; (b) 30, \$7.50.   5. Isosceles, with legs of $5\sqrt{2}$.
6. 16 cubic inches.   7. 6 inches by 9 inches.   8. $8/(\pi + 4)$.   9. $\theta \approx 1.3$ radians.
11. r.   12. (b) $60°$.   13. (b) $8(\sqrt[3]{9} + 1)^{3/2}$.
14. Where the perpendicular bisector of AB intersects the far shore.
16. Bill should walk on the paved roads all the way.

PROBLEMS 24,   p. 140

1. (a) $s = 16, v = r = 20, a = 16$; (c) $s = \frac{1}{2}, v = -\frac{1}{6}\pi\sqrt{3}, r = \frac{1}{6}\pi\sqrt{3}, a = -\frac{1}{18}\pi^2$;
(e) $s = 2, v = -2, r = 2, a = 6$.   4. 1, 0, $-\frac{16}{9}$.   5. (a) 256 feet; (b) 400 feet;
(c) 160 feet per second.   6. (a) 100 miles; (b) $400\sqrt{3}$ miles per minute;
(c) $\frac{1}{3}\pi$ minutes; (d) $400\sqrt{3}$ miles per minute.   7. $\dfrac{ds}{dt} = gT$.   8. $\frac{24}{5}$ feet per second.

9. $1\frac{3}{12}$ feet per second.   10. 60 cubic inches per minute, $\frac{1}{3}$ inch per minute.
11. $\frac{4}{5}$ feet per second.   12. 800 feet per second, 400 feet per second.
13. $8\frac{1}{3}$ feet per second, $3\frac{1}{3}$ feet per second.   14. (a) $2/\pi$ inches per second;
(b) $1/\pi$ inches per second.   15. Increasing.

PROBLEMS 25,   p. 145

1. (a) $x_2 = \frac{3}{4}$; (b) $x_1 = 1\frac{1}{24}$; (c) $x_1 = (\pi + 4)/(4\sqrt{2} + 4)$;
(d) $x_1 = (\pi + 4\sqrt{2} - 4)/(4\sqrt{2} + 4)$.   2. .5.   3. (a) .68; (b) 1.16.   5. (a) $2\frac{1}{6}$, $2\frac{11}{72}$;
(b) $2\frac{1}{6}$, $2\frac{235}{1521}$.   6. 3.14 feet.   7. Approximately (.59, .35).   8. $r - x_1 = \frac{1}{8}$,
$r - x_2 = \frac{1}{656}$.   10. $x_1 = [(n - 1)x_0^n + a]/nx_0^{n-1}$.

PROBLEMS 26,   p. 150

1. (a) $(1 + y^2 + y^4)^{-1}$; (c) $(2y - y^3)/(3xy^2 - 2x + 2y)$;
(e) $2\sqrt{1 + y}/(1 - 2\sqrt{1 + y})$; (g) $[\cos x - \cos (x + y)]/[\cos (x + y) - \cos y]$.
3. (a) $y = 3x + 6$; (b) $x - y = 4$; (c) $11x + 3y = 36$; (d) $3x - 4y = 10$.
4. $my_0(x - x_0) + nx_0(y - y_0) = 0$.   6. 90°.   8. (a) $-y^{-3}$; (c) $-4a^2y^{-3}$.
9. $2x + y = 12$ and $14x + y = 60$.   12. Since $\sin y \le 1 < 5 + x^2$,
the "curve" $\sin y = 5 + x^2$ contains no points.

REVIEW PROBLEMS, CHAPTER 3,   p. 152

1. $f(1) = f(-1) = 0$.   3. $(x, 0)$ is a critical point of g if $f(x) = 0$.   5. $[-\frac{3}{2}, 3]$.
3. .17 seconds after the initial instant.   4. $t = \sqrt{(2k - 1)\pi}$.   8. 3.02.
9. $2x_1^* - x_1 = x_0$.   10. $b^2 < 3ac$.   11. $10\sqrt{5}$.   12. (1, 1).   13. $\sqrt[3]{3}$.
14. (a) $\frac{1}{3}K$; (b) $K/(R^2 - 1)$ if $R > 2$.   15. (b) 12 by 12 by 24.

PROBLEMS 27,   p. 158

1. (a) $(x - 2)^2 + (y + 3)^2 = 25$; (c) $x^2 + y^2 = 2x \cos 3 + 2y \sin 3$.
2. (a) Center $(-2, 1)$, radius 3; (c) center $(\frac{1}{2}, 1)$, radius $\frac{1}{2}\sqrt{5}$.
3. (a) $(x - 2)^2 + (y + 3)^2 = 17$; (c) $(x + 1)^2 + (y - 2)^2 = 10$;
(e) $(x - 8)^2 + (y - 14)^2 = 100$; (g) $x^2 + (y - 5)^2 = 25$ or $x^2 + (y + 1)^2 = 25$.

4. (a) The exterior of the disk of radius 3 whose center is $(0, 0)$;
(c) the ring between the circles of radius 2 and 5 whose center is $(0, 0)$.
5. (a) Circles with center $(c, 0)$ and radius 3;
(b) circles with center $(-c, -2c)$ and radius 5;
(c) circles with center $(c, 0)$ and radius $|c|$;
(d) circles with center $(-c, -c)$ and radius $\sqrt{2}\,|c|$.   6. (a) $(4, -8), (0, -2)$;
(c) $(-3, -1), (-7, 5)$; (e) $(\frac{3}{5}, -\frac{16}{7}), (-\frac{17}{5}, \frac{26}{7})$; (g) $(\log 300, \log .002)$,
$(\log .03, \log 200)$.   7. (a) $(2, 1), \bar{y} = 2\bar{x}, \bar{x} = 0$; (c) $(2, -1), \bar{x} = 2\bar{y}, 3\bar{x} = -\bar{y}$.
8. $\bar{y} = y, \bar{x} = x - \frac{1}{2}\pi$.   9. (a) $\bar{x} = x - 2, \bar{y} = y + 1$; (c) $\bar{x} = x - 2, \bar{y} = y - 1$.
10. The circle of radius 3 whose center is $(0, 0)$.   11. The circle $x^2 + y^2 + x - 3y = 2$.
12. $a^2 + b^2 > 4c$.   13. Replace x with $x - h$ and y with $y - k$.

PROBLEMS 28,   p. 166

1. (a) $x^2/16 + y^2/9 = 1$; (c) $(x - 2)^2/9 + (y + 1)^2/4 = 1$.   2. (a) $x^2/9 + y^2/8 = 1$;
(c) $(x - 1)^2/4 + (y - 4)^2/3 = 1$.   3. (a) $x^2/25 + y^2/9 = 1$;
(c) $(x - 5)^2/25 + (y + 1)^2/4 = 1$.   4. (a) $(-4, 0), (4, 0)$; (c) $(-4, 0), (-2, 0)$.
5. (a) $(x - 4)^2/25 + (y - 2)^2/4 = 1$; (c) $(x - 1)^2 + (y - 2)^2/5 = 1$.
6. (a) Symmetric about both axes and the origin;
(c) symmetric about both axes and the origin; (e) symmetric about Y-axis.
7. (a) The curve is a circle, the given points are endpoints of a diameter.
9. Tacks should be placed in a line through the center of the paper and parallel to
the 11-inch edge and at points $\frac{1}{4}\sqrt{195} \approx \frac{7}{2}$ inches from the center. The string should
be about 18 inches long.   10. The upper half of the ellipse $x^2/25 + y^2/9 = 1$.
12. Area $= \pi ab$.   13. Slope is $-c/a$.

PROBLEMS 29,   p. 173

1. (a) Vertices: $(-2, 0), (2, 0)$, foci: $(-\sqrt{5}, 0), (\sqrt{5}, 0)$, asymptotes: $y = \frac{1}{2}x, y = -\frac{1}{2}x$;
(c) vertices: $(0, -1), (0, 1)$, foci: $(0, -\sqrt{5}), (0, \sqrt{5})$, asymptotes: $y = \frac{1}{2}x, y = -\frac{1}{2}x$;
(e) vertices: $(-5, 2), (3, 2)$, foci: $(-6, 2), (4, 2)$, asymptotes: $3x + 4y = 5$,
$3x - 4y = -11$; (g) vertices: $(1, -5), (1, 1)$, foci: $(1, -7), (1, 3)$,
asymptotes: $3x - 4y = 11, 3x + 4y = -5$.   2. (a) $3x^2 - y^2 = 3$;
(c) $5(x - 3)^2 - 4(y - 1)^2 = 20$.   3. (a) $(-7, 1), (3, 1)$; (c) $(-2, -4), (-2, 6)$.
4. (a) $x^2/36 - y^2/28 = 1$; (c) $(y - 5)^2/25 - (x - 1)^2/3 = 1$.   5. (a) $4x^2 - y^2 = 3$;
(c) $(y - 2)^2 - (x - 1)^2 = 3$.   6. (a) $7y^2 - 4x^2 = 28$.   7. No, no.
8. (a) Tangent: $y = 2x - 4$, normal: $x + 2y = 7$.   10. A hyperbola.
12. The second man is standing on a point of the hyperbola $3x^2 - y^2 = 3 \cdot 550^2$;
the first man is at a focus.   13. The union of an ellipse and a hyperbola.
15. Slope is c/a.

PROBLEMS 30, p. 178

1. (a) Focus: (4, 0), directrix: $x = -4$; (c) focus: (2, 3), directrix: $y = -9$;
(e) focus: $(-1, -2)$, directrix: $y = 0$. 2. (a) $y^2 = 16x$; (c) $(y - 2)^2 = 12(x + 2)$.
3. (a) $(x - 2)^2 = 16y$; (c) $(x + 1)^2 = -2(y - 3)$. 4. (b) 16, 12, 12, 16;
(c) $x^2 + y^2 = 5cx$. 5. Slopes are 1 and -1, tangents meet at right angles.
7. A parabola with the buoy as focus and shoreline as directrix.
10. Directrix: $x + y = -2$, parabola $(x - y)^2 = 8(x + y)$.
11. (a) $3x - 2y + 3 = 0, 2x + 3y = 11$; (b) $y = -x, y = x + 4$;
(c) $x = 4, y = -1$; (d) $x + y = 1, y - x = -11$. 13. $\frac{1}{8}x^2 + \frac{1}{4}y^2 = 1$.

PROBLEMS 31, p. 187

1. (a) $e = \frac{1}{2}$, directrices: $x = 4, x = -4$; (c) $e = \frac{5}{4}$, directrices: $x = \frac{16}{5}, x = -\frac{16}{5}$;
(e) $e = \sqrt{2}$, directrices: $x = \frac{1}{2}\sqrt{2}, x = -\frac{1}{2}\sqrt{2}$; (g) $e = \frac{5}{3}$, directrices: $y = \frac{9}{5}, y = -\frac{9}{5}$.
2. (a) $3x^2 - y^2 + 20x + 12 = 0$; (c) $3x^2 + 4y^2 - 20x + 12 = 0$; (e) $y^2 + 2x = 3$;
(g) $9x^2 + 8y^2 - 18x - 40y + 41 = 0$. 3. (a) $e = \frac{1}{2}$, directrices: $x = 5, x = -3$;
(c) $e = \sqrt{2}$, directrices: $x = 1 + \frac{1}{2}\sqrt{2}, x = 1 - \frac{1}{2}\sqrt{2}$. 4. $e = c/b$,
directrices: $y = b^2/c, y = -b^2/c$. 5. (a) An ellipse with eccentricity near 0 is
relatively circular; an ellipse with eccentricity near 1 is relatively flat or narrow;
(b) a hyperbola with eccentricity near 1 is relatively narrow; a hyperbola with a large
eccentricity is relatively open or wide. 6. $y^2 - x^2 + 6x + 9 = 0$ or
$y^2 - x^2 - 6x + 9 = 0$. 7. $2^{-1/2}$. 8. (a) $3x + 4y = -25$; (c) $3x + 4y = -50$.
9. (a) An arc of a parabola; (b) an arc of a hyperbola; (c) an arc of an ellipse.
10. (a) $|\sec c|$; (b) $|\sin c|$; (c) $|\sec c|$; (d) $|\csc c|$. 11. $\sqrt{2}$. 12. $\frac{1}{2}\sqrt{2}$.
14. (a) $\frac{3}{4}$; (b) $\frac{2}{5}$.

REVIEW PROBLEMS, CHAPTER 4, p. 189

3. You don't survive. 4. Plural. 5. Yes. 7. $(r \cos s/r, r \sin s/r)$.
8. $(4\pi, 2)$. 9. An ellipse with major diameter of 100 feet and minor diameter of $50\sqrt{3}$ feet.
10. There is an elliptical disk in which an attacker could get hit twice.
11. $(d_2 - d_1)/(d_2 + d_1)$.
12. Ellipse if a, b, and $c^2/a + d^2/b - e$ have the same sign, hyperbola if
$ab(c^2/a + d^2/b - e)^2 < 0$, parabola if $a = 0$ and $bc \neq 0$ or $b = 0$ and $ad \neq 0$.
15. $y' = -e$. 16. 10 units, (5, 0). 18. $y^2 = 4(c - a)(x - a)$.
19. $y^2 = 4(c - a)(x - a)$.

PROBLEMS 32, p. 199

1. (a) 4, 6, 8, 10; (c) $-\frac{7}{5}, \frac{1}{5}, \frac{9}{5}, \frac{17}{5}$.   2. Actual areas: (a) 152; (c) 40; (e) 44.
3. (a) 328, 284; (b) 168, 204; (c) 240, 242.   4. Actual areas: (a) $\frac{8}{3}$; (b) $\frac{4}{3}$; (c) $\frac{1}{4}\pi$;
(d) 4.   5. (a) 3, 1; (c) $\frac{3}{2}, \frac{1}{2}$; (e) 4, 3.   7. (a) 86; (c) $\frac{1}{2}(2 + \sqrt{2} + \sqrt{3})$;
(e) $F(5) - F(0)$.   8. (a) 205; (c) 164.

PROBLEMS 33, p. 204

1. (a) 2; (c) 17; (e) 2; (g) 4; (i) 9.   2. (a) $2 + \frac{1}{2}\pi$; (c) π.   4. (a) -36; (c) -36;
(e) 0.   5. (a) 16; (c) 2.   6. $\frac{17}{4}$.   9. $S(u) = \{c(b - a)\}$.   11. $S(1) = \{0, 1\}$.

PROBLEMS 34, p. 211

1. (a) Does not exist; (c) does not exist; (e) 11; (g) 4.   3. (a) $8 + \pi$; (c) $12 + 9\pi$.
5. $\pi, m = \pm\sqrt{16 - \pi^2}$.   6. $m_1 = \frac{1}{2}(d + c), f(m_1) = \frac{1}{2}m(d + c) + b$.   7. 0.
8. $\sqrt{2} + \sqrt{3} - 3$.   10. (a) No; (b) yes; (c) no.

PROBLEMS 35, p. 217

1. (a) 24; (c) 24; (e) $-\frac{26}{3}$; (g) $\frac{2}{3}$.   3. (a) $\frac{64}{3}$; (c) 30; (e) $8a^4$.   4. (a) 32; (c) 16;
(e) -6.   7. 18, $(2\sqrt{13} - 1)/3$ 8. $\frac{16}{3}$.   9. $a^{-1} - b^{-1}$.   11. (a) $\frac{1295}{4}$; (c) 15; (e) $\frac{17}{4}$.
12. $-\frac{25}{6}$.

PROBLEMS 36, p. 226

2. $\frac{7}{3}, \frac{75}{32}$, less than $\frac{1}{2}$ per cent.   3. (a) $\frac{31}{40}$; (c) $\frac{45}{8}$; (d) $\frac{9}{4}$; (e) $\frac{1}{4}\pi(\sqrt{2} + 1)$; (f) 0.
4. (c) $\frac{65}{12}$; (d) $\frac{13}{6}$; (e) $\frac{1}{6}\pi(2\sqrt{2} + 1)$; (f) 0.   5. .0789, .3964.   8. No.
11. (c) Midpoint Rule gives .3966, Trapezoidal Rule gives .3960.   12. (a) 3.00;
(b) 3.08; (c) 3.18.   13. (a) No; (b) no.

PROBLEMS 37, p. 231

1. (a) 64; (c) $\frac{1}{4}$; (e) 2. 2. (a) $-\frac{7}{15}$; (c) $\frac{32}{3}$; (e) 16; (g) $\frac{662}{15}$. 3. (a) 24; (c) $\frac{1783}{20}$; (e) $\frac{1}{9}(4^9 - 1)$. 4. (a) 16; (c) $\frac{2}{3}\sqrt{2}$. 6. (a) $\frac{256}{3}$; (c) $\frac{8}{3}a^5$. 7. $f(\frac{4}{3} + \frac{2}{3}\sqrt{3}) = 6$. 8. $x = 2$. 9. 6. 10. No. 11. (a) $3 - 2\sqrt{2}$; (c) 1.

PROBLEMS 38, p. 237

1. (a) 3; (c) 1; (e) 0. 2. (a) $3 - \cos 1$; (c) 0; (e) 0; (g) 0. 3. (a) 4; (c) $\frac{2}{3}$. 4. (a) 0; (c) 4; (e) 8. 5. $\frac{1}{6}\pi$. 6. (a) $\frac{1}{8}$; (c) 1; (e) $\frac{1}{3}$; (g) 3. 7. (a) $\frac{1}{8}$; (c) 1; (e) $\frac{1}{3}$; (g) 4. 8. 2.4.

PROBLEMS 39, p. 245

1. (a) 4; (c) $\frac{1}{2}$; (e) 2. 2. (a) $\frac{32}{3}$. 3. (a) $\frac{4}{3}$; (b) $\frac{4}{3}$. 4. (a) $\frac{8}{3}$; (b) $\frac{32}{3}$; (c) $\frac{4}{3}$; (d) $\frac{9}{2}$. 5. $\frac{5}{6}\sqrt{5}$. 6. (a) 9; (c) 18; (e) $\frac{71}{6}$. 7. $\frac{8}{3}\sqrt{2} - 2$. 8. (a) -4. 12. $n/(2n + 4)$, $\frac{1}{2}$. 13. $|m| = \frac{103}{100}$. 14. $\int_a^b |f(x)|\, dx$.

PROBLEMS 40, p. 251

1. (a) $\frac{128}{7}\pi$; (c) $\frac{1}{15}\pi a^3$; (e) $\pi\sqrt{3}$. 2. (a) $\frac{8}{3}\pi$; (b) $\frac{512}{15}\pi$; (c) $\frac{56}{13}\pi$; (d) $\frac{16}{5}\pi a^3$. 3. (a) $\frac{96}{5}\pi$; (c) $\frac{64}{7}\pi$. 4. (a) $\frac{1}{6}\pi$; (b) 8π. 5. (a) $\frac{16}{3}\pi$; (c) $\frac{1024}{35}\pi$. 6. $\frac{1}{2}\pi^2$. 7. $\frac{1}{2}\pi$. 8. $\frac{4}{3}\pi ab^2$. 9. $\frac{1}{3}\pi h(a^2 + ab + b^2)$. 10. $\frac{112}{15}\pi a^3$. 11. $2n$, $2/n$. 12. $2\pi^2 a^2 b$. 13. $4\pi ab(b^2 - a^2 + a\sqrt{a^2 + b^2})/3\sqrt{a^2 + b^2}$ if $a < b$. 15. (a) $\frac{1}{2}\pi$; (c) 4π.

PROBLEMS 41, p. 256

1. 10. 2. $4\pi r^3/3k$. 3. $20\pi\sqrt{15}$. 4. 6. 5. $6\sqrt{3}$. 6. $\frac{256}{3}\sqrt{3}$. 7. $\frac{32}{3}$. 8. π. 9. $2\sqrt{3}$. 10. (a) $64\sqrt{3}$; (b) 256. 11. $\frac{16}{3}ab^2$. 12. $\frac{4}{3}\pi abc$. 13. $\frac{1}{3}$. 14. $\frac{16}{3}r^3$.

PROBLEMS 42, p. 259

1. 50 inch pounds.   2. $\frac{200}{3}$ foot pounds.   3. 56.25 foot pounds.
4. (a) 14,400 foot pounds; (b) 10,800 foot pounds.   5. $1/\sqrt{2}$ of the way.
6. 250π foot pounds.   7. 875π foot pounds.   8. $6.9 \times 10^4\pi$ foot pounds.
9. $\frac{745}{3} \times 10^7$ foot pounds.   10. 112,500 foot pounds.   11. 460 foot pounds.
12. $54,000\pi$ foot pounds.   13. 1.6×10^7 mile pounds.   14. 5000 mile pounds.

REVIEW PROBLEMS, CHAPTER 5, p. 261

1. (a) $\frac{1}{2}x^2 - \frac{1}{3}t$; (b) $\frac{14}{3} - \frac{5}{3}\sqrt{2}$; (c) $\frac{51}{4}$; (d) $\frac{3}{2}$; (e) $2\cos 1 - 2 - \sin 1$;
(f) $\frac{3}{2} + \sqrt{2}$; (g) $-\frac{8}{3} - \frac{2}{3}\sqrt{3}$; (h) $4\sqrt{3} + 5\pi/3$.   2. (a) $p = q = -1$; (b) $p = -2$;
(c) $(p, q) = (\frac{4}{3}\sqrt{2}, \frac{3}{2})$; (d) $(p, q) = (1, -1)$; (e) $p = q = -1$; (f) $(p, q) = (-1, 2)$.
3. (a) The second; (b) the first; (c) the second; (d) the second; (e) the first;
(f) the second.   4. (a) $\frac{8}{3}$; (b) $\frac{48}{5}\pi$; (c) $\frac{24}{5}\pi$.   5. $\frac{1}{8}\pi$.   6. 2750 foot pounds.
7. About 12.   8. (a) $\{-1, 1\}$; (b) $\{-2\sqrt{2}, 3\}$; (c) $\{-1, 1\}$; (d) 0;
(e) $\{-2\sqrt{3}, 4\}$; (f) $\{-2, 2\}$; (g) $\{-2, 2\}$; (h) $\{-\sqrt{2}, \sqrt{2}\}$.   12. S_1.

PROBLEMS 43, p. 270

1. (b) $F(-2) = 0$, $F(-1) = \frac{3}{2}$, $F(0) = 2$, $F(1) = \frac{5}{2}$, $F(2) = 4$.   2. (b) $F(-2) = 0$,
$F(-1) = -1$, $F(0) = -2$, $F(1) = -1$, $F(2) = 0$; (e) $F'(x) = x/|x|$, $F''(x) = 0$;
(f) $F(x) = |x| - 2$.   3. Domain is $[-3, 3]$, range is $[0, \frac{9}{2}\pi]$, $F(0) = \frac{9}{4}\pi$, $F'(0) = 3$,
$F''(0) = 0$.   4. (a) 0; (b) maximum when $x = \pi$, minimum when $x = 0$;
(d) concave up in $(-\frac{1}{2}\pi, \frac{1}{2}\pi)$, and so on.   7. (a) $5^x + x^5$; (c) $-(1 + x^2)^{-1}$.
8. (a) $2x^5$; (c) $f(-x)$.   9. (d) $S(x) = \pi$.   11. Maximum when $x = 0$,
minimum when $x = 1$.   15. $F(x) = G(x)$.

PROBLEMS 44, p. 275

2. (a) 1.79; (c) 3.3; (e) .83.   3. (a) $(\sqrt{\ln x} - 1)/2x\sqrt{\ln x}$; (c) $-1/x$;
(e) $\cot x - (\cos \ln x)/x$.   4. (a) $-2/x$; (c) $1/x\ln x$; (e) $2\cot x$.
5. (a) 0 if $x \geq 1$, $2/x$ if $x \in (0, 1)$; (c) $\sec x$.   7. (a) $-.6144$; (c) .05.   8. $\frac{1}{3}\ln 2$.
9. $y = 3$, $y' = -2$.   12. $G(2) = .4$, $G(\frac{1}{2}) = .15$.   13. (a) $\ln \frac{1}{2}x$; (c) $\ln (x - 1)$.
15. (a) $\ln \frac{10}{7}$.

PROBLEMS 45, p. 281

1. (a) $f^{-1}(x) = \frac{1}{2}(x + 1)$; (b) $f^{-1}(x) = \frac{1}{4}(3 - x)$; (c) no inverse;
(d) $f^{-1}(x) = \sqrt[3]{x} - 1$.   2. (a) $f^{-1}(x) = (x + 1)^{1/3}$; (c) $f^{-1}(x) = f(x)$.
3. $f^{-1}(x) = \sqrt{x}$ if $x \geq 0$ and $f^{-1}(x) = -\sqrt{-x}$ if $x < 0$, yes, no, not for $x = 0$.
5. (a) $(-\infty, \infty)$; (c) $(-\infty, \infty)$; (e) no inverse; (f) $(-1, 1)$.   6. (a) $-x/y$, $-y/x$;
(b) $-(2x + y^3)/3xy^2$, $-3xy^2/(2x + y^3)$; (c) $(1 - 2xy \cos x^2 y)/(2y + x^2 \cos x^2 y)$,
$(2y + x^2 \cos x^2 y)/(1 - 2xy \cos x^2 y)$;
(d) $\dfrac{dy}{dx} = -\sqrt{y}(4x^{3/2} \sin x^2 + \sqrt{y})/\sqrt{x}(4y^{3/2} + \sqrt{x})$.   8. $b^2 < 3ac$.

9. (a) $y = \sqrt[3]{x} + 1$.

PROBLEMS 46, p. 287

1. (a) $2e^x \cos 2e^x$; (b) $2e^{2x} \cos e^{2x}$; (c) $e^x \sin 2e^x$; (d) $2x \exp x^2 \cos \exp x^2$;
(e) $\cos x \exp \sin x$; (f) $\sin 2x \exp \sin^2 x$.   2. (a) $2xe^{x^2} - 2e^{2x}$; (c) $e^x \exp e^x$;
(e) 0.   3. (a) 10.4986; (c) .0659.   4. (a) e; (c) $\frac{1}{25}$.
5. (a) Maximum when $x = 2$, minimum when $x = 0$.   6. (b) $(A, B) = (\frac{1}{2}, \frac{1}{2})$.
7. $(e^{-y} - 1)/(1 - e^{-x})$.   9. (b) $m = \frac{1}{2}$ or $m = -2$.   10. .57.   12. $\frac{1}{4}\pi$.
15. 1350, 23 years.   16. (a) $13 \exp \frac{1}{2}x^2$.

PROBLEMS 47, p. 291

1. (a) $\frac{1}{3}(e^3 - 1)$; (b) $\frac{1}{2}e^4 + 2e^2 - \frac{1}{2}$; (c) $\ln 2 + \frac{1}{2}$; (d) $\ln 2 - \frac{1}{2}$; (e) 1; (f) -1.
2. $\frac{1}{2}\pi(e^2 - 4e + 5)$.   3. (a) $\ln 2$; (c) $\frac{1}{3}e^3 + 2e - e^{-1} - \frac{4}{3}$; (e) 3.

5. (a) $\displaystyle\int xe^x \, dx = e^x(x - 1)$; (c) $\displaystyle\int xe^x(1 + x)^{-2} \, dx = e^x(1 + x)^{-1}$.

7. (a) $2e^3 + 1$; (b) 3; (c) $1 + (e - 1) \ln 3$; (d) $\frac{1}{2}(e^2 - 1) \ln 3$; (e) $\frac{1}{3}$;

(f) $(e - 1) \ln 3 - 1$.   8. $\displaystyle\int \ln ax \, dx = x \ln ax - x$.   9. $\frac{1}{2}\pi(e - e^{-1})$.   11. -48.
12. e^{-1}.

PROBLEMS 48, p. 298

1. (a) $3^x \ln 9$; (c) $-(\ln 3)/3^x$; (e) $\cos x \, 3^{\sin x} \ln 3$; (g) $x^\pi \pi^x (\ln \pi + \pi/x)$.
2. (a) $1/x \ln 3$; (c) $1/x \ln x \ln 3$; (e) $\log_3 5$.   3. (a) 1; (c) $\pi^e \approx 22.5$, $e^\pi \approx 23.1$.
4. (a) e^π; (c) $(\pi^e - 1)/e \ln \pi$; (e) $8/\ln 27$; (g) $10e$.   5. (a) 4; (b) 4.   6. (a) $3/\ln 4$.
7. (a) $x^{\sqrt{x}}(1 + \frac{1}{2} \ln x)/\sqrt{x}$; (c) $x^{\sin x} [(\sin x)/x + \ln x \cos x]$.   8. $e^{-1/e}$.
10. $-\ln 3/x (\ln x)^2$.

PROBLEMS 49, p. 304

1. (a) $\frac{1}{3}\pi$; (c) $\frac{1}{2}\pi$; (e) .2; (g) $-\pi$. 2. (a) .6; (c) $\frac{4}{5}$; (e) $\frac{1}{2}\sqrt{3}$. 3. (a) x; (c) $\text{Sin}^{-1} x$.
4. (a) Positive; (c) positive; (e) negative. 6. (a) $\frac{1}{2}\pi - 1$; (c) $\frac{1}{12}\pi + \frac{1}{2}\sqrt{3} - 1$.
7. $2^{-1/2}$. 8. $(\sqrt{\frac{1}{2}(\sqrt{5} - 1)}, \text{Sin}^{-1} \frac{1}{2}(\sqrt{5} - 1))$. 13. (a) Cos; (c) $\{(\frac{1}{4}\pi, \frac{1}{2}\sqrt{2})\}$;
(e) Tan.

PROBLEMS 50, p. 308

1. (a) $1/\sqrt{4 - x^2}$; (c) $e^x/\sqrt{1 - e^{2x}}$; (e) $1/(1 + x^2)$.
2. (a) $2x/(1 + x^4) - 2 (\text{Arctan } x)/(1 + x^2)$;
(c) $-1/(\text{Cos}^{-1} x)\sqrt{1 - x^2} + 1/x\sqrt{1 - (\ln x)^2}$; (e) $-\sin x/|\sin x| + x/\sqrt{1 - x^2}$.
3. $2 \text{Tan}^{-1} 2x$. 4. (a) $(\cos x)/|\cos x|$; (c) $(\sin x)/|\sin x|$. 5. (a) $-\sqrt{(1 - y^2)/(1 - x^2)}$;
(c) $-(1 + y^2)/\sqrt{1 - x^2}$. 6. $\dfrac{dy}{dx} = -(\sin x)/|\sin x| - (\cos x)/|\cos x|$.

8. $\text{Arctan } x^{-1} = \frac{1}{2}\pi x/|x| - \text{Arctan } x$. 9. (a) Yes; (c) if $x \in (-1, 1)$. 10. 12 feet.
11. $h/\sqrt{2}$. 12. 1. 13. (b) $\frac{5}{4}\pi^3$.

PROBLEMS 51, p. 312

1. (a) $\frac{1}{6}\pi$; (c) $\frac{1}{4}\pi$; (e) $\frac{1}{18}\pi$; (g) $\frac{1}{8}\pi$. 2. $\frac{10}{9}\pi$. 3. (a) $\frac{1}{3}\pi$; (c) $7\pi/6\sqrt{3}$; (e) .23.
4. (a) $\frac{1}{2}\sqrt{3} + \frac{1}{3}\pi$; (c) $\frac{1}{4}\sin 2 + \frac{1}{2}$. 5. Approximately 63.2. 6. (a) $\frac{1}{2}\pi - 2x$.
8. $\frac{1}{4}\pi^2$. 9. $4\pi - 3\sqrt{3}$. 12. $V \approx \frac{1}{4}\pi^2$.

PROBLEMS 52, p. 319

1. (a) $e^x \sinh e^x - e^{\cosh x} \sinh x$; (c) $\tanh x - \coth x$;
(e) $\sinh x \cos (\cosh x) + \cosh x \sin (\sinh x)$. 2. (a) 3.5256; (c) .43; (e) $\frac{3}{4}$.

8. (b) No; (c) $\displaystyle\int \text{sech } x \, dx = 2 \text{Tan}^{-1} e^x$, $\displaystyle\int \text{sech } x \, dx = \text{Tan}^{-1} (\sinh x)$.

10. (c) $\frac{1}{4}\pi(2 + \sinh 2)$.

PROBLEMS 53, p. 324

3. (a) $2x/\sqrt{x^4 + 1}$; (c) $-2 (\sinh x)^{-3} \cosh x$. 4. (a) $-\csc x$; (c) $-\text{csch } x$;
(e) y only defined when x is an integral multiple of 2π. 5. (a) .9; (c) .4; (e) $\ln 3$;
(g) .0933. 7. (a) $f = g$; (b) $f \subseteq g$. 8. (a) 1; (b) $\frac{1}{4}\sqrt{2}$.

REVIEW PROBLEMS, CHAPTER 6, p. 326

1. (a) $x^{e^e-1} + ex^{e-1}e^{x^e} + e^x e^{e^x}$; (b) $\ln 3 + 4/x$;
(c) $1/\sqrt{1-x^2} - x^{-2}\cos x^{-1} - \cos x \, (\mathrm{Sin}\, x)^{-2}$;
(d) $2 \, \mathrm{Tan}^{-1} x/(1+x^2) + 2x/(1+x^4) + 2 \, \mathrm{Tan}\, x \sec^2 x$; (e) $(\sinh \ln x)/x + \tanh x$;
(f) $e^x/\sqrt{e^{2x}+1} + \coth x$.   4. (a) $\{1, e^2\}$; (b) $\{0, 2\}$; (c) $\varnothing$; (d) $\varnothing$.   8. 1.
10. (a) $\frac{1}{3}\pi$; (b) $\ln \frac{1}{2}(3 + \sqrt{5})$; (c) $(1/b) \, \mathrm{Tan}^{-1} \frac{1}{2}$; (d) $(1/2b) \ln 3$.
11. $(A, B) = (3, -2)$.   12. $(99e^{101} - 100e^{100} + e)/(e-1)$.   13. $r < 1$.
16. A hyperbola.   17. 12, 6, 0, $3 + \sqrt{3}$.

PROBLEMS 54, p. 332

1. (a) $\frac{1}{2} \, \mathrm{Sin}^{-1} 2x$; (b) $\frac{1}{2} \ln |x/(x+2)|$; (c) $(x+1)^{-1} + \ln |x+1|$;
(d) $\frac{1}{9}x^3(\ln x^3 - 1)$.   2. (a) $\ln (1 + \sqrt{2})$; (b) $\frac{1}{8}\pi$; (c) $\frac{1}{4} \ln 3$; (d) $\frac{1}{2} \, \mathrm{Sin}^{-1} 2/\sqrt{5}$.
3. (a) $\frac{1}{6}$; (c) -1.   4. (a) $e^{x+2} + x^{e+3}/(e+3)$.
6. (a) $-1/4(3x-4) + \frac{1}{16} \ln |x/(3x-4)|$;
(c) $-\frac{1}{4}x(5-x^2)^{3/2} + \frac{5}{8}x(5-x^2)^{1/2} + \frac{25}{8} \, \mathrm{Arcsin}\, x/\sqrt{5}$; (e) $\frac{1}{5}x^5(\ln x - \frac{1}{5})$.
7. (a) $\dfrac{7 \cdot 5 \cdot 3 \cdot 1}{8 \cdot 6 \cdot 4 \cdot 2}\dfrac{\pi}{2}$; (c) $9e - 24$.   8. (a) $\frac{1}{1024}(8 + 3\pi)$; (c) $1 + \frac{1}{2}\pi$.   10. $\frac{1}{3}\pi$.
11. $24\pi \, \mathrm{Tan}^{-1} \frac{4}{3}$.   12. $\frac{3}{2} - 2 \, \mathrm{Sin}^{-1} \frac{3}{5}$.   13. $\ln \frac{4}{3}$.   14. $\pi(\ln \frac{8}{5} - \frac{3}{10})$.   15. No.

PROBLEMS 55, p. 338

1. (a) e^{x^3}; (c) $-\frac{1}{3}\cos x^3$; (e) $\frac{1}{6} \, \mathrm{Arctan}\, \frac{3}{2}x$; (g) $\frac{1}{4}(\ln x)^4$; (i) $\ln |\sec x|$.
2. (a) $-\frac{1}{2}\cos(x^2+1)$; (b) $-\frac{1}{2}\cot x^2$; (c) $-(x^2-4)^{-1/2}$; (d) $\frac{1}{2}e^{x^2}$;
(e) $-\frac{4}{3}(\cos \frac{1}{2}x)^{3/2}$; (f) $\frac{1}{6}\sin^3 2x$; (g) $-\frac{1}{24}\cos^4 6x$; (h) $\frac{1}{3}\sec^3 x$; (i) $\sinh^{-1} e^x$;
(j) $x - \ln(e^x + 1)$.   3. (a) $\exp \sin x$; (c) $\frac{1}{2}\tan(2x+3)$; (e) $\frac{2}{3}\sqrt{x^3-5}$;
(g) $\sec e^x$.   4. (a) $\frac{1}{3}\sec^3 x$; (c) $\ln [(\sec 2x + \tan 2x)^{3/2} \sec^2 2x]$; (e) $e^x - \ln(1 + e^x)$.

5. (a) $\frac{1}{3} \, \mathrm{Arctan}\, \frac{1}{3}(x+2)$; (c) $\frac{1}{2} \, \mathrm{Arcsin}\,(x-1)$.   6. (a) $u = e^x, \int f(u)\, du$;

(c) $u = \sqrt{x}, 2 \int f(u)\, du$; (e) $u = x^r, r^{-1} \int f(u)\, du$; (g) $u = mx + b, m^{-1} \int f(u)\, du$.
7. (a) $\frac{1}{2} \, \mathrm{Sin}^{-1} x^2$; (b) $\frac{1}{2} \, \mathrm{Tan}^{-1} x^2$; (c) $x(\sin \ln x - \cos \ln x)$;
(d) $\ln(\sin^2 x + \sin x + 1) - (2/\sqrt{3}) \, \mathrm{Tan}^{-1} [(2\sin x + 1)/\sqrt{3}]$.
8. $\tan x + \frac{2}{3}\tan^3 x + \frac{1}{5}\tan^5 x$.

PROBLEMS 56, p. 344

1. (a) $\frac{1}{4}\sqrt{3}$; (c) $\frac{1}{3}\ln 2$; (e) $\frac{3}{10}$; (g) $1 - \cos 1$; (i) $2e(e - 1)$; (k) $\frac{1}{4}(1 - \operatorname{sech}^4 1)$.
2. $\frac{1}{2}\ln 2$.   3. (a) $\frac{37}{3}$; (c) $\ln \frac{13}{5}$; (e) $\frac{1}{6}$.   4. $\frac{1}{24}\pi$.   5. (a) $e^7 - e^2$; (c) $\frac{1}{4}\pi$; (e) $-2/\pi$;
(g) $-\frac{1}{8}(\ln 2)^2$.   6. $\frac{32}{3}$.   9. 4.   10. $\frac{8}{3}\pi$.   11. $\frac{1}{5}\pi$.

PROBLEMS 57, p. 349

1. (a) $-x - \cot x$; (c) $\frac{1}{3}\sin^3 x - \frac{1}{5}\sin^5 x$; (e) $-\frac{2}{3}(\cos x)^{3/2}$.   2. (a) $\frac{1}{2}$; (c) $\frac{1}{2}\pi$.
4. (a) $\frac{1}{4}\sec^4 x$; (c) $\ln |\sec x + \tan x| - \sin x$.   5. (a) $\dfrac{7 \cdot 5 \cdot 3 \cdot 1}{8 \cdot 6 \cdot 4 \cdot 2}\dfrac{\pi}{2}$; (b) $\dfrac{8 \cdot 6 \cdot 4 \cdot 2}{9 \cdot 7 \cdot 5 \cdot 3}$;
(e) $(7 \cdot 5 \cdot 3 \cdot 1)^2\, \pi/2^9 \cdot 8!$; (g) $\frac{8}{3}$.   7. (a) $-\sqrt{x^2 + 25}/25x$; (c) $\sqrt{x^2 - 4}/4x$;
(e) $-\operatorname{Sin}^{-1}\frac{1}{3}x - \sqrt{9 - x^2}/x$; (g) $\sqrt{x^2 - 9} + 3\operatorname{Sin}^{-1} 3/x$.   8. (a) $\frac{1}{64}(\pi + 2)$;
(c) $(2 - \sqrt{2})/9$; (e) $\sqrt{3} - \frac{1}{3}\pi$.   9. $\frac{1}{8}(\pi + 4)$.   10. $\frac{1}{2}\pi^2$.   11. $\frac{1}{2}\pi$.

PROBLEMS 58, p. 356

1. (a) $x^2 - x + 1 - 1/(x + 1)$; (c) $1 + 1/(x - 1)$; (e) $(5x + 2)/(x^2 + 4)$;
(g) $2/(x + 2) + 3/(x - 2)$.   2. (a) $-1/(x + 1)$; (c) $x - \ln (x + 1)^2 - 1/(x + 1)$.
4. (a) $\ln (x + 2)^3/(x + 3)^2$; (c) $\frac{1}{2}x^2 - 2x + 3\ln |x + 1| + 1/(x + 1)$;
(e) $\ln |x| + \frac{1}{2}\operatorname{Tan}^{-1}\frac{1}{2}x$.   5. (a) $\ln (x + 1)^2\sqrt{x^2 + x + 1} - \sqrt{3}\operatorname{Tan}^{-1}(2x + 1)/\sqrt{3}$;
(c) $\ln |x - 1| + \operatorname{Arctan} x$; (e) $\ln \sqrt{(x - 1)/(x + 1)} - \operatorname{Tan}^{-1} x$.   6. (a) $4\ln \frac{4}{3} - \frac{3}{2}$;
(c) $\frac{1}{3}\ln 2 + \frac{1}{9}\sqrt{3}\pi$; (e) $\ln \frac{12}{5}(\frac{20}{13})^{3/2}$.   7. Let $u = x^3 + 1$.   11. (a) $-2/(\tan \frac{1}{2}x + 1)$;
(c) $\frac{1}{2}\operatorname{Tan}^{-1}\frac{1}{4}(5\tan \frac{1}{2}x + 3)$.

PROBLEMS 59, p. 363

1. (a) $\sin x - x\cos x$; (c) $\frac{1}{2}xe^{2x} - \frac{1}{4}e^{2x}$; (e) $(x + 25)^{101}(101x - 25)/101 \cdot 102$;
(g) $x\operatorname{Sin}^{-1} x + \sqrt{1 - x^2}$; (i) $\underline{x\sinh^{-1} x - \sqrt{x^2 + 1}}$.   2. (a) $\frac{32}{15}$; (c) $\frac{1}{2}\pi - 1$;
(e) $2(\ln 2 - 1)^2$.   3. (a) $x^2\sin x + 2x\cos x - 2\sin x$;
(c) $\frac{1}{4}(x^4 - 1)\operatorname{Tan}^{-1} x - \frac{1}{12}x^3 + \frac{1}{4}x$; (e) $\pi - 2$.   4. (a) $\frac{1}{13}x^{13}(\ln x - \frac{1}{13})$;
(b) $\frac{1}{2}x^2 \ln 12x - \frac{1}{4}x^2$; (c) $6x^2 \ln x - 3x^2$; (d) $\frac{1}{13}x^{13}(\ln 12x^{12} - \frac{1}{13})$.
7. (a) $\frac{1}{3}(e^3 - 1)\ln 2 + \frac{1}{9} - \frac{2}{9}e^3$; (b) $\frac{5}{27}e^3 - \frac{2}{27}$; (c) $\frac{1}{2}e^\pi + \frac{1}{2}$; (d) $-\frac{1}{2}e^\pi - \frac{1}{2}$.
8. (a) $\frac{1}{2}x^2 \sin x^2 + \frac{1}{2}\cos x^2$.   9. $\frac{4}{15}\pi$.   11. (c) $\frac{4}{5} \cdot \frac{2}{3}$ and $\frac{5}{6} \cdot \frac{3}{4} \cdot \frac{1}{2} \cdot \frac{1}{2}\pi$.   13. $\frac{3}{16}\pi a^4$.

PROBLEMS 60, p. 372

1. (a) $y = \frac{1}{2}(3 - \cos x^2)$; (c) $y = 3 \ln x + 3$.   2. (a) $\sin \ln x + 2$; (c) $x|x| + 4$.
3. (a) $\cos x + \sin y = 1$; (c) $y = x$; (e) $y = -\ln(2 - e^x)$; (g) $y = \text{Tan}^{-1} x^2 + \pi$.
4. (a) $\sin x + x$; (c) $\frac{1}{4}x^4 - x^2 + 5x + 1$.   5. (a) 3.24.   6. $x^2 + y^2 = 13$.
7. (a) $y = 4x^2 + 4$; (b) $y = 2^{x+2}$; (c) $y = 2^{x^2+2}$.   8. $y^2 - 3x^2 = 1$.
9. $16(1 + \sqrt{3})$.   10. No.   11. $v_{100} \approx 4000$ feet per second, $s_{100} \approx 130{,}000$ feet,
rocket reaches a height of about 360,000 feet.   12. $\sqrt{\frac{19}{2}}$.

PROBLEMS 61, p. 377

1. (a) $y = 1000\, e^{.05x}$; (b) about 14 years; (c) about 14 years.
2. (a) $y = -\frac{1}{2}(e^{-2x} - 3)$; (c) $y = (1 - x)e^x$; (e) $y = -3 \sec x$.   3. (a) $y = 2x^3$;
(c) $y = 2x^5(e^{x^2} - e)$.   4. $w(100) \approx 4$.   5. $y = x^x e^{1-x}$.
6. $(y - y_0)/(y_1 - y_0) = (x - x_0)/(x_1 - x_0)$.   7. $y = 3^{x^2+1}$.
8. $p = (A - 2.5)e^{-2t} + 2.5, 2.5$.   9. $A(t) = 10(1 - \exp(-t/10))$.   10. 272.

REVIEW PROBLEMS, CHAPTER 7, p. 378

1. (a) $\frac{1}{3}e^{x^3}$; (b) $\frac{1}{2}(x^2 - 1)e^{x^2}$; (c) $\frac{1}{6}x^3 + \frac{1}{8}x^2 + \frac{1}{8}x - \frac{63}{16}\ln|2x - 1|$;
(d) $\frac{1}{8}\ln[(x - 2)^2/(x^2 + 2x + 4)] + \frac{5}{12}\sqrt{3}\,\text{Tan}^{-1}\frac{1}{3}\sqrt{3}(x + 1)$; (e) $\frac{2}{9}(x^3 + 1)^{3/2}$;
(f) $\frac{1}{15}(x^2 + 1)^{3/2}(3x^2 - 2)$; (g) $\frac{1}{4}x^2 + \frac{1}{8}\sin 2x^2$; (h) $x^2 \sin x + 2x \cos x - 2 \sin x$.
2. $I_2 = \frac{1}{2}(e - 1), I_3 = \frac{1}{2}$.   3. $(e^{x^2} - 1)/x$ and $(e^t - 1)/2t$.   4. (a) $2 \exp(\sin^2 x) - 2$;
(b) $\cos x (1 - \ln \cos x) - 1$; (c) $1 - \cos x^2$; (d) $x^{-2} \sin x^2 - \cos x^2$;
(e) $x^{-1} \text{Tan}^{-1} x^{-1} \sin x$; (f) $x^{-1}(\text{Tan}^{-1} xe^x - \text{Tan}^{-1} x)$.   7. $\ln(1 + \frac{2}{3}\sqrt{3})$.
8. (a) 1; (b) $\frac{1}{2}\pi$; (c) $\frac{1}{4}\sqrt{2}\ln(3 + 2\sqrt{2})$;
(d) $\frac{1}{2}\sin 6 - \frac{1}{2}\sin 4 + \frac{1}{3}\sin 9$.   9. (a) $\omega = 120\pi$; (b) $\sqrt{2}\,115$.
14. $y = r - \ln|1 + as - sx|$.
16. Strictly speaking, we should stay in the interval $(-\infty, 0)$.   17. (a) $5u + 100$;
(b) $50u - u^2 + (u - 5)|u - 5| + 375$; (c) $10u + 5e^{-2u} + 205$;
(d) $10u + 10\displaystyle\int_0^u e^{-t^2}\,dt + 200$; (e) $80 + 5u - u\,\text{Tan}^{-1} u + \frac{1}{2}\ln(1 + u^2)$;
(f) $50 + 10u - 2\ln\cosh 3u$.

PROBLEMS 62, p. 387

1. (a) $(4\sqrt{3}, 4)$; (c) $(5\sqrt{2}, -5\sqrt{2})$; (e) $(30, 30\sqrt{3})$.   2. (a) $(3, \pi)$; (c) $(2, -\frac{1}{4}\pi)$;
(e) $(8, -60°)$.   3. (a) $x^2 + y^2 = 4x$; (c) $x = 3$; (e) $y = 2x$.   6. $(\frac{1}{2}, -\frac{1}{2}\pi), (2, \frac{1}{2}\pi)$.
7. $(\frac{3}{2}\sqrt{2}, \frac{3}{4}\pi)$.   8. $r(\sin \theta - 2\cos \theta) = 2$.   9. No.   10. (a) $(2, 30°)$ and 7 other points;
(c) $\{(0, 0), (1, \frac{1}{4}\pi), (1, \frac{3}{4}\pi)\}$; (e) $\{(0, 0), (3, \frac{2}{3}\pi), (3, \frac{4}{3}\pi)\}$.

PROBLEMS 63, p. 394

1. (a) $r \cos (\theta - 30°) = 6$; (c) $r \cos (\theta + \frac{1}{4}\pi) = 8$. 2. (a) 6; (c) 11.
3. (a) $r = 2\sqrt{2} \cos (\theta - 135°)$; (c) $r^2 - 2\sqrt{2}\, r \cos (\theta - 135°) = 11$.
4. (a) Hyperbola, vertices: $(2, \frac{1}{2}\pi)$, $(-8, \frac{3}{2}\pi)$, foci: $(0, 0)$, $(10, \frac{1}{2}\pi)$;
(c) parabola, vertex: $(2, \frac{1}{2}\pi)$, focus: $(0, 0)$; (e) parabola, vertex: $(\frac{1}{2}, \frac{1}{2}\pi)$, focus: $(0, 0)$.
5. (a) $r = 52/(9 - 13 \cos \theta)$; (c) $r = 12/(5 - 3 \cos \theta)$. 8. $2ep$.
12. $x^2(1 - e^2) + y^2 = 2pe^2 x + e^2 p^2$.

PROBLEMS 64, p. 399

1. $\psi \approx 81°$. 2. (a) Undefined; (c) $-1/\sqrt{3}$. 3. (a) $\theta = \frac{1}{4}\pi$, $\theta = -\frac{1}{4}\pi$;
(c) same as (a). 4. (a) -2; (c) $\sqrt{\pi} + 2\pi$. 5. (a) $\sec \theta$; (b) $\theta \ln \theta$; (c) $\cot \theta$;

(d) $\tan 2\theta$; (e) $\tan 3\theta$; (f) $\tan \theta \ln \sin \theta$; (g) $e^{-\theta}$; (h) $\csc \theta^2 \displaystyle\int_0^\theta \sin t^2\, dt$.

6. (a) $\theta = 0$; (b) $\psi = \theta$. 8. $(4, \pi)$, $(1, \frac{1}{3}\pi)$, $(1, -\frac{1}{3}\pi)$. 9. $(3, \frac{2}{3}\pi)$. 11. No.
13. (a) $30°, 30°, 90°$. 15. $\frac{1}{2}\pi$.

PROBLEMS 65, p. 403

1. (a) $\frac{1}{4}\pi$; (b) $\frac{1}{8}\pi$; (c) 1; (d) $\frac{1}{2}$; (e) $\frac{1}{32}\pi$; (f) $\frac{1}{36}\pi$; (g) $\frac{1}{12}\pi$; (h) $\frac{1}{3}$; (i) $\frac{3}{16}\pi$; (j) $\frac{3}{16}\pi$.
2. (a) $\frac{4}{3}\pi^3$; (b) $\frac{1}{2}(e^{4\pi} - 1)$; (c) $\frac{1}{2}\pi + \frac{1}{8} \sinh 4\pi$. 3. 25π. 4. $\frac{1}{32}\pi$. 5. (a) $\frac{1}{8}\pi$; (b) 2;
(c) 1. 6. $\pi - \frac{3}{2}\sqrt{3}$. 7. 11π. 8. 8 square units. 9. π. 11. $\frac{32}{3}$. 13. $8\pi/3\sqrt{3}$.
14. $\frac{1}{2}ab$.

PROBLEMS 66, p. 408

1. (a) $(3 - 2\sqrt{3}, 2 + 3\sqrt{3})$; (c) $(2 - 3\sqrt{3}, -3 - 2\sqrt{3})$; (e) $(4, -6)$; (g) $(-4, 6)$.
2. (a) $(1, \sqrt{3})$; (c) $(-1 - 2\sqrt{3}, 2 - \sqrt{3})$. 3. (b) $x = .8\bar{x} - .6\bar{y}$, $y = .6\bar{x} + .8\bar{y}$;
(c) about $37°$; (d) $\bar{x} = 2$; (e) $\bar{x}^2 + \bar{y}^2 = 16$. 4. (a) 2; (b) 17; (c) 5; (d) 1; (e) 1;
(f) 1. 5. (a) $\bar{x} = y$, $\bar{y} = -x$; (c) $\bar{x} = -y$, $\bar{y} = x$. 6. (a) $\bar{x} = \frac{1}{2}\sqrt{3}x + \frac{1}{2}y$,
$\bar{y} = -\frac{1}{2}x + \frac{1}{2}\sqrt{3}y$; (c) $(2\sqrt{3} - 3)\bar{x} - (2 + 3\sqrt{3})\bar{y} = 5$. 7. (b) .28 radian;
(c) -3.478; (d) $-.2876$; (e) 2.08 radians; (f) 2.64 radians.
10. $x = \bar{x} \cos \alpha - \bar{y} \sin \alpha + h \cos \alpha - k \sin \alpha$, $y = \bar{x} \sin \alpha + \bar{y} \cos \alpha + h \sin \alpha + k \cos \alpha$.

PROBLEMS 67, p. 415

1. (a) $3\bar{x}^2 - \bar{y}^2$; (c) $5\bar{x}^2$; (e) $-\frac{11}{2}\bar{x}^2 + \frac{39}{2}\bar{y}^2$. 2. (a) $4\bar{x}^2 - \bar{y}^2 = 4$;
(c) $4\bar{x}^2 - \bar{y}^2 = 4$. 3. (a) $4\bar{x}^2 + \bar{y}^2 = 16$; (c) $\bar{y}^2 = 4\bar{x}$. 4. Two intersecting lines.
5. (a) Circle; (b) ellipse; (c) hyperbola; (d) hyperbola. 6. (a) Ellipse, empty set;
(c) parabola, two lines. 7. (a) $p > \frac{9}{7}$ (we cannot choose p so as to get a parabola);
(c) $p \in (-\sqrt{40}, \sqrt{40})$. 8. (a) Parabola; (c) hyperbola. 11. $A = C, B = 0$.
12. The point $(0, 0)$, one line, two lines, or the whole plane. 13. No.

PROBLEMS 68, p. 421

1. (a) $(1, 2)$; (c) $(1, 3)$. 2. (a) $3\sqrt{5}$; (b) $(0, 4)$; (c) $(5, 14)$; (d) $(-4, -4)$.
4. (a) $2\mathbf{PQ}$; (c) $\mathbf{PR}$; (e) $\mathbf{QP}$. 5. $\mathbf{a} - \mathbf{b}$. 6. (a) $\sqrt{|\mathbf{a}|^2 + |\mathbf{b}|^2}$; (c) $\sqrt{9|\mathbf{a}|^2 + 16|\mathbf{b}|^2}$;
(e) $\sqrt{2}\,|\mathbf{a}|\,|\mathbf{b}|$. 7. **0**. 8. (a) $t = 2$; (b) no solution. 9. $t = \frac{1}{2}$, no.
11. $(1 - 2/\sqrt{5}, 1 + 1/\sqrt{5})$ or $(1 + 2/\sqrt{5}, 1 - 1/\sqrt{5})$. 12. (a) $r \geq 0$;
(c) $\mathbf{a} = r\mathbf{b}$, where $r \geq 1$. 13. The second expressions are meaningless.

PROBLEMS 69, p. 426

1. (a) $3\mathbf{i} + 4\mathbf{j}, 5$; (c) $-5\mathbf{j}, 5$. 2. (a) $(4, -6)$; (c) $(6, -2)$; (e) $(16, -3)$.
4. (a) $(4, 3)$; (c) $(19, 23)$; (d) $(\frac{3}{5}\mathbf{i} + \frac{4}{5}\mathbf{j})$; (e) $(-17, -25)$; (f) $(2, -2)$ or $(-12, -16)$
or $(12, 16)$. 5. $45°, 45°, 90°$. 6. $2x + 3y = 0$. 7. (a) The line $3x + 4y = 12$;
(c) the circle $(x + 3)^2 + (y + 4)^2 = 144$. 9. $r = -\frac{1}{10}, s = -\frac{7}{10}\sqrt{10}$.

PROBLEMS 70, p. 433

1. (a) Semicircle; (c) line segment; (e) branch of a hyperbola.
3. (a) $\mathbf{r} = (3t + 2)\mathbf{i} + (3 - t)\mathbf{j}$; (c) $\mathbf{r} = 2(t - 1)\mathbf{i} - 3t\mathbf{j}$;
(e) $\mathbf{r} = (t - 1)\mathbf{i} + (3t + 1)\mathbf{j}$. 4. (a) $m = -\frac{5}{3}, b = -\frac{2}{3}$; (c) $m = 2, b = 0$.
5. (a) $x = 1 - t, y = 10 - 3t$; (c) $x = 5, y = t$. 6. (a) $x = t^2, y = t^3$ and
$x = t^{2/3}, y = t$; (c) $x = \sec t, y = \tan t$ and $x = \coth t, y = \operatorname{csch} t$.
11. (a) $(3, 2)$ and $(2, 1)$. 12. $\mathbf{m}, \mathbf{n}$, and $\mathbf{b} - \mathbf{c}$ all parallel.
13. $\mathbf{m} = b\mathbf{i} - a\mathbf{j}, \mathbf{b} = -c(a^2 + b^2)^{-1}(a\mathbf{i} + b\mathbf{j})$.

PROBLEMS 71, p. 441

1. (a) $\mathbf{i} + \mathbf{j}$; (c) $\sin 2(-\mathbf{i} + \mathbf{j})$. 2. $\mathbf{i} - \mathbf{j}, m = -e^{-2}$. 3. (a) $\frac{3}{5}\mathbf{i} + \frac{4}{5}\mathbf{j}$;
(c) $(1 + e^2)^{-1/2}(e\mathbf{i} + \mathbf{j})$. 4. $\mathbf{r} = (t + 1)e\mathbf{i} + t\mathbf{j}$. 6. $8a$.
7. $\frac{1}{2}\sqrt{5} + \frac{1}{4}\ln(2 + \sqrt{5})$. 8. 6. 9. $6a$. 10. $\frac{61}{27}$. 11. $\frac{56}{27}$. 12. $\frac{17}{12}$.

PROBLEMS 72, p. 451

1. (a) 2; (c) $2^{-1/2}$. 2. (a) 2; (c) 0.
3. Radius of curvature a minimum when $x = \pm 45^{-1/4}$, minimum curvature (0) at
$(0, 0)$. 5. $y' \approx 0$. 7. (a) $b\mathbf{j}$; (b) maximum speed a, minimum speed b; (c) $\mathbf{a} = -\mathbf{r}$.
8. (a) $(1, -3)$; (b) $(71, 7)$; (c) no; (d) yes; (e) no; (f) after 3 seconds;
(g) $\sqrt{293}$ centimeters per second; (h) E; (i) parabola; (j) somewhat W of N;
(k) $\mathbf{a} = 6\mathbf{i}$. 9. (a) $\frac{1}{30}\sqrt{10}$ and $\frac{1}{10}\sqrt{10}$; (b) 0 and $8\sqrt{2}$ 10. (a) $\mathbf{v} = 8\pi\mathbf{i}$, $\mathbf{a} = -8\pi^2\mathbf{j}$.

PROBLEMS 73, p. 456

1. (a) $\frac{49}{4}\pi$. 2. (a) $\frac{208}{3}\pi$; (c) $\frac{1}{2}\pi(2 + \sinh 2)$; (d) $\frac{208}{9}\pi$. 3. (a) $\frac{8}{27}\pi(10^{3/2} - 1)$;
(b) $2\pi[\sqrt{2} + \ln(1 + \sqrt{2})]$. 4. $\frac{13}{3}\pi$. 5. $\pi(\frac{3200}{243}\sqrt{10} + \frac{128}{1215})$. 6. $\frac{2}{3}\sqrt{2}\pi(1 - 2e^{-\pi})$.
7. (a) About Y-axis; (b) about Y-axis; (c) about X-axis; (d) about X-axis.
9. (a) $18\pi + \frac{75}{2}\pi \operatorname{Arcsin} \frac{4}{5}$; (b) 36π; (c) $50\pi + \frac{45}{2}\pi \ln 3$.
11. $2\pi a^2(2 - \sqrt{2})$.

REVIEW PROBLEMS, CHAPTER 8, p. 458

1. $Q(r, \pi - \theta)$, $R(-r, \theta)$, $S(r, -\theta)$. 4. $r = 4/(1 - \cos(\theta - \frac{1}{4}\pi))$. 6. $r = 2\sqrt{2}e^\theta$.
7. Any integral multiple of $\frac{1}{2}\pi$. 8. $(4\sqrt{2} + 3\sqrt{3})x - (3\sqrt{2} - 4\sqrt{3})y = 7\sqrt{2} - \sqrt{3}$.
9. (a) $2\mathbf{j}$; (b) $3\mathbf{i} - 4\mathbf{j}$; (c) $\mathbf{r} = 3t\mathbf{i} + (2 - 4t)\mathbf{j}$; (d) $\frac{61}{27}$. 10. (b) No.
15. $\mathbf{r} = \mathbf{f}'(a)t + \mathbf{f}(a)$. 16. (a) $\frac{1}{2}\sqrt{5} + \frac{1}{4}\ln(2 + \sqrt{5})$; (b) $2(1 + 4t^2)^{-3/2}$;
(c) $\frac{1}{6}\pi(5^{3/2} - 1)$.

PROBLEMS 74, p. 464

1. (a) $m = \frac{1}{4}\pi$; (c) $m = \frac{1}{4}\pi$. 2. (a) 3; (c) $-\frac{5}{2}$; (e) $1/2\pi$; (g) $\frac{3}{5}$. 3. (a) 2; (c) $\frac{1}{2}$;
(e) $\frac{1}{4}$. 4. (a) 1; (c) 0; (e) $\frac{1}{2}$. 5. (a) 1; (c) e. 6. If $r = 1$, the limit is e.
7. (a) $\frac{1}{2}$. 8. 0. 9. $3a$ units to the left of a. 10. Et/L.
11. $\sin f(a)/f(a)$ if $f(a) \neq 0$, 1 if $f(a) = 0$. 12. (a) 1; (c) 1; (e) 0.

PROBLEMS 75, p. 470

1. (a) $\frac{3}{7}$; (c) $-\frac{1}{2}\pi$; (e) 1; (g) ∞; (i) $\frac{1}{2}\pi$. 2. (a) 1; (c) 0; (e) 0. 3. (a) 0; (c) 0.
4. (a) 0; (c) ∞. 5. (a) 1; (c) 1. 6. (a) 0; (c) $\frac{1}{2}$. 7. e. 8. About -16. 9. r.
10. 0.

PROBLEMS 76, p. 476

1. (a) 1; (c) $1/2e^{25}$; (e) diverges; (g) diverges. 2. (a) $\frac{1}{2}\pi$; (c) diverges.
3. (a) Diverges; (c) converges; (e) converges. 4. (a) $\frac{3}{2}$; (c) $\frac{1}{2}\pi$; (e) diverges.
5. (a) Diverges; (c) π. 6. $(r-1)^{-2}$. 8. e. 9. (a) Diverges; (b) proper.
10. (c) $\frac{1}{2}$. 11. 42.24×10^9 foot pounds. 12. kq/r.

PROBLEMS 77, p. 483

1. (a) Converges; (c) converges; (e) diverges; (g) converges. 2. (a) Yes; (c) yes.
3. (a) Converges; (c) converges. 4. (a) Converges; (c) converges; (e) diverges;
(g) converges. 5. (a) Converges; (c) converges; (e) diverges. 6. (a) 2π; (c) $\frac{1}{2}\pi$.
9. No.

REVIEW PROBLEMS, CHAPTER 9, p. 485

1. e^r. 2. (a) 1; (b) $\frac{1}{3}$; (c) 1; (d) 1. 3. **j.** 4. ∞. 7. -1. 10. (a) Converges;
(b) converges; (c) diverges; (d) diverges; (e) converges; (f) converges.
11. (a) Diverges; (b) diverges; (c) converges. 12. (a) Converges; (b) diverges;
(c) diverges; (d) diverges; (e) converges; (f) diverges. 15. ln 4.5.

PROBLEMS 78, p. 493

1. (fifth term, convergent?, limit, smallest closed interval): (a) $(-32, \text{no}, —, [-\infty, \infty])$;
(c) $(\frac{1}{32}, \text{yes}, 0, [0, \frac{1}{2}])$; (e) $(-32, \text{no}, —, [-\infty, -2])$; (g) $(0, \text{no}, —, [-1, 1])$;
(i) $(0, \text{no}, —, [0, 1])$; (k) $(\frac{1}{5}\sin 5, \text{yes}, 0, [\frac{1}{5}\sin 5, \sin 1])$; (m) $(\frac{9}{7}, \text{yes}, 2, [\frac{4}{3}, 2])$;
(o) $(\frac{1}{42}, \text{yes}, 0, [0, \frac{1}{3}])$; (q) $(\frac{63}{256}, \text{yes}, 0\text{—hard to show}, [0, \frac{1}{2}])$;
(s) $\left(\int_0^5 e^{-x^2}\,dx, \text{yes}, \frac{1}{2}\sqrt{\pi}, \left[\int_0^1 e^{-x^2}\,dx, \frac{1}{2}\sqrt{\pi}\right]\right)$. 2. (a) $c_n = 5n!$; (c) $c_n = 3 \cdot 2^{1-n}$;
(e) $c_n = 2^{2^{n-1}}$; (g) $c_n = 6$ if $n > 5$. 3. (a) $2^n - 1$; (c) $n^2 - 1$; (e) $\sin \frac{1}{2}n\pi$.
(a) 9; (c) 12. 6. (a) $\frac{1}{2}(3^{n-1} + 1)$. 7. $\frac{1}{2}n \sin 2\pi/n, \pi$. 8. $a_n = \frac{1}{2}(n+1)/n$.
11. (a) The positive integers, the integers greater than h.

PROBLEMS 79, p. 500

1. (a) $S_n = 3n$; (c) $S_n = \frac{1}{2}((-1)^n - 1)$; (e) $S_n = \sin(n+1) - \sin 1$;
(g) $S_n = \frac{1}{2}n(n+1)$. 2. (a) $a_k = 1$; (c) $a_1 = 0, a_k = 2(-1)^k$; (e) $a_1 = 1, a_k = 0$.
4. (a) 3; (c) $\frac{36}{5}$; (e) $\frac{1}{2}$; (g) $\tan^2 x$. 7. 1. 8. (b) $\frac{1}{4}$. 9. $S_n = \ln(1 + 1/n), S = 0$.
11. (a) $a_k = 2/k(k+1)$; (b) $S_n = 2n/(n+1)$; (c) 2. 12. (a) e^{-1}; (b) yes. 13. No.

PROBLEMS 80, p. 505

5. All converge.   7. $\frac{243}{32}$.   10. $4^{1,000,000}$ will be more than sufficient.

PROBLEMS 81, p. 511

1. (a) Converges absolutely; (b) converges conditionally; (c) converges absolutely;
(d) diverges; (e) converges absolutely; (f) diverges.   2. All converge absolutely.
3. (a) Conditionally convergent; (c) conditionally convergent;
(e) same series as in part (a).   4. (a) .139; (c) .9884.
5. Converges if $r > 0$, absolutely if $r > 1$.   6. $\frac{3}{5} \cdot (\frac{2}{3})^n$.   8. All diverge.   9. .83.
10. Diverges, converges.

PROBLEMS 82, p. 518

1. (h) and (j) diverge.   2. (a) and (b) diverge.   3. The series diverges.   5. $r < -1$.
7. 1.075.

PROBLEMS 83, p. 524

1. (b), (d) diverge.   3. (b), (h) diverge.   4. (a), (b) converge.   5. No, no.
6. (a) $x \in (-1, 1)$; (c) $x < 0$; (e) $|x| \neq 1$; (g) $x \in (1/e, e)$.   7. Yes.   8. (a) 1; (c) 0.
12. It is proved in more advanced courses that whenever Theorem 83-2 applies,
so does Theorem 83-1.

PROBLEMS 84, p. 531

1. (a) 2, $(-2, 2)$, $(-2, 2)$; (c) 3, $(-4, 2)$, $(-4, 2)$; (e) 1, $(3, 5)$, $[3, 5)$;
(g) 1, $(-1, 1)$, $(-1, 1)$; (i) 1, $(-1, 1)$, $[-1, 1)$; (k) 1, $(1, 3)$, $(1, 3)$.   2. (a) $(0, 4)$;
(c) $(-\infty, \infty)$; (e) $(-e^{-1}, e^{-1})$; (g) $(1, 3)$.   3. (a) $[-1, 1]$; (b) $[-1, 1)$;
(c) $(-1, 1)$.   4. (a) $1/(1 + x)$; (c) $-3/x$.   8. (a) $\displaystyle\sum_{k=0}^{\infty} x^k$ and $\displaystyle\sum_{k=0}^{\infty} -x^k$;
(b) $1 + 2x + 2x^2 + 2x^3 + \cdots$ and $1 - 2x + 2x^2 - 2x^3 + \cdots$; (c) 1 and $1 - x$.

PROBLEMS 85, p. 536

1. (a) $[-1, 1]$; (b) $-\frac{65}{144}$; (d) 0.   2. (a) $(-\infty, \infty)$; (b) $(-\infty, \infty)$;
(f) $f'(x) = 1 + \displaystyle\int_0^x f(t)\, dt$.   3. $(4x + 2)/(1 - x^2)$.   6. (a) 1; (b) $-.33$.
7. (c) $f(x) = e^{2x}$.   8. 2, ln 4.   12. $(-\infty, \infty)$, no.

PROBLEMS 86, p. 541

1. (a) $10!$; (c) $8!$. 2. (a) $-2 - 3x + x^3$; (c) $-3(x + 1)^2 + (x + 1)^3$.

3. $\sum_{k=0}^{\infty} (-1)^k(k + 1)(x - 1)^k$, radius of convergence 1. 4. -3^7.

8. $x + \frac{1}{3}x^3 + \frac{2}{15}x^5 + \cdots$.

PROBLEMS 87, p. 545

1. (a) $x + x^2 + \frac{1}{3}x^3 - \frac{1}{30}x^5 + \cdots$; (c) $1 + \frac{1}{2}x^2 + \frac{5}{24}x^4 + \frac{61}{720}x^6 + \cdots$.
2. $\sin(x - 3) = (x - 3) - \frac{1}{6}(x - 3)^3 + \frac{1}{120}(x - 3)^5 + \cdots = -\sin 3 + \cos 3\, x + \frac{1}{2}\sin 3\, x^2 + \cdots$. 3. $\sin x = \sin a + \cos a\,(x - a) - \frac{1}{2}\sin a\,(x - a)^2 + \cdots$.
4. $\sin^2 x = x^2 - \frac{1}{3}x^4 + \cdots$. 6. (a) 1.01; (c) .99; (e) .22. 7. (a) .764; (c) .079.
8. $x + 1 \cdot x^3/2 \cdot 3 + 1 \cdot 3 \cdot x^5/2 \cdot 4 \cdot 5 + \cdots$. 9. .659.
10. $1 - x + x^2/2! - x^3/3! + \cdots$

REVIEW PROBLEMS, CHAPTER 10, p. 546

1. $a_n = 2^{(2-2^n)} \to 0$. 2. $\ln \frac{1}{2}\pi$. 3. (b) and (f) are divergent. 4. $3^{533}/533!$.
6. (a) $[-2, 0)$. 9. $(-1, 1)$. 10. (a) $e/(e - 1)$; (b) $(e - 1)/(e + 1) = \tanh\frac{1}{2}$;
(c) $(e + 1)(e^e - 1)$.

PROBLEMS 88, p. 552

1. (a) $(-2, 1)$; (c) $(1, 1)$; (e) $(3, -4)$. 2. (a) $(1, 2, -1)$; (c) $(2, 0, 1)$;
(e) $(1, -1, 0, 2)$. 3. (a) $(0, 0, 0)$. 4. (a) $(\frac{1}{2}, -\frac{1}{3})$; (c) (e^2, e^{-1}).
5. $x = (ud - vb)/(ad - bc)$, $y = (va - uc)/(ad - bc)$.
6. 60% cheap and 40% expensive. 7. 58. 9. $80°, 85°, 15°$.
11. (a) $\{(x, y) \mid x \in [-1, 0), y \in [2, 3)\}$; (c) (e^{-1}, e^2); (e) no solution.

PROBLEMS 89, p. 559

1. (a) $(2, 1)$; (c) $(-1, 1)$; (e) $(4, 3)$. 2. (a) $(1, -1, 1)$; (c) $(0, 0, 0)$;
(e) $(2, 0, -1, 1)$. 3. (a) $(5t, 3t, -7t)$. 4. Yes. 5. (a) $(7t, 1 + 3t, -1 - 5t)$;
(c) no solution. 6. 374, no. 7. $(x, y) = (u \cos \alpha - v \sin \alpha, u \sin \alpha + v \cos \alpha)$.
8. (a) $\frac{120}{13}$ hours; (b) 24 hours. 9. No. 10. Inconsistent.
11. (a) $(a, b, c) = (-1, 2, 3)$. 12. $f(x) = \frac{2}{3}x^3 - \frac{1}{2}x^2 + x - \frac{1}{6}$.
13. (a) $(t, 7t - 6, 11t - 10)$; (c) no solution.

PROBLEMS 90, p. 564

1. (g) $(0, 0, -2)$. 2. (a) $-\frac{3}{2}$; (c) $(x, y) = (2, 1)$. 3. $\mathbf{b} = 4\mathbf{i} + 3\mathbf{k}$,
$(x, y, z) = (\frac{3}{8}, -\frac{1}{8}, \frac{3}{8})$. 4. $A\mathbf{a} = (9, 14, 1)$. 7. (a) $(1, 1, 1)$; (b) $(2, 0, 1)$;
(c) $(9, 2, -10)$; (d) 0. 10. Monday vector is $(.25, .75)$, probability that we eat at
Joe's on Tuesday is $\frac{13}{40}$.

PROBLEMS 91, p. 572

1. (a) $\begin{bmatrix} 0 & 1 & 0 \\ 0 & 0 & 1 \\ -2 & 0 & 0 \end{bmatrix}$; (c) $\begin{bmatrix} 0 & 0 & 0 \\ 0 & 0 & 0 \\ 0 & 0 & 1 \end{bmatrix}$; (e) $\begin{bmatrix} 0 & 2 & 0 \\ 0 & 0 & 3 \\ 1 & 0 & 1 \end{bmatrix}$; (g) 0; (i) A.

3. (a) $\begin{bmatrix} \frac{3}{2} & 0 & 0 \\ 0 & -1 & 0 \\ 0 & 0 & \frac{5}{3} \end{bmatrix}$; (c) $\begin{bmatrix} -5 & -1 & 6 \\ 1 & 0 & -1 \\ 9 & 2 & -10 \end{bmatrix}$; (e) $\begin{bmatrix} 9 & -2 & 0 & 0 \\ -4 & 1 & 0 & 0 \\ 0 & 0 & -3 & 1 \\ 0 & 0 & -8 & 3 \end{bmatrix}$.

4. $(x, y, z) = (-5u - v + 6w, u - w, 9u + 2v - 10w)$.

5. (a) $\dfrac{1}{ad - bc} \begin{bmatrix} d & -b \\ -c & a \end{bmatrix}$. 6. (a), (c), and (f) are true. 8. (b) 0. 9. (a) $3^9 S$.

PROBLEMS 92, p. 578

1. (a) -2; (c) ac; (e) 1; (g) 0. 2. (a) -7; (c) 0; (e) -24. 3. (a) 8; (c) $\{1, 6\}$.
4. (a) -2; (c) $\{2, 7\}$. 5. (a) 0; (b) the matrix has an inverse, no matter what k is;
(c) $\{0, 2, 5\}$; (e) $\{1, -2\}$; (f) $\{-1, 3 + 2\sqrt{3}, 3 - 2\sqrt{3}\}$. 6. -25. 7. (a) 1;
(b) $\mathbf{x} = c(e, -1)$, where c is arbitrary; (c) $\mathbf{x} = \mathbf{0}$. 9. $\{-2, 0, 1\}$.

10. (c) $x - y + 1 = 0$. 12. (c) $A = \begin{bmatrix} 0 & 1 \\ -1 & 0 \end{bmatrix}$, for example.

REVIEW PROBLEMS, CHAPTER 11, p. 580

1. (a) $-A$; (b) A; (c) 0; (d) $2^{1065}A$. 2. (a) $\frac{1}{10} \begin{bmatrix} 5 & 5 & -5 \\ -1 & -3 & 5 \\ -3 & 1 & 5 \end{bmatrix}$;

(c) $\frac{1}{7} \begin{bmatrix} 2 & 3 & -4 \\ 1 & -2 & 5 \\ -3 & -1 & 6 \end{bmatrix}$. 3. $(3, 7, 2, 6, 1, 5, 0, 4, -1)$. 5. (a) $\begin{bmatrix} 10 & 6 \\ 5 & 5 \end{bmatrix}$;

(b) $\begin{bmatrix} 2 & 1 & 1 \\ -2 & 2 & 4 \end{bmatrix}$; (c) $[32]$; (d) $[3\ 6\ 9]$. 8. (b) $\begin{bmatrix} 3 & 0 \\ 0 & 5 \end{bmatrix}$.

10. 4 nickels, 3 dimes, and 3 quarters or 1 nickel, 7 dimes, and 2 quarters.
12. Bet \$40 on A, \$48 on B, \$60 on C, and \$80 on D.   13. No, not always.
18. $c_3 = c_1 + c_2$.   19. (a) The 2 by 2 matrices $A = B = 0$ and $A = B = I$,

for example.   20. $X = \begin{bmatrix} 1 & 0 \\ 0 & 0 \end{bmatrix}$, $Y = \begin{bmatrix} 0 & 0 \\ 0 & 1 \end{bmatrix}$, $Z = \begin{bmatrix} 1 & 0 \\ 0 & 1 \end{bmatrix}$.

21. Yes, $C = B^{-1}A^{-1}$.   22. $C = A^{-1}B$.   23. $\det A^* = \det A$.

24. $A^{-1} = \begin{bmatrix} 0 & -2 & 1 \\ 0 & 1 & 0 \\ -\frac{1}{3} & -\frac{4}{3} & \frac{2}{3} \end{bmatrix}$.

PROBLEMS 93, p. 586

1. Terminal points are (a) $(2, 2, 2)$; (c) $(2, -4, 2)$.   2. (a) $12\mathbf{i} + 9\mathbf{j} + 6\mathbf{k}$;
(c) $(3, 5, -1)$; (e) $(23, 20, 9)$.   3. (a) $\frac{1}{3}$.   4. 3.   5. $(2, 0, 3)$.   6. (a) 3.
9. $\mathbf{a} + \mathbf{b}, \frac{1}{2}(\mathbf{a} + \mathbf{b})$.   12. One whose vertices are the terminal points of $\mathbf{0}$, $\sqrt{2}\,\mathbf{a}$ and $\sqrt{2}\,\mathbf{b}$.

PROBLEMS 94, p. 590

1. (a) 3; (c) $\sqrt{3}\ln 2$; (d) $2\cosh x$.   2. (a) $2 - \sqrt{14}$; (c) $9\mathbf{i} + 3\mathbf{j} + 6\mathbf{k}$;
(e) $\frac{1}{3}\mathbf{i} + \frac{2}{3}\mathbf{j} - \frac{2}{3}\mathbf{k}$.   3. $(0, 4, -5)$.   4. -4, no.   6. $\pi - \mathrm{Cos}^{-1}\frac{4}{9}$ radians.   7. No.
8. $3\mathbf{a} - 7\mathbf{b}$, for example.   9. $3\mathbf{a}, -3\mathbf{a}$.
10. (a) $x^2 + y^2 + z^2 - 2x - 2y + 2z - 6$; (b) a sphere with diameter PQ.
12. (b) if $|\mathbf{a}| = |\mathbf{b}|$; (c) $-6^{-1/2}$.

PROBLEMS 95, p. 596

1. (a) $-\mathbf{i} + 2\mathbf{k}$; (c) $5\mathbf{i} + 3\mathbf{k}$; (e) $2\mathbf{i} + \mathbf{j} - 3\mathbf{k}$; (g) 1.   2. 3.   3. (a) $\mathbf{k}$;
(c) $\frac{2}{3}\mathbf{i} + \frac{1}{3}\mathbf{j} - \frac{2}{3}\mathbf{k}$.   4. $(x, y) = (6, -9)$.   5. 13.   6. $\frac{3}{2}\sqrt{2}$.   7. (a) $r\mathbf{a} + \mathbf{i} - \mathbf{j} + 2\mathbf{k}$;
(b) $r(\mathbf{i} + 2\mathbf{j}) + s(3\mathbf{j} + \mathbf{k}) - \sqrt{3}\,\mathbf{j}$, where r and s can be any scalars.
8. $\sin(\gamma - \beta)\mathbf{i} + \sin(\alpha - \gamma)\mathbf{j} + \sin(\beta - \alpha)\mathbf{k}$.   11. (c) $(\mathbf{a} \cdot \mathbf{c})\mathbf{b} - (\mathbf{b} \cdot \mathbf{c})\mathbf{a}$.

PROBLEMS 96, p. 600

1. (a) $\frac{1}{3}(2\mathbf{i} + \mathbf{j} - 2\mathbf{k})$; (c) $3^{-1/2}(\mathbf{i} + \mathbf{j} + \mathbf{k})$.   2. (a) $2x - 3y + z = -1$;
(c) $x = 0$.   3. (a) $2x - 4y + 5z = -3$; (b) $\frac{1}{2}x + \frac{1}{3}y + \frac{1}{5}z = 1$; (c) $2x - 3z = -7$;
(d) $x - y - z = -4$; (e) $3x + 4y - z = 16$; (f) $y = 0$.   4. (a) $\mathbf{i} - \mathbf{j} - \mathbf{k}$;
(b) $x - y - z = 7$.   6. $\frac{1}{6}\sqrt{21}$.   7. $6x - 14y - 6z = 15$.   8. (a) 3;
(b) $4x - 2y + 4z = 3$.   9. $(-2, -2, -2)$.
10. $x - 7y + 5\sqrt{2}z = 0$ or $x - 7y - 5\sqrt{2}z = 0$.   11. 12.

PROBLEMS 97,   p. 605

2. (a) $9(x^2 + y^2) = 4(z - 3)^2, 4(x^2 + z^2) = 9(y - 2)^2$;
(c) $y^2 + z^2 = x^2, z|z| = x^2 + y^2$; (e) $z = 4 - x^2 - y^2, (4 - y^2)^2 = x^2 + z^2$;
(g) $x^2 + y^2 = \operatorname{Sin}^2 z, y^2 = \operatorname{Sin}^2 \sqrt{x^2 + z^2}$.
4. Ellipse rotated about a line parallel to the Z-axis.
5. $(x^2 + y^2 + z^2 - 34)^2 = 100(9 - x^2)$.   6. (b) $(x + 1)^2 + (y - 2)^2 = 25$;
(c) $(x + 1)^2 + (y - 2)^2 \le 25$.   7. The parabolic cylinder $z^2 = 8(2 - x)$.
8. $4(x^2 + y^2 + z^2) = (x^2 + y^2 + z^2 + 2x)^2$.   9. (a) 100π;
(b) $\frac{5}{2}\pi(20 + 9 \ln 3)$.   10. $25(x^2 + y^2) + 16z^2 = 400$.
11. An ellipse of major diameter $10\sqrt{2}$ and minor diameter 10 in the plane $x = y$.

PROBLEMS 98,   p. 611

6. (a) Parabola, $(0, 0, 0)$; (b) two lines intersecting in $(0, 0, 0)$;
(c) parabola, $(0, 3, -1)$; (d) hyperbola, $(0, 3, -1)$ and $(0, -3, -1)$;
(e) parabola, $(4, 0, 4)$.   7. (a) $x^2 - 2y^2 + 3z^2 = 2$, hyperboloid of one sheet;
(c) $12x^2 + 5y^2 - 27z^2 = 5$, hyperboloid of one sheet.   8. (a) $(1, -1, 0)$, 2.
9. (a) $x + 2y + 2z = 9$; (c) $-14x - 5y + 2z = 113$.   10. (a) Ellipsoid;
(c) hyperboloid of one sheet.   11. (a) $5x^2 + 5z^2 = 2|y|$.   12. (a) $8y = x^2 + z^2$;
(c) $4x^2 + 4z^2 + 3y^2 - 20y + 12 = 0$.

PROBLEMS 99,   p. 615

1. (a) $\frac{2}{3}\mathbf{i} - \frac{1}{3}\mathbf{j} + \frac{2}{3}\mathbf{k}, 3\pi$; (c) $\frac{3}{5}\mathbf{i} + \frac{4}{5}\mathbf{k}, 5\pi$; (e) $3^{-1/2}(\mathbf{i} - \mathbf{j} - \mathbf{k}), 3^{1/2}(e^\pi - 1)$;
(g) $- \operatorname{sech}\frac{3}{2}\pi\mathbf{i} + \tanh\frac{3}{2}\pi\mathbf{k}, \sinh 3\pi$.   2. $26^{-1/2}(\mathbf{i} + 3\mathbf{j} + 4\mathbf{k})$.
3. $\mathbf{r} = t\mathbf{i} + \sin t\mathbf{j} + \cos t\mathbf{k}$.   4. (a) $(25, 9, 22)$; (b) $(\frac{1}{4}, \frac{7}{2}, -\frac{11}{2})$ 5. $\frac{1}{4}\pi$.   6. $\frac{14}{3}$.
7. (a) $\mathbf{v} = 3t^2\mathbf{i} + \sin t\mathbf{j} + e^t\mathbf{k}$; (b) $\mathbf{r} = (t^3 + 1)\mathbf{i} - \cos t\mathbf{j} + (e^t + 1)\mathbf{k}$.
9. $(\frac{1}{2}\sqrt{2} - \frac{1}{8}\pi\sqrt{2}, \frac{1}{2}\sqrt{2} + \frac{1}{8}\pi\sqrt{2}, \frac{1}{8}\pi)$.   10. $\mathbf{x} = e^{2t}(\mathbf{i} - \mathbf{j} + \mathbf{k})$.   12. (a) $\mathbf{f}(t) \times \mathbf{f}''(t)$;
(c) $\mathbf{f}(t) \cdot \mathbf{f}'(t) \times \mathbf{f}'''(t)$.   13. (a) $\mathbf{r} = f_y(t)\mathbf{j} + f_z(t)\mathbf{k}$.
14. $\mathbf{r} = \cos t\mathbf{i} + \sin t\mathbf{j} + t\mathbf{k}, t \in [0, 2\pi]$.

PROBLEMS 100,   p. 620

1. (a) $\mathbf{r} = (3t + 2)\mathbf{i} - 2t\mathbf{j} + (7t + 3)\mathbf{k}$; (c) $\mathbf{r} = t\mathbf{j}$;
(e) $\mathbf{r} = \ln 2^{t+2}\mathbf{i} + \ln 5^{t+1}\mathbf{j} + \ln 6 \cdot 2^{-t}\mathbf{k}$.
2. (a) $\mathbf{r} = (3t + 2)\mathbf{i} + (2t - 1)\mathbf{j} + (3 - t)\mathbf{k}$; (c) $\mathbf{r} = 3t\mathbf{i} - 2t\mathbf{k}$.
3. (a) $3x + 2y - z = 1$; (c) $3x = 2z$.   4. (a) $\mathbf{r} = (2t - 2)\mathbf{i} + (t + 1)\mathbf{j} + (2t + 2)\mathbf{k}$;
(c) $\mathbf{r} = (2 - 5t)\mathbf{i} + (2t - 3)\mathbf{j} + (\frac{1}{2} - 2t)\mathbf{k}$.   5. (a) $\mathbf{r} = (1 - 2t)\mathbf{i} - t\mathbf{j} + (t + 3)\mathbf{k}$;
(c) $\mathbf{r} = t\mathbf{k} + 2\mathbf{j}$.   6. $\mathbf{r} = (t - 1)\mathbf{i} + (2 - 7t)\mathbf{j} + (1 - 11t)\mathbf{k}$.   8. Yes.
9. (a) $-1/\sqrt{11}, 3/\sqrt{11}, 1/\sqrt{11}$ and $7/\sqrt{59}, -3/\sqrt{59}, 1/\sqrt{59}$; (b) yes, at $(-1, 8, 3)$.
10. $\frac{1}{4}\pi$ radians.   11. (a) $\frac{1}{7}$; (c) -5.   12. (b) $\mathbf{j} - 4\mathbf{k}$ and $2\mathbf{i} + 9\mathbf{k}$;
(c) $9x - 8y - 2z = -29$.   13. The line segment joining the terminal points of **a** and **b**.
15. $\mathbf{r} = \mathbf{f}'(a)(t - a) + \mathbf{f}(a)$.

PROBLEMS 101, p. 625

1. (a) $(2\sqrt{2}, 30°, 135°)$; (c) $(3\sqrt{2}, 135°, 135°)$; (e) $(\sqrt{5}, \frac{1}{3}\pi, \frac{1}{2}\pi - \text{Tan}^{-1}\frac{1}{2})$;
(g) $(\pi\sqrt{2}, \pi, \frac{1}{4}\pi)$.   2. (a) Cone; (c) ellipsoid of revolution;
(e) hyperboloid of one sheet; (g) the curve $z = \sin y$ rotated about the Z-axis.
3. (a) Sphere of radius 3; (c) the plane $z = 2$;
(e) the sphere with center $(0, 0, 1)$ and radius 1; (g) the plane $3x - 2y = 5z$.
4. (a) $r^2 = z$; (c) $3r \cos \theta + 4r \sin \theta + 5z = 6$; (e) $\rho = 4 \cos \phi$;
(f) $\rho^4 = 4(1 - \rho^2 \cos^2 \phi)$.   5. (a) Cylinders of radius $0, 1, 2, \ldots$;
(b) spheres of radius $0, 1, \ldots$; (c) planes containing the Z-axis; (d) cones;
(e) the solid cylinder of radius 1 about the Z-axis;
(f) the ball of radius 1 about the origin.   7. A "mesh" of coordinate surfaces.
8. (a) $\frac{5}{2} \ln 3$.   9. $\rho = 2, \theta = t, \phi = \frac{1}{4}\pi, L = 2\pi\sqrt{3}$.   10. $\frac{1}{2}\pi\sqrt{1 + k^2}$.
11. The cartesian curve.   12. It is symmetric about the Z-axis.
14. (a) $\sqrt{r_1^2 + r_2^2 - 2r_1r_2 \cos (\theta_1 - \theta_2) + (z_1 - z_2)^2}$;
(b) $\sqrt{\rho_1^2 + \rho_2^2 - 2\rho_1\rho_2 \sin \phi_1 \sin \phi_2 \left[\cos (\theta_1 - \theta_2) + \cos \phi_1 \cos \phi_2\right]}$.

PROBLEMS 102, p. 633

4. No.   5. v, x, and y must be 0; u, w, and z must be either $+1$ or -1.   8. $30°$.

10. $U = \begin{bmatrix} 0 & 1 & 0 \\ 0 & 0 & -1 \\ 1 & 0 & 0 \end{bmatrix}$.   11. (a) $(-6, 3, 9)$; (c) $\bar{x} + 2\bar{y} - 11\bar{z} = 1$;

(e) $2\bar{x} - 2\bar{y} = \bar{z}$.   12. $(1, 1, 0)$.   13. $U\mathbf{n}$.
16. $\bar{\mathbf{r}} = \mathbf{r} + \mathbf{a}, \bar{\mathbf{r}} = U\mathbf{r} + \mathbf{a}, \bar{\mathbf{r}} = U(\mathbf{r} + \mathbf{a})$.

PROBLEMS 103, p. 640

1. (a) $\lambda^2 - 4\lambda + 3 = 0$; (b) 3, 1; (c) $(1, 1), (-1, 1)$; (d) $\frac{1}{2}\sqrt{2} \begin{bmatrix} 1 & -1 \\ 1 & 1 \end{bmatrix}$.

3. (a) $\lambda^2 - 6\lambda - 16 = 0$; (b) $-2, 8$; (c) $(1, -3), (3, 1)$; (d) $\frac{1}{10}\sqrt{10} \begin{bmatrix} 1 & 3 \\ -3 & 1 \end{bmatrix}$.

4. (a) $\begin{bmatrix} 5 & 0 \\ 0 & 3 \end{bmatrix}$; (c) $\begin{bmatrix} 1 & 0 \\ 0 & -1 \end{bmatrix}$; (e) $\begin{bmatrix} 2 + \sqrt{5} & 0 \\ 0 & 2 - \sqrt{5} \end{bmatrix}$.   5. (a) $\lambda^3 - 5\lambda^2 + 6\lambda = 0$;

(b) 3, 2, 0; (c) $(1, 1, 1), (0, 1, -1), (2, -1, -1)$; (d) $\frac{1}{6}\sqrt{6} \begin{bmatrix} \sqrt{2} & 0 & 2 \\ \sqrt{2} & \sqrt{3} & -1 \\ \sqrt{2} & -\sqrt{3} & -1 \end{bmatrix}$.

7. (a) $\begin{bmatrix} 2 & 0 & 0 \\ 0 & 1 & 0 \\ 0 & 0 & -2 \end{bmatrix}$; (c) $\begin{bmatrix} 3 & 0 & 0 \\ 0 & 1 & 0 \\ 0 & 0 & 1 \end{bmatrix}$; (e) $\begin{bmatrix} 6 & 0 & 0 \\ 0 & 3 & 0 \\ 0 & 0 & 2 \end{bmatrix}$.   8. $(4, -3)$.

9. $x = \pm 4$, $\lambda = 12$; if $x = 4$, characteristic vectors are $(2, -1)$, $(1, 2)$.

12. $\begin{bmatrix} 819 & 410 \\ 410 & 204 \end{bmatrix}$.   13. It has the form xI.   15. Yes, not necessarily, yes.

PROBLEMS 104, p. 645

1. (a) $3\bar{x}^2 + \bar{y}^2 = 1$; (b) $\frac{1}{2}\sqrt{2}\begin{bmatrix} 1 & -1 \\ 1 & 1 \end{bmatrix}$; (c) $45°$.   3. $\bar{y}^2 - \frac{1}{4}\bar{x}^2 = 1$;

(b) $\frac{1}{10}\sqrt{10}\begin{bmatrix} 1 & -3 \\ 3 & 1 \end{bmatrix}$; (c) $-\text{Tan}^{-1} 3$ radians.   4. (a) $\dfrac{x^2}{3} + \dfrac{y^2}{5} = 1$;

(c) $\bar{x}^2 - \bar{y}^2 = 1$; (e) $(2 + \sqrt{5})\bar{x}^2 + (2 - \sqrt{5})\bar{y}^2 = 9$.   5. (a) $3\bar{x}^2 + 2\bar{y}^2 = 1$;

(b) $\frac{1}{6}\sqrt{6}\begin{bmatrix} \sqrt{2} & \sqrt{2} & \sqrt{2} \\ 0 & \sqrt{3} & -\sqrt{3} \\ 2 & -1 & -1 \end{bmatrix}$.   7. (a) $2\bar{x}^2 + \bar{y}^2 - 2\bar{z}^2 = 2$; (c) $3\bar{x}^2 + \bar{y}^2 + \bar{z}^2 = 5$;

(e) $6\bar{x}^2 + 3\bar{y}^2 + 2\bar{z}^2 = 6$.   8. (a) Hyperbola; (c) hyperbola.
9. A parabola whose vertex is the point $(-3, -2)$.
10. (c) $\mathbf{x} = \frac{1}{25}(9e^{16t} + 16e^{-9t}, 12e^{16t} - 12e^{-9t})$.   12. The surface is a sphere.
13. A cylinder.

REVIEW PROBLEMS, CHAPTER 12, p. 647

1. $Q(3, 4, -3)$, $R(-7, 4, 5)$, $S(-1, 0, 1)$, $T(-4, -3, 5)$.   2. $(4, -5, -7)$.   3. $\frac{4}{7}$.
4. $a^{-1}x + b^{-1}y + c^{-1}z = 1$.   5. $x + y - 2z = 1$.   6. $(10, 18, 17)$.
7. $\mathbf{r} = t\mathbf{i} + (2t + 1)\mathbf{j} - 4(t - 1)\mathbf{k}$, $x + 2y - 4z + 14 = 0$.   8. $(a, \theta, a \cot \phi)$.
10. Any numbers that satisfy the equation $2r - s = 3t$; the planes cannot intersect in exactly one point.   11. Opposite.   12. $\frac{1}{4}\bar{x}^2 - \bar{y}^2 + \bar{z}^2 = 0$.
13. A circular cone.   14. No solution for $t > 0$.
15. (a) 1 and $-.3$ are characteristic values, and $(4, 9)$ and $(1, -1)$ are corresponding characteristic vectors.   16. Characteristic values are 2, $-1 - \sqrt{3}$, $-1 + \sqrt{3}$.
19. $|d - e|(a^2 + b^2 + c^2)^{-1/2}$.   25. $b = c$ or $b = -c$ and $a = d$.

PROBLEMS 105, p. 653

1. (a) $f(x, y, z) = xy^2z^3$; (c) true if $xyz = 0$; (d) $f(-\mathbf{x}) = f(\mathbf{x})$; (e) $(xyz)^6$;
(f) $\lim\limits_{h \to 0} q(h) = 12$.   2. (b) y can be any integer, for example; (c) 5; (d) yes; (f) no.
3. (a) Domain is R^2, range is R^1; (c) domain is R^2, range is $[-1, 1]$;

(e) domain is R^2; range is $[0, \infty)$.   4. (a) Domain is $\{(x, y, z) \mid x \ge y, z > 0\}$;
(c) domain is $\{(x, y, z) \mid x + y + z \ge 0\}$.   5. $f(x, y) = y/x$.   7. $R = L/25r^2$.
8. The planes $y = x$, $z = x$, and $z = y$.   9. (a) The plane $2x - 3y + 4z = 7$;
(b) the line $\mathbf{x} = \mathbf{m}t + \mathbf{b}$, where $\mathbf{b} = -\mathbf{i} - \mathbf{j} - \mathbf{k}$.   10. (a) $V = xyz$;
(b) $V = (5 - xy)yx/(x + y)$.   11. (b) $f(x) = 5x^2 + 4xy + 2y^2$.
12. $f(x, y, z) = f(y, z, x)$.

PROBLEMS 106,   p. 659

1. (a) 18; (c) -144; (e) $-\frac{9}{4}$; (g) $6e^{-2}$; (i) not defined; (k) 6; (m) $-\frac{2}{3}$; (o) -1.
4. Positive Y-direction.   6. (a) 0; (c) x; (e) $\sin (x^2 - y^2)$.
9. $\frac{1}{20}$ newton, -2 cubic meters.   10. True for some functions, false for others.
13. $f_1(0, 0) = f_2(0, 0) = 0$.

PROBLEMS 107,   p. 665

1. (a) 1, (c) 0; (e) $2x$; (g) $-e^x - e^{-x}$.   2. (a) $\sec^2 x$; (b) $\cos 2y \sec^2 (x - y)$;
(c) $\cos 2x \sec^2 (x - y)$.   3. $x^x(1 + \ln x)$.   4. $3x^2 \sin x^6 - 2x \sin x^4$.   5. $38e^{15}$.
6. $2x$.   7. (a) 3; (c) $-\frac{3}{2}$; (e) $-3/\ln 4$; (g) e^{-5}.
8. They make an angle of $135°$ with the positive X-axis.

9. (a) $f_1(x, y) \dfrac{dx}{du} + f_2(x, y) \dfrac{dy}{du}$;   (c) $f_1(g, h) \dfrac{dg}{dx} + f_2(g, h) \dfrac{dh}{dx}$.   11. (b) $4e^{2x} - x^{-2}$.

PROBLEMS 108,   p. 672

1. (a) $108\mathbf{i} - 48\mathbf{j}$; (c) $-3\pi\mathbf{i} + 2\pi\mathbf{j}$; (e) $3\mathbf{i} - 2\mathbf{j}$.   2. (a) $-\frac{6}{5}$; (c) 0; (e) $\frac{4}{5}$ or $-\frac{4}{5}$;
(g) $\pm 19\pi\sqrt{3}/54$ 3. (a) Ascending at 33 meters per meter;
(b) neither ascending nor descending; (c) northeast; (d) northwest or southeast.
4. (a) $\mathbf{k}$; (c) $\frac{1}{2}\sqrt{2}\,(\mathbf{i} + \mathbf{k})$.   5. (a) 5; (b) $36\mathbf{i} - 72\mathbf{j} - 12\mathbf{k}$; (c) 66;
(d) $-54\mathbf{j} + 3\mathbf{k}$; (e) 6; (f) $-3\mathbf{i} + 8\mathbf{j}$.   6. $2e^{-14}\sqrt{14}$.
10. Their ratio is $|\nabla f(\mathbf{x})|/|\nabla g(\mathbf{x})|$.

PROBLEMS 109,   p. 677

2. (a) $2\mathbf{i} - 3\mathbf{j} + 4\mathbf{k}$; (c) $4\mathbf{i} + \mathbf{j} + \mathbf{k}$; (e) $2\mathbf{i} + 2\mathbf{j} - \mathbf{k}$.   3. (a) $7x - 11y + z + 9 = 0$;
(c) $z = 0$; (e) $2x - z = 1$.   4. $(-1, 3, -15)$.
5. $\mathbf{m} = 7\mathbf{i} + 4\mathbf{j} + 10\mathbf{k}$, $\mathbf{b} = \mathbf{i} - 2\mathbf{j} + \mathbf{k}$.   6. $(2, 2, 0)$ and $(-2, -2, 0)$.
7. 7.   8. $3\mathbf{i} + 2\mathbf{j} - 3\mathbf{k}$.   9. $10/\sqrt{122}$.   10. No, it is above the panel.
13. $y = x + 3$.

PROBLEMS 110,   p. 683

1. (a) $(1, 2, 2)$; (b) $(2, 1, 1)$.   2. (a) $(8, 12, -64)$; (b) $(1, -2, -4)$.
3. (a) $(-1, 3, 10)$; (b) $(1, -2, 4)$.   4. 2.   5. $\frac{1}{2}\sqrt{2}$.
6. $(2, -1, 1)$ and $(2, -1, -1)$.   7. $2x + y + 2z = 6$.   9. 2 feet by 2 feet by 5 feet.
10. 4 inches, at an angle of $60°$.   11. $\frac{1}{3}\sqrt{3}$.   14. $\frac{1}{27}$ and 0.

REVIEW PROBLEMS, CHAPTER 13,   p. 684

1. 54.   2. (a) $3\mathbf{i} + e \cdot 2^{e-1}\mathbf{j}$; (b) $3 \cdot 2^e\mathbf{i} + 3e \cdot 2^{e-1}\mathbf{j}$; (c) $e(\ln 3)^{e-1}\mathbf{i} + e^2\mathbf{j}$;
(d) $e^3(\ln 3)^{e-1}\mathbf{i} + e^2(\ln 3)^e\mathbf{j}$; (e) $3e^2(\mathbf{i} + \mathbf{j})$; (f) $18\mathbf{i} + 9 \ln 3\mathbf{j}$.
3. (a) The ellipsoidal region $4x^2 + 36y^2 + z^2 \le 36$; (b) $[-\frac{1}{6}\pi, \frac{1}{2}\pi]$;
(d) $x + 3y - z = 6$; (e) $-\frac{14}{27}\sqrt{3}$.   5. $0, \sqrt{\frac{1}{2}\pi}$.   6. 3.   7. 114, 57, 0.
10. There is a number c such that $f(\mathbf{x}) = g(\mathbf{x}) + c$.   13. $\mathbf{a} \cdot \mathbf{u}$.   14. $\frac{1}{2}(\mathbf{a} + \mathbf{b})$.
16. (a) 17; (b) $(p, q) = (2, -1)$.
17. $(2 + 200/k, -3 - 240/k, 440 + 1/k)$, where $k = \sqrt{97{,}601}$.   18. $\frac{3}{2}\pi^2\sqrt{2}°$ per day.
19. $(2, -1, 3)$ or $(-2, 1, -3)$, whichever is on the sunny side.   20. $17/3\sqrt{41}$.
21. (a) $(1, -1, 2)$; (b) $5, 2, 2$; (c) 1.   22. (a) $(x, y) = (2, 3)$ and $(x, y) = -(2, 3)$;
(b) $90 \pm 18\sqrt{37}$ and $-90 \pm 18\sqrt{37}$.   23. (c) $25, (1, 2)$ and $-5, (2, -1)$;
(d) minimum point; (e) maximum point; (f) no.

PROBLEMS 111,   p. 692

1. (a) 6; (c) 6; (e) 2.   2. 20, 92.   3. (a) 4; (c) 4; (e) 2; (g) $4e^\pi$; (i) 8; (k) 2;
(m) 1.   4. (a) $\frac{4}{3}$; (c) 2; (e) $6 \ln 3 - 4$; (g) 4.   5. (b) $\frac{16}{3}\pi$; (c) $\frac{8}{3}\pi$.   7. $\frac{51}{2}$.   8. (a) π;
(c) 4; (e) $4 - \pi$.   9. (a) and (b).

PROBLEMS 112,   p. 699

1. (a) $\frac{7}{6}$; (c) $\frac{1}{2}$; (e) 4; (f) $\ln 81$.   2. (a) 1; (b) π; (c) 2; (d) 0; (e) $\frac{1}{8}(\pi^2 - 4)$.
3. (a) $\frac{2}{3}$; (c) $(e^2 - 2)\sin 1 + \sin e^2$.   4. (a) $e - 2$; (c) $e - 1$.
5. (a) $\frac{4}{15}(5^{5/2} - 33)$; (c) $\cos e^2 - \cos(e^2 + 1) + \cos 2 - \cos 1$.   6. (a) 4; (b) 1;
(c) $4(e - 1)^2$.   7. (a) $\ln 2$; (b) $1 - \sin 1$.   8. $\int_0^{\pi/2} x^{-1}(1 - \cos \frac{1}{2}\pi x)\, dx$.
11. $F(1, 1)$ and $F_1(1, 1)$ are positive.

PROBLEMS 113,   p. 706

1. (a) $\frac{64}{3}$; (c) $2/3\pi$; (e) $\frac{1}{4}e^4 - \frac{1}{3}e^3 + \frac{1}{12}$.   2. (a) $\frac{8}{3}$; (c) $\frac{1}{8}\pi^2$.   3. (a) 2; (c) $\frac{1}{3}$; (e) 0;
(f) 0.   5. (a) $4 \ln \frac{3}{2} - 1$; (c) 18; (e) $e^4 - 4e$.   6. (a) $\frac{1}{4}(e^{16} - 1)$; (c) $e - 1$.
7. $e^{2\pi} - \frac{1}{2}e^\pi + \frac{1}{2}e^{\pi/2}$.   8. $\ln 2 - \frac{1}{4}$.
10. $\int_0^a \int_a^b f(x, y)\, dx\, dy + \int_a^b \int_y^b f(x, y)\, dx\, dy$.   11. (a) π; (c) πc.

PROBLEMS 114, p. 711

1. (a) $\frac{9}{2}$; (c) $\frac{15}{2} - \ln 256$; (e) 1.   2. 20.   3. $\frac{1}{6}abc$.   4. (a) $\frac{1}{2}$; (c) $\frac{52}{3}$.   6. $\frac{4}{3}$.   7. $\frac{8}{15}$.
8. 15π.   9. $\frac{1}{2}\pi$.   11. 4π.   12. $\frac{2}{9}$.

PROBLEMS 115, p. 719

1. (a) $(4a/3\pi, 4a/3\pi)$; (c) $(\frac{8}{5}, \frac{3}{4})$; (e) $(\frac{1}{3}(a + c), \frac{1}{3}b)$; (g) $(\frac{9}{20}, \frac{9}{20})$.   2. $(1, \frac{1}{8}\pi)$.
5. (a) $-5\pi a^3$; (c) $-\frac{80}{3}$.   6. (a) $\frac{1}{3}ab^3, \frac{1}{3}a^3b, \frac{1}{3}ab(a^2 + b^2)$; (c) $\frac{1}{3}, \frac{1}{3}, \frac{2}{3}$.   8. (a) $\frac{1}{4}\pi a^3b$;
(b) $\frac{1}{4}\pi ab^3, \frac{1}{4}\pi ab(a^2 + b^2)$; (c) $\frac{1}{4}\pi a^4$; (d) $\pi ab(\frac{5}{4}a^2 - b^2)$; (e) $\pi a^3b(\frac{1}{4} + a^2c^{-2})$.
9. $I_O = 10\pi, I_L = 9\pi$.   10. No.

PROBLEMS 116, p. 724

1. (a) π; (b) $\frac{1}{2}\pi$; (c) π; (e) $\frac{2}{3}\pi$; (g) $\frac{1}{2}\pi$; (h) $\frac{1}{2}\pi$; (i) $\frac{1}{2}\pi$.   3. (a) $\frac{2}{3}\pi a^3$; (c) $\frac{5}{3}\pi$;
(e) $\frac{4}{9}$.   4. (a) $\frac{1}{4}\pi(1 - e^{-4})$; (c) $\frac{1}{8}\pi(1 - \cos 18)$.   5. (a) $\frac{4}{3}\pi a^3(8 - 3\sqrt{3})$.   6. (a) $\frac{1}{4}a^3s$;
(b) $\frac{4}{3}a^2s^{-1} \sin \frac{1}{2}sa^{-1}$; (c) $\frac{1}{16}a^4(2sa^{-1} - \sin 2sa^{-1})$.   8. (a) $(-\frac{5}{6}a, 0)$;
(c) $(\frac{512}{105}\pi^{-1}, \frac{512}{105}\pi^{-1})$.   9. (a) $\bar{x} = 4b/3\pi$; (b) $2b/\pi$; (c) yes, yes.

PROBLEMS 117, p. 731

1. (a) 3; (b) $\frac{3}{2}\pi - 3$; (c) $(e - 1)^3$; (d) 0; (e) $\frac{10}{3}$; (g) $\frac{3}{8}$;
(i) $3\pi - \frac{3}{2} - 3\sqrt{3}$.   2. (a) 36; (b) 0; (c) $\frac{1}{12}$; (d) $\frac{1}{3}$; (e) $\frac{1}{15}$; (f) $\frac{1}{4}$.
4. $\frac{1}{3}abc(a^2 + c^2)$.   5. $(\frac{1}{4}a, \frac{1}{4}b, \frac{1}{4}c)$.   6. $(0, \frac{3}{7}, \frac{2}{7})$.   7. $\frac{1}{30}$.   8. (a) $8 - \frac{4}{3}\pi$.
9. $I_z + (a^2 + b^2)V, I_z - (a^2 + b^2)V$.

PROBLEMS 118, p. 737

1. $\frac{4}{3}\pi a^3$.   2. $\frac{1}{35}2^3 \cdot 3^6$.   3. $\frac{243}{16}$.   4. (a) $\frac{1}{2}V(a^2 + b^2)$;
(b) $\frac{1}{12}V[3(a^2 + b^2) + L^2]$, approximations are Vc^2 and $\frac{1}{6}V[3c^2 + L^2]$.
5. $\frac{8}{15}\pi(b^5 - a^5)$.   6. $\frac{2}{15}\pi(64 - 33\sqrt{3})$.   7. (b) $\frac{8}{3}\pi$; (c) $\bar{z} = \frac{3}{4}$; (d) $\frac{16}{3}\pi$.   8. (b) $\frac{8}{9}\pi$;
(c) $(\rho, \theta, \phi) = (\frac{9}{4}\sqrt{2 - \sqrt{3}}, \frac{1}{12}\pi, \frac{1}{2}\pi)$; (d) $\frac{64}{45}\pi$.   9. $\frac{1}{10}\pi a^4h, \frac{1}{4}h$.
10. $\bar{z} = \frac{3}{16}(b^4 - a^4)/(b^3 - a^3)$.

PROBLEMS 119, p. 743

1. (a) 0; (b) 18; (c) 12; (d) $\frac{21}{2}$; (e) 4; (f) $\frac{1}{8}\pi + \frac{3}{2}\sqrt{2} - \frac{11}{4}$.   2. (a) $\frac{5}{4}$;
(c) $\frac{8}{3}\pi^{-1} + \frac{1}{2}$; (e) $\frac{3}{2}$.   3. 11.   4. (a) $\frac{2}{3}ab(a - b)$.   7. 2, no.   8. $a = \frac{1}{3}$.

PROBLEMS 120, p. 751

1. (a) 16; (c) 17; (e) $\frac{1}{2}\sqrt{3}$; (f) $\frac{3}{2}$. 2. (a) -16; (c) $\frac{1}{2}\sqrt{2}$. 3. (a) -7π; (b) -84;
(c) -42; (d) $-\frac{21}{2}$. 5. (a) $3xy^2$; (b) x^3. 7. (b) $-\pi$ along upper arc, π along lower.

PROBLEMS 121, p. 757

1. (a) $xe^{xy} - 1$; (c) $3xy^2 + 2x - 5$. 2. (a) $4x^2y^3$; (b) no solution;
(c) $\sin(x^2 + y)$; (d) $\sin^2 xy$. 3. (a) xe^{xy^2}; (b) ye^{xy^2}; (c) $\ln(x^2 + y^2)$;
(d) not a conservative force. 4. (a) $xy^2 + 4$; (b) $x^2y - 2$; (c) $-\cos xy - 1$;
(d) $\sin xy$. 5. $F(x, y) = 2x^3y + xy^2$. 6. $F_1(-1, 2) = -6$, $F_2(3, 0) = 2 \cdot 3^9$.

7. $F(x, y) = \displaystyle\int_0^{xy} e^{t^2}\, dt$. 8. $(x, y) = (-1, 2)$. 10. (a) $G(\mathbf{x}) = -F(\mathbf{x})$;

(b) $\nabla H(\mathbf{x}) = \nabla F(\mathbf{x})$; (c) $\nabla K(\mathbf{x}) = -\mathbf{f}(-\mathbf{x})$.

PROBLEMS 122, p. 762

1. (a) $y = \sqrt{4 + x^2}$; (c) $y = (2 + x)/(1 - 2x)$; (d) $y = (5x + 2)/(1 - 2x)$;
(e) $y = x + \frac{1}{2}\pi$. 2. (a) $y = 2x^{-2/3}$; (b) $y = -2\sqrt{2x^{-3}}$; (c) $y = (1 - x)e^{-x}$;
(e) $y = \text{Tan } x^2$. 3. $y = 2x^{3/2}$. 4. $y = \pi - \text{Sin}^{-1} x^3$. 5. $w(x) = -\ln(5 - e^x)$.
6. $y = cx$. 7. $y = \frac{1}{9}(x^{3/2} + 5)^2$. 8. $y = 3x^2$. 9. $\sqrt{3}$.

REVIEW PROBLEMS, CHAPTER 14, p. 764

1. (a) $\frac{1}{2}$; (b) e; (c) $\frac{1}{3}(e^6 - e^{-3})$; (d) $e^{e^2} - e^{e^{-1}} - e^2 + e^{-1}$; (e) $e^4 + e - 2$;
(f) $e^4 - 2e^2 + 2e^{-1} + 1$. 2. $I_3 < I_1 < I_2$. 3. Approximate value $\frac{5}{3}\pi$, true value $\frac{16}{3}$.

4. π. 5. $\frac{8}{3}$. 6. (a) $\displaystyle\int_0^1 \int_{\text{Sin}^{-1}y}^{\pi/2} f(x, y)\, dx\, dy$; (b) $\displaystyle\int_0^1 \int_0^{\text{Cos}^{-1}y} f(x, y)\, dx\, dy$;

(c) $\displaystyle\int_0^1 \int_{\text{Sin}^{-1}y}^{\pi-\text{Sin}^{-1}y} f(x, y)\, dx\, dy$; (d) $\displaystyle\int_{-1}^1 \int_0^{\text{Cos}^{-1}y} f(x, y)\, dx\, dy$. 7. 4π. 8. 8.
9. Negative. 10. (b) $\frac{5}{4}\pi$. 11. $\frac{3}{16}\pi^2$. 13. (a) $f(x)$. 14. $\pi - 2$. 15. (a) -1;
(b) -1; (c) -2; (d) $\frac{1}{2}$. 17. $\frac{5}{4} - 2\cos 1 + \frac{1}{4}\cos 2$. 18. $\frac{1}{3}a^4[\sqrt{2} + \ln(\sqrt{2} + 1)]$.
21. 24. 22. (a) $g_1(x, y) = 1 + f_2(x, y)$; (b) $x + 2xy$.
24. (b) $\nabla(\mathbf{x} \cdot M\mathbf{x}) = 2(px + qy)\mathbf{i} + 2(rx + sy)\mathbf{j}$.

Index

Derived functions:
on R^1, 61, 107
on R^n, 654
Determinant of a matrix, 574
Determinative system of linear equations, 557
Diagonalization:
of a matrix, 639
of quadratic expressions, 644
Diameter:
of an ellipse, 163
of a hyperbola, 169
Difference quotient, 67
Differentiability:
definition, 68
from the left, 71
Differential equation:
of the first order, 109
in general, 75, 365
linear, 374
partial, 658
of the second order, 109, 368
separable, 369, 761
Differential operator, 75
Differentiation:
definition, 68
general and specific formulas, 116, 117
implicit, 147
partial, 654
Directed arc, 739
Directed path, 741
Direction cosines, 619
Direction numbers, 619
Directional derivative, 670
Directrix:
of a conic, 183
of an ellipse, 182
of a hyperbola, 182
of a parabola, 175
Discontinuous function, 56
Discriminant, 415, 644
Distance between two points:
in cylindrical coordinates, 627
of the number scale, 3
in the plane, 12
in polar coordinates, 388
in space, 588
in spherical coordinates, 627
Divergence:
of improper integrals, 473, 475, 476
of infinite series, 496
of sequences, 490
Domain of a function, 22
Dominating series, 503
Dot product:
of elements of R^n, 589
of vectors in the plane, 424
of vectors in space, 588
Double integral:
definition, 689
in polar coordinates, 722

E

e:
definition, 285
as a limit, 297, 494
as the sum of a series, 539
value, 285
Eccentricity of a conic, 183
Echelon form of a system of linear equations, 561
Element of a set, 6
Ellipse, 160–165
center, 162
definition, 160
diameters, 163
directrices, 182
eccentricity, 183
equation, 161
foci, 160
focusing property, 186, 400–401
graph, 162
latus rectum, 167
polar coordinate equation, 392
vertices, 163
Ellipsoid, 606
Elliptic paraboloid, 608
Elliptical cone, 609
Empty set ($\emptyset$), 6
Endpoint, 7, 125
Equality of vectors, 417
Equation of a set, 17
Equilateral hyperbola, 171
Equivalent relations, 16
Equivalent systems of equations, 550
Error function (erf), 546
Euler's constant, 516
Euler's formula for e^{ix}, 548
Even function, 89
exp, 283
Exponential function, 286
Extended limit, 467
Extension of a function, 57, 288

F

Family of curves, 149, 159
Final segment of a sequence, 490
Finite interval, 7
First Derivative Test for maxima and minima, 126
First octant, 587
Focal length of a parabola, 175
Foci:
of an ellipse, 160
of a hyperbola, 167
Focus of a parabola, 174
Focusing property of conics, 185, 186, 399, 400
Frustum of a cone, area, 457